THERMAL CONDUCTIVITY 17

THERMAL CONDUCTIVITY 17

Edited by
J. G. Hust

Thermophysical Properties Division
National Bureau of Standards
Boulder, Colorado

Springer Science+Business Media, LLC

Library of Congress Cataloging in Publication Data

International Thermal Conductivity Conference (17th: 1981: Gaithersburg, Md.)
   Thermal conductivity 17.

   "Proceedings of the Seventeenth International Thermal Conductivity Conference, held
June 15–18, 1981, in Gaithersburg, Maryland" — T.p. verso.
   Includes bibliographical references and indexes.
   1. Heat — Conduction — Congresses. 2. Materials — Thermal properties — Congresses. I.
Hust, J. G. (Jerome G.) II. Title. III. Title: Thermal conductivity seventeen.
QC320.8.I58 1981   536′.2012   82-18051
ISBN 978-1-4899-5438-1

ISBN 978-1-4899-5438-1   ISBN 978-1-4899-5436-7 (eBook)
DOI 10.1007/978-1-4899-5436-7

Proceedings of the Seventeenth International Thermal Conductivity
Conference, held June 15–18, in Gaithersburg, Maryland

© 1983 Springer Science+Business Media New York
Originally published by Purdue Research Foundation in 1983
Softcover reprint of the hardover 1st edition 1983

FOREWORD

The International Thermal Conductivity Conference was started in 1961 with the initiative of Mr. C. F. Lucks and grew out of the needs of researchers in the field. From 1961 to 1973 the Conferences were held annually, and have been held biennially since 1975 when our Center for Information and Numerical Data Analysis and Synthesis (CINDAS) of Purdue University became the permanent Sponsor of the Conferences. These Conferences provide a broadly based forum for researchers actively working on the thermal conductivity and closely related properties to convene on a regular basis to exchange their ideas and experiences and report their findings and results.

The Conferences have been self-perpetuating and are an example of how a technical community with a common purpose can transcend the invisible, artificial barriers between disciplines and gather together in increasing numbers without the need of national publicity and continuing funding support, when they see something worthwhile going on. It is believed that this series of Conferences not only will grow stronger, but will set an example for researchers in other fields on how to jointly attack their own problem areas.

Of the first thirteen Conferences, only four published formal Proceedings. However, effective with the Fourteenth Conference, a policy of publishing formal Proceedings on a continuing and uniform basis has been established. Thus, including the present volume, the following formal Proceedings have been published:

Conference and Year	Title of Volume	Publisher and Year
7th (1967)	THERMAL CONDUCTIVITY Proceedings of the Seventh Conference	U.S. Government Printing Office (1968)
8th 1968)	THERMAL CONDUCTIVITY Proceedings of the Eighth Conference	Plenum Press (1969)

9th (1969)	NINTH CONFERENCE ON THERMAL CONDUCTIVITY	USAEC (1970)
13th (1973)	ADVANCES IN THERMAL CONDUCTIVITY Papers Presented at XIII International Conference on Thermal Conductivity	University of Missouri, Rolla (1974)
14th (1975)	THERMAL CONDUCTIVITY 14	Plenum Press (1976)
15th (1977)	THERMAL CONDUCTIVITY 15	Plenum Press (1978)
16th (1979)	THERMAL CONDUCTIVITY 16	Plenum Press (1982)
17th (1981)	THERMAL CONDUCTIVITY 17	Plenum Press (1982)

J. G. Hust, Chairman of the Seventeenth Conference, is to be congratulated for his excellent leadership in conducting the Conference and for his painstaking efforts which made the present volume possible. CINDAS looks forward to working with future host institutions to ensure that future Conferences continue to produce high-quality volumes of Proceedings in this important, specialized field.

This Foreword should have been written by Y. S. Touloukian, the founder and founding Director of CINDAS for the past 25 years. It was all owing to Y. S. Touloukian's great efforts that the Proceedings of recent and future Conferences have been and will be published formally by Plenum Press on a continuing and uniform basis, thus making them a continuing series of uniform volumes serving as the permanent major vehicle for the reporting of research results on thermal conductivity. I regret most deeply to report that Y. S. Touloukian died suddenly on 12 June 1981, three days before the opening of the Conference at which he was scheduled to deliver a welcoming speech. His passing away is a great loss to the field of thermophysical properties, to which he had devoted his entire life.

C. Y. Ho
Interim Director
Center for Information and Numerical
   Data Analysis and Synthesis
Purdue University

May 1982
West Lafayette, Indiana

PREFACE

     The Seventeenth International Thermal Conductivity Conference
was held at the National Bureau of Standards, Gaithersburg, Maryland.
It was held concurrently with the Eighth Symposium on Thermophysical
Properties and the Eighth International Thermal Expansion Symposium
under the auspices of the International Thermophysics Congress. This
was the first time these three conferences were held concurrently and
the conferees indicated it to be a very successful venture. The
conference chairman was J. G. Hust, National Bureau of Standards,
Boulder, Colorado. The arrangements committee consisted of:

          R. F. Martin
          J. A. Lorden
          K. C. Stang
          National Bureau of Standards
          Gaithersburg, Maryland

The Joint Conference Organizing Committee was:

          A. Cezairliyan, Chairman
          International Thermophysics Congress

          J. G. Hust, Chairman
          Seventeenth International Thermal Conductivity Conference

          T. A. Hahn, Chairman
          Eighth International Thermal Expansion Symposium

          J. V. Sengers, Chairman
          Eighth Symposium on Thermophysical Properties

CINDAS/Purdue University provided continuing sponsorship of the 17th
ITCC along with vital mailing assistance and made the arrangements
for the publication of these proceedings.

     A somber note is added in recalling that Y. S. Touloukian,
CINDAS, passed away while attending a CODATA meeting in the conference
hotel just prior to the start of the Joint Conference. He contributed

significantly to the organization of thermophysical properties data
and his departure is noted with sorrow. He was honored posthumously
at the conference banquet.

The Thermal Conductivity Conferences provide a unique forum for
the discussion of current thermal conductivity research and problem
areas. At this conference emphasis was directed toward discussions
of the following areas: a) Insulations – energy conservation, b)
Standard Reference Materials, and c) Composites and Aggregates.

The prefaces of past proceedings have usually listed pertinent
data about the organization for the Thermal Conductivity Conferences.
Several years ago it was decided that a history of these conferences
would be of interest to the readers. C. F. Lucks agreed to write this
history and it was completed in time for this proceedings. This paper
is included, in place of the usual documentation in the preface, as
the first paper of this proceedings. The chairman wishes to express
his appreciation to C. F. Lucks for this time consuming effort.

The thermal conductivity award was presented jointly to R.
Berman, Oxford University, England, and C. Y. Ho, CINDAS, Purdue
University.

The Eighteenth International Thermal Conductivity Conference
(1983) will be chaired by T. Ashworth at the South Dakota School of
Mines and Technology, Rapid City, South Dakota.

The chairman of the 17th ITCC acknowledges the assistance given
by various individuals and organizations to make this conference a
success. First, the excellent cooperation of the joint conference
chairmen is acknowledged: A. Cezairliyan, J. V. Sengers, and T. A.
Hahn. The arrangements committee is also commended for a job well
done. Thanks are given to all of the session chairmen for conducting
the sessions, reviewing the papers for technical content, and their
prompt responses to my requests. The support and assistance of the
NBS staff through the various stages of the conference is acknowledged. Appreciation is expressed to the staff of CINDAS, continuing
co-sponsor, for the publication arrangements and guidance in preparing
a quality publication. The advice of the ITCC governing board was
invaluable. The assistance given by Marilyn Steig, editorial reader,
and Edithann Allen, secretary and typist, is gratefully acknowledged.
Last, but not least, the assistance of my wife, Shirley Hust, in maintaining orderly files and performing numerous thankless tasks is noted
with appreciation.

Boulder, Colorado   J. G. Hust
May 1982   Chairman, 17th ITCC

CONTENTS

INTERNATIONAL THERMAL CONDUCTIVITY CONFERENCES 1961-1981:
A PROFILE   3
    C. F. Lucks

PLENARY SESSION TC-1

Keynote address by J. Kestin, Brown University, Thermophysical
Properties in Science and Technology (Not published)

SESSION TC-2

Theory, Review, and Correlation - I

Theory of Heat Conduction in Evacuated Metal Powders   25
    P.G. Klemens

A Local Composition Model for Thermal Conductivity in
Multicomponent Liquid Mixtures   31
    R.L. Rowley

Evaluation of Specific Heat Data for POCO Graphite and
Carbon-Carbon Composites   45
    M.S. Deshpande and R.H. Bogaard

Thermal Properties of POCO-Process Graphite   55
    L.L. Isaacs and W.Y. Wang

Phonon Conduction in Elastically Anisotropic Cubic
Crystals   63
    A.K. McCurdy

SESSION TC-3

Standard Reference Materials and Data - I

The Standard Reference Materials and Data Programs of the
CODATA Task Group on Thermophysical Properties   73
    M.L. Minges

Recommended Values of Electrical Resistivity and Thermal
Conductivity of Platinum   95
    G.K. White

Standard Reference Materials for Thermal Conductivity
Below 100 K   105
    R. Berman, N.D. Hardy, M. Sahota, J.G. Hust, and
    R.J. Tainsh

Thermal Diffissivity Measurements on Reference Materials
at IKE   117
    R. Brandt and G. Neuer

Standard Reference Materials and Low Thermal Conductivity
Experimental Methods   125
    F. Cabannes

SESSION TC-4

Standard Reference Materials and Data - II

Nondestructive Characterization of POCO AXM Graphite   137
    J.R. Koenig and C.D. Pears

Thermal Diffusivity of a POCO Graphite   147
    V.V. Mirkovich

The Thermal Transport Properties of a POCO AXM-5Q1
Graphite from 80 to 970 K   153
    J.P. Moore and R.S. Graves

Thermal Diffusivity of Candidate Standard Reference
Materials   163
    K. Maglic, N. Perovic, and Z. Zivotic

SESSION TC-5

Theory, Review, and Correlation - II

Attenuation of Seismic Waves by Heat Conduction   177
    P.G. Klemens

Illustrative Numerical Comparisons between Phonon Mean
Free Paths and Phonon Thermal Conductivity   185
    W.M. MacDonald and A.C. Anderson

Thermoelectric Power of Selected Binary Alloy Systems . . .   195
    C.Y. Ho, T.C. Chi, R.H. Bogaard, T.N. Havill,
    and H.M. James

SESSION TC-6

Metals and Alloys

Thermal Conductivity and Lorenz Number of Plutonium
and Plutonium-Gallium Alloys   209
    J.F. Andrew and P.G. Klemens

The Physical Properties of 9 Cr-1 Mo Steel from 300
to 1000 K   219
    R.K. Williams, R.S. Graves, F.J. Weaver, and
    D.L. McElroy

Thermal Diffusivity of Nitrided Steel   229
    H.J. Lee, S.K. Lee, and C.O. Lee

SESSION TC-7

Liquids and Gases - I

Thermal Conductivity of Liquids   241
    M.P. Saksena and Prabhuram

Thermal Conductivity Measurement of the Potassium
Nitrate-Sodium Nitrate System Using a Transient
Method with a Liquid-Metal Probe   251
    T. Omotani, Y. Nagasaka, and A. Nagashima

Thermal Conductivity of Normal Hydrogen   257
    H.M. Roder

The Thermal Conductivity of Solid N-Eicosane,
N-Octadecane, N-Heptadecane, N-Pentadecane, and
N-Tetradecane   265
    D.W. Yarbrough and C.N. Kuan

Thermal Conductivity Research of Gaseous and Liquid
Naphthenic Hydrocarbons over a Wide Range of State
Parameters   275
    Y.M. Naziev and A.N. Shakhverdiyev

Organic Liquid Mixtures: Comparison Between
Experimental and Predicted Thermal Conductivity
Values   285
    C. Baroncini, P. DiFilippo, G. Latini, and
    M. Pacetti

SESSION TC-8

Liquids and Gases - II

Thermal Conductivity of Carbon Monoxide in the
Temperature Range 350 - 2230 K   295
    R. Afshar, S. Murad, and S.C. Saxena

Absolute Measurement of the Thermal Conductivity
of Electrically Conducting Liquids by the
Transient Hot-Wire Method   307
    Y. Nagasaka and A. Nagashima

Measurements of the Conduction of Heat in Water
Vapor, Nitrogen and Mixtures of These Gases in a
Wide Knudsen-Number-Range   315
    A. Frohn and M. Westerdorf

SESSION TC-9

Theory, Review and Correlation - III

A Diffusion Model of Radiative Heat Transfer in
Scattering and Absorbing Media   329
    J.I. Berg

Thermal Conductivity Measurement of Materials
During Ablation--A Further Treatment on the
Moving Boundary Problem   341
    B. Zhou, Z. Wei, and J. Lin

A Method to Obtain the Analytical Solution of the
Thermal Conduction Equation with Moving Boundaries   349
    L. Dong, L. Bai, and B. Zhou

SESSION TC-10

Insulations - I

The Thermal Conductivity of Semitransparent
Materials   359
    H.A. Fine, S.H. Jury, D.L. McElroy, and
    D.W. Yarbrough

On the Thermal Conductivity of Ceramic Fibrous and
Rigid Insulations   369
    P. Brockerhoff

Correlation between the Structural Parameters and
the Thermal Resistance of Fibrous Insulants at
High Temperatures   381
    J. Boulant, C. Langlais, and S. Klarsfeld

Precision and Accuracy of Guarded Hot Plate Method   393
    M. Bomberg and K.R. Solvason

SESSION TC-11A

Insulations - II

Comparison of Results of Measurements Made on a
Line-Heat-Source and a Distributed-Heat-Source
Guarded-Hot-Plate Apparatus   413
    M.C.I. Siu

Development of Calibrated Transfer Specimens of Thick,
Low-Density Insulation   419
    B.G. Rennex, R.R. Jones, and D.G. Ober

SESSION TC-11B

Dielectrics - I

The Thermal Conductivity of Helium Crystals Containing
Neon Impurity   429
    R. Berman and D.M. Livesley

Thermal Conductivity of Ultrapure NaF Using Two-Fluid
Anharmonic Phonon Theory   437
    B.H. Armstrong

Heat Pulse Thermal Diffusivity Measurements on
Transparent Materials   447
    S.P. Howlett, R. Taylor, and R. Morrell

Thermal Conductivity and Diffusivity of Climax Stock
Quartz Monzonite at High Pressure and Temperature   459
    W.B. Durham and A.E. Abey

SESSION TC-12A

Insulations - III

An Experimental and Mathematical Study of the Effect
of Thickness in Low-Density Glass-Fiber Insulation   471
    M.A. Albers and C.M. Pelanne

Effective Thermal Conductivity of Glass-Fiber Board
and Blanket Standard Reference Materials   483
    D.R. Smith and J.G. Hust

Proficiency Testing for Thermal Insulation Materials
in the National Voluntary Laboratory Accreditation
Program   497
    D. Kirkpatrick and J. Horlick

Factors Affecting the Thermal Performance of a
Perlite Insulated System for Buildings   507
    R.P. Tye and S.C. Spinney

SESSION TC-12B

Dielectrics - II

Thermal Conductivity of Sodium-Conducting
Beta-Aluminas   519
    V.V. Mirkovich and T.A. Wheat

Thermal Conductivity of Orientationally-Disordered
Crystals under High Pressure   527
    P. Andersson, R.G. Ross, and G. Backstrom

Thermopower and Thermal Conductivity Measurements
on Intercalation Compounds   537
    J-P. Issi, J. Boxus, B. Poulaert, and J. Heremans

SESSION TC-13A

Techniques and Apparatus - I

A Rotating Annulus Device for Precise Thermal
Conductivity Measurements on Liquids; Values
for Ethanol, JP-4, and R-113   547
    R. Braun and A. Schaber

Automated Thermal Conductivity Measurements   559
    V.L. Morris and C. Haverlah

An Evacuated, Load-Bearing Powder Insulation
for a High Temperature Na/S Battery   569
    H. Reiss

Application of Parameter Estimation Techniques
to Thermal Conductivity Probe Data Reduction   587
    J.A. Koski and D.F. McVey

A Method to Measure the Thermal Diffusivity of
Low-Conductivity Materials with a Conventional
Twin-Plate Device for Measurement of the
Thermal Conductivity   601
    B.X. Wang, Z.P. Ren, and Z.H. Fang

SESSION TC-13B

Composites, Aggregates, and Geological Materials - I

Thermophysical Properties of Geologic Materials   611
    R.E. Taylor and H. Groot

Induced Convection During Cylindrical Probe
Conductivity Measurements on Permeable Media   619
    S.P. Fodemesi and A.E. Beck

Measurements on Thermal Conductivity and Thermal
Diffusivity of Alberta Oil Sands   635
    N. Seki, K.C. Cheng, and S. Fukusako

The Application of Thermal Conductivity and Diffusivity
Measurements in the Study of the Microstructure and
Phase Transition of Ceramic Materials   643
      T.G. Xi, B.M. Wang, Q.T. Chen, H.L. Ni,
      Z.W. Yin, and T.S. Yen

SESSION TC-14A

Composites, Aggregates, and Geological Materials - II

Thermal Conductivity of Concrete Mortar   655
      L.L. Sparks

Measurement of Soil Thermal Conductivity Using
an Infrared Scanning Technique   665
      C.K. Hsieh and X.A. Wang

Thermal Diffusivity of Silicon Carbide-Silicon
Composites   677
      M. Srinivasan, L.D. Bentsen, and
      D.P.H. Hasselman

Thermal and Electrical Conductivity of Metals
Embedded with Ceramic Granules   689
      W. Neumann

Nonstationary Thermal Behavior of Reinforced Composites:
A Better Evaluation of Wall Energy Balance for
Convective Conditions   699
      A.M. Luc and D.L. Balageas

SESSION TC-14B

Miscellaneous

The Thermophysical Properties of Coated Reinforced
Carbon-Carbon Composites   711
      R.P. Tye and A.O. Desjarlais

A Technique for Measuring the Apparent Thermal
Conductivity of Flat Insulations   727
      J.P. Moore, D.L. McElroy, and S.H. Jury

Thermal Conductivity of Carbonate Gneiss Under
Applied Uniaxial Pressure   737
      T. Ashworth, T.M. Alexander, and E. Ashworth

The Thermal Conductivity of Two Clathrate Hydrates   745
     J.G. Cook and M.J. Laubitz

Thermal Diffusivity of POCO Graphite and
Stainless Steel SRM 735-S.   753
     R. Taylor

The Development of Low-Density Glass-Fiber Insulation
as Thermal Transmission Reference Standards   763
     C.M. Pelanne

A Model of Apparent Thermal Conductivity for
Glass-Fiber Insulations   777
     L.J. Van Poolen, J.G. Hust, and D.R. Smith

Author Index   789

Subject Index   791

INTERNATIONAL THERMAL CONDUCTIVITY
CONFERENCES 1961-1981: A PROFILE

C.F. Lucks
Columbus, Ohio

INTERNATIONAL THERMAL CONDUCTIVITY CONFERENCES 1961 - 1981:

A PROFILE

C. F. Lucks*

Columbus, Ohio

INTRODUCTION

Since the time Fourier so concisely set forth the principles of heat conduction, the theoretical treatment of heat transmission through media has been and will continue to be important to many fields of science and engineering. In like manner, the passing of the "ice age" in the late 1920's and early 30's was the beginning of an increasingly wide interest in thermal-conductivity measurements. The advent of the mechanical refrigerator at that time led to considerable activity in thermal-conductivity measurements on low-conductivity materials for thermal insulation. This was followed by a high and continuing interest in thermal-conductivity measurements on insulating brick and refractories used in industrial applications. In the late 1930's and early 40's an increased need to know the thermal conductivity of metals was in evidence and was further accelerated by auto industry, nuclear energy, and aircraft developments. Additional impetus for this type of measurement was associated with the emphasis on semiconductors and aerospace materials. The need in 1960 for accurate experimental data was the impetus for the first conference on measurement methods.

The conferences on thermal conductivity stemmed from an informal discussion of methods by A. I. Dahl, General Electric Company; D. R. Flynn, National Bureau of Standards; D. L. McElroy, Oak Ridge National Laboratory; and H. W. Deem and C. F. Lucks, Battelle

---

*Retired Chief, Instrumentation Division, Battelle Memorial Institute, Columbus Laboratories (Historian of ITCC). Present address, 1858 W. Lane Ave., Columbus, OH 43221.

Memorial Institute, during the Fourth Temperature Symposium in
Columbus, Ohio, on March 30, 1961. The group believed it would be
very worthwhile to expand the exchange of ideas and experiences to
an informal conference on thermal conductivity methods. Because of
the breadth of materials and temperatures of interest, however, the
scope of the conference was to be limited, as much as possible, to
those methods in use for thermal conductivity measurements on solid
nuclear-energy and aerospace materials at elevated temperatures. By
request of the group mentioned, Battelle was very happy to assume
the responsibility of developing and sponsoring the first conference.

Each succeeding conference over the past two decades has, in
its own way, added breadth and depth to the subject of thermal conductivity. The conferences now serve as a "trading post" for both
the experimentalist and the theorist. C. Y. Ho and R. E. Taylor in
their preface to the eighth conference proceedings comment "The
(Thermal Conductivity) Conference proved so worthwhile that it has
been self-perpetuating even though there has been no professional
society, governmental agency, nor academic institution providing
continuing support. Each succeeding year an attendant has offered
to host the conference for the following year." And from the Foreword of the 14th conference proceedings, Y. S. Touloukian comments,
"The International Thermal Conductivity Conferences are an example
of how a technical community with a common purpose can transcend the
invisible artificial barriers between disciplines and gather in
increasing numbers without the need of national publicity and continuing funding sponsors, when they see something worthwhile going on."

The Profile is based primarily on the prefaces prepared by the
various conference chairmen for the abstracts or other publications
associated with their respective conference. It is a collection of
the major facets of conference prefaces augmented with other information of interest. The Profile is not intended to be a review of
the approximately eleven hundred papers presented during the seventeen conferences.

Appendixes A and B present a miniprofile of the seventeen conferences and the membership of the Board of Governors for 1972-1981,
respectively.

FIRST CONFERENCE

The first International Thermal Conductivity Conference was
held October 26-28, 1961, at Battelle Memorial Institute, Columbus,
Ohio. For the purpose of promoting discussions, attendance was by
invitation. The conference opened with most of the forty-five
attendants present to hear the address of welcome by H. W. Russell,
Battelle's Chief Physicist. Subjects covered in the first conference were steady-state methods with both longitudinal and radial

heat flows, transient methods and discussions on units and reference materials, (Armco iron round- robin already was in progress). The nineteen papers presented at the conference were authored by twenty-nine persons representing twelve organizations, including the National Research Council of Canada. A 309-page volume, "Conference on Thermal Conductivity Methods," was produced. The volume was not released as a formal publication, and for this reason, its distribution was limited to persons active in the field of thermal-conductivity measurements. It is now out of print. The conference chairman was C. F. Lucks. Assisting with the conference details were: H. W. Deem, E. S. Briich, E. A. Eldridge, and J. Matolich, Jr.

SECOND CONFERENCE

The gathering at the first conference of a group of people with the same interests and problems proved so useful and stimulating that it appeared highly desirable to hold a similar conferences in the future. The Division of Applied Physics, National Research Council of Canada, volunteered to sponsor the second thermal conductivity conference, which was held October 10-12, 1962, in Ottawa, Ontario. The conference chairman was M. J. Laubitz. Sixty-eight persons attended the conference.

The object of this conference was to bring together, informally, a small number of persons active in the field of thermal conductivity. Thus, the attendance was by invitation and all sessions were run very informally, including spontaneous evening sessions. The second conference extended the scope of the first conference from methods alone to discussions of techniques, mechanisms of heat transfer, and standards in greater depth with thirteen such materials suggested from the floor. As a follow up to information presented at the first conference, four organizations presented progress reports. Authors from Australia, Canada, and the U.S.A. provided twenty-one papers and four progress reports that were issued as a 378-page volume.

THIRD CONFERENCE

The Third Thermal Conductivity Conference was sponsored by the Oak Ridge National Laboratory with D. L. McElroy as conference chairman. The conference meetings were held at the Mountain View Hotel, Gatlinburg, Tennessee, October 16-18, 1963, and had an attendance of ninety. The conferees were welcomed by H. G. MacPherson of Oak Ridge National Laboratory. The scope of the third conference was broader and more detailed than the previous conferences.

Early sessions were devoted to thermal conductivity theory and thermal conductivity standards to facilitate more complete

discussions throughout the conference. M. L. Pickelsimer, Oak Ridge
National Laboratory, presented an invited paper on microstructural
examination to emphasize the need for techniques to better characterize specimens. One session was on those important solids termed
powders; another session was on the measurement of specific heat.
Two papers on calorimeters were presented by invitation; one by
E. E. Stansbury, University of Tennessee, on the adiabatic calorimeter and the other by T. B. Douglas, National Bureau of Standards,
on the drop-type calorimeter. Accurate specific heat values are
needed to calculate thermal conductivity from thermal diffusivity
data, because a great deal of thermal conductivity data are being
generated in this way.

Three especially interesting sessions were those on steady-
state measurements, non-steady-state measurements, and the progress
reports. In these sessions, as in the others, it was apparent that
all authors had a growing interest, not only in good experimental
results over extended temperature ranges, but also in the interpretation of the results through theory to gain insight into basic
heat-transfer mechanisms. This mechanistic approach led to the measurement of other related properties, such as electrical conductivity, Seebeck coefficient, and infrared optical properties. By invitation, R. W. Powell, National Physical Laboratory, England, presented a paper on thermal conductivity correlations. This trend was
expected to yield exciting results. In addition, it was expected
that more original theoretical thought would be introduced at future
conferences.

In addition to the introduction of invited papers and speakers,
the first banquet took place with Alvin M. Weinberg, Director, Oak
Ridge National Laboratory, addressing the conferees on the subject,
"Science, Government, and Information." The presence of R. W. Powell
at the conference enriched the international flavor; the 1964
National Physical Laboratory Conference on Thermophysical Properties
initiated the series of European Conferences on Thermophysical Properties.

The conference chairman, D. L. McElroy, and those who worked
with him, W. Fulkerson, T. G. Collie, T. G. Godfrey, J. P. Moore,
H. W. Godbee, and R. S. Graves, were most impressed in reading the
various papers and, in numerous places, found ideas that were unique
in conveying some of the best possible thinking about this complex
subject. On the lighter side, they were surprised when, during some
of the late night reading, they suddenly found themselves reading
about the thermal conductivity of moon dust and, still later, a
romantic reference to Shakespeare. The conference proceedings were
issued as an 835-page, two-volume set. The third conference proceedings were the first to include an author index.

FOURTH CONFERENCE

The U.S. Naval Radiological Defense Laboratory hosted the Fourth Thermal Conductivity Conference at its laboratory in San Francisco, October 13-16, 1964, with R. Rudkin as chairman. Eighty-eight persons attended the conference.

The first session was devoted to fundamentals and standards. The conference again included a session on measurements of heterogeneous mixtures. Although there was strong interest in all of the sessions, there appeared to be particular interest in the area of thermal contact conductance. There were papers on high temperature measurements and on transient techniques. Correlated properties were included, as well as a session on techniques and apparatus, that included two papers on thermocouple tests. The last session of the conference was made up of short, informal, progress reports given by representatives of various organizations actively engaged in the measurement of thermal conductivity. There appeared to be a great desire on the part of the participants to share their problems and experiences informally. The after-dinner talk by P. Butler on early scientists in San Francisco during the gold-rush era was especially enjoyable.

Thirty-two papers were issued as a single 760-page volume. The volume contained, for the first time, a data index.

FIFTH CONFERENCE

The Department of Metallurgy, College of Engineering, University of Denver, hosted the Fifth ITCC at the Denver Hilton Hotel, Denver, Colorado, October 20-22, 1965, with J. D. Plunkett as chairman. The first two sessions were devoted to papers on transient measurement methods. The invited paper of A. H. Klein, (NASA) Moffett Field, California, "Thermal Radiation - A Mechanism of Thermal Conduction," introduced the session on theory. Following sessions on steady-state apparatus, standards, and new data, the closing session's invited paper by J. C. Chato, University of Illinois, introduced a new subject: "A Survey of Thermal Conductivity Data on Biological Materials." Another very interesting paper in the closing session was by R. C. Birkebak and C. J. Cremers on the heat loss from animal systems under the influence of various environmental conditions as determined from thermal modeling techniques. In addition to the programmed sessions, three informal discussion sessions provided opportunity for a more detailed exchange on subjects of interest.

The conference was unique in that an ad hoc committee, appointed by D. L. McElroy, presented a summary covering the technical content, informal discussion groups, general remarks, and comments on the future of the conference.

The forty-one papers by seventy-seven authors were issued as a 990-page, two-volume set.

SIXTH CONFERENCE

The Sixth Thermal Conductivity Conference was hosted by the Air Force Materials Laboratory, Wright-Patterson Air Force Base, October 19-21, 1966. The conference meetings were held at the Statler Hilton Hotel, Dayton, Ohio, with 130 persons in attendance. The conference was cochaired by M. L. Minges and G. L. Denman. A. M. Lovelace, Chief Scientist, Air Force Materials Laboratory, Dayton, Ohio, welcomed the conferees with his speech, "Thermophysics Programs in the Air Force."

Fifty-nine papers were presented in seven sessions: gases and liquids; measurement techniques; ceramic materials; metallic materials, low temperatures; metallic materials, high temperatures; polymeric and graphitic materials; and general progress reports. In line with the recommendation of the Ad Hoc Committee of the previous conference (fifth), the chairman of each session presented an overview of the session subject as a practical method of assuring more time for discussion. P. G. Klemens, Westinghouse Research Laboratories, Pittsburgh, was invited to address the ceramic session on the "Theory of Thermal Conductivity in Disordered Solids." Although most of the sessions were concerned to some degree with measurement methods and results, there was considerable emphasis on standard reference materials, including the initial outline of the Air Force program in this field, which eventually extended over a ten-year period in cooperation with NBS, AGARD (NATO), and CODATA. A very informative address, "Materials Applications in Advanced Weapon System Development," was presented at the banquet by M. L. Minges, Air Force Materials Laboratory.

The papers of the conference were issued in a 1181-page volume.

SEVENTH CONFERENCE

The seventh conference was held October 13-16, 1967, at the Gaithersburg, Maryland, facilities of the National Bureau of Standards. Ninety papers by 149 authors were presented in thirteen sessions. The first session included state-of-the-knowledge presentations on metallic and nonmetallic elements, liquids, gases, the development of thermal methods and mathematical analysis, metals at very low, intermediate, and high temperatures, nonmetallics and graphites, nuclear materials, building elements, rocks and soils, gases, liquids, two-phase systems, and thermal contact conductance. L. Kushner, Acting Deputy Director, National Bureau of Standards, addressed the attendants at the conference banquet in the Sheraton Park Hotel in Washington, D.C.

The conference was cochaired by D. R. Flynn and B. A. Peavy, Jr., who were particularly appreciative of the assistance given them by W. L. Carroll. The cochairmen believed that, because the proceedings of the previous conferences were distributed as informal publications on a limited basis, a great deal of valuable information was not reaching the open literature. For this reason it was decided that the proceedings of the Seventh Conference on Thermal Conductivity would be published formally. National Bureau of Standards Special Publication 302, "Thermal Conductivity--Proceedings of the Seventh Conference" is an 801-page volume containing the papers of the conference and is available from the Superintendent of Documents, U.S. Government Printing Office, Washington, D.C. 20402.

EIGHTH CONFERENCE

The Eighth Thermal Conductivity Conference was hosted by the Thermophysical Properties Research Center (TPRC), which has contributed greatly to the organization of our knowledge on thermal conductivity. The meetings were held in the Memorial Center, Purdue University, West Lafayette, Indiana, October 7-10, 1968. The conference was cochaired by C. Y. Ho and R. E. Taylor.

Y. S. Touloukian, Director, TPRC and Distinguished Alcoa Professor of Engineering at Purdue University, welcomed the 144 conference attendants. Two sessions of the fifteen-session conference were required for each of two subjects: thermal conductivity of metals and alloys at low temperatures and thermal conductivity of gases. Single sessions were devoted to metals at intermediate and high temperatures, thermal contact conductance, nuclear materials, liquids, reference materials and low temperature insulators, thermal conductivity apparatus, semiconductors, thermal diffusivity methods for obtaining thermal conductivity, thermal diffusivity property measurements, and fibers, polymers, ice, and soil. During the early evening of October 8, the Thermophysical Properties Research Center had an open house for the attendants at their facilities located in the Purdue University Research Park. This was followed later in the evening with a banquet at the Morris Bryant Hotel at which H. Wooster, Director, Information Sciences, Air Force Office of Scientific Research, Arlington, Virginia, addressed the conferees on the subject, "Basic Research and the Department of Defense."

The cochairmen decided to publish the contributed papers formally using the referee system. It was recognized that such a policy for all papers presented at the eighth conference would seriously restrict the number and type of papers submitted and thus detract from the informal atmosphere that prevailed at previous conferences. For this reason, each person presenting his paper orally at the conference was given the option of not submitting his paper for review and subsequent inclusion in the proceedings of the conference. A

total of 112 papers by 197 authors appear in the 1169-page publication "Thermal Conductivity--Proceedings of the Eighth Conference,"
C. Y. Ho and R. E. Taylor, editors, Plenum Press, New York, 1969.

NINTH CONFERENCE

     The Ninth Thermal Conductivity Conference was held at Iowa
State University, October 6-8, 1969. Cosponsors of the conference
were: Ames Laboratory of the United States Atomic Energy Commission, Ames, Iowa; the Office of Naval Research, Washington, D.C.;
and Iowa State University, Ames, Iowa. The conference chairman was
H. R. Shanks of Iowa State University. The attendants numbered an
even 100 with nine countries being represented, most certainly an
international representation.

     The conference opened with a welcoming address by D. J.
Zaffarano, Chairman, Physics Department, Iowa State University.
Subjects covered in the nine conference sessions included low and
high temperature mea surements, gases, liquids, graphite, experimen-
tal methods, and low-conductivity methods. Three invited speakers
gave oral presentations during the conference. The invited speakers
and their subjects were: A. E. Wechsler, Arthur D. Little, Inc.,
Cambridge, Massachusetts, "Thermophysical Properties of Biological
Materials"; S. Legvold, Ames Laboratory, Ames, Iowa, "Transport
Properties of Rare Earth Metals"; and H. Plumb, National Bureau of
Standards, Gaithersburg, Maryland, "The International Temperature
Scale of 1968." R. S. Hansen of Iowa State University presented a
very interesting address at the banquet on the subject, "Reflections
on the Impact of Science on Our Society."

     The text of many of the papers presented at the conference and
several that were scheduled for presentation but, because of travel
problems were not presented, were published formally by the United
States Atomic Energy Committee, Division of Technical Information.
The 743-page volume, "Ninth Conference on Thermal Conductivity,"
containing seventy-six papers by 126 authors, is identified as CONF-
691002, Physics (TID-4500) and is available from the Clearinghouse
for Federal Scientific and Technical Information, National Bureau of
Standards, U.S. Department of Commerce, Springfield, Virginia 22151.

TENTH CONFERENCE

     The Tenth Thermal Conductivity Conference was held at the
Marriott Motor Hotel, Newton, Massachusetts, September 28-30, 1970.
The conference was cohosted by Dynatech R/D Company and Arthur D.
Little, Inc., both of Cambridge, Massachusetts, and was cochaired by
R. P. Tye, Dynatech R/D Company, and A. E. Wechsler, Arthur D.
Little, Inc. Welcoming addresses were given by W. M. Rohsenow,

were given by W. M. Rohsenow, Head, Heat Transfer Section, Department
of Mechanical Engineering, Massachusetts Institute of Technology,
and P. E. Glaser, Vice President, Engineering, Arthur D. Little, Inc.
One hundred and six persons attended the conference.

The conference had eight sessions, two of which were on fluids.
Subjects covered in the other six sessions were: theory and low
temperature, heterogeneous materials, measurement methods, nonmetallics, metals, alloys, semiconductors, and a group of miscellaneous,
related topics. Included in the program were tours of the facilities of Dynatech R/D Company and Arthur D. Little, Inc., in Cambridge.
The guest speaker at the banquet was C. Linde, President, National
Academy of Engineering. The cochairmen chose this occasion for
another conference first: the presentation of a traveling Thermal
Conductivity Award. Physically, the award is a solid right cylinder,
about one inch in diameter by four inches high, of the original lot
of ingot iron referred to in the thermal-conductivity literature as
Armco iron, that was used in a round-robin investigation among many
laboratories. This type of iron also had been studied and used as a
thermal conductivity reference material by a number of laboratories
for a period of a little more than forty years. The right cylinder
of Armco iron is mounted on a clear plastic base faced on the bottom
with a black-and-white laminated plastic on which the recipient's
name and year of award are engraved. The recipient is chosen to
receive the Thermal Conductivity Award "in recognition of his accom-
plishments in the field of thermal conductivity and as a mark of the
high esteem he is held by his fellow workers." The 1970 and first
recipient of the Thermal Conductivity Award was R. W. Powell, who
retired in November 1964 as Senior Principal Scientific Officer from
the National Physical Laboratory, Teddington, England, and subsequently continued his work in thermal properties as Senior Researcher,
Thermophysical Properties Research Center, Purdue University, West
Lafayette, Indiana, U.S.A.

This was the first conference to be hosted by independent commercial organizations. The cochairmen attempted to obtain greater
industrial participation with the hope that much of the technology
developed in achieving past national goals could be utilized in the
industrial community. In this respect, the horizons of interest
were extended to cover more areas of application and to include a
wide range of participants involved in real problems and commercial
materials. Despite the previous years' cut-backs in expenditures
for research and development in the field of thermal conductivity,
there were sixty-nine papers presented by 106 authors. In keeping
with the general wishes of the active participants, the cochairmen
decided to return to a more informal approach and for the first time
issued the 138-page conference publication as a volume of informal
extended abstracts.

ELEVENTH CONFERENCE

With the ever-widening international interest in this series of
conferences because of their scope and representation (the mailing
list now includes about a thousand names), the hosts for the 1971
conference decided that at the beginning of the second decade of
this series, the conference title should be modified to include the
word, "International"; a change they believe could have been made
some years past with equal justification. Almost thirty percent of
the papers presented at the 1971 Eleventh International Thermal Conductivity Conference were from countries other than the U.S.A. The
eighty-five papers presented at the conference included contributions
from twelve countries.

Hosts for the Eleventh International Thermal Conductivity Conference (ITCC) were: Los Alamos Scientific Laboratory, Los Alamos,
New Mexico and Sandia Laboratories and the University of New Mexico,
both of Albuquerque, New Mexico. The cochairmen of the conference
were R. U. Acton, Sandia Laboratories, P. Wagner, Los Alamos Scientific Laboratory, and A. V. Houghton III, University of New Mexico.
The technical sessions were held September 28-October 1 at the Western Skies Hotel in Albuquerque. The welcoming address by J. E.
McDonald, Director of Material Sciences, Sandia Laboratories, was
followed by a plenary session invited paper, "Thermal Conductivity
Implications," by D. L. McElroy, Oak Ridge National Laboratories,
Oak Ridge, Tennessee.

The conference had an attendance of 134 and the papers were
divided for presentation into ten sessions. Because of the number
of participants and range of subject matter, parallel sessions were
necessary in three instances. Topics covered were experimental
results, fluids and related topics, measurement techniques, theory
and analysis,gases, multiphase media, biological applications, and
thermal diffusivity. In addition to the papers of the ten sessions,
two additional papers were presented by invitation. These were:
"The Heat Pipe at Los Alamos" by J. E. Kemme of Los Alamos Scientific Laboratory and "Sensitivity Analysis" by T. Ishimoto of TRW.
The evenings were occupied with the conferees attending The Barn
Dinner Theater on September 28, an informal session of progress
reports and open discussion on the following night, and on September
30, a banquet at Summit House Restaurant atop 10,500-foot Sandia Peak.

The 1971 International Thermal Conductivity Award was presented
at the banquet. The recipient of the 1971 Award was D. L. McElroy,
Oak Ridge National Laboratory. After the presentation of the award,
D. Smylie of the U.S. Forest Service, Albuquerque, New Mexico,
addressed the conferees on "The Cibola National Forest."

The Eleventh ITCC continued the informality of the previous conference by issuing a 175-page volume of informal, extended abstracts.

The past conference cochairmen present at this conference discussed the possibility of organizing along more formal lines. R. E.
Taylor, as chairman of the Committee on Organization, addressed a
letter (December 9, 1971) to Y. S. Touloukian inquiring as to what
part the Thermophysical Properties Research Center of Purdue University might play in such an organization. Professor Touloukian
responded December 17, 1971, suggesting TPRC might become a titular
sponsor or cosponsor for future conferences.

TWELFTH CONFERENCE

The Twelfth ITCC was cohosted on September 12–15, 1972, by
Southern Research Institute and the University of Alabama–Birmingham,
both of Birmingham, Alabama. Cochairmen for the conference were:
W. T. Engelke and S. E. Bapat of Southern Research Institute and M.
Crawford of the University of Alabama–Birmingham. Assisting with
conference arrangements were: C. D. Pears, J. H. Strickland, W. F.
Brooks, M. Pennington, S. Lewis, and R. O'Donnell of Southern
Research Institute. The conference sessions were held at Parliament
House Motor Hotel, Birmingham, with sixty attendants.

The welcoming address for the conference was given by W. M.
Murray, Jr., President, Southern Research Institute, Birmingham,
Alabama. A plenary session followed with the invited paper, "Theory
of Thermal Resistivity of Insulators Due to Anharmonic Interactions,"
presented by P. G. Klemens, University of Connecticut, Storrs,
Connecticut. A second plenary session was held September 14, 1972,
with invited speaker G. Ruffino, Leeds and Northrup, Italina, S.P.A.,
presenting "Highlights of Third European Thermal Conductivity Conference." The fifty-one papers by seventy-four authors were presented in seven tandem sessions covering the subjects: theory and
analysis, experimental methods and results (three sessions), powders
and insulation, theory and low temperature, and liquids and gases.
The conference issued a 175-page volume of informal extended
abstracts.

The presentation of the Conference Award was made at the banquet held in the Crown Room of the Parliament House Motor Motel.
The recipient of the 1972 International Thermal Conductivity Award
was P. G. Klemens, University of Connecticut. Following the award
presentation, E. Stuhlinger, Associate Director for Science, George
C. Marshall Space Flight Center, NASA, Huntsville, Alabama, addressed
the conferees. The banquet was followed by a late evening, shirtsleeve session.

A Board of Governors, composed of past conference chairmen, was
formed at this conference. This informal incorporation was a mile-
stone in the history of the conferences. R. U. Acton was elected
chairman and W. T. Engelke, secretary of the Board of Governors.

The offer of Y. S. Touloukian for TPRC to act as a continuing titular cosponsor of future conferences was accepted. The drafting of
bylaws was assigned to P. Wagner and R. E. Taylor.

THIRTEENTH CONFERENCE

The Thirteenth ITCC was hosted November 5-8, 1973, by the
University of Missouri-Rolla, Rolla, Missouri, and was cochaired by
R. L. Reisbig, H. J. Sauer, Jr., T. R. Faucett, and C. R. Remington,
University of Missouri-Rolla. The conference meetings were held at
the Lodge of the Four Seasons, Lake Ozark, Missouri, with an atten-
dance of 113 persons. The conferees were welcomed by Bill Atchley,
Governor's Science Advisor, University of Missouri-Rolla.

The program of 105 papers by 186 authors contained fourteen
parallel sessions and five single sessions, two of which were plenary sessions. The invited speakers and their subjects for the plenary sessions were: J. Der Hovanesian, "Application of Coherent
Optics and Hologram Interferometry to Thermal Measurements," Oakland
University, Detroit, Michigan, and Y. S. Touloukian, "Numerical Data
of Science and Technology," Thermophysical Properties Research Center, Purdue University, West Lafayette, Indiana. The subjects
covered in the sessions were: gases, fluids and related topics,
theory and analysis, transient methods, laser and optical methods,
experimental methods and procedures, heterogeneous materials, soils
and rocks, solids theory and solids at low and moderate-to-high temperatures. Forty-three of the papers were published formally in the
458-page volume, "Advances in Thermal Conductivity," R. L. Reisbig
and H. J. Sauer, Jr., editors, University of Missouri-Rolla, Rolla,
Missouri. Abstracts of most of the papers presented at the conference are contained in an informal 345-page preconference volume.

On Monday evening, November 5, the conferees enjoyed an Ozark
Bar-B-Q with all its trimmings. Following the Bar-B-Q, a late evening, shirtsleeve session was held. On the following evening the
banquet was held at which K. D. Timmerhouse, National Science Foundation, Washington, D.C., addressed the group. The recipient of the
1973 International Thermal Conductivity Award presented at the banquet was C. F. Lucks (Retired), Battelle Memorial Institute, Columbus
Laboratories.

The conference marked the start of a stronger interaction among
the groups interested in thermophysical properties, both in North
America and overseas. The International Thermal Expansion Symposium,
ITES, was held immediately after the International Thermal Conductivity Conference. To avoid conflicts and encourage attendance at
both the ITCC/ITES and the next European Thermophysical Properties
Conference, the next conference was scheduled for 1975.

The Board of Governors held two meetings during the conference.
The previously distributed bylaws prepared by P. Wagner and R. E.
Taylor were discussed and adopted with minor changes. The adopted
bylaws appear in the above-mentioned volume, "Advances in Thermal
Conductivity," pages 454-458. The following committee chairmen were
appointed by Board Chairman, R. U. Acton:

> Awards - P. G. Klemens
> Bylaws - R. E. Taylor and P. Wagner
> Improvements - D. L. McElroy
> Sponsors and Sites - G. L. Denman and R. P. Tye
> Editorial and Publications - R. L. Reisbig
> Placement - R. E. Taylor
> Historian - C. F. Lucks

The Board reaffirmed the offer of TPRC to be a continuing titular co-sponsor of the International Thermal Conductivity Conferences,
as outlined in the correspondence of March 14, 1973, and March 27,
1973, between R. U. Acton and Y. S. Touloukian.

FOURTEENTH CONFERENCE

The Fourteenth ITCC was held June 2-4, 1975, at the University
of Connecticut at Storrs, Connecticut, with P. G. Klemens as conference chairman. Assisting with the conference were D. H. Damon, H.
Hilding, and F. P. Lipschultz of the university. The conference was
sponsored by the university along with the continuing sponsorship of
the Center for Information and Numerical Data Analysis and Synthesis
(CINDAS), Purdue University, West Lafayette, Indiana, formerly the
Thermophysical Properties Research Center (TPRC). The conference
was attended by 113 persons who were welcomed by L. V. Azaroff,
Director, Institute of Materials Research, University of Connecticut.
The conference program included 77 papers by 135 authors.

The Introductory Session included six papers on kinetic-theory
relation of thermal conductivity to other gas properties and thermal
conductivity of diamonds. It was followed by nine sessions that met
as four parallel sessions and one informal evening session. The
concluding single session was a first in that it was a joint session
with the ITES. The subjects covered in the ITCC were: solids at
low and high temperatures, reference and technical materials, gases
and liquids, and experimental methods and analysis. The papers were
published in the 566-page formal volume, "Thermal Conductivity 14,"
P. G. Klemens and T. K. Chu, editors, Plenum Press, New York, 1976.
This is the first conference publication to include a subject
index. The editors had restructured the subjects of the conference
sessions to a more compatible grouping. A review of the conference
by P. G. Klemens appeared in the Physics Bulletin, September 1975,
pages 392-393.

The conference banquet was held the evening of June 3 at the
Faculty Club of the University of Connecticut. C. Stern of the
University, School of Agriculture and Natural Resources, addressed
the attendants on the subject, "Energy and Environment." The recipient of the 1975 ITCC Award was M. J. Laubitz, NRC, Canada.

The Board of Governors met twice during the conference, June 1
and June 3. P. Wagner was elected secretary upon the resignation of
W. T. Engelke.

The Board of Governors resolved to have CINDAS/Purdue University as a permanent sponsor of the conferences and to establish a
policy of publishing the proceedings of future conferences on a continuing and uniform basis.

R. Berman of Clarendon Laboratory, Oxford, England, and R. P.
Tye, Dynatech R/D Company, Cambridge, Massachusetts, were elected
Fellows.

The elected officers and members of the Board of Governors for
1975-1977 are given in Appendix B.

FIFTEENTH CONFERENCE

The Fifteenth ITCC returned to the city of the second conference, Ottawa, Ontario, Canada. Meetings were held August 24-26,
1977, at the Holiday Inn, Ottawa Centre, with V. V. Mirkovich as
Conference Chairman. Assistance was provided locally by M. J.
Laubitz, J. G. Cook, and R. M. Buchanan. The conference was cospon-
sored by the Canadian Department of Energy, Mines and Resources;
Canada Centre for Mineral and Energy Technology, both of Ottawa; and
CINDAS/Purdue University, West Lafayette, Indiana, U.S.A. Six Cana-
dian companies: Aluminum Company of Canada, Fiberglass Canada, Gulf
Oil Canada, Canadian Refractories, Construction Materials, and
Imperial Oil (ESSO) provided much appreciated external financial
support for the conference.

The eighty-six conferees were welcomed by D. F. Coates, Director General, Canada Centre for Mineral and Energy Technology. The
conference consisted of two plenary and ten technical sessions.
Addressing the plenary sessions were invited speakers P. G. Klemens,
University of Connecticut, speaking on the basic motivations for
thermal conductivity studies and G. Ruffino, Leeds and Northrup
Italiana, Torino, Italy, discussing the "Progress in High Temperature Measurement in Laboratory." A. Cezairliyan, National Bureau of
Standards, Gaithersburg, Maryland, also an invited speaker, addressed
the conference on "International Activities and Interconference
Cooperation in Thermophysics." Subjects covered in the ten technical sessions were: solids at high and low temperatures, reactor

materials, techniques, data analysis, gases, liquids, rocks, soils, insulation, nuclear waste disposal, coal, diffusivity, and contact conductance. The ten sessions were made up of fifty-three papers authored by ninety-seven persons. The papers appear in the 493-page volume "Thermal Conductivity 15," Vladimir V. Mirkovich, editor, Plenum Press, New York, 1978.

In an unusual action, the ITCC Award was awarded to three candidates. Recipients of the 1977 Award were: R. E. Taylor, CINDAS and Properties Research Laboratory, Purdue University; Y. S. Touloukian, CINDAS and Purdue University; and R. P. Tye, Dynatech R/D Company.

The Board of Governors met in the evening of August 24. After extended discussion it was decided that the conference name should remain intact with its emphasis to be on thermophysics rather than just thermal conductivity. No Fellows were recommended by the Awards Committee. Board chairman, R. U. Acton, submitted "Guidelines to Organizations Who Serve as Host to ITCC," outlining responsibilities for future hosts of the conferences.

SIXTEENTH CONFERENCE

The Sixteenth ITCC and the Seventh ITES were held concurrently for the first time in Chicago, Illinois, November 7-9, 1979. Host for the conferences was IIT Research Institute with D. C. Larsen as general chairman assisted by C. Galassi, D. Lancaster, J. Adams, and Y. Haradal. CINDAS/Purdue University was cosponsor.

The Sixteenth ITCC consisted of two plenary and thirteen technical sessions. K. D. Maglic, Boris Kidrich Institute of Nuclear Sciences, Belgrade, Yugoslavia, presented the highlights of the scheduled Seventh European Thermophysical Properties Conference at the first plenary session. The speaker at the second plenary session, A. Cezairliyan, National Bureau of Standards, discussed the formation of the International Thermophysics Congress, which presently has representation from four conferences: The Thermophysical Properties Conference of ASME, ITCC, ITES, and the European Thermophysical Properties Conference. The thirteen technical sessions consisted of eighty-eight papers by 127 authors. It is worthy to note that two sessions each were required for theory of solids at low temperature and geologic materials. Other session subjects were: contact conductance, gases, liquids, thermal diffusivity, correlation of properties, solids at high temperature, and general. The papers appear in "Thermal Conductivity 16," D. C. Larsen, editor, Plenum Press, 1982.

The Board of Governors met November 7 and voted to continue having separate proceedings for the International Thermal Conductivity

Conferences. The resignation of H. Shanks as alternate chairman was accepted with regrets. D. L. McElroy was elected to be alternate chairman. By unanimous vote the 1979 Thermal Conductivity Award recipient was J. P. Moore, Oak Ridge National Laboratory. The award was presented at the banquet.

SEVENTEENTH CONFERENCE

The Seventeenth ITCC was held June 15-19, 1981, as part of the International Joint Conferences on Thermophysical Properties at the National Bureau of Standards, Gaithersburg, Maryland. J. G. Hust, National Bureau of Standards, Boulder, Colorado, was chairman of the conference.

This was a first, in that three thermophysics conferences were held concurrently. The three conferences were: 1) Eighth Symposium on Thermophysical Properties, 2) Seventeenth ITCC, and 3) Eighth ITES. This joint conference was highly successful with about 400 attendants and over 300 papers. Four parallel sessions were scheduled during most of the week. Eighty-five papers were presented at the eighteen sessions of the Seventeenth ITCC. Although numerous topics were included, emphasis was placed on a) theory and correlation, b) standard reference materials, c) insulations, and d) composites and aggregates.

The keynote address entitled, Thermophysical Properties in Science and Technology was presented by J. Kestin, Brown University.

The Board of Governors met on June 14 and again on June 17. Considerable attention was given to continued joint meetings and the possibility of merging one or more of these conferences. Since the site selection committee had made tentative recommendations for 1983 and 1985, a similar joint meeting was impossible prior to 1989. However, some interest continued to exist in a merger with the ITES, and the conference chairman agreed to pursue this possibility.

A somber note is added in recalling that Y. S. Touloukian, CINDAS and Purdue University, passed away while attending a CODATA meeting in the conference hotel just prior to the start of the joint conference. He contributed significantly to the organization of thermophysical properties data and his departure is noted with sorrow. He was honored posthumously at the banquet.

The Traveling Thermal Conductivity Award was presented jointly to R. Berman, Oxford University, England, and C. Y. Ho, CINDAS, Purdue University, at the banquet.

The Board of Governors accepted the offer of T. Ashworth, South Dakota School of Mines, to hold the 1983 International Thermal Conductivity Conference.

APPENDIX A. MINIPROFILE OF THE INTERNATIONAL THERMAL CONDUCTIVITY CONFERENCES 1961 – 1981

Conference and Year	Host Organization & Site	Chairmen	Number of Attendants				Numbers	
			U.S.A.	Canada	Other	Total	Papers	Authors
1st 1962	Battelle Memorial Inst. Columbus, OH	C. F. Lucks	41	2	–	43	19	29
2nd 1962	National Research Council (Canada) Ottawa, Canada	M. J. Laubitz	45	18	5	68	21	32
3rd 1963	Oak Ridge National Lab. Gatlinburg, TN	D. L. McElroy	80	5	5	90	35	66
4th 1964	U.S. Naval Radiological Defense Laboratory San Francisco, CA	R. L. Rudkin	87	1	–	88	32	76
5th 1965	University of Denver Denver, CO	J. D. Plunkett	100	6	5	111	41	77
6th 1966	A. F. Materials Lab. Dayton, OH	M. L. Minges & G. L. Denman	126	2	2	130	59	78
7th[a] 1967	Nat. Bureau of Stand. Gaithersburg, MD	D. R. Flynn & B. A. Peavy	170	6	18	194	90	149
8th[a] 1968	Thermophysical Prop. Res. Center/Purdue Univ. W. Lafayette, IN	C. Y. Ho & R. E. Taylor	123	6	15	144	112	197

(a) Formal publication issued.

(Cont.)

APPENDIX A. MINIPROFILE OF THE INTERNATIONAL THERMAL CONDUCTIVITY CONFERENCES 1961 – 1981 (Cont.)

Conference and Year	Host Organization & Site	Chairman	Number of Attendants				Numbers	
			U.S.A.	Canada	Other	Total	Papers	Authors
9th(a) 1969	Ames Lab. & Office of Naval Res., Ames, IA	H. R. Shanks	86	6	8	100	76	126
10th(b) 1970	Arthur D. Little, Inc. & Dynatech R/D Co. Boston, MA	A. E. Wechsler & R. P. Tye	88	10	7	105	69	106
11th(c) 1971	Sandia Labs., Los Alamos Scientific Lab. & Univ. of New Mexico, Albuquerque, NM	R. U. Acton P. Wagner & A. V. Houghton, III	76	7	13	96	85	134
12th 1972	Southern Res. Inst. & Univ. of Alabama Birmingham, AL	W. T. Engelke S. G. Bapat & M. Crawford	50	6	4	60	51	74
13th(a)(d) 1973	Univ. of Missouri-Rolla Lake of the Ozarks, MO	R. L. Reisbig & H. J. Sauer, Jr.	98	3	12	113	105	186
14th(a)(e) 1975	Univ. of Connecticut Storrs, CT	P. G. Klemens	86	14	13	113	77	135

(a) Formal publication issued.
(b) First Thermal Conductivity Award presented.
(c) First conference to modify the conference title to include the word International.
(d) Bylaws adopted at this conference.
(e) Board of Governors elect CINDAS/Purdue University as a permanent sponsor of the conferences.

Conference and Year	Host Organization & Site	Chairman	Number of Attendants				Numbers	
			U.S.A.	Canada	Other	Total	Papers	Authors
15th(a) 1977	Dept. of Energy, Mines & Resources, Ottawa, Canada	V. V. Mirkovich	35	18	17	86	53	97
16th(a) 1979	IIT Research Institute Chicago, IL	D. C. Larsen	91	7	15	113	88	127
17th(a) 1981	Nat. Bureau of Stand. Gaithersburg, MD	J. G. Hust	52	4	31	87	85	157

(a) Formal publication issued

APPENDIX B. BOARDS OF GOVERNORS OF THE INTERNATIONAL THERMAL CONDUCTIVITY CONFERENCES

	1972	1973/1975	1975/1977	1977/1979	1979/1981
Chairman –	R. U. Acton	R. U. Acton	R. U. Acton	R. U. Acton	R. U. Acton
Alternate					
Chairman –	–	H. R. Shanks	H. R. Shanks	H. R. Shanks	D. L. McElroy
Secretary –	W. T. Engelke	W. T. Engelke	P. Wagner	P. Wagner	R. P. Tye
Steward –	–	C. Y. Ho	C. Y. Ho	C. Y. Ho	C. Y. Ho
	S. G. Bapat	S. G. Bapat	S. G. Bapat	D. H. Damon	D. H. Damon
	G. L. Denman	G. L. Denman	D. H. Damon	P. G. Klemens	J. G. Hust
	C. Y. Ho	P. G. Klemens	G. L. Denman	D. C. Larsen	P. G. Klemens
	M. J. Laubitz	M. J. Laubitz	P. G. Klemens	M. J. Laubitz	D. C. Larsen
	C. F. Lucks	C. F. Lucks	M. J. Laubitz	C. F. Lucks	M. J. Laubitz
	D. L. McElroy	D. L. McElroy	C. F. Lucks	D. L. McElroy	C. F. Lucks
	M. L. Minges	M. L. Minges	D. L. McElroy	M. L. Minges	M. L. Minges
	J. D. Plunkett	J. D. Plunkett	M. L. Minges	V. V. Mirkovich	V. V. Mirkovich
	R. L. Reisbig	R. L. Reisbig	V. V. Mirkovich	H. J. Sauer, Jr.	H. J. Sauer, Jr.
	H. J. Sauer, Jr.	H. J. Sauer, Jr.	J. D. Plunkett	R. E. Taylor	R. E. Taylor
	H. R. Shanks	R. E. Taylor	R. L. Reisbig	R. P. Tye	
	R. E. Taylor	R. P. Tye	H. J. Sauer, Jr.		
	R. P. Tye	P. Wagner	R. E. Taylor		
	P. Wagner	A. E. Wechsler	R. P. Tye		
	A. E. Wechsler		A. E. Wechsler		

SESSION   TC-2

Theory, Review, and Correlation - I

CHAIRMAN

P. G. Klemens
University of Connecticut
Storrs, Connecticut

THEORY OF HEAT CONDUCTION IN EVACUATED METAL POWDERS

Paul G. Klemens

Department of Physics and
Institute of Materials Science
University of Connecticut, Storrs, CT 06268

ABSTRACT

The effective thermal conductivity is calculated for fine-grained metal powders, with the spaces between particles filled with gas at low pressure. There are three mechanisms of heat transfer: conduction between particles at the points of contact, gas conduction in the interstices, and radiative conduction. Their relative importance depends on particle size, gas pressure, and temperature. Contact conduction depends on the area of contact between particles, which in turn depends on the packing pressure and on the yield stress. It thus varies with depth below the powder surface and depends strongly on temperature through the temperature dependence of the yield stress. In addition, contact conduction will depend on the thermal history of the powder since its last mechanical disturbance. Gas conduction will be nearly independent of pressure, unless the pressure is low enough for the gas to be in the Knudsen region. Gas pressure is determined by the volume of powder that has reached outgassing temperature. Numerical values are estimated for a 50 micrometer diameter powder at 1000°C, where radiation is minor, but gas conduction can make a substantial contribution.

INTRODUCTION

In connection with the heating of large masses of metal powder prior to further treatment or compaction, one needs to know the thermal diffusivity to estimate the time required for temperature equilibrium to be established. There are three mechanisms of heat transfer between particles: (1) conduction across the regions of contact between the particles, (2) gas conduction, and (3) radiation. It will be assumed that the powder particles are opaque and that the thermal conductivity of the metal particles is much higher than that of the bulk powder, so that each particle is essentially isothermal. Hence, the thermal resistance arises in the interstices between the particles or in the regions of constricted heat flow at the contact points.

CONTACT CONDUCTION

The thermal conductance due to contact between the particles is determined by the area of contact at each contact point and the average number of contact points per particle. The contact area is governed by the local hydrostatic packing pressure of the powder and the yield stress of the material. Let the average number of contact points per particle be n; n is determined by the packing density ϱ/ϱ_0, where ϱ is the powder density and ϱ_0 that of the bulk metal.

The local hydrostatic pressure, P, causes an average force, F, between particles at each contact point. Let the contact points be randomly oriented, with a mean square direction cosine of 1/3 with respect to the direction of the temperature gradient. Equating the force component due to half the contact points with the product of P and the cross-sectional area of the average volume per particle, one finds

$$\frac{n}{2} F/\sqrt{3} = R^2 P(\varrho_0/\varrho)^{2/3} \tag{1}$$

where R is the particle radius. If A is the contact area at each contact point and σ_y the yield stress of the metal,

$$F = A\sigma_y \tag{2}$$

so that

$$A = 2\pi\sqrt{3}\ PR^2(\sigma_y n)^{-1}(\varrho_0/\varrho)^{2/3} \tag{3}$$

The temperature difference between two particles that touch at a randomly oriented contact point is

$$T = 2R\ \text{grad}T/\sqrt{3} \tag{4}$$

where grad T is the average temperature gradient in the powder. The heat flow across a contact region of area A is

$$B\lambda_0 \Delta T \sqrt{A} \tag{5}$$

where λ_0 is the bulk thermal conductivity and B is a numerical coefficient. If we assume the contact area to be a square and retain only the leading term in a Fourier expansion, we find $B = 4\sqrt{2}/\pi$. The heat flow component in the direction of the temperature gradient per contact point is $Q/\sqrt{3}$, and heat enters each particle through $n/2$ contact points. Since each particle has an effective cross-sectional area of $\pi R^2 (\rho_0/\rho)^{2/3}$ we obtain the following expression for the effective thermal conductivity due to particle contact:

$$\lambda_c = \sqrt{8n}\ 3^{-3/4}\ \pi^{-3/2}\ (\rho/\rho_0)^{1/3}\sqrt{P/\sigma_y}\ \lambda_0 \qquad (6)$$

Here n depends on the packing density. An approximate relation between n and the packing density, which fits a number of regular structures of uniform spheres, is

$$n = 2.5/(1-\rho/\rho_0) \qquad (7)$$

For a typical packing density of 65%, one obtains

$$\lambda_c = 1.43\sqrt{P/\sigma_y}\ \lambda_0 \qquad (8)$$

independently of particle size R.

In the context of equation (8), the hydrostatic pressure, P, arises from the weight of the powder itself, plus any external pressure that may be applied (e.g., by a press). However, gas in the interstices does not contribute to P, since it does not force the particles together. Of course, it provides an additional mechanism of heat conduction. The yield stress usually decreases with increasing temperature. The contact area depends on the highest value of the ratio P/σ_y which the powder has been subjected to since last disturbed mechanically, so that λ_c should show hysterisis. Furthermore, in the absence of externally applied pressure, λ_c should increase with depth below the powder surface.

GAS CONDUCTION

In addition to conduction at the points of contact, there will be heat transfer across the interstices both by radiation and by gas conduction. The latter is significant when the space between the powder particles is not evacuated to below the Knudsen range. Although each particle is practically isothermal, there is a temperature difference across each interstice. The average temperature gradient across the interstices is α grad T, where α depends on the particle shape and the packing density. If we assume that the thermal resistance presented by the interstices is proportional to the fractional volume of the interstices, the contribution of gas conduction to the overall thermal conductivity is

$$\lambda_g = \lambda_g^o / (1 - \rho/\rho_o) \tag{9}$$

where λ_g^o is the thermal conductivity of the gas in the inter-stices. We can write[1]

$$\lambda_g^o = \lambda_g^o(\text{bulk}) \ p/(p_o + p) \tag{10}$$

where λ_g^o (bulk) is the thermal conductivity of the gas in bulk, and p_o is the pressure at which the intrinsic mean free path equals ℓ_o, the average linear dimension of the interstices. The mean free path ℓ can be estimated from the known bulk thermal conductivity, the specific heat and the root mean square molecu-lar velocity. Taking, rather arbitrarily, $\ell_o = R/2$, one can estimate p_o.

For hydrogen at 1270 K, $\lambda_g^o(\text{bulk}) = 0.30$ W·m⁻¹·K⁻¹; thus if $\ell_o = 10 \mu$m, $p_o = 11$ kPa. For argon at the same temperature, $\lambda_g^o(\text{bulk}) = 4.24 \times 10^{-2}$ W·m⁻¹·K⁻¹; with the same value of ℓ_o, $p_o = 12$ kPa. Note that p_o varies inversely with ℓ_o or particle size. At pressures above p_o, λ_g is independent of particle size and depends only on packing density through equation (9).

RADIATION

The role of radiation is only minor at low temperatures, but increases rapidly with temperature. It also depends on particle size.

The rate of heat radiated per unit area is σT^4, where σ is the Stefan-Boltzmann constant (5.72×10^{-16}W·m⁻²·deg⁻⁴). The net heat transported between two surfaces of temperature difference ΔT is

$$Q = 4 \sigma T^3 \Delta T \tag{11}$$

Taking $\Delta T = 2$ R gradT, the radiative component of the thermal conductivity becomes

$$\lambda_{rad} = \frac{Q}{\text{gradT}} = 8 \sigma R T^3 \tag{12}$$

where R is the particle radius. The radiative component will thus be the larger the coarser the powder.

NUMERICAL EXAMPLE

Consider a high-strength nickel alloy at a temperature of

1270 K with an electrical resistivity of $100\mu\Omega\cdot$cm, so that (without lattice conductivity) $\lambda_o = 3.1$ W$\cdot$m$^{-1}\cdot$K^{-1}. Let it be in the form of a powder, having 70% of its bulk density, with an average particle diameter of $50\,\mu$m.   Let the powder be at a hydrostatic pressure of 8 kPa/cm^2,   corresponding to a depth of about 15 cm below the surface. Assume a yield stress of 200 MPa   at this temperature. We obtain the following values for the contact and radiative components from equations (8) and (17):

$$\lambda_c = 0.28 \text{ W}\cdot\text{m}^{-1}\cdot\text{K}^{-1}$$

$$\lambda_{rad} = 0.023 \text{ W}\cdot\text{m}^{-1}\cdot\text{K}^{-1}$$

Gas conduction depends, of course, on the nature of the gas and on the gas pressure. It thus contributes the greatest uncertainty to the estimate of the overall conductivity. For the above particle size one finds p_o to be about 9 kPa irrespective of the nature of the gas, so that

$$\lambda_g = 3.3 \sum_i \lambda_g^o(i) \, p_i/(p+9) \tag{13}$$

Here p_i is the partial pressure (in kPa) of component i, and p is the total gas pressure. Since many metal powders are prepared in an atmosphere of argon, one could expect argon to be the dominant gas in the interstices. Hydrogen is another possibility, because of the strong affinity of transition metals to hydrogen. Suppose the alloy contains 1 ppm (atomic) of hydrogen, leading to a partial pressure at 1270 K of about 400 Pa, and a monomolecular layer of argon, leading to a gas pressure of 25 kPa after it is driven off. The gas conduction would then be mainly due to argon, with hydrogen contributing only about 10% of the total. From (13) we find

$$\lambda_g = 0.11 \text{ W}\cdot\text{m}^{-1}\cdot\text{K}^{-1} \; .$$

Thus, if the powder were fully evacuated, the overall thermal conductivity would be about 0.3 W$\cdot$cm$^{-1}\cdot$K^{-1}; gas conduction could enhance this value by up to 30% if the dominant gas were argon. A greater increase by gas conduction could occur if the partial pressure of hydrogen were to approach or exceed that of argon.

Another quantity of interest is the thermal diffusivity, defined by

$$D = (\lambda_c + \lambda_{rad} + \lambda_g)/C \tag{14}$$

where C denotes the specific heat per unit volume of the powder.

For the bulk metal $C = 4.8$ MJ$\cdot$m$^{-3}\cdot$K^{-1}, and for the 70% dense powder $C = 3.4$ MJ$\cdot$m$^{-3}\cdot$K^{-1}. For a thermal conductivity of say 0.4 W$\cdot$m$^{-1}\cdot$K^{-1}, D is therefore about 1.2×10^{-7}m^2/s. Thus for a container of radius r = 0.1 m, typical times to reach thermal equilibrium, defined as r^2/D, are about 24 hours, a surprisingly large value.

FURTHER CONSIDERATIONS

The present calculations were done for a powder essentially at one temperature, with only a small temperature gradient. In actual practice, powder bodies are heated (or cooled) with large temperature differences. Not only is the thermal diffusivity a function of temperature, but the gas pressure depends on the thermal history of the entire powder body, since the gas can migrate from the hot to the cold regions.

Contact conductance is strongly temperature dependent because the yield stress, which occurs in equation (6) under the inverse square root, decreases with increasing temperature. Radiation is negligible at low temperatures. Thus when a powder is heated for the first time, the cold regions are heated mainly by gas conduction, the gas being supplied by those regions that are already hot and where the powder has been outgassed. At low pressures, the pressure distribution across the powder body will vary as $\sqrt{T}$, where T is the local absolute temperature, owing to the thermomolecular effect, whereas above p_0 the pressure will be independent of temperature and location. Gas conduction at low pressure will thus vary with position, being proportional to the local value of T, but it will also be proportional to the volume of powder that has reached or exceeded the outgassing temperature; it will thus change with time.

In coarser powders, radiative conduction is enhanced, and so is gas conduction at low pressures $(p < p_0)$, since p_0 itself is decreased.

In powders that are being heated for the first time, contact conduction, which is important at higher temperatures, depends on the hydrostatic pressure, P, and thus increases with depth below the powder surface. The region just below the surface is thus the last region to reach final temperature if heating or cooling is from below. If heating or cooling takes place both from the sides and from the top, the last region to attain final temperature is in the middle of the bed, at a depth comparable to the width of the powder bed.

1. C. Bankvall, Heat Transfer in Fibrous Materials, <u>Journal of Testing and Evaluation</u>, JTEVA1:235 (1973).

A LOCAL COMPOSITION MODEL FOR THERMAL CONDUCTIVITY

IN MULTICOMPONENT LIQUID MIXTURES

Richard L. Rowley

Department of Chemical Engineering
William Marsh Rice University
Houston, Texas

ABSTRACT

A local composition model for multicomponent, liquid mixture thermal
conductivity has been developed and tested. Only binary equilibrium
thermodynamic information is used in the model to obtain local compositions. No mixture thermal conductivity data are required and no
adjustable parameters are used; a single binary interaction parameter
is obtained from equation (15) based on an assumption for the local
compositional contributions to the thermal conductivity. Predictions
based on this model agreed, within experimental uncertainty, with the
experimental results for 18 different binary mixtures. An average
absolute percent deviation from experiment of 1.0% was obtained over
the entire composition range for the eighteen systems. The maximum
deviation at any of the tested compositions was 3.5%. Thermal conductivities for ternary systems have also been computed.

INTRODUCTION

Accurate methods for prediction of transport properties in
liquid mixtures have yet to be developed in spite of their importance
in science, engineering, and industry. The main problem in their
development is effective treatment of the liquid structure, which is
neither completely random nor totally structured. Statistical mechanical methods have not yet been fruitful. Generalization of the
Enskog theory becomes extremely complex even for a binary hard sphere
mixture.[1] Likewise, theories based either on the Kirkwood theory[2] or
the fluctuation-dissipation theory[3] have never been adequately solved
for other than very specific cases not applicable to liquid mixtures.
Recent progress has been made by Murad and Gubbins[4] in the use of a

corresponding states correlation for dense fluids. Unfortunately, the method is currently amenable only to nonpolar systems for which a suitable reference fluid thermal conductivity can be computed based on assumed mixing rules. Teja and Rice[5] have also used a corresponding states method for prediction of mixture thermal conductivities. This method is not suitable for systems containing water.

Binary mixture thermal conductivities are most often estimated from empirical data correlations. Although a few correlations do not require mixture thermal conductivity data,[6-8] most correlations require at least one adjustable parameter[9-11] to account properly for differences in interactions between molecules within quite different types of mixtures. Thus, for the same correlation to be applied to both polar-polar and nonpolar-nonpolar mixtures, an adjustable parameter must be introduced to "fit" the equation to the particular system. Worse yet, most existing data correlations cannot be extended to multicomponent systems.

The method for prediction of mixture thermal conductivities presented here is based on a local composition model for the liquid. It uses readily available thermodynamic equilibrium data for binary mixtures to evaluate local compositions in the multicomponent system. The mixture thermal conductivity is then evaluated from the local compositions. The net result is a technique for prediction of transport properties from a minimal amount of equilibrium data. The model requires neither mixture thermal conductivities nor system specific adjustable parameters to make it general, but does require an assumption about the local compositional contributions to the thermal conductivity. The input equilibrium thermodynamic information serves to "adjust" the equation for each specific system based on specific molecular interactions.

LOCAL COMPOSITIONS

The two liquid theory developed by Scott[12] was successfully used by Renon and Prausnitz[13] and by Prausnitz[14] to determine excess free energies. A binary fluid mixture can be represented as a mixture of two hypothetical fluids; fluid one contains molecule 1 at the center with its cell of appropriate nearest neighbors of molecules 1 and 2 while the second fluid contains molecule 2 at the center, again with nearest neighbors of components 1 and 2. The local mole fraction of component i molecules surrounding a central molecule of type j is represented by x_{ij}.† For an n-component system, the local mole fractions must obviously obey conservation equations,

$$\sum_{i=1}^{n} x_{ij} = 1 \quad ; \quad (j = 1,2 \ldots n) \ . \tag{1}$$

Local mole fractions deviate from overall mole fractions owing to

† Symbols are defined in a notation section at the end of the paper.

stronger or weaker 1-2 molecular interactions relative to the 1-1 and 2-2 counterparts. This deviation from random mixing is reflected in the excess properties of the mixture.

In the Nonrandom Two-Liquid (NRTL) model developed by Renon and Prausnitz,[13] the local mole fractions are assumed to be related to the overall mole fractions by a Boltzmann factor containing a nonrandom mixing parameter, α. Thus,

$$\frac{x_{ij}}{x_{jj}} = \frac{x_i}{x_j} \exp(-\alpha A_{ij}/RT) = \frac{x_i}{x_j} G_{ij} \tag{2}$$

where

$$G_{ij} = \exp(-\alpha A_{ij}/RT) \quad . \tag{3}$$

When the NRTL method is applied to the excess free energy, the A_{ij} are treated as adjustable parameters[11] with all $A_{ii} = 0$. The nonrandomness parameter, α, is indicative of the nonrandom mixing nature of the system. Thus, when $\alpha = 0$, $G_{ij} = 1$ and the local mole fractions are identical to the overall mole fractions. Renon and Prausnitz used local compositions to obtain a relation for the excess free energy, g^E, of a binary mixture,

$$g^E = x_1 x_2 \left(\frac{A_{21}G_{21}}{x_1 + x_2 G_{21}} + \frac{A_{12}G_{12}}{x_2 + x_1 G_{12}} \right) \quad . \tag{4}$$

Since activity coefficients are readily obtained from a known composition dependence of g^E, either vapor-liquid or liquid-liquid equilibrium data could, in principle, be used to obtain the NRTL parameters α, A_{21}, and A_{12}. In practice, the two sets of parameters obtained from liquid-liquid equilibrium and from vapor-liquid equilibrium data are not consistent.[15] This will be briefly discussed later. If equilibrium data are available, local compositions are calculable from equations (2) through (4).

THERMAL CONDUCTIVITY MODEL

Analogous to thermodynamic properties, an excess thermal conductivity, k^E, can be defined as the deviation between the mixture value and a mass fraction average of pure component values, k_i^o,

$$k^E = k - \sum_{i=1}^{n} w_i k_i^o \quad . \tag{5}$$

This is consistent with the concepts used by many authors[16,17] and is indeed the basis of many of the most widely used data correlations. Many correlations are, in fact, a mass fraction average plus a devia-

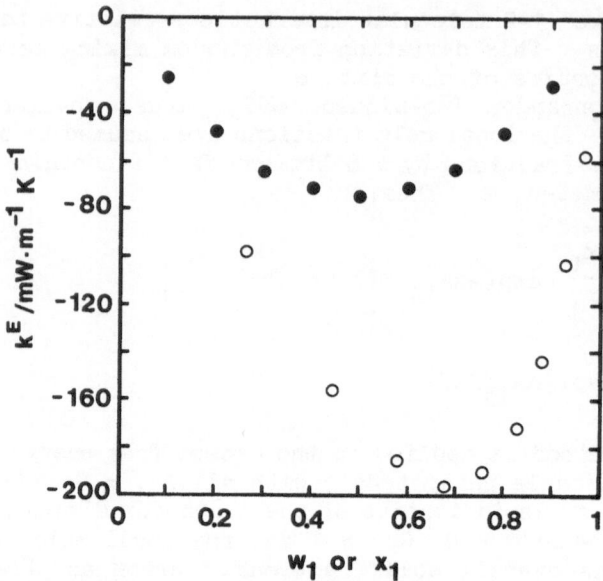

Figure 1.   Excess thermal conductivity of the water-acetone mixture.
Experimental data plotted vs. mass fraction, ●; vs. mole fraction, ○.

tion term (e.g., NEL equation,[10] Filippov equation,[17] and Jordan-
Coates equation[6]).

       This choice of mass fractions in the model for the excess
thermal conductivity is not derivable from theory.   It is, however, suggested by our model.   This can be seen from equation (4)
and the expectation that the excess thermal conductivity based on
this model will have a similar form.   Thus, a choice of composition
for equation (5) consistent with the model should yield an excess
thermal conductivity that is somewhat symmetric in composition for
nearly random mixtures.   For a random mixture ($\alpha = 0$), equation (4)
reduces to the two-suffix Margules expression, which is symmetric
with respect to mole fraction.   For nonrandom mixtures ($\alpha \neq 0$), g^E is
skewed from the symmetric Margules expression owing to the different
interactions between similar and dissimilar molecules.   When experimental excess thermal conductivities defined by equation (5) are plotted vs. mass fraction, a fairly symmetric representation is obtained
as is illustrated by the water-acetone system depicted in Figure 1.
Since this system is not entirely random, the experimental k^E is not
perfectly symmetric.   Figure 1 also shows the water-acetone excess
thermal conductivity that would be obtained if k^E were defined with
respect to   mole rather than mass fractions. Note that it is highly
skewed from symmetry.

       The thermal conductivity of a multicomponent liquid in this
model is a mass fraction average of the thermal conductivities of the
n hypothetical fluids that represent the n-component mixture,

$$k = \sum_{i=1}^{n} w_i k^{(i)} \quad . \tag{6}$$

Each of the n hypothetical fluid thermal conductivities is assumed to be related to the local mass fractions by

$$k^{(i)} = \sum_{j=1}^{n} w_{ji} k_{ji} \quad , \tag{7}$$

where the k_{ij} are yet unidentified parameters characteristic of j-i thermal conduction interactions and where $k_{ij} = k_{ji}$. The local mass fractions must, of course, obey conservation equations similar to equation (1),

$$\sum_{j=1}^{n} w_{ji} = 1 \quad ; \quad (i = 1,2 \ldots n) \quad . \tag{8}$$

Local mass and mole fractions are related through the identity

$$w_{ji} = M_j x_{ji} / (\sum_{\ell=1}^{n} M_\ell x_{\ell i}) \quad . \tag{9}$$

Substitution of equation (9) and the analogous identity between over-all mass and mole fractions into equation (2) yields,

$$\frac{w_{ji}}{w_{ii}} = \frac{w_j}{w_i} G_{ji} \quad . \tag{10}$$

The local mass fractions of equation (7) can now be reexpressed in terms of G_{ji} and overall mass fractions by application of equations (8) and (10). Thus,

$$w_{ii} = w_i / (\sum_{\ell=1}^{n} w_\ell G_{\ell i}) \quad ; \quad w_{ji} = w_j G_{ji} / (\sum_{\ell=1}^{n} w_\ell G_{\ell i}) \quad , \tag{11}$$

where all $G_{\ell\ell}$ are unity. Substitution of equation (7) into equation (6) and elimination of local composition variables by use of equation (11) yields

$$k = \sum_{i=1}^{n} w_i \sum_{j=1}^{n} k_{ji} w_j G_{ji} / (\sum_{i=1}^{n} w_\ell G_{\ell i}) \quad . \tag{12}$$

To obtain a nonparametric thermal conductivity equation of state, the k_{ji} interaction terms must be identified. This is easily done for all k_{ii} by evaluating equation (12) at each pure component limit. When $w_i = 1$, all other w_j ($j \neq i$) are zero. All terms in the summations of equation (7) vanish except for the $i = j = \ell$ term. In this pure component limit, $k = k_i^o$, $G_{ii} = 1$ and

$$k_{ii} = k_i^o \quad . \tag{13}$$

Unfortunately, mathematical manipulation of equation (12) does not elucidate a computational scheme for k_{ij} when $i \neq j$. Instead, an intuitive rule is assumed between the binary interaction term and the binary mixture thermal conductivity. Because equation (12) involves only binary interaction parameters, it only need be written for binary mixtures to evaluate the k_{ij}. For a mixture of components 1 and 2, the binary interaction parameter k_{21}, equal to k_{12}, is assumed to simply be the binary mixture thermal conductivity when the local mole fractions x_{12} and x_{21} are equal. This rule allows direct evaluation of k_{21} in terms of the input thermodynamic information G_{21} and G_{12}. When x_{12} and x_{21} are equal, there are the same number of 1-2 and 2-1 interactions. Likewise, there must also be the same number of 1-1 and 2-2 interactions at this composition according to equation (1).

The local mole fractions x_{12} and x_{21} become equal at only one composition,

$$w_1^* = M_1\sqrt{G_{21}} \;/\; (M_1\sqrt{G_{21}} + M_2\sqrt{G_{12}}) \quad , \tag{14}$$

where the relation between mass and mole fractions has also been used. The mass fraction at which $x_{12} = x_{21}$, denoted w_1^*, is seen to be determined entirely by the equilibrium thermodynamics of the mixture. Similarly, the local composition, w_{11}^*, related to w_1^* by equation (11), is also fixed. Setting the binary thermal conductivity in equations (6) and (7) equal to the independent interaction term, k_{21}, at the particular composition where $x_{12} = x_{21}$ yields

$$k_{21} = \frac{w_1^* w_{11}^* k_1^o + w_2^* w_{22}^* k_2^o}{(w_1^* w_{11}^* + w_2^* w_{22}^*)} \quad , \tag{15}$$

where equation (13) has also been used. Equation (15) is used to calculate all of the binary interaction terms involved in the multicomponent mixture of interest. There are no adjustable parameters involved. The interaction terms are completely defined in terms of the pure component thermal conductivities and the mixture equilibrium thermodynamic data inherent in G_{21} and G_{12}.

To complete the thermal conductivity model, equation (12) can be rearranged,

$$k = \sum_{i=1}^{n} w_i k_i^o + \sum_{i=1}^{n} w_i \left[\sum_{j=1}^{n} w_j G_{ji}(k_{ji} - k_i^o) / (\sum_{\ell=1}^{n} w_\ell G_{\ell i}) \right], \tag{16}$$

where all of the k_{ii} are directly evaluated from equation (15) using only pure component thermal conductivities and the thermodynamic parameters G_{ij} and G_{ji}. The excess thermal conductivity, k^E, is the second term of equation (16). As in the case of equation (4) for equilibrium properties, equation (16) reduces to a symmetric excess thermal conductivity contribution (similar to the two-suffix Margules expression for g^E) when random mixing occurs; i.e., $G_{ij} = 1$ for all values of i and j.

The procedure for use of the NRTL thermal conductivity model of a multicomponent liquid mixture is:

1. Obtain NRTL parameters α, A_{21}, and A_{12} for all pertinent binary mixtures. These are often tabulated[18,19] but can be easily obtained from equilibrium data using equation (4) and a regression program.

2. Compute G_{21} and G_{12} for each binary system at the desired temperature for the mixture.

3. Compute the binary interaction parameters for each binary system using equations (14) and (15).

4. Compute the mixture thermal conductivity at any desired composition from the final form of the model, equation (16).

DISCUSSION

The NRTL model for thermal conductivity contains no adjustable parameters; yet it should be quite general because information concerning binary intermolecular interactions is input via equilibrium thermodynamic data for the constituent binary systems. Equation (16) becomes a system specific equation each time new equilibrium thermodynamic information is used. Because only equilibrium (rather than mixture thermal conductivity) information determines the magnitude and shape of k^E, a comparison of predicted and measured mixture thermal conductivities provides a stringent test for the model and the assumed interaction rule for k_{21}.

Binary mixture thermal conductivities were computed for 18 systems that range from highly nonideal, nonrandom mixtures to quite ideal, random mixtures. The eighteen systems were chosen on the basis of NRTL parameter availability[18-20] and the corresponding availability of reliable mixture thermal conductivity data.[9] The results of this comparison are shown as a percent deviation of predicted from experimental values in Figure 2. Since the best experimental pure component thermal conductivities are assumed to be accurate to within 1-2%, whereas the best mixture thermal conductivities have an accuracy of only 3-5% and a precision of about 2%, the quite different types of mixtures illustrated in Figure 2 collapse down quite nicely to within experimental error.

Table I shows another test of the NRTL method's generality. An average absolute deviation (AAD) between experimental and predicted values was computed over the entire composition range of the 18 sys-

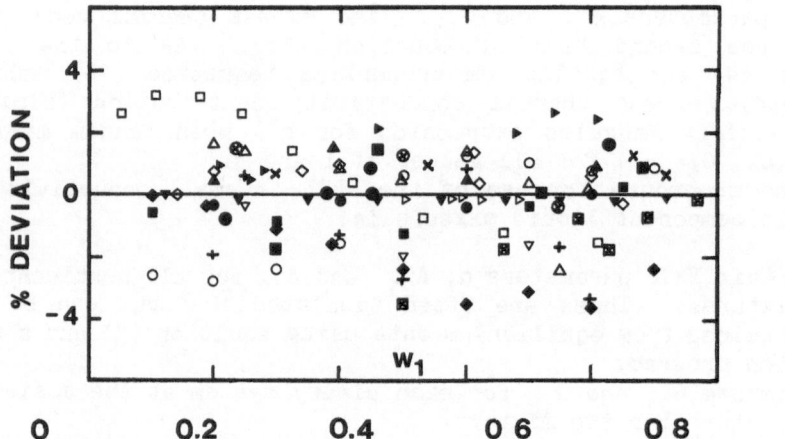

Figure 2.   Percent deviation of predicted thermal conductivities
from experimental values.

   ○ water-acetone
   ● carbon tetrachloride-benzene
   ■ ethanol-benzene
   □ acetone-chloroform
   ▲ benzene-1-propanol
   △ methanol-carbon tetrachloride
   ◆ ethanol-water
   ◇ acetone-benzene
   ▽ acetone-methanol
   ▼ acetone-ethanol

   ▶ benzene-toluene
   + diethylether-methanol
   ● methanol-benzene
   ⊗ acetone-isopropanol
   ✕ methanol-methylformate
   ■ carbon tetrachloride-
     1-propanol
   □ chlorobenzene-methanol
   ▷ carbon tetrachloride-
     cyclohexane [21]

tems. This is compared to the AAD obtained for several of the better
known data correlations. Adjustable parameters in the correlations
were fixed at the values suggested by their authors for nonparametric
estimation. Also shown in Table I are the AAD's obtained using only
the ideal or mass fraction average thermal conductivity. This is
included for comparison of the relative size of k^E and the improved
prediction expected from the local composition model.

Table I. A comparison of predicted thermal conductivity data to
experimental values for 18 binary systems.

Quantity	NRTL	NEL[10]	JC[6]	F[19]	Li[7]	k^{ID}
Overall AAD	1.0	1.3	1.5	1.6	3.3	4.0
Maximum AAD for any one system	2.0	3.5	3.0	5.3	9.9	18.6
Maximum % deviation at any one composition	3.5	6.0	5.5	7.9	14.3	27.2

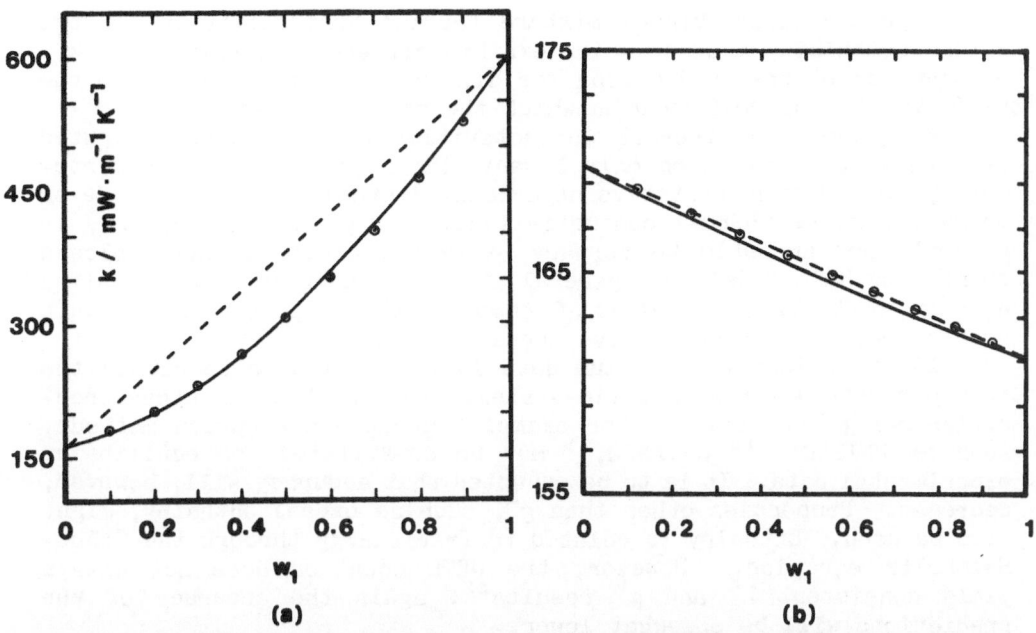

(a)   **(b)**

Figure 3. Thermal conductivity for (a) water(1)-acetone(2) and (b) acetone(1)-ethanol(2) binary mixtures at 25°C. Experimental, ———, ideal, ------, and predicted, o, values are shown.

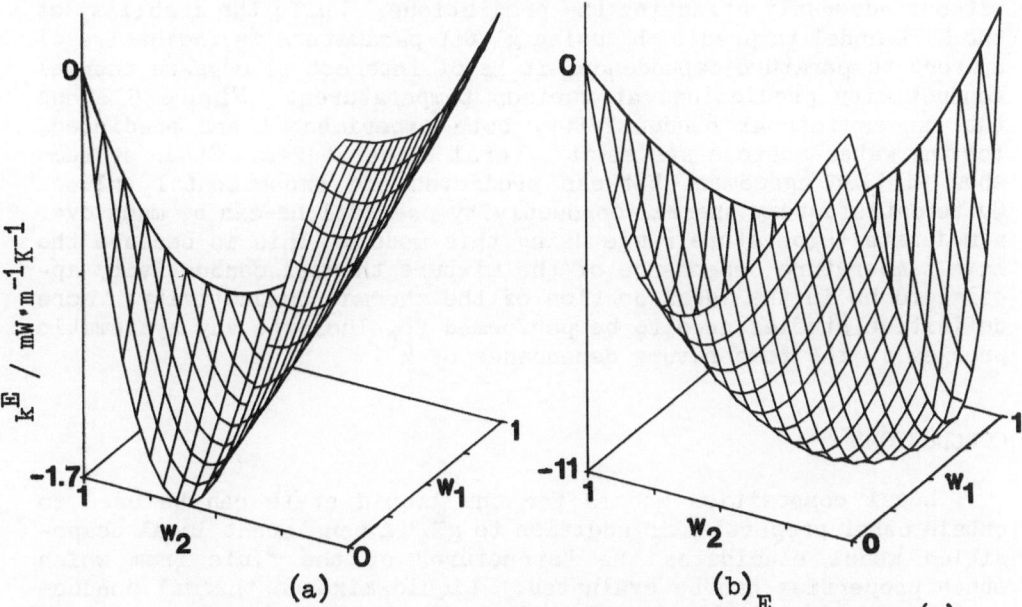

(a)   (b)

Figure 4. Predicted excess thermal conductivity, k^E, for the (a) acetone-benzene-ethanol and the (b) carbon tetrachloride-benzene-methanol systems at 25°C.

Figure 3 shows binary mixture thermal conductivities for two systems containing acetone which exhibit different magnitudes of k^E. The accuracy of the predictions for different mixtures is due to the thermodynamic information upon which the model is based.

Multicomponent thermal conductivities can be easily computed from equation (16) using only binary thermodynamic data. Unfortunately, very few quantitative experimental studies have been made on ternary mixture thermal conductivities. Data correlations are, in general, not amenable to ternary systems because of their closed form. The NRTL model is a general multicomponent model and applies equally well for any number of components. Figure 4 shows NRTL predictions for representative ternary systems.

If sufficient equilibrium data is not available to obtain the NRTL parameters for a specific system, other methods or other properties may prove useful. For example, group contribution methods, such as UNIFAC[11,15] or ASOG,[11] may be substituted for equilibrium experimental data. It is to be expected that accuracy will, however, decrease. Properties other than g^E, such as excess enthalpy, might also be used. Enthalpy is related to free energy through the Gibbs-Helmholtz equation. However, the NRTL equation does not always yield consistent h^E and g^E results[22]; again the accuracy of the predictions will be somewhat lower.

Vapor-liquid and liquid-liquid equilibrium data do not in general yield a unique set of NRTL parameters. Although vapor-liquid data were used for the computation of most of the mixtures illustrated in this paper, some liquid-liquid equilibrium data were also used without adversely affecting the predictions. While the inability of the NRTL model to predict h^E using g^E fit parameters is indicative of a wrong temperature dependence, it is of interest to compare thermal conductivity predictions at various temperatures. Figure 5 shows the excess thermal conductivity, both experimental and predicted, for the water-acetone system at several temperatures. Other systems show similar agreement between predicted and experimental values. Quite satisfactory thermal conductivity predictions can be made over a moderate temperature range using this model. This is because the main temperature dependence of the mixture thermal conductivity appears to be in the ideal portion of the thermal conductivity. More definitive studies need to be performed to elucidate any systematic problem in the temperature dependence of k^E.

CONCLUSIONS

Local composition models for the liquid state can be used to obtain other properties in addition to g^E. A consistent local composition model elucidates the "structure" of the fluid from which other properties can be evaluated. Liquid mixture thermal conductivity can be effectively related to local compositions using the NRTL model. Input of equilibrium data allows computation of local compositions from overall or bulk compositions. These, in turn, are

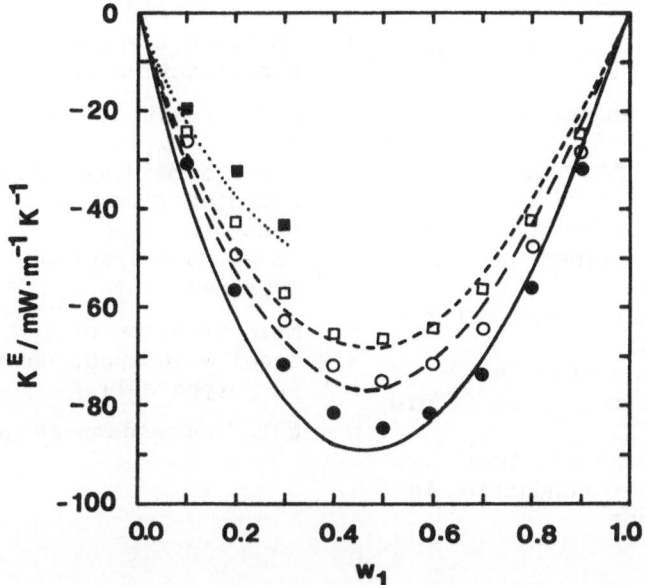

Figure 5. Predicted (curves) and experimental (points) excess thermal conductivity, k^E, for the water-acetone binary system at 233K (·······, ■), 273K (-----, □), 293K (— — —, ○), and 313K (———, ●).

used to predict mixture thermal conductivity data. Only binary equilibrium thermodynamic data are required regardless of the number of components in the actual mixture.

   Predicted values for binary systems generally agree with experimental data within experimental error. Moreover, the model is applicable to all types of mixtures without recourse to adjustable parameters. Instead, multicomponent predictions are easily made, but more experimental data are required to test these predictions. The model's temperature dependence also appears reliable over moderate temperature ranges.

   Studies are currently being made on the extension of local composition models to other transport properties. More sophisticated local composition models are also being tested. However, more and better quality liquid mixture transport data, particularly in ternary systems, are needed to refine and test local composition models for transport coefficients.

NOTATION

AAD	average absolute deviation	M_i	molecular weight of component i
A_{ij}	NRTL parameter	R	gas constant
g^E	excess molar Gibbs free energy	T	absolute temperature
		w_i	mass fraction of i
G_{ij}	NRTL parameter defined by equation (3)	w_i^*	mass fraction of i when $x_{12} = x_{21}$
k	mixture thermal conductivity	w_{ij}	local mass fraction of i surrounding a molecule of component j
k^E	excess thermal conductivity defined by equation (5)	w_{ij}^*	local mass fraction of i in j-type cell when $x_{12} = x_{21}$
k_i^o	pure component thermal conductivity	x_i	mole fraction of i
$k^{(i)}$	thermal conductivity of hypothetical pure fluid of type i	x_{ij}	local mole fraction of i in j-type cell
k_{ji}	characteristic term for j-i thermal conduction interactions	α	NRTL nonrandomness parameter.

REFERENCES

1. E. McLaughlin, The Thermal Conductivity of Liquids and Dense Gases, Chem. Rev. 64:389 (1964).
2. R.J. Bearman and V.S. Vaidhyanathan, Composition Dependence of the Thermal Conductivity of Regular Solutions, J. Chem. Phys. 34:264 (1961).
3. G.V. Chester, The Theory of Irreversible Processes, Rept. Progr. Phys. 26:411 (1963).
4. S. Murad and K.E. Gubbins, Corresponding States Correlation for Thermal Conductivity of Dense Fluids, Chem. Eng. Sci. 32:499 (1977).
5. A.S. Teja and P. Rice, A Generalized Corresponding States Method for the Prediction of Liquid Thermal Conductivity of Liquids and Liquid Mixtures, Chem. Eng. Sci., 36:417 (1981).
6. H.B. Jordan, M.S. Thesis, Louisiana State University, Baton Rouge (1961).
7. C.C. Li, Thermal Conductivity of Liquid Mixtures, AIChE J. 22: 928 (1976).
8. G.H. Schroff, Measurement and Correlation of Thermal Energy Transport in Non-Ideal Binary Liquid Ststems, in: "Proc. 8th Conf. Thermal Conductivity," C.Y. Ho and R.E. Taylor (eds.), Plenum, NY, (1969).
9. D.T. Jamieson, J.B. Irving and J.S. Tudhope, "Liquid Thermal Conductivity: A Data Survey to 1973," HMO, Edinburgh, (1975).

10. D.T. Jamieson and E.H. Hastings, The Thermal Conductivity of Binary Liquid Mixtures, in: "Proc. 8th Conf. Thermal Conductivity," C.Y. Ho and R.E. Taylor eds., Plenum, NY, (1969).

11. R.C. Reid, J.M. Prausnitz and T.K. Sherwood, "The Properties of Gases and Liquids, 3rd Edition," McGraw-Hill, (1977).

12. R.L. Scott, Corresponding States Treatment of Nonelectrolyte Solutions, J. Chem. Phys. 25:193 (1956).

13. H. Renon and J.M. Prausnitz, Local Compositions in Thermodynamic Excess Functions for Liquid Mixtures, AIChE J. 14:135 (1968).

14. J.M. Prausnitz, "Molecular Thermodynamics of Fluid-Phase Equilibria," Prentice-Hall, Englewood Cliffs (1969).

15. J. Prausnitz, T. Anderson, E. Grens, C. Eckert, R. Hsieh and J. O'Connell, "Computer Calculations for Multicomponent Vapor-Liquid and Liquid-Liquid Equilibria," Prentice-Hall, Englewood Cliffs, (1980).

16. N.V. Tsederberg, "Thermal Conductivity of Gases and Liquids," M.I.T. Press, Boston (1965).

17. L.P. Filippov, Liquid Thermal Conductivity Research at Moscow University, Int. J. Heat Mass Trans. 11:331 (1968).

18. I. Nagata, Prediction Accuracy of Multicomponent Vapor-Liquid Equilibrium Data From Binary Parameters, J. Chem. Eng. Japan 6:18 (1973).

19. J. Gmehling, and U. Onken, "Vapor-Liquid-Equilibrium Data Collection; Dachema Chem. Data Series, Vol. 1, part 2a," Verlag & Druckerel Friedrich Bischoff, Frankfurt, (1977).

20. E. Bender and U. Block, Thermodynamic Calculation of Fluid Extraction, Verfahrens Technik 9:106 (1975).

21. R.L. Rowley and F.H. Horne, The Dufour Effect. III. Direct Experimental Determination of the Heat of Transport of Carbon Tetrachloride-Cyclohexane Liquid Mixtures, J. Chem. Phys. 72: 131 (1980).

22. S. Skjold-Jorgensen, P. Rasmussen and Aa. Fredenslund, on the Temperature Dependence of the UNIQUAC/UNIFAC Models, Chem. Eng. Sci. 35:2389 (1980).

EVALUATION OF SPECIFIC HEAT DATA FOR POCO GRAPHITE AND

CARBON-CARBON COMPOSITES*

M. S. Deshpande and R. H. Bogaard

TEPIAC/CINDAS
Purdue University
West Lafayette, Indiana   47906

ABSTRACT

In the thermophysical characterization of carbon-carbon materials, properties such as the thermal diffusivity, specific heat, thermal conductivity, and thermal expansion are of central importance. Specific heat data are of interest for instance, in the conversion between thermal conductivity and thermal diffusivity. In this report, specific heat values are presented on different grades of graphites and different types of carbon-carbon composites over the temperature range 300 K to 3000 K. Recommended values are developed for the specific heat of POCO graphite over this range. The application of these values is made to a physical model for the specific heat of graphite at high temperatures. Finally, a comparison of the recommendations for POCO graphite is carried out with literature data on the specific heat of carbon-carbon composites, and the results are discussed.

INTRODUCTION

The wide range of applications of carbon-carbon composites at high temperatures has prompted the development of carbon-carbon (C-C) technology. New material and process developments are analyzed for their impact on the thermophysical behavior of C-C composites. The specific heat of such composites plays an important role in deriving thermal property information. Of particular interest is the conversion between thermal transport properties. The

---

*This work was performed under the auspices of the DOD carbon-carbon data-base program.

data generated for the specific heat of C–C composites show wide
variations. Most of the early efforts on property generation
argued that these materials be considered as graphitic and that
graphite specific heat be used for property conversion. An evaluation was, therefore, undertaken of the available information on the
specific heat of graphite with the hope of using it in thermal
transport property conversion.

The compilation and evaluation efforts are restricted to POCO
graphite and do not include all graphites accessible in the literature. The POCO AXM-5Q-1* graphite is of interest owing to its
possible use as a broad-temperature-range, international-reference
material for thermophysical properties. This interest stems, in
part, from desirable physical properties of POCO AXM-5Q-1, such as
high purity, small grain size with overall isotropy, and heat
treatment at a temperature of 2500°C which is within the range of
temperatures used for the processing of C–C composite materials.
Though the effects of graphitization temperature on specific heat
are not completely understood, similarity in high temperature
treatment (HTT) temperature is an essential factor for the characterization of graphites. Recommendations were, therefore, developed for POCO AXM-5Q-1 graphite. A physical model developed for
graphite was applied to the specific heat of POCO graphite. Finally, the recommendations are compared with specific heat data on
selected C–C composites to demonstrate their use for property
conversion.

SPECIFIC HEAT OF GRAPHITES

Specific heat data on various grade graphites is shown in
Figure 1 as a consolidated plot from 300 K to 3000 K. This compilation is somewhat selective in that widely divergent data sets are
not included. Generally, the listing is thought to be representative
of the numerous graphites for which data are available in the literature. The graphites presented here include 3447D, 7087, GBH, GBE,
CFZ, CCH, various nuclear grade graphites, pyrolytic graphite, ATJ,
POCO, and a few spectrally pure graphites. Most data sets are derived from experimentally measured enthalpy data, with the others
being directly measured specific heat obtained by differential adiabatic calorimetry, differential scanning calorimetry, or heat pulse
techniques. The accuracy of the measurements varies from ±3% to
±10%. The diversity observed in Figure 1 indicates the excursions
at higher temperatures occurring in some earlier work along with the
rather good agreement among more recent results. A separate evaluation for each type of graphite is beyond the scope of this paper.

---

*POCO graphite: Product of POCO Graphite, Inc., Decatur, Texas.
 Grade designation: AXM: medium grain fuel cell grade; 5Q: 2500°C
 graphitization temperature; 1—purified.

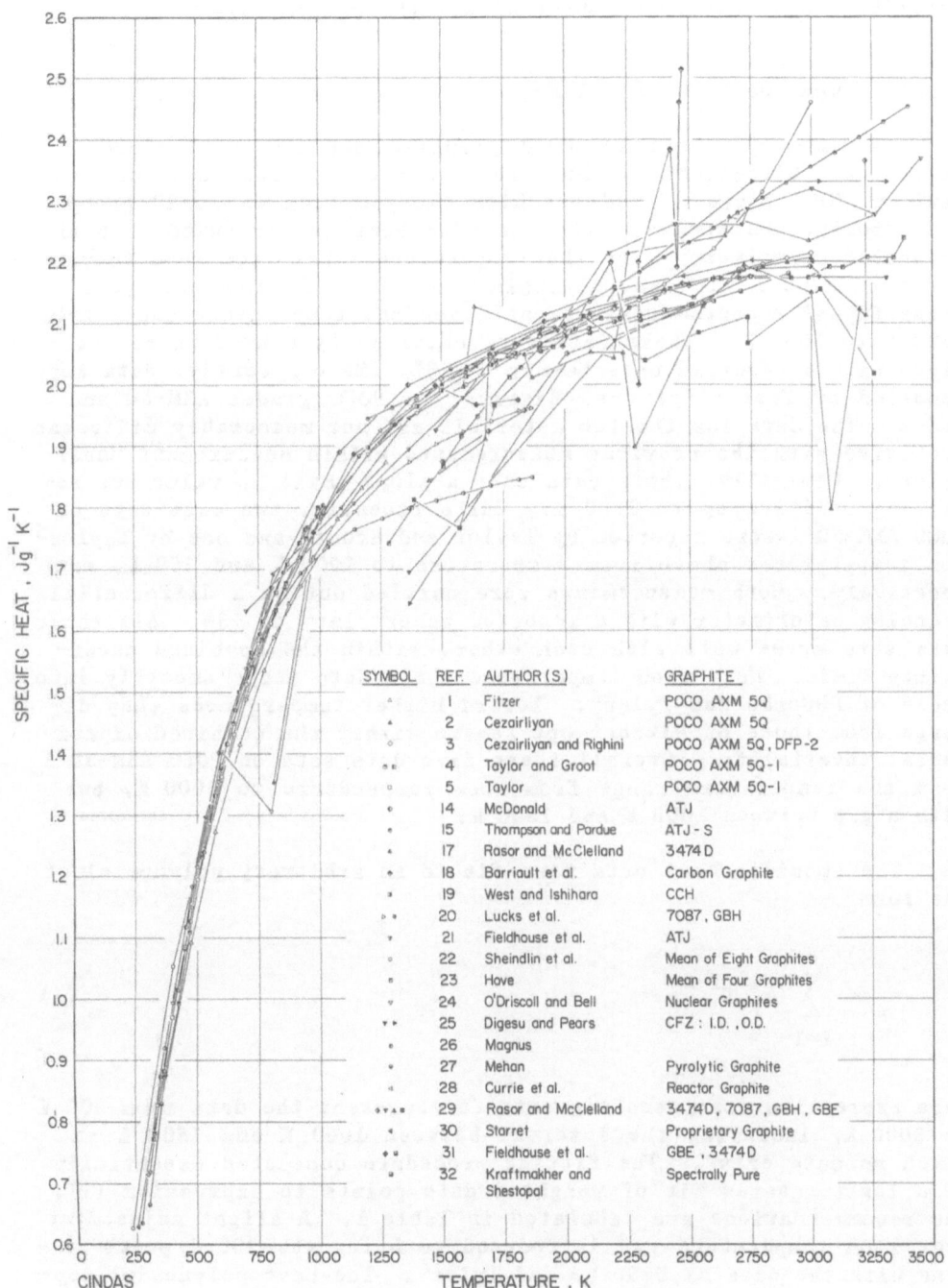

SYMBOL	REF.	AUTHOR(S)	GRAPHITE
	1	Fitzer	POCO AXM 5Q
	2	Cezairliyan	POCO AXM 5Q
	3	Cezairliyan and Righini	POCO AXM 5Q, DFP-2
	4	Taylor and Groot	POCO AXM 5Q-1
	5	Taylor	POCO AXM 5Q-1
	14	McDonald	ATJ
	15	Thompson and Pardue	ATJ-S
	17	Rasor and McClelland	3474D
	18	Barriault et al.	Carbon Graphite
	19	West and Ishihara	CCH
	20	Lucks et al.	7087, GBH
	21	Fieldhouse et al.	ATJ
	22	Sheindlin et al.	Mean of Eight Graphites
	23	Hove	Mean of Four Graphites
	24	O'Driscoll and Bell	Nuclear Graphites
	25	Digesu and Pears	CFZ : I.D. , O.D.
	26	Magnus	
	27	Mehan	Pyrolytic Graphite
	28	Currie et al.	Reactor Graphite
	29	Rasor and McClelland	3474D, 7087, GBH , GBE
	30	Starret	Proprietary Graphite
	31	Fieldhouse et al.	GBE , 3474D
	32	Kraftmakher and Shestopal	Spectrally Pure

Fig. 1. Specific Heat of Graphites.

The POCO AXM-5Q-1 graphite appears to be a very well-characterized material and is the focal material for the present study.

SPECIFIC HEAT OF POCO GRAPHITE

The specific heat of POCO graphite has received recent attention. Historically, the first reported specific heat measurement on POCO grade AXM-5Q graphite emerged from an AGARD group characterization study[1]. The specific heat was measured with an adiabatic calorimeter over the temperature range from room temperature to 1200 K with an uncertainty of 10%. In the same year, Cezairliyan[2] reported measurements over the temperature range from 1500 K to 3000 K. These data were obtained by a heat pulse technique with a reported uncertainty of ±3%. More recently, data were reported by Cezairliyan and Righini[3] on POCO grades AXM-5Q and DFP-2. The data for the two materials are not measurably different and agree with the previous measurements within measurement uncertainty. Generally, these data show a slope small in value but remaining positive up to 3000 K. Quite recently, two data sets on POCO AXM-5Q-1 were reported by Taylor and Groot[4] and one by Taylor[5] for temperatures above room temperature to 1000 K and 750 K, respectively. Both measurements were carried out on a differential scanning calorimeter with a reported uncertainty of ±3%. All three data sets agree well with each other, within the combined uncertainty limit. Near room temperature these data merge smoothly into those of DeSorbo and Tyler[6]. Toward higher temperatures they diverge from those of Fitzer[1] but remain within the combined experimental uncertainty. Overall these five data sets on POCO AXM-5Q-1 span the temperature range from room temperature to 3000 K, but with a gap between 1000 K and 1500 K.

The specific heat data were fit to an arbitrary polynomial of the form

$$c_p = \sum_{i=1}^{8} A_i T^{i-3} \tag{1}$$

This expression was found adequate to represent the data from 300 K to 3000 K, including the interval between 1000 K and 1500 K in which no data exist. The fitting procedure consisted essentially of a least-squares fit of weighted data points to expression (1). The recommendations are tabulated in Table I. A slight adjustment near room temperature was introduced to bring the 300 K point in line with the data of DeSorbo and Tyler[6]. The best polynomial representation of the recommendation is given as

$$c_p = 8.1353E+04T^{-2} - 6.2119E+02T^{-1} + 1.2892 + 2.6326E-03T - 2.5559E-06T^2 + 1.2376E-09T^3 - 2.9593E-13T^4 + 2.7904E-17T^5 \quad (2)$$

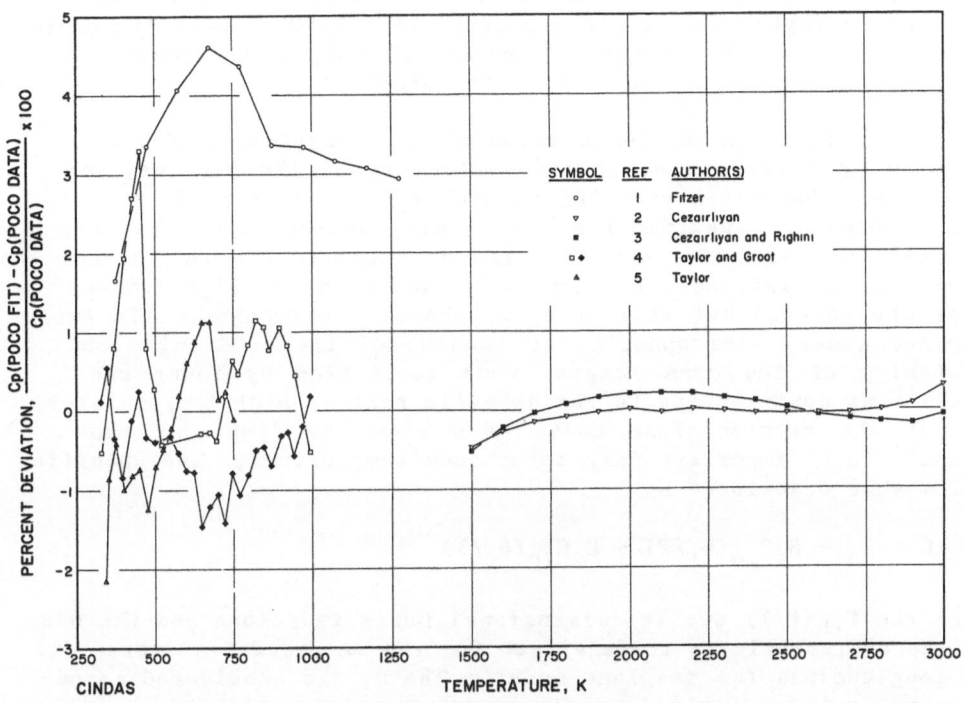

Fig. 2. Deviation of Specific Heat Recommendations for POCO Graphite from Experimental Data.

Table I. Recommended Values for the Specific Heat of POCO AXM-5Q-1 Graphite

[Temperature, T, K; Specific Heat, c_p, J $g^{-1}K^{-1}$]

T	c_p	T	c_p	T	c_p
300	0.713	1300	1.940	2300	2.142
400	0.961	1400	1.972	2400	2.154
500	1.187	1500	2.000	2500	2.165
600	1.370	1600	2.024	2600	2.174
700	1.516	1700	2.046	2700	2.184
800	1.632	1800	2.066	2800	2.193
900	1.723	1900	2.083	2900	2.202
1000	1.796	2000	2.100	3000	2.213
1100	1.854	2100	2.115		
1200	1.901	2200	2.129		

Figure 2 shows the deviation of the fit with respect to the experimental data. It is observed that with the exception of the reference 1 data, the deviations are generally less than 2%. The standard deviation of the fit is 0.019. These recommendations agree closely (within 2%) with the fitted values for POCO graphite reported by Taylor and Procter[7] except above 2600 K where the present values are 4% to 6% lower. The uncertainty in the recommendations is thought to be about 3% up to 2500°C.

An application of these results to a physical model was prompted by a concern about the continuity of the data between 1000 K and 1500 K. A simplified model for graphite applicable to high temperature specific heat is readily adapted from the work reported by Lutkov and Dymov[8]. The essentially two-dimensional nature of the lattice vibration modes important at high temperatures was pointed out earlier by Krumhansl and Brooks[9]. The two in-plane modes, corresponding to bending of the bond angle and stretching of the bond length, were identified by Young and Koppel[10] as contributing to the specific heat at high temperatures, whereas the between-plane modes were shown by Yoshimori and Kitano[11] to be important only below room temperature. The specific heat may be described[8] by

$$C = B_0 + B_1 C_{2D}(\theta_t/T) + B_2 C_{2D}(\theta_\ell/T) \tag{3}$$

where the $C_{2D}(\theta/T)$ are two-dimensional Debye functions and the θ's are the characteristic temperatures for the so-named transverse (t) and longitudinal (ℓ) in-plane modes. The B_i are considered as parameters to be determined by fitting the experimental data to expression (3). Corrections for c_p-c_v were neglected on the basis of estimates reported by Butland and Maddison[16], which amounted to only 3.2% at 1800 K.

The results of this study are:

1. The simplified model can be readily applied to POCO specific heat data. In the troublesome region from 1000 K to 1500 K, the values generated lie about 1% above those from the polynomial fit. Over the entire temperature range 300 K to 3000 K, the values generated from the model fit lie within about 2% of the measured data.

2. Definitive values for the pair of temperatures, θ_t and θ_ℓ, failed to emerge. Rather, a range of values for this pair was found from 950 K and 2700 K, respectively, up to 1500 K and 3500 K, respectively, with no real indication of a statistically 'better' choice. The parameter ratio B_2/B_1 took on values of 2.0 and 1.0, respectively, over this same range. The value of B_0 was always small, less than about 5% of the specific heat at 400 K.

3. The lack of success in Item 2 above appears to focus upon
θ_ℓ. The choice of θ_ℓ, of course, depends very sensitively
upon the slope of the specific heat at the highest temper-
atures. Further, c_p-c_v corrections should be applied before conclusions are drawn concerning θ_ℓ.

SPECIFIC HEAT OF CARBON–CARBON COMPOSITES AND CARBONACEOUS MATERIALS

Only limited efforts have been devoted in the past to
measurements of the specific heat of C–C composites and carbonaceous materials. Most of the measurements published on C–C composites are of recent origin. Data made available by Taylor[5],
Cezairliyan and Müller[12], Goetzel[13], and McDonald[14] are shown in
Figure 3. The data reported by Taylor[5] are on a fine-weave 2–2–3
composite graphitized at 2750°C and on a 3D cylindrical coarse-
weave C–C composite; the latter was tested at both inside and outside diameter locations. In spite of the differences in structure
and processing, these three data sets agree with POCO AXM-5Q-1
within measurement uncertainty. The data reported by Cezairliyan
and Müller[12] are on two specimens of a 2–2–3 C–C composite graphitized at 2750°C. A series of measurements involving pulse heating
to high temperatures a total of ten times resulted in no measurable
change in the specific heat. Goetzel[13] reported enthalpy measurements on flat, laminated C–C composites. Data were reported for
materials having three distinct density levels: 1.525 g/cm^3, 1.65
g/cm^3, and 1.85 g/cm^3. The specific heat reportedly correlates
with density. An explanation in terms of trapped gas and void
structure is suggested. McDonald[14] has communicated data on seven
different proprietary C–C composites. Of these, six are 3D whereas
one is a 7D C–C composite. The thermal histories of these materials are not available. These specific heat data are obtained from
enthalpy measurements by an adiabatic calorimetric technique and
cover the temperature range 300 K to 3000 K. It is noted that companion data on ATJ graphite[14] agree within 2% with measurements on
POCO graphite (see Figure 1). Also shown in Figure 3 are specific
heat data on another carbonaceous material, CVD felt. The density
of this graphitized CVD felt is in the range of 1.703 g/cm^3 to
1.749 g/cm^3 as reported by Thompson and Pardue[15]. These data,
generated by an aneroid drop-calorimeter technique, agree rather
well with companion data on ATJ-S graphite (see Figure 1).

As shown in Figure 3, the specific heat for composite
materials having densities between about 1.60 g/cm^3 and 1.85 g/cm^3
falls within ±5% of the recommendations for POCO AXM-5Q-1 graphite.
This confirms the contention that since the matrix of a C–C composite is graphitic, the specific heat of POCO graphite may be substituted. At present the available data are insufficient to carry
out an examination of effects due to HTT temperature, degree of

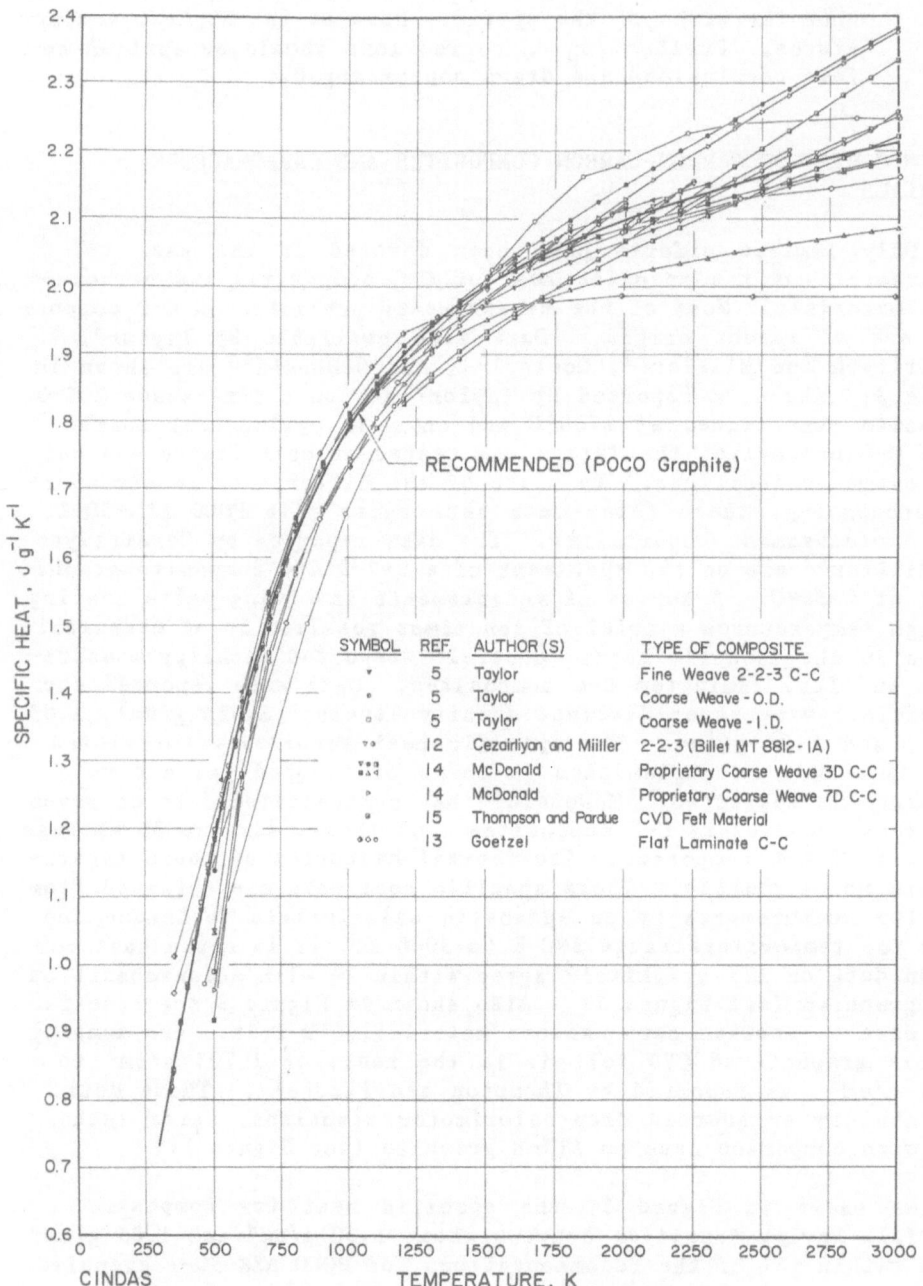

Fig. 3. Specific Heat of C–C Composites.

graphitization, and the trapped gas and void structure conjecture. Thus, the observed variations in the specific heat of C–C composites remain unexplained.

SUMMARY OF CONCLUSIONS

1. The evaluation and analysis of the experimental specific heat data for POCO graphite grade AXM–5Q–1 were carried out. A polynomial fit to the data, applicable to the temperature range from 300 K to 3000 K, was obtained.

2. The evaluation of composite–material specific heat data revealed that the recommendations for POCO AXM–5Q–1 graphite can be substituted for the specific heat of C–C composites, with the provision that the C–C composite material be fully graphitized at high temperatures.

REFERENCES

1. E. Fitzer, AGARD Advisory Rept. R–606, 107 pp. (1973). [AD 760 165]
2. A. Cezairliyan, in 'Proceedings of the Sixth Symposium on Thermophysical Properties' P. E. Liley, ed., ASME, New York, NY, 279–85 (1973).
3. A. Cezairliyan, and F. Righini, Rev. Int. Hautes Temp. Refract., 12(2):124 (1975).
4. R. E. Taylor and H. Groot, High Temp.–High Pressures, 12:147 (1980).
5. R. E. Taylor, Thermophysical Properties Research Lab. Rept. TPRL 235, Purdue Univ., W. Lafayette, IN, 11 pp. (1980).
6. W. DeSorbo and W. W. Tyler, J. Chem. Phys., 21(10):1660 (1953).
7. R. Taylor and R. N. Procter, Interim Scientific Report to AFOSR, EOARD, and AFML, 30 Sept. 1977–30 Sept. 1979, Contract No. AFOSR 77–3449, Univ. of Manchester, Inst. of Science and Technology, Manchester, England, 66 pp. (1979).
8. A. I. Lutkov and B. K. Dymov, Zh. Fiz. Khim., 49(12):3098 (1975); Engl. transl., Russ. J. Phys. Chem., 49(12):1832 (1975).
9. J. Krumhansl and H. Brooks, J. Chem. Phys., 21(10):1663 (1953).
10. J. A. Young and J. U. Koppel, J. Chem. Phys., 42:357 (1965).
11. A. Yoshimori and Y. Kitano, J. Phys. Soc. Jpn., 11(4):352 (1956).
12. A. Cezairliyan and A. Miiller, Int. J. Thermophys., 1(3):317 (1980).
13. C. G. Goetzel, High Temp.–High Pressures, 12(1):11 (1980).

14. J. R. McDonald, Southern Research Institute, Birmingham, AL,
    Private communication.

15. J. R. Thompson, Jr. and W. M. Pardue, Sandia Labs. Final Rept.
    FAO 82-5881, Serial No. 1760, Battelle, Columbus, OH, 31
    pp. (1971).

16. A. T. D. Butland and R. J. Maddison, Atomic Energy Establish-
    ment Rept. AEEW-R-815, Winfrith, England, 22 pp. (1972).

17. N. S. Rasor and J. D. McClelland, J. Phys. Chem. Solids, 15:17
    (1960).

18. R. J. Barriault, S. L. Bender, R. E. Dreikorn, T. H.
    Einwohner, R. C. Feber, R. E. Gannon, P. L. Hanst, M. E.
    Ihnat, J. P. Phaneuf, H. L. Schick, and C. H. Ward, U.S.
    Air Force Rept. ASD TR 61-260, 404 pp. (1962).
    [AD 278 633]

19. E. D. West and S. Ishihara, in 'Third ASME Symposium on Ther-
    mophysical Properties,' Purdue University, W. Lafayette,
    IN, 146-51 (1965).

20. C. F. Lucks, H. W. Deem, and W. D. Wood, Am. Ceram. Soc.
    Bull., 39(6):313 (1960).

21. I. B. Fieldhouse, J. I. Lang, and H. H. Blau, U.S. Air Force
    Rept. WADC TR 59-744, Vol. 4, 78 pp. (1960).
    [AD 249 166]

22. J. E. Hove, in 'Proceedings 1st Conference Industrial Carbon
    and Graphite, London, 1957,' Society of Chemical Indus-
    try, London, England, 501 (1958).

23. A. E. Sheindlin, I. S. Belevich, and I. G. Kozhevnikov,
    Teplofiz. Vys. Temp., 10(5):997 (1972); Engl. transl.,
    High Temp., 10(5):897 (1972).

24. W. G. O'Driscoll and J. C. Bell, Nucl. Engr., 11:479 (1958).

25. F. J. Digesu and C. D. Pears, Southern Research Institute,
    Birmingham, AL, Private communication.

26. A. Magnus, Ann. Phys. (Leipzig), 70:303 (1923).

27. R. L. Mehan, U.S. Air Force Tech. Rept., General Electric,
    Philadelphia, PA, 14 pp. (1963). [AD 342 002]

28. L. M. Currie, V. C. Hamister, and H. G. MacPherson, in 'Pro-
    ceedings of the International Conference on Peaceful Uses
    Atomic Energy,' Geneva, Switzerland, 451-73 (1955).

29. N. S. Rasor and J. D. McClelland, U.S. Air Force Rept.
    WADC-TR-56-400 (Pt. 1), 53 pp. + Appendix (1957).
    [AD 118 144]

30. H. S. Starret, Southern Research Institute, Birmingham, AL,
    Private communication.

31. I. B. Fieldhouse, J. C. Hedge, J. I. Lang, A. N. Takata, and
    T. E. Waterman, U.S. Air Force Rept. WADC-55-495 (Pt. 1),
    64 pp. (1952). [AD 110 404]

32. Ya. A. Kraftmakher and V. O. Shestopal, Zh. Prikl. Mekh. Tekh.
    Fiz., 4:170 (1965); Engl. transl., J. Appl. Mech.
    Technol. Phys., 4 (1965).

THERMAL PROPERTIES OF POCO-PROCESS GRAPHITE

L. L. Isaacs and W. Y. Wang

Department of Chemical Engineering
The City College of CUNY
New York, NY 10031

INTRODUCTION

The thermal properties of natural and synthetic carbons are of continuing scientific and technical interest. Carbons in the form of chars and cokes are products of processes designed to convert coals into liquid and gaseous fuels. The production of synthetic graphite is of technical importance. The understanding of the physical phenomena involved in the mode of carbon formation on catalytic surfaces is an ongoing concern of surface science and catalysis.

It is well-known that one cannot produce a synthetic graphite that will duplicate the thermal properties of natural graphite.[1] The closeness of approach to graphitic behavior is a function of the raw material from which the carbon is formed and of the thermal history of the carbon.[2]

Although our interests were mainly in the correlation of the thermal and physical properties of chars (soft carbon), we did measure the specific heat of the "hard" synthetic carbon, POCO-graphite. This measurement serves as a baseline for comparing the property variations of the chars.

SAMPLE DESCRIPTION

A rod of POCO-AXM 5Q1 (sample designation 103-2) 0.051 m long by 0.025 m diameter was supplied by J. G. Hust of the Thermophysical Properties Division of the National Engineering Laboratory. Part of this rod was powdered, dry, using a water-cooled analytical grinding mill. The amount of powder used in the calorimetric measurements was 4.5 g. Before calorimetry, the powder was vacuum dried at 110°C for one hour. After the completion of the calori- metric measurements, the powder was analyzed using a Per- kin-Elmer model 240-B elemental analyzer. The quantitative composition (w/o) of the sample was found to be:

$$C = 99.42 \pm 0.04$$
$$H = 0.83 \pm 0.04$$
$$N = 0.05 \pm 0.03$$
$$Ash = Nil$$

Instrument accuracy is $\pm$ 0.3 w/o for a given element. Hence, one gram of POCO graphite contains 0.091 mol of atoms.

CALORIMETRY

Data were collected in the 80 to 300 K temperature range. A cryostat was modified for using the adiabatic shield technique of calorimetry.[3]

A silicon-diode thermometer (Scientific Instruments, Inc. model Si-400 GG) was used for temperature measure- ments. The thermometer was calibrated against NBS refer- ence standards. Temperatures were calculated from a com- bination of a polynominal curve fit of the calibration data and a deviation plot.

$$T^* = \sum_{i=0}^{4} a_i V^i \qquad (1)$$

$$T = T^* + \Delta T \qquad (2)$$

V is the voltage drop, across the diode for 10 μamp exci- tation current; the a_is are the constants resulting from a least squares curve fit to the calibration data; T^* is the temperature calculated from the measured voltage drop;

ΔT is the correction obtained by plotting the difference between T* and the corresponding calibration temperature against T*; thus T is the corrected temperature accurate to 0.01K. Power inputs of 100 second duration to the sample were adjusted to give temperature increments of the order of 6 to 10 degrees. The heat capacity of the empty calorimeter (addendum) was measured in a separate experiment. The addendum heat capacity is approximately 30 percent of the total measured heat capacity. The same sample of the powdered POCO graphite was measure three times in separate experimental runs. The accuracy and reproducibility of the calorimetric data and the validity of the data reduction scheme were estimated by the determination of the heat capacity of a copper sample. This piece of copper was previously used in establishing the presently accepted low temperature heat capacity curve of copper.[4] Individual experimental points may scatter as much as 10 percent about a least squares curve fit of the heat capacity against temperature. However, copper heat capacities calculated from the least squares equation agreed with the values reported in the NBS tables.[5]

RESULTS

The actual data are presented in Table 1. The smoothed experimental data are shown in Figure 1. The smoothed curve was obtained by a least squares fitting of the data. The three-term polynominal

$$C(T) = 8.729 \times 10^{-4} T + 6.27 \times 10^{-6} T^2 + 6.309 \times 10^{-9} T^3 \pm 0.003 \quad (3)$$

gives an adequate representation of the data. C(T) is ·the heat capacity in J/g· K and T is the temperature.

DISCUSSION

As seen in Figure 1, the heat capacity of the POCO-graphite is greater than the heat capacity of natural graphite, and the temperature dependence is different. From a plot of log C versus log T, the overall temperature dependence of the POCO heat capacity varies as $T^{1.5}$ below 100 K. The heat capacity of natural graphite has a T^2 temperature dependence between 10 and 50 K.[6] This dependence changes toward lower powers of T with increasing

Table 1.   Specific Heat of POCO AXM-5Q1 Graphite

Run #	T (K)	C (J/g·K)	Run #	T (K)	C (J/g·K)
1	82.80	0.1172	3	172.88	0.3500
	89.96	0.1558		180.78	0.4192
	95.38	0.1473		187.52	0.4492
	111.24	0.1802		194.25	0.4594
	118.31	0.2016		209.14	0.5015
	188.02	0.4600		215.71	0.5730
	221.25	0.5498		232.12	0.6387
	228.75	0.5593		238.96	0.6587
	235.04	0.6183		245.67	0.6703
	271.08	0.8175		260.80	0.7600
				269.05	0.8342
2	116.14	0.1829		274.10	0.8180
	125.52	0.2227		279.56	0.8751
	133.89	0.2395		285.04	0.8899
	142.70	0.2654		289.42	0.9554
	152.06	0.2880			
	160.45	0.3077			
	168.02	0.3640			
	175.24	0.3696			
	181.90	0.4232			
	188.81	0.4053			

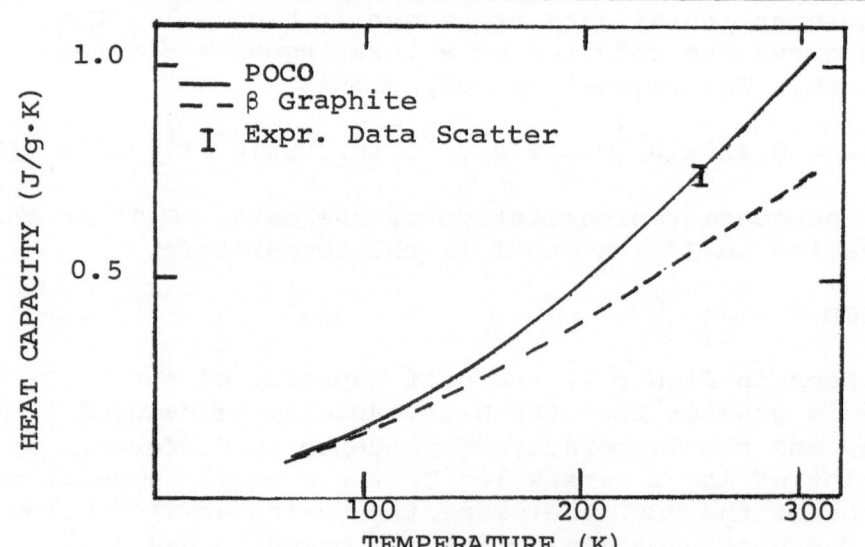

Fig. 1.   Heat Capacity versus Temperature.

temperature. The T^2 range has been explained in terms of the (planar) anisotropy of the interatomic forces of the material.[7] The POCO data must be extended to lower temperatures before definite conclusions can be drawn about the atomic force anisotropy of this material.

It is instructive to compare "effective" Debye temperatures, θ_D, of the POCO graphite with that of a natural graphite. Figure 2 was obtained by inverting the specific

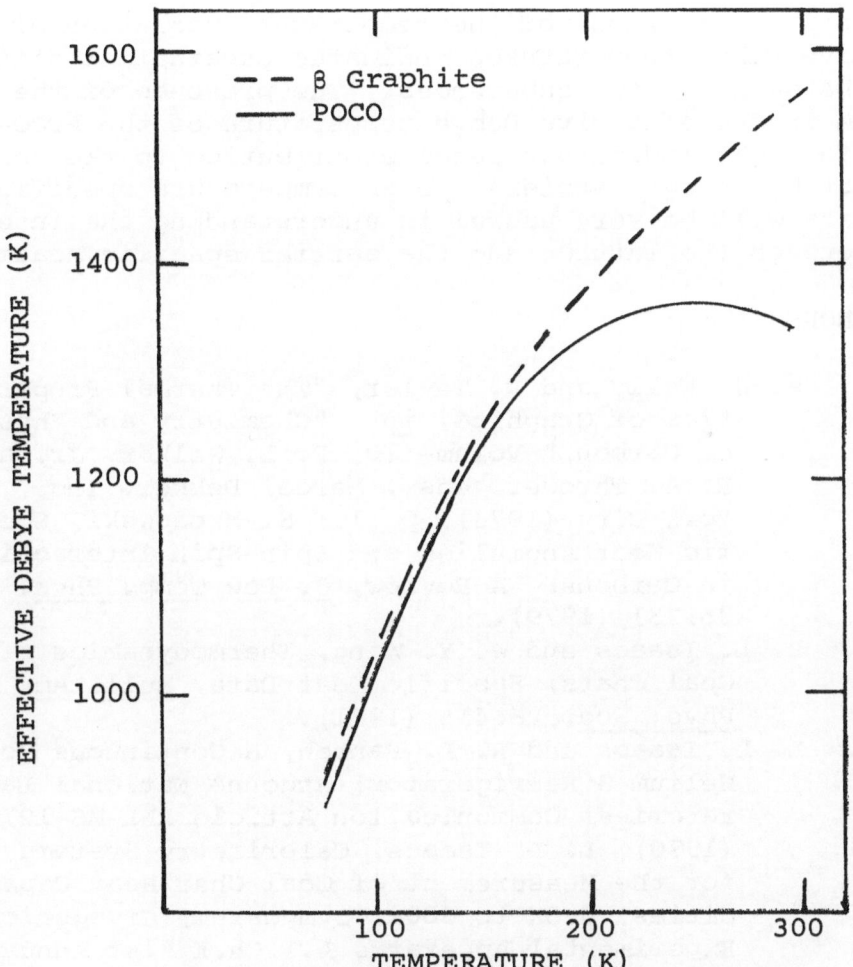

Fig. 2.   Effective Debye Temperature versus Temperature.

heat data with the use of the tabulated three-dimensional Debye integra.[8] This is theoretically incorrect because the specific heat of graphite and graphite-like materials should be given by[7]:

$$C_{v/mol} = f_2 \ (\theta_x/T) + 2f_2 \ (\theta_{xy}/T) \tag{4}$$

where for natural graphite $\theta_z = 950°$, $\theta_{xy} = 2500°$, and $f_2(y)$ is the two-dimensional Debye integral.

$$f_2(y) = \frac{2}{y^2} \int_o^y \frac{x^3 e^x}{(e^x-1)^2} \ dx. \tag{5}$$

Qualitative comparison of the temperature variation of the effective Debye temperatures indicates substantial differences between the two substances. The presence of the maximum in the effective Debye temperature of the POCO-graphite might indicate a large contribution to the specific heat by charge carriers. High temperature specific heat data will be very useful in understanding the interplay between the lattice and the carrier specific heats.

REFERENCES

1. B. J. Kelly and R. Taylor, "The Thermal Properties of Graphite, in: "Chemistry and Physics of Carbon," Volume 10, P. L. Walker, Jr. and R. A. Thrower, eds., Marcel Dekker, Inc., New York City (1973), p. 1.; S. Mrosówski, Specific Heat Anomalies and Spin-Spin Interactions in Carbons: A Review, J. Low Temp. Phys. 35:231 (1979).

2. L. L. Isaacs and W. Y. Wang, Thermodynamics of Coal Chars; Specific Heat Data, Bull. Am. Phys. Soc. 26:415 (1981).

3. L. L. Isaacs and R. L. Panosh, A Continuous Mode Helium-3 Refrigerator; Argonne National Laboratories' Communication Article ANL-MS-1270 (1970); L. L. Isaacs, Calorimetry Systems for the Measurement of Coal Char Heat Capacities; 0.5K to 300K; Symposium, Cryogenic Experimental Apparatus A.I.Ch.E 71st Annual Meeting (1978).

4. D. W. Osborne, H. E. Flotow and F. Schreiner,
    Calibration and Use of Germanium Resistance
    Thermometers for Precise Heat Capacity Mea-
    surements from 1 to 25°K. High Purity Cop-
    per for Interlaboratory Heat Capacity Com-
    parisons, <u>Rev. of Sci. Instr</u>. 38:159
    (1967).
5. G. T. Furakawa, W. G. Shaba and M. L. Reilly,
    NSROS—NBS—18, U.S. GPO, Washington, DC
    (1968).
6. W. Desorbo and W. W. Tyler, The Specific Heat of
    Graphite from 13° to 300°K, <u>J. Chem. Phys</u>.
    21:1660 (1953).
7. J. Krumhansl and H. Brooks, The Lattice Vibra-
    tion Specific Heat of Graphite, <u>J. Chem. Phys</u>.
    21:1663 (1953).
8. K. S. Pitzer, "Quantum Chemistry," Prentice Hall,
    New York City (1954), p. 502.

PHONON CONDUCTION IN ELASTICALLY

ANISOTROPIC CUBIC CRYSTALS

A. K. McCurdy

Worcester Polytechnic Institute

Worcester, Massachusetts 01609

INTRODUCTION

Thermal energy in dielectric solids is carried by phonons. At sufficiently low temperatures the phonons propagate ballistically so that in the absence of defect or impurity scattering, the mean-free path becomes limited by the linear dimensions of the sample.[1-4] A theory of the thermal conductivity applicable to this temperature range was first developed by Casimir.[5] Corrections to Casimir's theory have been derived for samples of finite length[6] and for samples in which a fraction of the phonons are specularly reflected from the end surfaces.[7]

Heat-pulse experiments, however, have shown striking differences (up to factors of 100) in the intensity of phonons propagating ballistically in an elastically anisotropic crystal.[8,9] These results were shown to arise from phonon focusing owing to the fact that in elastically anisotropic crystals the group velocity or direction of energy flow is not, in general, parallel to the phase velocity.[10] Phonon focusing occurs whenever the direction of the group velocity varies more slowly over some small solid angle with wave-vector direction than for an elastically isotropic solid, so that an isotropic distribution of wave vectors as emitted by an ideal heat source gives rise to an increased density in group-velocity space. Since energy flow is in the direction of the group velocity an enhanced energy flow occurs about any strongly focused direction.[10] The strongest phonon focusing occurs along directions where the group-velocity surface exhibits cuspidal edges or cuspidal onset.[10-14]

Calculations for silicon and calcium fluoride showed strong focusing of transverse waves in silicon about the <100> axes and in

63

all directions in {100} planes, and in calcium fluoride about the
<111> axes and in all directions in {110} planes.[15] This is
illustrated in Figs. 1a and b, respectively, where the dark areas
illustrating high intensity are bounded by cuspidal edges in the
group-velocity surfaces. In calcium fluoride the focusing about the
<111> axis is also due to narrow cusps around a collinear axis,
<θ_s>, with respect to the <100> axis in the {110} plane. Angle θ_s
is given in terms of the second-order elastic constants as:[14]

$$\tan \theta_s = [2(C_{11}+C_{12})/(C_{11}+3C_{12}+2C_{44})]^{\frac{1}{2}} \tag{1}$$

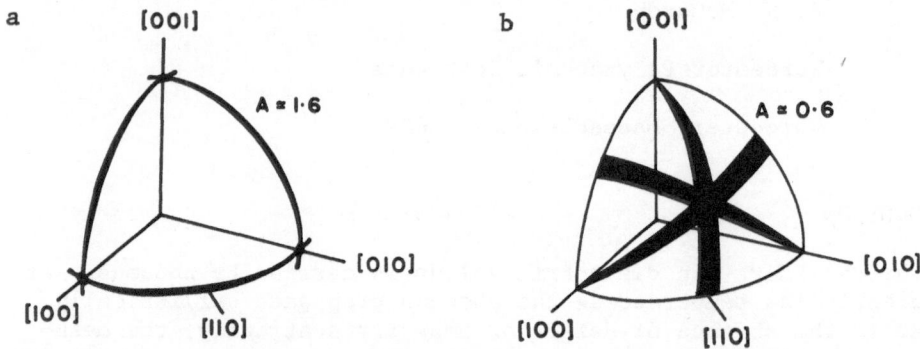

Fig. 1a,b. Simplified schematic diagrams illustrating phonon
            focusing for the two transverse modes of a cubic
            crystal with the contrasting elastic anisotropy factor,
            A, of (a) silicon, (b) calcium fluoride. Directions of
            high phonon intensity are given by constructing lines
            from the origin to the heavy dark areas.

     Subsequent measurements of the thermal conductivity of silicon
and calcium fluoride demonstrated clearly observable anisotropies
at all temperatures below the thermal conductivity maximum.[15,16]
The thermal conduction for samples of square cross section was found
to depend upon the crystallographic orientation of the rod axis,
the variation being as much as 50% for silicon and 40% in calcium
fluoride. For samples of rectangular cross section, the thermal conduction was found to depend, in addition, upon the orientation and
width ratio of the side faces, the variation being as much as 30%
for silicon samples with the same <110> rod axis.[15] The predictions
of Casimir's theory, end-corrected for finite thermal length and
generalized to include phonon focusing, gave quantitative agreement
with experimental results without any adjustable parameters. More
recent measurements of the thermal conductivity of lithium fluoride[17]
and diamond[18] were qualitatively consistent with those expected from
phonon focusing considerations. Similar anisotropies in the thermal
conductivity have been predicted in sufficiently defect-free super-
conducting lead and niobium at $T/T_c \ll 1$.[19]

Phonon focusing effects have also been studied in elastically anisotropic hexagonal,[13] tetragonal and orthorhombic,[14] trigonal and monoclinic crystals.[20] Calculations have recently been published for diamond and silicon,[21] aluminum oxide and α-quartz,[22] and gallium arsenide.[23,24] Effects of phonon focusing have been recently observed in heat-pulse experiments in solid ^{4}He,[25] diffuse and specular phonon reflection experiments in aluminum oxide,[26] imaging of ballistic phonons in germanium,[27,28] and the imaging of phonon distributions in silicon using the fountain pressure of superfluid ^{4}He films.[29] Calculations for silicon and calcium fluoride have recently been generalized to include elastically anisotropic materials of any crystal structure. Results have been obtained for a number of materials including Al_2O_3 and α-quartz.[30]

RESULTS

Calculations of the boundary-scattered phonon conductivity, κ, for elastically anisotropic crystals of thermal length, L, are described elsewhere.[15,16] The thermal length of the sample is defined as the length of the thermal gradient produced along the heat-flow axis in a conventional thermal conductivity measurement, and thus is usually less than the overall length of the sample.* Values of κ for cubic crystals depend upon the second-order elastic constants C_{11}, C_{44}, and C_{12} and thus upon the elastic anisotropy factor, A, defined as:

$$A = 2C_{44}/(C_{11}-C_{12}). \tag{2}$$

Results for samples of square cross section are given in Figs. 2a,b, and c. Values of κ for silicon and calcium fluoride samples of rectangular cross section and varying side-face-width ratio are given in Figs. 3a and b. All results are expressed in dimensionless units as $\kappa/\langle\kappa_c\rangle$ where $\langle\kappa_c\rangle$ is the Casimir thermal conductivity for infinitely long, elastically anisotropic crystals defined here as:

$$\langle\kappa_c\rangle = \frac{1}{3} \langle C_v\rangle\langle v_c\rangle \Lambda_c. \tag{3}$$

In Eq. 3 Λ_c is the Casimir length for rods of infinite thermal length; $\langle C_v\rangle$ and $\langle v_c\rangle$ are the approximate specific heat and Casimir velocity, respectively, for elastically anisotropic crystals. For rods of circular cross section with diameter, D:

$$\Lambda_c = D \tag{4}$$

but for rods of rectangular cross section[15,16] with side-face-width ratio, n, and sides, $D_1 = nD$ and $D_2 = D$:

---

*The thermal length in our original experiments was the distance between the centers of the heat source and heat sink, respectively.

Table 1.   Values of ρS^2 for the longitudinal (L) and the two trans-
            verse (T) modes for the principal directions in cubic
            crystals in terms of the second-order elastic constants.

Axis	ρS^2_L	ρS^2_T	ρS^2_T
<100>	C_{11}	C_{44}	C_{44}
<110>	$(C_{11}+C_{12})/2+C_{44}$	C_{44}	$(C_{11}-C_{12})/2$
<111>	$(C_{11}+2C_{12}+4C_{44})/3$	$(C_{11}-C_{12}+C_{44})/3$	$(C_{11}-C_{12}+C_{44})/3$

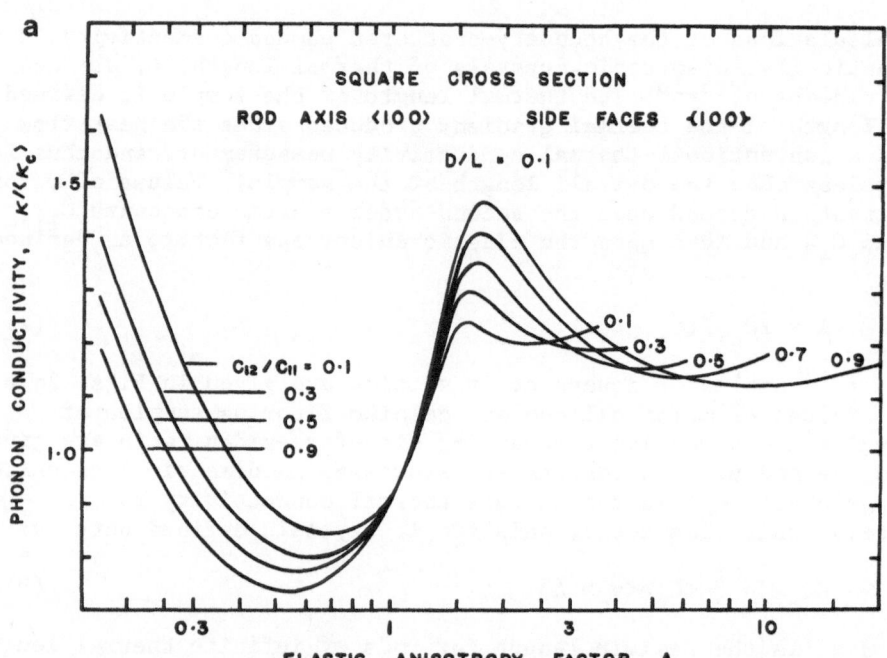

Fig. 2a.   Dimensionless boundary-scattered thermal conductivity,
            $\kappa/\langle\kappa_c\rangle$, for the <100> heat-flow axis as a function of A
            and the elastic constant ratio C_{12}/C_{11}.

$$\Lambda_c = (n^{\frac{1}{2}}D/4)\ \{3n^{\frac{1}{2}}\ \ln[n^{-1}+(n^{-2}+1)^{\frac{1}{2}}]\ +3n^{-\frac{1}{2}}\ \ln[n+(n^2+1)^{\frac{1}{2}}]$$
$$-(n+n^3)^{\frac{1}{2}}\ +n^{3/2}-(n^{-1}+n^{-3})^{\frac{1}{2}}\ +n^{-3/2}\}. \qquad (5)$$

For rods of square cross section, Eq. 5 reduces to:

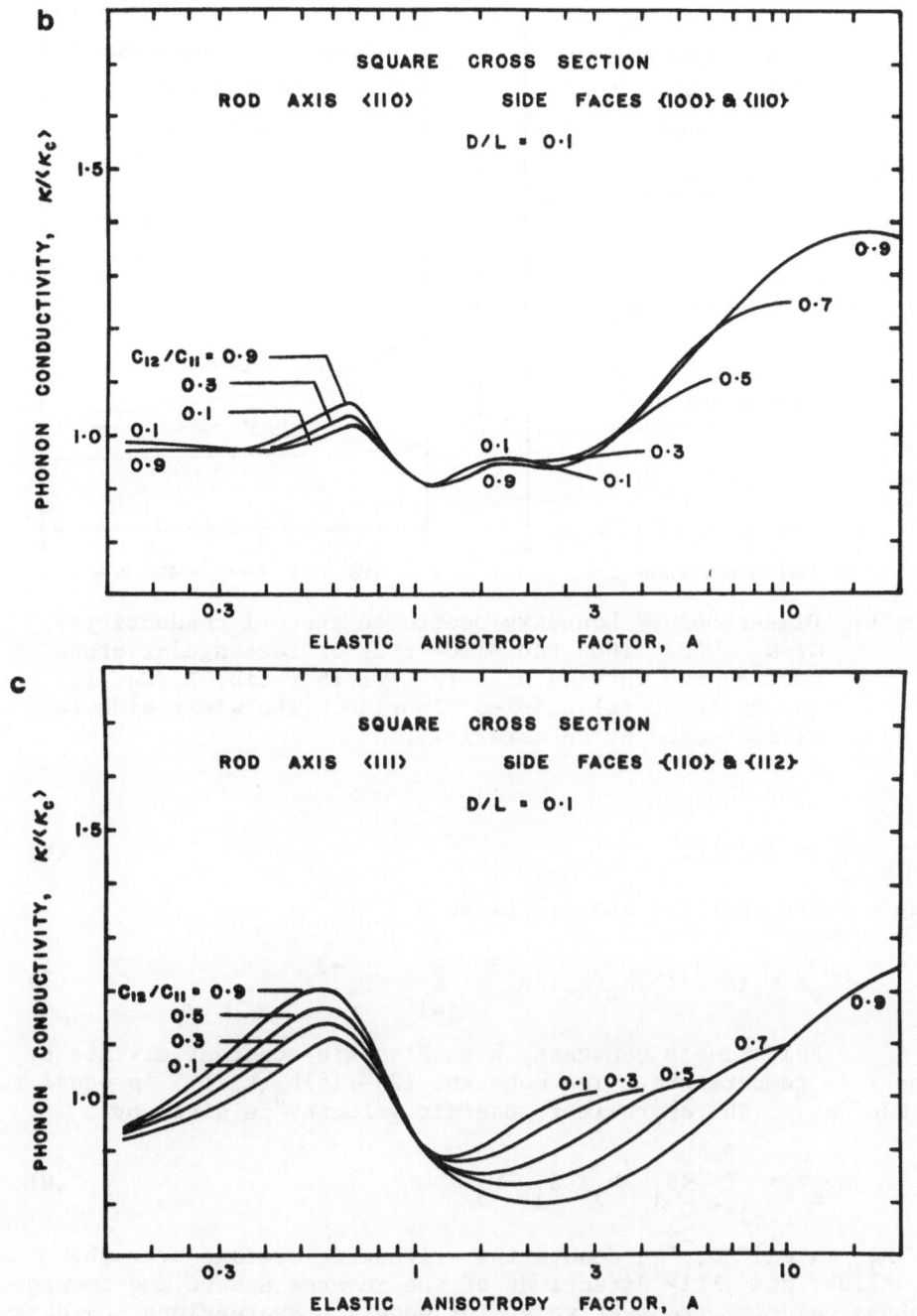

Fig. 2b,c. Dimensionless boundary-scattered thermal conductivity, $\kappa/\langle\kappa_c\rangle$, for the <110> and <111> heat-flow axes, respectively, as a function of A and the elastic constant ratio C_{12}/C_{11}.

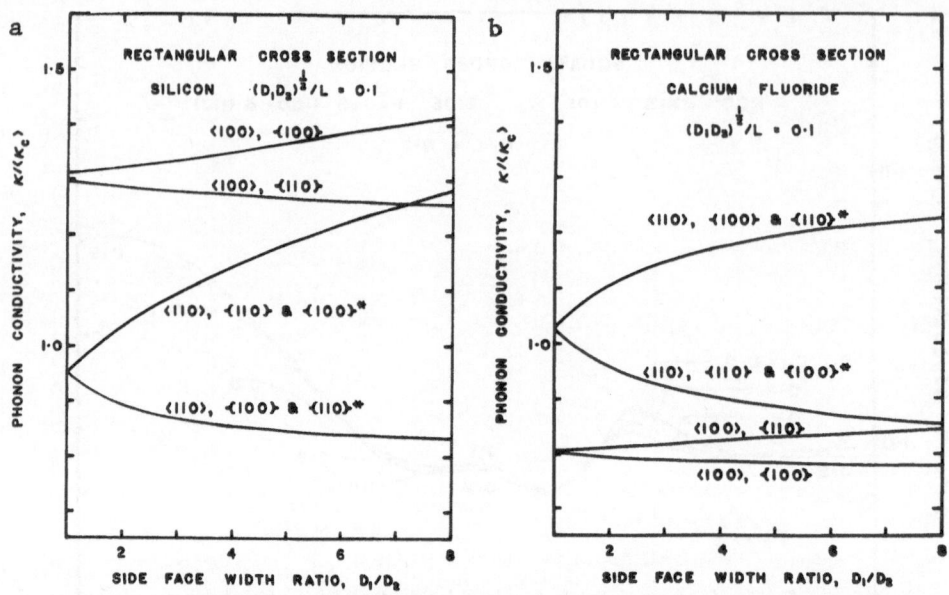

Fig. 3a,b. Dimensionless boundary-scattered thermal conductivity,
$\kappa/\langle\kappa_c\rangle$, for $\langle100\rangle$ and $\langle110\rangle$ rods of rectangular cross
section and varying side-face-width ratio, D_1/D_2, for
(a) silicon, (b) calcium fluoride. The wider side face
is indicated by an asterisk.

$$\Lambda_c = 1.115D. \qquad (6)$$

The approximate specific heat is given by:

$$\langle C_v\rangle = (2\pi^2/15)k_B(k_B T/\hbar)^3 \sum_{j=1}^{3} \langle S_j^{-3}\rangle_3 \qquad (7)$$

where k_B is Boltzmann's constant, $\hbar$ is Planck's constant divided by
2π, and T is temperature. The constant $(2\pi^2/15)k_B(k_B/\hbar)^3$ is equal to
$40.77 \text{ GJK}^{-4}\text{s}^{-3}$. The approximate Casimir velocity is given by:

$$\langle v_c\rangle = \sum_{j=1}^{3} \langle S_j^{-2}\rangle_3/\langle S_j^{-3}\rangle_3 \qquad (8)$$

where $\langle S_j^{-2}\rangle_3$ and $\langle S_j^{-3}\rangle_3$ denote the arithmetic average over the
$\langle100\rangle$, $\langle110\rangle$, and $\langle111\rangle$ directions of the inverse square and inverse
cube phase velocity, respectively, for mode j. Expressions for ρS^2,
where ρ is the density, are given for the three principal axes of
the cubic lattice in Table 1. For elastically isotropic crystals,
$A = 1$ and the end-corrected thermal conductivity $\kappa/\langle\kappa_c\rangle = 0.915$
for $D/L = 0.1$.

     Use of Figs. 2 and 3 and a hand calculator to determine $<\kappa_c>$
yields the correct boundary-scattered phonon conductivity for cubic
crystals. Note that there are no adjustable parameters; the thermal
conductivity, κ/T^3, depends only upon the linear dimensions D_1, D_2,
and L; the sample density, ρ; and the second-order elastic constants.

DISCUSSION

     For elastically isotropic crystals the only correction to the
boundary-scattered thermal conductivity is due to finite thermal
length. A slight increase in the elastic anisotropy factor (1.5>A>1),
however, gives a sharp rise in the thermal conductivity along a
<100> heat-flow axis as a result of the strong focusing of transverse
waves about the <100> direction (see Figs. 1a and 2a). For larger
values of A (20>A>1.6) an increase in A broadens the cusps about the
<100> axes and about the <100> planes decreasing the heat flow along
the <100> rod axis. A decrease in A (0.6<A<1) increases the conduction along a <111> rod axis due to strong focusing of transverse waves
about $<\theta_s>$ and in the {110} planes which intersect along the <111>
directions (see Figs. 1b and 2c). Samller values of A (0.16<A<0.6)
broaden these cusps decreasing the focusing and heat flow along the
<111> rod axis. Silicon (A = 1.566, C_{12}/C_{11} = 0.388) is very near the
maximum for <100> rods in Fig. 2a, and near the minimum for <111>
rods in Fig. 2c. Calcium fluoride, however, (A = 0.609, C_{12}/C_{11} =
0.322) is very near the maximum for <111> rods in Fig. 2c, and near
the minimum for <100> rods in Fig. 2a.

     Note also that for constant A an increase in C_{12}/C_{11} decreases
C_{44}/C_{11} and thus increases the relative contribution of the slower
modes to the heat flow. Since the slower modes are strongly focused
there is an increase in the thermal conductivity maximum with C_{12}/C_{11}
for <100> and <111> axis rods (see Figs. 2a and c). For small A and
C_{12}/C_{11} there is strong focusing of longitudinal waves about the
<100> axis. A decrease in C_{12}/C_{11} at constant A increases the
relative contribution of longitudinal waves giving a rapid rise in
thermal conductivity along a <100> heat-flow axis for smaller values
of A. For A>4 some focusing of longitudinal waves occurs about the
<110> and <111> directions, but the major contribution to the increase
in $\kappa/<\kappa_c>$, however, in Figs. 2b and c is due to corrections for $<C_v>$
and $<v_c>$ which become more significant for larger values of A.

     Phonon focusing occurs along the <110> directions in both silicon
and calcium fluoride; for silicon the focusing is concentrated in
{100} planes, for calcium fluoride in {110} planes (see Fig. 1). This
shows why silicon samples of rectangular cross section had greater
conductivity when the wider side face was oriented in the plane of
high phonon intensity.[15,16] Results for silicon and calcium fluoride
samples of rectangular cross section are given in Figs. 3a and b,
respectively. Higher conductivities are predicted for the same rod
axis when the wider side face is oriented in the plane of high phonon

intensity, the amount of anisotropy depending on the rod axis and the width ratio of the side faces. The greatest differences occur for rods aligned along the <110> directions.

REFERENCES

1. J.M. Ziman, "Electrons and Phonons," Oxford, London (1960).
2. P. Carruthers, Rev. Mod. Phys. 33:92 (1961).
3. P.G. Klemens, in:"Solid State Physics," Vol. 7, p. 1, F. Seitz and D. Turnbull, eds., Academic Press, N.Y., (1958).
4. P.G. Klemens, in:"Thermal Conductivity," Vol. 1, p. 1, R.P. Tye, ed., Academic Press, N.Y., (1969).
5. H.B.G. Casimir, Physica (Utr.) 5:495 (1938).
6. R. Berman, F.E. Simon, and J.M. Ziman, Proc. R. Soc. Lond. A220:171 (1953).
7. R. Berman, E.L. Foster, and J.M. Ziman, Proc. R. Soc. Lond. A231:130 (1955).
8. B. Taylor, H.J. Maris, and C. Elbaum, Phys. Rev. Lett. 23:416 (1969).
9. B. Taylor, H.J. Maris, and C. Elbaum, Phys. Rev. B3:1462 (1971.
10. H.J. Maris, J. Acoust. Soc. Am. 50:812 (1971).
11. M.J.P. Musgrave, Proc. Camb. Philos. Soc. 53:897 (1957).
12. M.J.P. Musgrave, "Crystal Acoustics," Holden-Day, San Francisco (1970).
13. A.K. McCurdy, Phys. Rev. B9:466 (1974).
14. C.G. Winternheimer and A.K. McCurdy, Phys. Rev. B18:6576 (1978).
15. A.K. McCurdy, "Ph.D. Thesis," Brown Univ. (1971), Univ. Microfilms, Inc., Ann Arbor, Michigan, Order No. 72-12,047.
16. A.K. McCurdy, H.J. Maris, and C. Elbaum, Phys. Rev. B2:4077 (1970).
17. A.C. Anderson and M.E. Malinowski, Phys. Rev. B5:3199 (1972).
18. J.W. Vandersande, Phys. Rev. B13:4560 (1976).
19. C.G. Winternheimer and A.K. McCurdy, Solid State Commun. 14:919 (1974).
20. A.K. McCurdy, in:"Phonon Scattering in Condensed Matter," p. 341, H.J. Maris, ed., Plenum, N.Y., (1980).
21. F. Rösch and O. Weis, Phys. B25:101 (1976).
22. F. Rösch and O. Weis, Phys. B25:115 (1976).
23. M. Lax and V. Narayanamurti, Phys. Rev. B22:4876 (1980).
24. J. Philip and K.S. Viswanathan, Phys. Rev. B17:4969 (1978).
25. V. Narayanamurti and R.C. Dynes, Phys. Rev. B12:1731 (1975).
26. P. Taborek and D. Goodstein, Phys. Rev. B22:1550 (1980).
27. G.A. Northrop and J.P. Wolfe, Phys. Rev. Lett. 43:1424 (1979).
28. G.A. Northrop and J.P. Wolfe, Phys. Rev. B22, 6196 (1980).
29. W. Eisenmenger, in:"Phonon Scattering in Condensed Matter," p. 303, H.J., Maris, ed., Plenum, N.Y., (1980).
30. A.K. McCurdy, at the Third International Conference on Phonon Scattering in Condensed Matter, Brown Univ., Aug. 28-31, (1979).

SESSION   TC-3

Standard Reference Materials and Data - I

CHAIRMAN

G. K. White
CSIRO
Sydney, Australia

THE STANDARD REFERENCE MATERIALS AND DATA PROGRAMS OF THE

CODATA TASK GROUP ON THERMOPHYSICAL PROPERTIES

Merrill L. Minges

Electromagnetic Materials Division
Materials Laboratory
Air Force Wright Aeronautical Laboratories
Dayton OH

INTRODUCTION

Three of the five principal CODATA Task Group program areas
shown in Figure 1 pertain to standard reference materials and data:
Areas I, II, and IV. The Area I program deals with an evaluation
of experimental methods over a very broad temperature range
(4-2500 K) and includes a number of projects on standard reference
materials (SRMs). In the geosciences field (Area IV), attention
has been recently focused on a survey of SRM candidates that
cover the lower thermal conductivity range. Finally, critical
data analysis is underway on six thermophysical properties of eight
different materials (Area II). Two SRMs are included in this
critically evaluated reference data project, although the criteria
for selection of the materials studied under the two areas are
different.

The objective of this paper is to summarize progress in these
three Task Group project areas that is pertinent to the Thermal
Conductivity Conference.

The Task Group on Thermophysical Properties of Solids was
established by the General Assembly of the International Council
of Scientific Unions Committee on Data for Science and Technology
(CODATA). Formal procedures for organizing the Task Group were
initiated during the Second International CODATA Conference in
St. Andrews, Scotland in 1970 by the Task Group's founding chairman, Prof. Y. S. Touloukian of Purdue University, who was Director of the Center for Information and Numerical Data Analysis and

AREA	PROJECT MANAGEMENT
I. EVALUATION OF EXPERIMENTAL METHODS AND TECHNIQUES/STANDARDS REFERENCE MATERIALS FOR THERMAL AND ELECTRICAL TRANSPORT PROPERTIES.	DR. R. BERMAN: CRYOGENIC TEMPERATURE DR. M.L. MINGES } ORDINARY AND HIGH DR. J.G. HUST } TEMPERATURES
II. PREPARATION OF INTERNATIONALLY ACCEPTED TABLES OF CRITICALLY EVALUATED REFERENCE DATA	PROF. Y.S. TOULOUKIAN DR. G.K. WHITE PROF. F. CABANNES DR. J.G. HUST
III. THE ORGANIZATION OF THERMOPHYSICAL PROPERTIES OF BIOLOGICAL MATERIALS WITH INITIAL EMPHASIS ON FOODS.	MR. R. JOWITT
IV. THE ORGANIZATION OF THERMOPHYSICAL PROPERTIES OF GEOLOGICAL MATERIALS	PROF. A.E. BECK
V. GUIDE PREPARATION	ALL MEMBERS

Figure 1 - Project Management Responsibility

Synthesis until his untimely death in June 1981. The Task Group was formally established by the CODATA General Assembly in Le-Creusot, France, in 1973 and held its first meeting in Warsaw, Poland in 1974.[1] Reflecting the broadened scope of Task Group projects in recent years, the title was changed in 1980 to encompass "Thermophysical Properties." The Task Group is charged with the responsibility of promoting, on an international basis, the availability of reliable reference data on the thermophysical properties of solids in the physical, earth, and life sciences. Formation of the Task Group was motivated, in large measure, by the needs in both science and engineering to increase the breadth of coverage and to improve the reliability of thermophysical information available on a worldwide basis. Implicit in the reliability goal were the needs of better organization and critical evaluation of the information already at hand and guidance in the generation of new information (e.g., optimized use of measurement techniques and the development and improvement of standard reference materials for calibration).

STANDARD REFERENCE MATERIALS

Background

Quantitative evaluation of experimental methods for property measurement is closely coupled to appropriate SRMs and often is successful only when conducted in conjunction with them. As described elsewhere,[2,3] Task Group Project I was organized to con-

sider a systematic, simultaneous study of measurement techniques
and SRMs. As outlined in Reference 2, attention has been focused
on materials with relatively high thermal conductivity. Recently,
at least partially motivated by needs in the geosciences area
(Task Group Project IV), consideration is being given to an evaluation of SRM candidates in a lower conductivity range. The
following section summarizes the current state of affairs on the
principal materials under study: iron, tungsten, austenitic
stainless steel and polycrystalline graphite. Then, briefly, the
near term plans in the lower conductivity area are summarized.

Iron

Iron has been studied as a reference material for nearly
fifty years. Most of the earlier work on a material in the medium

1934 - 1970

MANY THERMAL TRANSPORT PROPERTY MEASUREMENTS, CORRELATIONS AND
ROUND ROBINS ON INGOT IRON ("ARMCO" IRON)

TEMPERATURE: 100 TO 1600K   RRR: 9 TO 14

1966

ORNL MEASUREMENTS ON HIGH PURITY IRON
TEMPERATURE: 300 TO 1200K   RRR: 23

1971 - 1974

NBS ESTABLISHED ELECTROLYTIC IRON AS LOW TEMPERATURE THERMAL
CONDUCTIVITY (SRM 734) AND ELECTRICAL RESISTIVITY STANDARDS (SRM 797)
TEMPERATURE: 4 TO 300K   RRR: 21.0 TO 23.5

PROPOSED EXTENSION TO 1000K BASED ON ORNL MEASUREMENTS

CERTIFICATION: 4-280K: 2.5%, 280 - 1000K: 3%

1978 - 1981

MEASUREMENTS UNDER THE AUSPICES OF CODATA WITH THE FOLLOWING
OBJECTIVES:

1) CONFIRM AND BROADEN THE LOW TEMPERATURE CERTIFICATION BY
QUANTIFYING RRR - THERMAL CONDUCTIVITY RELATIONSHIP

2) MAKE SRM SPECIMENS AVAILABLE FOR MEASUREMENTS ABOVE ROOM
TEMPERATURE

Figure 2   Iron: Thermal Conductivity Chronology
of Investigations

- SRM CERTIFICATION ENCOMPASSES MATERIAL IN RRR RANGE: 20-24

- DIFFERENCES IN RRR BETWEEN NBS AND ORNL MATERIAL HAVE LESS THAN 0.2% EFFECT ON THERMAL CONDUCTIVITY AT ROOM TEMPERATURE AND ABOVE

- LOW TEMPERATURE MEASUREMENTS (4-90K) COMPLETED IN EUROPE AND AUSTRALIA CORRELATED WITH AND CONFIRM NBS CERTIFICATION

- AT LOW TEMPERATURES SRM SPECIMEN THERMAL CONDUCTIVITY IS A FUNCTION OF HEAT TREATMENT/ANNEAL HISTORY:

  - THIS DEPENDENCE IS ACCURATELY DESCRIBED VIA THE ELECTRICAL RESISTIVITY
  - FOR MATERIAL WITH RRR FROM 20 TO 24, PROCEDURE ESTABLISHED FOR CALCULATING THERMAL CONDUCTIVITY
  - THE IMPORTANCE OF THIS EFFECT DIMINISHES RAPIDLY ABOVE ABOUT 60K (WHERE IT IS APPROXIMATELY 1%)

Figure 3   Iron: Thermal Conductivity Current Status

purity range (Armco Iron) has been summarized by Lucks.[4] As indicated in Figure 2, in recent years data have been reported on three different types of iron most readily distinguished by their range of residual resistivity ratio (RRR): high purity iron (RRR 23), NBS Electrolytic Iron (RRR: 21 to 24), and Armco Iron (RRR 9 to 14). Task Group interest centered around key issues on the NBS certified SRM.[5] The main concerns were (1) confirming the certification below room temperature through a limited number of high accuracy measurements; this was judged to be necessary since the primary data set from which the certification was developed consisted of results from only one laboratory and (2) examining the influence of heat treatment and annealing history on the low temperature conductivity and determining whether electrical transport measurements could be used to describe these influences quantitatively. As summarized in Figure 3, these issues were successfully resolved. Details of this recent work are reported elsewhere in these proceedings by Berman, Hardy, Sahota, Hust, and Tainsh.[6]

Recommendations for further studies on the iron are of limited scope. Those considered to be of particular importance are:

1969 - 1972

EXTENSIVE MEASUREMENTS UNDER AGARD AUSPICES: 4 TO 1500K

1972 - 1974

EXCELLENT LOW TEMPERATURE FEATURES OBSERVED (THAT IS, INSENSITIVITY TO
MECHANICAL AND THERMAL HISTORY AND MATERIALS VARIABILITY EFFECTS ON
TRANSPORT PROPERTIES OF 1% OR LESS)
ESTABLISHED BY NBS AS A LOW TEMPERATURE THERMAL CONDUCTIVITY SRM
(1972) AND ELECTRICAL RESISTIVITY SRM (1974): 5 TO 280K

1975

BASED ON AGARD RESULTS, ESTABLISHED BY NBS AS A HIGH TEMPERATURE
THERMAL CONDUCTIVITY AND ELECTRICAL RESISTIVITY SRM: TO 1200K

1977 - 1981

MEASUREMENTS UNDER THE AUSPICES OF CODATA WITH THE FOLLOWING
OBJECTIVES:

   1) CONFIRM THERMAL CONDUCTIVITY CERTIFICATION ABOVE ROOM
      TEMPERATURE

   2) VERIFY PERFORMANCE OF LOW TEMPERATURE EXPERIMENTAL
      METHODS AND CONFIRM CERTIFICATION BELOW ROOM TEMPERATURE

   3) REMEASURE THERMAL DIFFUSIVITY IN THE TEMPERATURE RANGE WHERE
      THE SRM IS FULLY STABLE

   4) MAKE SRM SPECIMENS AVAILABLE INTERNATIONALLY FOR EXPERIMENTAL
      METHODS EVALUATION

Figure 4   SRM 735: Austenitic Stainless Steel --
Chronology of Investigations

1. Preparation of a brief but quantitative
guide in terms of RRR to show the accuracy
bounds expected when using different iron
specimens as standards. This is needed because, in practical terms, iron materials
other than the NBS Electrolytic Iron Standard have been and will continue to be used
as SRMs (especially Armco iron and possibly
other iron materials in Europe).

2. Consideration of electrolytic iron as a
thermal diffusivity SRM. As a first step,
this should include an evaluation of the historical thermal diffusivity data base on iron,
which is surprisingly broad, and a critical
analysis of the heat capacity of iron in order

to develop the necessary quantitative interrelationships with the thermal conductivity
SRM certification. Iron is a promising diffusivity SRM candidate from a number of views,
including possible certification across the
α-γ transition.

Stainless Steel

The particular austenitic alloy that was eventually certified
as an SRM_8 by NBS was evaluated initially under a program sponsored
by AGARD.[8] Both the AGARD and the NBS lots of material were
supplied by the German manufacturer Deutsche Edelstahlwerke, Kro-
feld, Federal Republic of Germany. Although difficulties were encountered with material stability at high temperatures (above
1400 K), the excellent low temperature homogeneity, interlot uniformity, and transport property insensitivity to mechanical working
and annealing motivated the studies that resulted in SRM certification. The chronology of developments is summarized in Figure 4.

Experimental measurements of the thermal conductivity both at
cryogenic temperatures and above room temperature were undertaken
as part of the Task Group program to confirm the certification
recommendation. Although some unexpectedly high data scatter

• AT VERY LOW TEMPERATURES (4 TO 30K) EXPERIMENTAL MEASUREMENT ERRORS
  LEAD VERY SURPRISINGLY TO THERMAL CONDUCTIVITY DATA SCATTER CONSIDERABLY
  OUTSIDE CERTIFICATION BOUNDS.

    -MEASUREMENTS IN EUROPE AND AUSTRALIA (ONE SPECIMEN/FOUR LABORATORIES):
        -DIFFERENCES OF UP TO ± 10% BELOW 30K
        -WITHIN CERTIFICATION LIMITS ABOVE 30K (TO 90K)
    MEASUREMENTS AT NBS (SEVERAL SPECIMENS/ONE APPARATUS):
        -DIFFERENCES OF UP TO ± 8% BELOW 10K
        -WITHIN CERTIFICATION BOUNDS ABOVE 10K

• BASED ON THE CODATA PROGRAM RESULTS, BOTH ABOVE AND BELOW ROOM TEMPER-
  ATURE, THE FOLLOWING THERMAL CONDUCTIVITY CERTIFICATION IS RECOMMENDED:
    4 TO 100K: 2% UNCERTAINTY
    100 TO 300K: 3% UNCERTAINTY
    300K TO 1000K: 5% UNCERTAINTY

• FURTHER THERMAL DIFFUSIVITY MEASUREMENTS (ALONG WITH CONFIRMATORY HEAT
  CAPACITY MEASUREMENTS) WOULD BE HELPFUL IN QUALIFYING THE PERFORMANCE
  OF THIS MATERIAL AS A DIFFUSIVITY SRM.

Figure 5   SRM 735: Austenitic Stainless Steel –
Current Status

was observed below room temperature, as indicated in Figure 5, these difficulties were traceable to instrumentation factors in the individual apparatus. Thus, no modification was suggested in the certification bounds at low temperatures. Further details on these measurements are reported elsewhere in these proceedings by Berman et al.[6]

Above room temperature only a very limited number of direct thermal conductivity measurements have been performed. Thus, the recent Oak Ridge National Laboratory results generated under the Task Group project by Graves and Moore should be added to the primary data set used to establish the SRM certification above room temperature. These new results fall within the SRM certification bounds shown in Figure 5; however, because they fall consistently in the lower part of the envelope, it is likely that above 400 K the least-squares representation of the certification curve will drop by about 0.5 to 1.0%.

Thermal diffusivity results reported earlier by seven different investigators under the AGARD program were scattered owing at least in part to material instability at high temperatures. If measurements were limited to a maximum of about 1200 K, it would be expected that this conductivity SRM would serve very well as a thermal diffusivity SRM also. As indicated in Figure 5, two tasks of relatively modest scope are required:

> 1. Heat capacity measurements to develop the quantitative interrelationship between the SRM conductivity certification and the diffusivity.
>
> 2. A limited number of direct thermal diffusivity measurements to no more than 1200 K to develop a primary data set and confirm the thermal stability of the austenitic alloy in this range.

Tungsten

Tungsten was selected for evaluation as an SRM because of its very broad temperature range of potential use. Although the focus of the experimental methods measurement program was developing the numerical thermal conductivity data base and determining its practical suitability in different types of measurement apparatus, there were other special areas of concern:

> 1. Ensuring that physically sound specimens could be obtained from either arc-cast or sintered billet stock; this was an especially important factor for this low-ductility, oxi-

1969 - 1972

> EXTENSIVE MEASUREMENTS UNDER AGARD AUSPICES: 300 TO 3600K
> MOST DATA ON A EUROPEAN SINTERED TUNGSTEN
> SOME DATA ON US-SUPPLIED ARC CAST TUNGSTEN

1975

> BASED ON AGARD, NBS AND ORNL PRIMARY DATA SETS ESTABLISHED BY
> NBS AS A BROAD TEMPERATURE RANGE SRM (4 TO 3000K)
> NBS SINTERED TUNGSTEN (3.2 - 6.4 mm D)
> ARC CAST TUNGSTEN (6.4 TO 12.7 mm D)

1977 - 1981

> MEASUREMENTS UNDER THE AUSPICES OF CODATA WITH THE FOLLOWING
> OBJECTIVES
> > 1) RESOLVE HIGH TEMPERATURE HEAT CAPACITY ISSUES
> > 2) CONFIRM SRM CERTIFICATION ABOVE ROOM TEMPERATURE
> > 3) CONFIRM CERTIFICATION BELOW ROOM TEMPERATURE AND EXAMINE
> >   CONDUCTIVITY / RRR RELATIONSHIP
> > 4) RESOLVE DIFFERENCES BETWEEN NUMEROUS RECOMMENDED VALUE
> >   CURVES AT HIGH TEMPERATURES
> > 5) EXPLORE FEASIBILITY OF USE AS A DIFFUSIVITY SRM

Figure 6   SRM 730:   Tungsten – Chronology
of Investigations

dation-susceptible, high-melting-point material.

2.   Describing the low temperature thermal
conductivity in terms of chemical and physical impurity effects, as for any pure metal
SRM.

3.   Obtaining high accuracy heat capacity and
electrical resistivity data for use in correlating the thermal transport property results and
evaluating the material as a potential thermal
diffusivity SRM.

Addressing the first special concern: three different
lots of material were evaluated, as summarized in Figure 6,
sintered material from Metallwerk Plansee in Austria[8] and the
National Bureau of Standards[10] and arc-cast material from the
Air Force Materials Laboratory.[11]   The material with the best

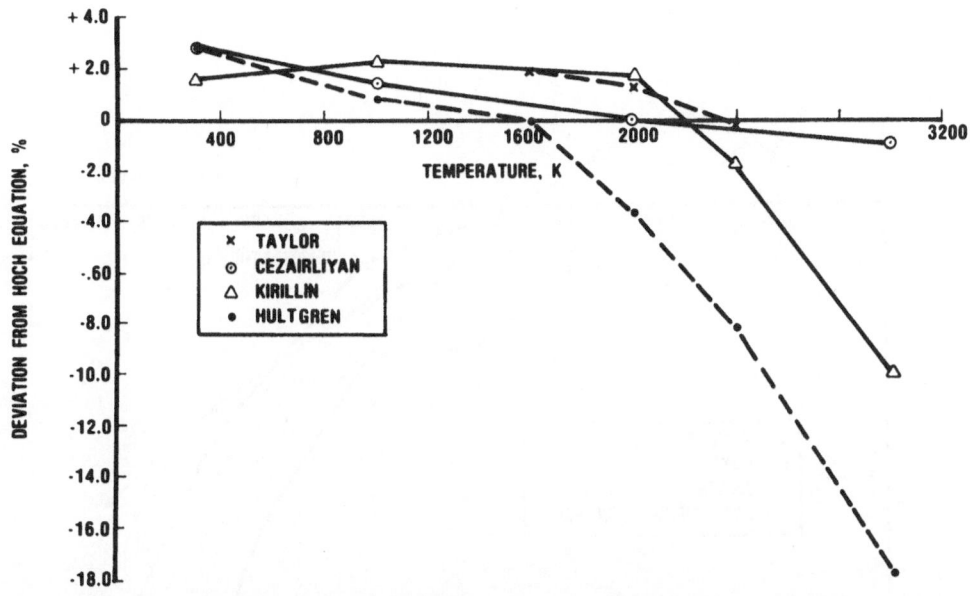

Figure 7   Heat Capacity of Tungsten

physical integrity was the NBS sintered tungsten. The AFML arc-cast tungsten was included in the certification since it was available in larger diameters than the NBS material.

The other two special concerns, together with thermal conduc-tivity certification confirmation, were the focus of activities initiated under the auspices of the Task Group, as summarized in Figure 6. Heat capacity data on tungsten above about 1600 K was found to fall into three groupings, as discussed in Reference 11 and as illustrated in Figure 7: the lowest values were represen-ted by extrapolation of the Hultgren formulation[12] to high temper-atures; the highest values were represented by the work of Cezair-liyan and McClure of NBS;[13] values between these were represented by the work of Kirillin et al.[14] Taylor[15] has recently confirmed Cezairliyan and McClure's work; thus these two sets are recom-mended as the baseline heat capacity relationship for intercompar-ison of thermal conductivity and diffusivity data. The analytical formulation of Cezairliyan and McClure and one given by Hoch[16] are both theoretically based. However, use of the Cezairliyan and McClure formulation is recommended since it is more consistent with the Hultgren expression at lower temperature (Figure 7).

Considering the thermal conductivity below room temperature, measurements from the Task Group project are described by Berman et al.[6] These data obtained below 100 K on specimens with

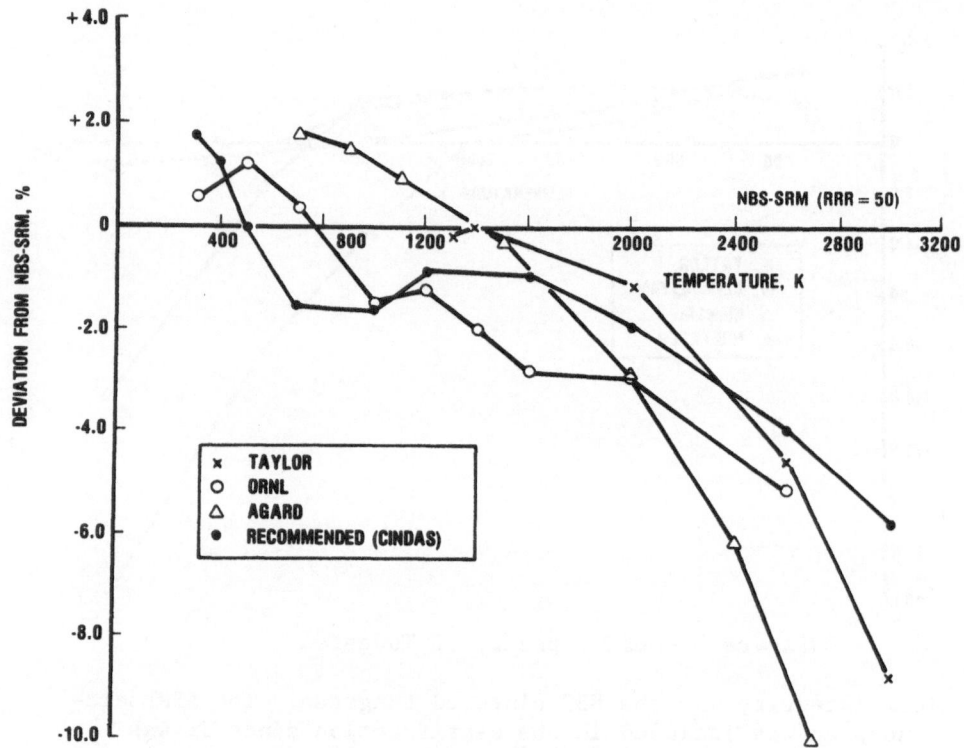

Figure 8   Thermal Conductivity of Tungsten

RRR = 72.6 and 75.4 confirm the NBS certification of Hust and
Giarratano.[10]   Both Berman and Hust give the relationship for
λ = f (RRR); although slightly different in form, either can be
used to develop λ values for particular specimens within the
bounds of the NBS certification. It should be noted that for very
high purity material, the Berman formulation is consistent with
Oak Ridge National Laboratory (ORNL) results on electron-beam-
melted tungsten (RRR = 4900),[17] whereas the CINDAS recommendation
of 1974 for material with RRR = 2906[18] is considerably different
from either expression.

     At high temperature, recent expansion-corrected measurements
by Taylor[15] to 3000 K suggest that above about 1200 K the SRM
certification should be revised downward. This would effectively
be accomplished if the recent Taylor data were included in the
primary data set and if the earlier converted thermal diffusivity
data from the AGARD program (which are now included in the primary set) were given less weight. This situation is shown in
the deviation plot, Figure 8. The rationale for assigning lesser
weight to the AGARD program results recognizes the difficulty in

HEAT CAPACITY
- **RECOMMENDED VALUES PROPOSED: RT TO 3200K**

THERMAL DIFFUSIVITY
- **NOT YET RECOMMENDED FOR SRM CERTIFICATION**

THERMAL CONDUCTIVITY (SRM 730)
- **ABOVE ROOM TEMPERATURE**
  CHANGE PRIMARY DATA SET, ELIMINATING CONVERTED
     DIFFUSIVITY DATA
  PRIMARY DATA SET MAY INCLUDE MATERIALS WITH
     FOLLOWING RRR RANGE: 30-80. SIGNIFICANT λ (RRR)
     DEPENDENCE ONLY BELOW 200K
  ABOVE 1200K NEW SRM CURVE SUGGESTED

- **BELOW ROOM TEMPERATURE**
  EXTENSIVE MEASUREMENTS AT NBS: 6 TO 280K
       SINTERED AND ARC CAST MATERIAL: RRR 40-76
  MEASUREMENT IN ENGLAND AND AUSTRALIA (4-90K) ON SAME
       SINTERED MATERIAL SPECIMEN: RRR 72.6-75.3
  PROCEDURE DEVELOPED FOR CALCULATING THERMAL
       CONDUCTIVITY FROM MEASURED RRR (50-100)
  CONFIRMATION OF NBS CERTIFICATION WITHIN 1-3%

Figure 9   SRM 730:   Tungsten – Current Status

obtaining high precision diffusivity data on tungsten, especially using the laser flash method, and the possible questionable quality of some of the specimens. Below about 1200 K, recent results do not suggest that the certification be altered.

Because there is reasonable confidence in the error bounds on the thermal conductivity and heat capacity of tungsten at high temperature as discussed above, a good estimate of the thermal diffusivity could be made. Thus, it is recommended that tungsten be reconsidered as a diffusivity SRM. A key action required would be the generation of additional diffusivity data on selected specimens where:

1. The physical integrity of the specimens can be assured (a historic problem) and

2. The measurement system is configured to
produce high precision and moderate to high
accuracy results.

The observations are summarized in Figure 9.

Polycrystalline Graphite

The AXM-5Q1 polycrystalline graphite, produced by POCO Graphite, is the candidate SRM material on which recent measurement
results are probably more extensive than on any other SRM. This
situation presents an important opportunity to determine exactly
where we stand on a graphite SRM; it also presents some perplex-
ity. Many of the important recent results are presented formally
for the first time elsewhere in this proceedings and thus a
full analysis bringing together all these new data cannot yet be
undertaken. Thus, this section must necessarily be brief and
unfortunately incomplete.

The chronology of activity is summarized in Figure 10. The
practical reasons for considering an isotropic but polycrystalline graphite as an SRM candidate have been summarized elsewhere.[3]
Polycrystalline graphites are really physical mixtures of graphitic
particles in a graphitized binder matrix, which may be homogeneous
and isotropic on one scale but heterogeneous on another. Thus an
important series of tasks in addition to transport and thermodynamic
property measurements has been the study of lot variability and
methods of quantifying it. The principal SRM issue is not that
variability exists but whether the thermal conductivity of a given
specimen can be determined in terms of relatively simple, easily
measured indices (e.g., density and electrical resistivity) and
whether the total lot variability (e.g., interbillet and intra-
billet) is within reasonable limits.

Progress in understanding and revolving these issues has been
good. Hust[19] at NBS has found a strong correlation between the
density and the Lorenz ratio of POCO specimens at room temperature.
This relationship, which gives the room temperature conductivity
with ±2%, covers a range of individual specimen resistivity of
about 20%, a density range of about ±7%, a thermal conductivity
range of about ±25%, and specimen annealing conditions to 3300 K.

High precision thermal property measurements, which have been
reported on the material at elevated temperatures (e.g., Taylor and
Groot[20]), will be used as the basis for extension of this correlation to high temperatures. These results exemplify one of the
important features of this candidate SRM: ease of measurement,
which is manifest as high precision data, allows the interspecimen
"fine structure" in transport property differences to be delineated
and described. Further progress will come from analyses of the

1968 - 1969
- 26 MATERIALS EVALUATED AS POTENTIAL HIGH TEMPERATURE SRMs: TWO GRAPHITE AMONG THESE; MEASUREMENTS CONDUCTED IN 6 LABORATORIES IN THE US

1970 - 1974
- INTERNATIONAL ROUND ROBIN UNDER THE AUSPICES OF AGARD
- 9 PARTICIPATING LABORATORIES

1975 - 1977
- OUTLINE OF SPECIFIC AREAS FOR FURTHER STUDY
- PURCHASE AND EVALUATION OF NBS LOT OF MATERIAL

1978 - 1981
- FURTHER MEASUREMENTS UNDER THE AUSPICES OF CODATA WITH THE FOLLOWING OBJECTIVES:
  1) EVALUATION OF PHYSICAL AND CHEMICAL STABILITY TO TEMPERATURES ABOVE 3000K
  2) DETERMINE USEFULNESS OF CLASSICAL NDI ANALYSES IN SPECIMEN CHARACTERIZATION
  3) EXPLORE THERMAL CONDUCTIVITY/THERMAL DIFFUSIVITY RELATIONSHIP
  4) EXPLORE THERMAL/ELECTRICAL TRANSPORT PROPERTY AND DENSITY RELATIONSHIPS
  5) MAKE SPECIMENS AVAILABLE INTERNATIONALLY FOR MEASUREMENT /SPECIMEN INTERCOMPARISONS

Figure 10   Polycrystalline Graphite SRM Candidate – Chronology of Investigations

results included in these proceedings [21-26] and from other program participants who are expected to complete their thermal conductivity and thermal diffusivity measurements in the near future.

Another important development has been the reconciliation of heat capacity data on POCO graphite based on relatively recent results of Taylor and Groot[20] and Cezairliyan and Righini.[27] Although the lower temperature results of the former obtained via differential scanning calorimetry do not overlay the high temperature results of the latter obtained using a pulse heating method, the reported data accuracies are excellent and the two sets of results join smoothly. Hence, there is confidence that the heat capacity of this material has been established within reasonable bounds. Specifically, on the basis of the Taylor-Groot inaccuracy estimate of 2-3% and the Cezairliyan-Righini estimate of 3%, it is projected that the heat capacity is known to within 3% over the full range from 300 to 3000 K. These results will be especially useful where quantitative comparisons of the thermal diffusivity and thermal conductivity data are undertaken in the near future.

Low Conductivity SRMs

Under the direction of Prof. F. Cabannes, CNRS, Orleans, France, a feasibility study was undertaken to explore whether the Task Group work with SRMs should be broadened to consider this conductivity range. A summary of this study has been prepared by Cabannes.[28]

The principal rationale for extending the experimental methods project can be summarized as follows: The previous project acti- vities, although covering a very broad temperature range (4-2500 K) considered methods and SRMs useful in transport property measure- ments on materials such as pure metals and some single crystal oxides in the high conductivity range ($\lambda \simeq 100$ to 25,000 W/m K) and the medium conductivity range ($\lambda \simeq 5$ to 500), such as metallic alloys, polycrystalline graphite, and certain other pure metals. Not included were the materials in the low conductivity range ($\lambda \simeq 0.5$ to 30), such as pure polycrystalline oxides, mixed oxide ceramics, and geological materials (e.g., minerals), and the very low conductivity range ($\lambda \simeq 0.001$ to 1.0), such as fibrous insulation, powders, and multilayer insulation. Some of the mater- ials in the lowest conductivity range are evacuated to eliminate gas phase conduction and convection ($\lambda_{air:RT} \simeq 0.1$).

The measurement methods, important temperature measurement issues and SRMs, for the low and very low conductivity ranges are fundamentally different from those that have been studied to date. These unique features could present considerably different pro- blems.

For the broad class of ceramic materials, there is consider- able practical interest in technology and industry for SRMs with conductivities in this range. SRMs will be required to facili- tate reliable, relatively high-accuracy thermophysical property measurements.

In the geological science field "ceramic-like" SRM materials are needed to extend the temperature range beyond about 100°C where vitreous silica ($\lambda \simeq 1$) and crystalline quartz ($\lambda \simeq 10$) are now used. This extension would be to higher temperature ($\simeq 1000^{\circ}$C) but still at nominally low pressures as contrasted to the high pressure- high temperature range where materials selection is considerably more limited owing to phase change problems. SRMs should include the porosity ranges normally encountered in geothermal and earth-mantle heat-transport studies.

In the very low-conductivity, high-temperature range associated with the several classes of thermal insulations, there are a variety of very important practical scientific and engineering applications. Reflecting this interest, there is currently an ongoing international activity organized under the auspices of ASTM, Federation Europeene des Fabricants de Produits Refractaires, and the National Bureau of Standards. Key representatives of this program have indicated that this Task Group of CODATA could perform a valuable function in the area.

On the basis of the survey by Cabannes, it has been decided to proceed with initial studies on the NBS lot of Pyroceram by conducting a controlled international series of measurements and analyses. This will include thermophysical property measurements as well as measurements of other properties (e.g., Young's modulus, acoustic attenuation), which may aid in specimen characterization and transport property correlations. In addition, further information will be obtained on a limited number of other promising candidates; specifically, stabilized zirconia and VICOR impregnated with opacifiers.

PREPARATION OF INTERNATIONALLY ACCEPTED TABLES OF REFERENCE DATA

Rationale

In contrast to the programs just described where new data are being generated, this project focuses on existing archival thermophysical properties information. The objective of the project is to generate internationally accepted recommended value tabulations as a function of temperature, including estimates of inaccuracy. Important ancillary formulation, such as material composition or fabrication parameters, and a concise summary of the data evaluation rationale are to be included. The technical approach was to form international teams of analysts and reviewers who conducted critical analysis and critiqued the results.

The specific materials and the thermophysical properties selected for initial attention in the project are summarized in Figure 11. The criteria for this selection was multi faceted: 1) scientific importance, 2) practical engineering importance, 3) existence of a sufficiently broad data base to make critical evaluation feasible. In addition, special attention was given to areas where basic correlation techniques had been demonstrated or appeared promising (e.g., resistivity rat) correlation of thermal transport in copper). This project of course, builds upon, updates, and broadens earlier critical evaluation predic-

PROPERTY MATERIAL	THERMAL CONDUCTIVITY	ELECTRICAL RESISTIVITY	HEAT CAPACITY	THERMAL EXPANSION	THERMAL DIFFUSIVITY	THERMO-ELECTRIC PROPERTIES
COPPER	X	X	X	X	X	X
IRON	X	X	X	X	X	
TUNGSTEN	X	X	X	X	X	
ALUMINUM	X	X				
PLATINUM	X	X				X
LEAD						X
SINGLE CRYSTAL ALUMINUM OXIDE			X	X		
SILICON	X	X		X		
PARTICIPATING COUNTRIES	AUSTRALIA, CANADA, FRANCE, GREAT BRITAIN, USA					

Figure 11   Program Summary: Project on Preparation of
Internationally Accepted Tables

tions in this field. Note that the selection criteria did not
explicitly include whether the material was used as an SRM, although some of these materials do appear on the list. The objectives and thus the material and property selection criteria for
this project were different from those applied in the measurements methods/SRM project.

Methodology

Another important output of this project will be a general
critical data analysis methodology for the thermophysical properties of solids.

After due consideration, primarily by G. White and J. Hust,
in connection with the preliminary investigation on the thermal
expansion and thermal conductivity data for copper, guidelines
have been developed for similar efforts on other properties and
materials. The chronology would be:

1. Literature search for data sources.

2. Data extraction and appraisal of precision and
accuracy, including specimen characterization.

3. Selection of data sets on which key values are to be based: the primary data set.

4. Estimation of uncertainty of key values, based upon scatter of data and expert opinion of analyst.

5. Weighted least-squares functional representation of this primary data set and reassessment of the set.

6. Comparisons with other data not in the primary data set.

Selection of primary data sets and estimates of uncertainty is the core of the critical analysis--the intellectual process of data evaluation. It is a process that necessarily contains subjective elements. Important criteria are summarized in Figure 12 and include:

a. Reported data inaccuracy and imprecision. In general, data characterized by inaccuracy in the 1-2% range, where this is satisfactorily substantiated, would merit their inclusion in the primary data set. Adequacy of the graphical, tabular, and functional format of the data representation is also important.

b. Internal empirical consistency and self-consistency: that is, agreement among data obtained in a given laboratory on a given specimen but using different measurement techniques. Adequate description and analysis of the equipment and measurement procedures or the agreement among quantitatively interrelated properties, such as thermal conductivity and thermal diffusivity, is implicit here. Adequate analysis of systematic error sources is also assumed.

- SCIENTIFIC STATURE

- REPORTED/VERIFIABLE ACCURACY

- INTERNAL CONSISTENCY

- THEORETICAL CONSISTENCY

- COMPLETENESS

    - SPECIMEN CHARACTERIZATION
    - MEASUREMENT PROCEDURES
    - FORMATTING OF NUMERICAL RESULTS

Figure 12 Critical Analysis Methodology

ARCHIVAL REFERENCE	USER SUMMARY

• SELECTION CRITERIA
• ANALYSIS METHODOLOGY
• NUMERICAL TABULATION
    INCLUDING UNCERTAINTIES OF THE
    RECOMMENDED VALUES
• GRAPHICAL DEVIATION DISPLAY
    PRIMARY DATA SET
        AND
    REJECTED DATA
• ANALYTICAL FUNCTIONAL FORM
    INCLUDING THEORETICAL BASIS
• GRAPHICAL DISPLAY
• MATERIALS CHARACTERIZATION
  INFORMATION
• BIBLIOGRAPHY
	• NUMERICAL TABULATION
	  $\left(\begin{array}{l}\text{LINEAR INTERPOLATION USAGE}\\ \text{WITH STATED UNCERTAINTY}\end{array}\right)$
	• GRAPHICAL DEVIATION DISPLAY
	• ANALYTICAL FUNCTIONAL FORM
	    (COMPUTATIONAL USAGE)
	• GRAPHICAL DISPLAY
	    (ENGINEERING USAGE)
	• BIBLIOGRAPHY

Figure 13   Presentation of Key Values

     c.   Completeness of specimen characterization information
and correlation of the experimental results with important specimen
parameters (e.g., purity and porosity).

     d.   Theoretical consistency; e.g., agreement with known
constraints on magnitude and temperature dependency of the measured
properties.

     e.   Scientific peer group stature.

Presentation of Results

   An important dichotomy must be addressed here. On the one
hand, the majority of users of the recommended numerical values
likely have little interest in or use for the details of the
analysis procedures as long as the uncertainties in the recommendations are clearly given. On the other hand, the analyst is really
obliged to document the important details of the procedures and rationale employed, especially the subjective elements. It is suggested that these different needs can be met by publication of two

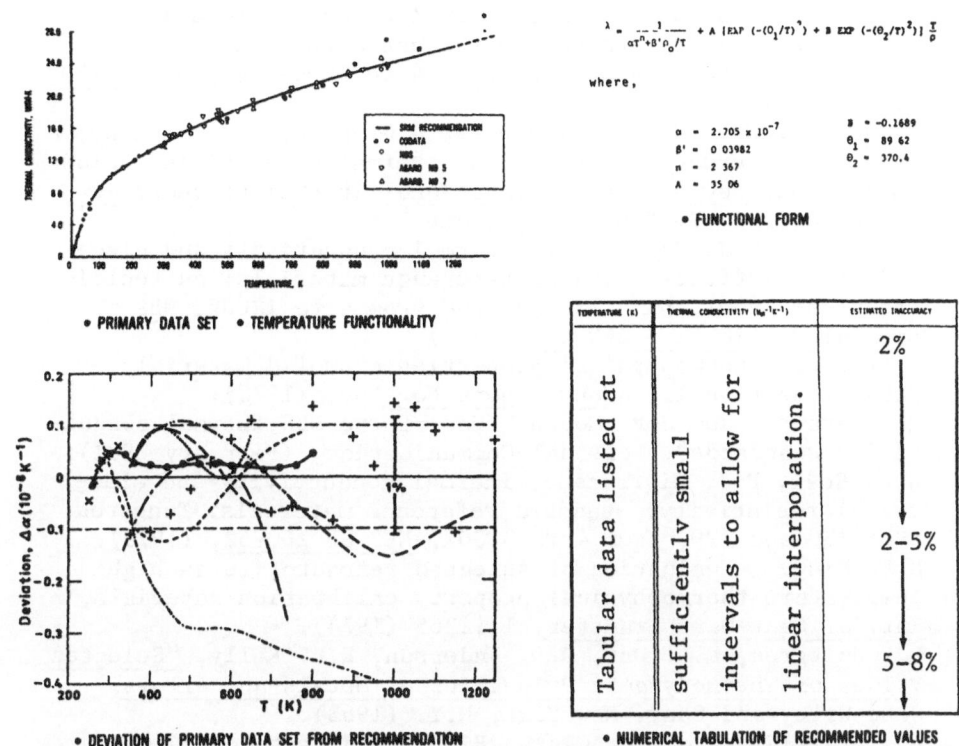

Figure 14   Critically Evaluated Reference Data:
Presentation Format

documents, an archival reference and a brief user summary. The
items comprising these documents are summarized in Figure 13. A
recommended format for the principal items in the user's summary
are illustrated in Figure 14.

REFERENCES

1. Report on first meeting of CODATA Task Group on Transport
    Properties, Warsaw, Poland, Thermophys. Electron. Newslett.,
    Vol 3, No 4, (July 1974).

2. M.L. Minges, High temperature thermophysical property refer-
    ence standards and the CODATA Task Group on Transport Proper-
    ties, High Temp. High Pressures, 8(4): 359-366 (1976).

3. M.L. Minges, The CODATA Task Group on Transport Properties,
    in: "Proc. Seventh Symposium Thermophysical Properties", ASME,
    New York. (May 1977)

4. C.F. Lucks, Armco Iron: New concept and broad data base
    justify its use as a thermal conductivity reference material,
    J. Test and Eval., 1(5): 422, (1973).

5.   J.G. Hust, P.J. Giarratano, Thermal conductivity and elec-
     trical resistivity standard reference materials: electro-
     lytic iron, SRMs 734 and 797 from 4 to 1000K, NBS SP 260-50,
     (June 1975).

6.   R. Berman, N.D. Hardy, M. Sahota, J.G. Hust, R.J. Tainsh,
     Standard reference materials for thermal conductivity below
     100K, Proc. Seventeenth International Thermal Conductivity
     Conference, To Be Published, (June 1981).

7.   J.G. Hust, P.J. Giarratano, Thermal conductivity and elec-
     trical resistivity standard reference materials: austenitic
     stainless steel, SRMs 735 and 798 from 4 to 1200K, NBS
     Special Publication 260-46, (1975).

8.   E. Fitzer, Thermophysical properties of solid materials -
     project section II, AGARD Report No. 606, (1973).

9.   R.S. Graves, and J.P. Moore, "Use of two NBS thermal conduc-
     tivity standards", Personal Communication, (February 1981).

10. J.G. Hust, P.J. Giarratano, Thermal conductivity and elec-
     trical resistivity, standard reference materials: Tungsten
     SRMs 730 and 799, from 4 to 3000K, NBS SP 260-52, (1975).

11. M.L. Minges, Evaluation of selected refractories as high
     temperature thermophysical property calibration materials,
     Int. J. Heat Mass Transfer, 17:1365 (1974).

12. R. Hultgren, R.L. Orr, P.D. Anderson, K.K. Kelly, "Selected
     Values of Thermodynamic Properties of Metals and Alloys,
     John Wiley and Sons, New York, N.Y. (1963).

13. A. Cezairliyan, J.L. McClure, High speed (subsecond) measure-
     ments of heat capacity, J. Res. NBS, 75A(4):283 (1971).

14. V.A. Kirillin, A.E. Sheindlin, V.Y. Checkovski, V.A.
     Petrov, Thermodynamic properties of tungsten, J. Phys. Chem.
     37:1212 (1963).

15. R.E. Taylor, Thermal properties of tungsten SRMs 730 and 799,
     J. Heat Trans., (2):330 (1978).

16. M. Hoch, The high temperature specific heat of body-centered-
     cubic refractory metals, High Temp. High Pressures, 1:531
     (1969).

17. R.K. Williams, W. Fulkerson, Separation of the electronic and
     lattice contributions to the thermal conductivity of metals
     and alloys, Proc. 8th Conf. Thermal Conductivity, 389 (1969).

18. C.Y. Ho, R.W. Powell, P.E. Liley, Thermal conductivity of the
     elements, J. Phys. Chem. Ref. Data, 3(1):691 (1974).

19. J.G. Hust, Homogeneity investigation on POCO graphite for use as
     NBS SRMs Personal Communication, (Nov. 1976).

20. R.E. Taylor, H. Groot, Thermophysical properties of POCO
     graphite, High Temp. High Pressure, 12:147 (1980).

21. A. Sugawara, M.L. Minges, Thermal conductivity of POCO graphite
     and stainless steel, To Be Published.

22. J.R. Koenig, and C.D. Pears, Nondestructive characterization of POCO AM graphite, Proc 17th Int'l Thermal Cond. Conf., To Be Published, (1981).

23. H.E. Schmidt, and H.A. Tasman, Thermal diffusivity data for POCO graphite as determined by a modulated electron beam technique, Proc. 17th Int'l Thermal Cond. Conf., To Be Published, (1981).

24. V. Mirkovich, Thermal diffusivity of a POCO graphite, Proc. 17th Int'l Thermal Cond. Conf., To Be Published (1981).

25. J.P. Moore, R.S. Graves, The thermal conductivity and electrical resistivity of a POCO AM-5Q1 Graphite, Proc. 17th Int'l Thermal Cond. Conf., To Be Published, (1981).

26. K.D. Maglic, N. Perovic, Z. Zivotic, Thermophysical characterization of candidate standard reference materials, Proc. 17th Int'l Thermal Cond. Conf., To Be Published (1981).

27. A. Cezairliyan, F. Righini, Measurement of heat capacity, electrical resistivity and hemispherical total emittance of two grades of graphite, Rev. Int. Hautes Temp. Refract., 12:124 (1975).

28. F. Cabannes, "Standard reference materials and low thermal conductivity experimental methods (A feasibility study for CODATA Task Group on Thermophysical Properties of Solids), Proc. 17th Int'l. Thermal Cond. Conf., To Be Published, (1981).

RECOMMENDED VALUES OF ELECTRICAL RESISTIVITY

AND THERMAL CONDUCTIVITY OF PLATINUM

Guy K. White

CSIRO Division of Applied Physics
Sydney, Australia   2070

INTRODUCTION

The CODATA Task Group on Thermophysical Properties of Solids is currently preparing "recommended" values for properties of some key solids used for calibration or reference purposes.[1]   The properties include thermal conductivity (λ), electrical resistivity (ρ), heat capacity (C_p), linear coefficient of thermal expansion (α), diffusivity (D), and absolute thermoelectric power (S).   The key materials include copper, iron, and tungsten for most properties, silicon (α), sapphire (C_p and α), and platinum (λ,ρ,S).   Platinum, Pt, is included because: (a) it is relatively stable and inert and available in high purity;   (b) the electrical resistance-temperature relations are well established from 13.8 to $\sim$ 1200 K through research on the International Practical Temperature Scale (IPTS);[2]   (c) measurements of λ(T) from 100 to 1200 K at a number of national laboratories agree well.[3]

I shall discuss first ρ(T) and then λ(T), explaining the selection of data, the fitting of these to polynomials, corrections for thermal expansion, deviations of input data from mean, estimated uncertainties, and tabulated values.

ELECTRICAL RESISTIVITY

The task of selecting electrical data for platinum is made easier because of its role in the realization of the temperature scale.   The IPTS is defined from 13.8 to 904 K (630.7°C) in terms of the resistance ratio, $W_T = R_T/R_{273}$, of suitably pure platinum, using polynomials to relate W(T) to T with defining fixed points that have themselves been determined thermodynamically by precision gas

thermometry.[4]   There have also been recent efforts to extend platinum
resistance thermometers (PRT) to the gold point (1337.6 K). We have
used these data on $W(T)$, corrected them where necessary for impurity
scattering ($W_o \leqslant 7 \times 10^{-4}$ for the PRT used), converted to resistivity by assuming that $\rho_{273.15}^i$ (ideally pure Pt) = 9.82 (± 0.01) $\times 10^{-8}$
Ωm (averaged [3,5,6,7]) and corrected for thermal expansion. These
dimensional changes $\Delta \ell / \ell$ are listed in Table 1 and are taken from the
NBS linear expansion data[8] above 293 K and from CINDAS recommendations[9] below 293 K. Values of $\Delta \ell / \ell$ from these two sources and the
American Institute of Physics Handbook (3rd Ed. 1972) agree within
0.3×10^{-4} up to 1600 K and 0.5×10^{-4} from 1600 K to 1900 K. The
corrections to ρ (and λ) for change in geometry reach 1% at 1250 K
and 2% at 2000 K.

2 to 13.8 K

At these temperatures, the impurity resistance, R_o, is generally
large compared with the ideal resistance, R^i, and departures from
additivity, $R_T = R_T^i + R_o$ (Matthiessen's Rule), are so significant that
we have not attempted to recommend values for ideal resistivity, ρ_T^i,
which could be misleading. Instead Table 1 lists values of $W_T - W_o$
and ρ_T^i for <u>one</u> PRT (No. 320) used by Berry.[10,11] His data on three
different PRT's show variations in $W_T - W_o$ of $\lesssim 20 \times 10^{-6}$ that are
considerable and far outweigh any discrepancies in the temperature
scale ($\sim$ 5 mK or 1 in 10^3) that he used.

13.8 to 278 K

Besley and Kemp[12] have calculated $Z_T = (R_T - R_4)/(R_{273} - R_4) =$
$(W_T - W_4)/(1 - W_4)$ for 51 PRT's from data taken at the National
Physical Laboratory and National Research Council of Canada and have
shown that deviations of Z values decrease from $\pm$ 1% at 20 K to $\pm$
0.01% at 150 K, while values of W_4 range from 3.4 to 6.8×10^{-4},
corresponding to an average resistance ratio (RRR) $\sim$ 2000. R. C. Kemp
has now reprocessed the values for 34 PRT's from NPL[13] (which are more
uniformly spaced in T) to obtain the smoothed values of Z_T and thence
ρ_T^i given in Table 1. The procedure follows from

$$\begin{aligned}
\rho_T^i &= (A/\ell)_T \ (R_T - R_o) \\
&= (1 + \Delta \ell / \ell) \ (1 - \delta) \ (Z + \delta) \ \rho_{273}^i \\
&= (1 + \Delta \ell / \ell) \ (0.999975) \ (Z_T + 0.000025) \ (9.82 \times 10^{-8}) \ \Omega m
\end{aligned}$$

where $\delta = W_4 - W_o = (25 \pm 1) \quad 10^{-6}$ and A/ℓ is the area/length.
Deviations of Z in Fig. 1 are from the polynomial fit

$$Z = \sum_{c=1}^{17} C_n \ [\frac{\ln \ (T/273.15 + 1.492}{1.492}]^{-1}$$

273 to 1300 K

McAllan[14] has fitted data on W_T to a fifth-order polynomial, after correcting for impurity resistance using $W_T^i = W_T + W_0(W_T - 1)$ and $W_0 = 200 (0.0039289-\alpha)^{11}$. From 273 to 733 K, he included the NBS gas thermometer data[15] and above this, values for 19 PRT's taken at four national laboratories at the following points: 903.89(Sb), 1052.97 K (Cu Ag), 1235.22 (Ag), and 1337.58 (Au). Uncertainties may reach 2 in 10^4 from the T scale and 1 in 10^4 from the scatter among the PRT's.

1300 to 2000 K

This range has not been used in platinum resistance thermometry and uncertainties in the temperature scale may reach 1 K. We have used the following input data, subtracted an impurity resistance, fitted a third-order polynomial to obtain smoothed values at 50 K intervals and corrected for expansion (Table 1):

1. Values of ρ_T measured at Turin[16] on 5 samples by a pulse technique from 1000 to 2000 K; these were of low electrical purity with $\rho_{273} = 10.24 \pm 0.03 \times 10^{-8}$ Ωm

2. Tabulation of W_T by Roeser[17] of NBS for thermometer grade wire up to 1500°C

3. Measurements by Kraftmakher[18] of W_T up to 2000 K. He does not quote electrical purity but values near 1300 K indicate thermometric grade platinum with $\rho_0 \approx 0.01 \times 10^{-8}$ Ωm

4. Values of ρ up to 1500 K from NRC[5] on rod having $W_0 = 5 \times 10^{-4}$. Deviation of experimental points from their fitted equation was less than 0.2%.

5. NBS measurements[6] extending up to 1373 K on annealed bar for which RRR = 393, i.e., $\rho_0 = 0.025 \times 10^{-8}$ Ωm.

The criterion for inclusion was that input data should not deviate significantly in the vicinity of 1200 K. The polynomial fitting was done over the range 1000 to 2000 K to overlap with the 273 to 1300 K data on PRT's. The deviations shown in Fig. 2 include input data from Martin *et al.*[7] on two samples of RRR = 5000 and 900.

Accuracy

For platinum of electrical purity or RRR in the range 1500 to 3000, values of ideal resistivity (assuming $\rho_{273}^i = 9.82 \times 10^{-8}$ Ωm) given in Table 1 should not be in error by more than 2% at 15 K, 1% (20 K), 0.1% (50 K), 0.02% (150 - 1000 K), 0.1% (1000 - 1300 K) and 0.3% (1300 - 2000 K). The choice of ρ_{273} could be in error by ± 0.01

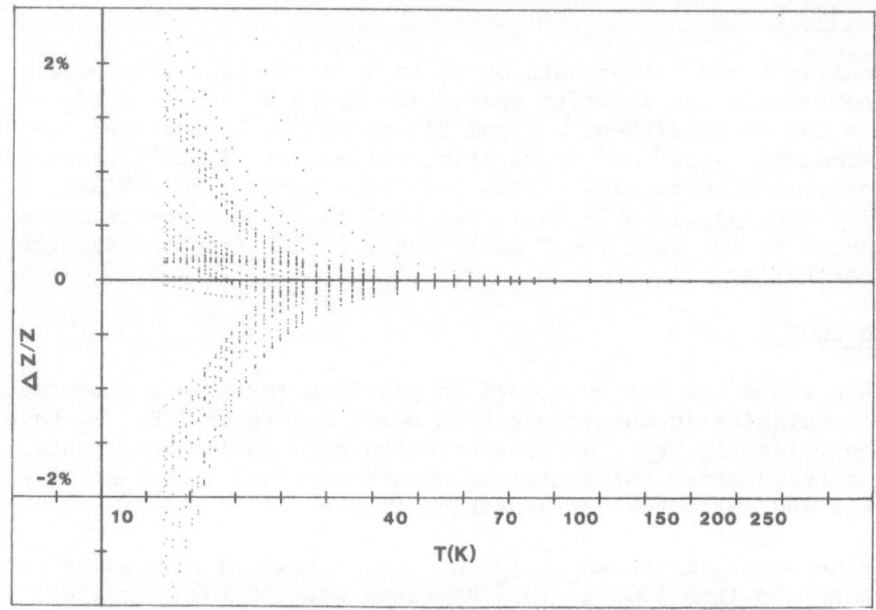

Fig. 1.   Spread of Z values on log T scale.

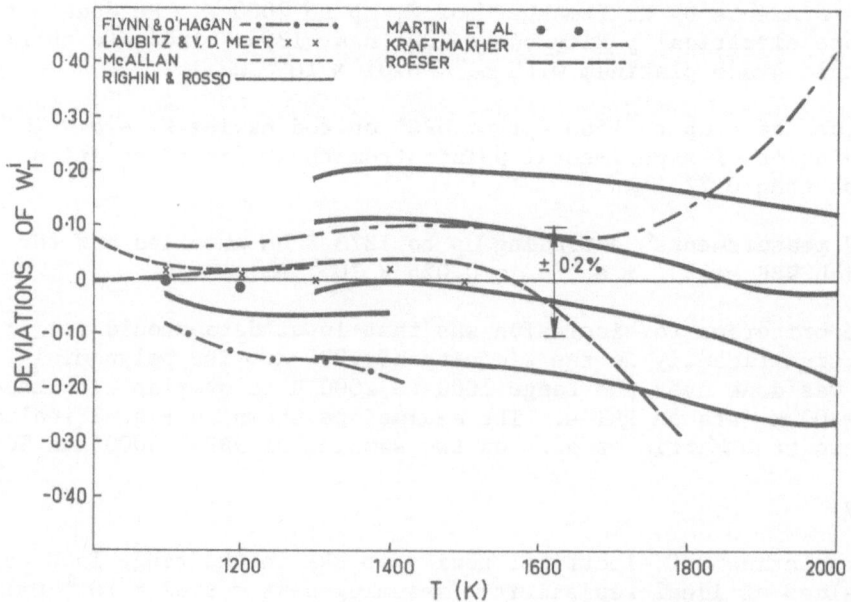

Fig. 2.   Spread of W_T^i values above 1000 K.

Table 1. Platinum. (a) Electrical Resistivity Ratios, W^i_T (2-12 K, >273K) or Z_T (14-273K), Uncorrected. (b) Expansion Correction. (c) ρ^i_T, Corrected for Expansion.

T (K)	W^i_T	$1 + \frac{\Delta \ell}{\ell}$	ρ^i_T (10^{-8} Ωm)	T (K)	Z_T	$1 + \frac{\Delta \ell}{\ell}$	ρ^i_T (10^{-8} Ωm)
2	0.000 005	0.998 26	0.000 05$_0$	260	0.947 44	0.999 89	9.302 9
3	12		11$_6$	273.15	1	1	9.82
4	22		21$_6$		W^i_T		
4.22	25		24$_5$	293.15	1.079 58	1.000 18	10.603
5	37		36	300	1.106 72	24	10.871
6	58		57	350	1.103 10	69	12.805
8	130		0.001 27	400	1.496 47	1.001 15	14.712
10	267		2 62	450	1.686 89	1 61	16.592
12	511		5 01	500	1.874 39	2 09	18.445
	Z_T			550	2.058 98	2 57	20.271
14	0.000 9082		0.009 15	600	2.240 67	3 05	22.070
16	0.001 540		0.015 34	650	2.419 4	3 54	23.843
18	2 455		24 31	700	2.595 3	4 04	25.588
20	3 718		36 69	750	2.768 1	4 54	27.306
22	5 378		52 96	800	2.937 9	5 04	28.996
24	7 475	0.998 27	73 52	850	3.104 7	5 56	30.658
26	0.010 037		98 64	900	3.268 5	6 08	32.292
28	13 081		0.128 48	950	3.429 1	6 61	33.897
30	16 613	0.998 28	163 10	1000	3.586 7	7 15	35.473
35	27 525	0.998 29	270 07	1050	3.741 2	7 69	37.021
40	41 167	30	403 81	1100	3.892 6	8 25	38.540
45	57 106	31	560 06	1150	4.041 0	8 81	40.03
50	74 859	33	734 11	1200	4.186 5	9 38	41.50
55	93 975	35	921 53	1250	4.329 1	9 96	42.94
60	0.114 07	37	1.118 6	1300	4.469 1	1.010 55	44.35
65	134 84	39	1.322 2	1350	4.606	11 16	45.73
70	156 05	41	1.530 2	1400	4.739	11 77	47.09
75	177 54	43	1.740 9	1450	4.871	12 39	48.42
80	199 17	46	1.953 1	1500	5.001	13 02	49.74
85	220 88	49	2.165 9	1550	5.128	13 66	51.05
90	242 60	51	2.379 0	1600	5.256	14 31	52.34
100	285 93	57	2.804 0	1650	5.382	14 98	53.63
120	371 79	71	3.646 2	1700	5.507	15 66	54.93
140	456 43	86	4.477 2	1750	5.632	16 36	56.21
160	540 02	0.999 02	5.297 9	1800	5.758	17 07	57.51
180	622 76	19	6.110 7	1850	5.883	17 81	58.79
200	704 81	36	6.916 9	1900	6.009	18 57	60.11
220	786 23	54	7.717 3	1950	6.136	19 34	61.43
240	867 09	72	8.512 5	2000	6.265	1.020 13	62.76

For convenience a 9-order polynomial ($\Sigma a^n T^n$) has been fitted to the values of ρ^i_T above 100 K with deviations of $\lessgtr 0.002 \times 10^{-8}$ (100 - 1000 K) and $\lessgtr 0.01 \times 10^{-8}$ Ωm (1000 - 2000 K). Coefficients are:

$a_0 = -1.621\ 733$

$a_1 = 4.681\ 197 \times 10^{-2}$

$a_2 = -3.258\ 075 \times 10^{-5}$

$a_3 = 8.554\ 023 \times 10^{-8}$

$a_4 = -1.594\ 242 \times 10^{-10}$

$a_5 = 1.837\ 342 \times 10^{-13}$

$a_6 = -1.316\ 886 \times 10^{-16}$

$a_7 = 5.678\ 222 \times 10^{-20}$

$a_8 = -1.340\ 980 \times 10^{-23}$

$a_9 = 1.329\ 896 \times 10^{-27}$

$\times 10^{-8}$ Ωm or more. If later determinations of this ice point resistivity give a more accurate value, then all the tabulated values
can be multiplied by an appropriate correction constant. For less
pure platinum, departures from Matthiessen's rule will make table 1
much less accurate below say 50 K, but at high temperatures the
inaccuracy should remain in the vicinity of 0.2%.

THERMAL CONDUCTIVITY

CINDAS (formerly TPRC) has made critical surveys of experimental
data on $\lambda(T)$ and produced recommended values for platinum of RRR $\approx$
1000 in 1966 (NSRDS-NBS No. 8), 1970[19], and 1972[20]. These differ considerably at the higher temperature due presumably to inclusion of
more recent data. We take here the 1972 recommendations (called λ_R)
and see how various data deviate including later values above 1000 K.
We also produce suggested values for platinum of RRR $\approx$ 500 and 2000
to illustrate changes to be expected at low temperatures. They are
only a guide since data are scarce in this region. Differences in λ
arising from improving RRR $\geqslant$ 500 are not significant above 100 K compared with experimental uncertainty.

2 to 100 K

Experimental data of Mendelssohn and Rosenberg, White and Woods
(see reference 19) are from two platinum samples, each having RRR $\approx$
1000. They can be represented roughly by

$$\lambda^{-1} = w = w_0 \text{ (defects)} + w_i^* = \rho_0/L_0 T + w_i^*$$

where w_i^* is similar to the ideal electronic resistivity, w_i, but reduced by the presence of lattice conductivity, λ_g. If $\lambda_g/\lambda = \Delta \ll 1$,
then $w \approx w_e (1-\Delta) = (w_0 + w_i)(1-\Delta)$. As $T \to 0$, $\lambda \to 0$ and $w \to w_0$.
For T of 10 to 100 K, we expect $\Delta \lesssim 0.1$ and $w_i^* \approx w_i (1-\Delta)$ to be fairly
constant for changes of RRR of 500 to 2000. We calculate mean values
of w_i^* at 10, 15, 20 K, from the above two sets of data and then values
of λ in Table 2 from $\lambda = [\rho_0/L_0 T + w_i^*]^{-1}$ for ρ_0 = 0.02, 0.01, and 0.005
$\times 10^{-8}$ Ωm, which correspond to RRR values of 500, 1000, and 2000.
These λ values may be in error by up to 10% and are intended as a
guide only. Other reference materials, e.g., iron, stainless steel,
and copper, are to be preferred in this range.

100 to 1200 K

Table 2 lists recommended values from 1972 CINDAS publication[20]
and Fig. 3 shows the deviation of measurements on high purity platinum. Most of these data are listed in CINDAS compilation[19]. Additions
are from Moore *et al.*[21] Duggin[22], Zinovev *et al.*[23] and Vertogradskii[24].

Table 2. Thermal Conductivity of Platinum. Values λ_R above 100 K
from[20] (RRR~1000) are corrected for expansion.

T	λ (W m^{-1} K^{-1})			T	λ_R
(K)	$10^8 \rho_0$ = 0.02	$10^8 \rho_0$ = 0.01	$10^8 \rho_0$ = 0.05	(K)	W m^{-1} K^{-1}
2	240	480	970	350	71.6
4	480	970	1950	400	71.8
6	680	1280	2300	450	72.0
8	820	1400	2190	500	72.2
10	850	1310	1800	600	73.0
15	670	820	920	700	74.0
20	430	475	500	800	75.2
25	290	305	315	900	76.6
30	210	216	219	1000	78.1
40	139	141	142	1100	79.9
60	94	95	95	1200	81.8
80	83	83	83	1300	83.9
100	78	78	78	1400	86.1
		λ_R		1500	88.4
150		74.0		1600	90.6 (90)
200		72.6		1700	92.7 (90)
250		71.8		1800	94.5 (90)
273		71.7		1900	96.0 (89)
300		71.6		2000	97.4 (88)

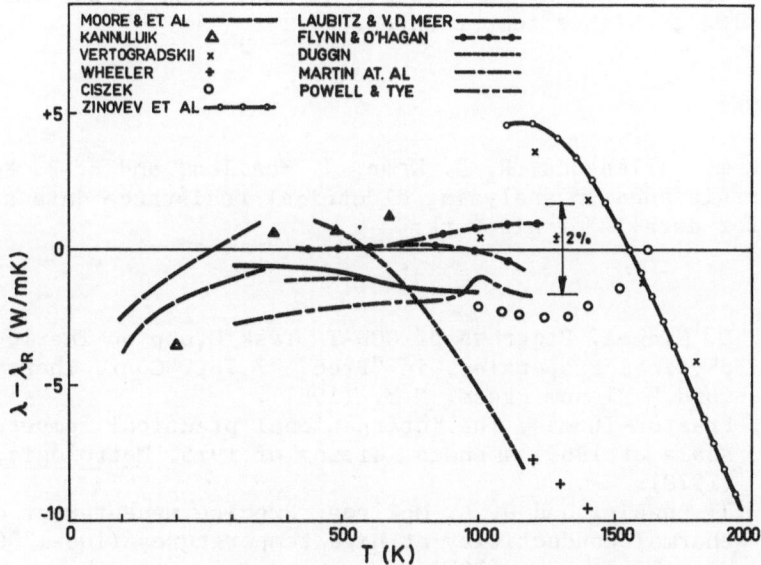

Fig. 3.   Departure of λ (T) from values of
          Ho et al.[20]

As Laubitz and McElroy[3] have shown, the data from ca. 300 to 1200 K taken by the national laboratories in North America agree within about 2%. It is unlikely that the values for λ_R[20] are in error by more than this. From 100 to 300 K the only comprehensive measurements were made on two samples (RRR of 426 and 600) at ORNL.[21] Confirmation is needed in this range.

1200 to 2000 K

In this range deviations are as large as ± 10%. The values λ_R recommended in 1972[20] (Table 2) increase with T. Subsequent values of Zinovev *et al.*[23] and Vertogradskii[24] are roughly constant above 1500 K giving $\lambda \approx$ 90 W/m.K at 1800 K compared with λ_R = 94.5 and earlier CINDAS recommendations of 95.7 W/mK[19] and 86.2 W/m.K in 1966. The value of 90 leads to a Lorenz ratio of L = 2.88 × 10^{-8} V^2/K^2 which is still surprisingly high at a temperature where the lattice conductivity, λ_g, should be less than 1% of the total. We give the lower values of λ in brackets in Table 2 as being more plausible. Further measurements are needed.

Accuracy

Below 100 K, paucity of data and possible departures from Matthiessen's rule limit the accuracy of suggested λ values. Uncertainties should not exceed 10% if platinum has an RRR greater than 500 and values are adjusted for change in ρ_0. From 100 to 300 K, data are limited but recommended values should not be in error by more than 3%. In the range 300 to 1200 K, uncertainty is less than 2% but increases to at least 10% at higher temperatures.

ACKNOWLEDGMENT

I thank my colleagues, R. C. Kemp, J. McAllan, and R. B. Roberts for expert assistance in analyzing electrical resistance data and F. Righini for details of his work.

REFERENCES

1.   M. L. Minges, Progress of CODATA Task Group on Thermophysical Properties, in "Proc. 17 Int. Conf. Thermal Cond." Plenum Press, N.Y. (1981).
2.   H. Preston-Thomas, The international practical temperature scale of 1968, amended edition of 1975, Metrologia, 12:7 (1978).
3.   M. J. Laubitz and D. L. McElroy, Precise measurement of thermal conductivity at high temperatures (100-1200 K), Metrologia, 7:1 (1971).
4.   T. J. Quinn and J. P. Compton, The foundations of thermometry, Rep. Prog. Phys. 38:151 (1975).

5.   M. J. Laubitz and M. P. Van der Meer, The thermal conduct-
     ivity of platinum between 300 and 1000 K, Can. J. Phys.
     44:3173 (1966).
6.   D. R. Flynn and M. E. O'Hagan, Measurements of thermal con-
     ductivity and electrical resistivity of platinum from
     100 to 900°C. J. Res. Nat. Bur. Stand. C71:255 (1967).
7.   J. J. Martin, P. H. Sidles, and G. C. Danielson, Thermal
     diffusivity of platinum from 300 to 1200 K. J. Appl.
     Phys. 38:3075 (1967).
8.   T. A. Hahn and R. K. Kirby, Thermal expansion of platinum
     from 293 to 1900 K, in "Thermal Expansion - 1971," A.I.P.
     Conference Proceedings No. 3. Amer. Inst. Phys. N.Y.
     (1972), p.87.
9.   Y. S. Touloukian, R. K. Kirby, R. E. Taylor and P. D. Desai,
     "Thermal Expansion, Metallic Elements and Alloys", Vol.
     12, Plenum Press, N.Y. (1977).
10. R. J. Berry, Platinum resistance thermometry below 10 K,
     Metrologia 3:53 (1967).
11. R. J. Berry, Ideal resistivity of platinum below 20 K, Can.
     J. Phys.45:1693 (1967).
12. L. M. Besley and R. C. Kemp, Two-point calibration of
     standard capsule platinum resistance thermometers for
     the range 13.81 K to 273.15 K, Cryogenics 18:497 (1978).
13. S. D. Ward and J. P. Compton, Intercomparison of platinum
     resistance thermometers and T_{68} calibrations, Metrologia
     15:31 (1979).
14. J. McAllan of Thermometry Section (C.S.I.R.O. Div. of Appl.
     Phys. Sydney), report to be presented to 1982 Symposium
     on Temperature at NBS, Washington, D.C.
15. L. A. Guildner and R. E. Edsinger, Deviation of individual
     platinum thermometers from thermodynamic temperature in
     range from 273.16 to 730 K. J. Res. Nat. Bur. Stand.
     A80:703 (1976).
16. F. Righini and A. Rosso, Measurement of Thermophysical
     properties by a pulse-heating method: platinum from 1000
     K to the melting point. High Temp. - High Press.12.335
     (1980).
17. W. Roeser in "Temperature, Its Measurement and Control in
     Science and Industry", Reinhold Book Corp. N.Y. (1941),
     p.1312.
18. Ya. A. Kraftmakher, The modulation method for measuring
     specific heat, High Temp. - High Press. 5.433 (1973).
19. Y. S. Touloukian, R. W. Powell, C. Y. Ho, and P. G. Klemens,
     "Thermal Conductivity, Metallic Elements and Alloys",
     Vol. 1, Plenum Press, N.Y. (1970), p.269.
20. C. Y. Ho, R. W. Powell, and P. E. Liley, Thermal conduct-
     ivity of the elements, J. Phys. Chem. Ref. Data 1:279
     (1972).

21. J. P. Moore, D. L. McElroy, and M. Barisoni, Thermal con-
     ductivity measurements between 78 and 340 K on Al, Fe,
     Pt, and W, in "Sixth Confererence on Thermal Conduct-
     ivity". M. L. Minges and G. L. Denman, ed. Air Force
     Materials Laboratory, Dayton (1966), p. 737.
22. M. J. Duggin, The thermal conductivities of Al and Pt,
     J. Phys. D.3:L21 (1970).
23. V. E. Zinovev, R. P. Krentsis, and P. V. Geld. Thermal
     conductivity and thermal diffusivity of Pt at high
     temperatures, Sov. Phys. - Sol. State 10:2228 (1969).
24. V. A. Vertogradskii, Thermal and electrical conductivity
     of platinum at high temperatures, High Temp. (USSR
     Transl.) 15:178 (1977).

STANDARD REFERENCE MATERIALS FOR THERMAL

CONDUCTIVITY BELOW 100 K

R. Berman

Clarendon Laboratory
University of Oxford
Oxford, England

N. D. Hardy and M. Sahota

Department of Physics
University of Leeds
Leeds, England

J. G. Hust

Thermophysical Properties Division
National Bureau of Standards
Boulder, Colorado, United States

R. J. Tainsh

CSIRO Division of Applied Physics
National Measurement Laboratory
Sydney, Australia

INTRODUCTION

Before the present investigation was undertaken, it had been
tacitly assumed by the low-temperature fraternity that anyone measuring thermal conductivity at temperatures low enough for radiation
not to be important (generally below ~100 K) would obtain values
correct to within perhaps 2 to 5%. However, an initial round-robin
conducted on a few specimens by several laboratories showed that
this assumption was far from valid; differences between measured
conductivities for the same specimen could be as large as 20% in
some cases, and very little, if any, of this difference could be

ascribed to real changes in conductivity due to handling of the specimens.

Therefore, it was evident that, just as for higher temperatures where accurate measurements are acknowledged to be difficult, low-temperature reference materials with accurately known thermal conductivities would be extremely useful. Anyone wishing to ensure the reliability of measuring equipment could then measure a suitably shaped specimen made out of such material and compare measured conductivities with the reference values. Because experimental problems depend on the range of conductivities to be measured, there would ideally be a number of standard reference materials (SRM's) covering different ranges of thermal conductivity. To study the extent to which certain materials could be designated as standards, a programme was undertaken by the Task Group on Thermophysical Properties of Solids set up by the Committee on Data for Science and Technology (CODATA).

Since a number of materials had already been studied in detail by Hust and his colleagues at the National Bureau of Standards (NBS) in Boulder, it was decided to work on some of the same materials and compare experimental results obtained in two other well-qualified laboratories. All measurements were made by the steady-state longitudinal heat flow method, the apparatus for which (or their ancestors) have already been described in the literature.[1]

An uncritical look at the experimental results suggests that the NBS measurements are less accurate because of the increased scatter of the points about a smooth curve, but this does not, in fact, necessarily reflect any differences between intrinsic accuracies. At NBS a long specimen is used and the temperature gradients at seven 25-mm sections along it are measured. One collective measurement thus gives the conductivities of seven sections, the mean temperatures of which differ progressively from one to the next. The scatter of results would then be expected to be roughly the same as if seven different specimens, with not necessarily identical conductivities, had been measured in seven separate apparatus of identical design, but with thermometers not having identical calibrations. The measurements are not quite so independent since of the eight thermometers, the inner six are common to two 'specimens' each. In some measurements a pattern can be seen with particular sections retaining their relative positions on either side of the mean curve through all the points. Below 10 K this spread can reach ±5%. In the other laboratories a single, shorter, specimen is measured and the scatter of points mainly reflects the precision with which all the instruments can be read, rather than indicating the accuracy with which the conductivity has been found. Repeat measurements may well give repeatedly wrong answers and suspicion only arises if definitely peculiar results are obtained. During the original round-robin, A.-M. de Goer at Centre d'Etudes Nucleaires,

Grenoble, made four separate sets of measurements on a single specimen of stainless steel using different thermometers, different methods of attaching them, and different lengths of specimen. Each set
of measurements gave results that show fairly small scatter about
the mean curve for that set; although the absolute values for two
sets are similar, the third set is ~3% higher, and the fourth set
is 4 to 5% higher than these two. This spread of values is thus
similar to that which has been observed at NBS. The procedures used
at NBS and by A.-M. de Goer, although resulting in greater scatter,
yield more information regarding the probable inaccuracy.

The three materials involved in the intercomparison so far are
stainless steel, tungsten, and electrolytic iron. It is only for
stainless steel that a direct comparison of results is possible.
Stainless steel is a sufficiently disordered solid for its thermal
conductivity to be dominated by its composition and to be only
slightly dependent on the thermal and mechanical treatment it
receives. For tungsten and iron the conductivity is sensitive to
thermal and mechanical treatment, and intercomparison can only be
made if the thermal conductivity is correlated with a measure of the
perfection of the actual specimen measured. This correlating parameter is obtained by measuring the electrical resistivity.

STAINLESS STEEL

This is an austenitic stainless steel described by Hust and
Giarratano[2] and referred to as SRM 735.

A comparison between the thermal conductivity results obtained
at NBS, The National Measurement Laboratory (NML)in Sydney, and the
University of Leeds is shown in figure 1, where $\Delta\lambda = \lambda - \lambda_{NBS}$, λ_{NBS}
being the tabulated NBS recommended values. Above 15 K, $\Delta\lambda/\lambda$ is not
more than ±3%. Below this temperature $\Delta\lambda/\lambda$ for the Leeds results
remains within this limit, but it increases for the NML results to
a maximum of 7 to 9% at 6 to 8 K. Below 8 K the NBS experimental
results themselves show appreciable scatter with points for different specimens showing some grouping together in the values of $\Delta\lambda/\lambda$,
the overall spread reaching ±5%. From the original round-robin, the
mean value of $\Delta\lambda/\lambda$ for all the Grenoble results below 25 K is approximately +3%, and for measurements at Nottingham by A. A. Ghazi below
20 K, it is approximately -2%.

It was felt that a great deal of effort would be required to
reduce the uncertainties in the conductivity at temperatures below
15 K to under 2 to 3% and that this extra accuracy would not at
present lead to any practical benefits. We thus see no compelling
reason to recommend values different from those proposed by Hust and
Giarratano, which for convenience are listed in table 1.

Table 1. NBS recommended values of thermal conductivity
          for SRM 735, austenitic stainless steel

Temperature (K)	Thermal Conductivity (W/m·K)
5	0.466
6	0.565
7	0.676
8	0.796
9	0.921
10	1.05
12	1.32
14	1.58
16	1.86
18	2.13
20	2.40
25	3.07
30	3.72
35	4.34
40	4.92
45	5.47
50	5.98
55	6.45
60	6.88
65	7.28
70	7.64
75	7.97
80	8.27
85	8.55
90	8.80
95	9.04
100	9.25

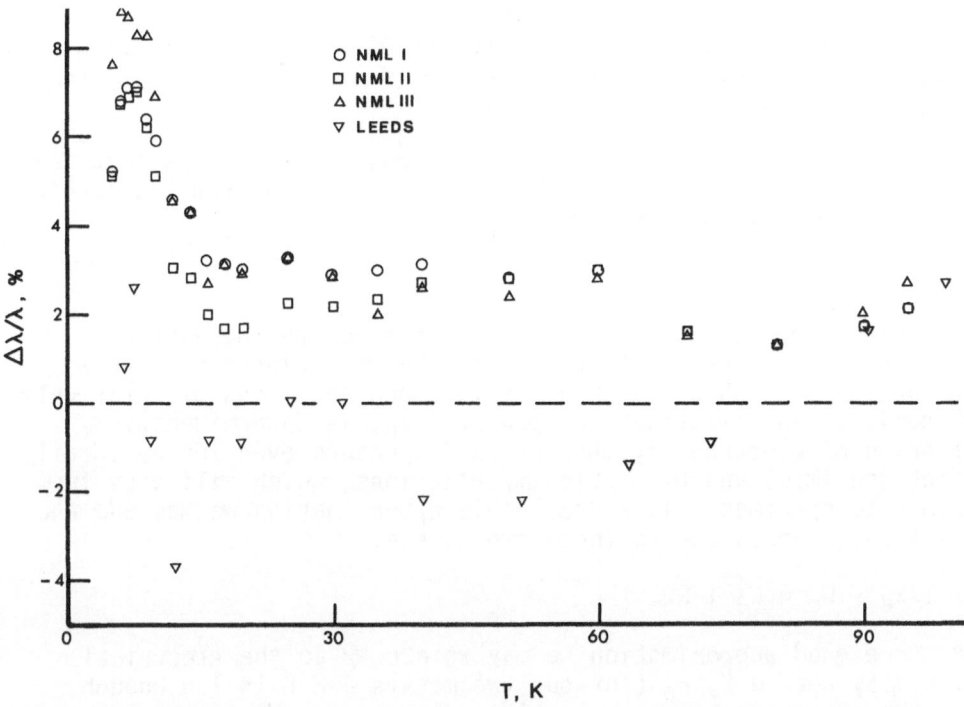

Figure 1. The relative difference between the measured thermal
conductivity, λ, of austenitic steel and the NBS recommended
values, λ_{NBS}. $\Delta\lambda/\lambda = (\lambda - \lambda_{NBS})/\lambda$.
NML I was also measured at Leeds. NML II had been measured at NBS.
NML III was specimen II reduced in length.

TUNGSTEN AND IRON

The thermal conductivity of a 'pure' metal is very dependent on
its physical state as well as on its chemical purity. It is not,
therefore, possible to make a meaningful direct comparison between
values obtained in different laboratories; transport and handling
may change the low-temperature conductivity by an appreciable
amount. Fortunately those factors that influence the thermal conductivity also show up in the electrical resistivity and affect the
ratio of the resistivity at the ice point to that at the boiling
point of helium (4.2 K). This ratio, the residual resistance ratio
(RRR), can be used as a measure of specimen perfection, and differences in RRR should correlate with differences in measured thermal
conductivities among specimens of the same metal. For our simple
method of estimating thermal conductivities from RRR, described

below, we assume that the electrical resistivity obeys Matthiessen's Rule, so that $\rho(T) = \rho^i(T) + \rho_0$, where $\rho^i(T)$ is the ideal resistivity at temperature T, assumed independent of specimen perfection, and ρ_0 is the residual resistivity at $T \rightarrow 0$ K. Then RRR $= \rho_{273}/\rho_0 = (\rho^i_{273} + \rho_0)/\rho_0$ and $\rho_0 = \rho^i_{273}/(RRR-1)$.

The thermal conductivity of a metal has contributions that can be identified as coming from the free electrons and from the lattice, and this is usually written as

$$\lambda = \lambda_e + \lambda_g.$$

The lattice component, λ_g, is mainly determined by the lattice itself and by scattering of phonons by the free electrons. This contribution to λ should change little among specimens of relatively high purity. The electronic components, λ_e, is determined by scattering of electrons by phonons (which occurs even for an ideally perfect specimen) and by static imperfections, which will vary from specimen to specimen. To a reasonable approximation we may add the thermal resistances due to these two causes:

$$1/\lambda_e = W_e = W_e^i + W_e^0.$$

Also, to a good approximation we may relate W_e^0 to the electrical resistivity near 0 K, ρ_0 (for our two metals 4.2 K is low enough for ρ there to be taken as ρ_0). According to the Wiedemann-Franz-Lorenz law, $W_e^0 = \rho_0/L_0T$, where L_0, the Lorenz number, has the value 2.44×10^{-8} W·Ω·K^{-2}.

In fact, all the scattering processes depend on one another, and this would have to be taken into account in a detailed investigation concerned with the theory of thermal conductivity. For our present purpose we only require a basis for comparison of measurements made on specimens of a metal of slightly different purities and for predicting to within no worse than 5% what the conductivity should be for a specimen with known RRR. To achieve this modest aim we assume that for each material $\lambda_g(T)$ and $W_e^i(T)$ are independent of RRR within the range of RRR measured for our specimens, so that only $W_e^0(T)$ is a function of RRR.

We assume that the thermal conductivity is dominated by λ_e and obtain $W_e^i(T)$ from each set of measurements by subtracting ρ_0/L_0T from the measured λ. We then fit polynomials to the mean of $W_e^i(T)$ for both materials. For both tungsten and iron, $W_e^i(T)$ computed in this way goes to zero at some finite temperature (see table 2), below which the calculated values would be negative. Since it is physically impossible for the true thermal resistivity, W_e^i, to be negative, the simplest explanation for the apparent negative values is that conduction by the lattice is not negligible, so that the overall thermal resistivity is less than that associated with electron

Table 2. W_e^i for tungsten and for electrolytic iron [†]

Temperature (K)	W_e^i tungsten* (10^{-4} $W^{-1} \cdot m \cdot K$)	W_e^i iron (10^{-4} $W^{-1} \cdot m \cdot K$)
10	0	
12	0.5$_5$	
14	1.0	
16	1.5	0
18	2.1	1.0
20	2.8	2.1
30	8.2	9.1
40	16.0	18.3
50	24.8	28.4
60	32.5	39.3
70	37.6	50.0
80	40.3	59.4
90	43.0	67.9

*$W_e^i = -1.50177 + 2.63090 \times 10^{-1}T - 1.90339 \times 10^{-2}T^2 + 1.13845 \times 10^{-3}T^3 - 1.68904 \times 10^{-5} \times T^4 + 7.67618 \times 10^{-8} T^5$ ($10^{-4}W^{-1} \cdot m \cdot K$)

[†]$W_e^i = -3.28095 - 1.02044 \times 10^{-1} T + 2.07585 \times 10^{-2} T^2 - 1.20589 \times 10^{-4} T^3$ (10^{-4} $W^{-1} \cdot m \cdot K$)

conduction alone. This, of course, implies that the positive calculated values of $W_e^i(T)$ are also not exactly the true $W_e^i(T)$ because
there must be some lattice conductivity at all temperatures.

It would be possible after more detailed measurements of the
relation between $\lambda(T)$ and RRR for a range of RRR values to disen-
tangle the separate contributions of λ_e and λ_g. However, as will be
shown, our simple-minded assumptions do allow us to derive prescriptions for predicting the thermal conductivity of iron and tungsten
that fit all the measurements below 100 K to within 5%, and in fact
to within 3% in most cases.

Tungsten

This material is a sintered tungsten; it is described by Hust
and Giarratano[3] and is designated as SRM 730. They give an expression that represents the thermal conductivity in terms of the electrical conductivity and covers the range 4 to 3000 K. It is derived
from measurements on two specimens with RRR ~40 and ~75. Because of
the wide temperature range to be covered, their expression is complicated, and we find that for the restricted range from 4 to 90 K
the conductivity can be represented in the simpler manner described
above.

Since the determination of RRR does not involve an accurate
measurement of specimen dimensions, it is this ratio that is normally measured, rather than the individual absolute resistivities.
It is thus necessary for our method of predicting λ to derive values
of ρ_0 from the measured RRR's. By considering a number of determinations ρ_{273K} and $\rho_{4.2K}$ we conclude that

$$\rho_0 = \frac{48.4}{RRR-1} \ n\Omega \cdot m,$$

48.4 $n\Omega \cdot m$ being the value taken for ρ^i_{273K}. The quantity ρ_0/L_0T is
subtracted from all the measured values of thermal resistivity
(interpolated to temperatures common to all sets) to obtain W_e^i.
Since the NBS recommended values for λ tabulated for three values of
RRR are not obtained independently of one another, the values of
W_e^i for these RRR's are averaged. The average values of W_e^i shown
in table 2 are obtained from this one set, which represents the NBS
measurements together with those derived from the NML and Leeds
results on the specimen with RRR 75. A fifth-degree polynomial was
used to fit the averages at each temperature; the coefficients are
given in table 2.

It can be seen that the polynomial gives $W_e^i(T) = 0$ at 10 K, and
it would be negative at lower temperatures. A simple interpretation
is that the lattice conductivity is becoming noticeable so that the

thermal resistance of the specimen is less than the sum of W_e^i and the true W_e^i. Over the short temperature interval below 10 K we can fit the measured conductivities by assuming that (1) $W_e^i = 0$, (2) the total electronic thermal conductivity is given by $L_0 T/\rho_0$, and (3) the lattice conductivity can be represented by a constant of value 3 $W \cdot m^{-1} \cdot K^{-1}$. Although the true lattice conductivity is, of course, temperature dependent and this constant is merely the value required to give the best fit to the measured conductivity, it is nevertheless of the expected order of magnitude for tungsten between 5 and 10 K.*

Figure 2 shows the fractional deviation of measured conductivities from values calculated according to the following prescription:

(1) Derive ρ_0 from $\rho_0 = \dfrac{48.4}{RRR-1}$ n$\Omega \cdot$m;

(2) At any temperature calculate $\rho_0/L_0 T$ and above 10 K add to this W_e^i read from table 2 or calculated from the polynomial given there;

(3) (a) Above 10 K invert the sum $\rho_0/L_0 T + W_e^i$ to give λ;

    (b) Below 10 K invert $\rho_0/L_0 T$ and add to it 3 $W \cdot m^{-1} \cdot K^{-1}$ to give λ.
The apparent discontinuity in λ at 10 K is considerably less than the uncertainties in measurement.

The experimental evidence suggests that the method should hold for RRR's between about 50 and 100.

Electrolytic Iron

Hust and Sparks[4] have reviewed the work on the thermal conductivity and electrical resistivity of electrolytic iron below 300 K. This material is designated as SRM 734 and its properties above 300 K are discussed by Hust and Giarratano.[5] They conclude that with the recommended heat treatment, the RRR and the thermal conductivity should be fixed to within about 1%. Measurements at NML and at Leeds on specimens that had been annealed according to the NBS recommendation gave values of both RRR and λ that differed by more than this amount from the NBS recommended values, so that a similar analysis was made to that described for tungsten.

---

*R. H. Bogaard has kindly collected together for us experimental evidence that helps one to see how our simple approach can be correlated with a more fundamental analysis.

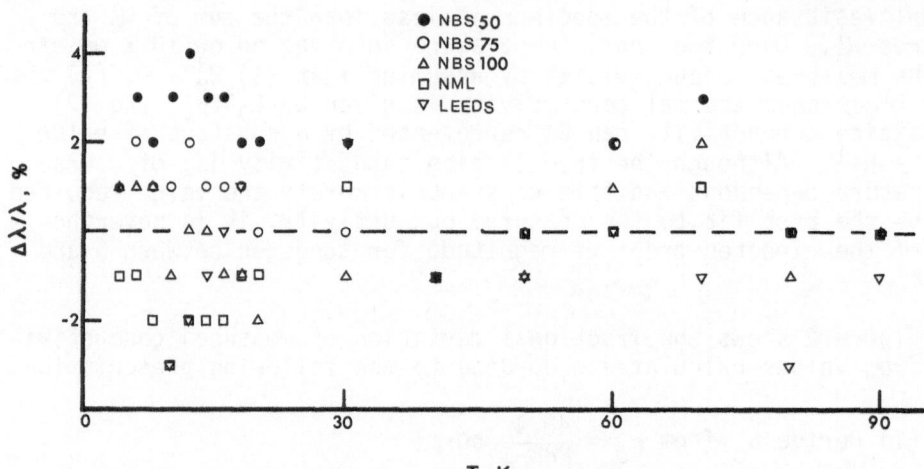

Figure 2. The relative difference (rounded to the nearest 1%) between various measured values of thermal conductivity, λ, of sintered tungsten and values predicted from the measured RRR, λ_{calc}. $\Delta\lambda/\lambda = (\lambda - \lambda_{calc})/\lambda$. The points for NBS represent the recommended values for the three values of RRR shown.

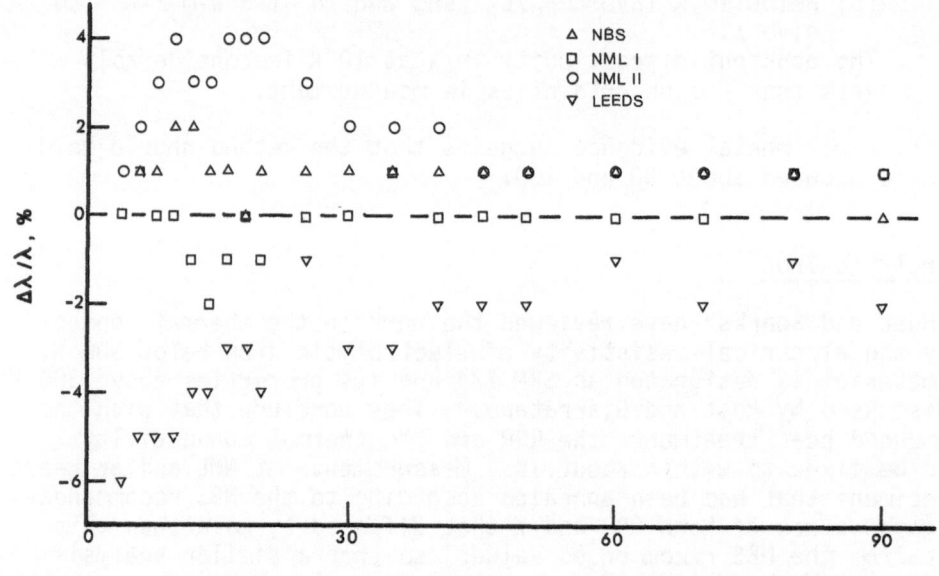

Figure 3. The relative difference (rounded to the nearest 1%) between various measured values of thermal conductivity, λ, of electrolytic iron and values predicted from the measured RRR, λ_{calc}. $\Delta\lambda/\lambda = (\lambda - \lambda_{calc})/\lambda$. The range of RRR is from 20.9 (NML II) to 23.4 (NBS).

The ρ_0 is taken as $\dfrac{87.5}{RRR-1}$ $n\Omega \cdot m$ and the average values of W_e^i are fitted with a third-order polynomial. As can be seen from table 2, this expression gives $W_e^i = 0$ at 16 K; our interpretation of this and for the negative values at lower temperatures is again as for tungsten. We obtain the best fit to the measured conductivities by adding 1 $W \cdot m^{-1} \cdot K^{-1}$ to the conductivities below 16 K calculated with $W_e^i = 0$, and our interpretation is that this number again represents a combination of the lattice contribution and a nonzero true W_e^i. Figure 3 shows the fractional deviation of measured conductivities from values calculated from the following prescription:

(1) Calculate ρ_0 from $\rho_0 = \dfrac{87.5}{RRR-1}$;

(2) At any temperature calculate $\rho_0/L_0 T$ and add to this W_e^i read from table 2 or calculated from the polynomial given there;

(3) (a) Above 16 K invert the sum of $\rho_0/L_0 T + W_e^i$ to give λ;

   (b) Below 16 K invert $\rho_0/L_0 T$ and add 1 $W \cdot m^{-1} \cdot K^{-1}$ to give λ.

Although the experimental evidence is derived from specimens with RRR between 20.9 and 23.4, it is likely that it would hold for values slightly outside this range, which might result from variation the effect of the recommended heat treatment.

CONCLUSIONS

We have measured the thermal conductivity below 100 K of an austenitic steel, sintered tungsten, and electrolytic iron. We believe that the values tabulated by Hust and Giarratano for stainless steel are accurate to within ~3% down to ~15 K. Below this temperature the uncertainty increases, but we do not consider that the considerable effort that would be required to reduce this uncertainty is, at present, warranted. For tungsten and electrolytic iron we present simple prescriptions for predicting the thermal conductivity from a measurement of the ratio, RRR, of the electrical resistivities at 273 and 4 K. Within the ranges of RRR involved, we think that the predicted conductivities are within 3 to 4% of the true values over the whole temperature range below 100 K.

ACKNOWLEDGMENTS

We are very grateful to the participants in the initial round-robin whose work opened our eyes to the difficulty of making accurate low-temperature measurements: A. A. Ghazi at Nottingham, A.-M. de Goer at Grenoble, and S. Kelham and M. Martinez at Oxford.

REFERENCES

1. e.g., J. G. Hust, R. L. Powell, and D. H. Weitzel, Thermal con-
    ductivity standard reference materials from 4 to 300 K, Armco
    Iron; including apparatus description and error analysis, <u>J. Res.
    NBS</u>, 47A:673 (1970).
    and
    G. K. White and S. B. Woods, Thermal and electrical conductiv-
    ities of solids at low temperatures, <u>Can. J. Phys.</u> 33:58 (1955).
2. J. G. Hust and P. J. Giarratano, "Thermal Conductivity and Elec-
    trical Resistivity Standard Reference Materials: Austenitic
    Stainless Steel, SRM's 735 and 798, from 4 to 1200 K," NBS
    Special Publication, 260-46, National Bureau of Standards,
    Boulder, Colorado (1975).
3. J. G. Hust and P. J. Giarratano, "Thermal Conductivity Electri-
    cal Resistivity Standard Reference Materials: Tungsten, SRM's
    730 and 799, from 4 to 3000 K," NBS Special Report, 260-52,
    National Bureau of Standards, Boulder, Colorado (1975).
4. J. G. Hust and L. L. Sparks, "Standard Reference Materials:
    Thermal Conductivity of Electrolytic Iron, SRM 734, from 4 to
    300 K," NBS Special Publication, 260-31, National Bureau of
    Standards, Boulder, Colorado (1971).
5. J. G. Hust and P. J. Giarratano, "Standard Reference Materials:
    Thermal Conductivity and Electrical Resistivity Standard Refer-
    ence Materials: Electrolytic Iron, SRM's 734 and 797, from 4 to
    1000 K," NBS Special Publication 260-5, National Bureau of
    Standards, Boulder, Colorado (1975).

THERMAL DIFFUSIVITY MEASUREMENTS ON REFERENCE

MATERIALS AT IKE

Rüdiger Brandt and Günther Neuer

Institut für Kernenergetik und Energiesysteme
(IKE), Universität Stuttgart

INTRODUCTION

International programs on measurements of thermophysical properties of reference materials have a two-
fold aim:
- to find materials suitable for manufacturing
  reference samples for testing equipment
- to compare different measurement techniques.
The requirements to be fulfilled by a reference material are
- reproducibility of different samples of the same
  material
- long term stability of the sample properties
- suitability for different specific measurement
  techniques to determine the reference property.
Two so-called round robin programs (AGARD and CODATA)
have been organized during the past years on thermal
conductivity and thermal diffusivity measurements. IKE
participated with thermal diffusivity measurements on
Poco (AXM-5Q1) graphite and arc-cast tungsten in the
temperature range 500-2400 K.

MEASUREMENT TECHNIQUE

The method and equipment used at IKE to measure
the thermal diffusivity has been described recently by
Brandt and Neuer.[1] A disc-shaped sample of 8 mm diameter and 1-2 mm thickness is heated using a sinusou-
dally modulated heating beam. The temperature oscilla-

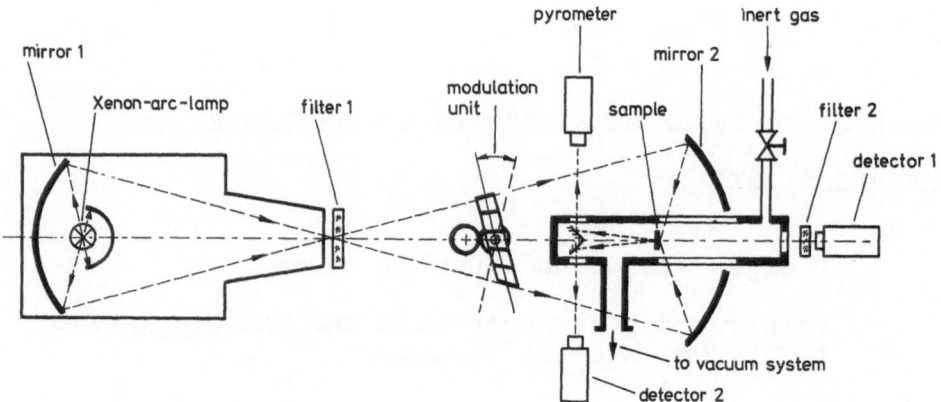

Fig. 1. Schematic of the thermal diffusivity
        measurement setup

tions propagate through the sample; the resulting phase
shift between the temperature oscillations at the
front-face and the rear face of the sample is measured
and is used to compute the thermal diffusivity. Attention is drawn to two salient characteristics of this
measurement technique:
  1. As shown in Figure 1, the xenon arc image is
     focused onto the sample by using two ellipsoi-
     dal mirrors. The precision of the adjustment is
     important for the resulting heating rate distri-
     bution and hence for the required one-dimen-
     sionality of the heat flux through the sample.
  2. The modulation frequency can be varied and the
     phase shifts both between front face and heating
     beam and between rear face and heating beam are
     measured. This enables us to correct for heat
     losses.
Before each reference measurement, the xenon lamp is adjusted in such a way that the phase shifts are measured
using a small target diameter of about 2 mm in the
center of the sample, and at four points towards the
edge on the right, on the left, and below and above the
center. The focus must be varied as long as the diffusivity values are equal everywhere and are not dependent
on the modulation frequency. To carry out the real
measurements, a measurement area of 5 mm is used.

All reference measurements were carried out at 4
different frequencies between 0.2 and 2 Hz. The mean
values are used as final results and the indicated
scattering range is due to the remaining frequency
dependence. All samples were heated in a vacuum.

MEASUREMENTS

Poco-AXM5Q1 Graphite

Results of diffusivity measurements on different
samples of Poco (AXM-5Q1) graphite are plotted in
Figure 2. The dashed line represents the values of the
sample used in the AGARD program.[2] It should be stated
that the same sample was used over a period of eight
years and the scatter of the comparative measurements
was within 4%.

The first CODATA measurements were made on sample
No. C-21 in September 1979. This sample was taken from
cylindrical rod No. 42F.

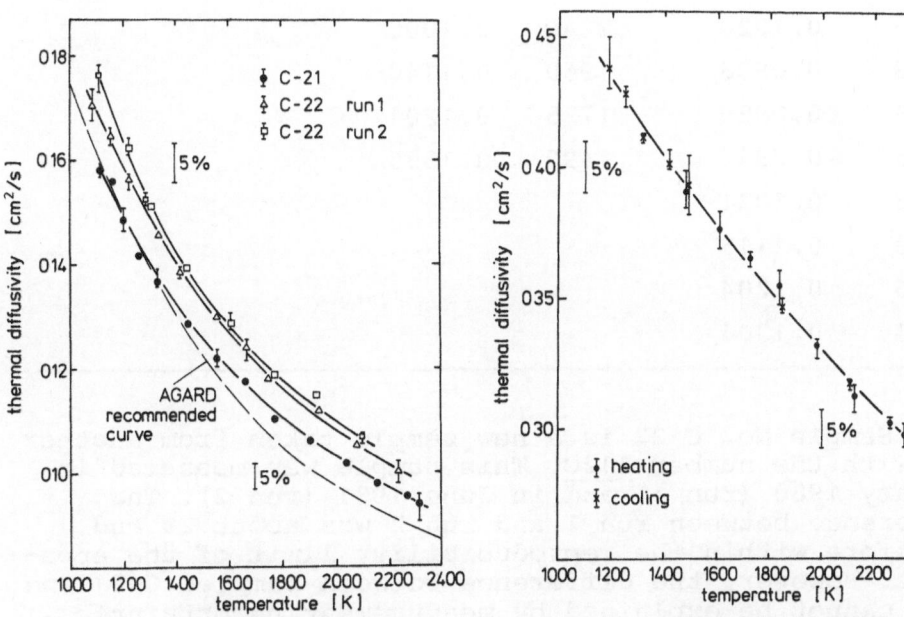

Fig. 2.   Thermal diffu-   Fig. 3.   Thermal diffu-
          sivity of Poco   sivity of arc-
          (AXM-5Q1) graphite   cast tungsten

Table 1 .   Thermal Diffusivity of Poco (AXM-5Q1)
             Graphite

Sample C-21 12.09.1979		Sample C-22 14.01.1980		Sample C-22 09.06.1981	
T, K	a, cm^2/s	T, K	a, cm^2/s	T, K	a, cm^2/s
1110	0.1582	1420	0.1392	1295	0.1513
1159	0.1564	1335	0.1459	1440	0.1394
1202	0.1491	1222	0.1564	1605	0.1291
1263	0.1421	1148	0.1651	1770	0.1189
1333	0.1372	1085	0.1708	1915	0.1152
1450	0.1291	1420	0.1395	2100	0.1073
1560	0.1226	1560	0.1303	1915	0.1160
1562	0.1234	1670	0.1244	1300	0.1530
1669	0.1183	1780	0.1185	1220	0.1628
1783	0.1105	1940	0.1124	1110	0.1769
1910	0.1065	2095	0.1062		
2045	0.1020	2240	0.1008		
2163	0.0983	1960	0.1140		
2275	0.0959	1725	0.1201		
2315	0.0944	1490	0.1355		
2045	0.1021				
1783	0.1115				
1556	0.1204				
1333	0.1304				

     Sample No. C-22 is a new sample taken from another
rod with the number 112C. This sample was measured in
January 1980 (run 1) and in June 1981 (run 2). The
difference between run 1 and run 2 was about 2% and
therefore within the reproducibility limit of the apparatus. However, the difference between samples C-21 and
C-22 cannot be explained by measurement uncertainties,
but must be caused by differences in the material.
Measurements on sample No. C-21 were also repeated in
June 1981 at a temperature of 1300 K. The thermal

diffusivity was 0.143 cm^2/s compared to 0.140 obtained
from the first measurement in 1979, indicating a
difference of about 2%. The values measured on C-22
agree with results of Taylor and Groot.[3] Compared with
the AGARD curve, the values of C-22 are 9-12% higher.
C-21 shows nearly the same values as the AGARD curve in
the temperature range 1100-1300 K, but with increasing
temperature the thermal diffusivity decreases slower,
and at 2000 K the values are 5% above the AGARD values.

The bars at the individual measurement points indicate the range of the results gained with different
heating frequencies of the Xenon lamp. This scattering
range does not exceed 3%.

Arc Cast Tungsten

The results of arc-cast tungsten are given in
Table 2 and are plotted in Figure 3. Between temperatures of 1200 and 2300 K the decrease of the values
is nearly linear versus temperature. Comparing the
scatter of measurements at different modulation frequencies, we see that the bars are in the range 1-5% and
the mean value of the reproducibility is about 4.5%
compared with that of graphite, which was below 3%.

Furthermore, it was noticed that the measurement
results of the tungsten sample were more sensitive to
the adjustments of the heating beam than those of
graphite. This is in accordance with results obtained
from a computational analysis published earlier by
Brandt and Neuer.[1]

In this analysis, nonuniform heating of a sample
(6 mm in diameter and 1 mm thick) was simulated by a
computer program, in which the phase shifts between
the temperature oscillations of small areas on the
front and rear faces (corresponding to the target diameter of the detectors) were computed as a function of
modulation frequency and of the location of the target
diameter on the sample faces. The simulations were
computed for a good thermal conductor (a = 0.3 cm^2/s)
and for a poor one (a = 0.03 cm^2/s). The results show
that the error in thermal diffusivity calculated from
these phase shifts is due to nonuniform heating and
is greater for the good conductor than for the poor
one.

Table 2.   Thermal Diffusivity of Arc-Cast Tungsten

Sample W-8	
T, K	$a, cm^2/s$
1482	0.391
1607	0.376
1723	0.366
1844	0.356
1981	0.332
2124	0.312
2265	0.302
2318	0.297
2124	0.317
1482	0.393
1414	0.402
1321	0.416
1256	0.428
1196	0.439

CONCLUSIONS

The thermal diffusivity of Poco (AXM-5Q1) graphite used for the CODATA Program seems to differ somewhat from the AGARD values. Since measurements on samples taken from different production lots and distributed within the frame of the CODATA Program lead to similar deviations, it can be concluded that the reproducibility of the manufacturing process is not negligible. The results of arc-cast tungsten lie within the range of scatter of the AGARD measurements, which, however, partly exceeded 10%. Hence, they do not satisfy the requirements for a reference material.

Our results showed that good conducting metallic materials are much more sensitive to the heating beam adjustment than poor heat conductors. Therefore, if a high accuracy of the thermal diffusivity of high conducting materials is to be ensured, a good conducting

reference sample is recommended to test the measurement
device. Despite the fact that Poco (AXM-5Q1) graphite is
an excellent reference material for the purpose of comparing different thermal diffusivity and conductivity
techniques, reliable metallic reference materials should
also be made available.

ACKNOWLEDGMENT

        This work was supported by Deutsche Forschungsge-
meinschaft (DFG).

REFERENCES

1.   R. Brandt and G. Neuer, Thermal diffusivity of
        solid—analysis of a modulated heating-beam
        technique. High Temp.-High Pressures, 11:
        59-68 (1979).

2.   E. Fitzer, "Thermophysical Properties of Solid
        Materials," AGARD Report No. 606, North
        Atlantic Treaty Organization (1973).

3.   R. E. Taylor and H. Groot. Thermophysical proper-
        ties of Poco graphite. High Temp.-High Pressures,
        12: 147-160 (1980).

STANDARD REFERENCE MATERIALS AND LOW THERMAL CONDUCTIVITY
EXPERIMENTAL METHODS (A feasibility study for CODATA Task
Group on Thermophysical Properties of Solids)

François Cabannes

Centre de Recherches sur la Physique des Hautes
Températures. CNRS, Université Orléans
45045 Orléans-cedex, France

INTRODUCTION

The measurement methods for the low conductivity range are fundamentally different from those for the high conductivity range. For
the broad class of ceramic materials with conductivity in the range
0.5 to 30 W/m.K there is a practical interest in technology and
industry for standard reference materials (SRM). SRM's will be
required to facilitate reliable thermophysical property measurements
of relatively high accuracy. However, we should keep in mind that a
misused technique will provide wrong results whatever SRM is used
for testing the technique.

When the radiation transport is effective, the measurement
techniques generally provide the thermal conductance instead of the
thermal conductivity. SRM's must be used with care. In some peculiar
problems, they are convenient for the measurement of the thermal
conductance; for instance, in the study of insulation systems for
buildings. But an SRM should be used with only one measurement
technique, with given sizes of the sample [1].

We may distinguish five temperature ranges:
1. Cryogenic: T < 100 K,
2. Low temperature: T < 500 K. Usually radiation transport is negligible, even in rather porous transparent materials,
3. Medium temperature; 500 - 1300 K. The radiation transport is
important in very low conductivity materials with high porosity,
4. High temperature: 1000 - 2000 K. The radiation transport is more
or less important, except in some dense ceramic materials,
5. Very high temperature: 2000 K < T. Only a few high refractory
compounds may be used.

Once SRM's have been chosen, the measurement techniques should be recommended. SRM's are useful to compare and to test different measurement techniques. At the same time, the best techniques are guides to choose one SRM.

We are concerned here with the low temperature range and partly with the high temperature range for the low conductivity materials. We do not emphasize the very low conductivity materials.

PREVIOUS STUDIES

The problem of low conductivity standards is not a new one. In 1963, a study of cordierite (9606 Pyroceram) was undertaken at the National Bureau of Standards, [2,3]. In 1966, the first selection of ceramic materials was investigated by A. E. Wechsler [4]. Five ceramics were studied: alumina, thoria, beryllia, zirconia, and calcium zirconate. The former two were chosen for continued studies because of their good chemical stability in vacuum and in contact with refractory metals. This means that the measurement techniques influence the selection of SRM's.

A few years later, among the same five ceramics, alumina and thoria were chosen again for an evaluation as high temperature thermophysical property calibration materials by M. L. Minges [5].

A lot of work has been done on uranium oxide, including a round-robin experimental program [6].

Beside the oxide ceramics, nitride and carbide materials have been studied, but much less extensiveley than the oxides. The thermal conductivity of carbides is too large, out of the range 0.5 to 30 W/m.K. Today considerable work is being done on materials development and characterization on silicon nitride and sialons for aerospace applications.

Results of the evaluation of selected refractories as high temperature thermophysical property calibration materials [5]

Alumina and thoria were selected, then studied in the "field-test" program of four laboratories. The main difficulties encountered in the use of alumina were occasional cracking of the samples during heating and probable radiation transport effect above 1400 K, despite the use of chromium doping (0.23% Cr).

For thoria, the data scatter was greater than that observed with alumina, amounting to 4 to 13% among the several investigations. However, it is worth studying more thoria because of the linear temperature dependence of the thermal resistivity. This dependence,

as derived from the least squares equations, was observed within 1%
to the highest test temperature, 2500 K. Such a relationship is quite
useful, implying that the radiation transport is negligible.

In alumina the radiation transport effect is probably 12% at
2100 K, thus pure alumina is not the most convenient material at
high temperatures. Chromium doping is not effective.

Results on uranium oxide

A lot of work has been devoted to uranium oxide, as shown in
figure 1! Because of the considerable interest in this material for
nuclear plants, a large amount of data clarify the problem, as illustrated in figure 2. Despite the effort of the international community,
the data scatter remains rather large. Either the materials or the
measurement techniques are responsible for the data scatter. The
results of a round-robin program demonstrate the influence of the
measurement techniques. Thus it is clear that difficulties arise in
selecting SRM's to be used with different measurement techniques.
A possible disadvantage of UO_2 is the influence of the O/U ratio on
the thermal conductivity.

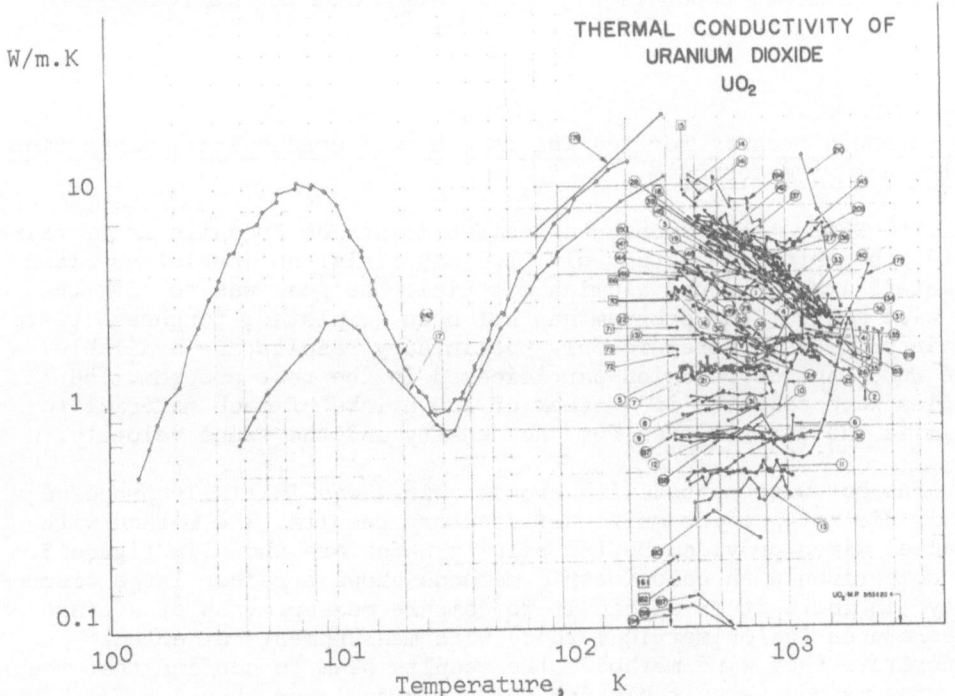

Fig. 1.   A lot of work has been done on uranium oxide!   Data published by TPRC, Purdue University.

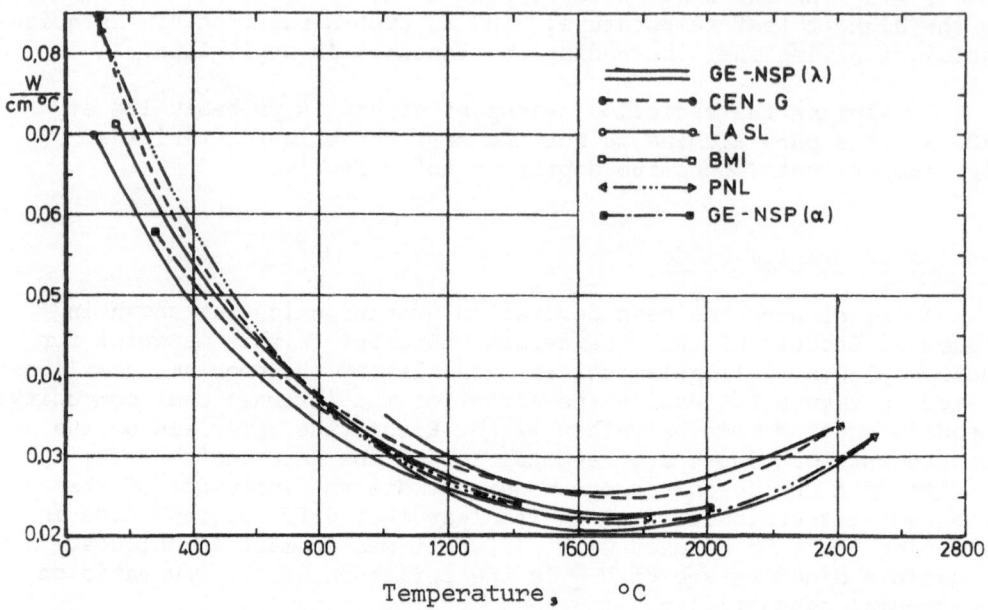

Fig. 2.   Thermal conductivity of uranium oxide UO_2; a round-robin
          program clarified the problem [6].

The European program for testing the thermal conductivity measurement by the hot wire method [7]

The Fédération Européenne des Fabricants de Produits Réfractaires
(PRE) has performed a study with two materials: an insulating class
28 material and a dense alumina material. The goal was to test the
hot wire method. The program has not been completed; toughness tests
are in progress, therefore only preliminary results are available.
Nine European laboratories participated in the test program. The
samples were selected in a stock of 100 bricks of each material to
minimize the data scatter for the density and the sound velocity.

The hot wire method with crossed wires was initially proposed,
but it did not provide quite satisfactory results. The method with
parallel wires provided better results which are shown in figure 3.
The comparison with calorimetric methods shows a rather large discrepancy. In fact, it is difficult to compare measurements at average
temperatures (calorimetric method) with measurements at actual
temperature (hot wire method). The results seem to confirm that the
hot wire method is suitable for conductivity lower than 1.5 W/m.K.
Nevertheless, the hot wire method with parallel wires is obviously
worthy of further tests, despite the difficulties encountered in the
thermal contacts in cases of conductivities higher than 1.5 W/m.K.

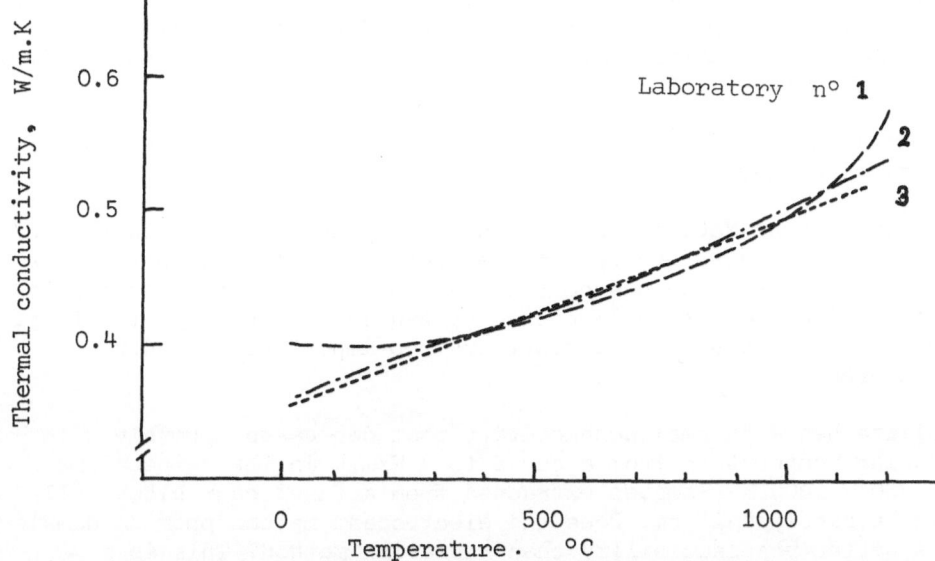

Fig. 3. Thermal conductivity of an insulating class 28 material;
measurements with the hot wire method (parallel wires) [7].

THE PRESENT FEASIBILITY STUDY FOR CODATA TASK GROUP ON THERMOPHYSICAL
PROPERTIES OF SOLIDS (Chairman: M. L. Minges)

To perform this feasibility study, 31 scientists from 9
countries were consulted. We received 15 answers from 7 countries.
Thirteen scientists are ready to participate when a task will be
undertaken; two would be able to participate in the future. Among
the 15 who replied, one is concerned with high-pressure field, one
with cryogenics, and two with very low conductivity SRM's. Unfortunately, all the laboratories active in the field were not consulted;
later we hope to get more information in order to extend the study.

The three topics of this study are:
1. Identification of potential SRM candidates and their state of
   development,
2. Establishment of criteria to be used in SRM characterization,
3. Identification of the measurement techniques used and recommendations for a comparative measurement program.

Materials

From results of previous studies, four materials may be selected
for further investigation: cordierite, alumina, thoria, and uranium
oxide. Some other materials were suggested in the answers we received
and are of interest:

1. Crystalline and fused quartz, which are commonly used as standards
for geological materials.

2. Single crystal NaCl and MgO could be proposed in low-temperature
high-pressure studies because of the simplicity of their structures
and the lack of a phase change, in the case of NaCl at pressures as
high as 30 GPa.They are well characterized materials. Unfortunately,
the effects of radiation transport are too important at medium
temperatures.

3. Mullite has a thermal conductivity that decreases smoothly with
increasing temperature from about 5 to 4 W/m.k in the temperature
range 300 - 1800 K. Samples extracted from a fused cast block will
likely be highly uniform. Does the electrocast method provide sample
with a better reproducibility than sintering method? This is a
worthy question. In the case of alumina-zirconia electrocast
materials, the answer is no, because of the phase change during the
cooling and further heat treatments. The answer could be different
in the case of mullite. By the way, we shall have to discuss the
characterization of SRM's.

4. Other materials must be kept in mind: partially stabilized
zirconia, silicon nitride, and sialons.

Some work has been carried out on partially stabilized zirconia,
but it seems too early to say that it is an SRM candidate. Silicon
nitride and sialons are the subject of considerable current materials
development and characterization work for various aerospace and
engine applications. Their oxidation susceptibility is a possible
disadvantage. As, for mullite, the thermal conductivity of RSSN
silicon nitride decreases smoothly with increasing temperature. The
effects of radiation transport are probably negligible, but the
porosity should be quite well controlled.

SRM CHARACTERIZATION

The criteria that the ideal thermal conductivity SRM should meet
were reviewed by D. R. Flynn et al. [3]. Moreover, the thermophysical
characterization of candidate SRM's was presented at this conference
by K. D. Maglic et al. [8]. Then we shall discuss the question briefly.
Of course, we should try to get samples with the most reproducible
characterization. Nevertheless, in a study of SRM candidates, all

samples should be characterized.

SRM's should have:
1. Isotropy,
2. Normal mechanical and thermal stresses,
3. Chemical inertness and chemical stability in temperature cycling
   and under conditions of use (air, inert gas, vacuum), and
4. High electrical resistivity offers an advantage.

The main features of the characterization are:
1. Porosity and type of pores (open or closed, preferably closed),
2. Infrared opacity,
3. Homogeneity and reproducibility of phase composition and of the
   microstructure. High purity is a disadvantage when the first traces
of impurity markedly affect the thermal conductivity.

Moreover SRM's should be:
1. Easy to fabricate in order to have stocks of materials,
2. Cheap, that means that the cost of the material is much lower
   than the cost of the measurement.

None of the SRM candidates will satisfy all of these features,
so that it will be necessary to select an SRM for a limited temperature range and a given conductivity range. Let us see how these
different features affect the choice of an SRM.

Isotropy. To compare different measurement techniques, such as radial
flow meters and the cut bar method, the samples have to be isotropic.
In porous and in dense materials, isotropy is easy to achieve.

Mechanical and thermal stresses. Some materials have a low density
and a high porosity (above 50%); therefore, their mechanical strength
is very low. These materials should be handled with care and the
measurement techniques adopted that will not destroy the samples,
nor change the sample sizes. One of the main problem is the achie-
vement of good thermal contacts. This problem is less drastic in the
case of dense materials, except sometimes in the high temperature
range.

Chemical inertness and chemical stability. The oxidation susceptibility of nitrides and carbides is a disadvantage. In the case of
uranium oxide the O/U ratio may change with different surroundings;
then it affects the thermal conductivity. The 9606 Pyroceram
(cordierite) is a mixture of several phases. The mixture may change
when heated through the range 900 - 1300 K. Then the material should
be preheated at the maximum working temperature before any conductivity measurement [3]. Except in the very high temperature range, the
chemical inertness of alumina, thoria, and mullite is very good, as
well as the chemical stability.

Porosity. The porosity is the most drastic property that affects the thermal conductivity. Glass-ceramic materials such as 9606 Pyroceram have an advantage since their porosity is zero. From this point of view, electrocast materials could be of interest. It seems preferable to fabricate, whenever it is possible, materials with closed porosity, but in this case the porosity is much more difficult to measure.

Opacity. The infrared opacity is very sensitive to the microstructure (grain size, porosity, absorbing impurities), so we should try to minimize the effects of radiation transport. Some studies on the doping of transparent materials, such as alumina, mullite, and cordierite, would be worthy. At 1300 K the radiation transport in 9606 Pyroceram seems to be few percent of the phonon conductivity.

IDENTIFICATION OF THE MEASUREMENT TECHNIQUES

The identification of a measurement technique, which could be recommended, is a difficult problem. SRM's should be a priori useful to compare different measurement methods and to test the apparatus. In fact, some techniques seem the most dependable and precise; therefore they were used to select SRM's in the previous studies.

For selection of SRM candidates, the absolute measurement techniques obviously should be used; they are:
1. Steady-state longitudinal heat flow meters (calorimetric ASTM method),
2. Steady-state radial heat flow meters, more suitable at high temperatures.

The former methods need large size samples; therefore SRM's must be cheap and easy to fabricate with good homogeneity. The latter methods need materials easy to machine.

Some people feel that other measurement techniques are worth investigating, for instance, monotonic heating techniques and hot wire methods. These methods were developed more recently and they present technical advantages. One of the advantages is that the conductivity is provided at the actual temperature in the hot wire methods, whereas the calorimetric methods provide the conductivity at average temperatures. Industry and technical centers are attracted by the hot wire methods, which seem easy to use. However, difficulties are encountered with thermal resistance contacts and the investigation of the technique with SRM's will give interesting answers.

Besides the thermal conductivity measurement techniques, we should mention the thermal diffusivity measurement methods. Usually they are easier to use, and they need smaller samples than the conductivity measurement techniques do.

The small size of the samples is an advantage for SRM characterization, except when the radiation transport may occur. The SRM would be selected in a different way for diffusivity measurements; however, we think it is better to select only one SRM for both diffusivity and conductivity measurements, even if the range of apparatus is more limited.

One question arises: does the use of an SRM with a measurement technique guarantee the results on other materials? We cannot say, for instance, that it is the case with the hot wire method and the heated probe technique. The key to success lies in accurate temperature time history measurements and obeying the boundary conditions. These are complex situations that may change from one run to another. Obviously, the recommended techniques must be the most dependable.

CONCLUSIONS

There is a considerable interest in industry and technology for the application of the broad class of ceramic materials and the knowledge of their thermophysical properties. SRM's are required to facilitate reliable thermophysical property measurements with relatively high accuracy. Therefore it is worth studying SRM candidates for the low thermal conductivity range.

From previous studies, four materials were selected: alumina, thoria, uranium oxide, and cordierite. For choosing SRM's, these studies need some further work. Thoria seems suitable for a large temperature range, and eventually, uranium oxide. Their conductivities change smoothly with increasing temperatures, that is an advantage. Alumina and cordierite may be good candidates, except at high temperatures (> 1400 K) because of the effects of the radiation transport. Mullite and stabilized zirconia could be candidates so far as they are very well characterized. A study of electrocast materials, in connection with their possibly good homogeneity and their low porosity, is of interest if the mechanical and thermal stresses are normal.

The laboratories participating in the SRM program should use the most dependable and accurate measurement techniques, such as steady-state longitudinal (and radial) heat flow meters, but other measurement methods have to be experienced simultaneously, for instance, the thermal diffusivity measurement methods and the monotonic heating regime. We must pay attention to the temperature range in which the SRM will be used. This temperature range is inevitably restricted and before selecting an SRM, it has to be chosen. This is not the least of the problems!

REFERENCES

1. F.J. Powell, Thermal conductance round-robin for guarded hot-
   plate and heat flow meters and status of ASTM C16 round-
   robin on hot boxes, Seventeenth Thermal Conductivity Conf.
   Gaithersburg (1981).
2. H. W. Flieger, The thermal diffusivity of pyroceram at high
   temperatures, Third Thermal Conductivity Conference,
   p. 769, ORNL (1963).
3. D. R. Flynn, H. E. Robinson, I. L. Martz, Present status of
   pyroceram code 9606 as a thermal conductivity reference
   standard, paper I-F, Fourth Thermal Conductivity Conference,
   USNRDL, San Francisco (1964).
4. A. E. Wechsler, High temperature thermal conductivity and
   diffusivity standards - Ceramics, p. 603, Sixth Thermal
   Conductivity Conference, MADAFML, Dayton (1966).
5. M. L. Minges, Evaluation of selected refractories as high
   temperature thermophysical property calibration materials,
   Tech. Rep. AFML - TR-74-96, (1975).
6. R. Brandt, G. Haufler, G. Neuer, Thermal conductivity and
   emittance of solid UO_2. A critical review and analysis of
   the literature, CINDAS/Purdue Univ. (1973).
7. Document ISO/TC 33/SC 2 105F, Nov. 1980, communicated by
   P. Lapoujade, PRE Secrétariat technique, Paris.
8. K. D. Maglic, N. Petrovic, Z. Zivotic, Thermal diffusivity
   of four candidate standard reference materials, Seventeenth Thermal Conductivity Conference, Gaithersburg
   (1981).

SESSION   TC-4

Standard Reference Materials and Data - II

CHAIRMAN

M. L. Minges
Air Force Materials Laboratory
Dayton, Ohio

NONDESTRUCTIVE CHARACTERIZATION OF POCO AXM GRAPHITE

J. R. Koenig and C. D. Pears

Southern Research Institute
Birmingham, Alabama   35255

INTRODUCTION

The establishment of international standard materials requires, perhaps first and foremost, consistency of the various samples of that material used by the various agencies for technique development, facility calibration, and scientific research. Some materials have an intrinsic homogeneity that allows them to be used as standards with only nominal screening beyond batch sampling and certification. Other materials, such as polycrystalline graphites, are inherently inhomogeneous, allowing the properties of interest to vary widely from sample to sample. When other considerations compel the use of such a material, other techniques must be utilized to assure minimal sample-to-sample variation. This is true of the Poco AXM-5Q1 grade graphite being discussed here, although among graphities, AXM-5Q1 has historically displayed superior consistency, which contributed to its selection as a candidate standard for high temperature uses.

The objective of this study was to utilize various nondestructive techniques to evaluate a respectable number of samples (28) to provide a verification of the ability to define acceptability of specimens as standards, and to establish the ability to select acceptable specimens for use as standards. Inherent in these objectives are the secondary objectives 1) to determine if nondestructive characterization monitors other than electrical resistivity could be used to distinguish batch differences that could be related to thermal conductivity and 2) to determine the restraints on specimen size and ultrasonic test frequency that would yield a consistent and accurate measure of ultrasonic velocity, should this monitor correlate with conductivity.

137

These tests were conducted "double blind" in that neither the pedigree of the specimens nor the eventual results of thermal conductivity studies on these specimens were known, with the exception of two samples in a round robin in which Southern Research was involved. The former is a limitation in that the statistical relevance of the results, as it is affected by batch to batch variation, is unknown. The latter leaves this study, to date, incomplete in developing criteria for acceptability.

The nondestructive tests selected for use in this study were:

1) visual inspection
2) radiography
3) density
4) open porosity
5) ultrasonic velocity

Electrical resistivity values were provided by the National Bureau of Standards (NBS). These techniques were selected on the basis of their prior successful utilization in sorting variations in other polycrystalline graphites (including other grades of Poco graphites) which were indicative of property changes.

VISUAL INSPECTION

Each of the specimens was notched for identification. The number and orientation of the notches on the ends identified the specimen number. Then each specimen was carefully inspected with the aid of low magnification. Specimen 500 had a 0.5 mm "hole" at 3 cm from the reference end. No other significant observation was noted.

RADIOGRAPHY

The objective of the radiography was to inspect for any "disparate zones" in the specimens. Examples of these would be cracks, pore clusters, flow lines, significant density variations, large pores, and flaws. The radiographs were made utilizing a radioflow 350 unit (Torr X-Ray Corporation), which is rated for operation between 0 to 120 kV at either 3 or 5 mA. A small focal spot size (0.35 mm) is used to provide high sensitivity and distortion-free imaging of small discontinuities. Two views 0 and $90°$ from the reference of each specimen were inspected. The radiograph of specimen 500 had a large, low-density spot corresponding to the "hole" seen visually. Five other specimens had radiographic indications, as follows:

Specimen Number	Indication
45	Several striations including a faint high density striation 1.2 cm from the unnotched end in the 90° view
220	Speckled (high density flakes)
300	Vague pore clusters
330	Speckled (high density flakes)
63	Lightly speckled
500	Large pore and several vague pore clusters

DENSITY AND OPEN POROSITY

The density monitors changes in both the green state processing (compacting pressure, variation in raw material, etc.) and pyrolization/graphitization changes. It was measured using standard gravimetric techniques, measuring the sample dimensions to the nearest 0.001 cm using calibrated micrometers and measuring the weight to the nearest ±0.0001 gm using an analytical balance. The open porosity, which is also a monitor of processing variations, was measured using a mineral spirits absorption technique, where dry, saturated, and suspended weights are used to calculate the open porosity. The results of the density and open porosity measurements are shown in Table 1.

The open porosity variations on these samples of AXM 5Q1 are reflecting the changes in the density, as shown in Figure 1, and are not detecting any independent variation between the pieces. This is also true if the different diameter materials are treated as separate groups.

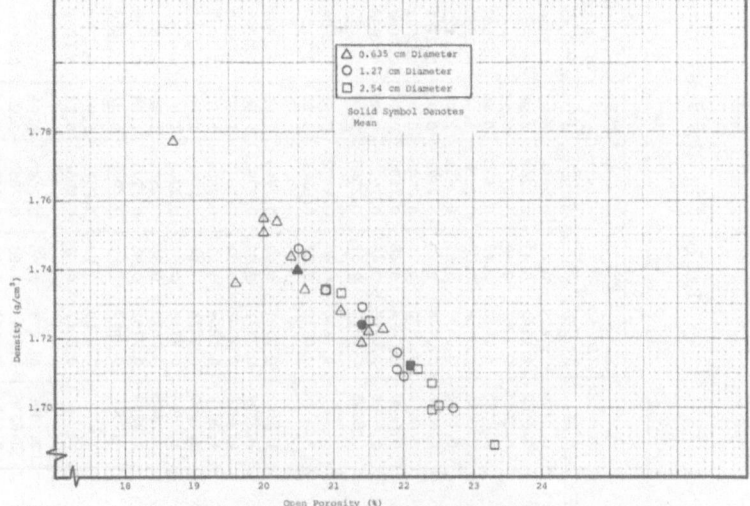

Figure 1.   Correlation of Open Porosity and Density on Specimens of Poco AXM 5Q1

Table 1. Nondestructive Monitors of NBS Poco Graphite Specimens.

Specimen	Diameter (cm)	0.2 MHz Velocity (cm/µs)	0.2 MHz D/λ	0.4 MHz Velocity (cm/µs)	0.4 MHz D/λ	1.0 MHz Velocity (cm/µs)	1.0 MHz D/λ	2.25 MHz Velocity (cm/µs)	2.25 MHz D/λ	5.0 MHz Velocity (cm/µs)	5.0 MHz D/λ	Open Porosity %	Electrical Resistivity (µΩm)	Length (cm)	Density (g/cm³)
5A	0.6459	0.2675	0.48	0.2664	0.97	0.2426	2.66	—	—	—	—	21.4	16.54	8.8951	1.719
5B	0.6462	0.2741	0.47	0.2738	0.94	—	—	—	—	—	—	20.2	15.03	10.0876	1.754
5C	0.6464	0.2761		0.2771	0.93	—	—	—	—	—	—	18.7	14.44	11.4193	1.777
103	0.6355	0.2621	0.48	0.2639	0.96	0.2604	2.44	0.2637	5.42	0.2624	12.11	21.7	15.49	4.9373	1.723
104	0.6439	0.2?8	0.50	0.2581	1.00	0.2558	2.52	0.2573	5.63	0.2573	12.51	20.0	17.31	4.7287	1.755
151	0.4672	0.2?13	0.53	0.2451	1.06	0.2431	2.66	0.2436	5.98	0.2410	13.42	21.1	16.61	4.9505	1.728
202	0.6386	0.2?48	0.46	0.2771	0.92	0.2741	2.33	0.2753	5.22	0.2746	11.63	20.0	13.27	4.9403	1.751
300	0.6421	0.2?3	0.50	0.2690	0.99	0.2593	2.48	0.2588	5.58	0.2598	13.82	20.6	15.67	4.4976	1.734
330	0.6449	0.2?4	0.48	0.2690	0.96	0.2664	2.42	0.2675	5.43	0.2687	12.00	21.5	22.01	5.0973	1.722
500	0.6469	0.2?8	0.45	0.2911	0.89	0.2898	2.23	0.2924	4.98	0.2898	11.16	19.6	15.08	4.7922	1.736
220	0.6452	0.2?2	0.51	0.2568	1.00	0.2537	2.54	0.2525	5.75	0.2520	12.80	20.4	18.95	4.8085	1.744
Avg.		0.2?4	0.48	0.2672	0.97	0.2606	2.48	0.2639	5.50	0.2631	12.43	20.5	16.40		1.740
S.D.		0.0130	0.02	0.0124	0.05	0.0150	0.14	0.0150	0.31	0.0147	0.89	0.9	2.40		0.018
13	1.2802	0.2761	0.93	0.2769	1.85	0.2764	4.63	0.2766	10.41	0.2758	23.20	21.0	12.38	10.1072	1.727
34A	1.2799	0.2771	0.92	0.2764	1.85	0.2769	4.62	0.2761	10.43	0.2776	23.05	21.4	15.56	8.7655	1.729
34B	1.2814	0.2809	0.91	0.2812	1.82	0.2802	4.57	0.2802	10.29	0.2802	22.87	20.5	14.80	11.4706	1.746
34C	1.2812	0.2804	0.91	0.2804	1.83	0.2807	4.56	0.2799	10.30	0.2804	22.84	20.6	14.35	9.9004	1.744
36	1.2807	0.2725	0.94	0.2718	1.88	0.2710	4.73	0.2723	10.58	—	—	22.7	14.43	10.0101	1.700
43	1.2807	0.2748	0.93	0.2743	1.87	0.2748	4.66	0.2748	10.48	—	—	21.9	16.43	10.1018	1.716
44	1.2809	0.2878	0.89	0.2880	1.78	0.2873	4.46	0.2880	10.01	0.2865	22.35	20.9	15.23	10.1910	1.734
46	1.2809	0.2761	0.93	0.2771	1.85	0.2769	4.63	0.2766	10.42	0.2766	23.15	21.9	17.03	10.1453	1.711
63	1.2812	0.2662	0.96	0.2675	1.92	0.2670	4.80	0.2672	10.79	0.2675	23.95	22.0	14.80	10.0736	1.709
Avg.		0.2769	0.92	0.2771	1.85	0.2769	4.63	0.2769	10.41	0.2779	23.06	21.4	15.00		1.724
S.D.		0.0061	0.02	0.0058	0.04	0.0058	0.10	0.0058	0.21	0.0058	0.48	0.7	1.33		0.016
32A	2.6096	0.2675	1.95	0.2690	3.88	0.2690	9.70	0.2664	22.04	0.2677	48.74	21.1	13.32	8.8854	1.733
32B	2.6215	0.2644	1.98	0.2652	3.95	0.2649	9.90	0.2631	22.42	—	—	22.5	13.68	11.2149	1.703
32C	2.6114	0.2654	1.97	0.2680	3.90	0.2662	9.81	0.2654	22.14	0.2680	48.73	21.5	13.97	10.0691	1.725
34	2.6119	0.2616	2.00	0.2614	4.00	0.2614	9.99	0.2601	22.59	0.2616	49.92	22.2	13.66	10.0698	1.711
42	2.5954	0.2545	2.04	0.2548	4.08	0.2555	10.16	0.2553	22.88	0.2560	50.68	22.8	14.73	9.9898	1.698
45	2.6083	0.2761	1.89	0.2769	3.77	0.2764	9.44	0.2756	21.30	0.2764	47.19	20.9	13.77	10.1943	1.734
52	2.6007	0.2598	2.00	0.2591	4.02	0.2609	9.97	0.2593	22.56	0.2604	49.95	22.4	13.39	10.0635	1.707
65	2.6020	0.2522	2.06	0.2515	4.14	0.2520	10.33	0.2507	23.35	0.2530	51.43	23.3	14.55	10.1257	1.689
Avg.		0.2626	1.99	0.2631	3.97	0.2634	9.91	0.2621	22.41	0.2634	49.52	22.1	13.88		1.712
S.D.		0.0076	0.05	0.0081	0.12	0.0076	0.27	0.0076	0.61	0.0079	1.41	0.8	0.52		0.017

Velocity values not provided were not obtained due to excessive signal attenuation.

ULTRASONIC VELOCITY

Sonic velocity measurements were made at frequencies of 0.2, 0.4, 1.0, 2.25, and 5.0 MHz using a technique where transmitting and receiving transducers are in contact with opposite ends of a specimen, coupled with alcohol. The measurements were made utilizing a Sperry UM 721 Reflectoscope as a pulser and a Tektronix 564 Oscilloscope as the signal measuring device. The reference point for determining the delay time was the first break of the rf signal. This technique gives consistent results by allowing precise measurement of the delay time. Table 1 lists the sonic velocity at each frequency for each specimen. For convenience, the values of open porosity, electrical resistivity (supplied by NBS), length, bulk density, and D/λ are also listed. The D/λ is the ratio of specimen diameter to wavelength of the ultrasonic energy.

Comparing the sonic velocities measured to the range of D/λ ratios existing within this data set shows very little change as a function of D/λ over the range 0.5 to 50. This is not in agreement with the model as presented by Bancroft[1] wherein, for an isotropic nondispersive material with a Poisson's ratio of approximately 0.2, a decrease in measured velocity of 40 percent would be anticipated over that range. Figure 2 shows the velocity trend for various D/λ ratios graphically. Also shown are the results of two other experiments, one on a carbon-carbon composite and the second on Union Carbide ATJ-S graphite. On the carbon-carbon the expected trend is seen. The carbon-carbon and ATJ-S experiments differ in that they were controlled experiments wherein a single specimen was machined in steps from a larger diameter to a small diameter while the length remained constant.

There is no evident trend in the velocity as a function of the length of the specimen. This is as would be expected since the diameter/length ratios for all of the specimens tested is below 0.15. Bancroft[2] shows, for homogeneous, nondispersive materials (which AXM 5Q1 is not, but it is as close as any graphite) that at ratios under 0.4 the error in velocity is less than 0.5 percent. When one considers only the specimens that are apparently from the same source billet (assumed based on I.D. numbers; 5 a to c, 32 a to c, and 34 a to c), there does appear to be a mild trend in the data as seen in Table 1. However, there is a tight correlation between the open porosity and the velocity for these specimens, indicating that the effect is probably due to the dispersive nature of these types of materials where the phase velocity is not equal to the group velocity/attenuation response of the material. The frequency/attenuation response is a function of the pore size distribution and grain size for graphitic materials.

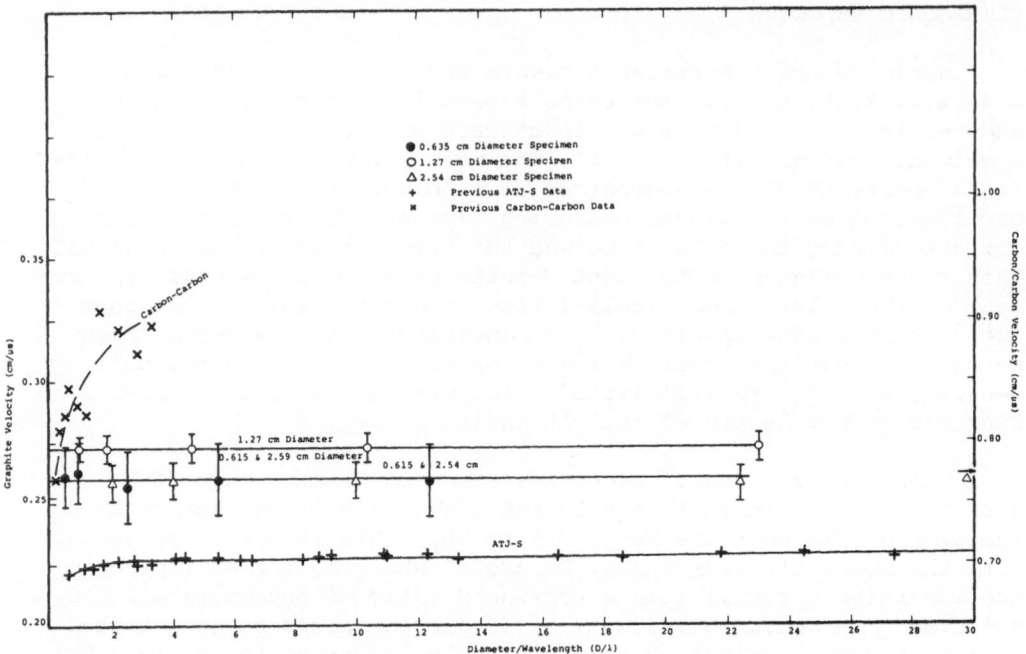

Figure 2.   Velocity Trends for Various Diameter to Wavelength Ratios

DISCUSSION

Recalling the primary objective of this effort, the work cannot be complete without the availability of consistent thermal conductivity on the set of samples evaluated. Some conclusions can be reached on the variability of the material, both specimen to specimen and by apparent lots. In addition, by making the assumption that the thermal conductivity of this material is, at least to some extent, governed by the Wiedemann-Franz law, then conclusions can be drawn by comparisons to the electrical resistivity.

Figure 3 shows a plot of density of electrical resistivity by specimen. The means of the specimens as grouped by diameter are gradually increasing, opposite of the expected trend; however, the data points within each set show the expected correspondence. The same is true (as would be expected from Figure 1) of the open porosity. The inverted correspondence of the means is illustrated in Figure 4. These data strongly indicate the existence of three separate populations among the samples evaluated.

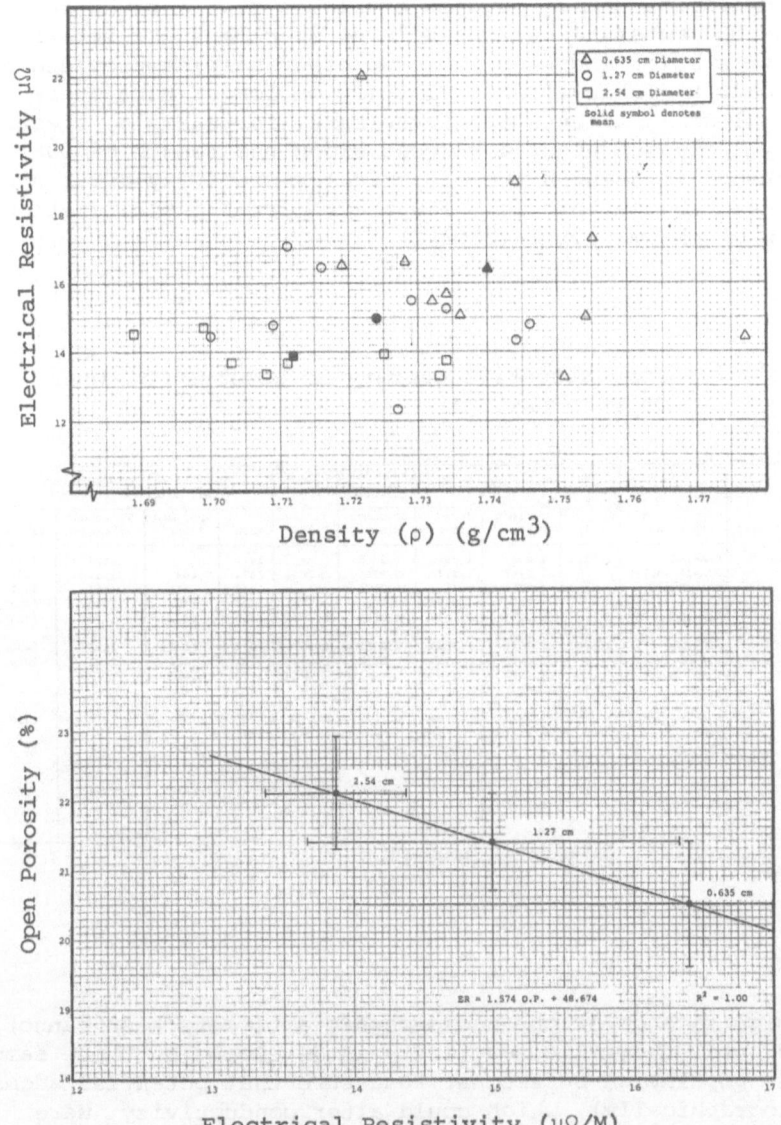

Figure 4. Electrical Resistivity Trends for Various Open
Porosity Values (Averages)

No strong correlation exists between sonic velocity and
electrical resistivity either for the entire population or for the
three sublets, even though a significant range of resistivity values
existed. Sonic velocity does have the potential, however, to sort
differences in processing for polycrystalline graphites, as it has
done in numerous previous studies.[3,4]

One additional set of data is available from a round robin on the NBS AXM 5Q1 material, which included specimens excised from two of the above samples and also from two other samples provided by NBS. These data (Table 2) show that (1) on longer bars there is the potential for gradients along the bars (for example, electrical resistivity increases about 4 percent along the length of 006) and (2) some change is seen in the electrical resistivity after exposure to 2200°C even though this is below the reported graphitization temperature of the material (2500°C).

Table 2.   Nondestructive Summary of Results Obtained on Poco AXM 5Q1

Material and Billet	Specimen	Density (g/cm³)			Velocity cm/μs Direction of Measurement = Q		Electrical Resistivity μΩ-cm (all axial except 32C)			
		Bulk	Specimen Blank	Final Values	Specimen Blanks	Final Values	Bulk	Specimen Blank	Finished* Specimen	Final* Values
POCO AXM-006	CRA-1-006	1.7103	1.6961	1.6960	0.2568	–	1415		1534	1567
	RIA-1A (0.7)		1.7165	1.7176	0.2565	0.2604	1472	1441	1511	1457
	RIA-1B (0.7)		1.7155	1.7147	0.2581	0.2576	1443	1414	1469	1421
	RIA-1C (0.7)		1.7176	1.7165	0.2581	0.2581	1436	1426	1442	1413
	RIA-1D (0.7)		1.7152	1.7135	0.2553	0.2560	1418	1402	1470	1423
POCO AXM-001	RIA-2C (0.5)	1.6900	1.7046	1.7053	0.2560	0.2525	1420	1372	1418	1434
	RIA-2D (0.5)		1.6932	1.6933	0.2537	0.2492	1435	1429	1463	1469
	TE-1 (Opt.)		1.6892	–	–	–	1420	1413	–	–
	TE-2		1.6973	–	–	–	1420	1404	–	–
POCO AXM-32C	RIA-2A (0.5)	1.725	1.7174	1.7138	0.2611	0.2548	~1397**	1391	1437	1436
	TE-4		1.7371	–	–	–	~1397**	1302	–	–
	CRA-2 (Across Rod)		1.7075	1.7260	0.2654	0.2642	~1397**	1325	1348	1330
POCO AXM-45	RIA-2B (0.5)	1.734	1.7411	1.7398	0.2764	0.2723	~1377**	1339	1384	1376
	TE-3 (Opt.)		1.7321	–	–	–	~1377**	1366	–	–

*Values not corrected for holes. Used for relative comparisons only.
**From NBS Data

CONCLUSIONS

The significance of the radiographic artifacts seen cannot be determined until thermal conductivity is developed on those samples. Nor was the population sufficient to assure that potential anomalies (seen radiographically), which could alter conductivity, were included. Once that population and correlation exists, it will have to be determined if those samples would be screened by other indicators before deleting this monitor.

Within the total population evaluated no distinctions were seen between the indications provided by density and open porosity. Both confirmed the existence of three populations within the materials evaluated, and within each of those sets, showed the expected correspondence with electrical resistivity, although the reverse correlation was seen in the population as a whole owing to differences in the populations.

The sonic velocity trends as a function of specimen dimensions did not follow expected theoretical trends, and in fact, were relatively insensitive to changes in D/λ or D/L. However, as a matter of good practice, the use of a constant D/λ for each measurement is suggested.

Sonic velocity does not correlate with electrical resistivity within the population tested; however, it does distinguish populations and has historically shown itself to be a good monitor of processing changes in polygraphites.

This effort is, as yet, incomplete in that the correlation to thermal conductivity is not available. However, a data base has been established for this one set of examples and these monitors have shown the ability to distinguish variations between apparent batches. This is illustrated in Table 3, which shows the means of the different subsets. Better tracking of batch and process for standards from this class of materials is indicated.

Table 3. Comparison of Mean Values by Subset

Subset	Diameter (cm)	Density (g/cm^3)	Open Porosity (%)	Sonic Velocity @ 0.4 MHz (cm/µs)	Electrical Resistivity (µohm/m)
A	0.64	1.74	20.5	0.267	16.4
B	1.27	1.72	21.4	0.277	15.0
C	2.54	1.71	22.1	0.263	13.9

ACKNOWLEDGMENTS

This work was conducted under the direction of H. E. Littleton who was the author of the initial report. The help of D. Van Wagoner in interpreting these data are gratefully acknowledged. The effort was funded by the Air Force Wright Aeronautical Laboratories/ Materials Laboratory under the technical direction of L. S. Theibert.

[1] The Velocity of Longitudinal Waves in Cylindrical Bars, Dennison Bancroft, Physical Review, Vol 59, 1941, pg 588-93.
[2] Ibid, pg 592.
[3] H. E. Littleton and C. D. Pears, Truncation of 994-2 Graphite, AFML-TR-76-204, August, 1977.
[4] J. R. Koenig, The Evaluation of Several Experimental Grades of Poco Graphite, AFML-TR-76-171, February, 1977.

THERMAL DIFFUSIVITY OF A POCO GRAPHITE

V.V. Mirkovich

Mineral Sciences Laboratories, CANMET
Energy, Mines and Resources Canada, Ottawa
K1A 0G1

ABSTRACT

Thermal diffusivity of a POCO graphite specimen was measured
in an apparatus based on the principle of an infinite cylinder and
radial symmetry. The measurements were performed in the temperature
range of 0° to 850°C. The specimen was obtained from the U.S.
National Bureau of Standards, from a batch used in a round-robin
cooperative program for evaluation of its suitability as a thermal
conductivity standard. Comparison with results of other studies
shows the data to be low by approximately 8 to 15%. Existence of
some anisotropism (grain orientation) congruent to the surface of
the cylindrical samples has been suggested.

MATERIAL

The material involved in this study is a specimen of POCO
graphite from a batch used in the CODATA round-robin cooperative
program to evaluate its suitability as a thermal-conductivity
standard. The specimen, 2.609 cm (1.027 in) in diameter and
approximately 7 cm long, was received from the U.S. National Bureau
of Standards (Boulder, Colorado) and was characterized as a product
of POCO Graphite Inc., Garland, Texas, grade AXM-5Q, medium grain
fuel cell grade, purified, 2500°C graphitization temperature.
Measured bulk density of the specimen was 1.718 g/cm^3.

147

EXPERIMENTAL PROCEDURE

The thermal diffusivity of POCO graphite was determined with an apparatus described previously in detail[1,2]. It is based on the principles of an infinite cylinder and radial symmetry. Heat pulses are generated on the cylindrical surface in either periodic or transient form. For these measurements only the periodic mode was used. The surface temperature in this case is a harmonic function of time and the progress of the resulting sinusoidal type heat wave is measured inside the cylinder between a point near the surface and, in the same radial plane, at the centre of the cylinder. The time interval needed for the heat wave to penetrate from the surface measuring point to the centre is related to the thermal diffusivity of the sample.

It is important in this measuring method that the time response of the temperature sensors (thermocouples in this case) be equal. Intrinsic thermocouples, where specimens are an integral part of the circuitry, are the most desirable. Unfortunately, intrinsic thermocouples could not be formed with the graphite, and thermocouples were therefore inserted in the specimen. In addition, the graphite required protection from oxidation during heating. The use of vacuum for this purpose had an adverse and uneven effect on the response time of the thermocouples measuring temperature changes. Consequently, an inert atmosphere was used.

RESULTS AND DISCUSSION

The thermal diffusivity results are shown in Fig. 1 The circles represent measured values. The solid curve was drawn by inspection. The standard deviation is   2.4%. The smoothed values are given in Table 1.

Table 1 – Thermal diffusivity of POCO graphite

Temp °C	Thermal diffusivity cm^2	Temp °C	Thermal diffusivity cm^2
0	0.725	400	0.257
50	0.581	500	0.224
100	0.487	600	0.197
200	0.376	700	0.176
300	0.303	800	0.161

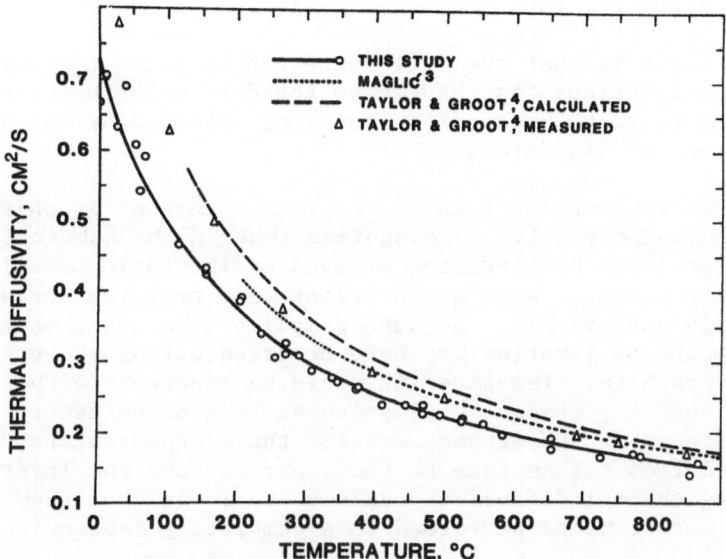

Fig. 1. Thermal Diffusivity of POCO graphite

For purposes of comparison, the thermal diffusivity values obtained by Maglić[3] and the recently published results by Taylor and Groot[4] are also included. The dashed curve is the thermal diffusivity calculated by Taylor/Groot from the measured values of thermal conductivity, specific heat and density, whereas the triangles represent their directly measured thermal diffusivities of the same sample. The dotted curve is the thermal diffusivity measured by Maglić, with which, incidentally, Taylor/Groot's experimental data generally coincide.

The present results are approximately 8 and 15% lower than that of Maglić and Taylor, respectively. These lower values could have resulted from unequal time-temperature response of the measuring thermocouples. On the other hand, if one assumes that the present results are accurate, then two other explanations may be offered.

One of these is that the difference can be accounted for by the physical variations that have been found to exist between nominally identical POCO graphite specimens obtained by the U.S. National Bureau of Standards.

The other explanation lies in the very nature of graphite, namely its anisotropy. It is recognized that in the fabrication of ceramic specimens by extrusion or even by isostatic pressing of anisotropic powders, some grain orientation near the surface is essentially unavoidable. Because graphite is a scaly material, a certain amount of layering may have occurred during the preparation of the POCO graphite. The layering would be congruent with the cylindrical surface, gradually diminishing toward the centre of the specimen. Since, in the method used for these measurements, heat pulses travel from the surface to the centre across the layering, the resulting thermal diffusivity values are lower than they would be were the measurements performed in a completely isotropic material.

In fact, the results obtained by Taylor/Groot could be interpreted to support the existence of anisotropy in POCO graphite. The authors measured electric resistivity of their specimen 001, originally at 2.54 cm diameter by 15.2 cm long rod, which was subsequently cut into 7 smaller specimens. Two of these, 001-A and 001-C, were machined in the form of thin rods (approximately 0.3 and 0.4 cm diameter, respectively, and 14.7 cm long). Both rods were cut equidistantly from the axis of the original rod. Their actual distance from the axis was not reported, but it must have been such that the thin rods contained less of the anisotropic structure than was the average for the orginal rod because their resistivity was found to be some 5% higher than that of the 2.54-cm diameter rod. It should also be noted that the difference between

the resistivities of the rods 001-A and 001-C was less than 0.5%,
which should not be surprising because their distance from the axis
of the original rod was the same and therefore they must have
contained equal quantities of layered graphite.

Furthermore, their thermal transport results are particularly
interesting. Direct thermal diffusivity measurements were made
using the laser-flash method. The 1.27 cm diameter by 0.38 cm
thick sample disc was machined from one end of the original rod, so
that its axis coincided with the axis of the rod[5]. That means that
the disc, and especially its central measuring section, contained
only isotropic material. Since rods 001-A and -B were cut at some
distance from the axis of the original rod, part of their graphite
must have been layered (parallel to the axes of the rods) and
therefore the thermal transport properties in the axial direction
should have been higher than that of the centrally located isotropic
material. The results confirm the correctness of this assumption:
as seen in Fig. 1, the calculated thermal diffusivity values for
rods 001-A and -B are indeed higher by some 5% than the measured
thermal diffusivity values for the disc.

CONCLUSION

It is concluded from the above analysis that layering congruent
with the cylindrical surface of POCO graphite specimens can account
for the apparent differences in the thermal diffusivity results:
the high and low values shown in Fig. 1 were obtained because
measurements were performed parallel and perpendicular to the
layering, respectively. On the other hand, the intermediate values,
obtained separately by Maglic and Taylor/Groot but using essentially
the same experimental methods, arose because the portion of the
specimen used was isotropic.

REFERENCES

1. V. V. Mirkovich, "Thermal diffusivity measurement of Armco iron
        by a novel method," Rev. Sci. Instum; 48(5)560-565 (1977).
2. V. V. Mirkovich, "An apparatus for measuring thermal diffusivity
        in air"; CANMET, Energy, Mines and Resources Canada; CANMET
        Report 77-21 (1976).
3. K. D. Maglić, Boris Kidrić Institute of Nuclear Sciences, Belgrade,
        Yugoslavia; Personal communication (1980).
4. R. E. Taylor and H. Groot, "Thermophysical properties of POCO
        graphite," High Temp. - High Pressures: 12,147-160 (1980).
5. R. E. Taylor, Thermophysical Properties Research Laboratory,
        Purdue University, West Lafayette, Indiana, U.S.A.; Personal
        communication (1981).

THE THERMAL TRANSPORT PROPERTIES OF A POCO AXM-5Q1 GRAPHITE

FROM 80 TO 970 K*

J. P. Moore and R. S. Graves

Metals and Ceramics Division
Oak Ridge National Laboratory
Oak Ridge, Tennessee 37830

ABSTRACT

The thermal transport properties of an AXM-5Q1 graphite with nominal room temperature electrical resistivity of 12.6 $\mu\Omega \cdot m$ and density of 1.77 Mg/m^3 were determined using three apparatuses. The thermal conductivity results display the normal broad maximum near 300 K that is attributed to phonon scattering at the crystallite grain boundaries. The experimental thermal resistivity is linear to within 0.7% at temperatures from 400 to 970 K.

INTRODUCTION

Graphite is used extensively in many applications because of the wide property ranges obtainable with slight differences in the fabrication process and because of its structural integrity at elevated temperatures. Its machinability into complex parts as compared to that of the refractory metal alloys has also been an attractive feature.

Graphite is of current interest as a potential high temperature thermal conductivity standard. This is because of the above features and because of its availability in large quantities and its high emissivity, which assists in high temperature thermometry. Unfortunately, there is justified concern about its use as a thermal conductivity standard because of its inhomogeneity and the tenuous

*Research sponsored by the Division of Materials Sciences, U.S. Department of Energy, under contract W-7405-eng-26 with the Union Carbide Corporation.

relationship between its thermal conductivity, λ, and more easily measured parameters, such as density and electrical resistivity. McElroy et al.[1] showed, however, that the room temperature λ of most graphites could be calculated to within ±8% using the electrical resistivity, ρ. In addition, Hust[2] has recently shown that the room temperature λ of AXM graphites can be calculated to within ±2% from the measured ρ and density.

The principal candidate for a high temperature λ standard is AXM-5Q1;* Taylor and Groot[3] have measured the thermophysical properties of this material from 300 to 2400 K, and Minges[4] has interpretated the transport properties.

We have measured the λ and the absolute Seebeck coefficient, S, of an AXM-5Q1 graphite from 80 to 900 K and the ρ from room temperature to 1000 K. These results are compared with previous experimental results and with the theoretical model described by Minges.[4]

APPARATUSES

The four different apparatuses used to measure ρ, λ, and S are described in Table 1. The knife-edge apparatus was used to determine the ρ at room temperature as accurately as possible, to determine the inhomogeneity of the rods by measuring ρ versus knife edge position, and to determine electrically the spacings between thermocouple wires attached to the specimens.

---

*AXM-5Q1 is a graphite fabricated by POCO Graphite, Inc., Garland, Texas. It is a medium grain isostatically pressed graphite with a graphitization temperature of 2800 K.

Table 1. Apparatuses Used for Measurements of the
Electrical Resistivity and Thermal Conductivity

Apparatus	Property Measured	Temperature Range (K)	Measurement Accuracy (%)		References
			λ	ρ	
Knife Edge	ρ	300	—	±0.2	7
LTL[a]	λ,S	80—400	±1.3	±0.4	6
HTL[b]	λ,ρ,S	300—1000	±1.5	±0.4	5
CL[c]	λ	300—360	±3	—	7

[a]Low Temperature Longitudinal.
[b]High Temperature Longitudinal.
[c]Comparative Longitudinal.

SPECIMEN DESCRIPTION

The specimens were two right circular cylinders of AXM-5Q1 graphite obtained from Hust,[2] who designated the specimens No. 102. Hust[2] reported nominal density, ρ at room temperature, and λ at room temperature of 1.77 Mg/m^3, 12.6 $\mu\Omega\cdot$m, and 112.6 W/m·K, respectively. These two cylinders were machined into 77.8-mm-long rods for measurements in the High Temperature Longitudinal, HTL,[5] and Low Temperature Longitudinal, LTL.[6] After the measurements in the LTL were completed, the specimen was machined into two small cylinders (6.35-mm diameter × 9.52-mm long) for measurements in the Comparative Longitudinal, CL.

RESULTS

The thermal conductivity results from the LTL and the CL are compared in Fig. 1. Since there were three thermocouples attached to the LTL specimen, the λ of two sections along the specimen were measured. The data from both sections are plotted in Fig. 1, and it is apparent that the two values of λ do not agree. The maximum difference was about 2%, which occurred at a temperature near 280 K. After completion of the LTL measurements, the two metering sections were cut from the rod, and the two sections were used as specimens in the CL. Although the results from the CL are higher than those from the LTL, the results agree to within the combined experimental uncertainties. Smoothed values for λ from 80 to 300 K were taken from the solid curve and listed in Table 2. This curve shows the broad maximum commonly observed in the λ of graphite near 300 K.

Experimental results from the HTL are shown in Fig. 2 as the reciprocal of the thermal conductivity, $R = \lambda^{-1}$. These experimental values were corrected for thermal expansion using expansion values from Taylor and Groot,[3] and the bars on each datum represent the ±1-1/2% measurement uncertainty. Above 450 K the R is linear with temperature, and all data are within ±0.7%* of

$$R = \lambda^{-1} = 0.4831 \times 10^{-2} + 0.10923 \times 10^{-4}\ T \qquad (1)$$

where R is in m·K/W and T is in K.

The solid curve in Fig. 3 represents the average thermal conductivity from the three apparatuses described in Table 1, and λ values at even temperature intervals were obtained from this curve

─────────────

*All data are within ±0.35% of the equation with one exception.

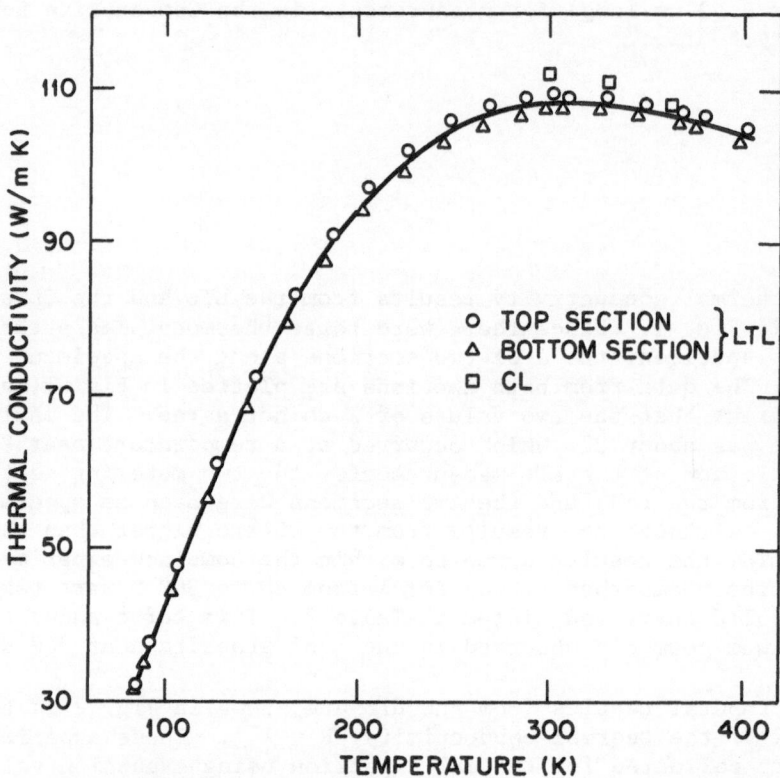

Fig. 1. Measured thermal conductivity values from the LTL and CL
         showing the good agreement of the results and a broad
         maximum near 300 K.

Table 2. Smoothed Values of the ρ, λ, and S of an AXM–5Q1
Graphite with a Room Temperature ρ of 12.5 $\mu\Omega$·m
and a Nominal Density of 1.77 Mg/m^3

Temperature (K)	ρ ($\mu\Omega$·m)	λ (W/m·K)	S (μV/K)
80	—	29.6	−2.83
100	—	43.0	−1.98
120	—	56.1	−1.17
140	—	68.0	−0.43
160	—	78.4	+0.23
180	—	87.1	0.81
200	—	94.1	1.30
220	—	99.5	1.71
240	—	103.5	2.09
260	—	106.2	2.29
280	—	107.8	2.47
300	12.51	108.4	2.58
320		108.3	2.62
340		108.2	2.61
360		107.5	2.54
380		106.5	2.42
400	10.92	105.3	2.25
500	10.01	98.3	0.99
600	9.36	88.3	−0.61
700	8.93	80.1	−2.03
800	8.66	74.6	−3.18
900	8.52	68.2	−3.54
950		65.9	
1000[a]		63.8	

[a]Since the highest measurement temperature was 970 K, this value was obtained by linear extrapolation.

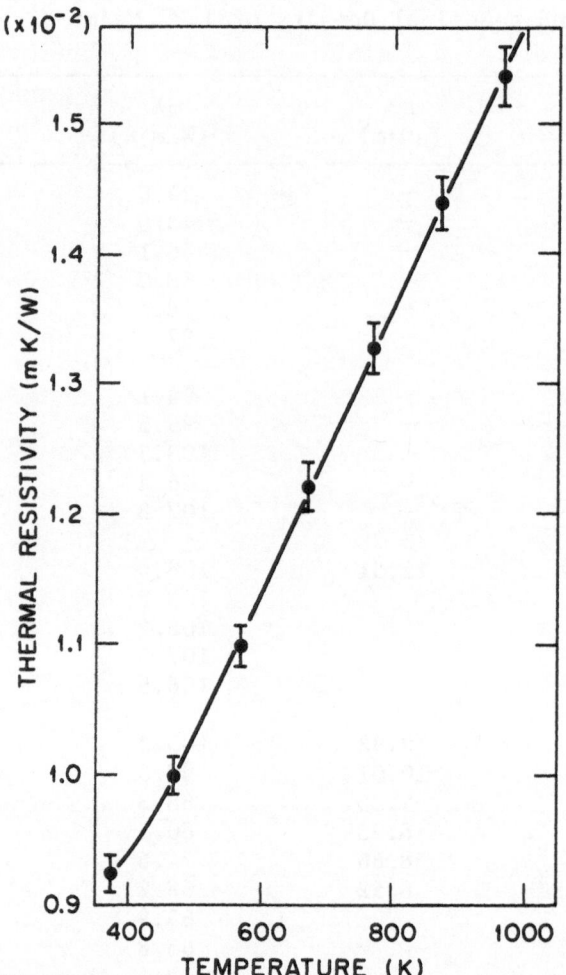

Fig. 2.   Thermal resistance results from 350 to 970 K showing the
          approximate linearity with temperature.

and are given in Table 2. In all cases, the difference between an individual data point and the solid curve was less than the experimental uncertainty of the apparatus used to obtain the data point. Data from Hust,[2] Taylor and Groot[3] and Minges[4] are shown for comparison. The datum by Hust near 300 K, which was obtained on a companion specimen to the one used in these measurements, is within 1% of the CL results. Taylor and Groot[3] measured the λ of several AXM-5Q1 graphites from 400 to 2400 K and their results on one specimen (3A-1) are below the present values and are shown in Fig. 3. High temperature λ values from Minges[4] are also below the present results.

The electrical resistivity values obtained at room temperature are in agreement with those obtained by Hust,[2] and the shape of the ρ-versus-temperature curve is similar to that noted by Taylor and Groot[3] and by Minges[4] on the graphites whose λ are plotted in Fig. 3. Smooth values of the ρ and S from this investigation are given in Table 2. The ρ was corrected for thermal expansion using Taylor and Groot's expansion results.[3] The Seebeck coefficient of this graphite has a broad maximum similar to that observed in the λ and occurring at the same temperature.

DISCUSSION OF RESULTS

Lattice Conduction from 80 to 1000 K

Minges[4] has described the thermal conductivity of AXM-5Q1 in detail over a wide temperature range. The total thermal conductivity is assumed to be due to heat conduction by phonons in the basal planes of the crystallites. Primary phonon scattering is by the crystallite boundaries and by three-phonon Umklapp processes. The thermal conductivity is given by

$$\lambda^{-1} = \alpha \ [\lambda_B^{-1} + \lambda_U^{-1}] \tag{2}$$

where λ_B^{-1} and λ_U^{-1} are the resistances to phonons by boundary scattering and by phonon scattering, respectively, and α is a tortuosity/porosity factor. The boundary scattering conductivity, λ_B, has been expressed by Kelly[8] and Taylor[3] as the sum of three components, each of which has a constant, L_a, that represents the "effective crystallite size." The λ_U component was assumed to be the same as the thermal conductivity of graphite in the basal plane, as measured by Taylor et al.[9] Minges calculated α as the ratio of the theoretically predicted λ_U to the experimental thermal conductivity at 1000 K. Using a calculated α of 6.51 for our specimens and a value of 0.15 μm suggested for L_a by Minges for his AXM-5Q1 graphite, we calculated a total thermal conductivity

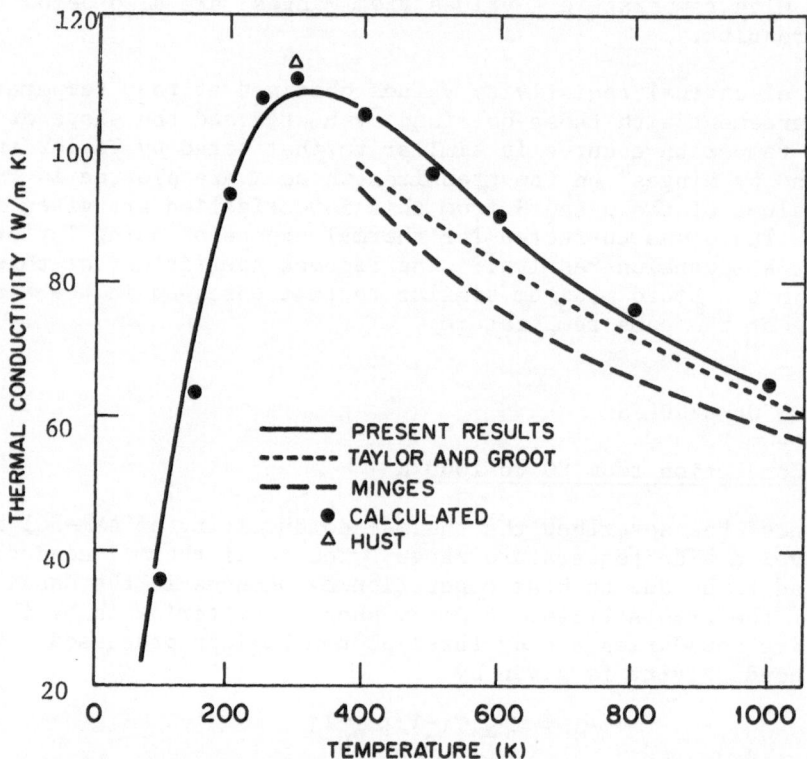

Fig. 3. Thermal conductivity results from this study compared with
results from other studies on AXM-5Q1 graphites with different densities and electrical resistivities. The calculated data were based on a model discussed by Minges.[4]

curve that covered our temperature measurement range. This curve deviated from experimental values by as much as 11% above 200 K. When L_a was shifted to 0.17 µm, the calculated results agreed with the experimental ones to within 3% at temperatures above 200 K, as shown in Fig. 3. Below 200 K, the agreement between theory and experiment is reduced, as can be seen by the 11 and 17% difference at 150 and 100 K, respectively.

Electronic Conduction in AXM-5Q1

Minges[4] has shown that the thermal resistance of AXM-5Q1 from direct measurements of the thermal conductivity and from measurements of the thermal diffusivity and specific heat is linear to within about ±10% from 300 to 3000 K. In detail, however, there appears to be systematic variations about the linear fit to the thermal resistance-versus-temperature results. It is possible that these systematic variations are caused by a nonnegligible high temperature electronic component, which would give a negative deviation from linearity and a higher order phonon scattering process that would give a positive deviation from linearity. If this were the case, the approximate linearity of the thermal resistance would be fortuitous.

CONCLUSIONS

1. The thermal conductivity, electrical resistivity, and Seebeck coefficient have been measured up to 970 K on an AXM-5Q1 graphite with a room temperature electrical resistivity of 12.6 µΩ·m and a density of 1.77 Mg/m^3.

2. The thermal conductivity results from three apparatuses agree to within ±3%.

3. The thermal conductivity results show a broad maximum near 300 K, which can be interpreted as boundary scattering of phonons with an effective crystallite size of 0.17 µm.

4. Although the thermal resistance of AXM-5Q1 is linear with temperature above about 450 K, this may be fortuitous.

ACKNOWLEDGMENTS

We gratefully acknowledge J. G. Hust of the National Bureau of Standards in Boulder, Colorado for sending us the specimens, D. L. McElroy for helpful suggestions, and S. J. Phillips for typing the text.

REFERENCES

1. D. L. McElroy, T. G. Kollie, W. M. Ewing, R. S. Graves, and
    R. M. Steele, Room Temperature Measurements of Electrical
    Resistivity and Thermal Conductivity of Various Graphites,
    ORNL-TM-3477, Oak Ridge National Laboratory, Oak Ridge,
    Tennessee (1971).
2. J. G. Hust, National Bureau of Standards, Boulder, Colorado,
    Second report on the homogeneity investigation on POCO
    graphite for use as NBS SRMs. Private communication
    (November 1976).
3. R. E. Taylor and Hans Groot, High Temp. High Pressures 12:147
    (1980).
4. M. L. Minges, Int. J. Heat Mass Transfer 20:1161 (1977).
5. J. P. Moore, D. L. McElroy, and R. S. Graves, A Technique for
    Determining Thermal and Electrical Conductivity and Absolute
    Seebeck Coefficient Between 300 and 1000 K, ORNL-4986, Oak
    Ridge National Laboratory, Oak Ridge, Tennessee (September
    1974).
6. J. P. Moore, R. K. Williams, and R. S. Graves, Rev. Sci. Instrum.
    45(1):87 (1974).
7. R. K. Williams, R. S. Graves, and J. P. Moore, A Study of the
    Effects of Several Variables on the Thermal Conductivity of
    2 1/4 Cr—1 Mo Steel, ORNL-5313, Oak Ridge National Laboratory,
    Oak Ridge, Tennessee (April 1978).
8. B. T. Kelly, Carbon 6:485 (1968).
9. R. Taylor, K. E. Gilchrist, and L. J. Poston, Carbon 6:537
    (1968).

THERMAL DIFFUSIVITY OF CANDIDATE

STANDARD REFERENCE MATERIALS

K. Maglić, N. Perović, and Z. Životić

Boris Kidrich Institute of Nuclear Sciences
Institute of Thermal Engineering and Energy Research
Belgrade, Yugoslavia

INTRODUCTION

In 1975 the CODATA Task Group on Transport Properties launched
an international project on the evaluation of experimental techniques
in the measurement of transport properties (CODATA Task Group,1975).
The program involved methods for the measurement of electrical and
thermal transport properties of a selected group of materials, in
the temperature range 0-2500 K. The assessment was envisaged through
experimental measurement of transport properties by different methods
in different research laboratories. The objectives of the program
were: (a) establishment of the optimal transport property measurement approach as a function of thermal conductivity level and temperature, and establishment and/or verification of the accuracy and
precision levels achievable with the various measurement techniques,
(b) experimental verification of the performance of certified
standard reference materials and improvement of the accuracy of the
certification, and (c) refinement of the theoretical interpretation
and understanding of energy transport in the selected reference materials used in the measurement program.

The project was divided in two temperature areas: cryogenic and
low temperature (up to 100 K) and intermediate and high temperature
(above 100 K). The program of intermediate and high temperature
area, to which belong the measurements contained in this paper, included measurements in 15 different laboratories in 10 countries.
Measurement methods covered four thermal conductivity and three
thermal diffusivity variants. Contribution of the Boris Kidrich
Institute Thermal Properties Laboratory consisted of the measurement
of thermal diffusivity of electrolytic iron, austenitic alloy, and
AXM-5Q graphite using the laser pulse technique. The results of the

163

measurement of austenitic alloy and AXM-5Q graphite were presented
at the 7th European Thermophysical Properties Conference by Maglić
et al (1980). For completeness of this report they are reproduced
here, together with new data on electrolytic iron measured in 1981.

EXPERIMENTAL PROCEDURES

All specimens were received from the National Bureau of Standards
in Washington (Gaithersburg) where the machining, annealing, and
physical and chemical characterization of specimen materials were
done. According to CODATA Task Group on Transport Properties Report
(1978), an extensive series of homogeneity and quality assurance
indice measurements on the polycrystalline graphite were carried out
at the National Bureau of Standards in Boulder, and further series
on nondestructive inspection evaluations of the graphite have been
initiated at Southern Research Institute in Alabama. References on
specimen materials and their characterization include Hust and
Giarratano (1975a, 1975b), Hust (1976) and Minges (1977), and the
results of nondestructive characterization of POCO AXM graphite at
Southern Research Institute were reported at the 17th International
Thermal Conductivity Conference (1981).

After thermophysical characterization the specimens were returned to the National Bureau of Standards for subsequent analysis.

Specimens were disc shaped, 1 cm in diameter and 2 to 3.5 mm
thick. Austenitic alloy and AXM-5Q graphite specimens were machined
at the National Bureau of Standards to 3.52 and 3.95 mm respecively.
Electrolytic iron was received in the shape of a disc, 3.18 cm in
diameter, from which 2 and 3 mm thick discs were turned and subsequently ground. Machining was done with much care to avoid damage
of the specimens and to provide plane parallel specimen faces.

The experimental and data reduction procedures have been
described before by Maglić et al (1980) and will not be given here.
It should be added only that the procedure described for the
austenitic alloy specimen was also used in the measurement of
thermal diffusivity of electrolytic iron.

EXPERIMENTAL RESULTS

In order to present experimental data in a more informative way,
they are given in tables which contain: "raw" values of "a" calculated with $t_{0.5}$ data based on the specimen thickness measured
at room temperature, given in the column I; values from the column
I corrected for thermal expansion given in the column II; values
from the column II corrected for thermal radiation losses (based
on dropping portion of the transient response according to Taylor

(1973)) given in the column III, and, for the AXM-5Q graphite specimen, values from the column II corrected for thermal radiation losses based on the rising portion of the transient response, which are given in the column IV (Table III only).

Electrolytic iron

Measurement of the thermal diffusivity of electrolytic iron was performed on two specimens of different thickness, machined from the same sample, designated NBS 1265. With the thicker specimen (3.00 mm thick), two runs of measurements were made and the maximum tempera- ture reached was 1060 K, when the maximum energy of the laser pulse became insufficient to ensure a favourable amplitude-to-noise ratio of the photodetector response. In the subsequent four runs a 2.00 mm thick specimen was used, and four different experiments were made in which the whole temperature range of investigation was traversed in different days. The intrinsic thermocouple as detector was used at room temperature only, as a check on possible changes in the specimen due to temperature cycling.

Table I lists all thermal diffusivity experimental results in the rising temperature sequence. Designation in the next column marks the thickness of the specimen (2 or 3) and number of the run (I to IV). Room temperature values measured with thermocouple are marked (TC).

Table 1. Thermal diffusivity of electrolytic iron

T (K)	Specimen Thickness (mm) and Run	$a \lfloor m^2 s^{-1} \rfloor \times 10^4$		
		I	II	III
296 (TC)	2-I	0.202	0.202	0.202
300 (TC)	3-I	0.229	0.229	0.229
300 (TC)	3-II	0.225	0.225	0.225
600	3-II	0.122	0.123	0.123
616	2-I	0.117	0.118	0.118
640	2-IV	0.112	0.113	0.113
656	2-I	0.111	0.112	0.112
663	3-III	0.109	0.110	0.110
678	2-III	0.106	0.106	0.106
708	3-II	0.0997	0.101	0.101
745	2-I	0.0929	0.0941	0.0941
776	3-III	0.0875	0.0887	0.0887
815	2-I	0.0805	0.0817	0.0815
839	3-III	0.0754	0.0766	0.0763
902	3-III	0.0662	0.0674	0.0669
905	2-IV	0.0643	0.0655	0.0650

(continued)

912	2-I	0.0634	0.0646	0.0641
946	3-III	0.0562	0.0573	0.0567
965	2-II	0.0520	0.0530	0.0524
967	2-I	0.0532	0.0543	0.0537
995	3-III	0.0473	0.0483	0.0477
1002	2-III	0.0449	0.0458	0.0452
1011	3-III	0.0405	0.0414	0.0408
1026	2-I	0.0385	0.0393	0.0387
1027	3-III	0.0366	0.0374	0.0369
1033	2-I	0.0361	0.0369	0.0363
1038	3-III	0.0343	0.0351	0.0346
1045	2-I	0.0335	0.0342	0.0337
1048	3-III	0.0389	0.0398	0.0392
1055	2-IV	0.0428	0.0428	0.0431
1057	2-I	0.0435	0.0445	0.0438
1060	3-III	0.0424	0.0434	0.0427
1072	2-IV	0.0474	0.0485	0.0477
1097	2-II	0.0492	0.0504	0.0495
1104	2-III	0.0499	0.0511	0.0501
1110	2-IV	0.0505	0.0517	0.0507
1142	2-II	0.0531	0.0544	0.0533
1151	2-III	0.0533	0.0547	0.0535
1161	2-III	0.0536	0.0550	0.0538
1170	2-II	0.0530	0.0544	0.0532
1172	2-III	0.0544	0.0558	0.0545
1183	2-II	0.0547	0.0558	0.0545
1186	2-IV	0.0647	0.0660	0.0645
1188	2-III	0.0645	0.0658	0.0643
1193	2-II	0.0616	0.0628	0.0613
1197	2-III	0.0625	0.0637	0.0622
1223	2-III	0.0642	0.0656	0.0639
1260	2-III	0.0652	0.0667	0.0649
1305	2-IV	0.0655	0.0671	0.0651
1334	2-III	0.0656	0.0673	0.0652
1373	2-IV	0.0654	0.0672	0.0650

Data for correction for thermal expansion taken from TPRC sourcebook (1967) were interpreted by the equation:

$$\frac{\Delta L}{L} = \frac{a}{100} (T - 300) + \frac{b}{100} \tag{1}$$

with coefficients: a = 0.00149 and b = 0 between 300 and 1183 K, and a = 0.002 and b = - 0.8 above 1183 K.

The 51 thermal diffusivity values in Table I and Fig.1 represent over 200 thermal diffusivity measurements: each experimental point in table or diagram is a mean value of 3 to 5 consecutive measurements at the same reference temperature (within 1-2 K) taken to account for statistical dispersion of individual measurements.

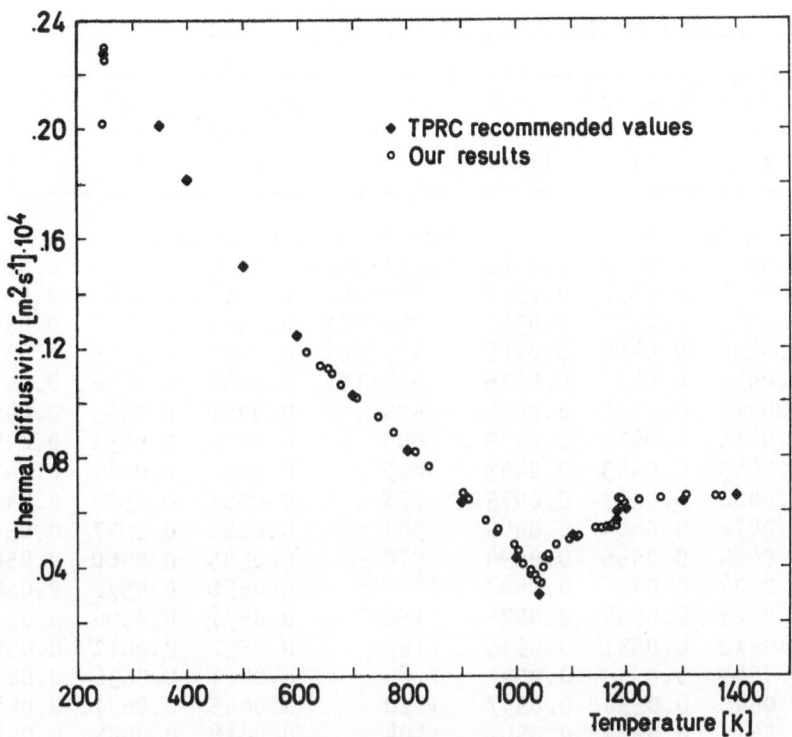

Fig. 1. Thermal Diffusivity of Electrolytic Iron

Phase transition α to γ occurred at a somewhat higher temperature than expected in the 2-III experimental run. At a temperature of 1187 K, three consecutive measurements gave continuously
rising values of thermal diffusivity (uncorrected values): 0.0563,
0.0572 and 0.0584 in 10^{-4} m^2/s to reach stable value in three
following measurements, which (at 1188 K mean temperature) amounted
to 0.0644, 0.0644 and 0.0647 in 10^{-4} m^2/s. As thermal expansion
data in the transition range were not known, the first three values
could not be corrected and entered in the table and the diagram. A
relatively high value at 1188 K was confirmed in the 2-IV run at
1186 K.

Thermal cycling had no visible effect on the measured thermal
diffusivity of electrolytic iron.

Austenitic alloy

The first set of measurements was performed on the austenitic

K. MAGLIC ET AL.

alloy specimen as received, namely a disc 1 cm in diameter and 3.52 mm thick. For the second set of measurements the specimen thickness was reduced to 2.50 mm. The results are given in Table II:

Table II. Thermal diffusivity of austenitic stainless steel

T (K)	$a\lvert m^2 s^{-1}\rvert \times 10^4$			T (K)	$a\lvert m^2 s^{-1}\rvert \times 10^4$		
	I	II	III		I	II	III
Specimen 3.52 mm thick				Specimen 2.50 mm thick			
286(TC)	0.0365	0.0365	0.0365	291(TC)	0.0356	0.0356	0.0356
291(TC)	0.0358	0.0358	0.0358	294(TC)	0.0344	0.0344	0.0344
380(TC)	0.0382	0.0383	0.0383	385(TC)	0.0381	0.0382	0.0382
484(TC)	0.0406	0.0409	0.0409	437(TC)	0.0383	0.0385	0.0385
507	0.0423	0.0426	0.0426	533(TC)	0.0444	0.0448	0.0448
603(TC)	0.0427	0.0432	0.0432	594	0.0450	0.0455	0.0455
603	0.0434	0.0439	0.0439	609	0.0448	0.0453	0.0453
639	0.0443	0.0449	0.0449	665	0.0456	0.0462	0.0462
703	0.0466	0.0473	0.0473	773	0.0491	0.0500	0.0498
751	0.0476	0.0484	0.0484	907	0.0525	0.0537	0.0530
806	0.0485	0.0495	0.0494	978	0.0545	0.0560	0.0549
880	0.0504	0.0516	0.0507	1013	0.0556	0.0572	0.0560
915	0.0523	0.0535	0.0524	1150	0.0577	0.0596	0.0578
920	0.0518	0.0531	0.0519	1160	0.0592	0.0612	0.0593
987	0.0544	0.0559	0.0541	1289	0.0611	0.0636	0.0610
1028	0.0542	0.0558	0.0537	1320	0.0645	0.0671	0.0643
1080	0.0572	0.0589	0.0563	1404	0.0628	0.0655	0.0623
1138	0.0580	0.0600	0.0569	1473	0.0659	0.0690	0.0653
1176	0.0598	0.0618	0.0585				
1221	0.0608	0.0630	0.0594				
1315	0.0621	0.0646	0.0606				
1401	0.0636	0.0664	0.0622				
1450	0.0645	0.0674	0.0631				

Thermal diffusivity values obtained for the specimen with the original thickness are plotted in figure 2. Discrete experimental points are again the means of three to four consecutive measurements at the same reference temperature (within 1-2 K). The scatter never exceeded $\pm$ 1% limits. Correction of the specimen reference temperatures were also made and is included in all of the results.

The scatter of values obtained for the specimen with reduced thickness is somewhat greater, and the experimental points tend to group 2 to 4% above the interpolated curve in figure 1, in the range of the optically measured transient response. This may be ascribed to a larger error caused by reduced characteristic time, $t_{0.5}$.

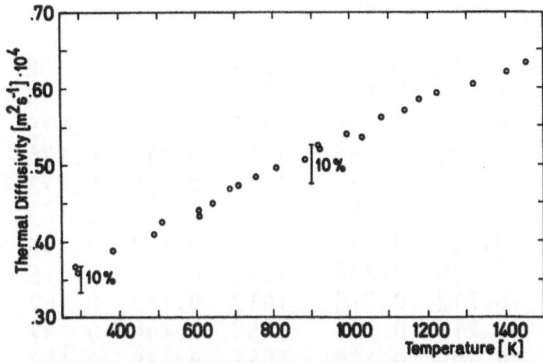

Fig. 2.   Thermal Diffusivity of the Austenitic Alloy

Thermal cycling had no influence on the measured thermal diffusivity of either specimen.

AXM-5Q graphite

Thermal diffusivity measurements were conducted on an AXM-5Q graphite specimen 3.95 mm thick, and the results are given in table III. The first three columns correspond to the respective columns for the austenitic alloy: the fourth contains values corrected for radiation heat losses with the use of an analysis of the rising part of the transient response curve (Clark and Taylor 1975). A comparison of the two methods of correction shows that the method based on the analysis of the rising part of the response curve gives somewhat lower values (2-4%) at high temperatures. Figure 3 shows the thermal diffusivity of AXM-5Q graphite. As in the case of the austenitic alloy, experimental points in the diagram represent means of three to four measurements, the scatter of individual measurements being within 1%.

Corrections for thermal expansion for both austenitic alloy and AXM-5Q graphite were made by using linear thermal expansion data for these two materials compiled by Fitzer (1971).

Table III. Thermal diffusivity of AXM-5Q graphite

T (K)	$a\lvert m^2 s^{-1}\rvert \times 10^4$				T (K)	$a\lvert m^2 s^{-1}\rvert \times 10^4$			
	I	II	III	IV		I	II	III	IV
480	0.413	0.415	0.415	0.415	1051	0.186	0.188	0.184	0.182
577	0.333	0.335	0.335	0.335	1077	0.183	0.185	0.181	0.179
605	0.317	0.319	0.319	0.319	1112	0.173	0.175	0.170	0.169
677	0.283	0.285	0.285	0.285	1149	0.174	0.176	0.171	0.170
708	0.270	0.271	0.271	0.271	1176	0.170	0.172	0.166	0.165
762	0.249	0.251	0.251	0.251	1224	0.162	0.165	0.159	0.157
777	0.246	0.247	0.247	0.247	1295	0.157	0.160	0.153	0.151
813	0.234	0.236	0.236	0.236	1339	0.152	0.155	0.148	0.146
866	0.218	0.220	0.217	0.215	1417	0.149	0.152	0.144	0.142
895	0.213	0.215	0.212	0.210	1458	0.144	0.147	0.139	0.136
897	0.211	0.213	0.209	0.207	1541	0.139	0.142	0.133	0.130
918	0.209	0.210	0.207	0.206	1558	0.136	0.139	0.130	0.127
996	0.202	0.194	0.190	0.198	1647	0.133	0.136	0.126	0.122
1000	0.193	0.196	0.191	0.190	1713	0.131	0.134	0.123	0.118

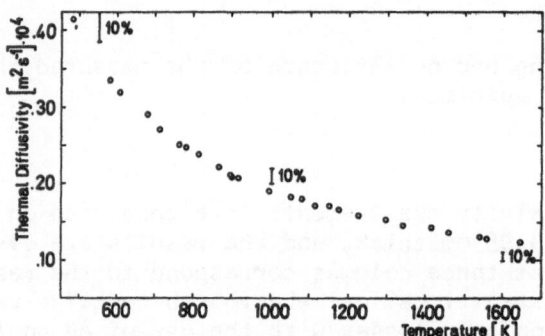

Fig. 3.   Thermal Diffusivity of AXM-5Q Graphite

Inaccuracy of thermal diffusivity measurements

Using the final expression for computing the thermal diffusivity from the measured values of the "half-time" and the specimen thickness

$$a = \frac{K_x L^2}{t_x} \tag{2}$$

the lower bound for the measurement inaccuracy is determined by the limits of the measuring instruments used, i.e., digital oscilloscope for time measurements and micrometer for length measurements. Assuming that the numeric parameter $k_x = 0.139$ is given exactly, this lower bound of inaccuracy (i.e. the resolution possible in the experiment) is given by the following expression:

$$\frac{\Delta a}{a} = \frac{\Delta \tau_{1/2}}{\tau_{1/2}} + 2 \frac{\Delta L}{L} \tag{3}$$

With the time resolution of the digital oscilloscope of 0.1 ms per point when the responses with a 50 ms half-time were reduced and with the measurement of the 2 mm thick specimen within 5 μm by means of the micrometer, expression (3) gives the minimum measurement error of 0.45%.

This ideal case of minimum error is never reached in real experiments. The measurement process is always affected by disturbances from different sources, causing changes in the results, with inaccuracy greater than that given by equation (3).

In the thermal diffusivity measurements dealt with in this paper, the greatest source of imprecision (as a part of the total inaccuracy) was presence of the noise on the temperature response trace at the digital oscilloscope. This trace width caused by electric and electronic disturbances (due to discharge of the laser capacitor bank, common mode voltage, ripple from electrically heated furnace, etc.) led to certain dispersion of measured values of the half-times and, consequently, to certain imprecision in computed thermal diffusivity values. To ensure the "best" values of thermal diffusivity, 3 to 5, and sometimes even more, measurements were performed at the same reference temperature, and the mean value of several measurement results was taken as "true". These values are presented in tables and diagrams. As an illustration of the measure of imprecision, standard deviation and probable error were calculated for measured electrolytic iron thermal diffusivity values (in three different temperature regions) in the form:

$$a = \bar{a} \pm 3\sigma \tag{4}$$

Typical thermal diffusivity values and standard deviations according to equation (4) were:

T = 656 K $a = (0.111 \pm 0.0048) \cdot 10^{-4} \ m^2/s$

T = 1045 K $a = (0.0335 \pm 0.00078) \cdot 10^{-4} \ m^2/s$

T = 1223 K $a = (0.0642 \pm 0.00156) \cdot 10^{-4} \ m^2/s$

which give the probable error (defined as $e_p = 0.674\sigma$) of $0.108 \cdot 10^{-6}$ m^2/s, $0.0175 \cdot 10^{-6}$ m^2/s and $0.035 \cdot 10^{-6}$ m^2/s respectively. Judging from these three typical values, the error was minimum at medium temperatures where the signal-to-noise ratio had its maximum due to the optimal combination of thermal diffusivity, specific heat capacity, total emissivity, and the sensitivity of the optical detector. In the lower and the higher temperature regions these conditions deviated from their optimum, and the smaller signal-to-noise ratio resulted in greater error.

Evaluation of the overall inaccuracy in the pulse diffusivity measurement involves consideration of (a) finite-pulse-time effect, (b) effect of heat losses, and (c) the effect of the nonuniformity of the laser beam.

Calculation of corrections for the finite-pulse-time effect showed that the uncertainty due to this effect in our measurements could be neglected.

Corrections due to radiation heat losses were necessary above 800 K. Using the recommended methods for calculation of heat losses by Cowan (1963) and Taylor (1973), the ratios V ten half-times/ /V half-time were measured at the cooling part of the temperature responses, and using the correction curve given by Taylor (1973). The correction factor, k_x, for equation (2) was obtained, and corrected thermal diffusivity values were calculated. In this procedure, due to noise present at higher temperatures, certain arbitrariness in the evaluation of the correction factor was present. This might explain the discrepancy between correction values obtained by two different methods for heat loss correction for AXM-5Q graphite.

The influence of the laser beam nonuniformity was observed at some thermal diffusivity values measured with the intrinsic thermocouple (electrolytic iron 2 mm thick). With the optical transient response measurement, where the detector integrated radiation from the whole specimen rear face, the effects of the laser beam nonuniformity were not observed.

Linear thermal expansion data for the three specimen materials investigated could be another source of the uncertainty in calculated thermal diffusivity values (maximum correction for thermal expansion of austenitic alloy amounted to 4.5%). System of presentation of experimental data, however, enables a simple correction when more accurate data become available.

CONCLUSIONS

The three investigated thermal diffusivity candidate standard reference materials proved stable in the course of our studies.They gave reproducible thermal diffusivity values after a number of temperature cycles, with different specimen thicknesses and over the entire range of measurement.

Electrolytic iron, with its pronounced variation of thermal diffusivity with temperature, characteristic and sharply defined phase transitions, and relatively simple machining of specimens seems an ideal SRM for the temperature range between 300 and 1400 K.

The tungsten specimen, which was also received from the National Bureau of Standards will be investigated after modification of our laser system, which will be effected in the near future.

Experimental results obtained in our study of three thermal diffusivity candidate standard reference materials have been presented in a form suitable for subsequent evaluation, in agreement with the recommendations of the CODATA, published in the Intl.J.Thermophys. (1980).

ACKNOWLEDGMENT

The authors wish to acknowledge support provided by the U.S. Yugoslav Joint Board on Scientific and Technological Cooperation, which facilitated realization of the research presented in this paper.

REFERENCES

Clark, L.M.III., Taylor, R.E., 1975, Radiation loss in the flash method for thermal diffusivity, J.Appl.Phys., 46:714.
CODATA Task Group on Transport Properties, 1975, Project on the evaluation of experimental techniques (Report to the 10th CODATA General Assembly).
CODATA Task Group on Transport Properties, 1978, Report to the 11th CODATA General Assembly.
CODATA Task Group on Thermophysical Properties of Solids, 1980, Criteria for the presentation in the primary literature of scientific and technical information on thermophysical properties of solids, Int.J.Thermophysics, 1, 1:135-140.
Cowan, R.D., 1963, Pulse method of measuring thermal diffusivity at high temperatures, J. Appl. Phys. 34:926-7.

Fitzer, E., 1971, Thermophysical properties of solid materials,
    Project Section II, Cooperative thermal expansion measurements
    up to 1000°C, AGARD Advisory Report 31, AGARD, Neuilly-sur-Seine,
    France.
Hust, J.G., Giarratano, P.J., 1975a, Thermal conductivity and elec-
    trical resistivity standard materials: Austenitic stainless
    steel SRM´s 735 and 798 from 4 to 1200 K, NBS Special Publi-
    cation 260-46, National Bureau of Standards, Boulder,Colorado.
Hust J.G., Giarratano, P.J., 1975b, Thermal conductivity and elec-
    trical resistivity standard reference materials: Tungsten SRM´s
    730 and 799, from 4 to 3000 K, NBS Special Publication 260-52,
    National Bureau of Standards, Boulder, Colorado.
Hust, J.G., 1976, Homogeneity investigation of POCO graphite for
    use as a NBS SRM, National Bureau of Standards, Boulder,
    Colorado (Personal communication to CODATA Task Group).
Koenig, J.R., Pears, C.D., 1981, Nondestructive characterization
    of POCO AXM graphite, in "Proceedings of the 17th Inter-
    national Thermal Conductivity Conference", J.G.Hust, ed.,
    (in press).
Maglić, K.D., Perović, N., Životić, Z., 1980. Thermal diffusivity
    measurements on standard reference materials, High Temps. -
    High Pressures, 12:55-60.
Minges, M.L., 1977, Analysis of thermal and electrical energy
    transport in POCO AXM-5Q graphite, Int. J. Heat Mass Transfer
    20:1161-72.
Taylor, R.E., 1973, Critical evaluation of flash method for mea-
    suring thermal diffusivity, Report PRF-6764, NTIS PB225
    591/7AS, National Technical Information Service, Springfield,
    VA U.S.A.
Taylor, R.E., 1978, Heat pulse diffusivity measurement, Report
    PRL 154, Purdue University, West Lafayette, IN U.S.A.
"Thermophysical Properties of Solid Materials", 1967, Vol.1:
    Elements, Y.S.Touloukian, ed.Macmillan Co., 589-90.

SESSION   TC-5

Theory, Review and Correlation - II

CHAIRMAN

M. J. Laubitz
National Research Council
Ottawa, Canada

ATTENUATION OF SEISMIC WAVES BY HEAT CONDUCTION

Paul G. Klemens

Department of Physics and
Institute of Materials Science
University of Connecticut, Storrs, CT 06268

ABSTRACT

Attenuation of seismic waves in the earth's mantle is described by Q factors ranging from 100 to 1000. Because of the low
frequency of these waves, it is difficult to account for this
relatively high attenuation. A heat conduction mechanism is proposed. Because rocks are inhomogeneous in thermal expansion, an
elastic wave causes adiabatic temperature changes that differ
from grain to grain or between regions of different composition
within the same grain. These temperature differences cause irreversible heat currents; the attenuation can be calculated from
the entropy generation. The loss can be expressed in terms of
the mean square variation of the Grüneisen parameter, the bulk
modulus, the absolute temperature, and a characteristic frequency, $f_o = \pi D/2L^2$, where L is the grain size and D the thermal
diffusivity. Typical Q values are around 300 at frequencies
near f_o for dry rocks at 1800 K, and about 100 for rocks containing 6% by volume of molten inclusions (asthenosphere). The
observed attenuations are somewhat higher and less frequency
dependent, but the heat conduction mechanism may well make a
significant contribution to the attenuation of seismic waves.

INTRODUCTION

The attenuation coefficient of seismic waves is difficult
to measure precisely, but it is of interest because the interpretation of seismic data requires mathematical models that in-
corporate attenuation as function of frequency and depth.[1] The

attenuation is described by the reciprocal of a quality factor,
$Q = \tau\omega$, where ω is the angular wave frequency and τ the decay
time. Present evidence suggests that Q is around 400 in the upper
mantle, decreases to a value around 80 in a highly absorbent low-
velocity layer about 100 km deep (asthenosphere), and then increases with depth, reaching a value in excess of 1000 in the
lowest mantle adjacent to the core.[1,2] It is believed that Q is
fairly insensitive to frequency in the seismographic range of frequencies. The evidence for this frequency independence is partly
derived from laboratory experiments and partly from the decay of
free oscillations of the earth. The former is perhaps not completely relevant, because at low pressures and high intensities
grain boundary sliding is an important loss mechanism that cannot
occur at significant depths. The decay of free oscillations
probes a large range of depths: it indicates a value of Q of
around 300 down to much lower frequencies[3] - as low as 0.0003 Hz.
There is evidence that above about 1 Hz, Q increases with
increasing frequency.[1]

Although attenuation can be described formally in terms of
anelastic parameters, it is difficult to account for these relatively high attenuation values at such low frequencies in terms
of physical processes. Furthermore, the relaxation mechanisms
that would be responsible for the attenuation are associated with
shear strain; recent evidence suggests that attenuation is also
associated with dilatational strain.[1,3]

In the present work, attenuation due to heat conduction is
considered. It is seen that it can be an important cause of
attenuation of seismic waves, though it will not be claimed to
be the only mechanism.

Heat conduction as a mechanism of attenuation has, in general, been discounted for seismic waves because the adiabatic
temperature differences produced by an elastic wave extend over
distances of the wavelength, i.e., several kilometers. Therefore
the temperature gradient and the consequent entropy production
are negligible. However, this argument assumes the material to
be homogeneous. The material of the earth's mantle, like all
rocks, is composed of a mixture of grains of different orientation
and composition. The individual grains differ in their response
to an elastic strain and exhibit different temperature changes
under adiabatic conditions. Since the grain size is much smaller
than the seismic wavelength, the temperature gradients and the
consequent entropy production are substantially enhanced. This
heat conduction mechanism was first discussed by Zener[4] in connection with ultrasonic attenuation in polycrystalline metals.
It will be shown that in the case of rocks in the mantle it can
yield attenuation values comparable to those that are observed.

THEORY

Consider a compressional wave with a local time-dependent dilatation, Δ. The rate of change of temperature, $T(\underline{r})$, a function of position, is given by

$$\frac{\partial}{\partial t} T(\underline{r}) = D\nabla^2 T(\underline{r}) - \gamma T \frac{\partial}{\partial t} \Delta \tag{1}$$

The first term on the right-hand side is due to heat conduction: D is the thermal diffusivity. The second term, derived from thermodynamics, describes the adiabatic temperature change due to dilatation (or compression): T is the absolute temperature and γ the Grüneisen constant. The latter is defined in terms of the volumetric coefficient of thermal expansion, β, the adiabatic bulk modulus, k, and the specific heat at constant pressure, C_P, per unit volume, V, as

$$\gamma = \beta k V / C_P \tag{2}$$

Now let the material be inhomogeneous, so that $\gamma(\underline{r})$ is a function of position. Therefore $T(\underline{r})$ is also a function of position, and in contrast to the case of a homogeneous material, it will reflect the spatial variation of $\gamma(\underline{r})$. Thus

$$\nabla^2 T(\underline{r}) = -q^2 \delta T \tag{3}$$

where δT is the excursion of the temperature from its mean value, T, and q is the wave number of the important Fourier components of $\gamma(r)$. In a granular material the important values of q are centered about

$$q \simeq \pi/L \tag{4}$$

where L^3 is the average grain volume. Writing $\delta T = T_o e^{i\omega t}$ and $\Delta = \Delta_o e^{i\omega t}$, equation (1) becomes

$$T_o = -\gamma(q) T\Delta_o (1 - i\omega_o/\omega)^{-1} \tag{5}$$

where

$$\omega_o = q^2 D = 2\pi f_o \tag{6}$$

defines a characteristic frequency f_o.

The entropy generation per unit volume is given by

$$\frac{dS}{dt} = \kappa (\text{grad } T/T)^2 = \frac{1}{2} \kappa q^2 T_o^2 / T^2 \tag{7}$$

where κ is the thermal conductivity. Note that $\kappa q^2 = \omega_o C_p/V$.
The rate of energy loss by the wave per unit volume is $T d\tilde{S}/dt$.
The energy content of the wave per unit volume is $\frac{1}{2}k\Delta_o^2$.
The decay rate thus becomes

$$\frac{1}{\tau} = \frac{2}{k\Delta_o^2} \; T \frac{dS}{dt} = \gamma^2 \omega_o \frac{C_p T}{V\,k} \left(1 + \frac{\omega_o^2}{\omega^2}\right)^{-1} \tag{8}$$

where γ^2 is the mean square fluctuation of $\gamma(\underline{r})$ about its average
value $\overline{\gamma}$. For a mixture of grains of different composition

$$\gamma^2 = \sum_i V_i (\gamma_i - \overline{\gamma})^2 \tag{9}$$

where γ_i is the value for each component of fractional volume, V_i.

APPLICATION TO THE EARTH's MANTLE

The specific dissipation function, given by equation (8) as

$$Q^{-1} = \frac{1}{\tau\omega} = \gamma^2 \frac{C_p\,T}{V\,k} \frac{f\,f_o}{f^2 + f_o^2} \tag{10}$$

depends on f_o, which in turn depends on grain size and the thermal
diffusivity. Neither of these quantities are well known for the
earth's mantle. If we adopt values typical of surface rocks, the
grain size, $L \simeq 0.3$ cm, and the thermal diffusivity, $D \approx 0.02$ cm^2/s,
so that f_o, given by equation (6), would be about 0.35 Hz. There
is, however, considerable uncertainty about this estimate of f_o.

It therefore seems better to estimate the maximum value of
Q^{-1}, which occurs when $f = f_o$. This maximum value is

$$Q_{max}^{-1} = \frac{1}{2} \; \gamma^2 \; C_p T/Vk \tag{11}$$

Since $C_p/V = 2$ MJ-m^{-3}-K^{-1} and $k = 50$ GJ-m^{-3}-K^{-1}, $Vk/C_p = 25,000$ K.
A representative value of the temperature at a depth of 400 km
is 1800 K.[5] Values of γ for the important minerals of the earth's
mantle are given by Anderson:[6] the average variability of γ
about its mean value is about 0.3, so that $\gamma^2 \approx 0.1$. Thus one obtains from equation (11) that $Q_{max}^{-1} \simeq 3.5 \times 10^{-3}$ or $Q_{min} \approx 300$.

At very high pressures, values of γ tend to decrease and k
tends to increase. One thus expects Q^{-1} to decrease with increasing depth in the lower mantle, which is in qualitative agreement with observations.

ASTHENOSPHERE

The low-velocity layer that shows high attenuation is believed to be a region in which there are inclusions of molten basalt, and to account for the low velocity these inclusions should occupy about 6% of the total volume.[7] The values of γ for molten materials are usually double those of the corresponding solids. The inclusions would therefore increase γ^2 by about 0.15 to a total value of perhaps 0.25. Hence one would expect in that region Q_{max}^{-1} to be about 9 x 10^{-3} or $Q_{min} \approx 110$. This attenuation is slightly less than what is observed ($Q \approx 80$). The thermal conduction model, however, agrees much better with the observed Q than a model based on Rayleigh scattering by the molten inclusions.[2]

SHARPNESS OF BOUNDARIES

So far we have assumed the inhomogeneous thermal expansivity, characterized by $\gamma(\underline{r})$, to have Fourier components of a single wave number, q, given by equation (4). If $\gamma(\underline{r})$ changes discontinuously at the grain boundaries, there will also be a series of harmonics (multiples of q), and the attenuation will be given by a sum of terms of the form of equation (8). The sum is essentially terminated when the overtone frequency equals ω, so that the number of terms in the sum is of order $(\omega/D)^{1/2} L/\pi$. Thus the attenuation still peaks at the frequency $f_o \sim \pi D/2L^2$, but at higher frequencies Q^{-1} falls off not as $1/f$, as equation (10) would demand for a single characteristic frequency, but as $1/\sqrt{f}$.

One can also understand this result physically. Note that Q^{-1} is the energy lost per cycle. At $f \sim f_o$, the entropy production takes place in the entire volume of each grain, but at high frequencies it takes place in a skin near each grain boundary of thickness of order $(D/\omega)^{1/2}$, so that the volume fraction of material involved in the entropy production is reduced by a factor $(f_o/f)^{1/2}$, and Q^{-1} falls off similarly.

However, if the composition in each grain changes gradually and merges into the composition of the next grain, a single resonance frequency, $f_o \sim \pi D/2L^2$, would be a better approximation, and equation (10) would describe the attenuation at all frequencies.

SHEAR WAVES

The attenuation of seismic shear waves is at least comparable to that of compressional waves. Since compressional waves also have a shear strain component, Knopoff[2] attributed all attenuation to anelasticity in the shear strain. Recent studies of the decay of free oscillations show a need for attenuation of oscillatory dilatational strains, which the present heat conduction model provides. However, this would not explain the attenuation of seismic shear waves, since these have no dilatational component.

A shear strain causes an adiabatic temperature change analogous to the second term in equation (1), provided the principal axes of strain do not coincide with the symmetry axes of the crystal and provided the crystal is noncubic. Hence the effective γ of each grain depends on orientation, not on composition, and the boundaries between grains must be regarded as sharp. Although individual values of γ are smaller than for dilatation, $\gamma=0$ and the mean square variability $(\gamma-\bar{\gamma})^2$ may well be of comparable magnitude. It is thus possible that heat conduction could cause attenuation of shear waves which is not much smaller than that of compressional waves. Values of f_o, which depend on thermal diffusivity and grain size, should be the same in both cases, and for frequencies $f>f_o$, Q^{-1} should decrease as $1/\sqrt{f}$, since boundaries between regions of different orientation are sharp.

DISCUSSION

The present theoretical model yields values of Q^{-1} that are comparable to observed attenuations, and which could perhaps be brought into better agreement with observations by adjusting some of the parameters involved. One must remember, however, that these values of Q^{-1} pertain only to one frequency, f_o. At lower frequencies, Q^{-1} falls off as f/f_o; at higher frequencies, as (f_o/f). The observed values of Q^{-1} seem to be essentially independent of frequency, particularly at the low frequencies of the earth's free vibrations.[1,3] It thus appears that the heat conduction mechanism makes a significant contribution, at least at some frequencies, but is not the only attenuation mechanism. At the present state of uncertainty, both in regard to attenuation values and in regard to a detailed physical picture of the earth's mantle, no firmer conclusion seems justified.

Two facts seem to give further support to the model, at least as a contribution to the attenuation: (1) Q^{-1} seems to decrease above about 1 Hz (see reference 1), a frequency not very different from the present rough estimate of f_o and (2) there is evidence for some attenuation associated with dilatational strain.[3]

B. H. Armstrong[8] has published a similar treatment. It differs mainly in taking a larger value of γ^2, but a lower temperature (300 K). He also considers two levels of inhomogeneity: granularity and variability of composition over larger distances (say 10 cm). This would weaken the frequency dependence of Q.

ACKNOWLEDGMENT

The author wishes to thank P. Dehlinger (University of Connecticut) and Sydney P. Clark, Jr. (Yale) for their helpful suggestions and discussions. B. H. Armstrong drew my attention to his work during the discussion following this paper.

REFERENCES

1. A. M. Dziewonski, Rev. Geophys. 17:303 (1979).
2. L. Knopoff, in "Mantle and Core in Planetary Physics," Academic Press, New York (1971), p. 146.
3. R. V. Sailor and A. M. Dziewonski, Geophys. J. R. Astron. Soc. 53:559 (1978).
4. C. Zener, "Elasticity and Anelasticity of Metals" University of Chicago Press, Chicago (1948).
5. A. E. Ringwood, in "Nature of the Solid Earth," McGraw-Hill, New York (1972), p. 67.
6. O. L. Anderson, in "Nature of the Solid Earth," McGraw-Hill, New York (1972), p. 575.
7. D. L. Anderson, C. Sammis, and T. Jordan, in "Nature of the Solid Earth," McGraw-Hill, New York (1972), p. 41.
8. B. H. Armstrong, Geophysics 45:1042 (1980).

ILLUSTRATIVE NUMERICAL COMPARISONS BETWEEN PHONON MEAN FREE PATHS AND PHONON THERMAL CONDUCTIVITY

W. M. MacDonald and A. C. Anderson

Department of Physics and Materials Research Laboratory
University of Illinois at Urbana-Champaign
Urbana, IL 61801

Measurements of thermal conductivity are often used as an interrogative technique to learn about phonon scattering processes in solids.[1] In general, the relationship between thermal conductivity, λ, and a phonon mean free path, ℓ, is complex, and it is therefore necessary to make some simplifying assumptions to make this relationship tractable. These assumptions when applied to experimental measurements may lead to erroneous conclusions, many of which have appeared in the published literature. The present paper attempts to provide an intuitive insight into the relationship between λ and ℓ.

In discussing the phonon thermal conductivity of a solid, we start with the usual series of simplifying assumptions. The more crucial of these will be discussed later. Considerable simplification is achieved if the solid is assumed to be <u>isotropic</u>. Thus the phonon thermal conductivity may be written in terms of phonon frequency, ν, as

$$\lambda = (2/3) \int C_t(\nu) v_t(\nu) \ell_t(\nu) d\nu + (1/3) \int C_\ell(\nu) v_\ell(\nu) \ell_\ell(\nu) d\nu \tag{1}$$

The subscript t refers to transverse acoustic phonons, and subscript ℓ refers to longitudinal acoustic phonons. In an isotropic medium the two transverse modes have the same velocity, v_t, hence both modes have been included in the first term of Eq. 1 by inserting the factor of 2. The quantity $C(\nu)d\nu$ can be thought to represent the contribution to the phonon specific heat by phonons within the frequency range ν to $\nu + d\nu$. It is given by

$$C_i(\nu) = (4\pi k \, \nu^2/v_i^3) x^2 \, e^x (e^x - 1)^{-2} \tag{2}$$

with x = hν/KT where h is Planck's constant and k is Boltzmann's
constant. Equation 2 is valid provided v_l is constant, that is
assuming there is no phonon dispersion. This is generally the
case for temperatures $\underline{T \lesssim 10 \text{ K}}$, and we now limit ourselves to this
region. Indeed, with the current availability of dilution refrig-
erators, many measurements are being made in the temperature range
0.01 to 10 K.

Assuming an isotropic material and $T \lesssim 10$ K, the only information needed to compute λ are the two velocities, v_t and v_l, measured at ultrasonic frequencies, and a knowledge of the mean free
paths, l_t and l_l. If $\underline{l_t = l_l = l}$, a constant, then Eq. 1 reduces
to

$$\lambda = 1.36 \times 10^{10} \ T^3 l (2v_t^{-2} + v_l^{-2}) \ (\text{W/m K}) \qquad (3)$$

Equation 2 can also be integrated to give the phonon specific heat,

$$C = 4.08 \times 10^{10} \ T^3 \ (2v_t^{-3} + v_l^{-3}) \ (\text{J/m}^3\text{K}) \qquad (4)$$

Thus it is possible to write

$$\lambda = \frac{1}{3.} C \ l [(2v_t^{-2} + v_l^{-2})/(2v_t^{-3} + v_l^{-3})] \qquad (5)$$

or

$$\lambda = C \ l \bar{v}/3 \qquad (6)$$

where the "average" $\bar{v}$ is defined by the square brackets in Eq. 5.
Thus if λ, C and $\bar{v}$ are measured,

$$l = 3\lambda/C\bar{v}. \qquad (7)$$

It should be stressed that the C used in Eq. 7 must be that contributed by phonons only. It is totally wrong, as is found occasionally
in the literature, to use a measured C that contains magnetic, electronic, or other contributions.

Equation 7 is a simple result, and is often applied or implied
when phonon mean free paths are quoted. But note the assumptions:
(i) isotropy, (ii) nondispersive, (iii) constant l.   The remainder of this paper will discuss the effects on λ of anisotropy and
of non-constant l while retaining the assumption of low temperature.

When l is a function of ν, we must use Eq. 1 explicitly. To
understand physically what occurs, it is useful to visualize C(ν).
This is plotted in Fig. 1 as a function of temperature, T, and of
frequency, ν. The plot shows the spectrum of phonons carrying the

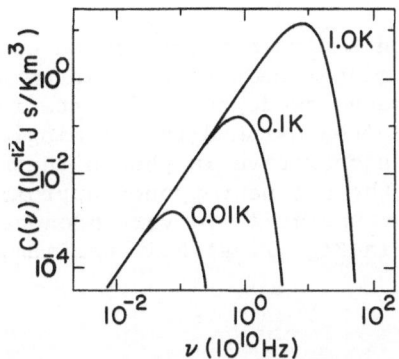

Fig. 1. The specific heat C(ν) contributed by phonons from the
frequency range ν to $\nu + d\nu$, as a function of frequency
at three temperatures.

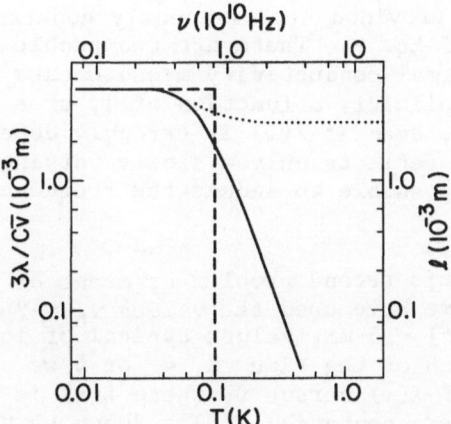

Fig. 2. Comparison of $\ell(\nu)$ versus ν (top and right coordinate),
the dashed line, and $\ell(T) = 3\lambda/C\bar{v}$ versus T (bottom and
left coordinates) represented by the solid line. The
dotted line, discussed in the text, occurs when only one
transverse mode is scattered according to $\ell(\nu)$.

heat at temperature T. The maximum in this spectrum may be varied
by changing T,

$$\nu_{max} = 3.83 \ kT/h = 7.98 \ x \ 10^{10} \ T \ (Hz) \qquad (8)$$

This corresponds to Wien's displacement law in optics, and it allows
a thermal conductivity measurement to be used as a phonon spectrometer to explore the frequency-dependent scattering of phonons.
For convenience, it is often assumed that a single phonon frequency
ν_{max}, given by Eq. 8, is operative in phonon thermal conduction.
This is referred to as the dominant-phonon approximation. But in
reality, our "phonon spectrometer" is very broadbanded. The full
width $\Delta\nu$ of the curves in Fig. 1, at half maximum, is

$$\Delta\nu = 1.3 \ \nu_{max} \qquad (9)$$

If $\ell(\nu)$ should vary slowly as a function of ν over the range
$\Delta\nu \approx 1.3\nu_{max}$ then, to a good approximation, $\ell(\nu)$ may be considered
constant in Eq. 1 and the integral performed at temperature T.
The value of the mean free path would be $\ell \approx \ell(\nu_{max})$ or, using
Eq. 8, $\ell \approx \ell(T)$. Hence

$$\lambda = C \ \bar{v} \ \ell(T)/3 \qquad (10)$$

provided $\ell_t(\nu) \approx \ell_\ell(\nu)$. Thus we again have the benefit of this
simplified equation, provided $\ell(\nu)$ is nearly constant over the
frequency range $\Delta\nu$ of Eq. 9. There are two problems in using Eq.
10. First, from thermal conductivity measurements alone, we cannot
determine if ℓ is explicitly a function of T, or a function of ν,
or both. And second, even if $\ell(\nu)$ is strongly dependent on ν, the
thermal conductivity reflects only a slowly varying function of T.
Thus, it may not be possible to deduce the frequency dependence of
$\ell(\nu)$ from λ.

We illustrate this second problem by means of Figs. 2 through
5. For simplicity, we have used the values v_t = 2 km/s, v_ℓ = 4 km/s,
and ℓ (where constant) = 5 mm, values typical of low-temperature
measurements. In each of the Figs. 2, 4, or 5 we show by the
dashed line a plot of $\ell(\nu)$ versus ν, where $\ell(\nu)$ is the phonon mean
free path used in the computation of λ. Shown by the solid line is
the relation $\ell(T) = 3\lambda/C\bar{v}$ from Eq. 10, where C has been calculated
using Eq. 4. The solid line thus represents the ℓ that would
be deduced from λ and C using the dominant-phonon approximation
of Eq. 8.

Figure 2 depicts an abrupt decrease in $\ell(\nu)$ at ν = 8 GHz.
This frequency was selected so that the related feature in λ would
occur at 0.1 K, well within the assumed temperature range of
T $\lesssim$ 10 K. Note from Fig. 2 that $3\lambda/C\bar{v}$ begins to decrease near

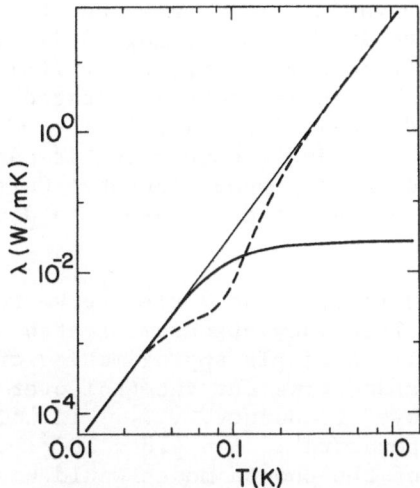

Fig. 3.   Thermal conductivity versus temperature, as discussed in
          the text. For reference, the light, straight line has a
          temperature dependence of $\lambda \propto T^3$.

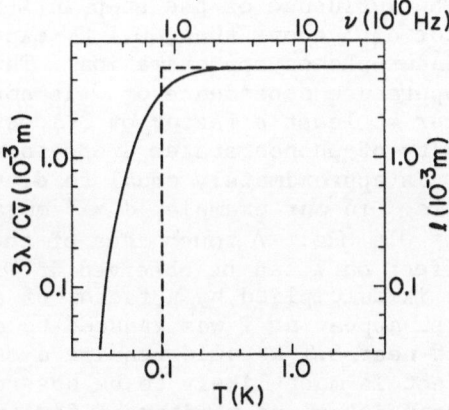

Fig. 4.   Comparison of $\ell(\nu)$, the dashed line, and $\ell(T) = 3\lambda/C\bar{v}$
          represented by the solid line.

0.033 K, a factor of 3 below the 0.1 K expected solely on the basis of the dominant-phonon approximation. This is a direct result of the width of the curves in Fig. 1 and reminds us that thermal conductivity is a broadband spectrometric technique. At temperatures well above 0.1 K, $3\lambda/C\bar{v}$ varies as T^{-3} meaning that $\lambda \propto T^0$. This is shown by the solid curve in Fig. 3. Figure 3 illustrates that no matter how strong the frequency dependence of $\ell(\nu)$, λ cannot decrease with increasing T. It is sometimes stated erroneously that scattering from point defects, for which $\ell(\nu) \propto \nu^{-4}$ (Rayleigh scattering) produces $\lambda \propto T^{-1}$. Such a conclusion is one example of the incorrect use of the dominant-phonon concept. Only an ℓ that is explicitly temperature dependent can cause λ to decrease with increasing T.

The solid lines in Figs. 2 and 3 also serve to demonstrate the effect of a finite frequency spectrum created by the atomic lattice of the material. A simple approximation of this fact is generally provided by truncating the integral over phonon frequencies in Eq. 1 at a "Debye" frequency, ν_D. In units of kelvin, this is the "Debye temperature": $\theta = 4.8 \times 10^{-11} \nu_D$. It is unlikely that all three of the phonon modes would have the same ν_D. However, for purposes of illustration, if $\nu_D = 8$ GHz for all three modes, the thermal conductivity would be represented by the solid curve in Fig. 3. Thus this temperature dependence, if observed experimentally, could be ascribed either to an abrupt decrease in $\ell(\nu)$, as shown by the dashed line in Fig. 2, or to an abrupt decrease in the density of phonon states at $\nu_D \simeq 8$ GHz.

Figure 4 shows the effect on $3\lambda/C\bar{v}$ of an abrupt increase in $\ell(\nu)$ at $\nu = 8$ GHz. The influence of the step in $\ell(\nu)$ has vanished at $T \gtrsim 0.33$ K, a factor of 3 above that, 0.1 K, expected solely on the basis of the dominant-phonon approximation. Thus, in general, the magnitude and temperature dependence of λ is not influenced by any features that occur at least a factor of 3 lower in temperature. For example, the density of phonon states drops to zero when the phonon wavelength, Λ, is approximately equal to d, where d is the dimension of the sample. In our example, $d \approx 5$ mm and $v_t = 2$ km/s. Hence, $\nu_{min} \approx v/\Lambda_{max} \approx 0.4$ MHz. A rough idea of the influence of this low-frequency effect on λ can be observed in Fig. 4 if the lower horizontal axis is multiplied by a factor of about 10^{-4}. The effect would thus first appear as T was reduced to about 10^{-5} K. To observe this effect near 0.1 K would require a sample of about 0.5 μm. Thus the effect is most likely to be observed during transport of heat in fine particles, as monitored, for example, by magnetic resonance measurements.

A resonant scattering of phonons of frequency ν from a defect that "vibrates" with frequency, ν_o, of 8 GHz can be approximated by the $\ell(\nu)$ shown in Fig. 5. This is simply a combination of curves like those in Figs. 2 and 4. The width of the resonant interaction is $\nu_h - \nu_\ell$. For Fig. 5a, $\nu_h/\nu_o = \nu_o/\nu_\ell = 1.10$, whereas for Fig. 5b, $\nu_h/\nu_o = \nu_\ell/\nu_\ell = 1.50$. The calculated $3\lambda/C\bar{v}$ shows three features of interest. First, as $\nu_h \to \nu_\ell$, the effect on λ is small, <u>independent</u> of the strength of the phonon scattering within the frequency interval $\nu_h - \nu_\ell$. Second, the width of the dip in $3\lambda/C\bar{v}$ is much larger than $\nu_h - \nu_\ell$ and is dictated by the width of the curves in Fig. 1. Third, the minimum in the dip does not fall at 0.1 K as expected solely on the basis of the dominant-phonon approximation, but rather at about 0.08 K. As $(\nu_h - \nu_\ell)$ becomes larger, the minimum does approach 0.1 K and, for $\nu_h/\nu_o = \nu_o/\nu_\ell \gtrsim 3.4$, falls at T > 0.1 K. The effect of a "resonance," having $\nu_h/\nu_o = \nu_o/\nu_\ell = 2$, on λ is shown by the dashed line in Fig. 3. Note again that for $T \gtrsim 0.5$ K, λ shows no influence of the resonance. The apparent phonon mean free path has returned to d = 5 mm, the size of the sample. This fact is also obvious in Fig. 5.

Often a mean free path is limited by more than one phonon scattering mechanism. The resultant mean free path, ℓ_{net}, is often approximated by the expression

$$\ell_{net}^{-1} = \ell_A^{-1} + \ell_B^{-1} + \ell_C^{-1} + \ldots \qquad (11)$$

The difficulty in obtaining the frequency dependence of, say, ℓ_B from the measured λ is demonstrated in Fig. 6 where, again, curves like those of Figs. 2 and 4 have been combined. Assume that we need a temperature range of at least a factor of 2 or 3 to determine the frequency dependence of ℓ_B from a measurement of λ. This is shown by the horizontal portion of the solid line in Fig. 6. It will be seen that ℓ_B, the dashed line, must therefore dominate ℓ_{net} over a frequency range of ≥ 20. Hence, in the entire temperature range of 0.01 to 10 K, it is difficult to deduce from λ the presence of more than 2 or 3 scattering mechanisms. Thus Eq. 11 should contain no more than 2 or 3 terms if a definitive analysis of λ data is to be expected.

The mean free paths, ℓ_t and ℓ_ℓ, need not be the same, and so the two terms in Eq. 1 may have to be computed independently. As an example, assume that one of the two transverse modes has a $\ell(\nu)$ like that of the dashed line in Fig. 2, while the longitudinal and other transverse modes continue to have $\ell = 5$ mm for all ν's. The result is shown by the dotted line in Fig. 2. Except for the region near 0.1 K, the temperature dependence is T^3. If $\ell = 5$ mm were caused by scattering from the surfaces of the sample (boundary scattering), then λ at both high and low temperatures would scale with the size of the sample. Hence the magnitude of λ above 0.1 K could be ascribed erroneously only to boundary scattering, whereas in reality it is a combination of boundary scattering for two modes plus zero conductance from the third, strongly scattered mode.

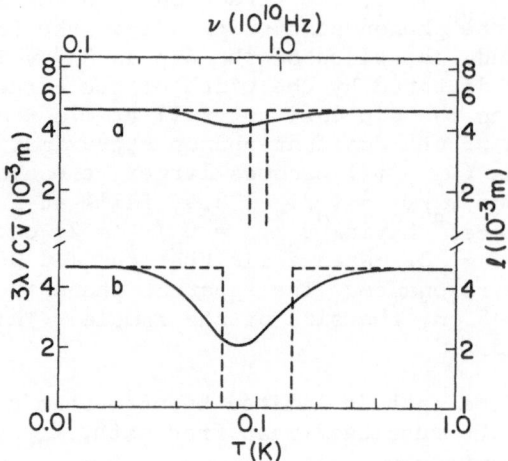

Fig. 5. Comparison of $\ell(\nu)$, the dashed lines, and $\ell(T) = 3\lambda/C\bar{v}$,
represented by the solid lines. For (a), $\nu_h/\nu_o =$
$\nu_o/\nu_\ell = 1.1$; for (b), $\nu_h/\nu_o = \nu_o/\nu_\ell = 1.5$. The symbols
ν_o, ν_h and ν_ℓ represent the center and the upper and
lower bounds of the drop in $\ell(\nu)$.

The interpretation of "λ" in the presence of boundary scattering is nasty. Phonons striking the surface may be scattered specularly or diffusively and may, in effect, change frequency or even
mode of propagation. Even the manner in which phonons scatter from
the ends of the sample may require consideration.[2]

Finally, all crystalline materials are anisotropic. As a result, phonon energy is channeled preferentially in certain directions.[3] Any calculation of the thermal conductivity of a crystal
that claims to be nearly accurate cannot avoid the significant computational complexities introduced by phonon "focusing."[4]

Problems associated with the limited resolution $\Delta\nu \stackrel{\sim}{} 1.3\,\nu_{max}$
(Eq. 9) could be avoided if a source of monochromatic phonons were
used. As yet, however, there is no fast and reliable experimental
method of inserting monochromatic phonons, having a frequency range
1 GHz to 1 THz, into a variety of materials. It is, therefore,
likely that thermal conductivity measurements will continue to provide an important source of information on phonon scattering.

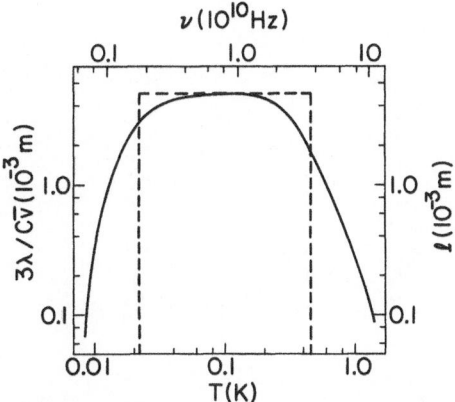

Fig. 6. A demonstration that $\ell(\nu)$ must be dominated by a single scattering mechanism over a frequency range a factor of 20 wide in order that the experimental data $3\lambda/C\bar{v}$ reflect the presence of this scattering mechanism over a temperature range of at least a factor of 2 or 3 in width.

ACKNOWLEDGMENT

This work was supported in part by the U.S. Department of Energy under Contract DE-AC02-76ER01198.

REFERENCES

1. R. Berman, "Thermal Conductivity in Solids" Clarendon, Oxford (1976).
2. R. Berman, E. L. Foster, and J. M. Ziman, Proc. R. Soc. Lond. 231A:130 (1955).
3. J. P. Wolfe, Phys. Today 33:44 (1980).
4. C. G. Winterherimer and A. K. McCurdy, Phys. Rev. B18:6576 (1978).

THERMOELECTRIC POWER OF SELECTED BINARY ALLOY SYSTEMS

C. Y. Ho, T. C. Chi, R. H. Bogaard, T. N. Havill,
and H. M. James

Center for Information and Numerical Data Analysis
and Synthesis
Purdue University
West Lafayette, Indiana   47906

ABSTRACT

A revised scale of the absolute thermoelectric power from 0
to 1300 K was developed to replace the old scale which was known
to be in serious error. The available experimental data exhaustively compiled from the literature for the thermoelectric power
of selected binary alloy systems were carefully converted to this
revised scale. The converted data were critically evaluated, correlated, analyzed, and synthesized, resulting in the recommended
values presented herein.

INTRODUCTION

Because the scale of the absolute thermoelectric power commonly
used throughout the world in the past 49 years had become known recently to be much in error, we developed a revised scale of the
absolute thermoelectric power from cryogenic to high temperatures.
This was done through critical evaluation, analysis, and synthesis
of (1) the available low-temperature data on the relative thermoelectric power of copper versus lead and (2) the available high-
temperature data on the Thomson coefficient of copper, and by accepting Roberts' values[1] for the absolute thermoelectric power of
lead up to 350 K. In the meantime, the available experimental data
on the thermoelectric power of selected binary alloy systems were
exhaustively compiled and carefully converted to our revised scale
of the absolute thermoelectric power. The resulting converted data
on the absolute thermoelectric power of the binary alloy systems

were critically evaluated, correlated, analyzed, and synthesized.
From the fragmentary and often conflicting data in which serious
gaps exist for both the composition and temperature dependences,
recommended values for the absolute thermoelectric power were
generated.

THE SCALE OF ABSOLUTE THERMOELECTRIC POWER

     The first scale of the absolute thermoelectric power was
established in 1932 by Borelius et al.[2] through a complex indirect
means of determination of the Thomson coefficients of lead, tin,
and a "silver-normal" alloy, which was an alloy of silver + 0.37
at.% gold + another (unknown) impurity. From the measured and derived Thomson coefficients in various temperature ranges, they
calculated the absolute thermoelectric power of lead and tin from
0 to 100 K and of the "silver-normal" alloy from 2 to 293 K and,
thus, established the first scale of absolute thermoelectric power.
This 1932 scale has since been in use as international standard,
though in 1958 Christian et al.[3] modified the values of Borelius
et al. for temperatures below 20 K.

     Christian et al.[3] made direct measurements relative to a Nb_3Sn
superconductor (mean transition temperature 17.92 K) of the absolute thermoelectric power of a zone-purified lead sample that was
"very much purer than the starting material which was already
99.999% chemically purified lead." Their determination of the absolute thermoelectric power of lead up to 18 K eliminated errors at
these low temperatures in the scale of Borelius et al., which had
resulted from the interpolation of the Thomson coefficient of tin
between 7.2 and 20 K. The modified scale of Christian et al.,
which was based on their own measurements below 20 K and on the
measurements of Borelius et al. from about 20 to 293 K, served as
the standard for low-temperature thermoelectric measurements from
1958 to 1977.

     In 1977 Roberts[1] shook the foundations of the older scale of
absolute thermoelectric power by reporting that the values for the
absolute thermoelectric power of lead as determined by Borelius et
al. and modified by Christian et al. were in error by as much as
0.3 $\mu V\ K^{-1}$, or nearly 50%. Roberts had undertaken the long overdue
low-temperature measurement of the Thomson coefficient of a 99.9999%
pure lead sample and calculated the absolute thermoelectric power
of lead up to 350 K from the measured Thomson coefficients.

     Accepting Roberts' values for the absolute thermoelectric
power of lead, we were able to generate the values for the absolute
thermoelectric power of copper up to 350 K through critical evaluation, analysis, and synthesis of the low-temperature data on the

relative thermoelectric power of copper versus lead available from the literature.

To generate values for the absolute thermoelectric power of copper above 350 K, the available high-temperature data on the Thomson coefficient of copper were critically evaluated, analyzed, and synthesized. The values of the absolute thermoelectric power of copper up to 1300 K were then calculated from the recommended values for the Thomson coefficient of copper. The recommended values for copper are shown in Figures 2 and 4.

ABSOLUTE THERMOELECTRIC POWER OF ALUMINUM-COPPER, COPPER-NICKEL, AND IRON-NICKEL ALLOY SYSTEMS

Most of the available data for the relative thermoelectric power of alloys were obtained from measurements against lead, copper, or platinum. Many authors also reported values for the absolute thermoelectric power of alloys, which were obtained by conversion from their measured relative values using incorrect values of the absolute thermoelectric power of lead, copper, and platinum.

We exhaustively compiled the available data on both the relative and the absolute thermoelectric power of alloys in the aluminum-copper, copper-nickel, and iron-nickel binary alloy systems. The compiled data were carefully converted according to the newly developed CINDAS revised scale of the absolute thermoelectric power. The resulting converted data for the absolute thermoelectric power were then critically evaluated, correlated, analyzed, and synthesized to generate a full field of internally consistent recommended values for the absolute thermoelectric power. The recommended values cover all the binary alloy compositions from one constituent pure element to the other and cover all the temperatures from absolute zero to the solidus temperature (melting starting point) or to 1300 K.

The recommended values for the aluminum-copper alloy system are presented in Figure 1 (for the temperature dependence of aluminum + copper alloys), Figure 2 (for the temperature dependence of copper + aluminum alloys), and Figure 3 (for the composition dependence of the entire aluminum-copper binary system). Similarly, the recommended values for the copper-nickel alloy system are presented in Figures 4 through 6 and those for the iron-nickel alloy system are presented in Figures 7 through 9. The uncertainties in most of these values are of the order of ±10 to ±20%, except for very small values in which uncertainties could be very large.

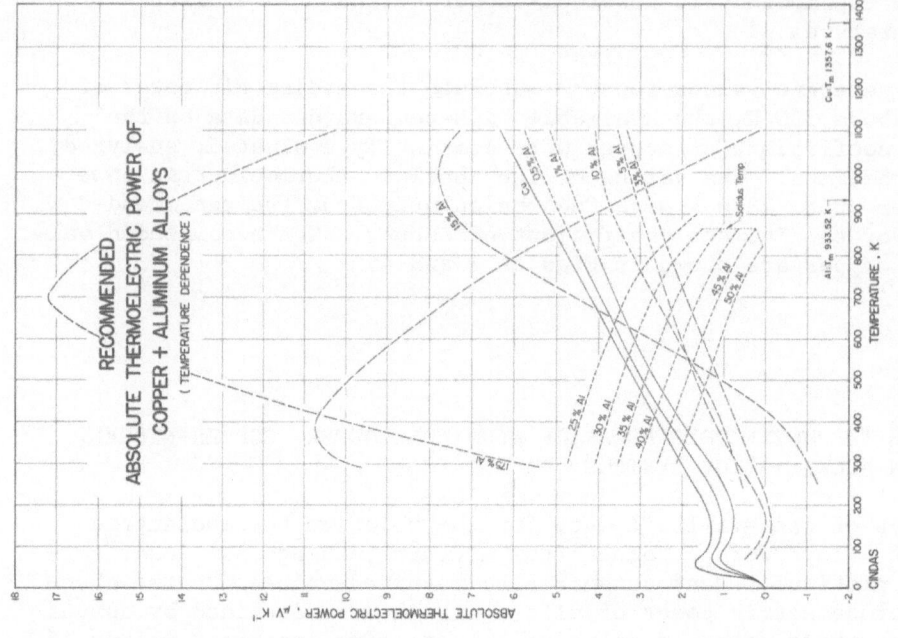

Fig. 2.   Recommended absolute thermoelectric power of copper + aluminum alloys.

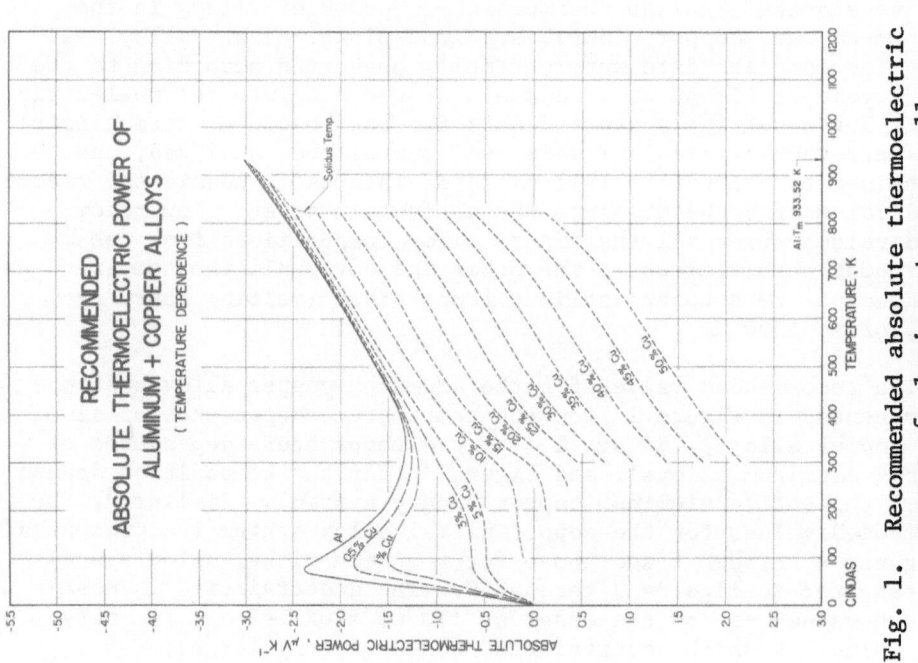

Fig. 1.   Recommended absolute thermoelectric power of aluminum + copper alloys.

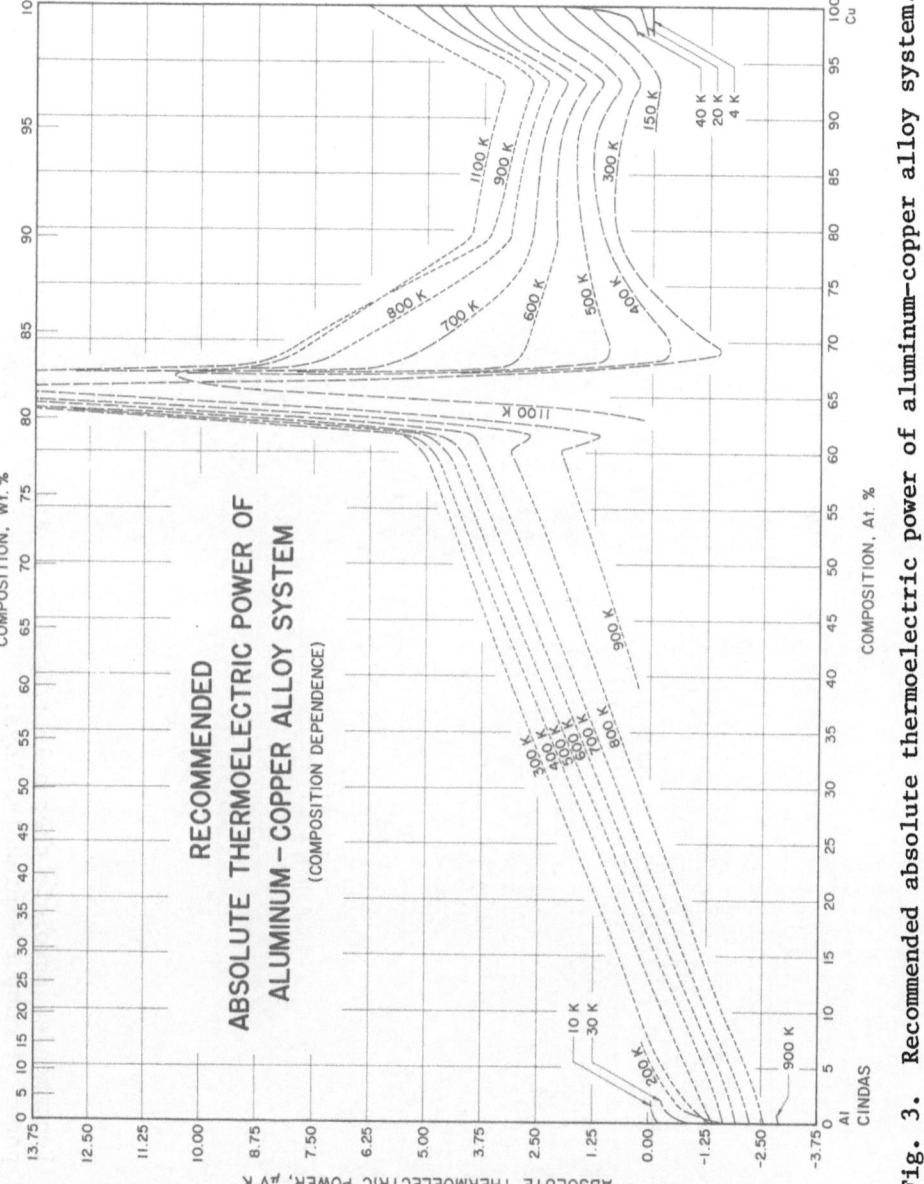

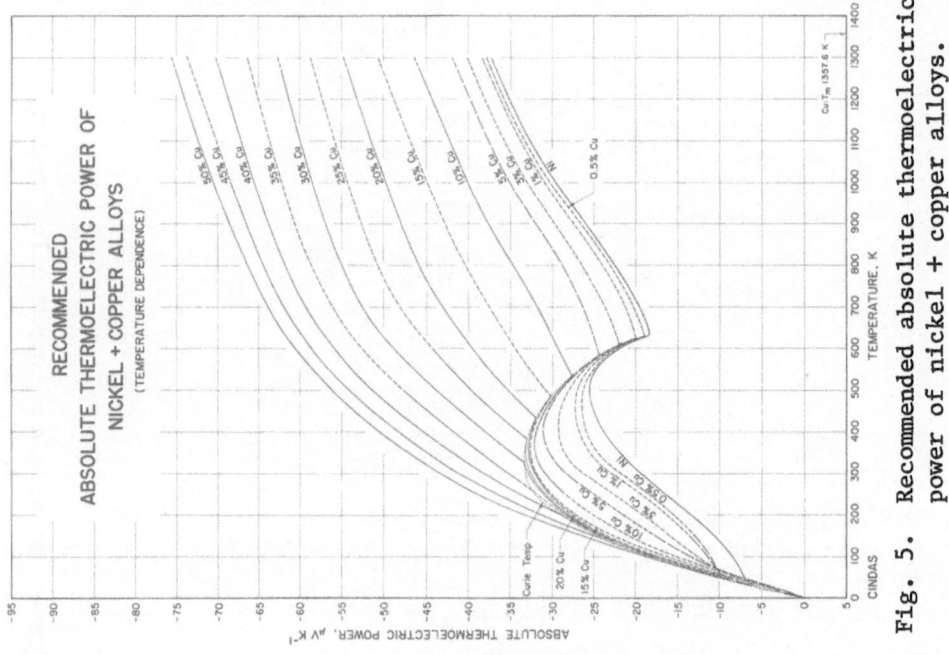

Fig. 5. Recommended absolute thermoelectric power of nickel + copper alloys.

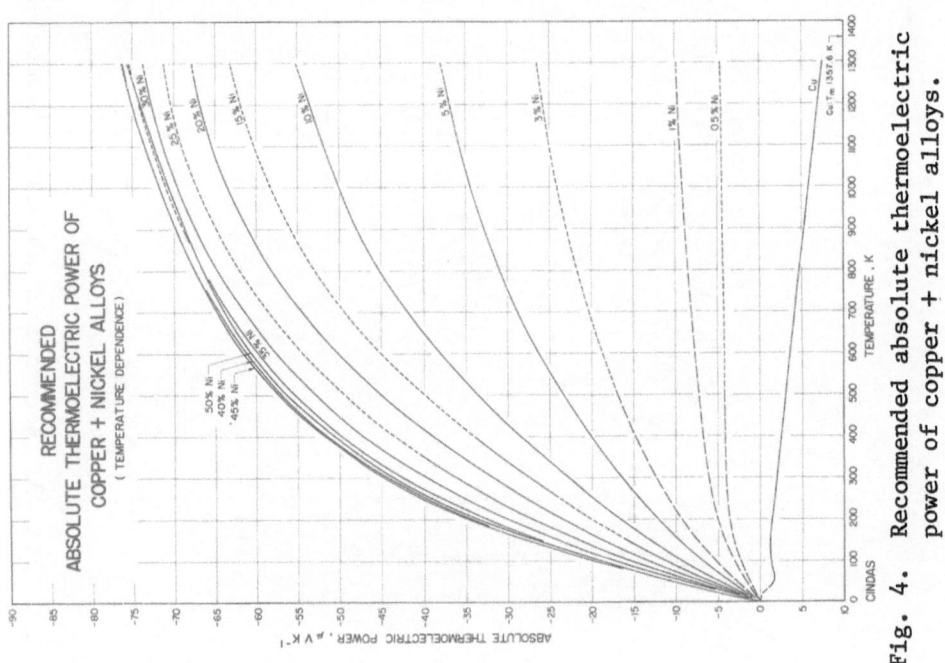

Fig. 4. Recommended absolute thermoelectric power of copper + nickel alloys.

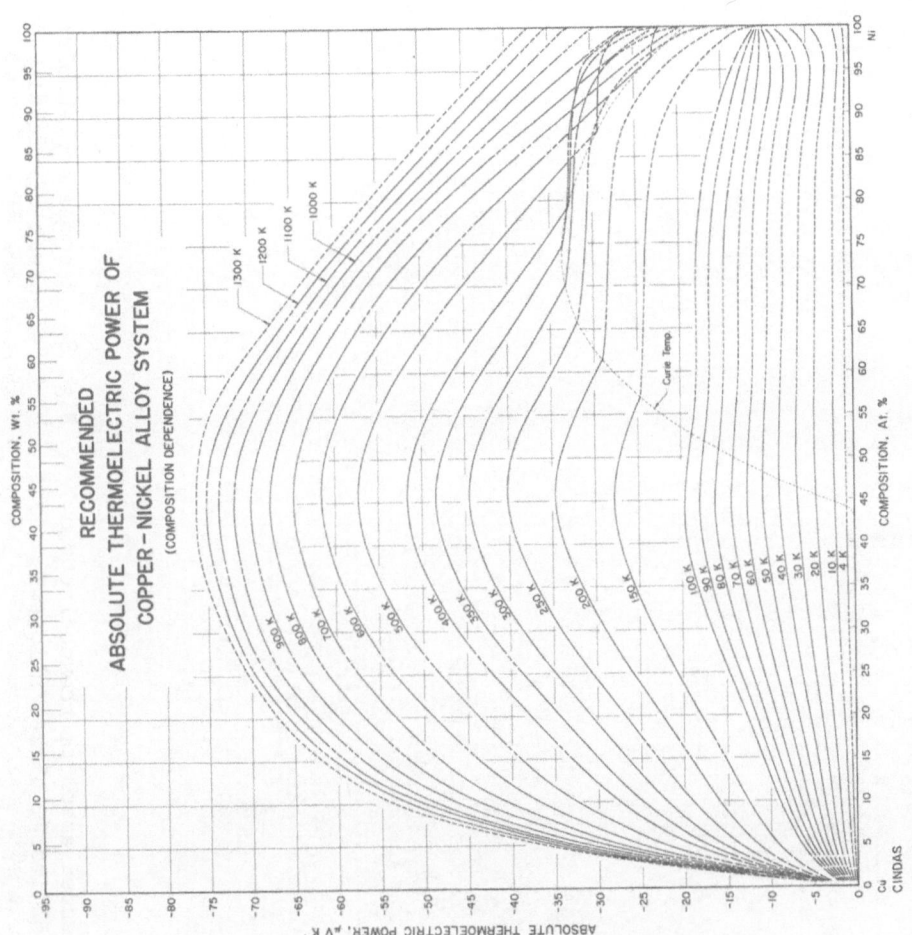

Fig. 6. Recommended absolute thermoelectric power of copper–nickel alloy system.

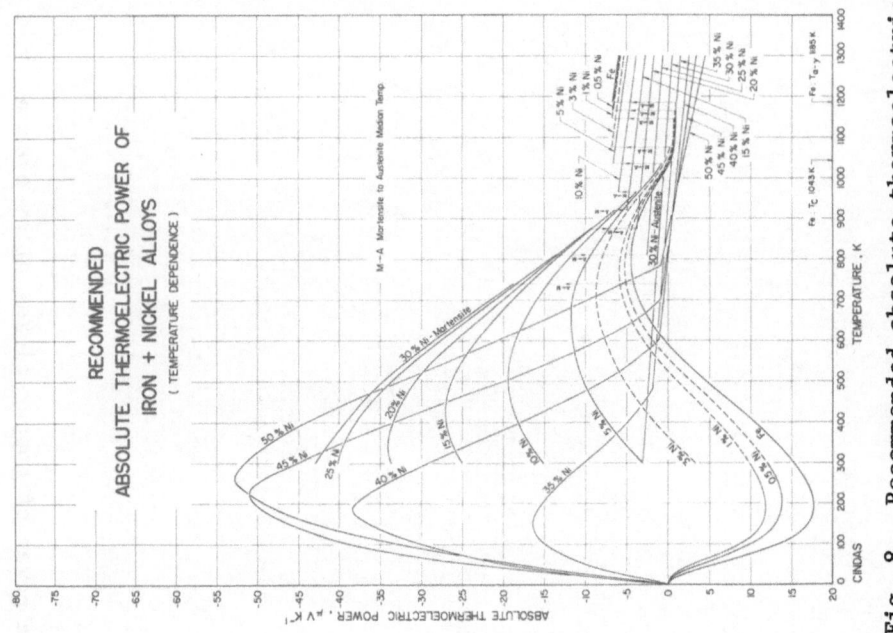

Fig. 8. Recommended absolute thermoelectric power of nickel + iron alloys.

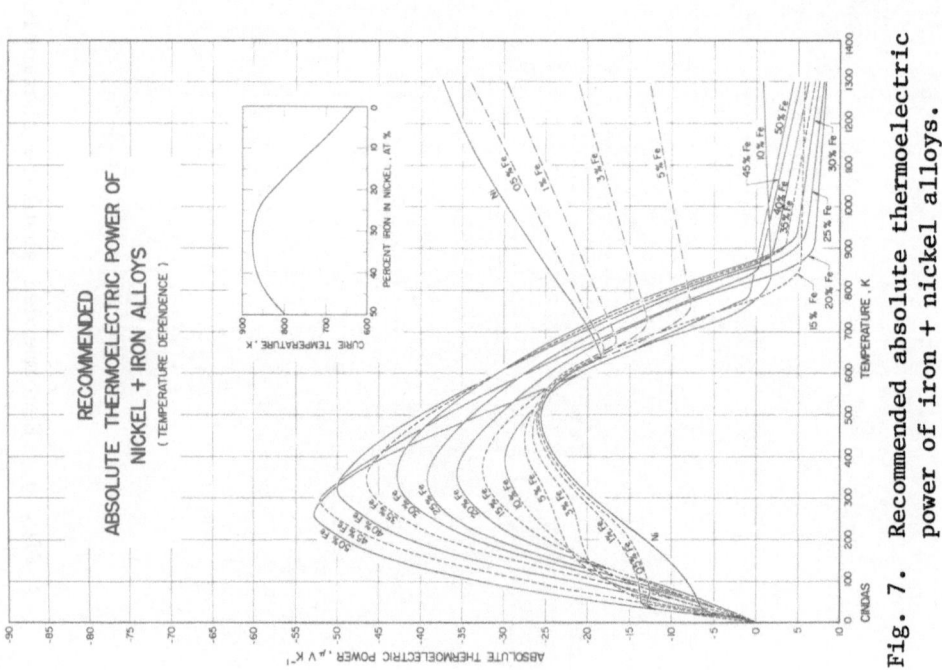

Fig. 7. Recommended absolute thermoelectric power of iron + nickel alloys.

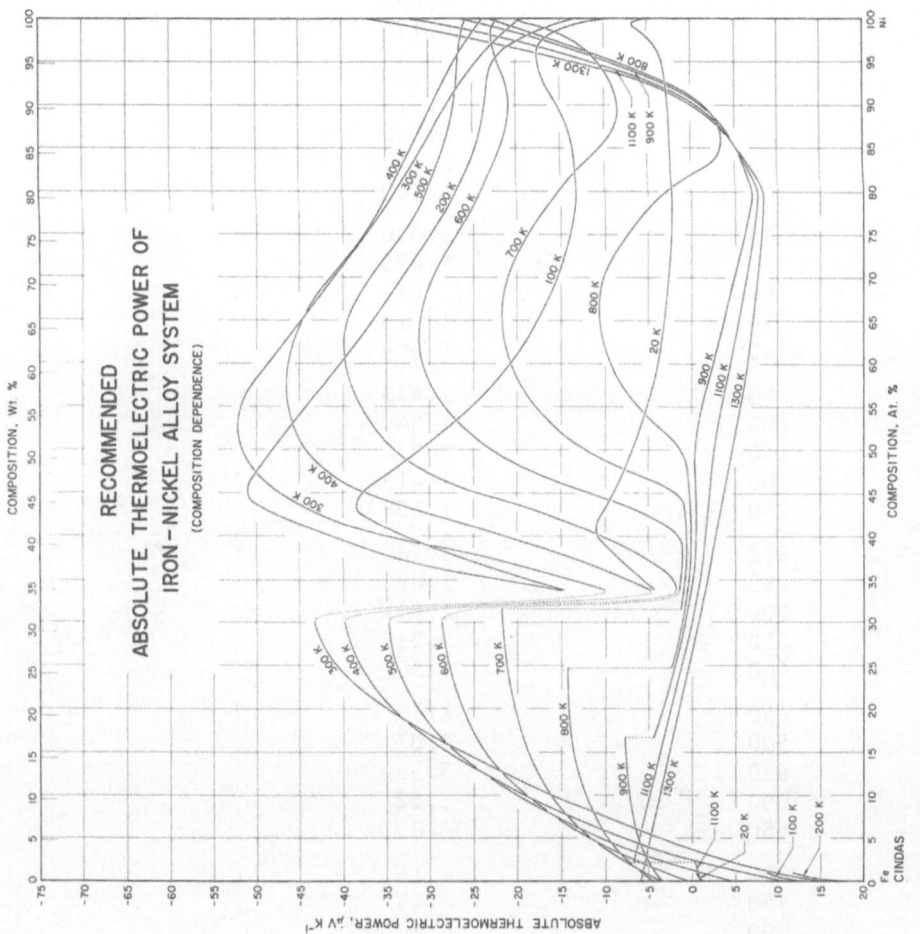

Fig. 9. Recommended absolute thermoelectric power of iron-nickel alloy system.

Table 1.   Recommended Absolute Thermoelectric Power of Copper

Temperature (K)	Absolute Thermoelectric Power ($\mu V \ K^{-1}$)
2	0.011*
4	0.023*
7	0.047*
10	0.082*
15	0.176*
20	0.330*
25	0.571*
30	0.895
35	1.186
40	1.400
45	1.540
50	1.625
60	1.674
70	1.616
80	1.509
90	1.413
100	1.349
150	1.278
200	1.413
250	1.658
273	1.785
293	1.890
300	1.931
350	2.216
400	2.50
450	2.78
500	3.07
550	3.35
600	3.62
650	3.89
700	4.16
750	4.43
800	4.70
850	4.96
900	5.23
1000	5.76
1100	6.30
1200	6.83
1300	7.37

*Values fitted to an idealized curve: $S = aT + bT^3$.

REFERENCES

1. R. B. Roberts, The absolute scale of thermoelectricity, <u>Philos. Mag.</u>, 36(1):91-107 (1977).
2. G. Borelius, W. H. Keesom, C. H. Johansson, and J. O. Linde, Establishment of an absolute scale for the thermoelectric force, <u>Proc. Roy. Acad. Sci. Amsterdam</u>, 35:10-4 (1932). (Note that Supplements 69b and 70c of Communications from the Kamerlingh Onnes Physical Laboratory of the University of Leiden are identical or nearly identical to one another and to this publication.)
3. J. W. Christian, J. P. Jan, W. B. Pearson, and I. M. Templeton, Thermoelectricity at low temperatures. VI. A redetermination of the absolute scale of thermoelectric power of lead, <u>Proc. R. Soc. Lond.</u>, A245:213-21 (1958).

ACKNOWLEDGMENT

   This work was supported by the Office of Standard Reference Data (OSRD) of the U.S. National Bureau of Standards (NBS), Gaithersburg, Maryland under the technical direction of Dr. Howard J. White, Jr., Program Manager of Industrial Process Data, OSRD.

NOTE ADDED AFTER PRESENTATION

   Immediately after this paper was presented at the Seventeenth International Thermal Conductivity Conference at NBS on 16 June 1981, we learned from Dr. G. K. White, a member of the CODATA Task Group on Thermophysical Properties, who was attending the Seventeenth International Thermal Conductivity Conference, that Dr. R. B. Roberts had recently measured the Thomson coefficient of copper, among other measurements, and that his calculated absolute thermoelectric power values over the temperature range 273 to 900 K were given in a manuscript which had been submitted for publication in Philosophical Magazine. A summary of Dr. Roberts' results were given in a letter to Dr. G. K. White, who kindly gave us a copy of his letter after our presentation of this paper.

   On comparison of Dr. Roberts' experimental data for the temperature range 273 to 900 K with our synthesized values for the absolute thermoelectric power of copper, we were very pleased to find that Dr. Roberts' data and our values were extremely close and almost identical, the differences being in the third significant digits. We chose to adjust the third significant digits of our values in conformity with his data over 400 to 900 K. Table 1 presents the recommended values for copper for the entire temperature range 2 to 1300 K.

SESSION   TC-6

Metals and Alloys

CHAIRMAN

J. P. Moore
Oak Ridge National Laboratory
Oak Ridge, Tennessee

THERMAL CONDUCTIVITY AND LORENZ NUMBER OF PLUTONIUM AND PLUTONIUM-GALLIUM ALLOYS

J. F. Andrew
Los Alamos National Laboratory
Los Alamos, New Mexico, 87545

and

P. G. Klemens
Dept. of Physics and Institute of Material Science
University of Connecticut
Storrs, Connecticut, 06268

The thermal diffusivities of Pu, Pu-2.7 at.% Ga, Pu-3.5 at.% Ga, and Pu-6.6 at.% Ga were measured from 25°C to around 500°C using a laser flash technique and electronic data acquisition. Although the Lorenz number, L, of pure Pu is well below the Sommerfeld value, L_0, except for the α-phase, L exceeds L_0 for the alloys at all temperatures and has a pronounced minimum around 200°C. At the lower temperatures we attribute the excess to lattice conduction and at temperatures above 200°C the excess is attributed to an electronic component. The negative deviation of L from L_0 for pure Pu is ascribed to a reduction of the electron mobilities in those energy ranges where the 5f bands overlap the conduction bands.

INTRODUCTION

This paper presents the results of thermal diffusivity measurements on a series of Pu-Ga alloys as a continuation of the diffusivity study begun by Lewis, et al., who reported results on 1 wt% Ga (3.3 at.% Ga).[1] Our study included high-purity unalloyed Pu and three stabilized delta-phase Pu-Ga alloys (2.7, 3.5, and

6.6 at.% Ga) in the temperature range 20-580°C. A number of elements, such as Al, Am, Ga, and Ce form solid solutions with Pu,
resulting in an fcc delta-phase alloy that is stable to below room
temperature. The measurements on the Pu + 3.5 at.% Ga alloy reported here are within 10% of the values reported previously by
Kruger and Robbins,[2] and along with the previous work by Andrew[3,4]
give a comprehensive picture of the thermal conductivity of Pu and
Pu-Ga alloys.

EXPERIMENTAL METHOD

The experimental method used in this study was similar to the
flash diffusivity measurement technique described by Parker et al.[5]
The front surface of a disk-shaped specimen was heated instantane-
ously by infrared radiation from a Nd-glass laser and the temperature rise on the back surface was measured by "Platinel 2" thermocouple wires contacting the Pu sample. The rear surface temperature was monitored on an oscilloscope operating in single-sweep,
memory mode and stored in an electronic data acquisition system.
An enlarged trace of the rear surface temperature-time history was
used to obtain the rise-time data for the diffusivity calculation.
The following equation, from Cowan,[6] was used to calculate the
thermal diffusivity:

$$\frac{T}{T_{max}} = 1 + 2 \sum_{n=1}^{\infty} (-1)^n e^{-n^2 \pi^2 \alpha t / z^2} \tag{1}$$

where T is rear surface temperature at time t, α is the thermal
diffusivity, and z is the thickness of the specimen disk. The
solutions of this equation for a series of T/T_{max} are given in
Table I. The heat loss corrections were included by calculating

Table I. Solutions to the Diffusi-
          vity Equations

T/T_{max}	$\alpha = \text{constant } z^2/t(x)$
0.1	$0.0661 \ z^2/t(0.1)$
0.2	$0.0843 \ z^2/t(0.2)$
0.3	$0.1012 \ z^2/t(0.3)$
0.4	$0.1190 \ z^2/t(0.4)$
0.5	$0.1388 \ z^2/t(0.5)$
0.6	$0.1622 \ z^2/t(0.6)$
0.7	$0.1919 \ z^2/t(0.7)$
0.8	$0.2331 \ z^2/t(0.8)$

the diffusivities at the T/T_{max} values given in Table I and extrapolating to zero time. This method of making the heat loss correction gives approximately the same results as the Cowan technique.[6]

The diffusivity apparatus and associated inert atmosphere
glovebox were previously described by Lewis et al.[1] Laser align-
ment and specimen temperature were checked by measuring the diffusivity of a sample of round-robin Armco Fe. Measured values of α
were within 5% of the results of Cody, Abeles, and Beers.[7]

Our Pu specimens were 10 mm in diameter and about 1.3 mm thick.
The unalloyed Pu specimen was 99.99 wt% Pu. The Pu-Ga alloys were
prepared from Pu stock with a purity greater than 99.98 wt% Pu. The
impurities were mainly Fe, Si, C, W, Ta, Al, and Ni; the overall concentration was, in all cases, less than 500 wt ppm. The Pu-Ga alloys
were annealed at 440°C for 200 h to insure delta-phase stability and
homogeneity.

Thermal conductivity values, λ, were calculated from the thermal diffusivities ($\lambda = \alpha\, D\, Cp$) using measured density values, D,
specific heat, Cp from Kay and Loasby[8] for Pu and from Rose et al.[9]
for the Pu-Ga alloys. The accuracy of the thermal conductivity
values are probably within 10% since the specific heat data was
taken from the literature and corrected for the Ga content.

RESULTS

The thermal diffusivity values are given in Table II for the
Pu-Ga alloys and these alloys along with the unalloyed Pu are shown
on Fig. 1. The diffusivity data were fitted to straight lines using
a linear regression technique. These equations are given in Table
III. Diffusivity values obtained from the equations in Table III,
along with values of specific heat and electrical resistivity used
to calculate the thermal conductivities and Lorenz numbers, are
presented in Table IV for all Pu specimens.

DISCUSSION

The thermal diffusivity, shown in Fig. 1 as a function of temperature, can be adequately represented by straight lines for all
phases of Pu. A small amount of Ga increases the thermal diffusivity, but further additions of Ga cause it to decrease. The same
behavior as a function of Ga content is shown by the thermal conductivity in Fig. 2. The data below 25°C are extrapolated from previous
results at -198°C.[3,4] The Lorenz number of pure Pu is well below the
Sommerfeld value, L_o (2.45 x 10⁻⁸ V²/K²) in the β, γ, δ, and ϵ phases
(see Table IV). However, the alloys have Lorenz numbers generally
above L_o (see Table IV).

Table II.   Measured Diffusivity Data for the Pu-Ga
            Alloys

2.7 at.% Ga		3.5 at.% Ga		6.6 at.% Ga	
T °C	α x10^6m^2/s	T °C	α x10^6m^2/s	T °C	α x10^6m^2/s
22	3.97	22	3.72	22	3.33
100	4.45	102	4.37	101	3.82
205	5.32	202	4.98	205	4.70
305	5.99	225	5.17	305	5.40
402	6.55	354	6.08	400	6.09
503	7.39	430	6.58	500	6.60
		503	7.10	550	6.88
				580	7.22

Table III.   Straight Line Approximation Equations for Thermal Dif-
             fusivity Data, where T is in °C and α is in m^2/s

Unalloyed Pu

monoclinic: α-phase	$\alpha = 1.69 \times 10^{-6} + 3.7 \times 10^{-9}$ T
b.c. monoclinic: β-phase	$\alpha = 2.03 \times 10^{-6} + 6.5 \times 10^{-9}$ T
f.c. orthorhombic: γ-phase	$\alpha = 1.95 \times 10^{-6} + 7.2 \times 10^{-9}$ T
f.c. cubic: δ-phase	$\alpha = 2.75 \times 10^{-6} + 5.0 \times 10^{-9}$ T
b.c. cubic: ε-phase	$\alpha = 3.02 \times 10^{-6} + 4.1 \times 10^{-9}$ T

f.c. cubic δ-phase Pu-Ga alloys

Pu + 2.7 at.% Ga:	$\alpha = 3.80 \times 10^{-6} + 7.1 \times 10^{-9}$ T
Pu + 3.5 at.% Ga:	$\alpha = 3.60 \times 10^{-6} + 7.0 \times 10^{-9}$ T
Pu + 6.6 at.% Ga:	$\alpha = 3.22 \times 10^{-6} + 6.9 \times 10^{-9}$ T

From L-L$_0$ we can calculate an excess thermal conductivity, $\Delta\lambda$,
by means of

$$\Delta\lambda/\lambda = (L-L_0)/L \qquad\qquad (2)$$

Values of $\Delta\lambda$ are shown in Fig. 3. At low temperatures, $\Delta\lambda$ has the
general behavior of a lattice thermal conductivity; it decreases with
increasing temperature and also decreases with increasing solute content. Above 200°C, however, $\Delta\lambda$ increases with temperatures, an unlikely behavior for lattice thermal conductivity. For comparison,
we have included a curve of the lattice thermal conductivity of pure

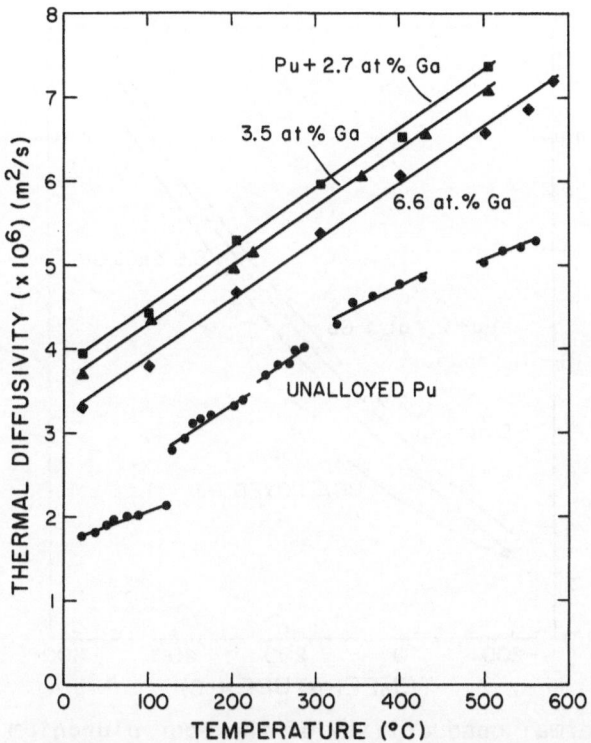

Fig. 1.   Thermal diffusivity of the four plutonium specimens as a function of temperature.

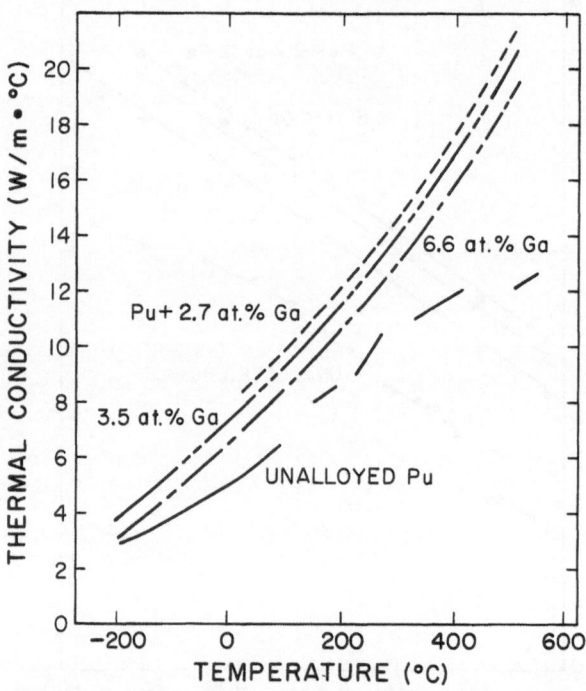

Fig. 2.   Thermal conductivity of the four plutonium specimens in Fig. 1 as a function of temperature.

Table IV. Calculated Values for Thermal Conductivity and Lorenz Numbers for High-Purity Unalloyed Pu

Pu Type $Dx10^{-3}kg/m^3$	T °C	α $x10^6 m^2/s$	Cp J/g°C	λ W/m°C	ρ $x10^8 \Omega m$	L $x10^8 V^2/K^2$
Unalloyed Pu						
α-phase	25	1.78	0.1481	5.20	142	2.48
19.71	100	2.10	0.1598	6.62	140	2.48
β-phase	150	3.10	0.1448	7.87	108	1.99
17.58	200	3.31	0.1490	8.67	108	1.98
γ-phase	225	3.53	0.1494	8.97	107	1.93
17.02	275	3.96	0.1552	10.46	107	2.04
δ-phase	325	4.40	0.1552	10.97	100	1.83
15.81	425	4.85	0.1577	12.10	100	1.73
ε-phase	500	5.04	0.1464	12.10	114	1.78
16.39	550	5.26	0.1464	12.62	114	1.75
Pu-Ga Alloys						
2.7 at.% Ga	25	3.98	0.1364	8.53	109	3.12
15.738	100	4.51	0.1389	9.85	108	2.85
	200	5.22	0.1439	11.81	107	2.67
	300	5.92	0.1540	14.33	107	2.68
	400	6.63	0.1669	17.40	107	2.77
	500	7.33	0.1845	21.26	108	2.97
3.5 at.% Ga	25	3.78	0.1372	8.18	110	3.02
15.713	100	4.30	0.1397	9.42	108	2.73
	200	4.99	0.1448	11.39	108	2.60
	300	5.69	0.1548	13.89	108	2.62
	400	6.38	0.1682	16.92	108	2.72
	500	7.08	0.1845	20.74	108	2.90
6.6 at.% Ga						
15.657	25	3.39	0.1406	7.46	113	2.83
15.610	100	3.91	0.1423	8.68	112	2.61
15.579	200	4.59	0.1481	10.50	110	2.46
15.533	300	5.28	0.1582	12.97	110	2.44
15.487	400	5.97	0.1724	15.94	110	2.61
15.471	500	6.66	0.1900	19.57	110	2.78
15.441	580	7.21	0.2050	22.82	110	2.94

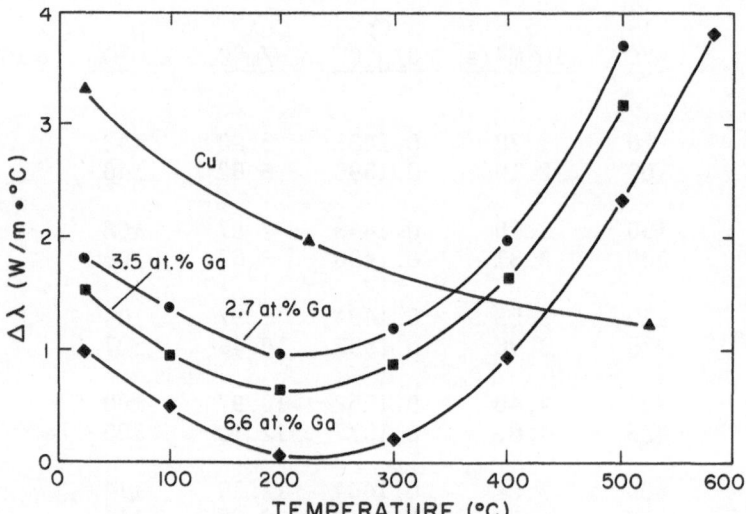

Fig. 3.   The excess thermal conductivity of the three Pu-
          Ga alloys and the lattice conductivity of Cu as
          a function of temperature.

Table V.   Calculated Lattice Conductivities for Pu-Ga
           Alloys from Scaling λ_g with $(cT)^{-\frac{1}{2}}$.   ($\Delta\lambda$ and
           λ_g units are W/m°C)

T, °C	2.7 at.% Ga		3.5 at.% Ga		6.6 at.% Ga	
	$\Delta\lambda$	λ_g	$\Delta\lambda$	λ_g	$\Delta\lambda$	λ_g
25	1.83	1.81	1.54	1.54	1.00	1.11
100	1.39	1.62	0.96	1.38	0.52	0.99
200	0.98	1.44	0.66	1.22	0.06	0.89
300	1.21	1.31	0.89	1.12	0.21	0.80
400	1.99	1.20	1.65	1.03	0.95	0.74
500	3.72	1.12	3.20	0.96	2.35	0.69

Cu,[10] obtained from the lattice thermal conductivities of dilute Cu alloys and extrapolated to higher temperatures as $1/T$. At low temperatures, $\Delta\lambda$ lies below the lattice thermal conductivity, λ_g, of Cu as one would expect for alloys. At high temperatures, however, $\Delta\lambda$ values for the alloys lie well above λ_g of Cu, this supports the belief that the excess thermal conductivity, $\Delta\lambda$, at higher temperatures is not due to lattice waves.

The lattice thermal conductivity, λ_g, should vary inversely as the square root of cT, where c is the solute concentration and T the absolute temperature.[11] At 25°C (298 K), values of $\Delta\lambda$ do indeed vary roughly as $c^{-\frac{1}{2}}$. If we identify $\Delta\lambda$ of the 3.5 at.% alloy at 25°C with λ_g, we can estimate λ_g at other temperatures and other concentrations: these estimates are given in Table V. The remaining discrepancy $\Delta\lambda_e = \Delta\lambda - \lambda_g$ is attributed to an electronic effect. We see that $\Delta\lambda_e$ is negative at 100 and 200°C, though this conclusion is somewhat uncertain in view of the uncertainty of estimating λ_g. However, at high temperatures $\Delta\lambda_e$ becomes clearly positive, and increases with temperature, and it seems to be relatively insensitive to composition at the highest temperatures.

Positive values of $\Delta\lambda_e$ are found in many transition metals; negative values of $\Delta\lambda_e$, or negative deviations of L from L_o at high temperatures seem to be a peculiarity of Pu. Negative deviations can be explained if the product of the density of states and the electron mobility has a maximum at the Fermi energy. In Pu the Fermi energy falls near a minimum between 5f electron subbands.[12] These electrons, however, do not necessarily contribute significantly to the electronic conductivity. The high density of states of the 5f electrons reduces the mobility of the electrons in the other bands (s, p, and d), so that the product of their density of states and mobility has a minimum at the energies where the density of states of the 5f electrons has a maximum, and a maximum near the Fermi surface, where the 5f electron density has a minimum. This could explain the low value of the Lorenz number in pure Pu. In the delta-phase alloys, the Lorenz number exceeds L_o at high temperatures; this is qualitatively similar to the behavior of the Lorenz number of many transition metals and indicates that the 5f electrons act quite differently in the alloys than in pure Pu. A complete description of the electronic transport properties must await a better understanding of the electronic structure.

ACKNOWLEDGMENTS

The authors wish to thank H. D. Lewis and D. R. Harbur for helpful discussions during the course of this investigation.

REFERENCES

1. H. D. Lewis, J. F. Kerrisk, and K. W. R. Johnson in: "Thermal
    Conductivity," Vol. 14, P. G. Klemens and T. K. Chu, eds,
    Plenum Publishing Company, New York (1976), pp. 201-8.

2. O. L. Kruger and J. L. Robbins, Heat Transport Properties of
    a Delta-Stabilized Pu-1 wt% Ga Alloy from Room Temperature to
    500 Degrees C., in: "Plutonium and Other Actinides," H. Blank
    and R. Linder, eds., North Holland Publishing Company,
    Amsterdam (1976), pp 547-556.

3. J. F. Andrew, Thermal Conductivity of Plutonium Metal, J. Phys.
    Chem. Solids 28:577-80 (1967).

4. J. F. Andrew, Thermal Conductivity of Some Delta-Plutonium
    Alloys, J. Nucl. Mat. 30:343-5 (1969).

5. W. J. Parker, K. J. Jenkens, C. P. Butler, and G. L. Abbot,
    Flash Method of Determining Thermal Diffusivity, Heat Capacity
    and Thermal Conductivity, J. Appl. Phys. 32(9):1679-84 (1961).

6. R. D. Cowan, Pulse Method of Measuring Thermal Diffusivity at
    High Temperatures, J. Appl. Phys. 34(4):926-7 (1963).

7. G. D. Cody, B. Abeles, and D. C. Beers, Thermal Diffusivity of
    Armco Iron Trans. Met. Soc. AIME 221:25 (1961).

8. A. E. Kay and R. G. Loasby, The Specific Heat of Plutonium at
    High Temperatures, Philos Mag. 9:43 (1964).

9. R. L. Rose, J. L. Robbins, and T. B. Massalski, Heat Content and
    Heat Capacity of Pu/1 wt% Ga Delta-Stabilized Alloy at Elevated
    Temperatures, J. Nucl. Mater. 36:99-107 (1970); and Enthalpy and
    Specific Heat of a Series of Pu/Ga Alloys at Elevated Tempera-
    tures, Ibid 75:98-104 (1978).

10. G. K. White in "Thermal Conductivity," Vol. I, R. P. Tye, ed.,
    Academic Press, London (1969), p. 104.

11. P. G. Klemens, Thermal Resistance due to Point Defects at High
    Temperature, Phys. Rev. 119:507 (1960).

12. E. A. Kmetko, The Band Structure, Specific Heat, Magnetic
    Susceptibility, and Anomalous Thermal Expansion of δ-Plutonium,
    in: "Plutonium 1965," A. E. Kay and M. B. Waldron, eds.,
    Chapman and Hall Ltd., London (1967), pp. 222-243.

THE PHYSICAL PROPERTIES OF 9 Cr—1 Mo STEEL FROM 300 TO 1000 K

R. K. Williams, R. S. Graves, F. J. Weaver, and
D. L. McElroy

Metals and Ceramics Division
Oak Ridge National Laboratory
Oak Ridge, Tennessee 37830

ABSTRACT

The thermal conductivity and expansion of an improved
9 Cr—1 Mo steel were studied. The results show that the electron
and phonon thermal conductivity components are both quite sensitive
to Si content. The thermal expansion of standard commercial material
is indistinguishable from that of the improved alloy.

INTRODUCTION

The mechanical properties of 9 Cr—1 Mo steel can be significantly improved by heat-treating material that contains small additions of the carbide-forming elements V and Nb.[1] The properties of
the modified alloy not only exceed those of the commercially available material, but also exceed those of type 304 stainless steel up
to about 850 K. The ferritic alloy also has considerably better
resistance to thermal shock and fatigue.[2] These improvements, in
addition to the lowered requirement for the strategic element Cr,
have led to interest in tests to qualify the modified alloy for
various load-bearing applications. As part of this effort, we have
measured the thermal expansion behavior of commercial and modified
9 Cr—1 Mo steel up to 1000 K and have begun a study of the thermal
conductivity, λ. Both steels were studied in the normalized
(1310 K, 1 h; air cool) and tempered (1030 K, 1 h; air cool) condition, and some data on the heat-to-heat variation of the modified
alloy are also presented.

SAMPLE CHARACTERIZATION

Table 1 shows chemical analyses of the eight heats studied. Most of the data are for two heats that were chosen to represent commercial (15965) and modified (30176) material. The melting and fabrication practice has been described elsewhere.[3] To provide baseline data on the α-Fe solid solution, a ternary alloy sample (1271) was also produced by arc-melting high purity Fe, Cr, and Mo stock and cold-swaging the casting to rod form. Extensive investigations[1,3] have shown that normalized material contains martensite, and this product decomposes to a bcc solid solution containing dislocations and carbide precipitates during the tempering heat treatment. Additional aging effects, involving changes in the structure, size, and composition of the carbide phase, have also been observed.[4] X-ray diffraction showed that water-quenched samples of the ternary alloy did not contain martensite.

The chemical analyses shows that the modified alloy contains significant amounts of the carbide-forming elements V and Nb, and the commercial alloy contains more Si. Additions of Si improve the fluidity of the steel and add corrosion resistance, but it lowers the impact resistance of the steel. Silicon also significantly lowers the λ of the alloy.

EXPERIMENTAL DATA

The thermal expansion data are shown in Table 2. The experimental technique and method used to smooth the data have been described elsewhere.[5] The four-probe, dc electrical resistivity, ρ, measurements on eight heats are shown in Table 3 and the λ data are presented in Table 4. Experiments on the high-temperature ρ and λ behavior are now in progress. These experimental techniques have also been described previously,[6] and a small (∿2%) negative correction was applied to the λ data because tests have shown that the results from this comparative apparatus tend to fall above those from absolute λ measurements.[6]

Table 1. Chemical Analyses of Commercial and Modified 9 Cr–1 Mo Steel

Element (wt. %)	Heat No.							
	15965 (Commercial)	30176 (Modified)	1271 (Experimental)	5349 (Modified)	30383 (Modified)	30394 (Modified)	10148 (Modified)	30182 (Modified)
Cr	9.38	8.41	9.37	8.83	8.43	8.35	9.48	8.46
Mo	0.99	0.90	1.00	0.94	1.03	1.03	0.96	0.89
C	0.11	0.075	0.004	0.10	0.088	0.084	0.089	0.087
Si	0.71	0.19	0.01	0.36	0.50	0.45	0.38	0.21
Mn	0.59	0.40	0.02	0.43	0.47	0.47	0.49	0.37
V	0.054	0.204	<0.001	0.208	0.202	0.202	0.210	0.225
Nb	0.010	0.072	0.004	0.058	0.08	0.080	0.061	0.75
Ni	0.07	0.10	<0.01	0.12	0.09	0.09	0.16	0.10
Ti	0.002	0.004	0.004	0.010	0.005	0.005	0.005	0.001
Co	0.081	0.016	0.009	0.023	0.061	0.061	0.025	0.019
Cu	0.03	0.03	<0.01	0.090	0.03	0.03	0.07	0.03
Al	0.001	0.005	<0.001	0.001	0.003	0.025	0.001	0.012
S	0.008	0.004	0.004	0.016	0.006	0.004	0.006	0.004
P	0.017	0.009	0.001	0.010	0.011	0.011	0.017	0.012
N	0.021	0.054	0.001	0.011	0.055	0.054	0.037	0.054
O	0.011	0.005	0.014	—	0.005	0.004	0.009	0.008

Table 2. Thermal Expansion Values for Standard and Modified
9 Cr—1 Mo Steels in the Normalized and Tempered Condition

T (K)	Commercial (15965)		Modified (30176)	
	$\alpha_o^* \times 10^6$	$\alpha_m^\dagger \times 10^6$	$\alpha_o^* \times 10^6$	$\alpha_o^\dagger \times 10^6$
293	10.37	10.37	10.47	10.47
373	11.05	10.72	11.12	10.81
473	11.83	11.12	11.86	11.20
573	12.53	11.50	12.52	11.58
673	13.14	11.85	13.12	11.92
773	13.66	12.18	13.63	12.24
873	14.11	12.47	14.07	12.54
973	14.47	12.74	14.44	12.81

$*\alpha_o$: instantaneous thermal expansion coefficient, K^{-1}.
$\dagger\alpha_m$: mean thermal expansion coefficient, K^{-1}.

Table 3. Electrical Resistivity Values at 300 K^* (10^{-8} Ωm)

Heat Number	Normalized Condition[†]	Normalized and Tempered Condition[b]
15965	61.27	58.05
30176	49.16	44.42[‡]
1271	39.17	38.92
5349	53.23	50.06
30383	57.04	52.46
30394	57.39	51.91
10148	55.90	51.38
30182	49.70	45.02

*Data were corrected to 300 K by using a temperature
 coefficient of 8.28×10^{-10} Ωm/deg.
†See text for descriptions of the heat treatments.
‡Average of measurements on four samples. Range of values
 ±0.5%.

Table 4.   Thermal Conductivity Values for Four Heats
of 9 Cr—1 Mo

Heat Identification	Thermal Conductivity* (W/m·K) at		
	300 K	330 K	360 K
15965	24.43	24.83	25.22
30176	29.49	29.64	29.79
1271	32.09	32.06	32.03
5349	26.87	27.06	27.25

*Measured by a comparative method, corrected by ∿2%
to force agreement with an absolute method and
smoothed using a linear least-squares fit.

DISCUSSION

The electrical resistivity data for seven commercial heats
(Table 3) show a 14% variation about the mean and a consistent drop
of about 7% when normalized material is tempered. This information
is of interest because it indicates the heat-to-heat variation of
the electronic thermal conductivity component. The heat-treatment
sensitivity could be due to either an inherent structural effect
(martensite vs α-Fe) or the fact that carbide precipitation
selectively removes some elements from solution during tempering.
An attempt to resolve this point was not successful because the
ternary alloy did not form martensite.

As shown in Fig. 1, the heat-to-heat variation of ρ is closely
correlated with Si content for both heat treatments. This is not
surprising because Si is known to produce a relatively large effect
on the ρ of electrical steels[7] and there was a relatively large
variation in the concentration of this element. The room-temperature
data also show that the ρ of 9 Cr—1 Mo steel is about 2.3 times more
sensitive to Si content than the ρ of Si containing electrical
steels.[7]

These ρ values, along with an experimental estimate of the
electronic Lorenz function of iron[8] and the assumption that the
Sommerfeld value[9] applies to the impurity contribution to ρ, can

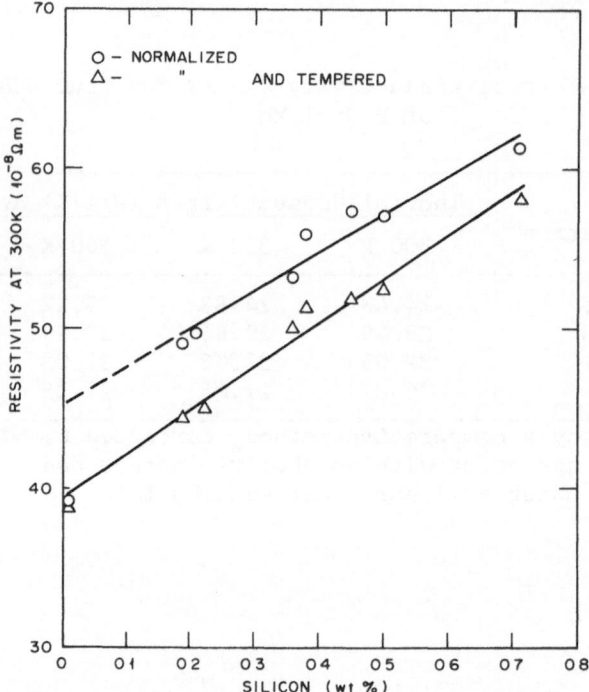

Fig. 1.   Effect of Si content on the room-temperature electrical
          resistivity of 9 Cr—1 Mo steel.

be used to estimate the electronic thermal conductivities, λ_e, of
the seven commercial heats. The values have a range of 26%, and
this variation should decrease at higher temperature. The λ measurements on three commercial heats, shown in Table 4, are considerably larger than the calculated λ_e values and the difference is
attributed to phonon conduction, λ_p. On the basis of this analysis
λ_p contributes 45—50% of the total λ at room temperature and this
relative contribution should continue at higher temperatures because
λ_e would be expected to decrease with increasing temperatures.[8]

It is interesting to try to use the λ_p estimates to identify
terms associated with particular solutes. This is useful because
λ_p is relatively large, and compositional variations can be used to
optimize this property or reach strength–conductivity compromises.
The analysis of data on binary and ternary α–Fe solutions[8] indicates
that

$$\lambda_p^{-1} \ (300 \ \text{K}) = 0.057 + 3.6 \times 10^{-4} \ (\text{at. \% Cr})$$

$$+ \ 1.9 \times 10^{-2} \ (\text{at. \% Mo}) \ (\text{m K/W}) \ . \qquad (1)$$

This shows that the effect of Mo on the λ_p of α-Fe is about 50 times greater than Cr. Presumably this is mostly due to mass difference scattering; the $(\Delta M/\bar{M})^2$ values associated with Cr and Mo have a ratio of 110.

The λ_p^{-1} values for the three commercial heats of 9 Cr-1 Mo steel are larger than the values predicted by Eq. (1), and Fig. 2 shows that the extra phonon thermal resistance is correlated with the Si content of the heats. The room-temperature coefficient for Si is about 50% of the value for Mo, and this also is about what would be expected for mass-difference scattering of phonons.

The thermal expansion coefficients agree within ±2% for standard and modified material in the normalized and tempered condition. This is not a surprising result, since an extensive review[10] of the literature indicates that thermal expansion is a slowly varying function of composition. These data are in good agreement with the results of Conway[11] and the Boiler Code,[12] but 10% below values suggested by Gilchrist et al.[13] The latter values were based on data for several steels and did not rely on any direct measurements on 9 Cr-1 Mo.

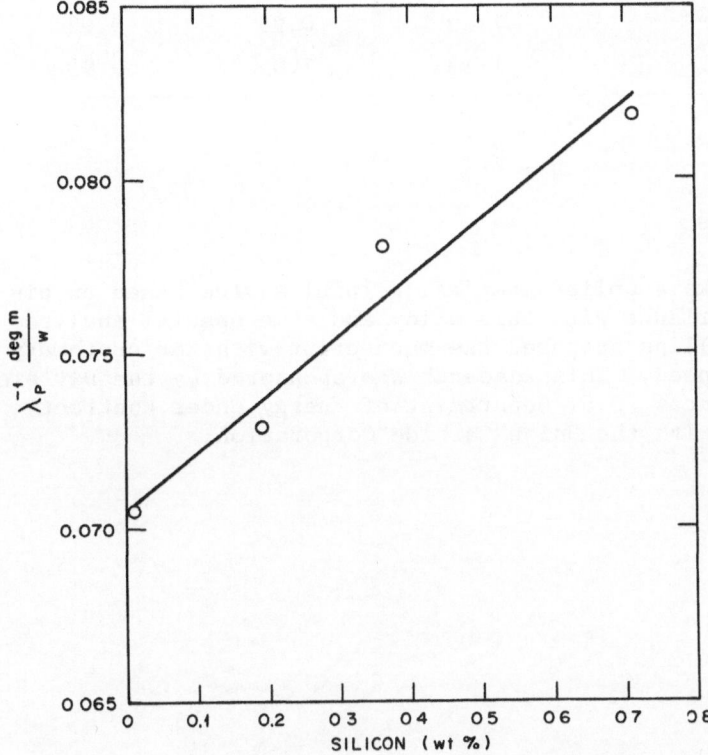

Fig. 2. Effect of Si content on the phonon thermal conductivity of 9 Cr-1 Mo steel at room temperature. All samples were in the normalized and tempered condition.

Knowledge of the ratio α_o/λ and high-temperature mechanical property data can indicate thermal stress resistance. Table 5 shows the α_o/λ value for modified 9 Cr—1 Mo steel is two to three times lower than that of Inconel Alloy 718 or type 316 stainless steel. The values for 9 Cr—1 Mo were obtained by assuming that λ is independent of temperature.

Table 5. Values of the α_o/λ Ratio for Several Alloys as a Function of Temperature

	α_o/λ, μm/W		
	300 K	600 K	900 K
9 Cr—1 Mo Steel			
Commercial (15965)	0.43	0.51	0.58
Modified (30176)	0.36	0.42	0.48
Alloy 718[14]	1.20	0.95	0.92
SS 316[15]	1.13	1.05	0.93

ACKNOWLEDGMENTS

V. K. Sikka supplied samples, helpful advice based on his extensive experience with this alloy and some special analyses for Si. S. J. Phillips prepared the manuscript with her customary accuracy and speed. This research was sponsored by the Division of Materials Sciences, U.S. Department of Energy under contract W—7405—eng—26 with the Union Carbide Corporation.

REFERENCES

1. G. C. Bodine, Jr., B. Chakravarti, C. M. Owens, B. W. Roberts, D. M. Vandergriff, and C. T. Ward, "A Program for the Development of Advanced Ferritic Alloys for LMFBR Structural Application," TR-MCD-015 (September 1977).
2. J. Y. Chang and W. E. Ray, Westinghouse Advanced Reactors Division, Waltz Mill Site, Pennsylvania, Personal Communication (June 22, 1981).
3. G. C. Bodine, Jr., B. Chakravarti, S. D. Harkness, C. M. Owens, B. Roberts, D. Vandergriff, and C. M. Owens, "The development of a 9 Cr steel with improved strength and toughness," in: "Ferritic Steels for Fast Reactor Steam Generators," Vol. 1, S. F. Pugh and E. A. Little, eds., British Nuclear Society, London (1978), pp. 160–164.
4. V. K. Sikka, "Substitution of Modified 9 Cr-1 Mo Steel for Austenitic Stainless Steel," Workshop on Conservation and Substitution Technology for Critical Materials, Nashville, Tennessee, June 1981, to be published by American Society for Metals, Metals Park, Ohio.
5. T. G. Kollie, D. L. McElroy, J. T. Hutton, and W. M. Ewing, AIP Conf. Proc. 17:129 (1974).
6. R. K. Williams, R. S. Graves, and J. P. Moore, "A Study of the Effects of Several Variables on the Thermal Conductivity of 2 1/4 Cr–1 Mo Steel," ORNL 5313, Oak Ridge National Laboratory, Oak Ridge, Tennessee (April 1978).
7. "Metals Handbook," 8th ed., Vol 1, American Society for Metals, Metals Park, Ohio (1960), p. 788.
8. R. K. Williams, D. W. Yarbrough, J. W. Masey, T. K. Holder, and R. S. Graves, J. Appl. Phys. 52:5167 (1981).
9. J. M. Ziman, "Electrons and Phonons," p. 385, Oxford at the Clarendon Press, London (1960).
10. D. L. McElroy, "Estimated Physical Properties of 2 1/4 Cr–1 Mo Steel," ORNL report in preparation, Oak Ridge National Laboratory, Oak Ridge, Tennessee.
11. J. B. Conway, "Revised Expansion Coefficients for 9 Cr–1 Mo Alloy Steel, Metcut Research Associates Inc.," private communication to P. L. Rittenhouse, Oak Ridge National Laboratory (1975).
12. "Rules for Construction of Nuclear Power Plant Components," ASME Boiler and Pressure Vessel Code, Section III, Table I-5.0, ASME, New York (1971), p. 413.
13. K. E. Gilchrist, S. D. Preston, and J. E. Brocklehurst, "Physical Property Data for 2 1/4 Cr/1 Mo and 9 Cr/1 Mo Steels," ND-M-346(s), Springfield Nuclear Power Development Laboratories (May 1978). Not for publication (commercial).

14. D. L. McElroy, R. K. Williams, J. P. Moore, R. S. Graves, and
     F. J. Weaver, The physical properties of Inconel alloy 718
     from 300 to 1000 K, in: "Thermal Conductivity," Vol. 15,
     V. V. Mirkovich, ed., Plenum Press, New York (1978),
     pp. 149–151.
15. R. A. Moen and E. H. Novendstern, Proposed revisions to physi-
     cal properties for austenitic stainless steels, Letter 25,
     4-11-77, in: Nuclear Systems Materials Handbook," (1977)
     Committee Correspondence.

THERMAL DIFFUSIVITY OF NITRIDED STEEL

Hung Joo Lee, Sang Kil Lee, and Chung Oh Lee*

Dept. of Ordnance Engr.   *Dept. of Mech. Science

Korea Military Academy   Korea Advanced Institute
Seoul, Korea   of Science and Technology
                                  Seoul, Korea

The thermal diffusivity of a diffusion layer of ion-nitrided steel was experimentally determined by the flash method. In the experiment, a high intensity heat pulse was impinged on the front surface of a nitrided, thin, disk-shaped metal and the transient temperature rise of the opposite surface was recorded. The thermal diffusivity of the sample material was then determined from the recorded temperature transient versus time curve by the computer data reduction with mathematical analysis.

From this study, it was found that the thermal diffusivity of a diffusion layer of nitrided steel was up to 10 % greater than that of nonnitrided steel.

INTRODUCTION

To satisfy the real needs of industry using various steels, an ion-nitriding[1-3] surface treatment on steel is most important because of its enhancing material properties in wear, erosion and corrosion resistance protection. It is, then, an important and interesting problem to predict and measure the thermophysical properties, including thermal conductivity and thermal diffusivity. In this work, the thermal diffusivity of ion-nitrided steel is measured by the flash method.

In the flash method for measuring the thermal diffusivity,[4-9] a high-intensity, short-duration heat pulse of radiant energy from an optical flash lamp or laser is used to irradiate one of the two flat, parallel surfaces of a homogeneous sample and the subsequent

229

effect at the opposite surface is monitored with a thermocouple or a photomultiplier tube. The pulse raises the average temperature of the sample only a few degrees above its initial value. The thermal diffusivity of the sample material may then be deduced from the shape of the resulting temperature transient if the heat diffusion equation has been solved for the particular boundary conditions of the experiment. It is assumed that heat flow is one dimensional, that material properties are temperature independent, and that the heat pulse is uniform over the sample surface.[4-9]

The effect of a noninstantaneous heat pulse on the shape of the rear-face transient is important. Cape and Lehman[7] have shown that this transient is retarded with pulses of finite duration, referring to the retardation as the "finite pulse time effect."[10-13] Larson and Koyama[10] have shown that one can completely eliminate those errors due to the finite pulse time effect, even for samples as thin as $1 = (\alpha\tau)^{\frac{1}{2}}$, by applying an empirical function that more closely describes the pulse shape. Here 1 is the thickness of the sample, α is the thermal diffusivity, and τ is the time in seconds representing the duration of the heat pulse. The thickness of the disk-shaped samples used in the experiment is less than $4(\alpha\tau)^{\frac{1}{2}}$. The samples are made of a homogeneous diffusion layer of an ion-nitrided steel.

MATHEMATICAL ANALYSIS

In the usual flash technique, the front surface of the homogeneous sample is subjected to a heat pulse, and the resulting temperature rise of the rear surface is recorded. From this temperature rise, the thermal diffusivity of the material is obtained by computer data reduction with mathematical analysis.[10] The normalized amplitude of heat pulse fired from the xenon flash tube employed in the experiments can be represented as

$$q(t) = (q_o/t_p^2)t \exp (-t/t_p), \quad t > 0, \quad (1)$$

where q_o is the total energy absorbed by the unit area of the front surface, t_p is the pulse peak time, and t is the time in seconds.

To solve the heat diffusion equation with the appropriate boundary conditions, the following assumptions are made:
(1) Heat flow is one dimensional.
(2) There is no heat loss from the sample surfaces.
(3) Heat pulse is uniformly absorbed on the front surface.
An appropriate model corresponding to these assumptions takes the form of a slab of infinite extent in radial direction, with the parallel and planar front and rear surface at z=0, and z=1, respectively.
At time t=0 when the slab is at temperature T_o, the front surface is uniformly irradiated by a flash tube, which results in energy absorption of the sample with temporal behavior described by the

equation (1). The subsequent temperature history is recorded at the rear surface. With these assumptions heat flow in the sample is described as follows:

$$-kU_{zz}(z,t) + RCU_t(z,t) = 0, \quad 0<z<1 \ , \ t>0 \tag{2}$$

$$U(0 <z< 1, \quad t = 0+) = 0 \tag{3}$$

$$-kU_z(z=0, \ t>0) = (q_0/t_p^2) \ t \ \exp(-t/t_p) \tag{4}$$

and

$$U_z(z = 1, t > 0) = 0 \tag{5}$$

Here R is density, C is heat capacity, and k is the thermal conductivity. In equations (2) through (5), the temperature excursion is denoted by $U(z,t) = T(z,t) - T_0$ where $T(z,t)$ is the temperature in the slab at location z and time t. The subscripts of U indicate the differentiation by the parameters.

By using Laplace and inverse Laplace transforms, the dimensionless temperature rise of the rear face can be obtained as follows:

$$V(t:y,t) = 1+2r^{\frac{1}{2}} \sum_{n=1}^{\infty}(-1)^n \frac{\exp(-n^2\pi^2 t/y)}{(r-n^2\pi^2)^2} - \frac{r^{\frac{1}{2}}}{2 \sin r^{\frac{1}{2}}} \exp(-t/t_p)$$

$$(1+2 \ \frac{t}{t_p} + r^{\frac{1}{2}} \cot r^{\frac{1}{2}}) \tag{6}$$

where y is the heat diffusion time and r is the pulse shape parameter. These are defined as follows:

$$y = 1^2/A \tag{7}$$

$$r = y/t_p \tag{8}$$

where
    1 is the thickness,
    A is the thermal diffusivity.

Equation (6) was used to compute the thermal diffusivity.

EXPERIMENTAL PROCEDURE

The flash method is shown schematically in Figure 1 using a flash lamp as an energy source and an oscilloscope trace as a recording device.

The success of the flash technique depends upon adequately meeting the boundary conditions of theory. The front surface of the sample must be uniformly irradiated. The thermocouple must measure the
actual rear surface temperature, and heat losses must be minimized.
Furthermore, the signal must be large enough to be well above the
noise level in the recording system, and the bandwidth of the amplifier and recorder must be wide enough to pass the signal without disturbances.

The xenon flash tube used for the energy source in the experiment is a commercially available unit (IL LI IND. CO. SS-1200) consisting of a ring-type quartz tube dissipating 400 J of energy in
each flash.

In the present experiment, the reflecting mirror, which is made
of aluminum foil, was attached to improve the irradiation effect. The
front faces of the sample were uniformly blackened with camphor black
to increase the amount of energy absorbed. These samples were mounted in a sample holder at a short distance (less than 5 cm) from the
flash tube. The front surfaces of the samples were parallel to the
axis of the xenon flash quartz tube and the back surfaces were welded
with a Chromel-Alumel thermocouple wire with a 0.1 mm diameter. The
thermocouple wires were junctioned at the center of the rear surface
of the samples to reduce the disturbing effects caused by heat loss
at the edge of the samples.

The output voltage from the thermocouple was shown on a storage
oscilloscope (Tektronix 7603) and photographed with a Polaroid Land
camera. The maximum voltage was usually 200 to 320 μV, indicating
a rear surface temperature rise of 5° to 8°C.

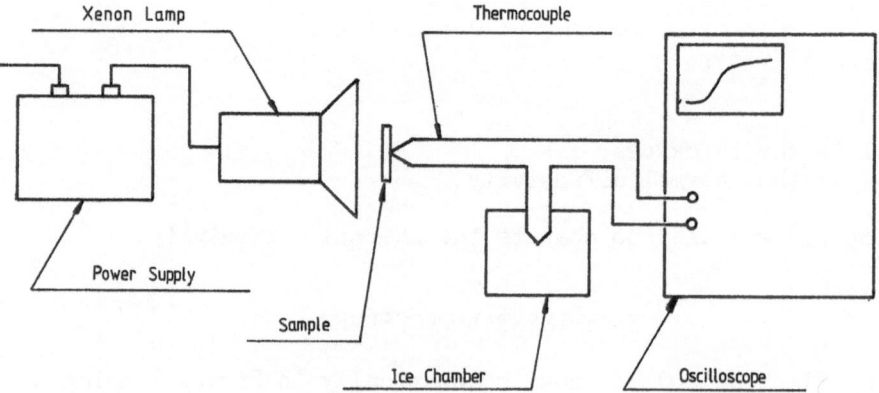

Figure 1. Outline of the Flash Diffusivity Apparatus

The sample holders must be opaque to prevent irradiation of the back surface of the sample. To reduce the heat transfer from the sample to the sample holder by conduction, the contact area is minimized. The sample thickness should also be minimized to prevent convective heat transfer.

If the final temperature indicated by the thermocouple is reached very slowly after an initial fast rise, it may indicate non-uniform heating on the front surface. If the signal gradually increases to the maximum temperature from the beginning to the end, it also represents the heat loss from the sample surfaces.

It is necessary to find the peak time (time to reach the maximum intensity) of the heat pulse to obtain better experimental results. The empirical function for the heat pulse can be represented as follows:

$$Q(t) = (t/t_p) \exp (1 - t/t_p) : t > 0, \qquad (9)$$

where $Q(t)$ is an empirical intensity function of the heat pulse, t is a time in seconds, and t_p is the time to reach the maximum intensity.

In the present experiment, a Korean Photoelectronics SP-1KL planar-type silicon photodiode was used. The photodiode detector has a spectral response at $0.8 \mu m$ wavelength and has substantial sensitivity in the infrared region.

The oscilloscope trace of the photodiode output for the xenon flash tube is shown in Figure 2.

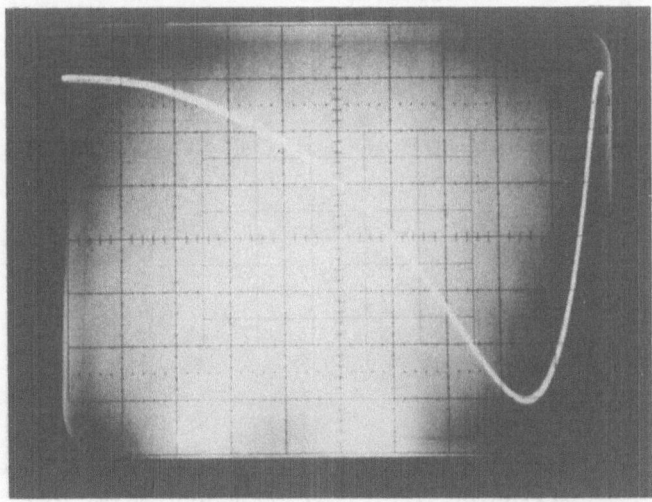

Figure 2. Photodiode Output of the Xenon Flash Tube

From Figure 2, the peak time t_p can be easily determined as 1.4 ms.

DISCUSSIONS AND CONCLUSIONS

It has been known that the nitrided material would have the various compound and diffusion layers according to the nitriding conditions. The specimens were obtained from homogeneous regions of the nitrided layer. The inhomogeneous regions of these layers were ground off. The metallographic structures of the nonnitrided sample are shown in Figure 3.

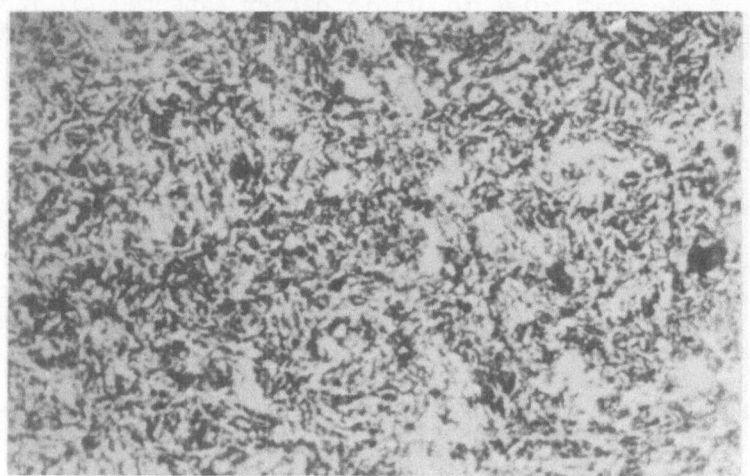

Figure 3. Metallographic View of SCM 3

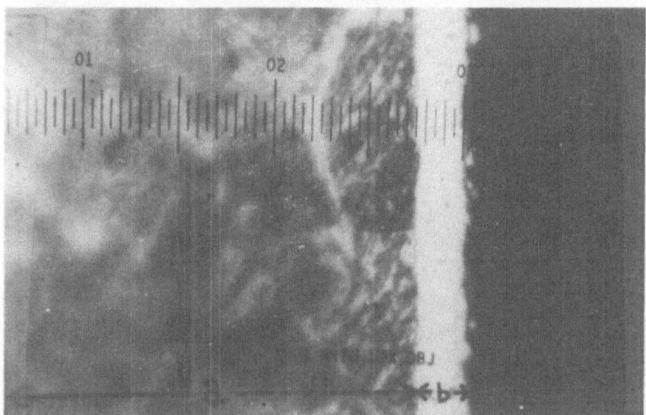

Figure 4. Metallographic View of Ion-Nitrided Steel
Diffusion Layer (a)+Compound Layer (b)
SCM 3: 667 N/m^2, 550°C, 6 hrs(400x)

The microscopic structures of a sample made of SCM 3 steel can be compared with the compound and diffusion layers of an ion-nitrided SCM 3 steel in Figure 4 and 5.

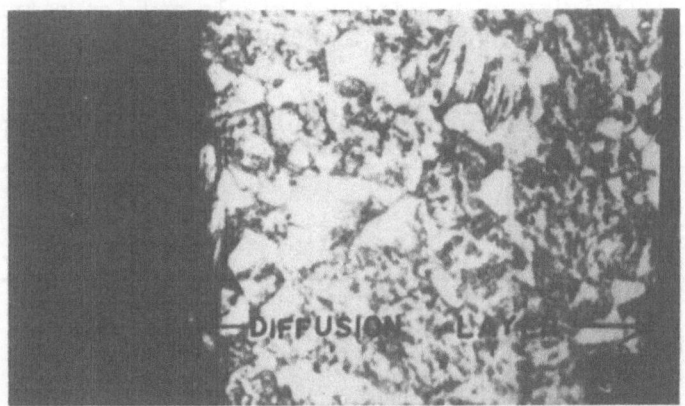

Figure 5.   Metallographic View of Ion-Nitrided Steel
            (Diffusion Layer) SCM 3: 667 N/m^2, 550°C,
            12 hrs (400x)

Theoretical temperature rise with a constant value of α fits the experimental temperature rise very well, as shown in Figure 6.

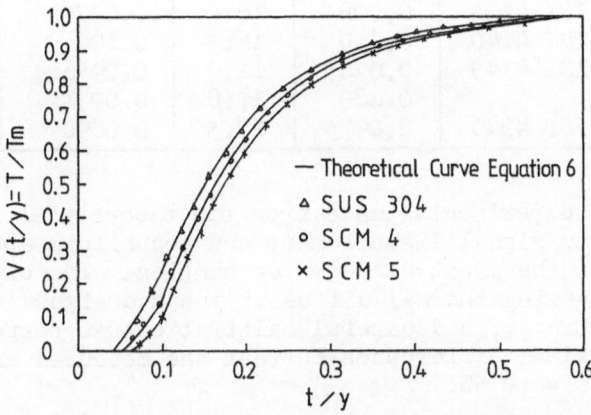

Figure 6.   Theoretical Rear Face Transients and Experimental
            Oscilloscope Trace Points for Various Samples

Therefore, it may be concluded that the influences tending to distort the shape of the experimental temporal transient away from the theoretical shape is almost negligible. The experimental results of the thermal diffusivity of nonnitrided steels and ion-nitrided steels at 22°C are listed in Table I and Table II, respectively.

Table I. Thermal Diffusivity of Steels Before Ion-Nitriding

Sample		L (cm)	$t_{\frac{1}{2}}$ (ms)	+at 22°C (cm^2/s)	(other sources) (cm^2/s)
JIS No.	AISI No.				
SCM 2	AISI 4130	0.0875	14.0	0.096	
SCM 3	AISI 4135	0.0980	17.0	0.0947	
SCM 4	AISI 4140	0.0990	18.5	0.0873	
SCM 5	AISI 4145	0.0795	13.1	0.0864	
SCM 22		0.085	13.8	0.0860	
SNCM 8	AISI 4340	0.0807	13.8	0.0873	0.091*
SUS 304	AISI 304	0.0950	38.0	0.0356	0.035^x

+ Obtained from Eq. (6).
* From reference 16.
x Taken from reference 17.

Table II. Thermal Diffusivity of Ion-Nitrided Steels

Sample		L (cm)	$t_{\frac{1}{2}}$ (ms)	at 22°C (cm^2/s)
JIS No.	AISI No.			
SCM 2	AISI 4130	0.0955	15.6	0.0100
SCM 3	AISI 4135	0.0985	16.8	0.0970
SCM 4	AISI 4140	0.099	16.5	0.100
SCM 5	AISI 4145	0.0785	11.9	0.0956
SCM 22		0.080	17.8	0.0931
SNCM 8	AISI 4340	0.0975	17.5	0.0890

Deviation of experimental data from the theoretical shape may come from noise-signal disturbances and nonuniform absorption of a heat pulse at the front surfaces of samples. The experimental apparatus for measuring thermal diffusivity was designed to minimize all of the disturbances, and careful calibration was performed. Therefore, the maximum differences between the measured and reported diffusivity values were 5%.

From this study, we conclude that the thermal diffusivity of the ion-nitrided surface region of the steel may be as much as 10% greater than that of the steel.

REFERENCES

1. Metals Handbook, 8th Edition, Vol. 2, American Society for Metals, Metals Park, Ohio, p. 149 (1964).
2. H. Bennek and O. Rudiger, Arch. Eisenhuettenw., 18 (10), (1944).
3. C.T. Jones, S.W. Martin, D.J. Struges, and M. Hudis, Ion Nitriding, Met. Prog. p. 72 (1972).
4. W.J. Parker, et al., Flash Method of Determining Thermal Diffusivity, Heat Capacity, and Thermal Conductivity, J. Appl. Phys. 32: 1679 (1961).
5. R.L. Rudkin, R.J. Jenkins, and W.J. Parker, Thermal Diffusivity Measurements on Metals at High Temperatures, Rev. Sci. Instrum. 33: 21 (1963).
6. R.D. Cowan, Pulse Method of Measuring Thermal Diffusivity, J. Appl. Phys. 34: 926 (1963).
7. J.A. Cape and G.W. Lehman, Temperature and Finite Pulse Time Effects in the Flash Method for Measuring Thermal Diffusivity, J. Appl. Phys. 34: 1909 (1963).
8. H.W. Deem and W.D. Wood, Flash Thermal Diffusivity Measurement Using a Laser, Rev. Sci. Instrum. 33: 1107 (1962).
9. R.J. Jenkins and W.J. Parker, A Flash Method for Determining Thermal Diffusivity Over a Wide Temperature Range, U.S. Naval and Radiological Defense Laboratory, WADD Technical Report pp. 61-95, (1961).
10. K.B. Larson and Karl Koyama, Correction for Finite Pulse Time Effects in Very Thin Samples Using the Flash Method of Measuring Thermal Diffusivity, J. Appl. Phys. 32 (9): pp. 1679-84 (1960).
11. R.C. Heckman, Finite Pulse Time and Heat Loss in Pulse Thermal Diffusivity Measurements, J. Appl. Phys. 44 (4): pp. 1415-1460 (1973).
12. R.E. Taylor and J.A. Cape, Finite Pulse Time Effects in the Flash Diffusivity Technique, Appl. Phys. Lett. 5: pp. 212-13 (1964).
13. D.A. Watt, Theory of Thermal Diffusivity by Pulse Technique, Br. J. Appl. Phys. 17: pp. 231-40 (1966).
14. H.S. Carslaw and J.C. Jaeger, Conduction of Heat in Solids, Oxford Univ. Press. 2nd Edition, p. 302 (1959).
15. F.B. Hildebrand, Advanced Calculus for Engineers, Chap. 10, Prentice-Hall Inc., New York, p. 524 (1949).
16. C.P. Buttler and E.C.Y. Inn in Thermodynamic and Transport Properties of Gases, Liquids and Solids, American Society of Mechanical Engineers, New York, (1959).
17. R.J. Jenkins and Westover. R.W., The Thermal Diffusivity of Stainless Steel Over the Temperature Range 120°C to 1000°C, UNSRDL TR-484, (AD 249578) pp. 1-13 (1960).

SESSION TC-7

Liquids and Gases - I

CHAIRMAN

S. C. Saxena
University of Illinois
Chicago, Illinois

THERMAL CONDUCTIVITY OF LIQUIDS

M.P. Saksena and Prabhuram

Department of Physics
University of Rajasthan
Jaipur, India

ABSTRACT

    An analysis of the radial distribution function of liquids has indicated that the liquid state consists of clusters of distorted microcrystallites. From this picture, the partition function of a liquid has been developed, and by minimizing the Helmholtz free energy of the system, an expression for the equilibrium number of clusters at a given temperature and volume has been obtained. This model was then used to investigate the problem of heat conduction in a liquid. By assuming a predominant role of structure scattering for the phonon a relation for the thermal conductivity of a liquid was obtained. Calculations of thermal conductivity of argon and nitrogen for different temperatures and pressures revealed the success of the theory developed.

    The theory was further extended to liquid metals. The total thermal conductivity of a liquid metal was assumed to be the sum of cluster and electronic thermal conductivities. The calculated values of thermal conductivity of metallic liquid mercury were found to be in excellent agreement with the experimental data over a wide range of temperature.

INTRODUCTION

During recent years several models[1-7] for the structure of the liquid state have been suggested and used to
predict the macroscopic properties of liquids. Of all
these models, the significant structure theory given by
Eyring et al.[2] has been most successful in explaining the
thermodynamic properties of liquids. According to this
theory, the liquids are treated as a mixture of crystals
and gases. This theory contains several ad hoc assumptions, and the assumption of the simultaneous existence
of two phases, crystalline and gaseous, appears untenable.
Recently Bagchi[8-9] through the analysis of radial distribution functions of liquids has suggested that the liquid
state consists of clusters of distorted microcrystallites.
Similar models have also been used by several other workers.[10-14] Since the liquid exhibits isotropic properties, these small mosaic blocks of distorted crystallites
must be randomly distributed. Further, as the temperature rises the size of the clusters decreases, and due
to the availability of larger space, the clusters as a
whole gain translational energy. Thus, in effect, this
model is equivalent to the two phase model proposed by
Eyring et al. However, it should be noted that in this
model an independent gaseous state does not exist along
with the solid state, but the gaseous behaviour is exhibited by the crystallite itself and it increases with
rise in temperature. We have attempted here to develop
a theory for the thermal conductivity of metallic and nonmetallic liquids.

DEVELOPMENT OF THEORY

According to Bagchi[8-9] a liquid consists of clusters of distorted microcrystallites randomly oriented
and distributed in the volume of the liquid. Inside the
clusters the molecules behave like solids, while the
cluster as a whole may have gas-like random translatory
motion. Thus, the partition function of the liquid can
be written as

$$Z_L = (Z_{cl}^s)^n (Z_{cl}^{tr})^n /n!$$

$$= [\exp(-\beta E_s n'/N) (T/\theta)^{3n'}]^n$$

$$\times [(2\pi mn'kT/h^2)^{3/2} (V-V_s/n)]^n/n!$$

where $\beta = 1/(kT)$, k is Boltzmann constant and T is temperature. The average number of molecules in a cluster is n' and n is number of clusters in a liquid of volume V at temperature T. These are related to each other by the condition that number of molecules, N, in a liquid is constant, i.e., nn' = N. The partition function for the translational motion of the cluster is represented by Z_{cl}^{tr}, and Z_{cl}^s represents the partition function of the cluster for solid like motion of the molecules. The Einstein temperature and sublimation energy of the solid are Θ and E_s, respectively. Thus, mn' is the average mass of a cluster, and $(V-V_s/n)$ is volume available to the cluster for translatory motion. Due to the large size of the clusters and the hinderance caused by other clusters, the movement of a cluster is limited to the free space available around it. In other words, the surrounding clusters create a cage around the given cluster, which keeps it confined in a given region[1].

By using the Sterling formula for n!, the equation (2) reduces to

$$Z_L = \left\{ \exp(-\beta E_s n'/N) \ (T/\Theta)^{3n'} \ (\frac{2\pi \ mn'kT}{h^2})^{3'/2} \right._n$$
$$\left. x \frac{(V-V_s)e}{n^2} \right\}$$

$$(3)$$

The number of clusters, n, at a given temperature and volume is now calculated from the equation (3) using the variational method. The Helmholtz free energy of the system is given by

$$(4)$$

$$A = -kT \ln Z_L$$

Minimizing A for the equilibrium, the average number of clusters at a temperature T and volume V is given by[15]

$$n = (\frac{2\pi \ mNk}{h^2})^{3/7} \ T^{3/7} \ (V-V_s)^{2/7} \ e^{-5/7} \qquad (5)$$

By considering the scattering of phonons owing to the mosaic blocks of microcrystallites, the mean free path can be related to the wave vector, $\vec{K}$, as[16]

$$l(\vec{K}) = Ma \ (1/aK)^2 \qquad (6)$$

where a is the lattice constant and M is a constant dependent on the size of the cluster and is inversely proportional to the equilibrium number of clusters at a given temperature and volume. Then one can write

$$M = \frac{A'}{T^{3/7} (V-V_s)^{2/7}} \tag{7}$$

where A' is a constant.

Further, the phonon exchange is limited by vacant sites in between microcrystallites. The energy exchange as well as the number of interactions is proportional to the concentration of the solid like molecules present in a given volume. Thus, the cluster contributions to the thermal conductivity of liquid are given by

$$K_{cl} = \frac{1}{3} x^2 \int_0^\infty C(\vec{K}) \; v \; \underline{1} \; (\vec{K}) \quad d\vec{K} \tag{8}$$

where $x = (V_s/V)$. V_s is the molar volume in the solid phase, $C(\vec{K})$ is the heat capacity of a mode having a wave vector $\vec{K}$, v is the velocity associated with this mode, and $1(\vec{K})$ is the mean free path. Using the values of $C(\vec{K})$ and $1(K)$

$$K_{cl} = \frac{1}{3} x^2 \int_0^\infty \frac{M}{a} \frac{2h^2 v^3 K^2}{(2\pi)^4 kT^2} \frac{e^{hvK/2\pi kT}}{(e^{hvK/2\pi kT}-1)^2} dK \tag{9}$$

$$= B \frac{T^{4/7}}{(V-V_s)^{2/7}} x^2 \tag{10}$$

where B is a constant.

The free electron contribution to the thermal conductivity of a metallic liquid is given by[17-19]:

$$K_{el} = \pi^2/3 \; (k/q)^2 \; \sigma T \tag{11}$$

where q represents charge on an electron, and σ is electrical conductivity. The total thermal conductivity of a metallic liquid is, thus, given by

$$K_L = K_{cl} + K_{el}$$

$$= B \frac{T^{4/7}}{(V-V_s)^{2/7}} \quad x^2 + \frac{\pi^2}{3} \left(\frac{k}{q}\right)^2 \sigma T \tag{12}$$

CALCULATION OF K_L AND DISCUSSION OF RESULTS

Three representative liquids, argon, nitrogen, and mercury, were chosen for the calculation of thermal conductivity. The temperature range considered was from the normal melting point to the critical temperature for these liquids. The pressure effects on the liquid thermal conductivity were also investigated for argon.

The calculation of thermal conductivity using relation (12) involves the evaluation of constant B for each liquid. The constant B is temperature and pressure independent, and this fact has been clearly established by the calculated results. Theoretical relations do not exist for the calculation of such a constant, even for crystalline solids. This, therefore, makes it essential to obtain the value of B empirically through the knowledge of an experimental value of K_L at a given temperature. The values of B so determined are; argon: B = 0.47 x 10^{-4}; nitrogen: B = 0.64 x 10^{-4}; and mercury: 0.69 x 10^{-4}. It may be noted that the significant structure theory[20-21] involves two such constants that have to be determined empirically.

By using the above values of B for the three liquids considered, the liquid thermal conductivity values obtained through relation (12) are shown in Figures 1, 2, and 3 as a function of temperature and pressure. In these calculations the same value of B is used for the different temperatures and pressures considered.

Figure 1 shows the comparison of theoretical and experimental thermal conductivity values for liquid argon. In this figure the solid line represents the theoretically calculated vlaues. The experimental data[20] at the pressure of 48 atm is depicted by open circles and the data at 72 atm by open squares. The comparison of theoretical and experimental thermal conductivity values of liquid nitrogen is shown in Figure 2. The solid line represents the theoretically

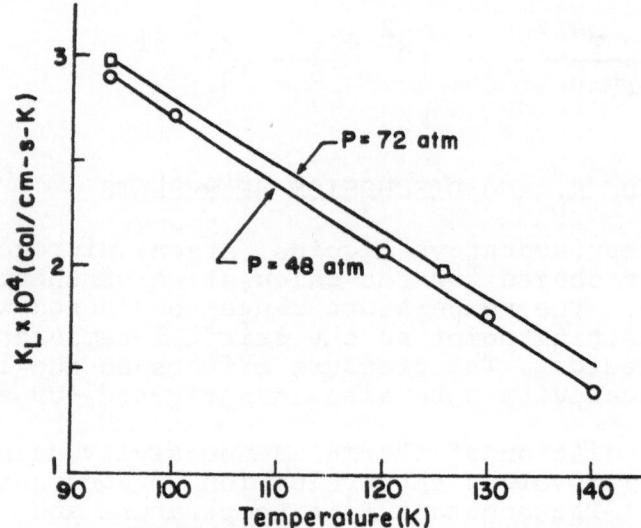

Fig. 1   Thermal conductivity of liquid argon

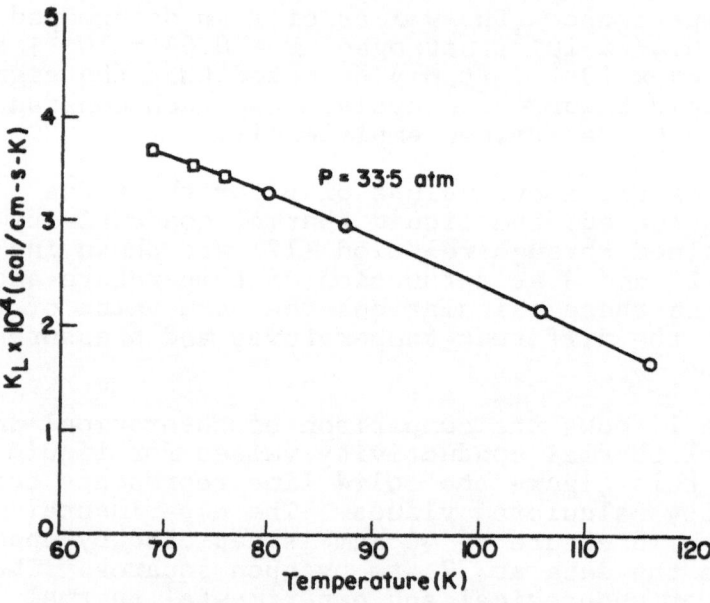

Fig. 2   Thermal conductivity of liquid nitrogen

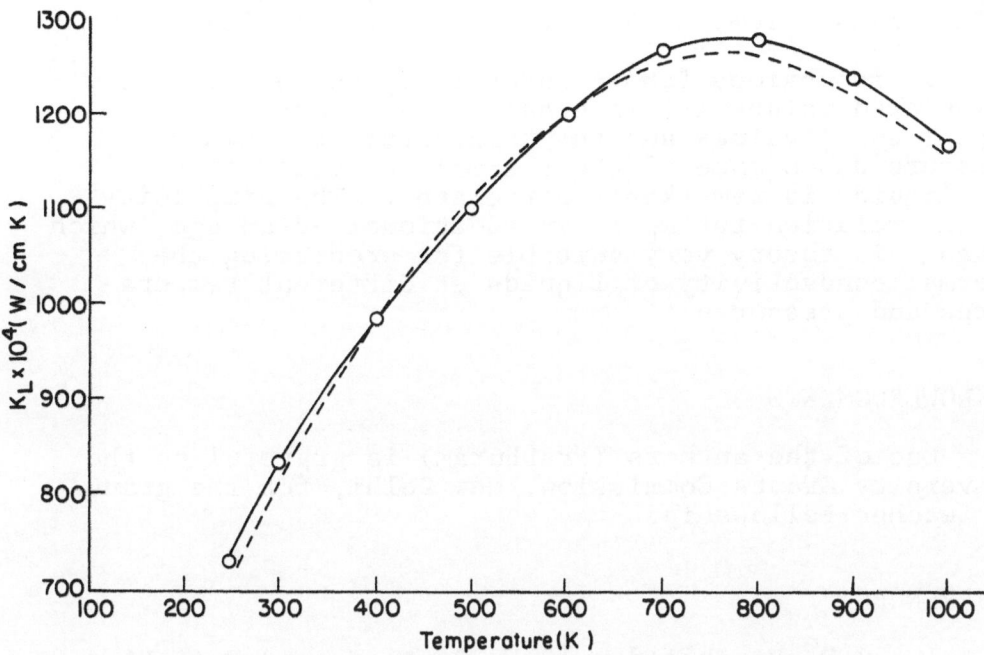

Fig. 3. Thermal conductivity of liquid mercury

calculated values. The experimental data taken from
Lin et al.[20] is shown by open circles and the data from
Powers et al.[21] by open squares. The temperature
dependence of the thermal conductivity of liquid mercury is depicted in Figure 3, where the solid curve
represents the calculated values using the relation (12)
and dashed curve represents the calculated values of
Groose[22-23]. The experimental data[24], taken from the
TPRC series volume 1, are shown by open circles.

It is obvious from Figures 1, 2, and 3 that the
calculated values are in excellent agreement with the
experimental values and the temperature as well as
pressure dependence of the thermal conductivity of all
the liquids is remarkably correlated. The simplicity
of the relation for K_L is an additional advantage, which
makes this theory very suitable for predicting the
thermal conductivity of liquids at different temperatures and pressures.

ACKNOWLEDGMENTS

One of the authors (Prabhuram) is grateful to the
University Grants Commission, New Delhi, for the grant
of Teacher Fellowship.

REFERENCES

1.   J.O. Hirshfelder, C.F. Curtiss, and R.B. Bird,
     "Molecular Theory of Liquids and Gases,"
     John Wiley & Sons, Inc., New York (1954).
2.   H. Eyring, T. Ree, and N. Hirai, Proc. Nat.
     Acad. Sci. (US), 44:683 (1958).
3.   H. Eyring and M.S. John, "Significant Liquid
     Structures," John Wiley & Sons, Inc.,
     New York (1969).
4.   S.C. Misra, Ind. J. Pure Appl. Phys., 7:296
     (1969).
5.   S.C. Misra, Ind. J. Pure Appl. Phys., 8:448
     (1970).
6.   "Physical Chemistry - An Advanced Treatise,
     Volume VIII A, Liquid State," D. Henderson,
     ed., Academic Press, New York (1971).
7.   A. Munster, "Statistical Thermodynamics,"
     Volume II, Academic Press, New York (1974),
     p. 321.
8.   S.N. Bagchi, Adv. Phys., 19:119 (1970).

9.   S.N. Bagchi, Acta. Crystallogr., A28:560 (1972).
10. J.J. Burton, J. Chem. Phys., 52:345 (1970).
11. J.J. Burton, Surf. Sci., 26:1 (1971).
12. J. McGinty, J. Chem. Phys., 55:580 (1971).
13. J. McGinty, J. Chem. Phys., 58:4733 (1973).
14. J.J. Gilman, Philos. Mag., B37:577 (1978).
15. J.J. Kozak and S.A. Rice, J. Chem. Phys.,
     48:1226 (1968).
16. I. Pomeranchuk, J. Phys. (U.S.S.R.), 6:237
     (1942).
17. N.E. Cusack, Rep. Prog. Phys., 26:361 (1963).
18. C. Kittel, "Introduction to Solid State
     Physics," 4th Edition, John Wiley & Sons,
     Inc., New York (1974), P. 262.
19. F. Seitz, "The Modern Theory of Solids,"
     McGraw Hill, New York (1940), P. 174.
20. S.H. Lin, H. Eyring, and W.J. Davis,
     J. Phys. Chem., 68:3017 (1964).
21. R.W. Powers, R.W. Mattox, and H.L. Johnston,
     J. Am. Chem. Soc., 76:5968 (1954).
22. A.V. Groose, J. Inorg. Nucl. Chem., 28:7950
     (1966).
23. A.V. Groose, J. Inorg. Nucl. Chem., 28:803 (1966).
24. Y.S. Touloukian, R.W. Powell, C.Y. Ho, and
     P.G. Klemens, "Thermophysical Properties of
     Matter, The TPRC Data Series," Volume 1,
     IFF/Plenum Data Publishing Co., New York
     (1970).

THERMAL CONDUCTIVITY MEASUREMENT OF THE POTASSIUM NITRATE-SODIUM

NITRATE SYSTEM USING A TRANSIENT METHOD WITH A LIQUID-METAL PROBE

T. Omotani, Y. Nagasaka, and A. Nagashima

Keio University
Yokohama, Japan

INTRODUCTION

In recent years, the transient hot-wire method has been developed as one of the most precise methods to measure the thermal
conductivity of fluids. The most important feature of this method
is that it can eliminate errors due to convection. Thus, this
method is attractive for measurement of high temperature melts,
such as molten salts and liquid-metals, in which convection is most
likely to occur. But since high temperature melts are electrically
conducting substances, a conventional hot-wire method connot be used
for them. Coating or sheathing of the wire with insulating materials
is not easy at high temperatures, because the differences in thermal
expansion coefficients or solid-to-solid surface contact causes
problems.

In the present study, the wire heat source in a conventional
hot-wire cell was replaced by a mercury thread confined in a thin
quartz glass capillary. This modification has made it possible to
apply the transient method to electrically conducting liquids at
high temperatures. Also, the transient hot-wire method is considered
to be good for mixtures, since errors due to thermal diffusion can
be avoided.

Measurements of mixtures of the KNO_3-NaNO_3 system were performed
from their melting points to 593 K. Temperature and composition dependences of the thermal conductivity of these mixtures were studied.

Among the various salt mixtures, the KNO_3-NaNO_3 system has been
one of the most controversial as far as thermal conductivity is concerned. Two previous measurements are available for the composition

251

dependence of the thermal conductivity. One of them showed a nearly linear dependence, whereas the other showed strong a nonlinear dependence with one minimum point at a certain ratio.

The intention of this study was to discover which measurement is the correct one.

EXPERIMENTAL APPARATUS AND PROCEDURE

The principle of the ordinary transient hot-wire method has been studied and written up by many researchers[1]. Details of the present apparatus and procedure have been described in our reports 2, 3 and only the outline is given here.

If a thin metallic wire in an infinite fluid layer starts to generate constant heat flux, the thermal conductivity of the fluid can be determined from the heat flux and temperature rise of the wire using the solution of the heat conduction equation for the present system, which is

$$\lambda = \frac{1}{4\pi l_e} \cdot \frac{dR}{dT} \cdot \frac{R\,S}{R+S} \cdot I^3 / (\frac{d\Delta V}{d\ln t}) \qquad \ldots\ldots (1)$$

where λ is the thermal conductivity of the fluid, R the resistance of the wire, T the temperature, S the resistance in a double bridge

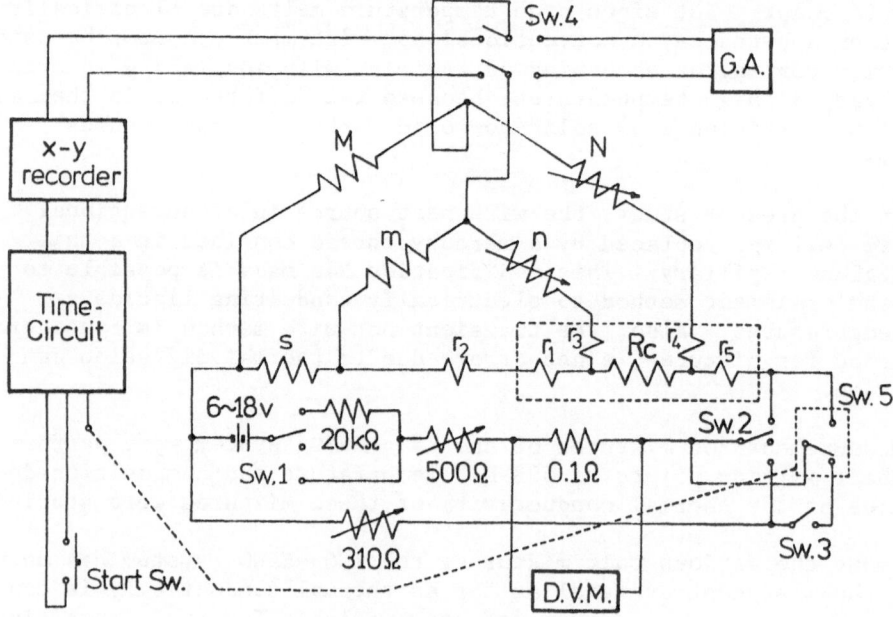

Fig. 1. Electrical circuit.

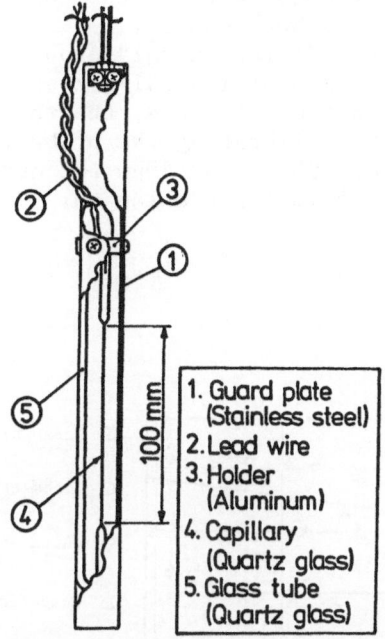

1. Guard plate (Stainless steel)
2. Lead wire
3. Holder (Aluminum)
4. Capillary (Quartz glass)
5. Glass tube (Quartz glass)

Fig. 2 Probe

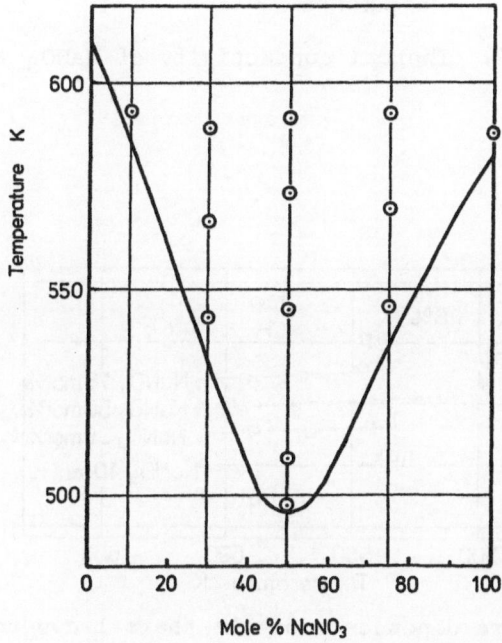

Fig. 3. Measured points on phase diagram.

circuit shown in Fig.1, I the current, ΔV the unbalanced voltage and t the time. For the wire heat source, l_e is the length of the wire, but in this study, the "liquid-metal" probe shown in Fig.2 was used and l_e was the experimentally calibrated "effective length" of the test section. As in the reference, the thermal conductivity of the toluene[4] was used in calibrating the probes. The actual dimensions of the probe were 90 μm and 45μm for outer and inner diameters, respectively, and 100mm for the length. In our first appa-

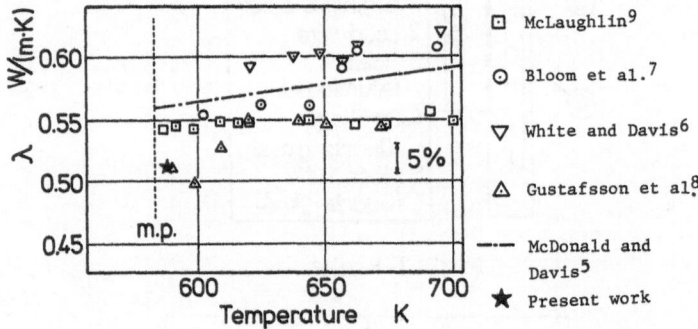

Fig. 4.   Thermal conductivity of $NaNO_3$

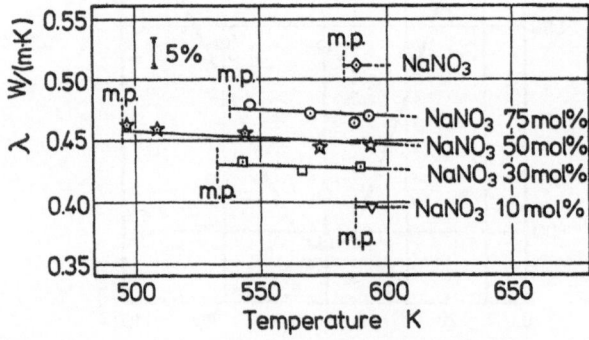

Fig. 5.   Temperature dependence of the thermal conductivity of mixtures

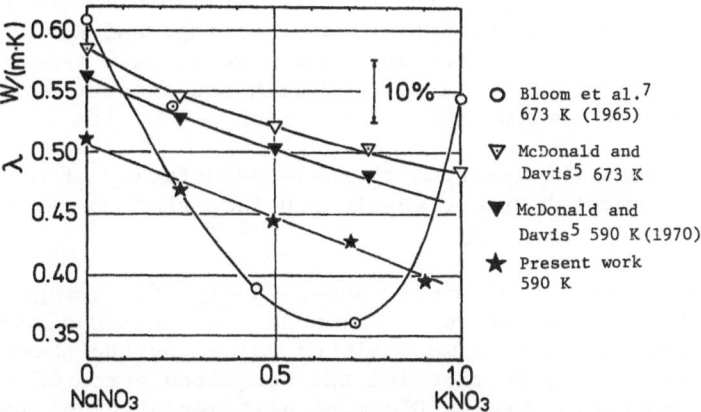

Fig. 6. Composition dependence of the thermal
conductivity of the KNO_3-$NaNO_3$ system.

ratus, a Pyrex glass capillary was used, but later it was replaced
by a quartz glass capillary, since the electrical resistivity of
Pyrex glass was found to be insufficient at high temperatures[3].

One measurement run lasted about 9 s. For each condition, 5 to
19 independent measurements were performed, and the average value
was adopted as the final result. The deviation of each measurement
from the average value did not exceed 2.1%. Mixings were performed
with reagent-grade KNO_3 (99.9% pure) and $NaNO_3$ (99.9% pure) by weighing on a balance. Before being poured into a container, the salts
were heated to 473 K, and then nitrogen gas was allowed to pass
through them for two hours.

The experimental error of the present measurement was estimated
to be 3%. A large portion of the error came from the uncertainties
in calibrated constant, l_e, the temperature coefficient of the resistivity of mercury, dR/dT, and the recording of unbalanced voltage,
$d\Delta V/d1nt$.

RESULTS AND DISCUSSION

Measurements were performed at the points shown in the phase
diagram of Fig.3. Mixtures of 10, 30, 50, and 75 mol% of $NaNO_3$ and
also pure $NaNO_3$ were measured. Figure 4 shows experimental data
for $NaNO_3$ carried out by the present authors and others.

Values by McDonald and Davis [5], White and Davis [6], and Bloom, Doroszkowski, and Tricklebank[7] are higher and show positive temperature dependences. At 593 K, the value established by the present authors agrees with those by Gustafsson, Halling, and Kjellander [8]. McLaughlin's values[9] are higher but show a very small temperature dependence, which agrees qualitatively with the present results on mixtures as shown below.

Figure 5 shows the measured thermal conductivity of mixtures. Within the experimental error, we can conclude that no temperature dependence exists.

The composition dependence is shown in Fig. 6. The gradient of the present study agrees very well with that of McDonald and Davis [5], although they gave a much higher value for the thermal conductivity. The deviations exceeded the estimated error of the present measurements. Data by Bloom et al.[7] deviated considerably, both qualitatively and quantitatively, from the present results.

CONCLUSIONS

Using a newly developed apparatus in combination with a modified transient hot-wire method, the thermal conductivity of salt mixtures of the KNO_3-$NaNO_3$ system was measured in the temperature range from each melting point to 593 K. Results showed that the temperature dependence of the thermal conductivity of this system was very small and that the composition dependence was almost linear. Although the applicable temperature range of the present method is not so high, the precision is better than methods used in earlier studies. The present method can be used to check the reliability of other instruments used for a wider temperature range.

REFERENCES

1. For example, Takizawa, S., Murata, H. and Nagashima, A., Bull. J. Am. Soc. Mach. Eng., 21-152(1978), 273.
2. Hoshi, M., Omotani, T. and Nagashima, A., Rev. Sci. Instrum., 52-4(1981).
3. Omotani, T. and Nagashima, A., (to be published).
4. Nagasaka, Y. and Nagashima, A., I&EC Fundamentals, 20-3 (1981), 216.
5. McDonald, J. and Davis, H. T., J. Phys. Chem., 74-4(1970) 725.
6. White, L. R. and Davis, H. T., J. Chem. Phys., 47-12(1967), 5433.
7. Bloom, H., Doroszkowski, A., and Tricklebank, S. B., Aust. J. Chem., 18(1965), 1171.
8. Gustafsson, S. E., Halling, N.-O. and Kjellander, R. A. E., Z. Naturforsch., 23a(1968), 44.
9. McLaughlin, E., Chem. Rev., 64(1964), 389.

THERMAL CONDUCTIVITY OF NORMAL HYDROGEN*

Hans M. Roder

Thermophysical Properties Division
National Bureau of Standards
Boulder, Colorado 80303

ABSTRACT

The paper presents new experimental measurements of the thermal conductivity of normal hydrogen for eight isotherms at temperatures from 78 to 300 K with pressures to 70 MPa and densities from 0 to 40 mol/L. The data are represented with an equation that is based in part on an existing correlation of the dilute gas. The data are compared with the experimental measurements of others through the new correlation. It is estimated that the overall uncertainty of both experimental and correlated thermal conductivity is 1.5 percent.

INTRODUCTION

A search of the literature reveals a relative abundance of papers on the thermal conductivity of hydrogen.[1] However, measurements that cover a wide range in both temperature and density or pressure are rare,[2,3] and as we shall see, differ considerably. It is, therefore, not surprising that efforts to correlate the thermal conductivity surface of hydrogen[4] are beset with difficulties, and that the results are of doubtful accuracy. In this paper new experimental measurements are presented that cover a large range in density for every isotherm, i.e., 0 to 19 mol/L for 300 K and 0 to 40 mol/L for 78 K. The results overlap our earlier measurements[2] and extend them to 300 K. The earlier measurements were done primarily at temperatures below 100 K. The new results are combined with an existing correlation for the dilute gas[5] to describe the

---

*This work was carried out at the National Bureau of Standards under the sponsorship of the National Aeronautics and Space Administration (C-32369-C). Not subject to copyright.

thermal conductivity surface of normal hydrogen between 78 and 300 K
for pressures up to 70 MPa.

EXPERIMENTAL SECTION

The measurements were made with a new transient hot wire thermal
conductivity apparatus, which is described elsewhere.[6] In the
transient hot wire technique, a thin platinum wire, immersed in the
fluid and initially in thermal equilibrium with it, is subjected at
time t = 0 to a step voltage. The wire behaves as a line source of
heat with constant heat generation. The physical arrangement closely
models an ideal line source, and the working equation for the temperature increase in the wire, ΔT is given by

$$\Delta T = \frac{q}{4\pi\lambda} \, \ell n \left(\frac{4K}{a^2 C} t \right) \tag{1}$$

where q is the heat generated per unit length of wire of radius a,
$K = \lambda/\rho \, C_p$ is the thermal diffusivity of the fluid at the reference
temperature, and $\ell nC = \gamma$, where γ is Euler's constant. K is nearly
constant since the fluid properties do not change drastically with a
small increase in temperature. Corrections to equation (1) have been
fully described elsewhere,[7] the most important one being the effect
of the finite heat capacity of the wire.

Use of a Wheatstone bridge provides end effect compensation,
while the voltages are measured directly with a fast response digital
voltmeter. The DVM is controlled by a minicomputer, which also
handles the switching of the power and the logging of the data. The
measurement of thermal conductivity for a single point is accomplished by balancing the bridge as close to null as is practical at
the cell or reference temperature. The lead resistances, the hot
wire resistances, and the ballast resistors are read first; then the
power supply is set to the desired power and the voltage developed
across the bridge as a function of time is read and stored. The
basic data form a set of 250 voltage readings taken at 3 ms intervals. The other variables measured include the applied power, the
cell temperature, and the pressure. All of the pertinent data are
written by the minicomputer onto a magnetic tape for subsequent
evaluation.

For each run, the data on the magnetic tape are processed on a
large computer. In addition to the reduction of the raw data, i.e.,
the conversion of bridge offset voltages to resistance changes and
then to temperature changes, the large computer also handles the wire
calibration data and evaluates the best straight line for the
ΔT-ℓn(t) data and determines the thermal conductivity. For the wire
calibration some 1800 values were collected for each wire in the

temperature range 77-320 K during an extended set of measurements on liquid oxygen.[8]

The samples used are research grade hydrogen stated by the supplier to be a minimum of 99.999 mol percent hydrogen. We used a small diaphragm compressor as a pressure intensifier.

RESULTS

Approximately 585 points were measured. The measurements are distributed among eight pseudo isotherms where the nominal isotherm temperatures are 78, 100, 125, 150, 175, 200, 250, and 300 K. There are roughly 72 points per isotherm taken at 24 different pressure levels, with three different power levels at each pressure. The pressure, temperature, and applied power are measured directly, the thermal conductivity and the associated regression error are obtained through the data reduction program, while the density is calculated from an equation of state[9] using the measured pressure and temperature. Since the temperature of measurement varies with the applied power, each experimental point has to be adjusted slightly to obtain values on isotherms. The shift in temperature is accomplished through use of the correlating equation given in the next section. The entire set of data is shown in thermal conductivity vs. density coordinates in figure 1. Shown as lines in figure 1 are the isotherms calculated from the correlating equation.

CORRELATION OF THE DATA

It is generally accepted that the thermal conductivity should be correlated in terms of density and temperature rather than temperature and pressure because over a wide range of experimental conditions the behavior of thermal conductivity is dominated by its density dependence. This preferred technique requires an equation of state[9] to translate measured pressures into equivalent densities. The dependence of thermal conductivity on temperature and density is normally expressed as

$$\lambda(\rho,T) = \lambda_o(T) + \lambda_{excess}(\rho,T) + \Delta\lambda_{critical}(\rho,T) \tag{2}$$

We neglect the critical contribution because it is not yet clear if it is present in the data. In detail, the functional form used to describe the thermal conductivity surface is

$$\lambda(\rho,T) = A + (B_1 + B_2T)\rho + (C_1 + C_2T)\rho^2 + D\rho^5 \tag{3}$$

Extrapolation of the present results to zero density shows an average

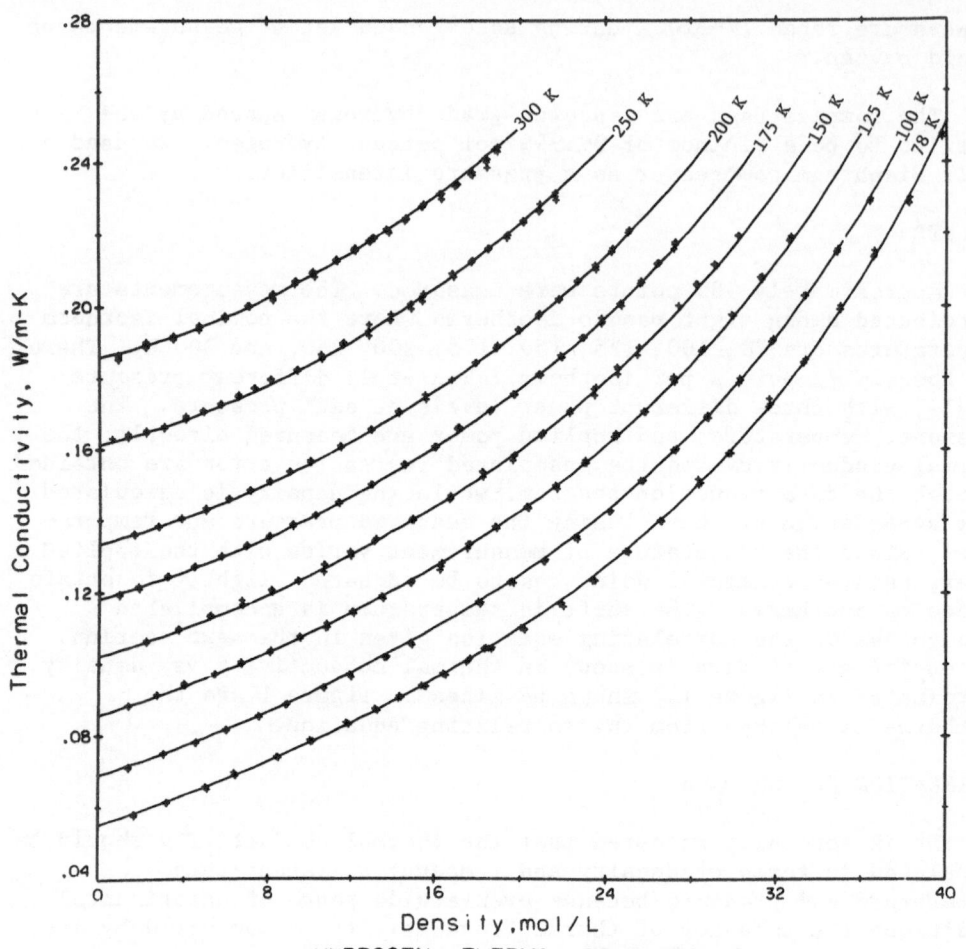

n-HYDROGEN, THERMAL CONDUCTIVITY

Fig. 1.   Isotherms of the thermal conductivity of normal hydrogen.
          + experimental results. ——— correlation.

offset for all isotherms of 0.6 percent when compared with the
theoretical values given by Hanley et al.[5] This discrepancy is
within the error estimate of the latter. However, for convenience we
take $A = \lambda_0(T)/1.006$ where $\lambda_0(T)$ is taken from reference 5.   An
abbreviated table of $\lambda_0(T)$ is given in the appendix.   The values
determined for the coefficients of equation (3) are:

$$
\begin{aligned}
B_1 &= 0.20695\text{E-}02 \pm 0.44\text{E-}04 \ (2 \text{ sigma}) \\
B_2 &= 0.23738\text{E-}05 \pm 0.21\text{E-}06 \\
C_1 &= 0.20555\text{E-}04 \pm 0.23\text{E-}05 \\
C_2 &= 0.22606\text{E-}06 \pm 0.11\text{E-}07 \\
D &= 0.58213\text{E-}09 \pm 0.17\text{E-}10
\end{aligned}
$$

with λ in W/m·K and ρ in mol/L.

The deviation of each experimental point from the correlated surface, equation (3), along isotherms is shown in figure 2 where the spacing between the horizontal isotherm lines is equal to 2 percent. The standard deviation for equation (3) is 0.0007 W/m·K, which is equivalent to 0.5 percent for a value in the middle of the range of thermal conductivities, or 1.5 percent at the 3 sigma level. Errors at the very lowest temperature are seen to increase because the experimental temperatures are quite close to the boiling point of the refrigerant, liquid nitrogen, leaving very little thermal control for the high pressure cell.

DISCUSSION

The precision of the measurements can be established from several considerations. These are the linear regression statistics for a single point, the variation in the measured thermal conductivity with applied power, and the variation obtained in a curve fit of the thermal conductivity surface considering different densities and different temperatures. All of these lead to a value of precision or reproducibility of a single measurement between $\pm$ 0.6 and $\pm$ 1.0 percent. We have compared the correlation to the data presented in this paper in figure 2. The accuracy of the measurement can be established, in principle, from the measurements and certain theoretical considerations, i.e., the Eucken factor for the rare gases. Accuracy can also be estimated by comparison to the results of others. These comparisons run between 1 and 2 percent for nitrogen, helium, argon, oxygen, and propane.[6] The comparison of the present results to those of others[2,3,10] for normal hydrogen is given in figure 3. The deviations are seen to range from + 4 to - 6 percent, a full 10 percent over the surface. The agreement between our earlier results,[2] using a different method and apparatus with an uncertainty of somewhat less than 2 percent, and the present ones is within the combined uncertainty of both measurements, especially at the lower temperatures, as seen in figure 3. The measurements of Golubev and Kal'sina[3] cover the same range of temperature and nearly the same range of densities or pressures as ours but were presented only as a table of smoothed values. The deviations between the two correlations of data are larger than the combined uncertainties and are, furthermore, systematic. The systematic differences arise because the present measurements reveal a distinct dependence on temperature of the λ_{excess}, whereas Golubev and Kal'sina used a single, temperature independent curve to correlate their results. The difference between the present results and those of Clifford et al.,[10] who also use a transient hot wire method, is a nearly constant 1.5 percent, the present results being lower.

Additional analysis of these results and the thermal conductivity surface describing them is planned after a second set of measurements over the same range of temperatures and pressures has been completed on parahydrogen.

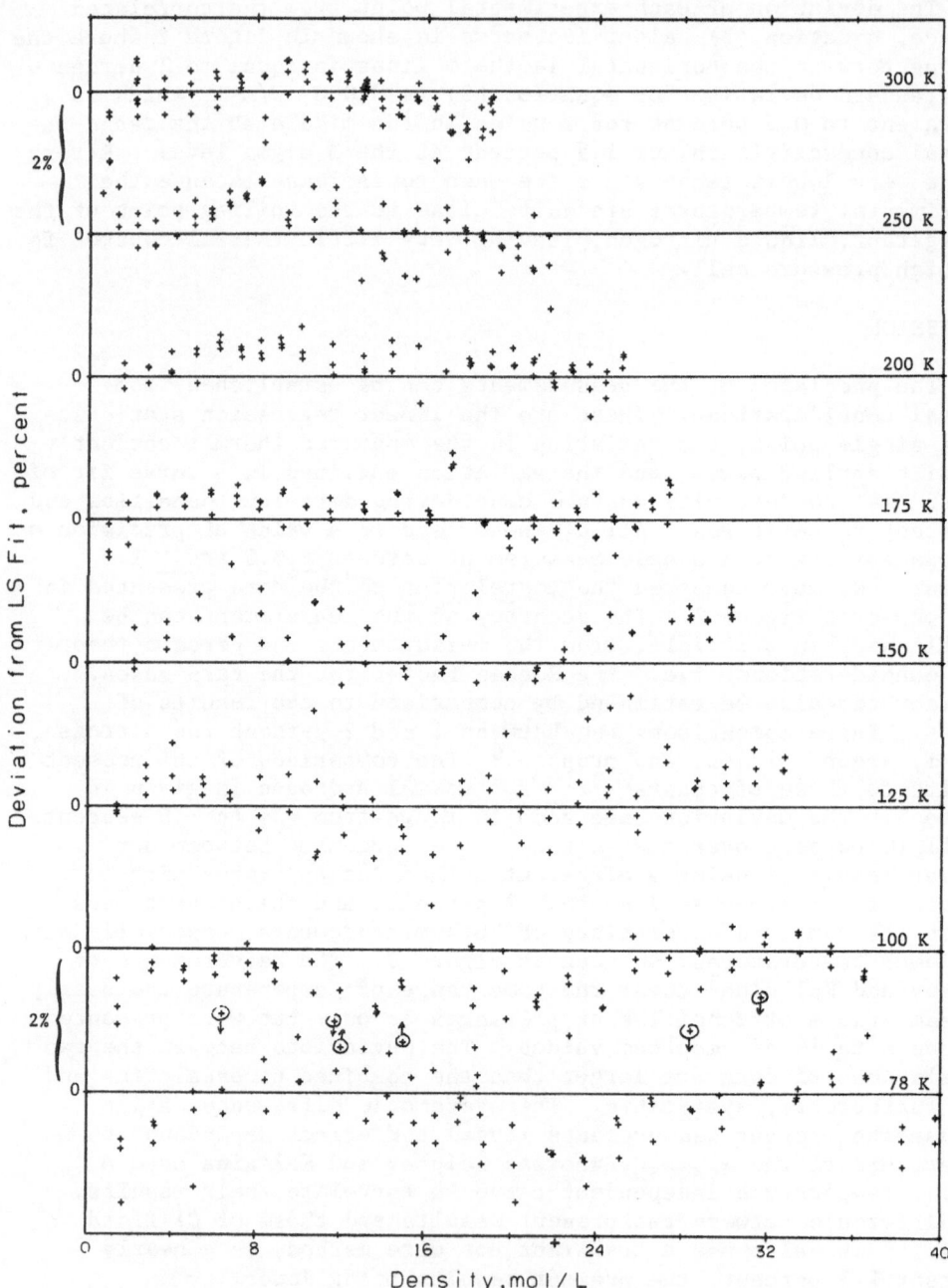

Fig. 2. Deviations of the experimental measurements from the
         correlated thermal conductivity surface along isotherms.
         The spacing between the horizontal isotherms is equal to
         2 percent.

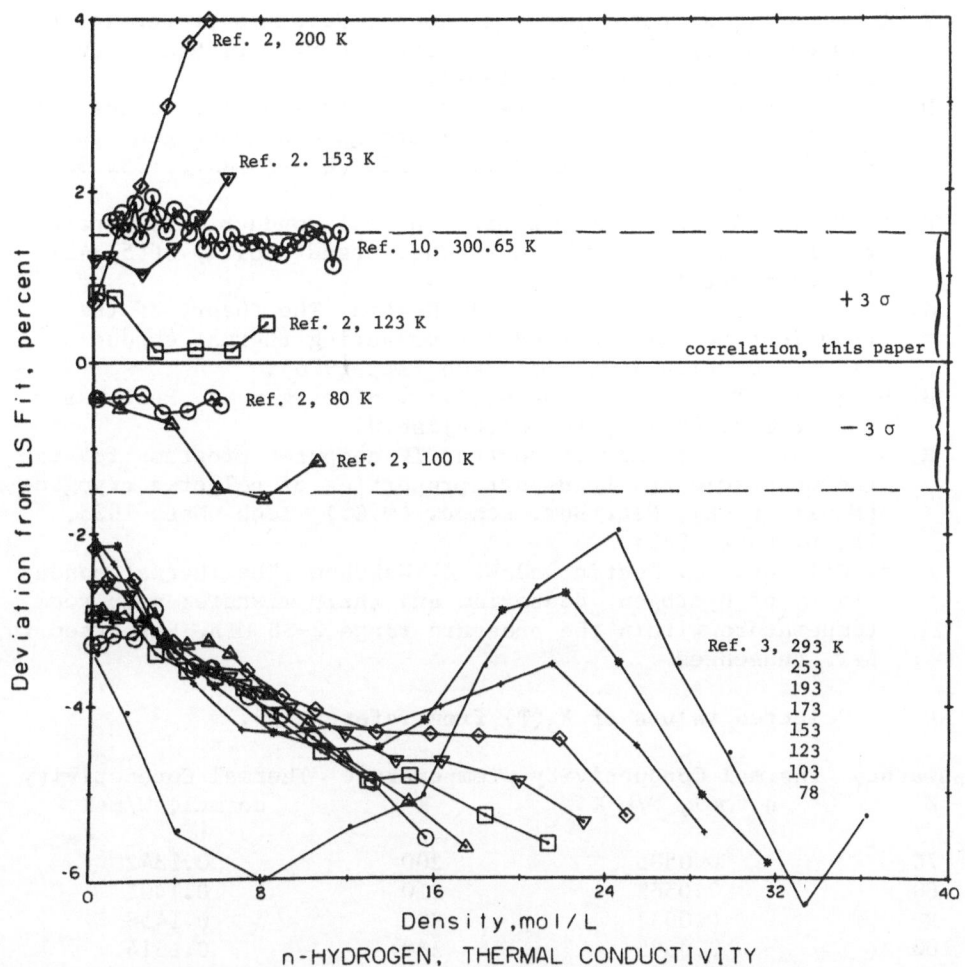

Fig. 3. Comparison of the present results with those of others for normal hydrogen.

REFERENCES

1. See for example appendices 6.1 and 6.2 in R. D. McCarty, Hydrogen-Technological Survey – Thermophysical Properties. National Aeronautics and Space Administration, NASA SP-3089, 530 p. (1975).

2. H. M. Roder and D. E. Diller, Thermal conductivity of gaseous and liquid hydrogen, J. Chem. Phys. 52(11):5928-5949 (Jun 1, 1970).

3. I. F. Golubev and M. V. Kal'sina, Thermal conductivity of nitrogen and hydrogen at temperatures from 20 to -195 C and pressures from 1 to 500 atm, Gazov. Prom. 9(8):41-43 (1964).

4. R. D. McCarty, J. Hord and H. M. Roder, Selected properties of
    hydrogen (Engineering Design Data), Nat. Bur. Stand. (U.S.),
    Monogr. 168, 523 p. (Feb 1981).
5. H. J. M. Hanley, R. D. McCarty and H. Intemann, The viscosity
    and thermal conductivity of dilute gaseous hydrogen from 15
    to 5000 K. J. Res. Nat. Bur. Stand. (U.S.) 74A(3):331–353,
    (May–Jun 1970).
6. H. M. Roder, A transient hot wire thermal conductivity appara-
    tus for fluids, J. Res. Nat. Bur. Stand. (U.S.) 86(5):457–
    493 (Sep–Oct 1981).
7. J. J. Healy, J. J. deGroot and J. Kestin, The theory of the
    transient hot-wire method for measuring thermal conduc-
    tivity, Physica 82C(2):392–408 (Apr 1976).
8. H. M. Roder, The thermal conductivity of oxygen, J. Res. Nat.
    Bur. Stand. (U.S.), to be published.
9. R. D. McCarty, Interactive fortran IV computer programs for the
    thermodynamic and transport properties of selected cryogens
    [Fluids Pack], Nat. Bur. Stand. (U.S.), Tech. Note 1025,
    112 p. (Oct 1980).
10. A. A. Clifford, J. Kestin and W. A. Wakeham, The thermal conduc-
    tivity of hydrogen, deuterium and their mixtures near room
    temperature within the pressure range 2–36 MPa, Submitted to
    Ber. Bunsenges.

APPENDIX: Selected values of $\lambda_0(T)$ from reference 5.

Temperature K	Thermal Conductivity normal, W/m·K	Temperature K	Thermal Conductivity normal, W/m·K
70	0.0505	200	0.1342
80	0.0568	210	0.1401
90	0.0632	220	0.1458
100	0.0695	230	0.1514
110	0.0763	240	0.1569
120	0.0829	250	0.1622
130	0.0896	260	0.1674
140	0.0962	270	0.1725
150	0.1026	280	0.1774
160	0.1092	290	0.1823
170	0.1157	300	0.1870
180	0.1220	310	0.1917
190	0.1282		

THE THERMAL CONDUCTIVITY OF SOLID N-EICOSANE, N-OCTADECANE, N-HEPTADECANE, N-PENTADECANE, AND N-TETRADECANE

David W. Yarbrough and Chih-Ning Kuan

Department of Chemical Engineering
Tennessee Technological University
Cookeville, Tennessee 38501

ABSTRACT

The thermal conductivities, k, of five solid, long-chain alkanes have been determined as a function of temperature from about $10^{\circ}C$ below their melting points to about $30^{\circ}C$ below their melting points. A long, unguarded, radial heat-flow apparatus was used for the measurements. HEATING5 was used to model the apparatus thermally and show end guards to be unnecessary. Experimental uncertainty was estimated to be less than ±6% for a majority of the k measurements.

Values for k at $0^{\circ}C$ were determined to be 0.260 $W/m^{\circ}C$ for n-tetradecane, 0.343 $W/m^{\circ}C$ for n-octadecane, 0.422 $W/m^{\circ}C$ for n-eicosane, 0.182 $W/m^{\circ}C$ for n-pentadecane, and 0.210 $W/m^{\circ}C$ for n-heptadecane. The k values decreased with temperature for the five solids studied.

INTRODUCTION

Published thermal conductivity data for solid n-tetradecane ($C_{14}H_{30}$), n-hexadecane ($C_{16}H_{34}$), n-octadecane ($C_{18}H_{38}$), n-nonadecane ($C_{19}H_{40}$), and n-eicosane ($C_{20}H_{42}$) are limited to a few sources. Reliable data, however, are needed for the design of thermal controllers and phase-change heat-storage systems that use the above compounds. Analyses of results[7,8] obtained in phase-change heat-storage experiments suggest that the k of solid n-hexadecane, n-nonadecane, and n-octadecane are up to three times greater than previously reported values. The apparent need for additional k data for the compounds mentioned above motivated this work.

EXPERIMENTAL APPARATUS AND PROCEDURE

Thermal conductivities of the five solids were measured using an unguarded radial heat-flow apparatus. Specimens of the compounds were placed in the annular space formed by centering a 0.635-cm o.d. stainless steel tube inside of a 1.994-cm i.d. copper tube. The stainless steel tube was held in position by two nylon caps machined to fit inside of the copper tube. The stainless steel tube extended through the caps to permit power leads to be attached. The copper tube was 80 cm in length; the stainless steel tube was approximately 84 cm in length. The concentric cylinders were mounted horizontally in a constant temperature bath constructed to permit the ends of the cylinders to be outside the bath. Thus, the outside curved surface of the copper tube was in contact with the bath solution, while the ends of the apparatus and the power leads were in contact with the surrounding air. The bath was insulated with mineral fiber insulation to reduce heat gain.

Temperatures on the inside and outside surfaces of the solid specimens were determined by copper-constantan thermocouples. Eleven thermocouples were located on the copper tube in a 13.5-cm-long region centered at the midpoint of the tube. Three thermocouples were placed 120° apart at the midpoint, three thermocouples were placed 120° apart at ±2.5 cm from the midpoint, and two thermocouples were placed ±6.25 cm from the midpoint. Two thermocouples were used to determine the inside surface temperature. The inside thermocouples were imbedded 180° apart in an acrylic plug machined to fit inside of the stainless steel tube. The thermocouples were held against the inside surface of the stainless steel tube by the plug. The plug assembly was moved axially to measure temperatures at positions corresponding to the fixed locations of the outside thermocouples. A preliminary set of measurements was made to compare temperatures sensed by the inside plug-mounted thermocouples and the three thermocouples attached to the outside surface of the stainless steel tube. The results indicated that the outside surface temperature averaged 0.15°C less than the temperature on the inside tube surface. All measured inside temperatures were consequently reduced 0.15°C and averaged to obtain an inside surface temperature for the specimen. Inside temperatures were measured at five locations (10 points). The 11 temperature measurements on the copper tube were averaged without adjustment. All of the thermocouples were read relative to an ice-water reference with a Leeds and Northrup K-5 potentiometer.

Electrical power was supplied to the stainless steel tube by a constant current dc power supply.[a] Electrical current to the

---

[a]Constant current dc power supply manufactured by the OPAD Electrical Company, New York, New York.

stainless steel tube was measured with a dc ammeter in series with the heater tube.[b] The voltage drop across the stainless steel tube was measured with the K-5 potentiometer. Calculations of k were made by assuming that the electrical resistance of the stainless steel tube did not vary with position in the range of temperatures observed.

The materials used in the study were nominal 99% pure compounds that were obtained commercially.[c] The compounds were used as received. Specimens of the compounds were poured into the space between the tube as a liquid, and additional liquid was added as the specimen solidified to produce a specimen free of voids.

A HEATING5[9] analysis showed that radial heat flow can be assumed in the apparatus for a 50-cm region centered in the cylindrical system. The HEATING5 computer program was used to calculate steady-state temperature distributions in a cylindrical system like the experimental apparatus. The program was run for an apparatus length of 70 cm and a specimen k of 0.4 W/m^{o}C. The results indicated negligible error in k resulting from axial heat transfer to the unguarded ends if measurements are restricted to the central 50 cm of the apparatus.

EXPERIMENTAL RESULTS

Thermal conductivities were calculated using

$$k = (Q/2 L T) \ln(r_o/r_i) \qquad\qquad (1)$$

were k is the thermal conductivity (W/m^{o}C), r_i is the inside radius of the specimen (cm), r_o is the outside radius of the specimen (cm), T is the difference between the inside specimen surface temperature and the outside specimen surface temperature (^{o}C), and Q/L is the heater power (W/m). Measured thermal conductivities for the five solids are given in Table 1 as a function of the mean specimen temperature. The variation of k with temperature is described by Equation (2) with values for the coefficients A and B determined by the method of least squares. Five sets of coefficients, correlation coefficients, and average values for the percent difference between the data and results calculated using Equation (2) are given in Table 2.

$$k = A + BT \qquad\qquad (2)$$

[b]Type DP-9 ammeter manufactured by the General Electric Company.

[c]Humphrey Chemical Company, North Haven, Connecticut 06473.

DISCUSSION OF RESULTS

Figures 1 and 2 show the thermal conductivity data as a function of temperature with solid lines representing Equation (2). The data are divided into two groups on the basis of the number of carbon atoms in the molecule (carbon number). In both groups, odd and even carbon number, k decreases with temperature and increases with carbon number. No attempt has been made in the present study to take into account possible solid-phase transitions. Values for k at 0 C calculated from Equation (2) are shown in Figure 3 with a few additional points from Griggs and Yarbrough and Mehlon. The additional data support the observation that the compounds with an even number of carbon atoms have k values above those with an odd carbon number.

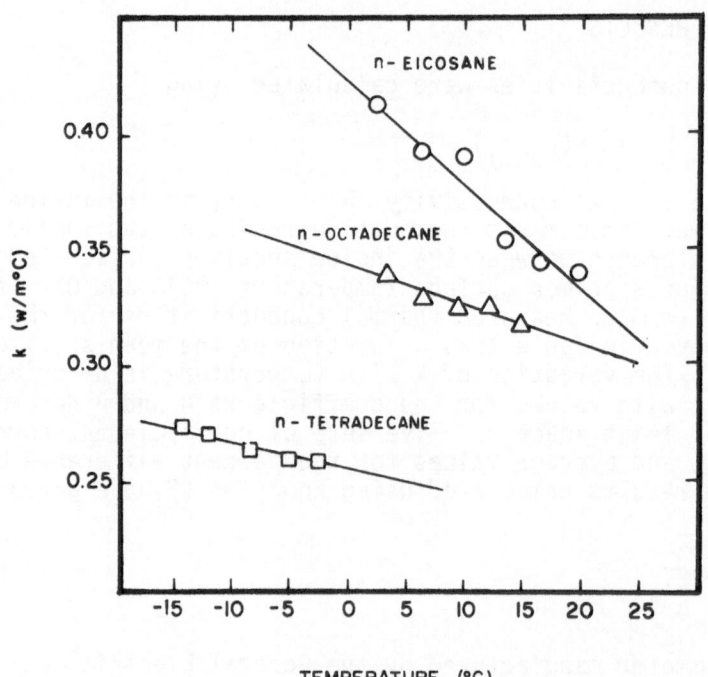

Figure 1. The thermal conductivities of n-tetradecane, n-octadecane, and n-eicosane as a function of temperature.

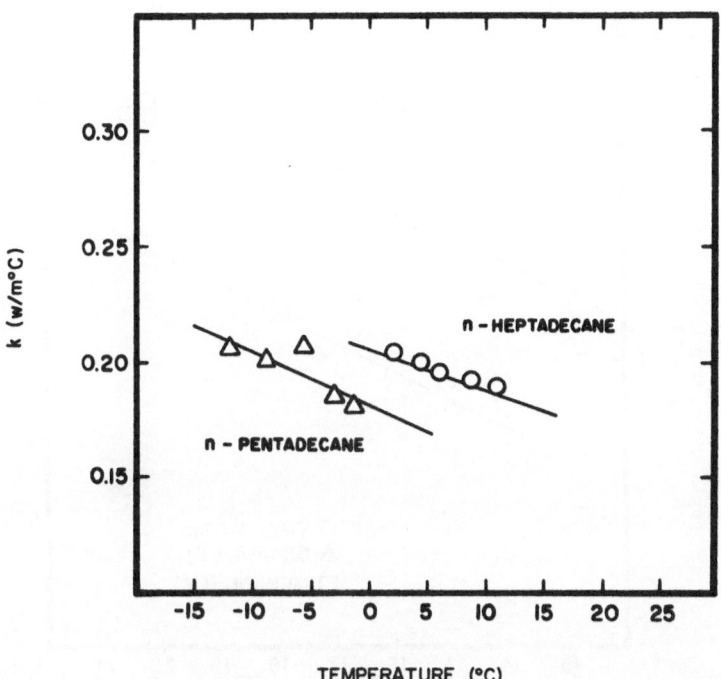

Figure 2. The thermal conductivity of n-pentadecane and
n-heptadecane as a function of temperature.

A detailed analysis by Kuan of the uncertainty associated with
the measurement of k resulted in a value of ±3.8% exclusive of the
temperature measurements. The total experimental uncertainty depends
on the temperature difference uncertainty, which is conservatively
estimated to be ±20/ T%. Seventeen of the 26 k values were determined with temperature differences greater than 9 C with a resultant
uncertainty of less than ±10%, whereas two measurements were made
with temperature differences below 3 C with a worst case uncertainty
of ±14%. The data were equally weighted in the determination of A
and B in Equation (2).

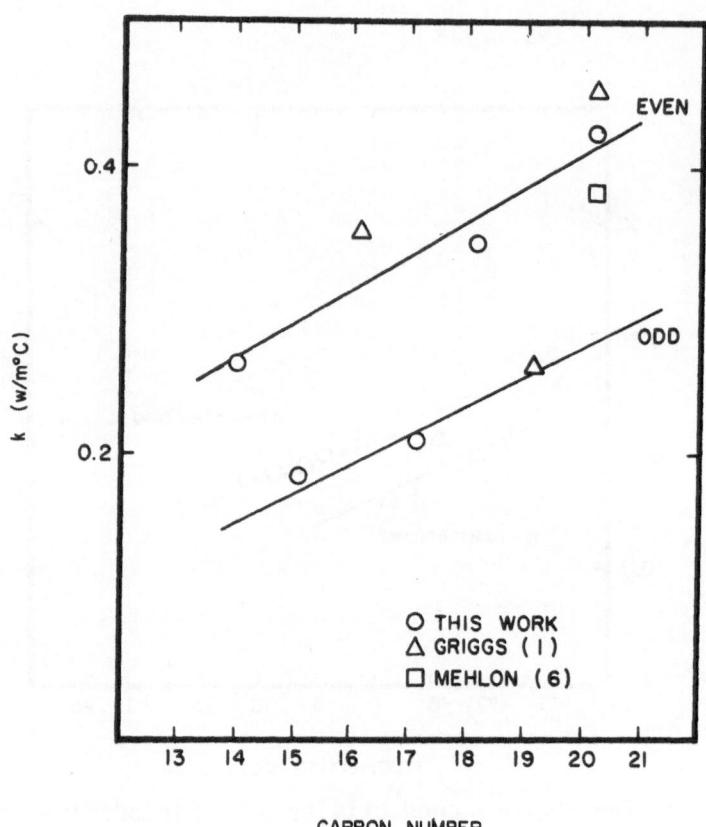

Figure 3. Thermal conductivity at 0°C as a function of carbon number.

CONCLUSIONS

Solid-phase thermal conductivities for n-eicosane, n-octadecane, n-heptadecane, n-pentadecane, and n-tetradecane have been measured as a function of temperature for a limited temperature interval below the melting points of the compounds and are described by functions that are linear in temperature. The five compounds show a decrease in k with temperature. The compounds with an even number of carbon atoms show increasing k with carbon number. The odd-carbon-number compounds also show increasing k with carbon number, but the k values are lower than those for the even-numbered compounds.

Table 1. Thermal Conductivity Data for Solid n-Eicosane, n-Octadecane, n-Heptadecane, n-Pentadecane, and n-Tetradecane.

Heater Power (W/m)	Average Inside Temperature (^{0}C)	Average Outside Temperature (^{0}C)	Average Specimen Temperature (^{0}C)	Thermal Conductivity (W/m^{0}C)
		n-eicosane		
6.748	3.586	0.613	2.10	0.413
17.128	9.665	1.740	5.70	0.393
29.924	16.599	2.692	9.65	0.392
37.126	22.292	3.264	12.78	0.355
46.545	28.457	3.862	16.16	0.344
55.681	34.351	4.673	19.51	0.341
		n-octadecane		
8.046	5.205	0.894	3.05	0.340
17.807	11.446	1.590	6.52	0.329
24.881	16.033	2.239	9.14	0.328
31.717	20.584	2.869	11.73	0.326
39.310	25.938	3.423	14.68	0.318
		n-heptadecane		
4.470	4.305	0.263	2.28	0.201
8.260	8.317	0.760	4.54	0.199
10.764	11.110	1.108	6.11	0.196
15.772	16.725	1.320	9.02	0.186
18.120	20.247	1.735	10.99	0.178

(con't)

Table 1 (cont.). Thermal Conductivity Data for Solid n-Eicosane, n-Octadecane, n-Heptadecane, n-Pentadecane, and n-Tetradecane.

Heater Power (W/m)	Average Inside Temperature (°C)	Average Outside Temperature (°C)	Average Specimen Temperature (°C)	Thermal Conductivity (W/m°C)
		n-pentadecane		
4.643	- 9.662	-13.768	-11.72	0.206
7.330	- 5.573	-12.192	- 8.88	0.202
12.356	- 0.091	-10.831	- 5.46	0.209
11.802	2.639	- 8.956	- 3.16	0.185
15.547	6.549	- 9.090	- 1.27	0.181
		n-tetradecane		
3.022	-13.571	-15.489	-14.53	0.287
5.071	-10.166	-13.464	-11.82	0.280
8.634	- 5.819	-11.549	- 8.68	0.274
12.713	- 0.992	- 9.531	- 5.26	0.271
13.666	1.899	- 7.525	- 2.81	0.264

Table 2. Coefficients for k = A+BT Obtained from the Five Sets of Thermal Conductivity Data

Compound	A	B	Average Error*(%)	Correlation Coefficient
n-tetradecane $(C_{14}H_{30})$	0.25951	-0.00182	0.4	0.99
n-octadecane $(C_{18}H_{38})$	0.34328	-0.00168	0.5	0.96
n-eicosane $(C_{20}H_{42})$	0.42194	-0.00446	1.8	0.96
n-pentadecane $(C_{15}H_{32})$	0.18220	-0.00236	2.8	0.79
n-heptadecane $(C_{17}H_{36})$	0.20991	-0.00272	1.0	0.97

* $(k_{calc}-k_{exp}) \times 100/k_{exp}$

REFERENCES

1. E.I. Griggs, and D.W. Yarbrough, Thermal conductivity of
    solid unbranched alkanes from n-hexadecane to n-eicosane,
    in: "Proceedings of the Southeastern Seminar on Thermal
    Sciences," North Carolina University, Raleigh, Apr. 6-7
    (1978) pgs. 256-267.
2. R.D. Sutherland, R.S. Davis, and W.F. Seyer, Heat transfer
    effects- molecular orientation of octadecane, Ind. Eng.
    Chem. 51(4):585 (1959).
3. I.L. Lyons and L.D. Russell, Phase change materials and
    thermal design, ASHRAE J. 19:47 (1977).
4. R.W. Powell and A.R. Challoner, Thermal conductivity of
    n-Octadecane, Ind. Eng. Chem. 53(7):581 (1961)
5. C.N. Kuan, "Determination of Thermal Conductivity of Heavy
    Hydrocarbons Using an Unguarded Concentric Cylinder
    Apparatus," Master of Science Thesis, Tennessee
    Technological University, Cookeville, Tennessee (June
    1979).
6. R.J. Mehlon, "The Thermal Conductivity of Heavy
    Hydrocarbons," Master of Science Thesis, Tennessee
    Technological University, Cookeville, Tennessee (August 1979).
7. J.A. Bailey and C. Liao, "Thermal Capacitor Design Rational;
    Part 1," Final Report to NASA-MSFC under agreement NCA-52
    with North Carolina State University (January 1974).
8. J.D. Dyer and E.I. Griggs, "Study of the Solidification and
    Melting of Nonadecane and Hexadecane Including Solid
    Phase Thermal Conductivity Measurements," Report ME-49,
    Tennessee Technological University, Cookeville, Tennessee
    (June 1975).

THERMAL CONDUCTIVITY RESEARCH OF GASEOUS AND LIQUID NAPHTHENIC HYDROCARBONS OVER A WIDE RANGE OF STATE PARAMETERS

Ya.M. Naziev and A.N. Shakhverdiyev
The Azerbaijan Ch.Ildrym Polytechnical
Institute, Baku, USSR

ABSTRACT

This paper deals with an experimental investigation of the thermal conductivities of three important representatives of naphthenic hydrocarbons over a wide temperature and pressure range. The investigation was carried out in a modified experimental unit by the tri-calorimeter method. The experimental data obtained are described analytically.

A relation of thermal conductivity to thermal pressure and dynamic viscosity has been established.

MAIN RESULTS OF THERMAL-CONDUCTIVITY MEASUREMENTS

The investigation of thermal-conductivity coefficients of gases and liquids at various pressures and temperatures is of particular practical and theoretical interest. Thermal-conductivity data of naphthenic hydrocarbons are actually lacking in published literature. This generates a need for measuring the thermal-conductivity coefficients of naphthenic hydrocarbons, in particular those of cyclohexane and methyl and ethyl

275

cyclohexane, which have found extensive use in the chemical industry recently.

Measurements were carried out in an experimental
unit by the method of a modified cylindrical tricalo-
rimeter developed by Ya.M.Naziev. A detailed description of the experimental unit is presented in earlier
publications [1,2].

When measuring the thermal conductivity coefficients of the above hydrocarbons, the pressure range
was 0.1 to 50 MPa and the temperature ranged from
293.15 to 633.15K including the critical and supercri-
tical region and the saturation line.

Studies were conducted on chemically pure cyclohexane, as well as on methyl and ethyl cyclohexane.

Factors that could cause convection were exclud-
ed in the course of experimentation.

The experiments showed that the thermal conductivity increases with rising pressure both for the gaseous and liquid state, but with the gas the effect of
pressure is greater than with the liquid. As the temperature is increased, the thermal conductivity drops
in the case of the liquid state and shows a marked increase with the gaseous state.

The experimental values of the thermal conductivity of cyclohexane, methyl cyclohexane, and ethylcyc-
lohexane are presented in Tables 1, 2, and 3, whereas
Table 4 gives thermal conductivity values at the saturation line both on the liquid and vapour side. In
the course of investigation, an anomalous behaviour of
the thermal conductivity coefficient was observed
near the critical point. However the thermal conductivi-

ty "maxima" of the materials studied are not presented in the tables, inasmuch as they have not been investigated comprehensively. So far, the phenomenon of the thermal-conductivity "maxima" remains a subject of discussion.

An analysis of all kinds of errors has shown that the aggregate mean-square relative error inherent in the cylindrical-tricalorimeter method is not in excess of 1.5%, with the exception of the critical region where the error of the experimental data reaches a value of 3% and even higher.

The Ryvkin method (Fig. 1) was used to process and describe analytically the experimental data obtained with cyclohexane. If the excessive heat conductivity $\Delta\lambda = \lambda_{P,T} - \lambda_{0,T}$ is shown in the P and T coordinate system, one will obtain, up to certain pressure and temperature values, straight lines described by a linear equation of the following form:

$$P = A(\Delta\lambda) + B(\Delta\lambda)T \qquad (1)$$

where A and B are coefficients depending on the excessive thermal conductivity ($\Delta\lambda$) and given by functional dependences in the forms of

$$A = \sum_{i=0}^{3} a_i (\Delta\lambda)^i \quad \text{and} \quad B = \sum_{i=0}^{3} b_i (\Delta\lambda)^i$$

These coefficients were found by the method of least squares with the use of a "Minsk-32" electronic computer.

The values of the a_i and b_i coefficients are given below:

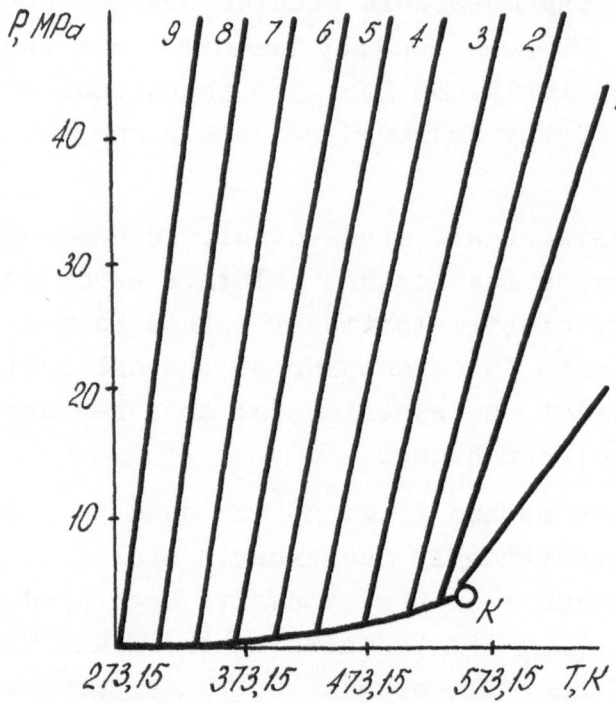

Fig. 1. Dependence of constant values of excessive thermal conductivity of cyclohexane
on temperature and pressure. $1 - \Delta\lambda = 400\frac{W}{m\cdot deg.}$;
$2 - 500$; $3 - 600$; $4 - 700$; $5 - 800$; $6 - 900$;
$7 - 1000$; $8 - 1100$; $9 - 1200$.

$a_0 = - 286.5357$ $b_0 = 4.8394$

$a_1 = 55.9941$ $b_1 = - 1.5007$

$a_2 = - 5.0719$ $b_2 = 0.1661$

$a_3 = 0.1951$ $b_3 = - 0.0058$

Compressed gases are known to be very close to
liquids in properties. On the basis of this fact, a
check has been carried out to make sure that the fol-

lowing equation relating thermal conductivity to thermal pressure is applicable to liquid hydrocarbons

$$\lambda_{\rho,T} = \lambda_T + \alpha_\lambda \left(\frac{P_T}{T}\right)^{n\lambda} \tag{2}$$

This equation was derived by Naziev [3] on the basis of the Enskog equation.

The thermal pressure is determined by graphical differentiation of P–V–T data. It is very convenient to use equation (2) in the logarithmic form. If log $(\lambda_{\rho,T} - \lambda_{0,T})$ is plotted along one of the coordinate axes and log $\frac{P_T}{T}$ is laid off along the other, we obtain straight lines given by the following equations (Fig. 2):

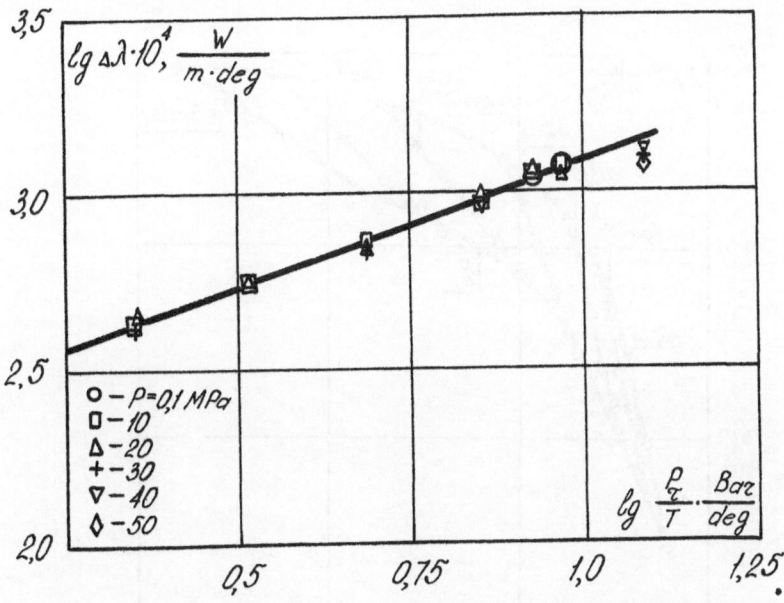

Fig. 2. Relation of excessive thermal conductivity of cyclohexane to thermal pressure.

for cyclohexane

$$\log (\quad \lambda_{e,T} = \lambda_{0,T} \quad) = 2.55 + 0.51 \log \frac{p_T}{T} \qquad (3)$$

for methyl cyclohexane

$$\log (\quad \lambda_{e,T} - \lambda_{0,T} \quad) = 1.96 + 1.259 \log \frac{p_T}{T} \qquad (4)$$

Equations (3) and (4) give thermal conductivity data within the experimental accuracy.

Mamedov et al. [4] revealed a relationship between thermal conductivity and viscosity for ordinary water· and toluene over a wide range of state parameters. The procedure used in the above study was checked by us with cyclohexane as an example, and in doing so we obtained in coordinates (Fig. 3) the following relation-

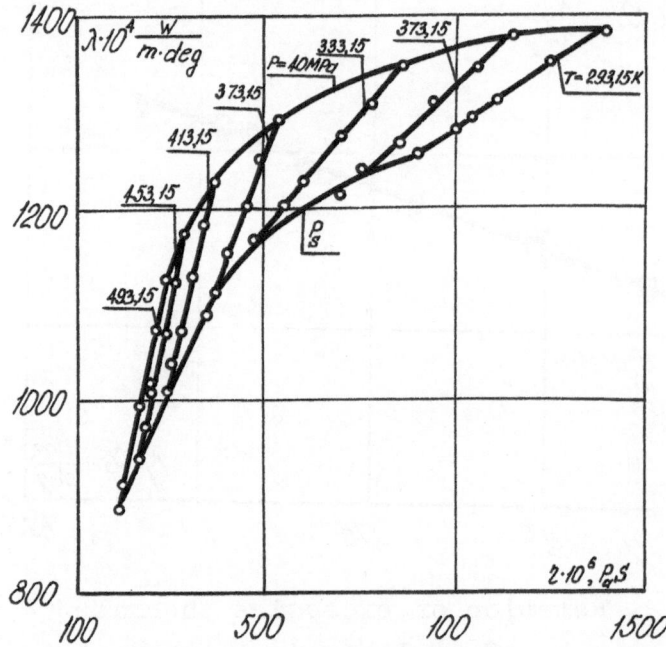

Fig. 3. Relation of thermal conductivity to dynamic viscosity of cyclohexane.

ship:
$$\lambda = A(T) + B(T)l \qquad (5)$$
Up to a pressure of 40 MPa, the experimental values of
cyclohexane thermal conductivity and dynamic viscosity
from Ref. 5 fall accurately enough on a straight line
in isotherms.

The values of coefficients A and B of relationship (5), which are a function of temperature, are given below

$$A = \sum_{i=0}^{4} a_i T^i \qquad \text{and} \quad B = \sum_{i=0}^{4} b_i T^i$$

when $x \ 10^4 \frac{W}{m \cdot deg.}$ and $x \ 10^6$ MPa·s

$a_0 = 1043 \cdot 3694$ $\qquad\qquad$ $b_0 = 0.2349$

$a_1 = \quad 1.7827$ $\qquad\qquad$ $b_1 = -0.1813 \times 10^2$

$a_2 = -0.0946$ $\qquad\qquad$ $b_2 = 0.9562 \times 10^4$

$a_3 = 0.5391 \times 10^3$ $\qquad$ $b_3 = 0.3665 \times 10^6$

$a_4 = -0.6762 \times 10^6$ $\qquad$ $b_4 = 0.3125 \times 10^8$

Thus, relationship (5) has been obtained, relating thermal conductivity to dynamic viscosity and describing well enough the experimental data on the thermal conductivity of cyclohexane up to a pressure of
40 MPa and a temperature of 493.15 K with a maximum
error of 1.7%.

NOMENCLATURE

λ = thermal conductivity, W/m·deg.

T = absolute temperature, K

P = pressure MPa

$\Delta\lambda$ = excessive thermal conductivity, W/m·deg.

$\lambda_{P,T}$ = thermal conductivity at specified pressure and
temperature, W/m·deg.

$\lambda_{g,T}$ = thermal conductivity in gaseous state at atmospheric pressure and various temperatures, W/m·deg.

a_{λ} and Π_{λ} = coefficients depending weakly on density

P_T = thermal pressure, MPa

l = dynamic viscosity, Pa·s

P_s = saturation pressure, MPa

REFERENCES

1. Ya.M. Naziev, Thermal conductivity of olefine hydrocarbons, in: Proceedings of the Fifth Symposium on Thermophysical Properties, The Amer.Soc.of Mech. Eng., New York (1970).

2. Ya.M. Naziev, A.N. Shakhverdiyev, and A.A. Abbasov, Heat conductivity investigation of gaseous and liquid methyl and ethyl cyclohexane at high pressures and temperatures. Izv.Vyssh.Uchebn.Zaved.Neft.Gaz., 5:54-57 (1979).

3. Ya.M. Naziev. Heat conductivity investigation of hydrocarbons at high pressures and some peculiarities of the methods used to measure it. Abstract of the Doctor's Thesis, Energet.Inst., Moscow (1970).

4. A.M. Mamedov, T.S. Akhundov, and S.T. Ismailov, Relationship between transfer properties of liquids Inzh.Fiz.Zh., 24 (4): 761-765 (1973).

5. S.O. Guseinov, Ya.M. Naziev, and A.K.Akhmedov, Viscosity of cyclohexane at high pressures. Izv. Vyssh.Uchebn.Zaved.Neft.Gaz., 2:65-67 (1973).

Table 1. Experimental Values of Thermal Conductivities
of Cyclohexane ($\lambda \times 10^4$ W/m·deg.)

P, MPa / T, K	0.1	2	5	10	20	30	40	50
296.25	1248	1257	1270	1291	1331	1369	1408	1446
323.35	1188	1198	1212	1234	1279	1324	1368	1413
349.25	1133	1141	1155	1179	1228	1275	1323	1369
373.35	181	1092	1106	1131	1182	1233	1284	1331
400.15	216	1037	1056	1082	1137	1189	1244	1296
424.15	249	991	1014	1042	1097	1152	1208	1263
449.25	284	944	972	1002	1061	1118	1176	1235
473.05	317	901	936	968	1027	1086	1145	1205
496.55	354	853	898	936	999	1058	1118	1177
522.65	387	407	848	901	969	1028	1087	1147
548.15	423	443	789	869	945	1004	1064	1124
571.65	459	479	713	843	926	987	1048	1109
598.15	494	511	680	818	911	973	1033	1095
630.15	539	555	669	792	891	954	1016	1079

Table 2. Experimental Values of Thermal Conductivities
of Methyl Cyclohexane ($\lambda \times 10^4$ W/m·deg.)

P, MPa / T, K	0.1	1	5	10	20	30	40	50
294.15	1117	1120	1133	1149	1183	1215	1247	1280
324.15	1051	1054	1067	1085	1159	1152	1186	1219
359.15	974	978	991	1009	1044	1077	1111	1145
376.15	177	941	953	972	1008	1043	1077	1112
393.95	183	902	916	934	970	1005	1039	1075
424.55	209	833	848	866	903	940	976	1014
460.65	246	755	770	790	828	866	904	941
476.75	264	717	734	754	794	833	873	905
493.45	279	289	701	722	763	803	843	876
525.65	319	327	623	654	701	739	778	816
551.25	352	360	563	597	649	691	731	770
573.05	388	395	520	561	611	651	692	732
593.65					577	616	657	698
612.15					552	593	634	675
624.15					541	582	624	665

Table 3.   Experimental Values of Thermal Conductivities
of Ethyl Cyclohexane (λ x 10^4 W/m·deg.)

P, MPa T, K	0.1	1	5	10	20	30	40	50
293.65	1152	1157	1170	1189	1228	1266	1303	1342
321.55	1097	1102	1117	1136	1175	1214	1254	1293
345.15	1055	1060	1076	1094	1134	1173	1213	1253
369.65	1008	1012	1027	1047	1087	1126	1166	1206
403.15	954	957	972	991	1031	1071	1111	1151
422.75	207	923	939	959	998	1037	1078	1117
462.65	244	858	874	894	933	973	1012	1052
493.85	276	798	821	847	886	927	966	1005
524.65	304	312	765	796	837	879	920	961
549.96	331	341	727	765	810	853	895	937
571.25	356	365	691	733	779	825	869	914
593.35	379	389	643	694	749	790	848	896
608.65	397	409	599	673	727	780	832	884
621.25	411	422	562	653	712	766	819	872

Table 4.   Experimental Values of Thermal Conductivities
of the Investigated Hydrocarbons at the Saturation Line (λ x 10^4 W/m·deg.)

Cyclohexane					
Liquid				Vapour	
T,K	λ	T, K	λ	T, K	λ
296.25	1248	473.15	891	373.35	181
323.35	1198	496.55	850	400.15	216
349.25	1134	522.85	810	424.15	249
373.35	1082	548.15	728	449.25	288
400.15	1024	571.65	620	473.05	325
424.15	981			496.55	374
449.25	930			522.85	420
				548.15	521
				571.65	538
Methyl Cyclohexane					
294.15	1118	476.75	716	376.15	177
324.15	1060	493.45	677	393.95	183
359.15	974	523.65	600	424.55	212
376.15	938	551.25	542	460.65	252
393.95	898	573.05	461	476.75	271
424.55	831			493.45	292
460.65	754			525.65	338
				551.25	384
				573.05	434
Ethyl Cyclohexane					
293.65	1152	524.65	738	422.75	207
321.55	1096	549.96	694	462.65	244
345.15	1054	571.25	653	493.85	280
369.65	1008	593.35	604	524.65	317
403.15	954			549.96	348
422.75	918			571.25	380
462.65	855			593.35	418
493.85	796				

ORGANIC LIQUID MIXTURES: COMPARISON BETWEEN EXPERIMENTAL AND PREDICTED THERMAL CONDUCTIVITY VALUES*

C. Baroncini, P. Di Filippo, G. Latini, and M. Pacetti

Istituto di Fisica Generale e Fisica Tecnica

Università di Ancona 60100 Ancona, Italy

ABSTRACT

A general correlation, proposed and checked for organic liquids in previous papers, is modified and fitted to predict the thermal conductivity of liquid mixtures. The comparison between experimental and estimated thermal conductivity values is performed also with data obtained by the authors using the steady hot-wire method. The mean general deviation between experimental and predicted thermal conductivity values is less than 2% for 30 nonaqueous mixtures.

INTRODUCTION

Thermal conductivity of organic mixtures is a very important physical property and its value is necessary in several engineering problems, but, because of the complexity of laboratory measurements, the experimental data cover a small part of the enormous variety of the substances possible. On the other hand, the correlations existing in literature[1-4] and proposed to predict the thermal conductivity, λ_m, of the binary mixtures require the knowledge of the thermal conductivity values of the mixture components λ_1 and λ_2, which is not always possible at every temperature.

The aim of this work is to propose a correlation that relates λ_m directly with the temperature and not with λ_1 and λ_2: this result is obtained because the binary mixture components are completely characterized, with respect to the thermal conductivity, by the respective critical temperatures, T_{c1} and T_{c2}, and by the respective temperature independent factors, A_1 and A_2.

*Work supported by Consiglio Nazionale delle Ricerche - Italy

The check of the proposed correlation is developed by taking
into account thermal conductivity values measured at the National
Engineering Laboratory (NEL)[5] and at Ancona University. This choice
was made to avoid difficult and often impossible comparisons among
experimental data published by various authors, who use different
techniques and sometimes do not specify the errors; moreover the
technique and the apparatus used at Ancona University[6] are quite
similar to those used at NEL, so that the experimental thermal conductivity values can be assumed to be accurate to within ± 3%.[1]

EXPERIMENTAL APPARATUS

A relative method was adopted employing the hot-wire principle.
The conductivity cell consists of a thin platinum wire (0.1 mm diameter and 81 mm in length), axially disposed along a Pyrex glass
tube (9 mm internal diameter) and welded to thicker platinum leads
(0.6 mm diameter). The apparatus is used in the four leads arrangement so that the wire electrical resistance can be measured by a
Kelvin Bridge; with this arrangement, a cell is mechanically and
electrically more reliable, even if an increase of the end thermal
losses must be taken into account and evaluated analytically.[6] The
cell is immersed in a constant temperature bath, whose temperature
could be controlled during the experiments better than ± 0.001°C.
The experimental technique consists essentially in passing
various known currents through the wire and measuring the corresponding temperatures when the temperatures are steady. Moreover, the
cell must be calibrated because it is not possible to measure the
cell geometric dimensions accurately; the calibration fluids were
water (λ = 0.606 W/m K at 25°C), toluene (λ = 0.1296 W/m K at 25°C)
and n-Butanol (λ = 0.1480 W/m K at 25°C) and the cell constants
were optimized by the Powell Method.[7]

CORRELATION PROPOSED

In a previous paper,[8] the following correlation was proposed
for pure organic liquids:

$$\lambda = A \cdot \frac{(1 - T_r)^{0.38}}{T_r^{1/6}} \tag{1}$$

were T_r is the reduced temperature and A is a factor that is practically temperature independent and characteristic of each liquid. S.
I. units are used throughout.
Equation (1) was checked for 144 organic liquids in large temperature ranges (generally from the melting point to the normal boiling point and higher) and the mean general deviation between experimental and predicted thermal conductivity values was found to be

Δ = 1.5%, while the maximum deviations exceed 8% (but are less then 11%) for only seven compounds.

The factor A can be evaluated if a few selected experimental thermal conductivity data are available, and it is interesting to point out that only one experimental value of λ at $T_r \simeq 0.55$[8] can be used in the calculation of A; moreover equation (1) shows that A is equal to the value of the thermal conductivity at $T_r = 0.36$[9].

If selected experimental thermal conductivity values are not available, simple correlations were proposed[8] that relate A with the molecular weight, M, the normal boiling point, T_b, and the critical temperature, T_c, of the members of 10 different families.

The accuracy of equation (1) and the extent of its use, for both the number of the investigated compounds and the reduced temperature ranges explored, lead one to look for a generalization of the cited correlation.

The following considerations can be put forth:
1) A and T_c characterize completely, with respect to the thermal conductivity, the behaviour of each pure liquid;
2) If the mixture is considered in the liquid state, it is reasonable to expect, for given concentrations of the components, a corresponding factor, A_m, that is temperature independent and dependent only on the concentrations of the components.

At this point, before the problem of A_m is solved, another difficulty must be overcome: the expression of the critical temperature, T_{cm}, of a mixture. The mixture critical temperature usually is not experimentally known and the existing prediction methods[4] were conceived and checked almost exclusively for hydrocarbon mixtures or for mixtures of hydrocarbons with CO_2, H_2S, CO, and the permanent gases; in conclusion the estimation of T_{cm} generally leads to errors difficult to be appraised, so that in this work a linear mole-fraction average of the pure-component critical temperatures is used. Actually, the true critical temperature is usually not a linear mole-fraction average:

$$T_{cm} = \sum_i x_i T_{ci} \tag{2}$$

and the expression (2) is a pseudocritical temperature, that is, a quantity to be used when the corresponding states theory is applied.

With all this oversimplification, equation (2) is the most simple means to surmount the obstacle. Also, errors of 15 K in Tcm produce variations of $(1 - T_r)^{0.38}/T_r^{1/6}$ less than 2%. Now reasonable hypotheses have to be proposed in order to give an expression for A_m; the solution of this problem was pursued as follows:
1) The thermal conductivity of organic liquid mixtures generally is less than those predicted with a mole-(or weight-) fraction average;
2) The factor A_m should depend only on the factors A_1, A_2 of the mixture components and on the respective mole fractions x_1, x_2.

Starting from the proposed "quadratic mixing law" proposed by Mc Laughlin:[10]

$$\lambda_m = \lambda_1 \, x_1^2 + \lambda_2 \, x_2^2 + 2 \, \lambda_{12} \, x_1 \, x_2 \tag{3}$$

considered the most logical form of the thermal conductivity of a binary mixture,[1] if the temperature dependence is fully represented by the expression $(1 - T_r)^{0.38}/T_r^{1/6}$, the factor A_m may be written as follows:

$$A_m = A_1 \, x_1^2 + A_2 \, x_2^2 + A_{12} \, x_1 \, x_2 \tag{4}$$

where the "cross-terms coefficient" A_{12}, which has the same dimensions as A_m, A_1, and A_2, should depend only on A_1 and A_2.

According to the dimensions and to the form of equation (4), the more probable expression for A_{12} is:

$$A_{12} = 2\sqrt{A_1 \cdot A_2} \tag{5}$$

and effectively this happens for several of the investigated mixtures, but unfortunately not for all the mixtures. This complication probably can be ascribed to the errors made using the correlation (2) for the T_{cm} estimation, but it cannot be overcome; therefore, the best choice for equation (4) is:

$$A_m = A_1 \, x_1^2 + A_2 \, x_2^2 + n \cdot \sqrt{A_1 \cdot A_2} \, x_1 \, x_2 \tag{6}$$

where n changes from mixture to mixture and generally varies from 1.5 to 2.0.

In conclusion, the following correlation is proposed for organic liquid mixture thermal conductivity prediction:

$$\lambda_m = (A_1 \, x_1 + A_2 \, x_2 + n \cdot \sqrt{A_1 \cdot A_2} \, x_1 \, x_2) \cdot \frac{(1 - T_r)^{0.38}}{T_r^{1/6}} \tag{7}$$

where T_r is calculated as a "reduced temperature" using the pseudo-critical temperature given by equation (2).

Table I offers an example of the results obtained above and summarized. This Table, concerning the mixture carbon tetrachloride/toluene, contains the mole fractions of the mixture components and the A_m factors obtained by using the λ_m experimental data in equation (6) at different temperatures and shows that the factor A_m does not vary appreciably with the temperature; it depends only on the mole fractions of the mixture components. Table II contains the experimental thermal conductivity values obtained at Ancona University; the thermal conductivity values obtained at NEL are

Table I – Values of A_m obtained using the experimental thermal conductivity data[5] in equation (7) for the mixture carbon tetrachloride (A=0.123 W/m K, T_c=556.3 K)/Toluene (A=0.153 W/m K, T_c=594.0 K).

Mole Fractions	C.Tetr.	Toluene	C.Tetr.	Toluene	C.Tetr.	Toluene	C.Tetr.	Toluene	C.Tetr.	Toluene
	0.0	1.0	0.1665	0.8335	0.3746	0.6254	0.6425	0.3575	1.0	0.0
A_m exp (t= 0°C)	0.1530		0.1450		0.1348		0.1255		0.1230	
A_m exp (t=50°C)			0.1434		0.1319		0.1247			
A_m exp (t=65°C)			0.1424		0.1315		0.1248			

Table II – Experimental thermal conductivity at t=25°C obtained at the Ancona University for 4 binary mixtures; w is the weight-fraction of the first component, λ_m is the thermal conductivity of the mixture in W/m K.

w	λ_m	w	λ_m	w	λ_m	w	λ_m
Acetone/Ethanol		n-Butanol/Toluene		Benzene/Diethyl Ether		Acetone/Carbon Tetrachloride	
0	0.1603	0	0.1296	0	0.1278	0	0.0982
0.25	0.1617	0.25	0.1338	0.25	0.1312	0.25	0.1122
0.50	0.1602	0.50	0.1376	0.50	0.1384	0.50	0.1303
0.75	0.1622	0.75	0.1414	0.75	0.1412	0.75	0.1417
1	0.1600	1	0.1480	1	0.1433	1	0.1600

Table III – General table containing the mean $\Delta\%$ and the maximum $\Delta_{Max}\%$ deviations between experimental and estimated thermal conductivity data. The subscripts 1 and 2 in the factor A (W/m K) and in the critical temperature T_c (K) refer to the first and to the second component of each mixture.

Mixture	A_1	A_2	T_{c1}	T_{c2}	n eq.(6)	Temp. expl.°C	$\Delta\%$	$\Delta_{Max}\%$
Acetone/Aniline	0.203	0.173	508.7	766.4	2	0	1.2	2.1
Acetone/n-Butanol	0.203	0.181	508.7	536.0	2	0	2.1	2.8
Acetone/Carbon Tetrachloride	0.203	0.123	508.7	556.3	1.5	0/25	1.6	3.8
Acetone/Ethanol	0.203	0.206	508.7	516.2	2	25	0.9	1.5
Acetone/Methanol	0.203	0.250	508.7	513.2	2	0/50	1.4	2.3
Acetone/Toluene	0.203	0.153	508.7	594.0	2	0	1.4	1.8
Aniline/Nitrobenzene	0.173	0.151	766.4	782.0	2	0/75/150	2.3	4.4
Benzene/Diethyl Ether	0.172	0.179	562.1	465.8	2	25	1.1	2.6
Benzene/Methanol	0.172	0.250	562.1	513.2	2	0	1.5	4.7
Benzene/Toluene	0.172	0.153	562.1	594.0	2	0/50	1.1	2.3
n-Butanol/Carbon Tetrachloride	0.181	0.123	563.0	556.3	1.5	0/50/65	1.2	3.2
n-Butanol/Toluene	0.181	0.153	563.0	594.0	2	0/25	2.4	4.0
n-Butyl Acetate/Diethyl Ether	0.162	0.179	579.0	465.8	2	0	1.3	2.9
n-Butyl Acetate/Toluene	0.162	0.153	579.0	594.0	2	0	2.7	4.7
Carbon Tetrachl./di-n-But.Eth.	0.123	0.145	556.3	580.0	2	0	3.2	6.5
Carbon Tetrachl./Dichlorometh.	0.123	0.178	556.3	510.0	1.7	0	1.1	3.3
Carbon Tetrachl./Diethyl Ether	0.123	0.179	556.3	465.8	1.5	0	1.5	2.4
Carbon Tetrachl./n-Heptanol	0.123	0.169	556.3	633.0	1.7	0/50/65	1.6	4.3
Carbon Tetrachl./n-Hexanol	0.123	0.178	556.3	610.0	1.5	0/50	1.8	6.4
Carbon Tetrachl./Toluene	0.123	0.153	556.3	594.0	1.7	0/50/65	0.9	2.3
Chloroform/di-n-Butyl Ether	0.142	0.145	536.4	580.0	1.8	0	1.8	4.7
Chloroform/Diethyl Ether	0.142	0.179	536.4	465.8	1.6	0/-50	1.8	3.9
di-n-Butyl Ether/Methanol	0.145	0.250	580.0	513.2	1.0	0	3.9	8.3
Diethyl Ether/1,2-Dichloroeth.	0.179	0.161	465.8	561.0	2	0	1.7	4.1
Diethyl Ether/Methanol	0.179	0.250	465.8	513.2	1.6	0	1.8	4.1
Diethyl Ether/Nitrobenzene	0.179	0.151	465.8	782.0	2	0	1.4	4.2
Diethyl Ether/Toluene	0.179	0.153	465.8	594.0	2	0	0.9	1.6
n-Heptanol/Toluene	0.169	0.153	633.0	594.0	2	0	1.2	2.3
n-Hexanol/Toluene	0.178	0.153	610.0	594.0	2	0	3.5	6.4
Methanol/Toluene	0.250	0.153	513.2	594.0	1.5	0/50	2.6	7.4

Mean general deviation (on 260 experimental values at various temperatures and concentrations = 1.8%)

published in reference 5. Table III contains the comparison between experimental and predicted thermal conductivity values for all the 30 organic liquid mixtures investigated.

CONCLUSIONS

As can be observed in Table III, the mean deviations between predicted and experimental thermal conductivity values are generally less than 2% and the maximum deviations are generally less than 5%; therefore, the accuracy of the equation (7) is more than acceptable for engineering purposes. The main criticisms of equation (7) concern its empirical nature and the simplification assumed in the T_{cm} estimation, but some advantages can be emphasized:
1) Equation (7) does not need values of the thermal conductivity of the mixture components, but contains the temperature-independent factors A_1 and A_2;
2) Equation (7), if A_1 and A_2 are known (and suggestions were given above in this paper), can be used at every temperature. Finally, the obtained results lead one to believe that, if the true critical temperature is known, the factor n in equation (7) can be assumed constant and equal to 2.0.

REFERENCES

1. D. T. Jamieson, and E. H. Hastings, "Thermal Conductivity - Proceedings of Eighth Conference," Plenum Press, New York (1969), pp. 631-641.
2. L. P. Filippov, Int. J. Heat Mass Transfer, 11:331 (1968).
3. G. H. Shroff, "Thermal Conductivity - Proceedings of Eighth Conference," Plenum Press, New York (1969) pp. 643-657.
4. R. C. Reid, J. M. Prausnitz, and T. K. Sherwood, "The Properties of Gases and Liquids," McGraw-Hill, New York (1977), pp. 531-537.
5. D. T. Jamieson, J. B. Irving, and J. S. Tudhope, "Liquid Thermal Conductivity, a Data Survey to 1973," Her Majesty's Stationery Office, Edinburgh (1975).
6. C. Baroncini, P. Di Filippo, G. Latini, and M. Pacetti, "Memorie del XXXV Congresso Nazionale A.T.I., CELID," La Grafica Nuova, Turin, Italy, (1980), pp. 217-228.
7. D. T. Jamieson, and J. B. Irving, NEL Report N. 609, National Engineering Laboratory, (1976).
8. C. Baroncini, P. Di Filippo, G. Latini, and M. Pacetti, Int. J. Thermophys. 2:21 (1981).
9. G. Latini, "Abacus for Organic Liquids Thermal Conductivity Estimation in the Reduced Temperature Range 0.3 to 0.8," Report N. 1 T, Ancona University, (1981).
10. E. McLaughlin, Chem. Rev., 64(4):390 (1964).

SESSION   TC-8

Liquids and Gases - II

CHAIRMAN

H. M. Roder
National Bureau of Standards
Boulder, Colorado

THERMAL CONDUCTIVITY OF CARBON MONOXIDE IN THE TEMPERATURE RANGE

350–2230 K

R. Afshar, S. Murad, and S.C. Saxena

Department of Energy Engineering
University of Illinois at Chicago Circle
Chicago, Illinois 60680

ABSTRACT

The thermal conductivity of research-grade pure carbon monoxide is measured by employing a hot-wire type thermal diffusion column at pressures of 88.0, 66.7, and 33.0 kPa and over the temperature range of 350 to 2230 K. This conductivity column method leads to values that have a maximum probable error of 0.95 percent at 2207 K, and it increases in magnitude to 2.85 percent as the temperature decreases to 388 K. The experimental data are correlated by the following cubic polynomial in temperature on the basis of least-squares analysis.

$$k(T) = -1.351 \times 10^{-2} + 9.733 \times 10^{-4}T - 4.358 \times 10^{-7}T^2 + 1.239 \times 10^{-10}T^3$$

Here k is in mW/cm.K and T is in K.

The experimental thermal conductivity values are compared with the predictions based on three kinetic theory expressions and a Lennard-Jones (12-6) intermolecular potential.

INTRODUCTION

There is considerable interest in the thermophysical properties of carbon monoxide because of its presence in combustion gases. Carbon monoxide is a weakly polar compound having a dipole moment[1] of 0.112×10^{-18} esu.cm and a quadrupole moment[2] of -2.2×10^{-26} esu.cm^2. However, since the dipole moment of carbon monoxide is rather small, it is generally regarded as a spherical molecule while describing its transport properties. Touloukian, Liley , and Saxena[3] have examined the four sets of available experimental data

295

for carbon monoxide and have recommended values for the range
80-1250 K. They have reported the estimated error in their recommended values to be up to 7 percent in the temperature range
1100-1250 K. Data above 1250 K are nonexistent for carbon monoxide.

In this paper we present the thermal conductivity of gaseous
carbon monoxide in the temperature range 350-2230 K and at pressures of 88.0, 66.7, and 33.3 kPa, where the thermal conductivity
is found to be independent of pressure. The experimental values
were represented by a cubic polynomial in temperature. In addition,
we examined several currently used theories for thermal conductivity and compared our experimental results with values calculated
using these theories. Finally, we employed the theory of Ahtye to
estimate the vibrational diffusion coefficient on the basis of our
experimental results.

EXPERIMENTATION

The measurements were made on 99.99 percent pure carbon monoxide supplied by Matheson Gas Products. The conductivity column
employed was similar in design to that of Chen and Saxena[4]. It was
provided with a 62-cm-long, 0.508-mm (20 mil)-diameter tungsten
wire. The tungsten wire, in addition to heater, was also used as
its own thermometer and was carefully calibrated according to the
procedure outlined by Chen and Saxena[4]. It involved the measurement of electrical resistance of a given sample of annealed wire
at the ice point (0.00°C), and the melting points of tin (231.9°C),
zinc (419.51°C), and aluminum (660.37°C) in a Leeds and Northrup
8411 fixed temperature facility. The tungsten wire was thermally
annealed in vacuum by heating in steps to a maximum temperature of
2400 K. Before each experiment the wire was heated to 2400 K for
fifteen minutes. The outer wall of the column was maintained at
316 K by circulating constant temperature water.

Measurements were made at three pressures, viz., 88.0, 66.7,
and 33.3 kPa as a function of temperature. The electrical power
required to heat the gas was found to be independent of pressure,
which conclusively establishes that convective heat transfer is
insignificant for these operating conditions and column geometry.
We also calculated the thermal conductivity at each pressure level
as a function of temperature from the heat transfer rate data and
found that the three sets of conductivity values do not differ
significantly from each other.

These raw experimental conductivity values were corrected for
the small temperature drop across the glass column wall, the change
in wire diameter due to changes in its length, and fluctuations in
the cold wall temperature along its length.[4] The total magnitude
of these three corrections was found to be always less than 0.76
percent.

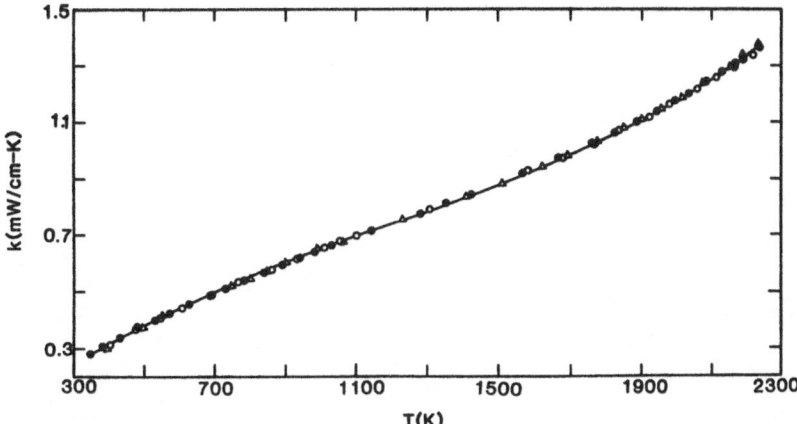

Figure 1. Thermal conductivity of carbon monoxide as a function
of temperature. 0 88.0, Δ 66.7, 33.0 kPa. The solid
line is based on eq. (1).

The three sets of experimental data are plotted in Figure 1.
A detailed error analysis gave a maximum uncertainty of 1.0 to 2.9
percent in this temperature range. Specifically, the errors are
2.85, 1.72 and 0.95 percent at 388, 1034, and 2207 K, respectively.
The experimental values were finally fitted to a cubic polynomial in
temperature using the method of least squares and giving equal
weight to all the data points. The final result is:

$$k(T) = -1.351 \times 10^{-2} + 9.7328 \times 10^{-4}T - 4.3576 \times 10^{-7}T^2 + 1.2385 \times 10^{-10}T^3$$

(1)

Here k is in mW/cm.K and T is in K. This correlation reproduces the
experimental results within an average absolute deviation of 0.28
percent and a maximum deviation of 1.8 percent. Carbon monoxide
does not dissociate at these temperatures. The deviations of the
three sets of data from predictions based on equation (1) are shown
in Figure 2. These deviations are always less than the estimated
errors in the data also shown by curve 4 in Figure 2.

The present conductivity data, as represented by equation (1)
are compared in Figure 3 with similar data of Geier and Schäfer,[5]
Saxena and Gupta,[6] Gardiner and Schäfer,[7] and the values recommended
by Touloukian et al.[3] The data of Geier and Schäfer,[5] and Touloukian
et al.[3] are within the estimated errors of our measurements.
However, the data of Gardiner and Schäfer[7] underestimate the thermal

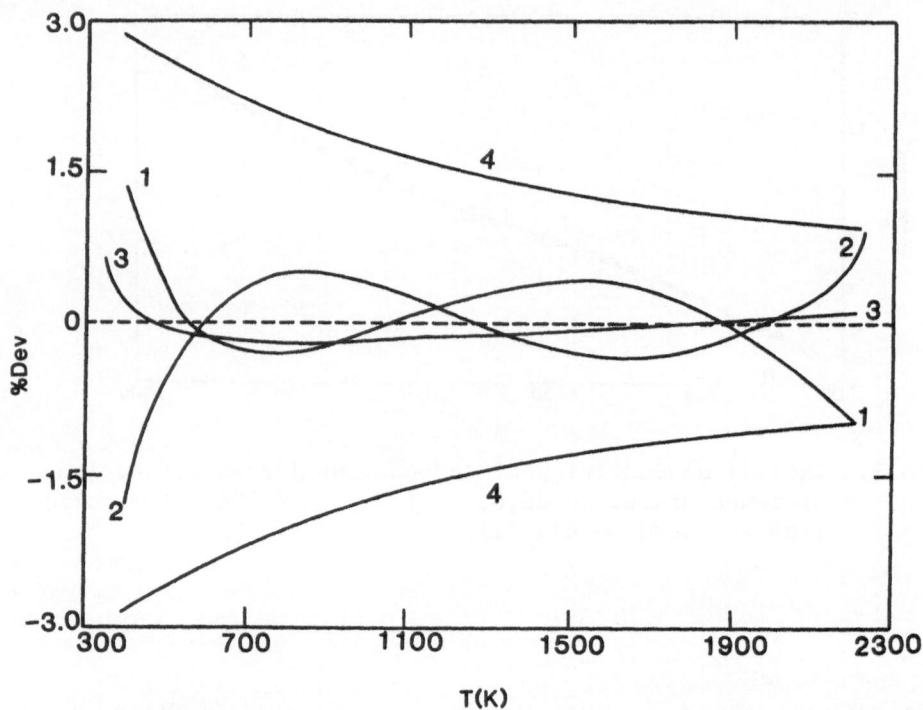

Figure 2.   Deviation curves for the thermal conductivity of carbon
monoxide from the predictions based on eq. (1). % Dev =
$100 \left[k_{exp} - k \text{ (eq. 1)} \right] / k_{exp}$. Curves 1, 2, and 3 refer
to pressures of 33.0, 66.7, and 88.0 kPa respectively.
Curve 4 represents the probable errors in our measurements.

conductivity by approximately 5 percent when compared with our
present data.   The reported values of Saxena and Gupta[6] are in
agreement with our experimental results up to about 700 K, but
above this temperature they deviate from our present data by as
much as 8 percent.   We believe the data of Gardiner and Schäfer[7]
and Saxena and Gupta[6] (over 700 K) to be in error.   This belief is
confirmed by the analysis of Touloukian et al.[3]

THEORETICAL INTERPRETATION

        Several theories have been proposed in recent years for cal-
culating the thermal conductivity of polyatomic gases.   In this
section we examine the validity of some of these theories by comparing them with our experimental results.

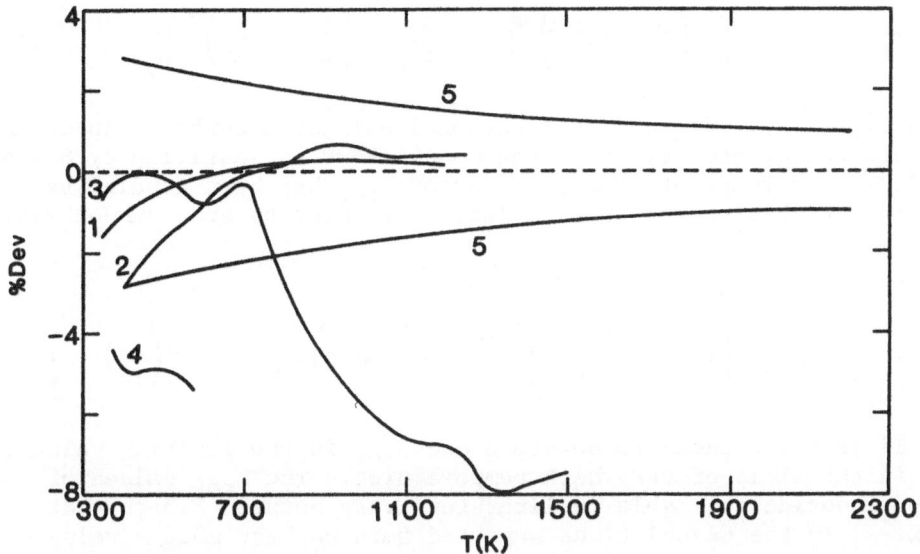

Figure 3. Comparison of the present data (eq. 1), with the
           measured values of (1) Geier and Schäfer;[5] (2) Touloukian
           et al;[3] (3) Saxena and Gupta;[6] (4) Gardiner and Schäfer.[7]
           Curve 5 represents the probable error in our measurements.

The thermal conductivity of gases is generally discussed in
terms of a dimensionless ratio, f, defined as

$$f = k \, M/\eta C_v \tag{2}$$

where M is the molecular weight, η is the viscosity, and C_v is the
constant volume specific heat. Eucken[8] proposed that f is given by,

$$f_E = 1 + (9R/4C_v) \tag{3}$$

Here R is the gas constant. Hirschfelder[9] later suggested the form

$$f_H = \frac{\rho D}{\eta} + \left(\frac{3R}{2C_v}\right)\left(\frac{5}{2} - \frac{\rho D}{\eta}\right) \tag{4}$$

where D is the self-diffusion coefficient and ρ is the density.
Mason and Monchick[10] separated the translational and internal contributions to thermal conductivity and suggested the following
expression for f[11]

$$f_M = \frac{\rho D}{\eta} + \frac{3}{2}\left(\frac{5}{2} - \frac{\rho D}{\eta}\right)\frac{R}{C_v} - \frac{3}{\pi Z_{rot}}\left(\frac{5}{2} - \frac{\rho D}{\eta}\right)^2 \frac{R}{C_v} \qquad (5)$$

where Z_{rot} is defined as the rotational collision number. In eq. 5 D replaces D_{int} used by Mason and Monchick,[10] as justified by Saxena et al.[12] The temperature dependence of Z_{rot} has been studied by Parker[13] and his expression, as later corrected by Brau and Jonkman,[14] is,

$$Z_{rot} = Z_{rot}^{\infty}\left[1 + \frac{\pi}{2}^{3/2}\left(\frac{1}{T*}\right)^{1/2} + \left(\frac{\pi^2}{4} + 2\right)\frac{1}{T*} + \pi^{3/2}\left(\frac{1}{T*}\right)^{3/2}\right]^{-1} \qquad (6)$$

Here $T*$ is the reduced temperature and Z_{rot}^{∞} is the limiting value of Z_{rot} in the limit of very high temperatures. The Z_{rot} values of carbon monoxide available in literature vary between 3.5 to 4 at 400 K.[15] In the calculations reported here we have used a value of 4.0. The thermal conductivity values are, however, quite insensitive to the value of Z_{rot}. A 50 percent change in Z_{rot} changes the thermal conductivity values by no more than 1.5 percent.

To employ the theories outlined above, values of η, D, and C_v are needed. Values for C_v are given by Svehla,[16] and η and D are calculated on the basis of Lennard-Jones (12-6) potential and the kinetic theory expressions.[17] The parameters of the Lennard-Jones potential for carbon monoxide were obtained by matching the calculated viscosity with the experimental values, as recommended by Touloukian et al.[18] The parameters obtained are: ε/k_B = 94.1 K, and σ = 3.685 x 10^{-10} m. Here k_B is the Boltzmann constant, ε is the well depth of the intermolecular potential, and σ is the molecular separation at which the potential is zero. The potential is tested by using it to calculate the self-diffusion coefficient between 150 and 2100 K and comparing these results with the available experimental values for the self-diffusion coefficient.[19] As can be seen from Figure 4, the agreement with the experimental results is satisfactory in the temperature range where experimental values are available. The calculated values of η and D are then used to calculate the thermal conductivity of carbon monoxide using the three theories outlined above. The results are shown in Figure 5. As can be seen, systematic deviation between theory and experiment remain, although the theory of Mason and Monchick[10] appears to be relatively more successful in reproducing these experimental data. This could be because the simple Lennard-Jones potential is unsuitable for carbon monoxide or because of the deficiencies that exist in these theories as a result of simplifying assumptions employed in their development.

More recently Ahtye[20] proposed a theory to overcome some of the

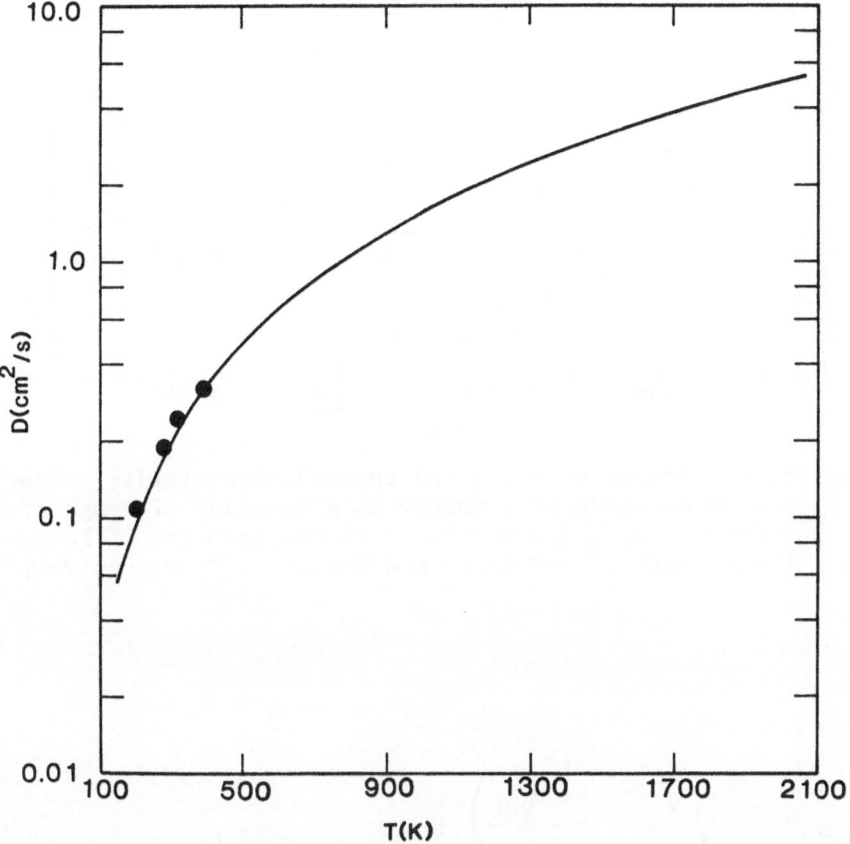

Figure 4. The self-diffusion coefficient, D, as a function of
temperature, T.   experimental data of Amdur and Shuler.[19]

shortcomings of the previous theories.[10,21] He suggested separating
the internal thermal conductivity as:

$$k = k_{tr} + k_{rot} + k_{vib} \tag{7}$$

$$= \frac{5\eta\, C_{v,tr}\, h_{tr}}{2M} + \frac{\rho\, D_{rot}\, C_{v,rot}\, h_{rot}}{M} + \frac{\rho\, D_{vib}\, C_{v,vib}}{M} \tag{8}$$

where

$$h_{tr} = 1 - \frac{10}{3\pi}\left(1 - \frac{2\rho\, D_{rot}}{5\eta}\right) \frac{C_{v,rot}}{R\, Z_{rot}} \tag{9}$$

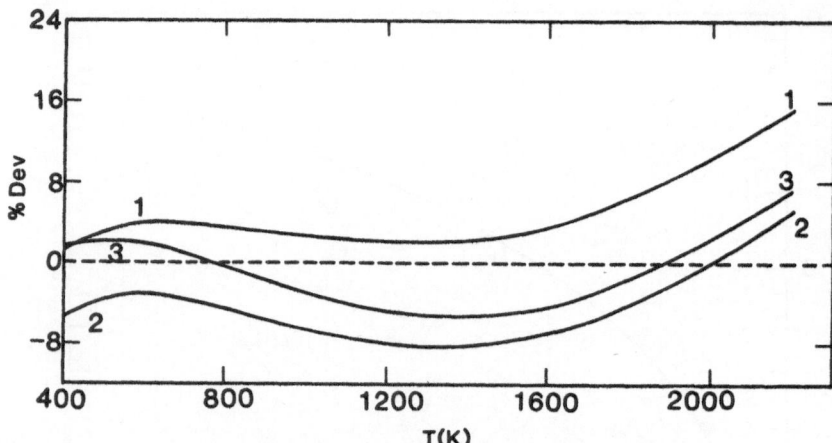

Figure 5. Comparison of experimental thermal conductivity values of
carbon monoxide with theory as a function of temperature.
Curves 1, 2, and 3 are based on the theories of Eucken,[8]
Hirschfelder,[9] and Mason and Monchick,[10] respectively.

and

$$h_{rot} = 1 + \frac{5}{\pi} \left(1 - \frac{2\rho \, D_{rot}}{5\eta} \right) \frac{1}{Z_{rot}} \qquad (10)$$

The subscripts tr, rot, and vib refer to the translational, rotational, and vibrational contributions, respectively. In addition,
in the present calculations we have assumed $D_{rot} = D$, $C_{v,rot} =$
R, $C_{v,vib} = C_v - (5/2)R$, and $C_{v,tr} = (3/2)R$. Experimental values
of k for carbon monoxide and equations (8) to (10) can then be
manipulated to calculate the vibrational diffusion coefficient,
D_{vib}. In Figure 6, $\rho \, D_{vib}$ and D_{vib}/D have been plotted as a function
of temperature. The minimum value of ρD_{vib} is around 1000 K, and
the ratio D_{vib}/D has a minimum value around 1200 K. Similar calculations for several other gases, such as carbon dioxide,[22] sulfur
dioxide,[23] and ammonia,[11] have been reported by us earlier.

The Prandtl number ($C_p\eta/k$, where C_p is the constant pressure
specific heat) and the Schmidt number ($\eta/\rho D$) are widely used in
engineering correlations for mass and heat transfer. In Table 1
we have listed values of Schmidt and Prandtl numbers in the range
150 - 2200 K. Values of viscosity and self-diffusion coefficients
are obtained from kinetic theory expressions, while the experimental
thermal conductivity values are used.

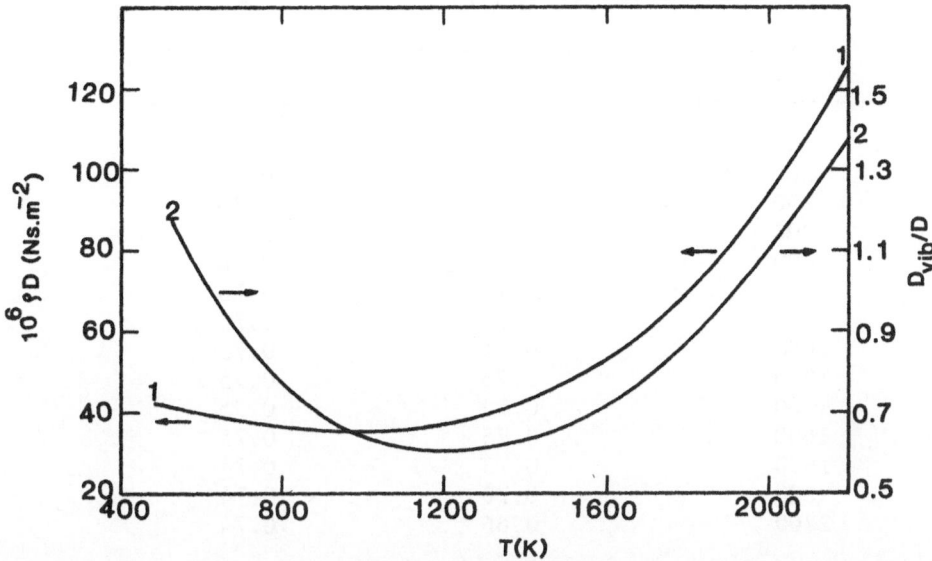

Figure 6. Variation of ρD_{vib} and D_{vib}/D with temperatures for carbon monoxide as obtained from Ahtye's theory.[20]

CONCLUSIONS

We have presented here accurate measurements of the thermal conductivity of carbon monoxide over a wide temperature range, 350 - 2230 K. It is found that none of the kinetic theories can accurately predict the thermal conductivity of carbon monoxide over the entire temperature range with one set of molecular potential parameters. There is, therefore, a need for improving the current theories for thermal conductivity while still keeping them tractable. The concept of separating the internal thermal conductivity part into rotational and vibrational components with a characteristic vibrational energy diffusion coefficient appears to be an attractive one and has been substantiated by our recent work on several polyatomic gases.

ACKNOWLEDGMENTS

The authors are thankful to National Science Foundation (Grant No. ENG-7917620) and to American Chemical Society (Petroleum Research Fund) for partly supporting this research.

Table 1.   Schmidt and Prandtl Numbers for Carbon
           Monoxide

Temperature (K)	Prandtl Number	Schmidt Number
150		0.76
200		0.76
250		0.76
300		0.76
400	0.74	0.76
600	0.73	0.76
800	0.73	0.75
1000	0.74	0.75
1200	0.75	0.75
1400	0.75	0.75
1600	0.75	0.75
1800	0.73	0.74
2000	0.70	0.74
2200	0.66	0.74

REFERENCES

1.   D.E. Stogryn and A.P. Stogryn, Molecular Multipole Moments,
     Mol. Phys. 11: 371 (1966).
2.   W.H. Flygare and R.C. Benson, The Molecular Zeeman Effect in
     Diamagnetic Molecules and the Determination of Molecular
     Magnetic Moments (g values), Magnetic Susceptibilities, and
     Molecular Quadrupole Moments, Mol. Phys. 20: 225 (1971).
3.   Y.S. Touloukian, P.E. Liley and S.C. Saxena, "Thermal Con-
     ductivity: Nonmetallic Liquids and Gases," Plenum, New York
     (1970).
4.   S.H.P. Chen and S.C. Saxena, Experimental Determination of
     Thermal Conductivity of Nitrogen in the Temperature Range 100 -
     2200 °C, High Temp. Sci. 5: 206 (1973).
5.   H.Geier and K. Schäfer, Heat Conductivity of Pure Gases and
     Gas Mixtures, Alg. Warm. 10: 70 (1961).
6.   S.C. Saxena and G.P. Gupta, The Column Method of Measuring
     Thermal Conductivity of Gases: Results on Carbon Monoxide and
     Oxygen in the Temperature Range 350° to 1500 °K, Prog.
     Astronaut. Aeronaut. 23: 45 (1970).
7.   W.C. Gardiner and K. Schäfer, Transportphänomene in Gasen und
     Zwischenmolekulare Kräfte, Z. Elektrochem. 60: 588 (1956).
8.   A. Eucken, The Thermal Conductivity, the Specific Heat and the
     Viscosity of Gases, Phys. Z. 14: 324 (1913).
9.   J.O. Hirschfelder, Heat Conductivity in Polyatomic or

Electronically Excited Gases II, J. Chem. Phys. <u>26</u>: 282 (1957).

10. E.A. Mason and L. Monchick, Heat Conductivity of Polyatomic and Polar Gases, J. Chem. Phys. 36: 1622 (1962).

11. R. Afshar, S. Murad and S.C. Saxena, Thermal Conductivity of Gaseous Ammonia in the Temperature Range 358 - 925 K, Chem. Eng. Commun. 10: 1 (1981).

12. S.C. Saxena and J.P. Agrawal, Thermal Conductivity of Polyatomic Gases and Relaxation Phenomena, J. Chem. Phys. 35: 2107 (1961).

13. J.G. Parker, Rotational and Vibrational Relaxation in Diatomic Gases, Phys. Fluids 2: 449 (1959).

14. C.A. Brau and R.M. Jonkman, Classical Theory of Rotational Relaxation in Diatomic Gases, J. Chem. Phys. 52: 477 (1970).

15. N.C. Petrellis and T.S. Storvick, Rotational Relaxation Numbers and Heat Conductivities from Thermal Transpiration Measurements on Argon, Carbon Monoxide, Carbon Dioxide, Oxygen and Sulfur Dioxide to 1240 K, Thirteen Int. Conf. Thermal Cond., Program and Abstract, p. 49, 1973, Sponsored by University of Missouri, Rolla, Missouri.

16. R.A. Svehla, Estimated Viscosities and Thermal Conductivities of Gases at High Temperatures, NASA TR, R-132 (1962).

17. J.O. Hirschfelder, C.F. Curtiss and R.B. Bird, "Molecular Theory of Gases and Liquids," Wiley, New York (1954).

18. Y.S. Touloukian, S.C. Saxena and P. Hestermans, "Viscosity," Plenum, New York (1975).

19. I. Amdur and L.M. Shuler, Diffusion Coefficients of the System CO-CO and CO-N_2, J. Chem. Phys. 38: 188 (1963).

20. W.F. Ahtye, Thermal Conductivity in Vibrationally Excited Gases, J. Chem. Phys. 57: 5542 (1972).

21. S.C. Saxena, M.P. Saksena and R.S. Gambhir, The Thermal Conductivity of Non-Polar Polyatomic Gases, Br. J. Appl. Phys. 15: 843 (1964).

22. S.H.P. Chen, P.C. Jain and S.C. Saxena, Thermal Conductivity and Effective Diffusion Coefficient for Vibrational Energy: Carbon Dioxide (350 - 2000 K), J. Phys. B. 8: 1962 (1975); ibid., 9: 1839 (1976).

23. R. Afshar, A. Alimadadian and S.C. Saxena, Thermal Conductivity of Sulfur Dioxide and Thermal Accommodation Coefficient for Sulfur Dioxide on a Gas-Covered Platinum Surface as a Function of Temperature, High Temp. Sci. 11: 79 (1979).

ABSOLUTE MEASUREMENT OF THE THERMAL CONDUCTIVITY OF ELECTRICALLY CONDUCTING LIQUIDS BY THE TRANSIENT HOT-WIRE METHOD (THERMAL CONDUCTIVITY OF AN AQUEOUS NaCl SOLUTION AT HIGH PRESSURE)

Y. Nagasaka and A. Nagashima

Department of Mechanical Engineering
Keio University
Yokohama, 223, JAPAN

ABSTRACT

An instrument for precise and absolute measurement of the thermal conductivity of electrically conducting liquids using the transient hot-wire method has been developed. A platinum wire coated with a thin polyester layer has been used as a line heat source. The thermal conductivity of an aqueous NaCl solution has been measured over the temperature range 0 - 80°C, pressure range 0.1 - 40 MPa, at a concentration of 1 molality, together with pure water. The experimental results have an estimated accuracy of $\pm$ 0.5%.

INTRODUCTION

Precise data on the thermodynamic and transport properties for the geothermal brines and sea water are required in the development and utilization of geothermal and ocean thermal energies. Thermophysical properties data for aqueous NaCl solutions, which can be considered as model substances of geothermal brines and sea water in engineering, such as density and viscosity data, have been reported in wide temperature, pressure, and concentration ranges[1, 2]. In case of the thermal conductivity, however, there exist only a few reliable data in a restricted temperature and concentration range. Besides, no data are available on the effect of pressures on the thermal conductivity of aqueous NaCl solutions[3].

We have developed a precise method for the measurement of the thermal conductivity of electrically conducting liquids based on the principle of the transient hot-wire technique[4]. The purpose

307

of the present study is to obtain precise data for the thermal conductivity of aqueous NaCl solutions in wide temperature, concentration, and especially pressure ranges using this method. In this
paper we report measurements of the thermal conductivity of an
aqueous NaCl solution over the temperature range 0 - 80°C, pressure
range 0.1 - 40 MPa, at a concentration of 1 molality and pure water
as a part of the complete concentration range.

PRINCIPLE OF MEASUREMENT

    To measure the thermal conductivity of an electrically conducting
liquid by the transient hot-wire method, a metallic wire coated with
a thin electrical insulation layer has been used both as a heating
element and a resistance thermometer instead of a bare metallic wire.
The effects on the measurement caused by the thin insulation layer

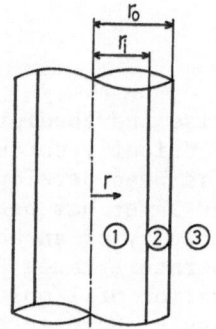

Figure 1.
The coordinate system
of an insulated wire.
(1) Metallic wire;
(2) insulator;
(3) sample liquid.

can be analyzed by solving the Fourier equations according to the
coordinate system shown in figure 1 using appropriate initial and
boundary conditions. The detailed analysis has been described
elsewhere[4]; we therefore introduce here only the final form of the
analysis.

    The average temperature rise of the metallic wire, $\Delta \bar{T}_1$, as a
function of time, t, is expressed as:

$$\Delta \overline{T}_1 = \frac{q}{4\pi\lambda_3}\left[\ln\frac{4\kappa_3 t}{r_o{}^2 C} + \frac{2\lambda_3}{\lambda_2}\ln\frac{r_o}{r_i} + \frac{\lambda_3}{2\lambda_1} + \frac{1}{t}\left\{\frac{r_i{}^2}{8}\left[\left(\frac{\lambda_3 - \lambda_2}{\lambda_1}\right)\left(\frac{1}{\kappa_1} - \frac{1}{\kappa_2}\right)+ \right.\right.\right.$$

$$\left. \frac{4}{\kappa_2} - \frac{2}{\kappa_1}\right] + \frac{r_o{}^2}{2}\left(\frac{1}{\kappa_3} - \frac{1}{\kappa_2}\right)+ \frac{r_i{}^2}{\lambda_2}\left(\frac{\lambda_2}{\kappa_2} - \frac{\lambda_1}{\kappa_1}\right)\ln\frac{r_o}{r_i} + $$

$$\left.\left. \frac{1}{2\lambda_3}\left[r_i{}^2\left(\frac{\lambda_2}{\kappa_2} - \frac{\lambda_1}{\kappa_1}\right)+ r_o{}^2\left(\frac{\lambda_3}{\kappa_3} - \frac{\lambda_2}{\kappa_2}\right)\right]\ln\frac{4\kappa_3 t}{r_o{}^2 C}\right\}\right] \qquad (1)$$

Where λ is the thermal conductivity, κ the thermal diffusivity, r the radial coordinate measured from the center of the wire, q the heat generated per unit length of the wire, r_i the radius of the metallic wire, r_o the radius of the insulated wire, and $C = \exp\gamma = 1.781\cdots$. The suffixes denote each material according to figure 1. Equation (1) is rewritten using A, B, and C, which become constant, fixing the insulated wire and measured substance as follows:

$$\Delta \overline{T}_1 = \frac{q}{4\pi\lambda_3}\left[\ln t + A + \frac{1}{t}(B\ln t + C)\right] \qquad (2)$$

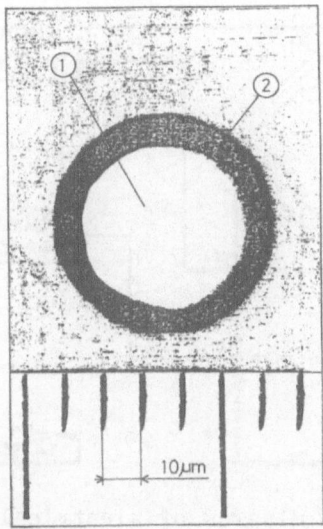

Figure 2. Cross section of the insulated wire.
(1) Platinum wire; (2) polyester layer.

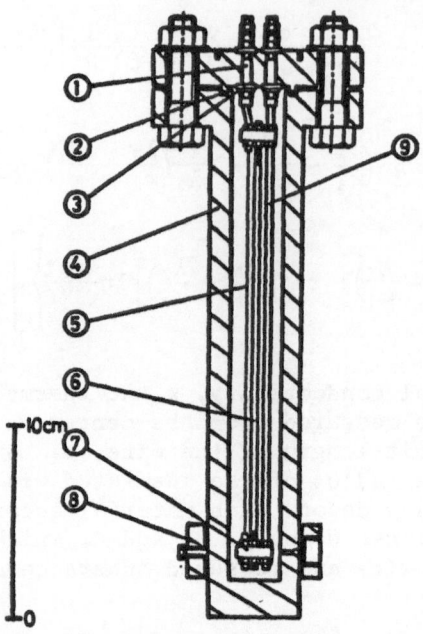

Figure 3. The hot-wire cell assembly. (1) Terminal; (2) Cu packing; (3) Teflon packing; (4) pressure vessel; (5) insulated Pt wire; (6) SUS rod; (7) ABS disk; (8) grand retaining ring; (9) insulated Cu rod.

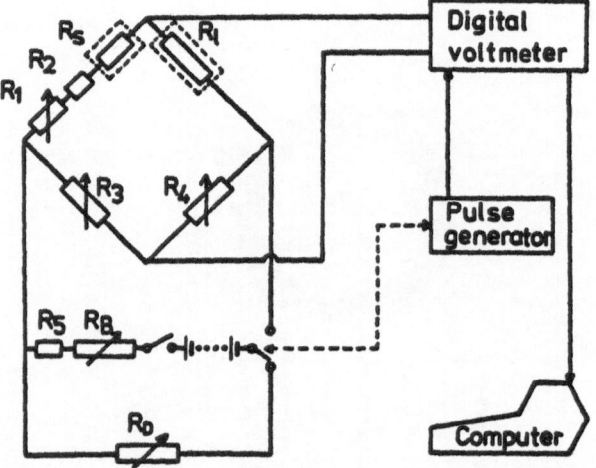

Figure 4. Block diagram of electrical system.

In the present measurement, the deviation from a linear relationship of $\Delta\bar{T}_1$ vs ln t, owing to the 1/t term in equation (2), is small enough (less than 0.05% of $\Delta\bar{T}_1$) that the thermal conductivity of the liquid, λ_3 is calculated by the following equation.

$$\lambda_3 = \frac{q}{4\pi} \bigg/ \frac{d\Delta\bar{T}_1}{d\ln t} \tag{3}$$

EXPERIMENTAL APPARATUS

Figure 2 shows the cross section of the insulated wire used in the present apparatus. This wire consists of platinum, 40 μm ± 1.5% in diameter①, and a thin polyester electrical insulation layer②, 7.5 μm ± 20% in thickness.

The hot-wire cell assembly is given in figure 3. The cell is designed for high pressure measurements up to 50 MPa. Two cells of this type differing only in their length have been constructed to compensate the wire-end effects. The length of the wires is about 200 mm and 100 mm, respectively. The inside portion of the terminals and other metallic parts, exposed to sample liquid are painted with thin silicon rubber for insulation.

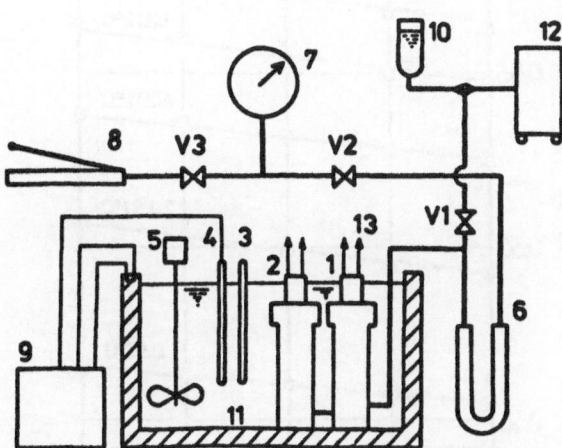

Figure 5. Pressurizing system with thermostat. (1) Cell (short wire); (2) cell (long wire); (3) standard thermometer; (4) thermocouple; (5) stirrer; (6) separator; (7) pressure gauge; (8) hand pump; (9) automatic temperature controller; (10) filling buret; (11) thermostatic bath, (12) vacuum pump; (13) to electrical circuit.

Figure 4 shows a block diagram of the electrical system. The temperature rise of the wire, $\Delta \overline{T}_1$, is calculated from transient voltages which are measured as the out-of-balance of the bridge circuit. Measured transient voltages are recorded and reduced to the thermal conductivity values by a desk-top computer.

The pressurizing system with thermostat is shown in figure 5. Both cells are immersed in a liquid thermostatic bath attaining temperature stability and a vertical temperature difference of no more than 0.05 K. Details of the filling and pressurizing system are omitted since they are quite standard in nature.

RESULTS

The aqueous NaCl solutions used for the present measurements were prepared by weight with reagent-grade NaCl (stated purity of 99.9%) and twice-distilled, ion-exchanged water. The density of a sample solution was measured before and after the measurement at room temperature by a precision pycnometer. The agreement between

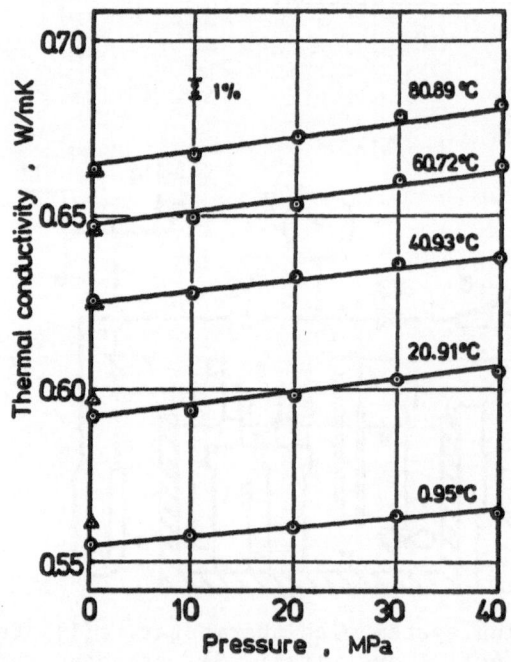

Figure 6. Pressure dependence of the thermal conductivity of aqueous NaCl solution (0.9995) molality). ⊙ = present study △ = reference [3].

measured values of the density before and after the measurements was good and the influence of corrosion on the density was checked.

The experimental results are listed in Tables 1 and 2. Each datum in the tables is the average value of three measurements at the same temperature and pressure. The available measuring time in one run was 1 - 5 s, and the temperature rise of the surface of the insulation layer was less than 1 K. The accuracy of the present measurements was estimated to be ±0.5% over the entire temperature

Table 1. Experimental results (H_2O).

T (°C)	P (MPa)	λ (W/mK)
0.94	0.10	0.5598
0.93	10.00	0.5644
0.93	20.00	0.5689
0.93	30.00	0.5721
0.93	40.00	0.5788
20.91	0.10	0.5997
20.90	10.00	0.6030
20.90	20.00	0.6078
20.89	30.00	0.6133
20.89	40.00	0.6171
40.93	0.10	0.6335
40.93	10.00	0.6378
40.91	20.00	0.6431
40.91	30.00	0.6459
40.90	40.00	0.6506
60.71	0.10	0.6548
60.74	10.00	0.6582
60.70	20.00	0.6630
60.67	30.00	0.6670
60.68	40.00	0.6733
80.86	0.10	0.6723
80.85	10.00	0.6760
80.86	20.00	0.6810
80.83	30.00	0.6855
80.85	40.00	0.6899

Table 2. Experimental results (aqueous NaCl solution, 0.9995 molality).

T (°C)	P (MPa)	λ (W/mK)
0.96	0.10	0.5556
0.95	10.00	0.5575
0.95	20.00	0.5603
0.95	30.00	0.5637
0.94	40.00	0.5667
20.92	0.10	0.5916
20.91	10.00	0.5943
20.91	20.00	0.5987
20.90	30.00	0.6034
20.90	40.00	0.6068
40.95	0.10	0.6253
40.94	10.00	0.6283
40.93	20.00	0.6327
40.93	30.00	0.6357
40.92	40.00	0.6390
60.74	0.10	0.6473
60.73	10.00	0.6492
60.72	20.00	0.6535
60.71	30.00	0.6595
60.71	40.00	0.6638
80.89	0.10	0.6634
80.88	10.00	0.6682
80.89	20.00	0.6724
80.88	30.00	0.6780
80.88	40.00	0.6807

and pressure range. The details of this estimation are given in ref. [4].

Figure 6 shows the pressure dependence of the thermal conductivity of the aqueous NaCl solution (0.9995 molality). Average deviation of the present results is ±0.3% while that of reference [3] is ±2%. As shown in this figure, the rate of increase of the thermal conductivity of this solution is almost equal to that of H_2O.

A further report on experiments of higher concentrations is reserved for a future data.

ACKNOWLEDGMENT

The authors wish to thank H. Okada for his assistance in carrying out these experiments.

REFERENCES

1. Rowe, A. M. and Chou, J. C. S., J. Chem. Eng. Data 15:61 (1970).
2. Kestin, J., Khalifa, H. E., Abe, Y., Grimes, C. E., Sookiazian, H. and Wakeham, W. A., J. Chem. Eng. Data 23:328 (1978).
3. Ozbek, H. and Phillips, S. L., J. Chem. Eng. Data 25:263 (1980).
4. Nagasaka, Y. and Nagashima, A., J. Phys. E: Sci. Instrum. (to be published).

MEASUREMENTS OF THE CONDUCTION OF HEAT IN WATER VAPOR, NITROGEN AND MIXTURES OF THESE GASES IN A WIDE KNUDSEN-NUMBER-RANGE

A. Frohn and M. Westerdorf

Institut für Thermodynamik der Luft- und Raumfahrt
Universität Stuttgart
Stuttgart, Federal Republic of Germany

ABSTRACT

The conduction of heat was measured in water vapor, nitrogen, and in mixtures of water vapor and nitrogen in the pressure range from 10 kPa to 1 Pa and for temperatures between 300 K and 700 K. Results of these experiments are important for the theoretical investigations of the transport properties of polyatomic gases.

The experiments were performed with a cylindrical heat-transfer cell having an inner diameter of 2 R_I = 0.1 mm and an outer diameter of 2 R_{II} = 40 mm. The length of the measuring section is H = 402 mm. In the first series of experiments for temperatures up to 450 K, an oil thermostat was used to keep the temperature of the outer cylinder constant, pumping the oil through a stainless steel jacket of the heat-transfer cell. To obtain higher temperatures, a new heat-transfer cell was built, whose outer cylinder was heated with electrical heating tapes. To maintain constant temperatures in the measuring section, the heating tapes were divided into three sections. A set of seven thermocouples and one resistance thermometer were used as input for the electronic regulating system and for a PDP 11 computer that controlled the temperature stability in the wall and the heat flux in the heating tapes. The temperature variations in the wall of the measuring section were less than 0.1 K. This small tolerance is needed to obtain the thermal conductivities with an accuracy of 1%. An automatic measuring system controlled by the PDP 11 computer has been developed for continuous sampling, processing, and storing of the data.

For this heat-transfer cell, the Knudsen number is defined as Kn = $\ell/(2R_I)$, where ℓ is the mean free path at the inner cylinder. In the experiments, a Knudsen number range of 10^{-2} < Kn < 10^2 was covered, so that measurements could be made over the entire range, from continuum conditions to the free molecule conditions.

Results of the heat conduction measurements in the pure gases and in the mixtures are presented. The experimental results for the heat flux of pure water vapor can be described throughout the range between continuum conditions and free molecule conditions by a simple mathematical relation. Thermal conductivities have been determined at continuum conditions for water vapor, nitrogen, and mixtures of these gases. The results of these measurements are in very good agreement with the tabulated data for the pure gases and the theoretical results of Chapman and Cowling for the mixtures.

INTRODUCTION

The conduction of heat in water vapor, nitrogen and mixtures of water vapor and nitrogen was measured in a cylindrical heat-transfer cell for pressures ranging from 10 kPa to 1 Pa. This corresponds to the Knudsen number range from $Kn = 10^{-2}$ to $Kn = 10^2$. Here the Knudsen number is defined as $Kn = \ell/(2R_I)$, where ℓ is the mean free path at the inner cylinder and R_I is the radius of the inner cylinder. The results of this paper are important for vacuum driers, for the heat-transfer of small droplets in fogs or cooling towers, for fuel injection into combustion chambers, and for high altitude flights.

BASIC EQUATIONS

Continuum Conditions

It is well known that the heat flux is independent of the density, ρ, for small Knudsen numbers. In this case with $Kn \ll 1$ the heat flux is given by the solution of the heat conduction equation, which for cylindrical geometry and small temperature difference is

$$Q_\infty = 2\pi H \bar{\lambda} (\theta_I - \theta_{II})/\ln(R_{II}/R_I) \tag{1}$$

Here the index I refers to the inner cylinder and the index II to the outer cylinder, R designates the radius of the cylinders, H is the length of the measuring section, $\theta_I - \theta_{II}$ is the temperature difference between inner and outer cylinders, and $\bar{\lambda} = (\lambda_I + \lambda_{II})/2$ is the mean heat conductivity. Equation (1) has been used to determine the heat conductivities.

Free Molecule Conditions

Free molecule conditions are obtained at large Knudsen numbers. For $Kn \gg 1$ the heat flux is given by Knudsen's formula, which Dushman and Lafferty (1962) have written in the form

$$Q_{FM} = 2\pi HR_I \left\{(\kappa+1)/(\kappa-1)\right\} \cdot (\alpha_I/4) \left\{2k/(\pi m\theta)\right\}^{1/2} \cdot p \cdot (\theta_I - \theta_{II}), \quad (2)$$

where k is the Boltzmann constant, m is the mass of a molecule, α_I is the thermal accommodation coefficient of the inner cylinder, p is the pressure, $\kappa = c_p/c_v$, c_p is the specific isobaric heat capacity, c_v is the specific isochoric heat capacity, and $\theta^{-1/2} = (\theta_I^{-1/2} + \theta_{II}^{-1/2})/2$. This equation can be used to determine the accommodation coefficient, α_I, when the heat flux, Q_{FM}, has been measured as a function of p. Introducing the viscosity, η, and the definition of the mean free path

$$\ell = \left\{\pi k\theta/(2m)\right\}^{1/2} \eta/p \quad (3)$$

the pressure, p, can be eliminated from eq. (2). Using the relation

$$\lambda = (15/4)(k/m) \cdot \eta \quad , \quad (4)$$

which is strictly valid for Maxwell molecules, and eq. (1), one obtains the heat flux in dimensionless form for $\kappa = 5/3$.

$$Q_{FM}/Q_\infty = \alpha_I R_I \ln(R_{II}/R_I)/(15\ell/4) \quad . \quad (5)$$

It has previously been shown in different papers that this equation is correct for monatomic gases (Dybbs and Springer, 1965, Yeh and Frohn, 1973). Vines and Bennet (1954) have suggested the following modification of eq. (4) for water vapor

$$\lambda = (16.9/4)(k/m) \cdot \eta \quad . \quad (6)$$

So for water vapor with $\kappa = 4/3$, eq. (5) becomes

$$Q_{FM}/Q_\infty = (7/4)\alpha_I R_I \ln(R_{II}/R_I)/(16.9 \cdot \ell/4) \quad . \quad (7)$$

Temperature Jump Region

For slightly rarefied gases, the heat conduction equation can be used in combination with temperature-jump boundary conditions

$$T_I - \theta_I = g_I(\partial T/\partial r)_{R_I} \quad \text{and} \quad \theta_{II} - T_{II} = g_{II}(\partial T/\partial r)_{R_{II}} \quad (8)$$

where T_I and T_{II} are the temperatures of the gas at the inner cylinder and at the wall. The solution is

$$Q/Q_\infty = \ln(R_{II}/R_I)/\left\{\ln(R_{II}/R_I) + g_I/R_I + g_{II}/R_{II}\right\} \quad . \quad (9)$$

Following Kennard (1938), the temperature-jump distances, g_I and g_{II} can be written as

$$g = (2-\alpha)/\alpha \cdot 4c/(\kappa+1) \cdot \lambda\ell/(\eta c_v) \quad (10)$$

with c = 0.4909. The viscosity, η, and the heat conductivity, λ, in this equation can be eliminated using eq. (4) for monatomic gases or eq. (6) for water vapor. The mean free path has been computed with the relation $\ell = kT/(\sqrt{2}\pi\sigma^2 p)$. Values for the gas kinetic diameter, σ, have been taken from D'Ans-Lax (1970), and values for the specific heat capacities have been taken from Schmidt and Grigull (1979).

Transition Region

A different approach to determine the heat flux for a large Knudsen number range was used by Lees and Liu (1962), who solved Maxwell's moment equations in combination with a two-sided Maxwellian distribution function for Maxwell molecules. For small temperature differences between inner and outer cylinders and for complete accommodation at the inner cylinder, their result can be put in the form

$$Q/Q_\infty = 1/\left\{1 + 15\ell/(4R_I \ln(R_{II}/R_I))\right\} \qquad . \qquad (11)$$

This result is plotted in Fig. 1. One recognizes clearly that the heat flux, Q, is independent of the density in the continuum region and proportional to the density in the free molecule region. The solution has been extended to arbitrary accommodation coefficients of the inner cylinder by Hurlbut (1964). His result is

$$Q/Q_\infty = 1/\left\{1 + 15\ell/\left(4\alpha_I R_I \ln(R_{II}/R_I)\right)\right\} \qquad . \qquad (12)$$

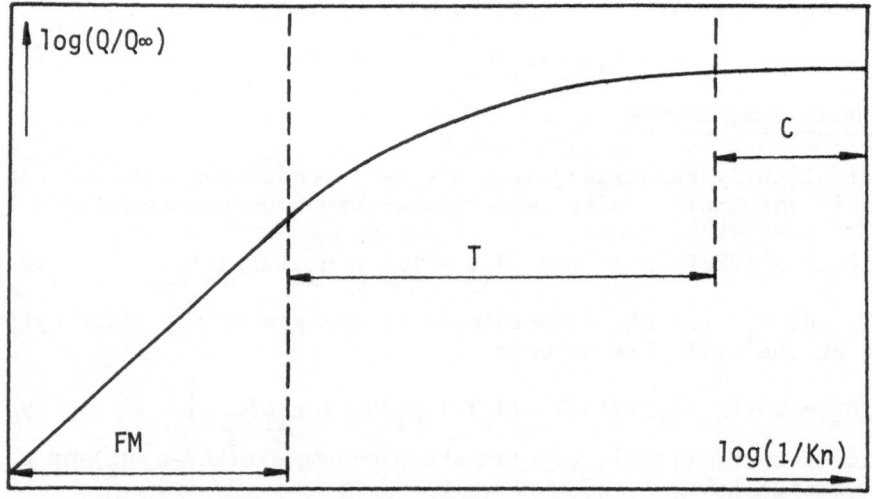

Fig. 1. Heat-transfer in gases from eq. (11) as a function of the inverse Knudsen number. C: Continuum conditions, T: Transition region, FM: Free molecule flow conditions

Modified Solution of the Momentum Equations

Equation (11) is strictly valid for Maxwell molecules. The correct solution for water vapor must give eq. (1) for Kn → 0 and eq. (7) for Kn → ∞. A simple modification of eq. (11) that fulfills both requirements is

$$Q/Q_\infty = 1/\left\{1 + f_{H_2O}15\ell/\left(4R_I\ln(R_{II}/R_I)\right)\right\} \qquad . \qquad (13)$$

with $f_{H_2O} = (4/7)\cdot 16.9/15$.

EXPERIMENTAL SECTION

The experimental setup consisted mainly of an ultrahigh-vacuum pumping station, heat-transfer cell, electronic measuring and control system, and a PDP 11 computer. A bright-drawn stainless steel tube having an inner diameter of $2R_{II}$ = 40 mm served as outer cylinder of the heat-transfer cell. A thin Pt wire with the diameter $2R_I$ = 0.1 mm formed the inner cylinder. The wire was tightened by a small weight attached at the lower end. To avoid end effects, the measuring section of the wire was limited by two potential leads. In the first series of experiments, an oil thermostat was used to keep the temperature of the outer cylinder constant (cell A). The oil was pumped through the stainless steel jacket of the heat-transfer cell. Because of the temperature boiling of the oil, it was not possible to have temperatures above 450 K. To obtain higher temperatures, a new heat-transfer cell was built whose outer cylinder was heated with electrical heating tapes (cell B). The electrical heating system allowed temperatures up to 725 K. Table 1 shows important data of cell A and B. Figure 2 shows a schematic view of the heat-transfer cells.

Table 1. Dimensions of Heat-Transfer Cells

	cell A	cell B
Length of heat-transfer cell, mm	750	930
Length of measuring section, mm	330	402
R_I, mm	0.05	0.05
R_{II}, mm	20	20
R_{II}/R_I	400	400
Maximum temperature, K	450	725
Pressure range (both cells), Pa	$5\cdot 10^{-7}$ up to	10^{+5}

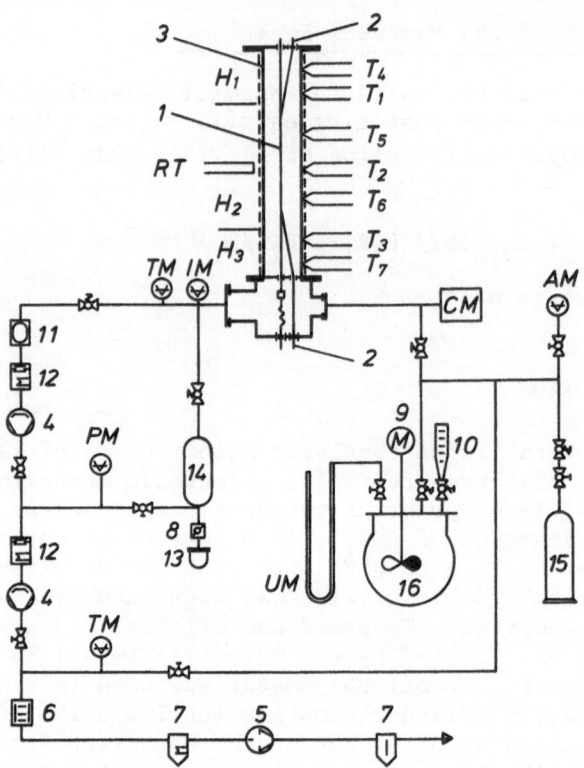

Fig. 2. Schematic view of the heat-transfer cell.

Nomenclature for Figures 2 and 3

1 Pt wire	8 Valve	14 Storage vessel for
2 Potential leads	9 Motor	water vapor
3 Outer cylinder	10 Water inlet	15 Storage vessel for
4 Diffusion pump	11 Liquid nitrogen trap	nitrogen
5 Mechanical pump	12 Water cooled baffle	16 Storage vessel for
6 Sorption trap	13 Water evaporator	mixtures
7 Oil trap	—⋈— Valves	—⋈— Leak valves

AM Aneroidmanometer	IR Automatic ice	RT Resistance
CM Capacitance	point reference	thermometer
manometer	MS Measuring section	RO Normal resistor
CR Current reverser	PM Penning gauge	T Thermocouple
DV Digital voltmeter	PS Power supply	TM Pirani gauge
H Heating tapes	RA Adjustable	UM U-tube manometer
HC Heating control	resistor	VM Measuring voltage
IM Ionization gauge	RS Range selector	multiplexer

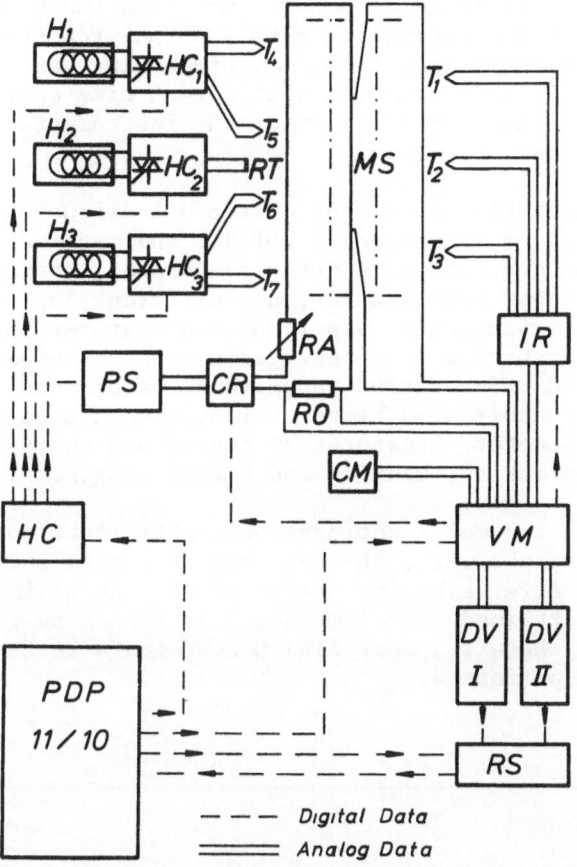

Fig. 3. Schematic diagram of instrumentation and experimental control

The temperature of the outer cylinder of the heat-transfer cell A was controlled by two NiCr-Ni thermocouples at the inlet and the outlet of the jacket. The temperature difference between inlet and outlet was less then 0.2 K for temperatures of the outer cylinder $T_{II} < 330$ K; for $T_{II} > 400$ K, the temperature difference reached 2 K. For cell B, temperature variations from the lower end to the upper end of the outer cylinder were continuously checked with seven thermocouples in the wall. Four of these thermocouples and an additional resistance thermometer were used to control the temperature of the outer wall by the electronic heating control, the other thermocouples were used as input for the PDP 11 computer. The maximum temperature variation along the tube was less than 0.1 K.

To adjust the temperature of the inner cylinder, the section between the two potential leads was used as resistance thermometer.

The electrical measurements were performed with two Data Precision
digital voltmeters in combination with a normal resistor of 1 Ω for
the current. Continuous data sampling, processing, and storing was
performed on a PDP 11/10 computer with a real-time operation system.
Figure 3 shows a schematic diagramm of the instrumentation and
experimental control.

The pumping station consisted of two oil diffusion pumps in
series, a cryotrap, a water-cooled baffle, and metal-sealed valves.
The inlet system consisted of a water evaporator; storage vessels for
nitrogen, water vapor and mixtures; and different leak valves. The
storage vessels for the water vapor and the mixtures could be heated
electrically. Pressures of the test gases were measured with Granville Phillips and MKS-Baratron capacitance manometers, which were
compared with a Consolidated Vacuum Corporation McLeod gauge and
with a U-tube manometer. Pressures at the vacuum station could be
measured with ionization, Penning and Pirani gauges.

Reproducible thermal accommodation coefficients could be
obtained by baking the wire, the heat-transfer cell and the vacuum
system at 725 K. After a baking period of 48 h an ultimate vacuum of
$5 \cdot 10^{-7}$ Pa was obtained. To eliminate heat losses by radiation and
conduction along the wire, heat flux measurements in the free molecule region were performed.

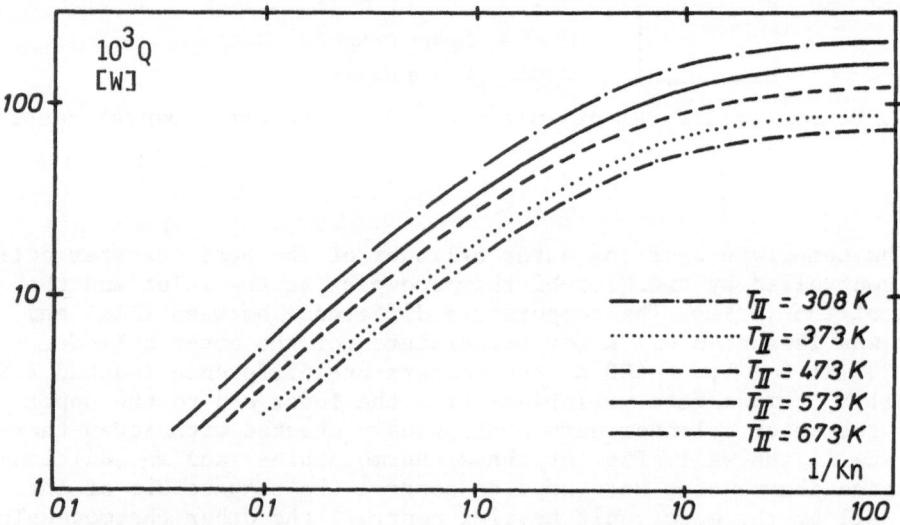

Fig. 4. Heat flux in water vapor as a function of the inverse
        Knudsen number for various temperatures of the outer
        cylinder at constant temperature difference of 10 K.
        The measuring length, H, in eq. (1) is reduced to 300 mm.

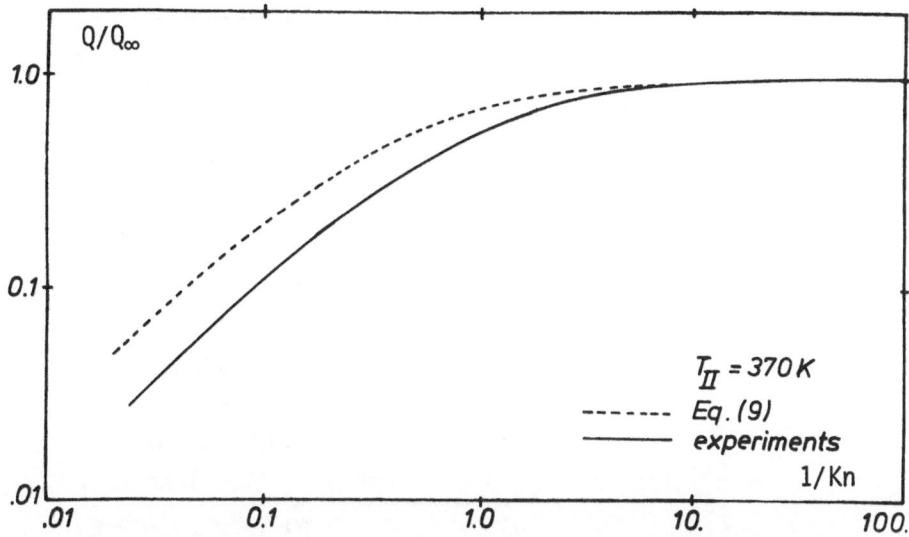

Fig. 5. Heat-transfer in water vapor as a function of the inverse
Knudsen number for a temperature difference of 10 K in
comparison with numerical results obtained from eq. (9).

RESULTS

Heat-Transfer and Thermal Conductivity of Water Vapor

Experimental results for water vapor in the pressure range from 4 kPa
to 1 Pa and for temperature between 300 K and 700 K are shown in
Fig. 4 as a function of the temperature of the outer cylinder. The
temperature of the inner cylinder is 10 K higher than the temperature
of the outer cylinder in all experiments. In Fig. 5, the experimental
results are compared with the theoretical results obtained with the
temperature-jump conditions obtained from eq. (9). Good agreement
between theory and experiment was obtained for the continuum region
and for slight deviations from the continuum region. Appreciable
deviations between theory and experiments occurred for Knudsen numbers >0.1. For Kn = 0.22, the deviation was 5% at a temperature of
T_{II} = 370 K. For Knudsen numbers >1, the error was larger than 50%
and for Knudsen numbers >10, the error was larger than 100%. In
Fig. 6 the experimental results for water vapor are compared with
eq. (13). Careful comparison of experimental and theoretical results
reveals that f_{H_2O} in eq. (13) decreases slightly with increasing temperature. For the temperature range of the present experiments, good
agreement was obtained for the mean value f_{H_2O} = 0.65, as can be seen
in Fig. 6.

From the present results at continuum conditions, the heat
conductivity of water vapor was determined in the temperature range

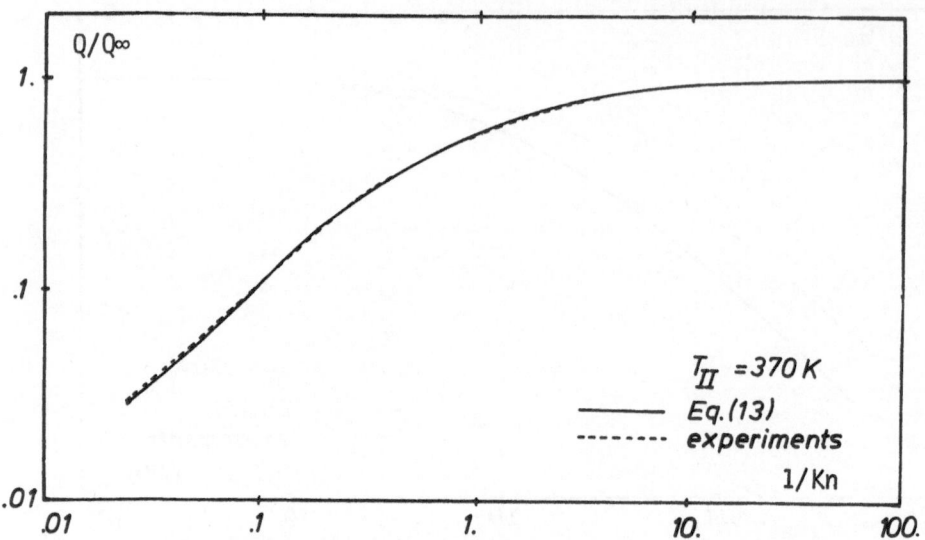

Fig. 6. Heat transfer in water vapor as a function of the inverse
       Knudsen number in comparison with numerical results obtained
       from eq. (13) with f_{H_2O} = 0.65

300 K to 680 K. The agreement with the new Skeleton Tables of Schmidt
and Grigull (1979) is always within the given tolerance of their
data. The results of the present paper can be approximated by the
empirical formula

$$10^3 \cdot \lambda_{H_2O} = 2.195 + 0.05719 \cdot T + 4.034 \cdot 10^{-5} \cdot T^2 \tag{14}$$

Thermal Conductivity of Nitrogen

Heat transfer measurements for nitrogen were performed at continuum
conditions for the temperature range from 300 K to 680 K. The heat
conductivity could be approximated by the empirical formula

$$10^3 \cdot \lambda_{N_2} = 3.789 + 0.08409 \cdot T - 3.0 \cdot 10^{-5} \cdot T^2 \quad . \tag{15}$$

which agree within 1% with the tabulated data of Landolt-Börnstein
(1969)

Thermal Conductivity of Water Vapor-Nitrogen Mixtures

Figures 7 and 8 represent experimental results for mixtures of water vapor and nitrogen together with the theoretical results of
Chapman and Cowling (1970)

$$\lambda = \lambda_1 x_1 / (x_1 + \alpha_{12} x_2 D_{11}/D_{12}) + \lambda_2 x_2 / (x_2 + \alpha_{21} x_1 D_{22}/D_{12}) \quad . \tag{16}$$

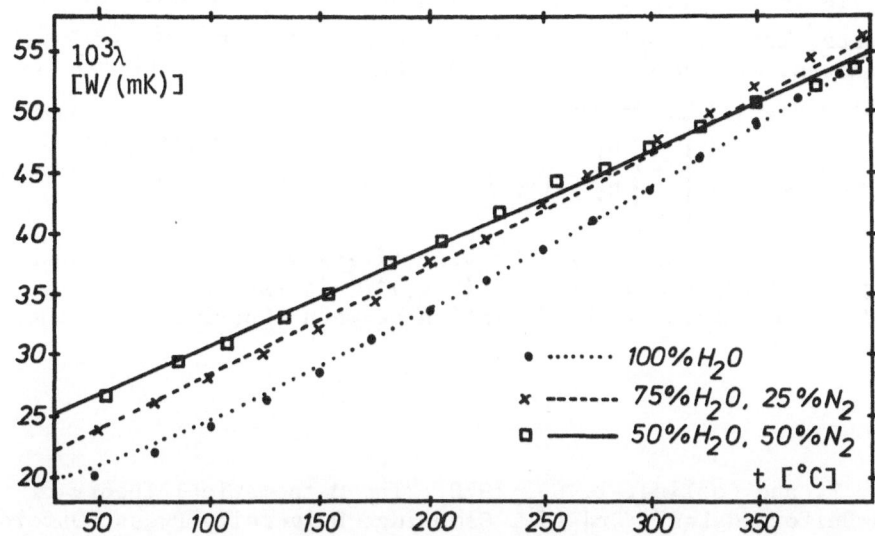

Fig. 7. Thermal conductivities of mixtures of water vapor and nitrogen as a function of the temperature in comparison
with results obtained from eq. (16).

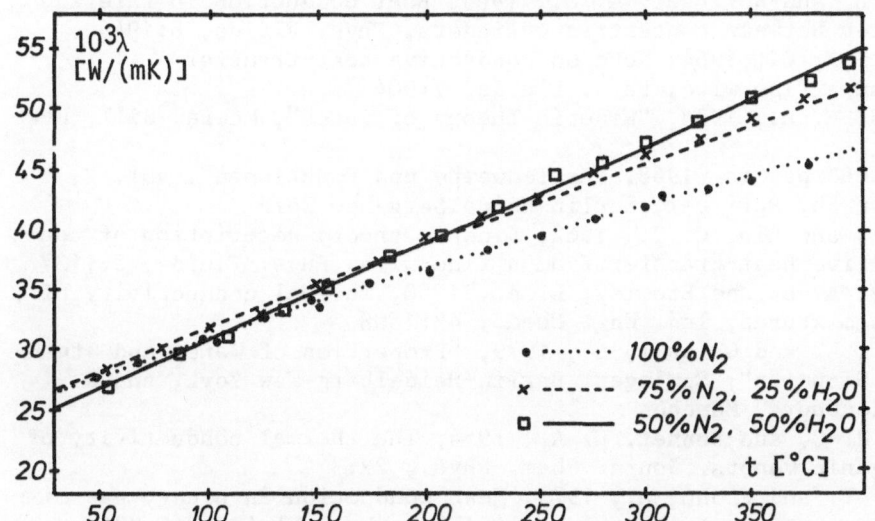

Fig. 8. Thermal conductivities of mixtures of water vapor and
nitrogen as a function of the temperature in comparison
with results obtained from eq. (16).

In this equation λ_1 and λ_2 denote the heat conductivities of the pure
gases; D_{11} and D_{22}, the coefficients of self-diffusion; and D_{12}, the
coefficient of mutual diffusion. The quantity, x_1, is defined by

$x_1 = p_1/(p_1+p_2)$, where p_1 and p_2 are the partial pressures of the components. The constants, α_{12} and α_{21}, are functions of the molecular masses, m_1 and m_2. Following Lindsay and Bromley (1950) the expression $\alpha_{12}D_{11}/D_{12}$ is replaced by

$$\alpha_{12}D_{11}/D_{12} = \frac{1}{4} \cdot \left[1 + \left\{ \frac{\eta_1}{\eta_2} \cdot \left\{ \frac{m_2}{m_1} \right\}^{3/4} \cdot \frac{1+S_1/T}{1+S_2/T} \right\}^{1/2} \right]^2 \cdot \frac{1+S_{12}/T}{1+S_1/T} , \quad (17)$$

where η_1, η_2, S_1 and S_2 are the viscosities and Sutherland constants of the pure gases, and $S_{12} = 0.733(S_1 S_1)^{1/2}$ is the Sutherland constant of the mixture. The deviation between experiment and theory is less than 2%.

REFERENCES

Chapman, S. and Cowling, T. G., 1970, "The Mathematical Theory of Non-Uniform Gases", 3rd ed., Cambridge University Press, Oxford

D'Ans-Lax, 1970, "Taschenbuch für Chemiker und Physiker", Vol. III, Springer, Berlin-Heidelberg-New York

Dushman, S. and Lafferty, J. M., 1962, "Scientific Foundation of Vacuum Technique", 2nd ed., John Wiley Sons Inc., New York-London

Dybbs, A. and Springer, G. S., 1965, Heat conduction in rarefied gases between concentric cylinders, Phys. Fluids, 8:1946

Hurlbut, F. C., 1964, Note on conductive heat-transfer from a fine wire, Phys. Fluids, 7:904

Kennard, E. H., 1938, "Kinetic Theory of Gases", McGraw-Hill, New York

Landolt-Börnstein, 1968, "Zahlenwerte und Funktionen", vol. 2, part 5b, Springer, Berlin-Heidelberg-New York

Lees, L. and Liu, C. Y., 1962, Kinetic theory description of conductive heat-transfer from a fine wire, Phys. Fluids, 5:1137

Lindsay, A. L. and Bromley, L. A., 1950, Thermal conductivity of gas mixtures, Ind. Eng. Chem., 42:1508

Schmidt, E. and Grigull, U., 1979, "Properties of Water and Steam in SI-Units", Springer, Berlin-Heidelberg-New York, and R. Oldenbourg, München

Vines, R. G. and Bennet, L. A., 1954, The thermal conductivity of organic vapors, Journ. Chem. Phys., 22:3

Yeh, B. T. and Frohn, A., 1973, Heat conduction in binary gas mixtures between concentric cylinders, Phys. Fluids, 16:801

SESSION   TC-9

Theory, Review, and Correlation - III

CHAIRMAN

D. L. McElroy
Oak Ridge National Laboratory
Oak Ridge, Tennessee

A DIFFUSION MODEL OF RADIATIVE HEAT TRANSFER

IN SCATTERING AND ABSORBING MEDIA

James I. Berg
R&D Division
Owens-Corning Fiberglas Corporation
Technical Center
Granville, OH 43023

INTRODUCTION

Thermal insulations scatter, absorb, and thermally emit infrared radiation. Scattering results from the refractive index discontinuities characteristic of the heterogeneous structure which may be
fibrous, particulate or foamed. Absorption by atomic mechanisms
nearly always occurs over some spectral regions within the very broad
thermal radiation band.

In the development, manufacturing, and improvement of insulation,
it is desirable to know how physical, chemical, and structural factors
affect the radiative conductivity. Regardless of the nature of the
insulation, radiative phenomena can be described in terms of coefficients for absorption and scattering and a phase function giving
the directional dependence of scattered radiation relative to the
incident direction. The primary objective of this paper is to give
a simple physical explanation of how these fundamental properties
affect the radiative portion of the conductivity. The properties
themselves could either be calculated from knowledge of the insulation
structure and composition or determined experimentally.

To retain mathematical simplicity, yet display the effects of
the above parameters and their wavelength dependence, diffusion conditions are assumed. The flux in response to a temperature gradient
is calculated at an internal point, removed by several mean free
paths from either boundary. Also, the gradient is assumed constant
over a distance of several mean free paths from the point. Since
mean free paths for insulations are commonly the order of a few milli-
meters, it is not surprising that diffusion conditions are apparently
approached rather closely for insulation layers a few inches thick.
This point will be dealt with in later sections.

There are two fundamental approaches to finding the radiation intensity in interacting media. The first, and most common in recent years, is to attempt to solve the equation of transfer.[1] This integro-differential equation expresses conservation of energy within infinitesimal volume and solid angle elements at each point. Several authors[1-8] have applied the equation of transfer to scattering and absorbing media, but solutions generally follow either through simplifying approximations to the scattering problem or through numerical methods. Neither are compatible with the objectives set forth in this study.

The second fundamental approach is to apply the emission, propagation and interaction laws directly. This is somewhat analogous to applying Coulomb's law rather than solving Poisson's Equation in an electrostatic problem. The approach is not as generally applicable as use of the equation of transfer, but can be easily applied under the diffusion conditions assumed here. Some of the earlier treatments of heat transfer in insulation[9-11] were based on this approach, but they were limited to absorption and thermal emission, with scattering neglected. The analysis which follows includes scattering, absorption, and emission in an emitter-receptor diffusion model. The medium is assumed to be uniform, but can be anisotropic. All quantites should be regarded as functions of wavelength, even though this dependence is not explicit in the symbols used.

EMISSION AND INTERACTION LAWS

The intensity $I(\vec{r},\Theta,\phi)$ is the power flowing at position $\vec{r}$, per unit area and solid angle (and wavelength interval) in the (Θ,ϕ) direction; where Θ is the polar angle and ϕ the azimuthal angle.

The properties characterizing the medium are independent of position and are introduced as follows. The fractional amount of extinction per unit distance in the (Θ,ϕ) direction is given by the extinction coefficient, $\alpha(\Theta,\phi)$. It is the sum of absorption and scattering parts;

$$\alpha(\Theta,\phi) = \alpha_a(\Theta,\phi) + \alpha_s(\Theta,\phi). \tag{1}$$

The power intercepted by a unit volume of material per unit solid angle is

$$E(\Theta,\phi) = \alpha(\Theta,\phi)I(\Theta,\phi), \tag{2}$$

with similar relationships holding for absorption and scattering separately.

Energy is emitted at each point both by thermal excitation and scattering. The power emitted thermally per unit volume and solid angle is given by

$$J_t(\Theta,\phi) = \alpha_a(\Theta,\phi)I_B, \tag{3}$$

where I_B is the radiation intensity in a blackbody cavity at the local temperature. Equation (3) can be derived from Equation (2) by equilibrium considerations applied to a specimen of the material in a blackbody cavity.

The scattering portion of emitted power is determined by the intensity incident on the emitting volume element and the redistribution in a direction according to the scattering phase function. Here, the phase function, $K(\Theta,\phi,\Theta_i,\phi_i)$, gives the portion of the unabsorbed intercepted energy incident in the (Θ_i,ϕ_i) direction which is redirected into a unit solid angle about (Θ,ϕ). Thus, the power emitted because of scattering is

$$J_s(\Theta,\phi) = \int K(\Theta,\phi,\Theta_i,\phi_i)\alpha_s(\Theta_i,\phi_i)I(\Theta_i,\phi_i)d\Omega_i \tag{4}$$

$$= K(\Theta,\phi)*\alpha_s(\Theta,\phi)I(\Theta,\phi).$$

Here $d\Omega_i$ is an incident solid angle element, $d\Omega_i = \sin\Theta_i d\Theta_i d\phi_i$, and the integration is over a complete sphere. Equation (4) defines the notation $K*$. By definition, K is normalized with respect to the scattering angles, or,

$$\int K(\Theta,\phi,\Theta_i,\phi_i)d\Omega = 1. \tag{5}$$

A homogeneous medium is symmetric upon linear reversal, so that

$$\alpha(\Pi-\Theta,\Pi+\phi) = \alpha(\Theta,\phi), \tag{6}$$

with similar relations holding for α_s and α_a. For the phase function symmetry implies that

$$K(\Pi-\Theta,\Pi+\phi,\Pi-\Theta_i,\Pi+\phi_i) = K(\Theta,\phi,\Theta_i,\phi_i). \tag{7}$$

The net power emitted is the sum of Equations (3) and (4), or

$$J = \alpha_a I_B + K*\alpha_s I, \tag{8}$$

where angular dependence is made implicit.

It is convenient to define the quiescent condition as that in which no net energy is transferred locally, so that $J = E$. Thus, from Eqs. (2) and (8), the quiescent state intensity, I_0, must satisfy the equation

$$\alpha I_0 = \alpha_a I_B + K * \alpha_s I_0. \qquad (9)$$

It follows from Eq. (9) and the reversal relationships Eqs. (6) and (7) that I_0 is symmetric under linear reversal. This must be so for there to be no net flux, or energy transfer by radiation. Integrating Eq. (9) over all solid angles, using the normalization condition, Eq. (5), gives

$$\overline{\alpha_a I_0} = \overline{\alpha_a} I_B, \qquad (10)$$

where bars denote averages over a sphere. A consequence of Eq. (10) is that, for an isotropic structure, $I_0 = I_B$, or the quiescent state intensity is the blackbody intensity in this case.

RADIATIVE FLUX IN THE DIFFUSION APPROXIMATION

To give a net flux, the angular distribution of intensity must not be completely symmetric, but have a non-zero antisymmetric component. The intensity at any point results from emission from other points, and only emission from internal points need be considered for receptor points several mean free paths from the boundaries. It is only when J is independent of position that no flux results. Since by Eq. (8), J depends on both I_B and I, any variation in the magnitudes of either can cause a flux. Except when there is no scattering, or the scattering is symmetric, there will be an additional contribution to the flux from the nonsymmetric angular dependence of I itself, through J.

Consider a planar sample with faces normal to the $\Theta=0$ or x-axis. The radiative flux in the x direction is

$$F_R = 4\pi \overline{I\cos\Theta}, \qquad (11)$$

where the bar again denotes the average over an entire sphere. A flux through the sample may arise from a temperature differential between the boundaries, or a difference in incident radiation intensity. The effect of a temperature variation appears through the temperature dependence of I_0, and thus of J. Since thermal emission depends only on temperature, the thermal emission portion from a point will be the same as in the quiescent state for the same local temperature.

Although the magnitude of I must vary, and I must be nonsymmetric in its angular dependence, there is no apparent reason why the angular distribution must vary with x, and here we attempt to derive distribution which, under diffusion conditions, is independent of x. The contribution to the intensity at point P from another point at a location relative to P specified by distance vector $\vec{r}$ is given in Fig. 1. Therefore, the net intensity at P is given by the integral

$$I(\Theta,\phi) = \int_0^\infty J(-\vec{r},\Theta,\phi)\exp(-\alpha(\Theta,\phi)r)dr, \qquad (12)$$

where $-\vec{r}$ refers to a point a distance r in the direction opposite to the (Θ,ϕ) direction, or the $(\Pi-\Theta, \Pi+\phi)$ direction. The integration limit of infinity implies that the point P is much greater than α^{-1} away from any boundary. Equations (8) and (12) may be combined to give an integral equation for intensity,

$$I = \alpha_a \int_0^\infty I_B(-\vec{r})\exp(-\alpha r)dr + K*\alpha_s \int_0^\infty I(-\vec{r})\exp(-\alpha r)dr, \qquad (13)$$

where the angular dependence in Eq. (13) is again made implicit. It may be verified that under constant intensity, Eq. (13) reduces to Eq. (9).

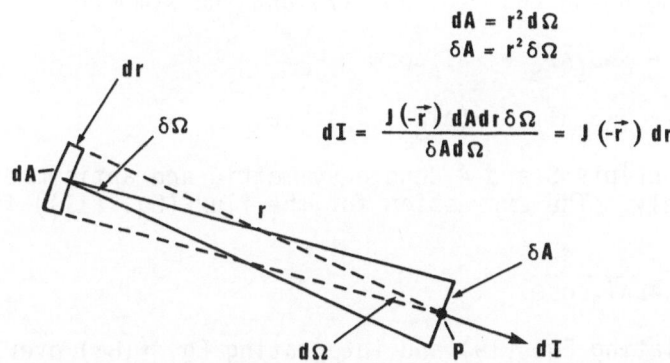

Fig. 1.   The intensity contribution $dI(0,\Theta,\phi)$ at point P from a point at $(r,\Pi-\Theta,\Pi+\phi)$.

The first-order diffusion approximation is as follows: I_B and I are expanded in Taylor's series in x about point P. It is assumed that, within a radial distance several times α^{-1}, only the linear terms in the expansion need be included. Equation (13) then becomes

$$\alpha I = \alpha_a I_B + K*\alpha_s I - (\cos\Theta/\alpha)(\alpha_a I_B' + K*\alpha_s I'), \qquad (14)$$

where the prime denotes the derivative with respect to x. An integral equation like (12) can be written for I' as well as I, and similar integration gives, keeping only linear terms,

$$\alpha I' = \alpha_a I'_B + K*\alpha_s I'. \tag{15}$$

Defining

$$\Delta I = I - I_0, \tag{16}$$

using Eq. (15) and subtracting Eq. (9) from Eq. (14) gives, finally,

$$\alpha \Delta I - K*\alpha_s \Delta I = -(I'_0 + \Delta I')\cos\theta. \tag{17}$$

Since I_0 gives zero flux, ΔI may be substituted for I in Eq. (11). Equation (17) can be solved for ΔI, the flux-carrying portion of the intensity, once the quiescent state portion, I_0, is known. Equation (17) is consistent with the equation of transfer,[1] although it applies here under the restrictive conditions that the position and directional dependences are separate.

The intensity, ΔI, is not necessarily symmetric or antisymmetric, but may have portions with either character. Writing Eq. (17) for the (θ,ϕ) and the $(\pi-\theta, \pi+\phi)$ directions and successively adding and subtracting using Eqs. (6) and (7) and the symmetry of I_0 gives

$$\alpha \Delta I_S - K*\alpha_s \Delta I_S = -\Delta I'_A \cos\theta \tag{18a}$$

$$\alpha \Delta I_A - K*\alpha_s \Delta I_A = -(\Delta I'_S + I'_0)\cos\theta, \tag{18b}$$

where subscripts S and A denote symmetric and antisymmetric parts, respectively. The expression for the flux (Eq. (11)) finally simplifies to

$$F_R = 4\pi\overline{\Delta I_A \cos\theta}. \tag{19}$$

Differentiating Eq. (19) and integrating Eq. (18a) over a sphere gives

$$F'_R = -4\pi\overline{\alpha_a \Delta I_S}. \tag{20}$$

A consequence of Eq. (20) is that if F_R is constant, ΔI is antisymmetric.

THE RADIATIVE CONDUCTIVITY

Heat transfer in insulation generally occurs by conduction and radiation, although convection may contribute.[12] We assume that convection is either neglegible or its contribution can be included in the conductive portion. Conduction is a diffusion phenomenon because of the very small mean free path. Under radiation diffusion conditions then, the total flux is

$$F = -(k_C + k_A)(dT/dx), \tag{21}$$

where the meaning of the subscripts is obvious. Since the total
flux must be constant, the radiative portion

$$F_R = -k_R(dT/dx) \tag{22}$$

must be constant also. Here we derive k_R.

Under constant F_R, $\Delta I_S = 0$, and Eq. (18b) reduces to

$$\alpha \Delta I_A - K*\alpha_s \Delta I_A = -I_o \cos\Theta. \tag{23}$$

Because of the antisymmetry of ΔI_A, the $0 \leqslant \Theta \leqslant \Pi/2$ and $\Pi/2 \leqslant \Theta \leqslant \Pi$
hemispheres contribute equally to the flux, and the two hemispheres
may be associated with emission from regions on opposite sides of
the plane through P and normal to the x-axis. Also from the
antisymmetry it follows that parts of the phase function symmetric
in incident angle, i.e. $K(\Theta,\phi,\Pi-\Theta_i,\phi_i+\Pi) = K(\Theta,\phi,\Theta_i,\phi_i)$, contribute
nothing to the second term in Eq. (23). But since both symmetric
and antisymmetric parts contribute to α along with absorption,
that term may be regarded as a correction for antisymmetric
scattering. This was referred to earlier.

The remainder of the discussion in this section is restricted
to isotropic structure. In this case, α and α_s are independent of
angle, and K depends only on the angle ψ between incident and
scattered directions. The intensity is then a function of Θ alone,
and, as noted, $I_o = I_B$.

The general form of the flux may be found after multiplying
Eq. (23) by $\cos\Theta$ and integrating, giving

$$F_R = -(4\Pi/3)I_B^!/\alpha(1-\Upsilon), \tag{24}$$

where $(1-\Upsilon)$ is the nonsymmetric scattering correction to the
extinction coefficient with Υ given by

$$\Upsilon = \eta \ \overline{H\Delta I_A}/\overline{\Delta I_A \cos\Theta}. \tag{25}$$

Here η is the scattering albedo, α_s/α, and H is the forward scattered
flux distribution function defined as

$$H(\Theta_i,\phi_i) = \int K(\Theta,\phi,\Theta_i,\phi_i)\cos\Theta d\Omega \ = H(\Theta_i). \tag{26}$$

The dependence of H on Θ_i alone follows from the above-mentioned
properties of the phase function, and this is clear from Fig. 2.

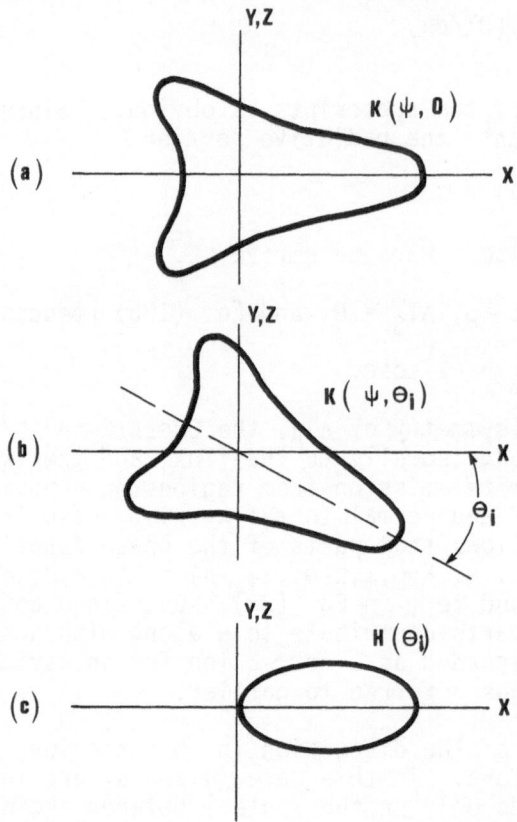

Fig. 2.   The function, H, derived from K; (a) scattered intensity
          for forward incidence, (b) for oblique incidence at angle
          Θ_i and (c) the result, H.

        The correction γ does depend on the functional form of
$\Delta I_A(\Theta)$, and thus it is necessary to solve the integral equation
(Eq. (23)) for accurate evaluation. However, the functional form
of ΔI_A is independent of I_B, so the proportionality in Eq. (24)
is correct. In accordance with the result already stated, symmetric
portions of phase function do not contribute to H. It follows from
the symmetry and normalization conditions for K that H is antisymmetric and less than or equal to unity. Thus, $H\Delta I_A$ is symmetric,
and the integrals implied in Eq. (25) need be carried out only over
one hemisphere.

The correction γ ranges between $-\eta$ and η, approaching the limiting values in the cases where the phase function consists of a strong backward ($\Theta=\Pi$) or forward ($\Theta=0$) lobe, respectively. In either of these cases, Eq. (23) reduces to an algebraic equation in $\Delta I_A(\Theta)$ as it does in the case of completely symmetric scattering. If the phase function contains a backward or forward lobe and a symmetric portion, only the lobe contributes, and γ is reduced to the appropriate fraction of η. For completely symmetric scattering, γ is of course zero.

The conductivity, or negative ratio of flux to temperature gradient is, from Eq. (24),

$$K_R = (4/3)\widetilde{W}_B/\alpha(1-\gamma),\qquad\qquad(27)$$

where W_B is the blackbody hemispherical radiant flux density, and $\sim$ denotes partial differentiation with respect to temperature. Equation (27) resembles the Rosseland formula[13] with n = 1, except that it applies to a medium that both absorbs and scatters with arbitrary phase function. It is interesting to note that, although the mean free path for diffusion conductivity is effectively increased by nonsymmetric scattering, the conditions for validity of the diffusion approximation (see Eq. (12)) are in terms of total extinction alone, independently of the details of the phase function.

The bulk diffusion conductivity, k_R, actually applies to internal regions in the diffusion limit, and does not take boundaries into account. The linearity must be disturbed in boundary regions because the flux within a few mean free paths of the boundary depends upon the emissivity, ε. It is possible, under pure radiation conditions at least[1], to have a temperature discontinuity at the boundary. To give a crude accounting for boundary effects, we assume here that the bulk temperature line extends to the boundaries, with steps at each boundary, as shown in Fig. 3. The temperature steps may be found by requiring the net flux to be the same at the boundaries as in the bulk. Radiation emitted by the walls is assumed to penetrate without reflection, but that emitted from the medium undergoes reflection with coefficient 1-ε. The intensity is taken to be independent of direction. The flux emitted by the medium is the sum of thermal equilibrium and gradient-induced portions, the latter being half the internal flux because there is emission from regions on only one side of the boundary plane. The net radiative flux crossing the first boundary to the right in Fig. 3 is

$$F_R = \varepsilon_1\left[W_B(T_1) + F_R/2 - W_B(T_1 - \delta T_1)\right].\qquad\qquad(28)$$

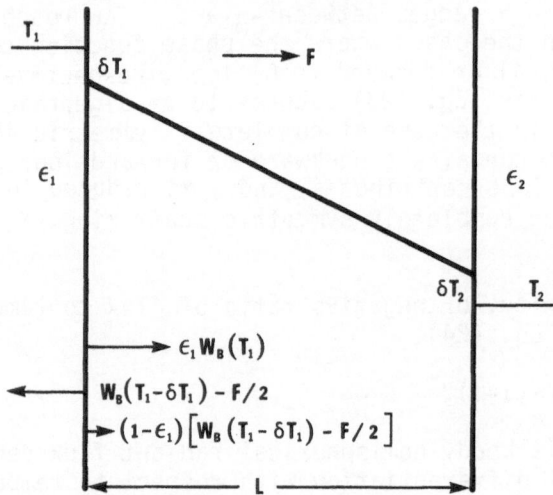

Fig. 3. Temperature variation in simplified model and boundary
        flux contributions.

Expanding W_B in a Taylor's series to first order in T and
solving for the temperature differential gives, for either boundary,

$$\delta T_{1,2} = F_R(1/\varepsilon_{1,2} - 1/2)/\widetilde{W}_B. \qquad (29)$$

The effective conductivity is the flux times the thickness, L,
divided by the net temperature difference between the boundaries,
which includes the bulk contribution from Eq. (27) and the two
boundary contributions from Eq. (29). The result is

$$k_R^{eff} = \widetilde{W}_B / \left((3/4)\alpha(1-\gamma) + (1/\varepsilon_1 + 1/\varepsilon_2 - 1)/L \right) . \qquad (30)$$

Equation (30) is a familiar and useful formula for description of
conductivity in fibrous and foamed insulations. Here it applies
to each wavelength separately, but the greybody form is usually
used. The formula was derived from a two-flux scattering approximation to the transfer equation by Larkin and Churchill,[2] and from
a diffusion and isotropic scattering model by Siegel and Howell.[1]
It was used successfully by Bhattacharyya[14] in an empirical description of fibrous wool conductivity for varied density and thickness.
The derivation here appears to be the most general in that it includes
absorption and scattering with arbitrary phase function. The result
shows that a diffusion approach gives a reasonably accurate description of heat flow in insulation, both in terms of basic physical
mechanisms and practical application.

NONTHERMAL RADIATION TRANSMISSION

By varying the intensity of incident radiation on a sample through pulsing or chopping, or by cooling the sample, it is possible to create conditions in which thermal emission can be ignored. The intensity is not constant, so that ΔI has both symmetric and anti-symmetric components.

Considering again an isotropic structure, Eq. (18b) reduces to a formula like Eq. (23), with I_B replaced by ΔI_S. Similar integration gives

$$F_R = -4\pi\overline{\Delta I'_S \cos^2\theta}/\alpha(1-\gamma). \tag{31}$$

Differentiating Eq. (31) and comparing with Eq. (20) gives an equation with solutions of the form

$$F_R(x) = C_1\exp(\mu x) + C_2\exp(-\mu x), \tag{32}$$

where C_1 and C_2 are constants, and

$$\mu = (g\alpha_a\alpha(1-\gamma))^{1/2}. \tag{33}$$

Here

$$g = \overline{\Delta I'_S} / \overline{\Delta I'_S \cos^2\theta}. \tag{34}$$

For $x \gg \mu^{-1}$, the negative exponential behavior prevails, and this is confirmed by transmittance vs. thickness experiments reported, for example, by Larkin and Churchill.[2] Since the quantity $\alpha(1-\gamma)$ appears in both the conductivity and the attenuation coefficient, μ, the diffusion model provides a possible way of correlating transmittance with conductivity. For spherical ΔI_S, $g = 3$. This may be a reasonable approximation since the antisymmetric part, not the symmetric part, is related to the flux, and may have to be more highly peaked in the forward direction. Only α_a must then be determined separately. For this, photoacoustic spectroscopy, or even estimation from homogeneous bulk material absorption are possibilities.

SUMMARY

It has been shown that the inclusion of scattering and absorption in a diffusion model of radiative heat transfer results in a conductivity formula similar to the Rosseland[13] formula except that there is a correction resulting from the antisymmetric portion of the scattering phase function. This correction increases the effective mean free path relating to conductivity, but does not affect the conditions for validity of the diffusion approximation in terms of net extinction. Through an approximate treatment of boundary effects, the familiar conductivity formula of Larkin and Churchill[2] is derived, although here it applies without restriction regarding scattering and absorption. Finally, a formula for the transmitted radiation attenuation coefficient is derived and related to the conductivity.

REFERENCES

1. R. Siegel and J. R. Howell, "Thermal Radiation Heat Transfer," McGraw-Hill, New York (1972).
2. B. K. Larkin and S. W. Churchill, Heat transfer by radiation through porous insulations, A.I.Ch.E. 5:467 (1959).
3. J. C. Chen and S. W. Churchill, Radiant heat transfer in packed beds, A.I.Ch.E. 9:35 (1963).
4. R. M. F. Linford, R. J. Schmitt and T. A. Hughes, Radiative contribution to the thermal conductivity of fibrous insulations, in "Heat Transmission Measurements in Thermal Insulations," ASTM STP 544, American Society for Testing and Materials, Philadelphia (1972) pp. 68-84.
5. G. J. Kowalski and J. W. Mitchell, An analytical and experimental investigation of the heat transfer mechanisms within fibrous media, ASME Winter Annual Meeting Paper No. 79-WA/HT-40.
6. R. Viskanta and R. J. Grosh, Heat transfer in a thermal radiation absorbing and scattering medium, in "International Developments in Heat Transfer," Part IV, ASME, New York (1961) pp. 820-828.
7. R. L. Houston, "Combined Radiation and Conduction in a Nongray Participating Medium that Absorbs, Emits and Anisotropically Scatters," Ph.D. Dissertation, The Ohio State University, Columbus, Ohio (1980).
8. T. W. Tong and C. L. Tien, Analytical models for thermal radiation in fibrous insulations, J. Therm. Insul. 4:27 (1980).
9. J. D. Verschoor and P. Greebler, Heat transfer by gas conduction and radiation in fibrous insulations, Trans. Am. Soc. Mech. Eng. 74:961 (1952).
10. H. M. Strong, F. P. Bundy, and H. P. Bovenkerk, Flat panel vacuum thermal insulation, J. Appl. Phys. 31:39 (1960).
11. N. E. Hager, Jr. and R. C. Steere, Radiant heat transfer in fibrous thermal insulation, J. Appl. Phys. 38:4663 (1967).
12. C. M. Pelanne, Thermal insulation: What it is and how it works, J. Therm. Ins. 1:223 (1978).
13. S. Rosseland, "Theoretical Astrophysics; Atomic Theory and Analysis of Stellar Atmospheres and Envelopes," Clarendon Press, Oxford (1936).
14. R. K. Bhattacharyya, Heat transfer model for fibrous insulations, in "Thermal Insulation Performance," ASTM, STP 7/8, American Society for Testing and Materials, Philadelphia (1978) pp. 272-286.

THERMAL CONDUCTIVITY MEASUREMENT OF MATERIALS DURING ABLATION

---A FURTHER TREATMENT ON THE MOVING BOUNDARY PROBLEM

Benlian Zhou, Zhen Wei, and Junheng Lin

Institute of Metal Research
Academia Sinica
Shenyang, China

Thermal diffusivity, α, and thermal conductivity, K, have been calculated from the temperature-increasing curves of the specimen, during ablation. It is discussed mathematically that there exists an equivalence between the moving boundary model of constant temperature and the fixed boundary model of varying temperature; thus the latter may be used to treat the case with ablation.

INTRODUCTION

The structure of coatings changes continuously during ablation; therefore, it is necessary to measure the continuous variation of the thermal conductivity of coatings. An ablation experiment by arc plasma makes it possible to obtain the temperature-increasing curves of the inner and outer surfaces of the specimen during its size changing process. Thermal diffusivities and thermal conductivities can thus be calculated. The experimental details and the first calculation may be found in references 1 and 2.

EXPERIMENTAL DATA AND CALCULATING METHODS

Experiments were divided into two groups: with and without ablation.

Moving Boundary with Constant Temperature

In Fig. 1, the surface temperatures of the specimens rise quickly (14.1s) to more than 2400°C and fluctuate within 4% generally, with 7.5% as the maximum. Therefore the outer surface may be treated as a boundary with constant temperature. The

341

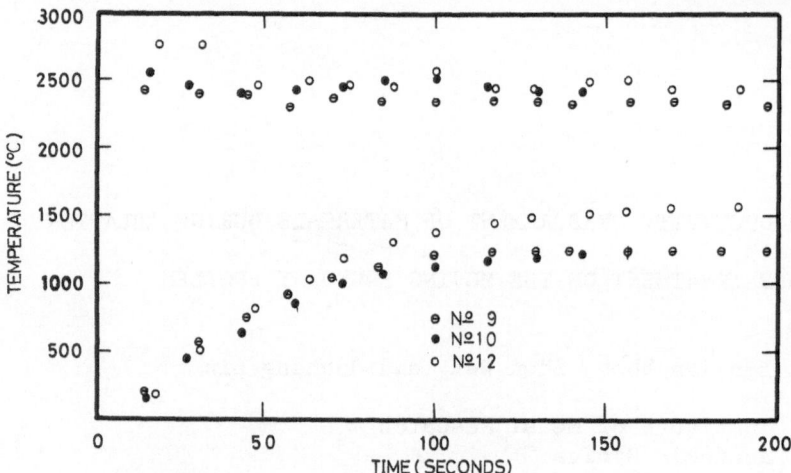

Figure 1.   Temperature process of inner and outer surfaces of
             specimens (moving boundary with constant temperature).

tubular specimen rotated during heating, so the temperatures of
the different parts of its inner surface were nearly the same,
thus the heat loss was negligible, because of radiation balance.
Therefore, the inner surface (x = o) can be treated as a heat-
insulated boundary[3], but the outer boundary is moving. According
to Carslaw and Jaeger[4], if the initial temperature of a plate with
thickness R is f(x), no heat flow at x=o, and the temperature at
x=R is $\phi(\tau)$, thus the temperature t at x and time $\tau(\lambda=o \to \tau)$ is

$$t(x,\tau) = \frac{2}{R} \sum_{n=o}^{\infty} e^{-\alpha(2n+1)^2\pi^2\tau/4R^2} \cdot \cos\frac{(2n+1)\pi x}{2R} \left\{ \frac{(2n+1)\pi\alpha(-1)^n}{2R} \right. \cdot$$

$$\left. \cdot \int_{o}^{\tau} e^{\alpha(2n+1)^2\pi^2\lambda/4R^2} \cdot \phi(\lambda)d\lambda + \int_{o}^{R} f(x')\cos\frac{(2n+1)\pi x'}{2R}dx' \right\} \qquad (1)$$

On the basis of this equation, we may treat the following:

Moving Boundary. The diagram is shown in Fig. 2(a); its
governing equation and limiting conditions are:

$$\frac{\partial t_1(x,\tau)}{\partial \tau} = \alpha \frac{\partial^2 t_1(x,\tau)}{\partial x^2} , \quad \left[\tau>0, 0<x<R(\tau)\right] \qquad (2)$$

$$I: \begin{cases} t_1(x,o)=f(x), & (3) \\ t_1\left[R(\tau),\tau\right]=t_c, & (4) \\ \dfrac{\partial t_1(0,\tau)}{\partial x} = 0. & (5) \end{cases}$$

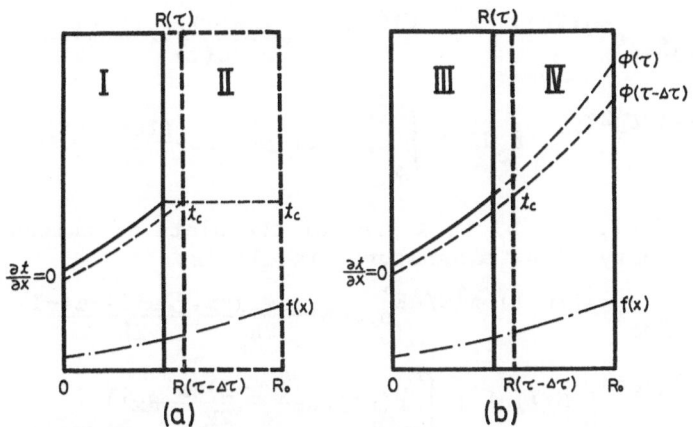

Figure 2. The temperature profiles of moving boundary with constant temperature (a) and fixed boundary with varying temperature (b).

$$t_2(x,\tau)=t_c,\ \left[\tau>0,\ R(\tau)\leq x\leq R_o\right] \tag{6}$$

II: $\left\{\begin{array}{ll} t_2(x,o)=f(x), & (7)\\ t_2[R(\tau),\tau]=t_1[R(\tau),\tau]=t_c & (4') \end{array}\right.$

Fixed Boundary. The diagram is shown in Fig. 2(b); its governing equation and limiting conditions are:

$$\frac{\partial t_3(x,\tau)}{\partial\tau}=\alpha\frac{\partial^2 t_3(x,\tau)}{\partial x^2},\ \left[\tau>0, 0<x<R(\tau)\right] \tag{8}$$

III: $\left\{\begin{array}{ll} t_3(x,o)=f(x), & (9)\\[2mm] t_3[R(\tau),\tau]=t_c, & (10)\\[2mm] \dfrac{\partial t_3(0,\tau)}{\partial x}=0. & (11) \end{array}\right.$

$$\frac{\partial t_4(x,\tau)}{\partial\tau}=\alpha\frac{\partial^2 t_4(x,\tau)}{\partial x^2},\ \left[\tau>0, R(\tau)<x<R_o\right] \tag{12}$$

IV: $\left\{\begin{array}{ll} t_4(x,o)=f(x), & (13)\\[2mm] t_4[R(\tau),\tau]=t_3[R(\tau),\tau]=t_c & (10')\\[2mm] \dfrac{\partial t_4[R(\tau),\tau]}{\partial x}=\dfrac{\partial t_3[R(\tau),\tau]}{\partial x} & (14) \end{array}\right.$

The temperature profile of part I in Fig. 2(a) at time τ is a development of the temperature profile at $\tau-\Delta\tau$ after an interval $\Delta\tau$. Because the variation of $R(\tau)$ in this short interval is small, we get from eq. (1):

$$t_1(x,\tau) = \frac{2}{R(\tau)} \sum_{n=0}^{\infty} e^{-\alpha(2n+1)^2\pi^2\tau/4R^2(\tau)} \cdot \cos\frac{(2n+1)\pi x}{2R(\tau)} \{ \frac{(2n+1)\pi\alpha(-1)^n}{2R(\tau)} \cdot$$

$$\cdot \int_{\tau-\Delta\tau}^{\tau} e^{\alpha(2n+1)^2\pi^2\lambda/4R^2(\tau)} \cdot t_c d\lambda + \int_0^{R(\tau)} t_1(x',\tau-\Delta\tau) \cdot \cos\frac{(2n+1)\pi x'}{2R(\tau)} dx' \} \quad (15)$$

Part III in Fig. 2(b) is a part of the plate of thickness R_o with fixed boundary, its temperature profile is:

$$t_3(x,\tau) = \frac{2}{R_o} \sum_{n=0}^{\infty} e^{-\alpha(2n+1)^2\pi^2\tau/4R_o^2} \cdot \cos\frac{(2n+1)\pi x}{2R_o} \{ \frac{(2n+1)\pi\alpha(-1)^n}{2R_o} \cdot$$

$$\cdot \int_0^{\tau} e^{\alpha(2n+1)^2\pi^2\lambda/4R_o^2} \cdot \phi(\lambda)d\lambda + \int_0^{R_o} f(x') \cos\frac{(2n+1)\pi x'}{2R_o} dx' \} \quad (16)$$

where $\phi(\tau)$ is the temperature function at boundary, R_o. But from another viewpoint, $t_3(x,\tau)$ is also a development of $t_3(x,\tau-\Delta\tau)$ after an interval $\Delta\tau$ and can be described as:

$$t_3(x,\tau) = \frac{2}{R_o} \sum_{n=0}^{\infty} e^{-\alpha(2n+1)^2\pi^2\tau/4R_o^2} \cdot \cos\frac{(2n+1)\pi x}{2R_o} \{ \frac{(2n+1)\pi\alpha(-1)^n}{2R_o} \cdot$$

$$\cdot \int_{\tau-\Delta\tau}^{\tau} e^{\alpha(2n+1)^2\pi^2\lambda/4R_o^2} \cdot \phi(\lambda)d\lambda + \int_0^{R_o} t_3(x',\tau-\Delta\tau) \cdot \cos\frac{(2n+1)\pi x'}{2R_o} dx' \} \quad (17)$$

Now, let's do the transformation: $x/R(\tau)=x_1/R_o, \tau/R^2(\tau)=\tau_1/R_o^2$,

thus $\tau_1 = R_o^2\tau/R^2(\tau) \equiv F(\tau)$ and the inverse function is

$\tau = G(\tau_1)$, so $R(\tau) = R[G(\tau_1)] \equiv R_1(\tau_1)$, then

$$x = \frac{R_1(\tau_1)}{R_o} x_1, \quad \tau = \frac{R_1^2(\tau_1)}{R_o^2} \tau_1,$$

therefore $\left[\text{ for certain time, } \tau, \text{ or } \tau_1 \right]$

$$\int_0^{R(\tau)} \ldots \cos\frac{(2n+1)\pi x'}{2R(\tau)} dx' = \frac{R_1(\tau_1)}{R_o} \int_0^{R_o} \ldots \cos\frac{(2n+1)\pi x_1'}{2R_o} dx_1',$$

and eq. (15) becomes

$$t_1\left[\frac{R_1(\tau_1)}{R_o}x_1, \frac{R_1^2(\tau_1)}{R_o^2}\tau_1\right] = \frac{2}{R_1(\tau_1)} \sum_{n=0}^{\infty} e^{-\alpha(2n+1)^2\pi^2\tau_1/4R_o^2} \cdot \cos\frac{(2n+1)\pi x_1}{2R_o} \cdot$$

$$\cdot \{ \frac{(2n+1)\pi\alpha(-1)^n}{2R_o} \cdot \frac{R_o}{R_1(\tau_1)} \cdot \int_{\tau_1-\Delta\tau_1}^{\tau_1} e^{\alpha(2n+1)^2\pi^2\lambda_1/4R_o^2} \cdot t_c \cdot d\left(\frac{R_1^2(\lambda_1)}{R_o^2}\lambda_1\right) \} +$$

$$+ \frac{R_1(\tau_1)}{R_o} \int_o^{R_o} t_1 \left[\frac{R_1(\tau_1)}{R_o} x_1', \frac{R_1^2(\tau_1)}{R_o^2}(\tau_1 - \Delta\tau_1) \right] \cdot \cos\frac{(2n+1)\pi x_1'}{2R_o} dx_1'\}$$

$$= \frac{2}{R_o} \sum_{n=o}^{\infty} e^{-\alpha(2n+1)^2\pi^2\tau_1/4R_o^2} \cdot \cos\frac{(2n+1)\pi x_1}{2R_o} \{\frac{(2n+1)\pi\alpha(-1)^n}{2R_o} \cdot$$

$$\cdot \int_{\tau_1-\Delta\tau_1}^{\tau_1} e^{\alpha(2n+1)^2\pi^2\lambda_1/4R_o^2} \cdot t_c \cdot \frac{R_1^2(\lambda_1)}{R_1^2(\tau_1)}\left[1 + \frac{2R_1'(\lambda_1)}{R_1(\lambda_1)} \lambda_1\right] d\lambda_1 +$$

$$+ \int_o^{R_o} t_1 \left[\frac{R_1(\tau_1)}{R_o} x_1', \frac{R_1^2(\tau_1)}{R_o^2} (\tau_1-\Delta\tau_1)\right] \cdot \cos\frac{(2n+1)\pi x_1'}{2R_o} dx_1'\} \qquad (18)$$

By changing the variables (x,τ) in eq. (17) into (x_1,τ_1), we have

$$t_3(x_1,\tau_1) = \frac{2}{R_o} \sum_{n=o}^{\infty} e^{-\alpha(2n+1)^2\pi^2\tau_1/4R_o^2} \cdot \cos\frac{(2n+1)\pi x_1}{2R_o} \{\frac{(2n+1)\pi\alpha(-1)^n}{2R_o} \cdot$$

$$\cdot \int_{\tau_1-\Delta\tau_1}^{\tau_1} e^{\alpha(2n+1)^2\pi^2\lambda_1/4R_o^2} \cdot \phi(\lambda_1) d\lambda_1 + \int_o^{R_o} t_3(x_1', \tau_1-\Delta\tau_1) \cdot \cos\frac{(2n+1)\pi x_1'}{2R_o} dx_1'\} (19)$$

Comparing eq. (18) and eq. (19) we know that, if

$$t_1 \left[\frac{R_1(\tau_1)}{R_o} x_1, \frac{R_1^2(\tau_1)}{R_o^2} (\tau_1-\Delta\tau_1)\right] = t_3(x_1,\tau_1-\Delta\tau_1)$$

for all $o \leq x_1 \leq R_o$ at time $\tau_1-\Delta\tau_1$, then the necessary condition for the realization of

$$t_1 \left[\frac{R_1(\tau_1)}{R_o} x_1, \frac{R_1^2(\tau_1)}{R_o^2} \tau_1 \right] = t_3(x_1,\tau_1) \qquad (20)$$

at time τ_1, is (when $\Delta\tau_1$ approaches zero)

$$\phi(\tau_1) = \left[1 + \frac{2R_1'(\tau_1)}{R_1(\tau_1)} \tau_1\right] t_c$$

as the mean value theorem of integration has been taken into account. By changing τ_1 to τ, we obtain

$$\phi(\tau) = \left[1 + \frac{2R_1'(\tau)}{R_1(\tau)} \tau\right] t_c \qquad (21)$$

Substitute eq. (21) into (16) and change τ_1 in both sides of eq. (20) to τ, then

$$t_1\left[\frac{R_1(\tau)}{R_o}x, \frac{R_1^2(\tau)}{R_o^2}\tau\right] = t_3(x,\tau) = \frac{2}{R_o}\sum_{n=o}^{\infty}e^{-\alpha(2n+1)^2\tau/4R_o^2}\cdot\cos\frac{(2n+1)\pi x}{2R_o}\cdot$$

$$\cdot\left\{\frac{(2n+1)\pi\alpha(-1)^n}{2R_o}\cdot\int_o^\tau e^{\alpha(2n+1)^2\pi^2\lambda/4R_o^2}\cdot\left[1+\frac{2R_1'(\tau)}{R_1(\tau)}\tau\right]t_c\cdot d\lambda +\right.$$

$$\left.+\int_o^{R_o}f(x')\cos\frac{(2n+1)\pi x'}{2R_o}dx'\right\} \tag{22}$$

Now we discuss eq. (22) as follows:

(a) So long as the first derivative $R_1'(\tau)$ exists, there is no restriction on the form of function $R_1(\tau)$.

(b) When the initial temperature is t_o (i.e., $f(x)=t_o$), the boundary does not move (i.e., $R_1(\tau)=R_o$, $R_1'(\tau)=o$) and its temperature remains constant as t_c, eq. (22) becomes: (j=1,3)

$$\tag{23}$$

$$t_j(x,\tau) = t_c^{*)} + (t_o-t_c)\cdot\frac{4}{\pi}\sum_{n=o}^{\infty}e^{-\alpha(2n+1)^2\pi^2\tau/4R_o^2}\cdot\frac{(-1)^n}{2n+1}\cos\frac{(2n+1)\pi x}{2R_o}$$

This is the expression for the case of fixed boundary of constant temperature[4].

(c) From eq. (22) it may be seen that, when the boundary moves as $R(\tau)$, it seems as if the coordinate, x, of the medium expands continuously to the boundary with a factor of $R_o/R_1(\tau)$. Meanwhile the time, τ, speeds up gradually in the ratio of $R_o^2/R_1^2(\tau)$ to realize the heat conduction effect. Therefore, we have found a way to treat the moving boundary of constant temperature with the model of fixed boundary of varying temperature. To calculate in this way is much more simple and convenient than the method of recurrence[2].

Fixed Boundary with Constant Heat Flow

As shown in Fig. 3, the temperature of the inner and outer surfaces rise together with time, but ablation did not happen, so it cannot be treated as constant temperature. The temperature of plasma is above 6000 K, but the surface temperature of specimen is below 2500 K, and the ratio of radiation heat is about $2.5^4/6^4 \doteq 3\%$. So the radiative heat of the specimen is negligible, while the inward radiative heat flow, q_c, remains almost constant and the other conditions are the same as mentioned above, the solution is[3]:

---

*) The Fourier expansion of constant 1 in the open region (-R,R) is[5]:

$$1 = \frac{4}{\pi}\left(\cos\frac{\pi x}{2R_o} - \frac{1}{3}\cos\frac{3\pi x}{2R_o} + \frac{1}{5}\cos\frac{5\pi x}{2R_o}\ldots\ldots\right)$$

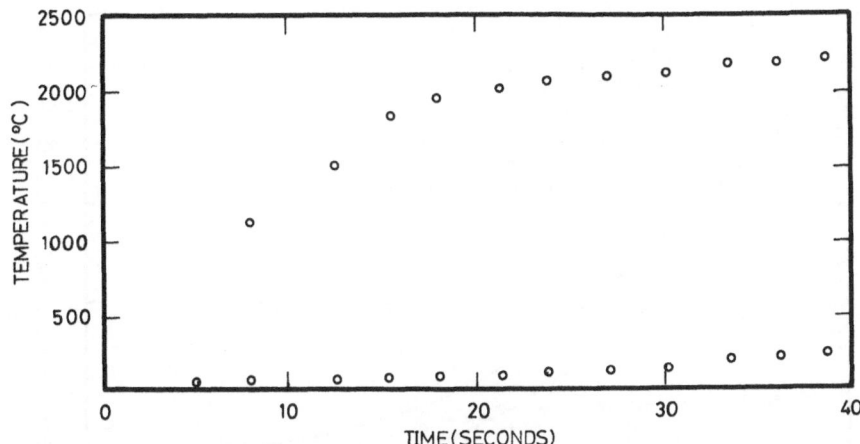

Figure 3. Temperature process of inner and outer surfaces of a specimen (fixed boundary with constant heat flow).

$$t(x,\tau)-t_o=\frac{q_c}{K}\left[\frac{\alpha\tau}{R} - \frac{R^2-3x^2}{6R} + R\sum_{n=1}^{\infty}(-1)^{n+1}\frac{2}{\mu_n^2}\cos\mu_n\frac{x}{R}\cdot e^{-\mu_n^2 Fo}\right] \quad (24)$$

where $\mu_n=n\pi$, $Fo=\alpha\tau/R^2$ is Fourier criterion. We surmount the difficulty of unknown q_c and K by the following method:

Let x=R and x=o respectively, since $(-1)^{n+1}\cos\mu_n=-1$, so the ratio of the relative temperature increases of the outer and inner surfaces is:

$$\frac{\theta_{ou}}{\theta_{in}} = \frac{t_{ou}-t_o}{t_{in}-t_o} = \frac{F_o + \frac{1}{3} - \sum_{n=1}^{\infty}\frac{2}{\mu_n^2}\cdot e^{-\mu_n^2 F_o}}{F_o - \frac{1}{6} + \sum_{n=1}^{\infty}(-1)^{n+1}\frac{2}{\mu_n^2}\cdot e^{-\mu_n^2 F_o}} \equiv \phi(F_o) \quad (25)$$

The unknown quantities q_c and K have already been cancelled.
From the value θ_{ou}/θ_{in} for different τ in Fig. 3, we may find the corresponding value of F_o, so that, α and K are obtained[2].

Results of Calculation

Different values of α for two groups of specimens were calculated almost the same way as the first treatment[2], and the values of K calculated from $\bar\alpha$ are shown in Fig. 4. The results calculated by the two methods mentioned above link up smoothly as shown.

DISCUSSION

1. Melting and vaporization of composite coatings occurred during ablation, so the data measured are equivalent α and K. It would be very difficult to measure these by the ordinary method.

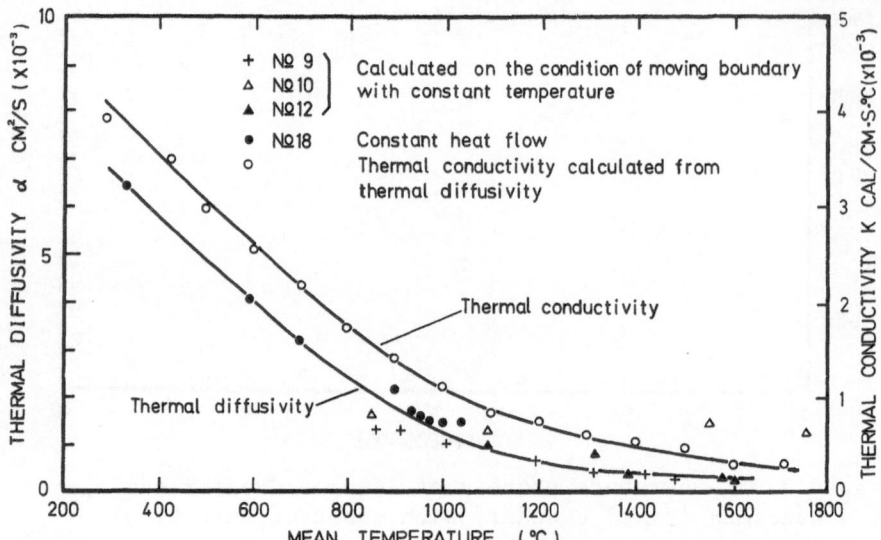

Figure 4. Thermal diffusivity α and thermal conductivity K of high temperature heat insulating composite coatings versus temperature (calculated from the temperature curves of ablation test).

    2. The value for q_c may be calculated from eq.(24) as α and K were obtained.

    3. The thermal conductivities of materials during ablation were calculated from differential equations with boundary and initial conditions; therefore these can be used for some other nonanalogous conditions.

ACKNOWLEDGMENT

    The authors are very grateful to Prof. M. Wachman and Dr. X. H. Ji of the Department of Mathematics of the University of Connecticut for their enthusiastic support, patient verification and genuine discussion. Thanks also to Mr. L. K. Dong and Mr. L. Y. Bai for their advantageous suggestions.

REFERENCES

1.   Institute of Metal Research, High Temperature Test with Low Power Arc Plasma, May, 1974. (in Chinese).
2.   B. L. Zhou, Z. Wei and J. H. Lin, J. Eng. Thermophysics, Vol. 2. May, 1981, p. 166-172. (China).
3.   A. V. Lekov, "Theory of Thermal Conductivity," GITTL, 1952. (in Russian).
4.   H. S. Carslaw and J. C. Jaeger, "Conduction of Heat in Solids," 2nd ed. Oxford, 1959, P. 104, p. 100.
5.   Y. C. Fan, "Teaching Materials of Higher Mathematics, "PEPH, Vol.2, p. 69. (in Chinese).

A METHOD TO OBTAIN THE ANALYTICAL SOLUTION OF THE THERMAL CONDUCTION EQUATION WITH MOVING BOUNDARIES

Lianke Dong, Liangyu Bai and Benlian Zhou

Institute of Metal Research

Academia Sinica, Shenyang, China

ABSTRACT

The right surface of a plate moves continuously with time during ablation, sublimation or erosion. In these cases, the thermal conduction process may be described by the thermal conduction equation with moving boundaries. Using a compressing (or expanding) axis system, this equation is transformed to a homogeneous parabolic type of differential equation with variable coefficients, while the outer boundary becomes fixed; by further transformations and derivations, the analytical solution is obtained.

INTRODUCTION

Let $t(x, \tau)$ be the temperature distribution in an infinite plate. B.-L.Zhou et al[1] had suggested a kind of thermal conduction equation with moving boundary in studying the determination of thermal conductivity during ablation, i,e,

$$(I) \begin{cases} \dfrac{\partial t(x,\tau)}{\partial \tau} = \alpha \dfrac{\partial^2 t(x,\tau)}{\partial x^2}, \quad (0 < x < R(\tau), \ \tau > 0) \\[2mm] t(x, 0) = f(x), \ (0 < x < R_0) \\[2mm] t(R(\tau), \tau) = t_c \end{cases}$$

$$\left| \frac{\partial t(0, \tau)}{\partial \tau} \right| = 0$$

where the function $t(x, \tau)$ has a continuous derivative of the first order with respect to τ and a continuous derivative of the second order with respect to x; $f(x)$ is a continuous function; $R(\tau)$ is a time-dependent function describing the moving boundary and has a continuous derivative of the first order. B.-L.Zhou et al solved this problem by making use of the equivalence between the moving boundary model of constant temperature and the fixed boundary model of varying temperature.

In this paper, we suggest a compressing (or expanding) transformation with respect to x; then problem (I) will transform to a homogeneous parabolic type of equation with variable coefficients. We will prove the equivalence of the two models in reference 1; hence, the existence and uniqueness of the solution of problem (I) will be proved.

By further transformations for the homogeneous parabolic type of equation with variable coefficients, an analytical solution will be given. Within the limit of experimental error, we find an approximate solution for problem (I) and compare it with experimental data.

THE DETERMINACY OF PROBLEM (I)

Let the temperature function $t(x, \tau)$ be defined in the region $D: \{(x, \tau) \mid 0 \le x \le R_0, \tau \ge 0\}$, where R_0 is the original thickness of the plate and τ is time. The coordinate system $S: \{x, \tau\}$ is an Ordinary Cartesian System.

Now we make a transformation like this

$$S^* : \begin{cases} x^* = x R_0 / R(\tau) \\ \tau^* = \tau \end{cases} \tag{1}$$

hence

$$S : \begin{cases} x = x^* R(\tau^*) / R_0 \\ \tau = \tau^* \end{cases} \tag{2}$$

Obviously the following relations hold;
1. If $\tau = 0$, then

$$R(\tau) = R_0, \quad x = x^* \tag{3}$$

It shows that the metric dimensions are the same for both systems.

2. The origins of both the fixed and the compressing

coordinates are zero, i.e.,

$$\chi^* = 0 \quad \text{and} \quad \chi = 0 \tag{4}$$

respectively. It means that the origins of the two coordinate systems coincide with each other.

3. If $\chi = R(\tau)$, we have

$$\chi^* = R_0 \tag{5}$$

at any time. Therefore, the outer boundary becomes fixed for the system S^*.

Consequently, system S is compressed and system S^* is expanded.

Theorem 1 (Theorem of equivalency of the physical model). Making use of transformation (1), problem (I) turns to the following problem:

$$(\text{II}) \begin{cases} \dfrac{\partial t(\chi^*, \tau^*)}{\partial \tau^*} = \dfrac{\alpha R_0^2}{[R(\tau^*)]^2} \dfrac{\partial^2 t(\chi^*, \tau^*)}{\partial \chi^{*2}} + \dfrac{\chi^* R'(\tau^*)}{R(\tau^*)} \dfrac{\partial t(\chi^*, \tau^*)}{\partial \chi^*}, & (0 \leq \chi^* \leq R_0, \tau^* > 0) \\[2mm] t(\chi^*, 0) = f(\chi^*) \\[2mm] t(R_0, \tau^*) = t_c \\[2mm] \dfrac{\partial t(0, \tau^*)}{\partial \chi^*} = 0 \end{cases}$$

Proof. In fact, from the differential principle of multivariable compound functions, we have

$$\frac{\partial t(\chi, \tau)}{\partial \chi} = \frac{R_0}{R(\tau^*)} \frac{\partial t(\chi^*, \tau^*)}{\partial \chi^*}$$

$$\frac{\partial^2 t(\chi, \tau)}{\partial \chi^2} = \frac{R_0^2}{[R(\tau^*)]^2} \frac{\partial^2 t(\chi^*, \tau^*)}{\partial \chi^{*2}}$$

and

$$\frac{\partial t(\chi, \tau)}{\partial \tau} = \frac{\partial t(\chi^*, \tau^*)}{\partial \tau^*} - \frac{\chi^* R'(\tau^*)}{R(\tau^*)} \frac{\partial t(\chi^*, \tau^*)}{\partial \chi^*}$$

Substituting the above relations into the governing differential equation of problem (I), we obtain the governing differential equation of problem (II).

$$\frac{\partial t(\chi^*, \tau^*)}{\partial \tau^*} = \frac{\alpha R_0^2}{[R(\tau^*)]^2} \frac{\partial^2 t(\chi^*, \tau^*)}{\partial \chi^{*2}} + \frac{\chi^* R'(\tau^*)}{R(\tau^*)} \frac{\partial t(\chi^*, \tau^*)}{\partial \chi^*}$$

From (3), (5) and (4) may prove the limiting condition of problem (II).

The physical meaning of the theorem of equivalency is that, under transformation (1), problem (I) of heat conduction with a moving boundary transforms equivalently to problem (II) of a parabolic type of equation with a fixed boundary, with a moving heat source term being added to the heat conduction equation.

Problem (II) is a homogeneous parabolic type of equation with variable coefficients and a fixed boundary. If the temperature function $t(x^*, \tau^*)$ and the ablation function $R(\tau^*)$ satisfy certain conditions, it is shown[2] that the solution of problem (II) not only exists, but is also unique. Therefore, we know that from transformations (1) or (2), the solution of problem (I) exists and is unique.

AN ANALYTICAL METHOD TO SOLVE PROBLEM (II)

According to the ablation experiment, we may suppose that

$$R(\tau^*) = R_0 e^{-\beta \tau^*} \tag{6}$$

here β can be determined by experiment. Now problem (II) becomes

$$(\mathrm{III}_1) \quad \begin{cases} \dfrac{\partial t(x^*, \tau^*)}{\partial \tau^*} = \alpha e^{2\beta \tau^*} \dfrac{\partial^2 t(x^*, \tau^*)}{\partial x^{*2}} - \beta x^* \dfrac{\partial t(x^*, \tau^*)}{\partial x^*} \\[2mm] t(x^*, 0) = f(x^*) \\[2mm] t(R_0, \tau^*) = t_c \\[2mm] \dfrac{\partial t(0, \tau^*)}{\partial x^*} = 0 \end{cases}$$

We make the following transformation

$$t(x^*, \tau^*) = t_c + u(x^*, \tau^*) \tag{7}$$

Substituting (7) into problem (III_1), we obtain the following

$$(\mathrm{III}_2) \quad \begin{cases} \dfrac{\partial u(x^*, \tau^*)}{\partial \tau^*} = \alpha e^{\beta x^*} \dfrac{\partial^2 u(x^*, \tau^*)}{\partial x^{*2}} - \beta x^* \dfrac{\partial u(x^*, \tau^*)}{\partial x^*} \\[2mm] u(x^*, 0) = f(x^*) - t_c \\[2mm] u(R_0, \tau^*) = 0 \end{cases}$$

$$\frac{\partial u(0,\tau^*)}{\partial x^*} = 0$$

Now, we take the transformation

$$u(x^*,\tau^*) = \exp\left[\frac{\beta x^{*2}}{4\alpha} e^{-2\beta\tau^*}\right] \cdot V(x^*,\tau^*) \qquad (8)$$

So we have the following theorem:

Theorem 2. Under the transformation (8), problem (III_2) becomes

$$(\text{IV}) \quad \begin{cases} \dfrac{\partial V(x^*,\tau^*)}{\partial \tau^*} = \alpha e^{2\beta\tau^*} \dfrac{\partial^2 V(x^*,\tau^*)}{\partial x^{*2}} + C(x^*,\tau^*)V(x^*,\tau^*) \\[2mm] V(x^*,0) = e^{-\beta x^{*2}/4\alpha}[f(x^*) - t_c] \\[2mm] V(R_0,\tau^*) = 0 \\[2mm] \dfrac{\partial V(0,\tau^*)}{\partial x^*} = 0 \end{cases}$$

here

$$C(x^*,\tau^*) = \frac{\beta}{2} + \frac{\beta^2 x^{*2}}{4\alpha} e^{-2\beta\tau^*} \qquad (9)$$

According to the method of З.И.Халилов[3] first of all, we must find the eigenvalue and eigenfunction for the following definite problem:

$$(\text{V}) \quad \begin{cases} \dfrac{\partial V(x^*,\tau^*)}{\partial \tau^*} = \alpha e^{2\beta\tau^*} \dfrac{\partial^2 V(x^*,\tau^*)}{\partial x^{*2}} \\[2mm] V(x^*,0) = e^{-\beta x^{*2}/4\alpha}[f(x^*) - t_c] \\[2mm] V(R_0,\tau^*) = 0 \\[2mm] \dfrac{\partial V(0,\tau^*)}{\partial x^*} = 0 \end{cases}$$

It is well known that we may obtain the solution by the method of separation variables for problem (V) We find the eigenvalue and eigenfunction as follows

$$\lambda_n = (2n+1)\pi/2R_0, \qquad n = 0,1,2,\cdots\cdots \qquad (10)$$

and

$$V_n(x^*) = X_n(x^*) = \cos[(2n+1)\pi x^*/2R_0], \quad n=0,1,2,\cdots\cdots \qquad (11)$$

respectively. Then the general solution of problem
may be written as

$$V(x^*, \tau^*) = \sum_{n=0}^{\infty} A_n(\tau^*) \cos \frac{(2n+1)\pi}{2R_0} x^* \qquad (12)$$

here the coefficients $A_n(\tau^*)$ are to be determined by the
initial condition. Substituting (12) into governing equation of problem ($\mathbb{V}$), we obtain the expression of $A_n(\tau^*)$.

Finally, we obtain the analytical solution of problem (I) as follows:

$$t(x, \tau) = t_c + e^{\beta x^2/4\alpha} \sum_{n=0}^{\infty} A_n(\tau) \cos \frac{(2n+1)\pi}{2R(\tau)} x \qquad (13)$$

COMPARISON WITH THE EXPERIMENTAL CURVE

Within the limit of experimental error, we may give
an approximate solution of problem (I) and compare it
with the experimental data.

Let n=o, we may give an approximate solution of
problem (I),

$$\hat{t}(x, \tau) = t_c + a_0 \exp\left[\frac{\beta R_0^2 x^2}{4\alpha [R(\tau)]^2} e^{-2\beta\tau}\right] \cdot \left[1 - \left(\frac{2\beta}{\alpha\pi^2} + \frac{1}{4R_0^2\pi^2}\right) e^{-2\beta\tau}\right.$$

$$+ \left(\frac{2\beta}{\alpha\pi^2} + \frac{1}{4R_0^2\pi^2}\right) e^{-(2\beta+\bar{a})\tau} - \frac{\beta^2 R_0^2 (\pi^2-6)}{3\pi^2\alpha(8\beta R_0 - \alpha\pi^2)} e^{-4\beta\tau}$$

$$+ \left. \frac{\beta^2 R_0^2 (\pi^2-6)}{3\pi^2\alpha(8\beta R_0 - \alpha\pi^2)} e^{-(2\beta+\bar{a})\tau}\right] \cdot \cos \frac{\pi x}{2R(\tau)}$$

here

$$a_0 = \frac{2}{R_0} \int_0^{R_0} e^{-\frac{\beta R_0^2 x'^2}{4\alpha [R(\tau)]^2}} \left[t\left(\frac{R_0 x'}{R(\tau)}\right) - t_c\right] \cos \frac{\pi x'}{2R(\tau)} dx'$$

and

$$\bar{a} = \alpha\pi^2/4R_0^2$$

Let x=0, an approximate solution of temperature distribution at the inner boundary is obtained:

$$\hat{t}(0,\tau) = t_c + a_0 \left[1 - \left(\frac{2\beta}{\alpha \pi^2} + \frac{1}{4R_0^2 \pi^2} \right) e^{-2\beta\tau} \right.$$

$$+ \left(\frac{2\beta}{\alpha \pi^2} + \frac{1}{4\alpha \pi^2} \right) e^{-(2\beta+\bar{a})\tau} - \frac{\beta^2 R_0^2 (\pi^2 - 6)}{3\pi^2 \alpha (8\beta R_0 - \alpha \pi^2)} e^{-4\beta\tau}$$

$$\left. + \frac{\beta^2 R_0^2 (\pi^2 - 6)}{3\pi^2 \alpha (8\beta R_0 - \alpha \pi^2)} e^{-(2\beta+\bar{a})\tau} \right]$$

From the second theorem of integral mean value, we have

$$a_0 = 2 \left[f\left(\frac{R_0 \xi}{R(\tau)} \right) - t_c \right] \cdot \left[1 - \frac{2p + \xi^2}{6} R_0^2 + \frac{12p^2 + 12p\xi + \xi^2}{120} R_0^4 \right.$$

$$- \frac{120p^3 + 120p^2\xi^2 + 30p\xi^4 + \xi^6}{5040} R_0^6 + \frac{1680p^4 + 3360p^3\xi^2 + 840p\xi^4 + 56p\xi^6 + \xi^8}{362880} R_0^8$$

$$\left. - \frac{840p^4\xi^2 + 280p^3\xi^4 + 28p^2\xi^6 + \xi^8}{443520} R_0^{10} \right]$$

here

$$p = \frac{\beta R_0^2}{4\alpha [R(\tau)]^2}, \quad \xi = \frac{\pi}{2R(\tau)}, \quad 0 < \xi < R_0$$

According to the data of specimen 9 of reference 1, we have calculated the temperature of its inner surface, which is shown in Fig.1. It can be seen that the experimental data coincides very well with the theoretical curve.

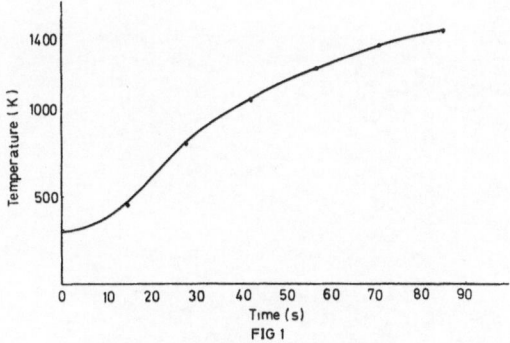

FIG 1

NOMENCLATURE

$R(\tau)$: The moving boundary of the specimen.
$t(x,\tau)$: Function of a temperature distribution.
x: Space coordinate in the system
x^*: Space coordinate in the system S^*.
α: Thermal diffusivity.
β: Ablation coefficient.
τ: Time in the system S.
τ^*: Time in the system S^*.

REFERENCES

1. Benlian Zhou, Zhen Wei and Junheng Lin, Journal of Engineering Thermo-physics, 2(2), 166-172 (1981, Chinese).
2. Tong Wang, Thermal Potential Theory, Science Press, Peking, 174-175, (1980, Chinese).
3. З. И. Халилов, ДАН, 83, 659-662, (1952).

SESSION TC-10

Insulations - I

CHAIRMAN

F. J. Powell
National Bureau of Standards
Gaithersburg, Maryland

THE THERMAL CONDUCTIVITY OF SEMITRANSPARENT MATERIALS[*]

H. A. Fine[1], S. H. Jury[2], D. L. McElroy[2] and
D. W. Yarbrough[2]

[1]Department of Metallurgical Engineering and
 Materials Science
 University of Kentucky
 Lexington, KY

[2]Metals and Ceramics Division
 Oak Ridge National Laboratory
 Oak Ridge, TN

ABSTRACT

    The three-region approximate solution for coupled conductive
and radiative heat transfer and an exact solution for uncoupled
conductive and radiative heat transfer in a grey semitransparent
medium bounded by infinite parallel isothermal plates are employed
to establish the dependence of the apparent thermal conductivity
of semitransparent materials on other material properties and
boundary conditions. An application of the analyses, which uses
apparent thermal conductivity versus density data to predict the
dependence of apparent thermal conductivity on temperature is demonstrated. The predictions for seven sets of R-11 fiberglass and
rock wool insulations agree with published measured values to within the limits of experimental error (±3%). Agreement for three
sets of R-19 fiberglass insulations was, however, not good.

INTRODUCTION

    Because of the presence of radiative heat transfer within
semitransparent (diathermanous) materials, the apparent thermal

---

[*]Research sponsored by the Division of Conservation, U. S. Department of Energy, under contract W-7405-eng-26 with the Union Carbide
Corporation.

conductivity of these materials is dependent upon the properties of the bounding surfaces, e.g., temperature and emittance, and other properties of the material, e.g., size (thickness), phonon conductivity, extinction coefficient, and albedo. In an earlier paper, the three-region approximate solution for coupled conductive and radiative heat transfer in a semitransparent medium bounded by infinite parallel black isothermal plates was developed.[1] This approximate solution and an exact solution to the uncoupled conductive and radiative heat transfer problem were then used to establish the thickness effect for fibrous building thermal insulation.[1,2]

The two solutions are briefly described in the current work. Their application for the analysis of apparent thermal conductivity data is then demonstrated using k_a versus ρ data (see Table 1 for nomenclature) to predict the relationship between k_a and temperature. Data sets for three fiberglass R-11*, three fiberglass R-19*, and four rock wool R-11* insulations, which have previously been published[3], are used to validate the predictions.

Table 1. Nomenclature

a, b, c, a', b', a'', b''	Constants in appropriate relationship
dL_i	Thickness of region i
k_a	Apparent thermal conductivity
k_C	Phonon conductivity
L	Sample thickness
q_t	Total heat flux
R_i	Thermal resistance of region i
T	Absolute temperature
T_m	Modified mean absolute temperature defined in Eq. (10)
T_H	Hot-plate absolute temperature
T_C	Cold-plate absolute temperature
T_H^*	Absolute temperature at interface between regions I and II
T_C^*	Absolute temperature at interface between regions II and III
x	Position
α	Absorption coefficient
ϵ	Emittance
ρ	Sample density
σ	Scattering coefficient
$\bar{\sigma}$	Stefan-Boltzmann constant

---

*Nominal Thermal Resistance Value in Customary Units, $kr \cdot Ft^2 \cdot °F/Btu$.

THEORETICAL CONSIDERATIONS

Absorption (and re-emission) or scattering of photons of radiant energy, or both, can occur within a semitransparent material. The pure absorption limiting case has been shown to be accurately described by the three-region approximate solution for optical thicknesses (αL) greater than four.[1] The pure scattering limiting case yields uncoupled conductive and radiative heat fluxes and has a closed-form solution.[1,4,5] These solutions are briefly described. A detailed development is presented in Ref. 1.

Three-Region Approximate Solution

An infinite slab of semitransparent material bounded by parallel black isothermal surfaces at T_H and T_C may be assumed to consist of an optically thin region immediately adjacent to the hot surface (Region I), an optically thick central region (Region II), and an optically thin region immediately adjacent to the cold surface (Region III) (see Fig. 1). Furthermore, if dL_1 and dL_2 are set equal to the distance at which half of an incident beam of radiation has been attenuated and the material is assumed at steady state, the series heat flow problem yields[1]

$$T_H^* = T_H - (T_H - T_C)\frac{R_I}{R_I + R_{II} + R_{III}}, \tag{1}$$

$$T_C^* = T_C + (T_H - T_C)\frac{R_{III}}{R_I + R_{II} + R_{III}}, \tag{2}$$

where

$$R_I = \frac{1}{(k_c\alpha/0.69315) + \bar{\sigma}(T_H^2 + T_H^2{}^*)(T_H^* + T_H)}, \tag{3}$$

$$R_{II} = \frac{\alpha L - 2(0.69315)}{k_c\alpha + \frac{4}{3}\bar{\sigma}(T_H^2{}^* + T_C^2{}^*)(T_H^* + T_C^*)}, \tag{4}$$

and

$$R_{III} = \frac{1}{(k_c\alpha/0.69315) + \bar{\sigma}(T_C^2{}^* + T_C^2)(T_C^* + T_C)}. \tag{5}$$

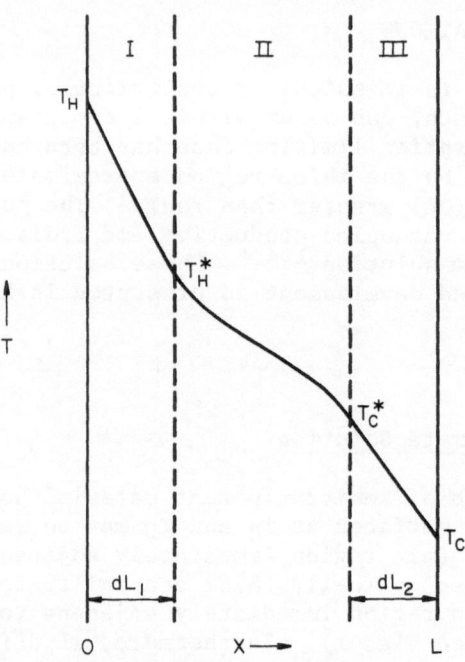

Fig. 1.   Schematic representation of the three-region approximation for a planar object.

Equations (1) through (5) may be solved iteratively, and the
total heat flux through the slab and apparent thermal conductivity
of the slab are calculated as follows:

$$q_t = \frac{T_H - T_C}{R_I + R_{II} + R_{III}}$$ (6)

and

$$k_a = \frac{q_t \cdot L}{T_H - T_C} \cdot$$ (7)

It is seen from this analysis that

$$k_a = F(k_C, \alpha, L, T_H, T_C).$$ (8)

Pure Scattering Solution

A closed-form solution for the uncoupled problem that results for the pure scattering case was developed by Heaslet and Warming.[5] For optical thickness (αL) greater than two, combination of this solution with Eq. (7) yields[1]

$$k_a = k_C + \frac{\frac{16}{3} \bar{\sigma} T_m^3}{\frac{4}{3}(\frac{2}{\varepsilon} - 2) + \alpha L + 1.42089} L \qquad (9)$$

where

$$T_m = \left\{ \frac{T_H^4 - T_C^4}{4(T_H - T_C)} \right\}^{1/3} \qquad (10)$$

It is seen from this analysis that

$$k_a = F(k_C, \sigma, L, T_H, T_C, \varepsilon). \qquad (11)$$

Furthermore, if σ, L, and ε are constant and k_C is given by

$$k_C = a + bT_m, \qquad (12)$$

then

$$k_a = a + bT_m + cT_m^3, \qquad (13)$$

where

$$c = \frac{\frac{16}{3} \bar{\sigma} L}{\frac{4}{3}(\frac{2}{\varepsilon} - 2) + \alpha L + 1.42089} . \qquad (14)$$

PRACTICAL APPLICATIONS

A demonstration of the use of the limiting case solutions for the analysis and extrapolation of apparent thermal conductivity measurements is presented in this section. In this demonstration, k_a versus ρ measurements at a fixed mean test temperature, $(T_H + T_C)/2$, and test temperature difference, $T_H - T_C$, and the limiting case solutions are used to predict the dependence of k_a

on $(T_H + T_C)/2$ and $T_H - T_C$. The validity of the predictions is then checked by comparison with measurements available in the literature for three R-11 fiberglass, three R-19 fiberglass, and four R-11 rock wool insulations.[3]

The initial step in the analysis was to use the k_a versus ρ data[3] (see Table 2) and the limiting case solutions to obtain α versus ρ and σ versus ρ data for each insulation. The σ value for a given ρ, L, ε, and k_a was obtained using Eq. (9).* The value of α for a given ρ, L, and k_a was found using a computer that performed the iterative solution for the three-region approximation. In this instance, values of α and L were supplied and the values of k_a were calculated. The value of α that gave the desired k_a was then used in further analyses.

The values of α and ρ or σ and ρ were then fit by the method of least squares to straight lines of the form

$$\alpha = a' + b'\rho \tag{15}$$

and

$$\sigma = a'' + b''\rho \tag{16}$$

Finally, Eq. (15) and the three-region approximation or Eqs. (9) and (16) were used to predict values of the apparent thermal conductivity at different densities, thicknesses, and test temperatures (see Tables 3 and 4).

The predictions for k_a at mean test temperatures of 255, 297, and 399K (0, 75, and 150°F) are in excellent agreement with the measured values for the seven R-11 insulations. As can be seen in Table 3, the largest discrepancy is 4.4%. Since the estimated measurement error[3] was ±3%, this agreement for all 21 cases is excellent.

The agreement for the R-19 fiberglass insulations was, however, only good at mean test temperatures of 297K (75°F). As can be seen in Table 3, the extrapolated values for other temperatures showed errors as large as 10.9%.

In an earlier paper, the dependence of k_a on L was examined using other data in the overall data set from which these three R-19 data sets were taken. In the previous analyses, excellent agreement was obtained between the predicted recovered-thickness

---

*The emittance for the surfaces in the guarded hot-plate apparatus[3,6] was 0.82 at 297K(75°F).

Table 2. Apparent Thermal Conductivity, Scattering Coefficient, and Absorption Coefficient Data for Ten Fiberglass and Rock Wool Insulations* - SI Units

Sample Code	Density (kg/m^3)	Thickness (cm)	Temperature Difference (K)	Apparent Thermal Conductivity (w/mK)	Scattering Coefficient (cm^{-1})	Absorption Coefficient (cm^{-1})
			R-11 Fiberglass			
1101-1	11.5	8.00	27.4	.0444	4.07	4.20
	13.0	7.11	28.2	.0423	4.59	4.79
	14.8	6.22	28.1	.0398	5.45	5.69
1107-1	9.62	8.00	28.1	.0459	3.74	3.87
	10.8	7.11	27.3	.0438	4.17	4.33
	12.4	6.22	27.4	.0414	4.86	5.02
1111-1	8.54	8.00	27.4	.0444	4.07	4.20
	9.62	7.11	28.3	.0424	4.56	4.72
	11.0	6.22	26.9	.0404	5.22	5.38
			R-19 Fiberglass			
1202-1	10.4	13.7	27.4	.0472	3.61	3.71
	11.7	12.2	28.4	.0451	4.00	4.07
	13.4	10.7	27.6	.0431	4.46	4.56
1204-2	10.4	13.7	28.1	.0428	4.56	4.65
	11.7	12.2	28.1	.0411	5.09	5.18
	13.3	10.7	27.4	.0394	5.74	5.84
1207-2	7.73	13.7	27.2	.0500	3.15	3.25
	8.70	12.2	28.1	.0482	3.41	3.51
	9.94	10.7	28.6	.0464	3.71	3.81
			R-11 Rock Wool			
1302-1	28.0	8.00	28.3	.0490	3.22	3.31
	31.6	7.11	28.1	.0470	3.51	3.64
	36.1	6.22	27.4	.0441	4.07	4.23
1304-1	34.4	8.00	28.1	.0415	4.86	4.99
	38.8	7.11	28.1	.0398	5.48	5.61
	44.3	6.22	27.3	.0382	6.17	6.37
1306-1	35.0	6.86	27.8	.0418	4.72	4.89
	40.4	6.10	27.4	.0401	5.32	5.48
	46.2	5.33	27.4	.0382	6.14	6.33
1309-2	29.0	8.00	27.8	.0436	4.27	4.40
	32.6	7.11	28.2	.0417	4.79	4.92
	37.3	6.22	27.9	.0402	5.25	5.45

*Measurements[3] at a mean test temperature of 297K (75°F).

Table 3. Apparent Thermal Conductivity versus Mean Test Temperature Data for Ten Fiberglass and Rock Wool Insulations -
SI Units

Sample Code	Density (kg/m³)	Thickness (cm)	Temperature (K) Mean	Difference	Apparent Thermal Conductivity Measured	Three-Region	(w/mK) Scattering
R-11 Fiberglass							
1101-1	10.2	9.04	255	27.8	.0369	.0366 (-0.8)[a]	.0366 (-0.8)
			297	28.3	.0469	.0475 (1.2)	.0475 (1.2)
			339	27.4	.0617	.0607 (-1.6)	.0607 (-1.6)
1107-1	9.66	7.98	255	27.8	.0316	.0356 (2.9)	.0356 (2.9)
			297	27.7	.0457	.0460 (0.6)	.0459 (0.3)
			339	28.2	.0587	.0586 (-0.2)	.0583 (-0.7)
1111-1	8.51	8.03	255	27.7	.0342	.0348 (1.7)	.0398 (1.7)
			297	27.6	.0444	.0444 (0.0)	.0444 (0.0)
			339	27.6	.0588	.0562 (-4.4)	.0564 (-4.2)
R-19 Fiberglass							
1202-1	10.4	13.7	255	27.8	.0352	.0363 (3.3)	.0365 (3.7)
			297	27.4	.0472	.0470 (-0.3)	.0472 (0.0)
			339	27.4	.0650	.0603 (-7.3)	.0603 (-7.3)
1204-2	9.62	14.8	255	28.2	.0325	.0345 (6.2)	.0345 (6.2)
			297	27.3	.0437	.0440 (0.7)	.0440 (0.7)
			339	28.0	.0624	.0557 (-10.9)	.0557 (-10.9)
1207-2	6.80	15.6	255	27.5	.0382	.0394 (3.0)	.0394 (3.0)
			297	27.6	.0521	.0519 (-0.3)	.0519 (-0.3)
			339	28.2	.0708	.0672 (-5.1)	.0672 (-5.1)
R-11 Rock Wool							
1302-1	25.2	8.89	255	27.4	.0391	.0397 (0.7)	.0394 (0.0)
			297	27.9	.0518	.0521 (0.6)	.0516 (-0.3)
			339	27.5	.0668	.0675 (1.1)	.0669 (0.2)
1304-1	33.1	8.33	255	27.4	.0330	.0333 (0.9)	.0332 (0.9)
			297	27.4	.0424	.0421 (-0.7)	.0421 (-0.7)
			339	27.7	.0519	.0529 (1.9)	.0527 (1.7)
1306-1	30.9	7.98	255	27.4	.0349	.0346 (-0.8)	.0347 (-0.4)
			297	27.6	.0449	.0443 (-1.3)	.0446 (-0.6)
			339	27.6	.0561	.0554 (-1.3)	.0564 (0.5)
1309-2	26.3	8.81	255	28.1	.0358	.0350 (2.0)	.0349 (-2.4)
			297	28.0	.0459	.0451 (-1.6)	.0499 (-2.2)
			339	27.7	.0583	.0573 (-1.7)	.0568 (-2.5)

[a]Values given in parentheses are the percentage difference between measured and predicted values.

Table 4. Apparent Thermal Conductivity versus Test Temperature
Difference - SI Units

| Temperature (K) | | Apparent Thermal Conductivity (w/mK) | | |
Mean	Difference	Measured	Three-Region	Scattering
297.1[a]	27.7	.0449	.0449[b] (0.0)	.0449[c] (0.0)[d]
297.9	59.2	.0454	.0451 (−0.6)	.0451 (−0.6)
297.2	39.8	.0453	.0449 (−1.0)	.0450 (−0.6)
297.3	9.45	.0447	.0447 (0.0)	.0449 (0.3)
297.5	3.77	.0449	.0449 (0.0)	.0450 (0.3)

[a]This specimen (1204-3) had a density of 10.5 kg/m^3 and was
tested at 7.32 cm.

[b]According to the three-region approximation, $\alpha = 4.10$ cm^{-1}.

[c]According to the scattering solution, $\sigma = 3.94$ cm^{-1}.

[d]Values given in parentheses are the percentage difference between
measured and predicted values.

R value and the measured recovered-thickness R value. Because
density measurements were not made on the metered section and because the previous agreement was excellent, it must be concluded
that differences in density measurement of the order 3% produced
these discrepancies between the predicted and measured values for
k_a for the R-19 insulations.

Finally, the predicted variations of k_a with test temperature difference are in excellent agreement with the measured values
and show, as does the original data, that the effect of $T_H - T_C$ is
negligible (see Table 4). Furthermore, it must be stated that each
set of k_a versus T_m data fits a linear line to within an error of
$\pm 1\%$ and the relationship shown in Eq. (13) to within 0.05%. It is
therefore not surprising that a dependence on $T_H - T_C$ was not found,
since the measurements of any thermal conductivity data that have a
linear temperature dependence will not show a dependence on the
test temperature difference.

CONCLUSIONS

1. The three-region approximate solution and the pure scattering solution to conductive and radiative heat transfer may be
used as models for the analysis and extrapolation of the apparent
thermal conductivity of semitransparent materials.

2. Using measured k_a versus ρ data as a starting point, the limiting case solutions are excellent predictors of k_a at other values of T_m and $T_H - T_C$ for R-11 fiberglass and rock wool insulation. The predictions for R-19 fiberglass insulations are not as good. However, it is believed that this results from improper density measurement and is not the fault of the models or the k_a measurements.

3. The apparent thermal conductivity of R-11 fiberglass and rock wool insulations is linearly dependent ($\pm 1\%$) on T_m between 255 and 339K (0 and 150°F).

4. The apparent thermal conductivity of R-11 fiberglass and rock wool insulations is not dependent upon test temperature differences up to 61K (110°F).

REFERENCES

1. H. A. Fine, S. H. Jury, D. W. Yarbrough, and D. L. McElroy, "Analysis of Heat Transfer in Building Thermal Insulation", ORNL/TM-7481, Oak Ridge National Laboratory, Oak Ridge, Tennessee (December, 1980).
2. H. A. Fine, S. H. Jury, D. W. Yarbrough, and D. L. McElroy, Heat transfer in building thermal insulation-- the thickness effect, ASHRAE Trans., in press.
3. R. P. Tye, A. O. Desjarlais, D. W. Yarbrough, and D. L. McElroy, "An Experimental Study of Thermal Resistance Values (R-Values) of Low-Density Mineral-Fiber Building Insulation Batts Commercially Available in 1977", ORNL/TM-7266, Oak Ridge National Laboratory, Oak Ridge, Tennessee (April 1980).
4. E. M. Sparrow and R. D. Cess, "Radiation Heat Transfer", Brooks/Cole Publishing Co., Belmont, California (1970).
5. M. A. Heaslet and R. F. Warming, Radiative transport and wall temperature slip in an absorbing planar medium, Int. J. Heat Mass Transfer 8:979 (1965).

ON THE THERMAL CONDUCTIVITY OF CERAMIC FIBROUS AND RIGID INSULATIONS

P. Bröckerhoff

Institut für Reaktorbauelemente
Kernforschungsanlage Jülich GmbH
D 517 Jülich/West Germany

ABSTRACT

Investigations on the effective thermal conductivity of insulating materials have been carried out. In the first step, small specimens were installed between electrically heated plates. The stationary method of Pönsgen was applied. The thermal conductivities of fibrous and rigid materials were determined for various pressures, temperatures, and coolant media. Full-size systems were tested in a high pressure wind tunnel in a horizontal position at different pressures, temperatures, media, and axial pressure gradients.

INTRODUCTION

Contrary to normal technical applications, the insulations of gas-cooled nuclear reactors are placed on the inner walls of the vessels and the hot gas ducts. Thus, these insulating systems are in direct contact with the hot gas. Their main object is to protect the pressure bearing walls from high temperatures. Furthermore, the thermal fluxes have to be low because of the efficiency of the power plant.

Since the insulating systems are filled with gas at high pressure, natural convection is a greater problem than in case of outer insulations, e.g., buildings that are filled with air at ambient pressure. However, free convection within the structures must be avoided or held at a low level. Additionally, differences of static pressure due to pressure gradients, which are caused by the flowing medium, may lead to an increase of thermal fluxes, especially in bends.

The insulating structures must satisfy severe requirements. Usually damages cannot be repaired, because the insulations are installed in areas of difficult access. In the case of rapid depressurization, e.g., during operation or after rupture of a hot gas duct, pressure differences between the spaces of the insulations and the inner free cross section occur, which may damage the insulation.

AIM OF THE WORK

The design data of hot duct insulations depend upon the reactor type. In a High Temperature Reactor for direct cycle (HHT), a gas temperature of 850 $^\circ$C is provided; in the case of the project Prototype Nuclear Process Heat (PNP) a temperature of 950 $^\circ$C is promised. The corresponding pressures are 7×10^6 Pa and 4×10^6 Pa, respectively. Helium is used as coolant. The mean velocities are about 120 m/s for HHT and 60 m/s in case of PNP. For the vessel insulations, the design conditions are less severe.

The selection criteria applied to the materials and insulation structures are: low effective or apparent thermal conductivities, compatability with other materials (e.g., with the coolant), long-term stability (e.g., mechanical resistance over the lifetime of the power station), and chemical and physical resistance. Thus, a very comprehensive program has been agreed upon by the partners of the above-mentioned projects. In addition to the investigations on the behaviour during rapid depressurization, KFA was responsible for thermal conductivity measurements of materials and for the thermal performance tests of complete insulations. An extensive experimental program was performed to study the influence of pressure, temperature, and medium on the thermal conductivities of different samples. Investigations on full-size systems were necessary, because the homogeneity of these structures is disturbed for constructional reasons. Thus, there exist gaps through which hot gas may pass leading to inadmissibly high temperatures of the metallic bearing tube and causing additional thermal fluxes and apparent thermal conductivities.

The effective thermal conductivity of porous insulating materials is a complex property. It is composed of four parts: The first contribution is heat conduction through the solid. The second item is heat conduction through the gas within the pores and cells of the structure. The third part is radiation between the surfaces of the insulations. Radiation becomes important at high temperatures, temperature differences, and highly porous materials. The fourth part is convection, which also occur under certain conditions. Free convection is caused by differences in coolant density due to nonuniform temperature distribution and high static pressure. Forced convection occurs in the case of pressure gradients due to the velocity of the medium. Convection within the insulations can be suppressed or reduced by increasing the flow resistance. Thus, the permeability is diminished. That could mean a disadvantage in case of depressurization.

The complicated heat transfer mechanism can be described by the effective thermal conductivity, λ_{eff}, which is defined by the Fourier equation

$$q_t = \lambda_{eff} \cdot \frac{\Delta T}{d} \qquad (1)$$

Here q_t is the total heat flux, d is the thickness of the insulation, and ΔT is the temperature difference across the insulation structure. In the literature, the apparent conductivity often is defined as sum of the particular contributions, e.g., conduction through the solid and the spaces filled with gas, radiation, and free and forced convection. Sometimes more or less complicated expressions have been derived. Under certain conditions, gas conduction is the most important mechanism of heat transfer. This is the case for low pressure and in stagnant gas.

DESCRIPTION OF EXPERIMENTAL APPARATUS

For the screening tests on small samples, measurements were carried out in two environments by means of the guarded hot plate method of Pönsgen (1912). This method was chosen instead of the hot wire method, which is also often used. The guarded hot plate measurement is either stationary or nonstationary, according to the time dependence of the temperatures within the samples. The nonstationary method is not appropriate for testing the influence of radiation, as mentioned by v.d. Held (1952) and Vos (1957). The nonstationary hot wire method is based upon the measurement of a temperature increase of a line shaped heat source that is embedded in the specimen. The electric power produced is constant over the time and the length of the test sample. A detailed description is given in DIN 51046 (1973). The standard of the guarded hot plate technique is described in the U.K., U.S.A., and Germany by the British Standard 874 (1973), ANSI/ASTM Standard C 177-76 (1976), and DIN 52612 (1972), respectively. A technique often used in laboratories requires a carefully calibrated sample for reference purposes. In this case the dimensions of the test samples, however, are very small. Thus, this method did not seem appropriate for the provided measurements. The flash method was eliminated, because fibrous materials would also be tested. As pointed out by Corsan and Williams (1980), these steady-state techniques are very simple. However, the accuracy of the measurements depends

mainly on the quality of the equipment and care taken in fitting
the thermocouples.

A scheme of the test facility is shown in fig. 1. A total of
four heaters was used. The main heater was square with a length of
250 mm. The square heater reommended by Cammerer (1960) was chosen
since it produces a more uniform heat flux than a round heater.

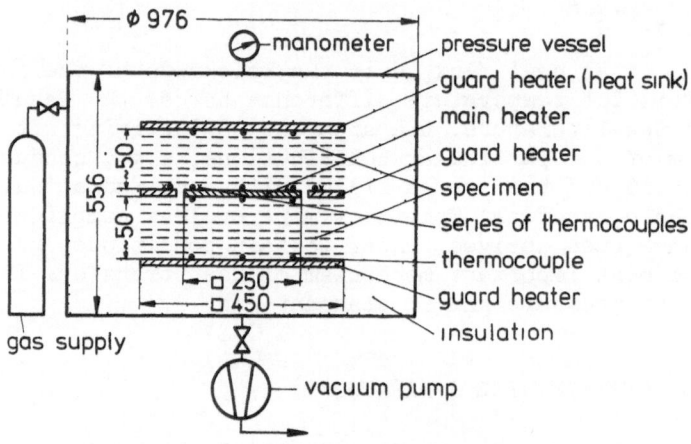

Fig. 1. Test facility.

The main heater was surrounded by a guard heater to avoid radial
heat losses. To ensure that the temperature difference over the
gap between both heaters was nearly zero, a series of thermocouples consisting of 16 single elements was installed. The heat produced by the main heater was allowed to flow only perpendicular to
the main heater surface. If the heat was flowing upwards, the lower
heating system was held at the same temperature level as the main
heater. The positions of the thermocouples that were fixed to the
heaters are marked in fig. 1. In case of samples with high thermal
conductivities, a water cooling system was used as the heat sink.
To get temperatures as high as possible inside the specimens, the
temperature differences across the insulations were kept mostly at
50 K. The samples were situated between the main heater and the
heaters of the top and the bottom. The heating system was completely insulated by Kaowool to avoid excessive heat losses to the
surroundings. A vacuum pump at the bottom of the vessel allowed a
vacuum of about 0.01 Pa. Helium was supplied from a gas tank.

The heaters consisted of metallic plates and wires that were
bent in a meandering shape. Glimmer and Kaowool papers were used as

electrical insulators between plates and wires. The thermocouples were positioned in such a manner that the temperatures in the middle of the main system and near the edges and those of the guard systems could be measured. They were placed on the surfaces of the metallic plates, if the conductivities of the samples were low. In case of elevated conductivities, they were installed in small grooves in the rigid ceramic samples because of the temperature difference between the heaters and the specimens. If the regulating systems were working well, the temperature differences of the thermocouples at the main heater and the surrounding guard heater should be as low as possible.

The effective thermal conductivity of the samples was calculated from

$$\lambda_{eff} = \frac{U \cdot I}{S} \cdot \frac{d}{\Delta T} \tag{2}$$

where U and I are the voltage and current of the main heater and S is the surface area of the main heater.

Nearly full-size systems were tested in a closed high pressure wind tunnel at pressures up to 4×10^6 Pa and temperatures up to $400 \ ^\circ C$. Both air and helium were used as coolant media. A test facility at KFA in which helium temperatures of $850 \ ^\circ C$, pressures of 5.1×10^6 Pa, and velocities of about 120 m/s has been reached was successfully operated. Two tubes were declared test sections. They are insulated with a metallic foil and a ceramic fibrous insulation.

RESULTS

At the beginning of the experimental research, ceramic fibrous materials were tested. In fig. 2 the dependence of the apparent thermal conductivities upon the mean temperature within Kaowool samples is given. The main components of Kaowool are SiO_2 and Al_2O_3. Gulf General Atomic uses Kaowool for the insulation of the concrete reactor vessel of the Fort St. Vrain reactor. In Great Britain it is also used in the nuclear power stations in Hunterston and Hinkley. The presented results are valid for experiments in vacuum and in air and helium at atmospheric pressure. The bulk densities of the samples were 95 kg/m³ and 165 kg/m³, respectively. In the lower density sample, the fibers become more transparent for heat radiation. Thus, the thermal conductivity is higher in vacuum and in air than the higher density samples. This is not true when using helium, at least in the low temperature region. For comparison the results of MITI and ERANS (1979) are shown. The latter values hold for a bulk density of 192 kg/m³. Our own results are in relatively good agreement with those mentioned in literature, which lie in a broad scatter band.

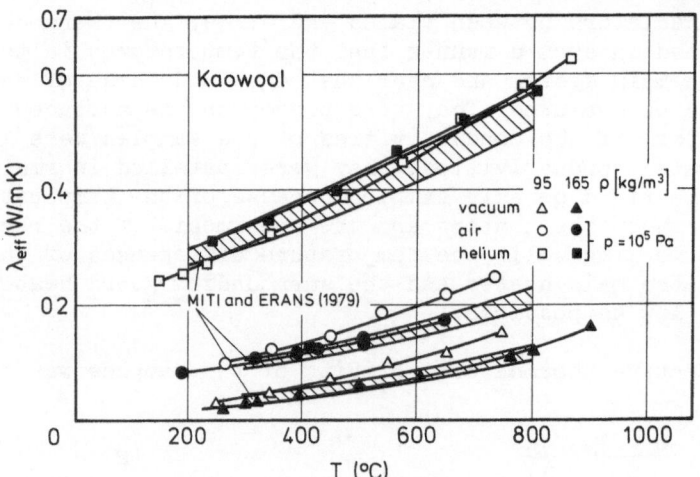

Fig. 2. Thermal conductivities of Kaowool versus
        mean temperature.

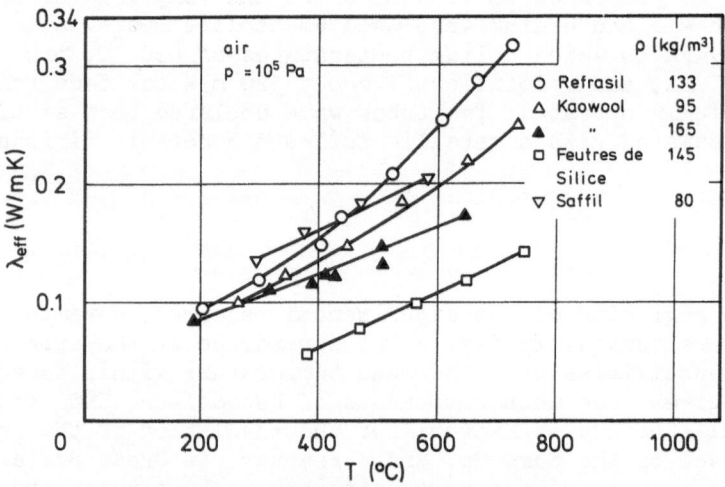

Fig. 3. Effective thermal conductivities of fibrous
        materials versus mean temperature.

    As shown in fig. 3, the apparent thermal conductivities in-
crease over the entire temperature range for the fibrous materials.
The results hold for air at a pressure of 10^5 Pa. Four different
materials have been investigated. The bulk densities are given.
Although the porosity of the samples and thus the density normally
are a measure for heat losses, the lowest conductivities were

gained with Feutres de Silice whose density is 145 kg/m^3. The density of the Refrasil sample is only 12 kg/m^3 lower. Nevertheless, the
measured values are nearly two times higher over the entire temperature range. The structure of the blankets, e.g., the direction
of fibers and their diameter, seems to be more effective than density. The conductivities of the Saffil sample, whose density was
only 80 kg/m^3, are slightly higher in the low temperature region
than those of Refrasil and Kaowool.

When using helium as environment the thermal conductivity of
the gas becomes more effective (see fig. 4). In comparison with
the previously discussed figure, there is an increase of conductivity to about 0.7 W/mK at a mean temperature of 800 $^\circ$C for Refrasil. Also in case of Feutres de Silice, the apparent conductivities

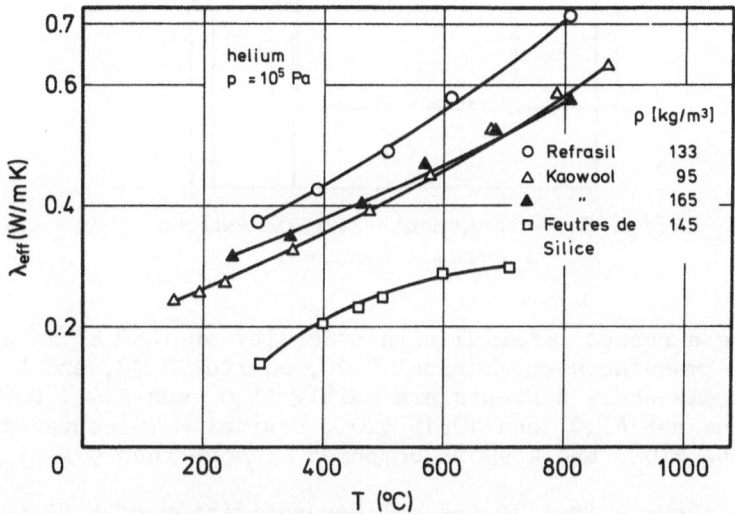

Fig. 4. Effective thermal conductivities of fibrous
        materials versus mean temperature.

are nearly two times higher than those gained with air. As discussed in fig. 2, the influence of bulk densities is less dominant
for Kaowool. The values differ only by 10% at a mean temperature
of 300 $^\circ$C. This confirms the effect due to the high contribution
of gas, which is dominant in comparison with radiation and conduction through the pure material.

Further investigations were carried out on rigid materials.
The samples were subdivided in the same way as the metallic heaters.
Figure 5 shows the subdivision. The specimen is situated in the
middle of the whole arrangement. The length of the square plate
was 250 mm. The four guard systems are of rectangular shape with a

length of 352 mm and a width of 98 mm. The gaps between the central
plate and the surrounding pieces were filled with Kaowool paper to
minimize heat losses.

Investigations on Si_3N_4, Durital E 90, and MPK 34/90 have
been carried out with air and helium at pressures up to 4×10^6 Pa.

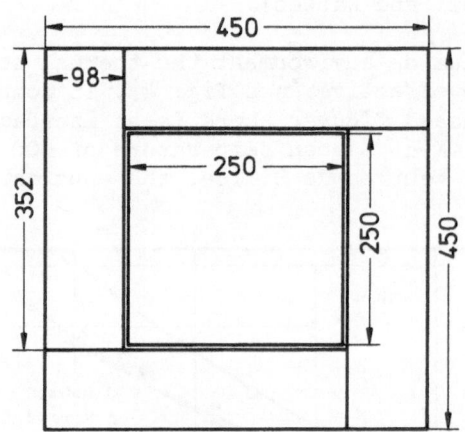

Fig. 5. Arrangements and dimensions
of ceramic samples.

MPK 34/90 is a porous material with a density of 1450 kg/m^3 as
given by the manufacturer. Durital E 90, Durital M 70, and K 99 are
dense materials whose contents are mainly Al_2O_3 and SiO_2. Durital
E 90 contains 89% Al_2O_3 and 10.1% SiO_2, Durital M 70 contains 73%
Al_2O_3 and 25% SiO_2, and K 99 is composed of more than 99% Al_2O_3.

Fig. 6 shows effective thermal conductivities of K 99 versus
mean temperature. The results hold for air at 10^5 Pa and 2×10^6 Pa,
respectively. The thickness of the distance, d [see equation (1)],
was 43 mm. There is relatively good agreement between the results
given by the manufacturer and those obtained at ambient pressure.
In contrast to the thermal conductivities of fibrous materials,
the values decrease with increasing temperatures. Also, the results
are much higher. In the lower temperature range, the values at
elevated pressures are much higher than those at atmospheric pressure. At temperatures of about 500 $^\circ$C, all the presented results
are nearly identical.

In fig. 7 the dependence of apparent thermal conductivities
on mean temperature are represented for four different rigid materials. The environment was helium at 10^5 Pa. The slopes of the
curves are almost zero for Si_3N_4, Durital E 90, and Durital M 70
at temperatures above 600 $^\circ$C. Because of the high thermal conduc-

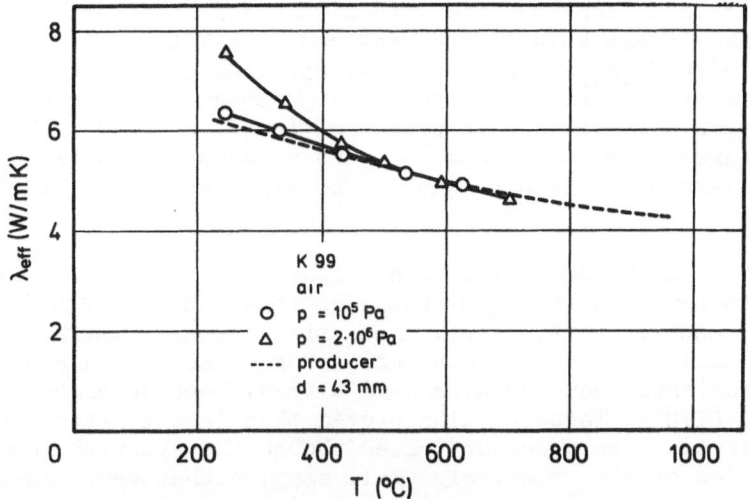

Fig. 6. Effective thermal conductivities of K 99 versus mean temperature.

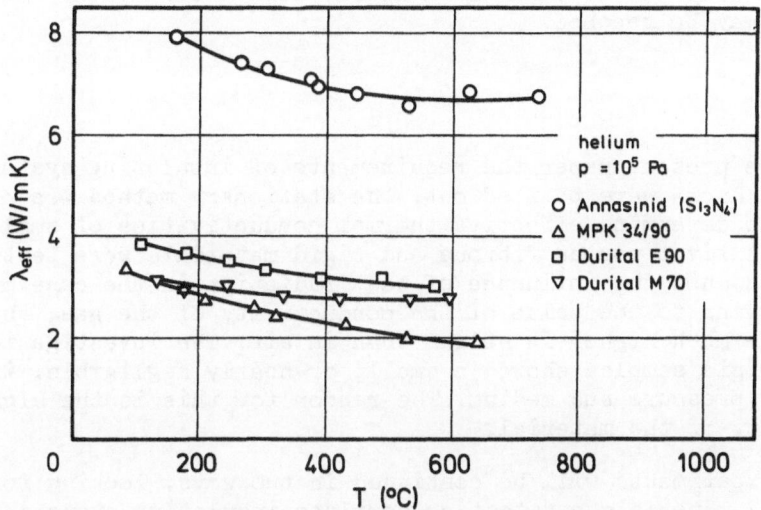

Fig. 7. Effective thermal conductivities of rigid materials versus mean temperature

tivity of the materials, only a maximum mean temperature of 630 °C could be reached. When comparing the results in air and helium one observes but a small influence of medium. This is due to the high amount of solid conduction of the pure material. With increasing porosity and lower density, the effective thermal conductivities decrease. The lowest values were measured with MPK 34/90. In comparison with Si_3N_4 and both Durital materials, the MPK 34/90 is much more appropriate for insulating purposes. Further experiments on rigid ceramics are in progress, (e.g., on K 99 in helium and at different pressures) and more are planned for the future. A Glasrock sample has been prepared. The experiment will start soon.

In the high pressure wind tunnel four insulations have been tested up to now in a straight tube. In some cases free and forced convection occurred within gaps. Thus, the effective thermal conductivities normally increased with increasing pressure and hot side pressure gradients. Some results have already been described by Bröckerhoff (1980). These results proved that investigations on full-size systems are necessary. Even though the apparent thermal conductivities of the relatively small samples that were tested in stagnant gas are well known, it is impossible or difficult for engineers to design complete structures to satisfy the mentioned requirements. At present, experiments are being conducted on an inner insulated bend. Two test pieces are prepared for investigations in a test facility in which nearly design conditions for a HHT plant may be reached.

CONCLUSIONS

In the present paper the requirements of insulating systems of hot gas ducts were pointed out. The stationary method was employed to determine effective thermal conductivities of small samples. At first porous fibrous and rigid materials were tested. The results show the influence of heat radiation in the case of Kaowool. Owing to the value of the conductivity of the gas, the results are much higher in helium than in air. The investigations on dense rigid samples showed a small, or nearly negligible, influence of pressure and medium. The reason for this is the high conductivity of the material.

The experiments will be continued in two ways: looking for appropriate materials and testing complete insulating systems. The knowledge of the thermal conductivities in different environments at elevated pressures and temperatures is not sufficient for designers and engineers. Full-size systems must be tested since they often consist of different materials. Thus, structures are provided with inner shrouds or cover plates composed of ceramic materials. Behind these components, materials of low thermal conductivities are needed. Final judgements on the thermal performance and efficiency of complete insulations can only be done after testing them at temperatures and pressures that are expected in a nuclear power plant.

REFERENCES

Bröckerhoff, P., 1980, The behaviour of thermal insulations for hot
    gas ducts of high temperature reactors, in "Proceedings
    of the ANS/ASME/NRC International Topical Meeting on
    Nuclear Reactor Thermal Hydraulics", Vol. 3, Office of
    the Nuclear Regulatory Research U.S. Nuclear Regula-
    tory Commission, Washington, pp. 2337-2351.
Cammerer, W.F., 1960, Die Messung der Wärmeleitfähigkeit von Iso-
    lierstoffen bei tiefen Temperaturen, Kältetechnik 12:
    107-110.
Corsan, J.M. and Williams, I., 1980, "Errors Associated with Imper-
    fect Surfaces in Standard Hot-Plate Thermal Conducti-
    vity Measurements", NPL report QU 57, National Physi-
    cal Laboratory, Teddington.
v.d. Held, E.F.M., 1952, The contribution of radiation to the con-
    duction of heat, Appl. Sci. Res. A,3:237-249
Pönsgen, R., 1912, Ein technisches Verfahren zur Ermittlung der
    Wärmeleitfähigkeit plattenförmiger Stoffe, Z.Ver.dt.
    Ing., Bd. 56:1653-1658.
Vos, B.F., 1957, Determination of the thermal conductivity of in-
    sulating materials by a non-steady state method down
    to approx. -200°C, Supplement Bull. Inst. Int. Froid,
    Annexe 1:47-54.
MITI, (Agency of industrial science and technology), ERANS (Engi-
    neering research association of nuclear steelmaking),
    1979, "Progress Report (Japan), National Research &
    Development Program on Direct Steelmaking Using High
    Temperature Reducing Gas".
ANSI/ASTM Standard C 177-76, 1976, "Standard Test Method for
    Steady-State Thermal Transmission Properties by Means
    of the Guarded Hot Plate", American Society for Test-
    ing and Materials, Philadelphia.
British Standard 874, 1973, "Methods for Determining Thermal Insu-
    lating Properties with Definitions of Thermal Insula-
    ting Terms", British Standards Institution, London.
DIN 51046, 1973, "Bestimmung der Wärmeleitfähigkeit bei Tempera-
    turen bis 1600 °C nach dem Heißdraht-Verfahren",
    Fachnormenausschuß Materialprüfung (FNM) im Deutschen
    Normenausschuß (DNA).
DIN 52612, 1972, "Bestimmung der Wärmeleitfähigkeit mit dem Plat-
    tengerät", Fachnormenausschuß Materialprüfung (FNM)
    im Deutschen Normenausschuß (DNA).

CORRELATION BETWEEN THE STRUCTURAL PARAMETERS AND THE THERMAL

RESISTANCE OF FIBROUS INSULANTS AT HIGH TEMPERATURES

Jean Boulant, Catherine Langlais, and Sorin Klarsfeld

Isover Saint Gobain, Centre de Recherches Industrielles

Rantigny, France

ABSTRACT

The principle and the technical characteristics of an apparatus
for measuring thermal conductivity are described. This apparatus,
of the "guarded hot-plate" type, has been specially designed for
high temperatures and is mainly intended for fibrous insulants.
Because of a significant decrease in the measurement time, repeti-
tive measurements can be made on a large number of samples. This
apparatus has been used to investigate further and more thoroughly
the correlation between the thermal resistance and the structural
parameters of anisotropic insulants, such as porosity or specific
surface. The experimental data are compared with computational results obtained from optical parameters determined on each type of
insulant under study.

INTRODUCTION

The use of rapid apparatus for the measurement of thermal conductivity at temperatures near ambient temperature has enabled us
to considerably increase the number of measurements and consequently to have a better knowledge of fibrous insulants. In particular,
it has been possible :

. to show the statistical nature of the measurement data obtained on products of the same lot (and to determine the governing
parameters),
. to investigate more thoroughly the correlation between the
thermal conductivity and the structure parameters of the insulants
[1,2].

These results concerned mainly building applications, i.e, low-density insulants used at temperatures near ambient. Little attention has been given so far to fibrous insulating products used at high temperatures. This type of product generally has a greater density (100 kg/m^3 or more), but at ambient temperature, it is not strongly affected by variations of structure parameters.

All this led us to undertake a study similar to the one done at ambient temperature on low-density products but this time at high temperatures and on high-density products. This study required an apparatus that could measure thermal conductivity at high temperatures far more rapidly than existing apparatus.

DESCRIPTION OF THE APPARATUS

The general principle of the apparatus is similar to the one previously described for temperatures near ambient in Refs. 3-6. A biguarded hot plate, in which the central part is supplied with stabilized direct current, generates a heat flow through a single sample towards an auxiliary hot plate ensuring the required level of cold temperature.

Several requirements are necessary to allow rapid measurement of the thermal conductivity of insulating products at high temperatures in stationary conditions. The apparatus must be capable of more measurements on the same type of product in order to obtain statistical data. These requirements suggest an apparatus that has :

. a "hot" and a "cold" plate constantly maintained at a given temperature,
. a continuously working change-of-sample system that eliminates a direct contact between the high temperature plates and the ambient air and permits preheating of the samples (fig. 1)
. automatic controllers that minimize the time necessary to reach a stationary condition, ensure constant temperature (T_C and T_H) by a simple display, and simultaneously allow a precise measurement of the power dissipated in the central part of the hot plate (fig. 2).

To meet these requirements, the apparatus has the capability of simultaneously accepting three samples of the same thickness in a measurement tunnel. In fig. 1 the central sample is under measurement, the left sample is to be measured, and the last one has already been measured and is ready to be taken out of the apparatus once a new sample is introduced:the entire system works from left to right. This "three-sample" design limits the lateral losses, but it also should impede the thermal shock caused by the contact between the ambient air and the surfaces of the plates (temperature difference > 350°C).

The tunnel must permit the movement of the samples without
friction. The "cold" plate at the top center of the tunnel can be
slightly displaced vertically to ensure a good contact, under controlled pressure, between the plates during the measurement ("low"
position) and to allow the exit of the sample ("high" position)
when it must be replaced.

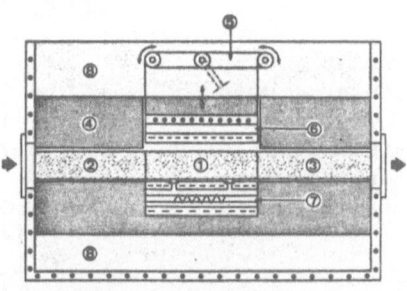

1. Sample under measurement
2. Sample to be measured
3. Measured sample
4. Tunnel in refractory bricks
5. Displacement system
6. Cold plate
7. Hot plate
8. Insulation

Fig. 1.   High temperature bi-guarded hot plate apparatus.

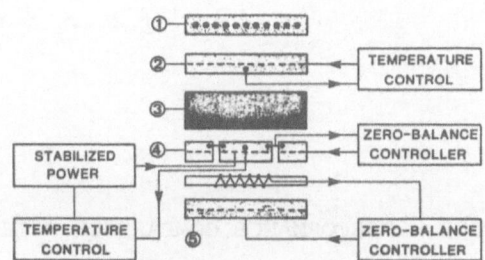

1. Refrigerated heat sink
2. Auxiliary heater unit
3. Sample
4. Heating unit : central and guard ring
5. Lower guard-plate

Fig. 2.   High temperature bi-guarded apparatus.
Temperature and zero-balance controllers.

Table 1.   High temperature biguarded hot plate apparatus.
           Technical characteristics and performances

SPECIMEN
   . surface   300 by 300   mm
   . thickness   38 to 40   mm
   . metering area   150 by 150   mm

GUARDED HP + COLD PLATE
   . total area: center + guard   300 by 300   mm
   . metering area   150 by 150   mm
   . range of possible temperature   100 to 700   °C
   . temperature heterogeneity   < ± 1   °C
   . temperature stability   ± 0,25   °C
   . planeity   0,1   mm
   . parallelism   0,1   mm

THICKNESS MEASUREMENT   ± 0,1   mm

MEASUREMENT DEVICES
   . digital voltmeter   200.000 digits
   . control thermocouples, power   resolution:1μV

TIME NEEDED TO REACH STEADY-STATE   3 to 10 hours
(fibrous insulants)

REPRODUCIBILITY   ± 2 ... 3 %

ACCURACY   ± 2 ... 5 %

TECHNICAL CHARACTERISTICS - PERFORMANCE COMPARISON WITH PREVIOUS
RESULTS

     The technical characteristics and the performance of the appa-
ratus for a mean temperature of 400°C are given in Table 1.

     The time needed to reach steady state, varies between 3 and
10 h depending on the type of insulant.

     Under steady-state conditions, the stability of the hot and
cold temperatures is about ± 0.25°C. The temperature differences
from one point on the surfaces to another are less than 1°C.

     The reproducibility of the apparatus is about ± 2 - 3%.

The accuracy of the apparatus lies somewhere between ± 2% and ± 5%. This order of magnitude has been determined from comparative measurements with other types of apparatus working in a common or a complementary range of temperature (fig. 3, 4 and 5).

Three types of fibrous insulants were used as reference samples. Among those three materials, we find the reference panels RBGFB (resin-bonded glass-fiber board) from the B.C.R. (Community Bureau of Reference, Commission of the European Communities) are generally used between 170 and 370 [7-9].

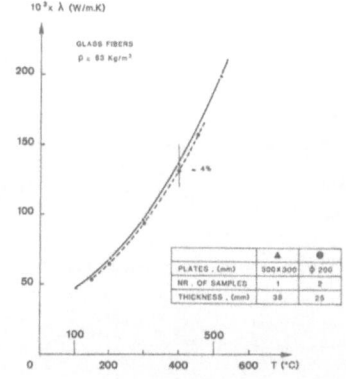

Fig. 3. Thermal conductivity, λ, vs. mean temperature, measured for a glass fiber reference specimen

(▲) high temperature bi-guarded hot plate apparatus.
(●) Dynatech "TCFG" guarded hot plate apparatus.

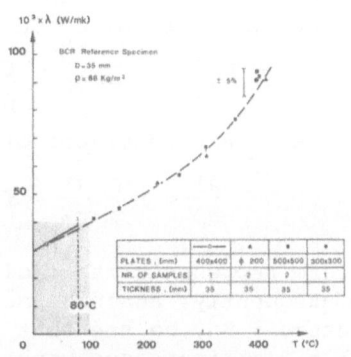

Fig. 5. Thermal conductivity, λ, vs. mean temperature measured for BCR reference specimen
□ CRIR "standard" bi-guarded hot plate apparatus (for measurements near ambient temperature)
▲ Dynatech "TCFG" guarded hot plate apparatus
■ CRIR, 500 x 500 mm guarded hot plate apparatus
● CRIR new "high temperature" bi-guarded hot plate apparatus.

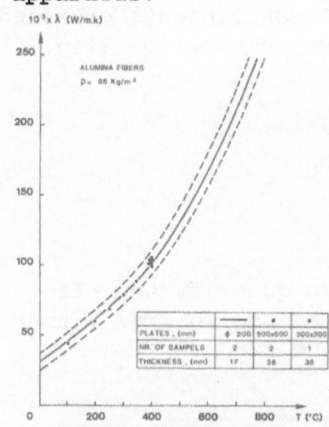

Fig. 4. Thermal Conductivity, λ, vs. mean temperature, measured for an alumina fiber reference specimen. (—) Dynatech "TCFG" guarded hot plate apparatus (mean curve and total statistical dispersion zone) (■) CRIR, 500x500mm, guarded hot plate apparatus, (●)CRIR, new"high temperature" bi-guarded hot plate apparatus.

CORRELATION THERMAL CONDUCTIVITY STRUCTURE

At ambient temperature, we know that the thermal conductivity of a fibrous insulant is mainly influenced by two structure parameters:

- its density, ρ (i.e, its porosity, ε)
- its specific surface, S_v, quotient of the area of the fluid-solid interface by the total volume of the solid matrix

It is of interest to compare the relative influence of these two parameters as a function of temperature.

Influence of Density

We know that for given fiberizing conditions and at a fixed temperature the thermal conductivity can be expressed as [2]

$$\lambda_T = A + B\rho + \frac{C}{\rho}$$

where A, B, C are functions of temperature. Thus, we can write:

$$\frac{\partial \lambda}{\partial \rho} = B - \frac{C}{\rho^2} = f(T, \rho)$$

The $\frac{C}{\rho}$ term represents the radiative component of the heat transfer and, consequently, an increase with temperature (C increases when T increases). So, we see that for a given density, the slope of the curve $\lambda = f(\rho)$ increases with temperature. This means that the influence of density on thermal conductivity is even greater at higher temperatures.

To quantify this influence in a given configuration, we have represented $\partial \lambda / \partial \rho = f(\rho)$ using the B and C coefficients obtained by regression on a fiberglass product. This representation (fig. 6) shows that:

$$(\frac{\partial \lambda}{\partial \rho})_{\substack{400°C \\ 100 \text{ kg/m}^3}} \simeq (\frac{\partial \lambda}{\partial \rho})_{\substack{24°C \\ 30 \text{ kg/m}^3}}$$

Influence of Specific Surface

As was done for density, we can try to quantify the effect of S_v on λ for the products under study. We can see in fig. 7 that there is a linear relationship between λ and $1/S_v$.

If we compute the slopes, we get:

$$\frac{\partial \lambda}{\partial (1/S_v)}\bigg|_{24°C} = 620 \text{ W/m}^2 \text{ K}$$

$$\frac{\partial \lambda}{\partial (1/S_v)}\bigg|_{400°C} = 19310 \text{ W/m}^2 \text{ K}$$

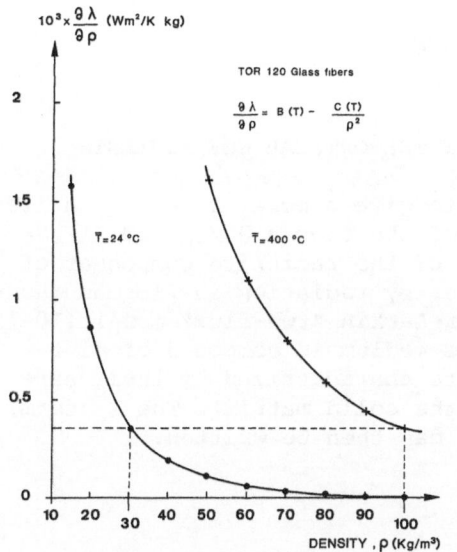

Fig. 6. $\frac{\partial \lambda}{\partial \rho}$ vs. density for TOR 120 glass fibers at two reference temperatures.

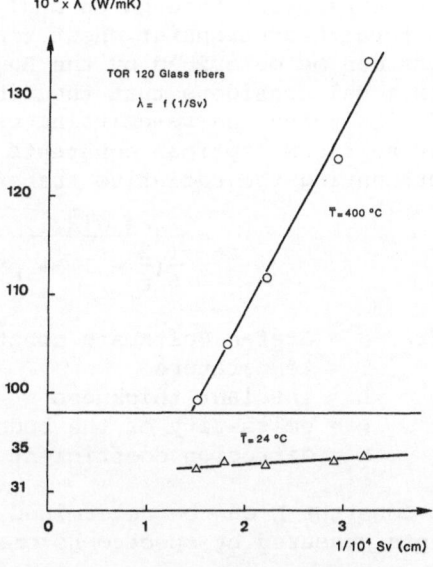

Fig. 7. Thermal conductivity vs. $1/S_v$ for TOR 120 glass fibers at two reference temperatures.

which gives on a "standard" product

$$(-\frac{\partial\lambda}{\partial S_v})_{24°C} = \frac{620.10^{-4}}{(4306)^2} = 3.34.10^{-9} \text{ W/K}$$

$$(-\frac{\partial\lambda}{\partial S_v})_{400°C} = \frac{19.31.10^{-4}}{(4306)^2} = 1.041.10^{-7} \text{ W/K}$$

which confirm that

$$(-\frac{\partial\lambda}{\partial S_v})_{400°C} >> (-\frac{\partial\lambda}{\partial S_v})_{24°C}$$

INTERPRETATION OF THE RESULTS THROUGH THE OPTICAL COEFFICIENTS

In this section, we would like to give a more or less quantitative explanation of the variations of the slopes $\partial\lambda/\partial\rho$ and $\partial\lambda/\partial S_v$ with temperature through an analysis of the radiative component of the total heat transfer. Heat transfer by radiation in fibrous insulants can be described by the Hamaker-Larkin "two-flux" model [10-12]. This model considers that the fibrous medium is composed of diffusing-absorbing and re-emitting centers characterized by their effective surfaces (optical constants of the solid matrix). The λ_r term, representing the radiative transfer, can then be written:

$$\lambda_r = \frac{4 \sigma T^3}{\frac{1}{L}(\frac{2}{\epsilon} - 1) + \rho N}$$

where σ = Stefan Boltzmann constant
        T = temperature
        L = insulant thickness
        ϵ = emissivity of the boundaries
        N = diffusion coefficient (m^2/kg ; independant of ρ)

The constant N can be determined from infrared transmission coefficients measured by spectrophotometry.

If we assume $\epsilon = 1$ and $\frac{1}{L} << \rho N$, we get $\lambda_r \sim 4\sigma T^3/\rho N$ and $\lambda = (A + B\rho + 4\sigma T^3)/\rho N$. Knowing the N coefficients of a few materials from measurements performed in our laboratory we can compute the slopes $\partial\lambda/\partial\rho$ and $\partial\lambda/\partial S_v$ from this last expression of λ

Influence of Density

We have $\frac{\partial\lambda}{\partial\rho} = B - \frac{4\sigma T^3}{\rho^2 N}$

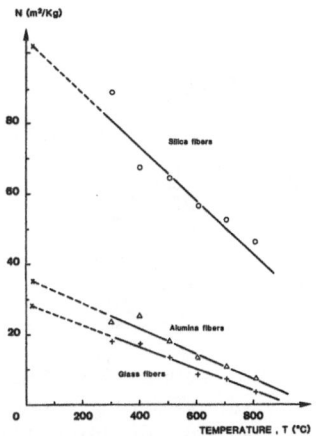

Fig. 8. Diffusion coefficient, N, vs. temperature for three types of mineral fibers.

In fig. 8, we can see that the N coefficient measured on different types of insulating products decreases quasi-linearly with temperature. Therefore, it is clear that for a given density, the term T^3/N increases with temperature. This fact confirms that the influence of density is greater at high temperatures. We can also see that

$$(\frac{\partial \lambda}{\partial \rho})_{\substack{\rho_1 \\ T_1}} = (\frac{\partial \lambda}{\partial \rho})_{\substack{\rho_2 \\ T_2}} \rightarrow T_1^3/\rho_1^2 N_1 = T_2^3/\rho_2^2 N_2$$

For T_1 = 24°C, T_2 = 400°C and ρ_2 = 100 kg/m^3, we have

$$\rho_1 = 100 \sqrt{(297/673)^3} \times \sqrt{N_2/N_1} \sim 29\sqrt{N_2/N_1}$$

Extrapolating the results of fig. 8, we can take $\sqrt{N_2/N_1} \sim 0.8$ (mean value), which gives

$$(\frac{\partial \lambda}{\partial \rho})_{\substack{400°C \\ 100 \text{ kg/m}^3}} \sim (\frac{\partial \lambda}{\partial \rho})_{\substack{24°C \\ 23 \text{ kg/m}^3}}$$

i.e, a result very close to the one previously obtained

Influence of Specific Surface

Using again the expression of λ that we derived, we can write:

$$\frac{\partial \lambda}{\partial S_v} \simeq - \frac{4\sigma}{\rho 2} \times \frac{T^3}{N^2} \times \frac{\partial N}{\partial S_v}$$

or
$$\frac{(\frac{\partial \lambda}{\partial S_v})_{T_1}^{\rho}}{(\frac{\partial \lambda}{\partial S_v})_{T_2}^{\rho_2}} = \frac{(\frac{T_1}{N_1})^3}{(\frac{T_2}{N_2})^3} \times \frac{(\frac{\partial N}{\partial S_v})_{T_1}}{(\frac{\partial N}{\partial S_v})_{T_2}}$$

In first approximation, we will assume that $\partial N / \partial S_v$ does not vary much with temperature. A few measurements (shown in fig. 8) allow us to make this assumption ; we can see that the slope of the curve $N = f(S_v)$ is almost independent of temperature. However, this result should undoubtedly be confirmed. We can write:

$$\frac{(\frac{\partial \lambda}{\partial S_v})_{T_1}}{(\frac{\partial \lambda}{\partial S_v})_{T_2}} = \frac{\frac{T_1^3}{N_1^2}}{\frac{T_2^3}{N_2^2}} = \frac{(\frac{T^3}{N^2})_1}{(\frac{T^3}{N^2})_2}$$

As we have seen before, the term T^3/N^2 increases with temperature and, consequently, if $T_1 > T_2$,

$$(\frac{\partial \lambda}{\partial S_v})_{T_1} > (\frac{\partial \lambda}{\partial S_v})_{T_2}$$

Thus, we confirm that the influence of the specific surface is greater at high temperatures. For $T_1 = 400°C$ and $T_2 = 24°C$,

$$(\frac{N_{400}}{N_{24}})^2 \simeq 0.6 \text{ (mean value from fig.8)}$$

we get

$$\frac{(\frac{\partial \lambda}{\partial S_v})_{24°C}}{(\frac{\partial \lambda}{\partial S_v})_{400°C}} \simeq 5\%$$

i.e, a result close to the one obtained

By taking into account the radiative component of the total heat transfer, we have been able to give a satisfactory qualitative physical interpretation to the correlation of thermal conductivity and structure parameters found experimentally. However, we must not forget the assumptions made in our computations, especially, that the optical diffusion coefficient of our materials is greater than the absorbtion coefficient and that the nature of the solid matrix (i.e, the radiation absorbtion) is negligible. This cannot be considered except as a first approximation that should probably be revised later.

CONCLUSIONS

This new apparatus allows us, to confirm the statistical nature of the results obtained on high-density products of given nominal characteristics and therefore to have a better knowledge of the properties of the products $\overline{\lambda} \pm \sigma_\lambda$.

A single or a few measurements are in no way sufficient to determine the thermal conductivity; by means of the new apparatus, we confirmed that the correlation $\lambda - \varepsilon$, S_v existing at ambient temperature remains valid at higher temperatures but with completely different slopes, $\partial\lambda/\partial\rho$ and $\partial\lambda/\partial S_v$. From a practical point of view, the determination of these slopes is essential, since it is the basis for improving the performances of the products.

Finally, we would like to emphasize that a complete understanding of the correlations found experimentally can only be achieved through knowledge of the radiative heat transfer mechanisms and determination of the optical parameters of the studied materials.

REFERENCES

1. D. Fournier, G. André and S. Klarsfeld, Etudes récentes sur la conductivité thermique des isolants fibreux effectuées à l'aide d'un appareil de mesure rapide de la conductivité thermique en régime permanent, Commission 2 de l'I.I.F., Trondheim 1966. Supplément au Bulletin de l'Institut International du Froid, Annexe 1966-2, p. 163-174.
2. S. Klarsfeld, La conductivité thermique des isolants fibreux en fonction de leur structure, cr du IIe Colloque CSTB/CFI, Cahiers du CSTB 1669, n° 213, (October 1980).
3. D. Fournier and S. Klarsfeld, Mesures de conductivité thermique des matériaux isolants par un appareil orientable à plaque chaude bigardée, Commission 2 et 6 de l'I.I.F., Liège 1969 - Supplément au Bulletin de l'Institut International du Froid, Annexe 1969-7, p. 321-331.

4. C. G. Bankvall, Guarded hot plate apparatus for the investiga
    tion of thermal insulations. National Swedish Building
    Research, Stockholm, Rapport D 5, (1972).

5. D. Fournier and S. Klarsfeld, "Some Recent Experimental Data
    on Glass Fiber Insulating Materials and Their Use for a
    Reliable Design of Insulations at Low Temperature",
    ASTM STP 544, American Society for Testing and Materials,
    Philadelphia (June 1974).

6. M. Degenne, S. Klarsfeld and M-P. Barthe, "Measurement of the
    Thermal Resistance of Thick Low Density Mineral Fiber
    Insulation", ASTM STP 660, American Society for Testing
    and Materials, Philadelphia (1978).

7. R. Doussin, Méthode et matériaux de référence pour la mesure
    de la conductivité thermique, Bull. BNM, 5 (18):(1974).

8. M. Bertasi, G. Bigolaro and F. De Ponte, "Fibrous Insulating
    Materials as Standard Reference Materials at Low Tempera-
    ture", ASTM STP 660, American Society for Testing and Ma-
    terials, Philadelphia (1978).

9. M. Ziebland, Certification report on a reference material for
    the thermal conductivity of insulating materials between
    170 K and 370 K. Resin-bonded glass fibre board. Commis-
    sion of the European Communities BCR Information, 1981,
    (to be published).

10. B. H. Larkin, A study of the rate of thermal radiation through
    porous insulating materials. University of Michigan,
    Ph. D, 1957, 169 p.

11. F. Cabannes, M. Hyrien, S. Klarsfeld and J-C Maurau, Radiative
    heat transfer in fiber-glass insulating materials, in
    connection with their optical properties. 6ème Conféren-
    ce européenne sur les propriétés thermophysiques des ma-
    tériaux, 26 au 30 Juin 1978, Dubrovnik, Yugoslavia.
    High Temperatures, High Pressure, vol. 11, (1979).

12. M. Hyrien, C. Langlais, G. Guilbert and S. Klarsfeld, Corre-
    lation between thermal resistance and structure parame-
    ters of fibrous and cellular media in relation with the
    heat transfer by radiation. Presented at the 7th Euro-
    pean Thermophysical Properties Conference 30 June -
    4 July 1980, Antwerpen (Belgium) (to be published).

PRECISION AND ACCURACY OF GUARDED HOT PLATE METHOD

M. Bomberg and K.R. Solvason

Thermal Properties Section
Division of Building Research
National Research Council of Canada
Ottawa, Canada K1A OR6

ABSTRACT

This paper presents a summary of an investigation of the precision and accuracy achievable with the NRCC 450 mm square GHP with a 300 mm main heater and a 75 mm guard ring heater and a 600 mm GHP with 300 mm main and a 150 mm guard ring heater. Only the uncertainties that may be attributed to the instrument are discussed here.

INTRODUCTION

This paper deals with instrument errors as calculated or measured on "ideal" specimens and illustrates the precision to which the thermal resistance of a stable specimen can be determined at the National Research Council of Canada (NRCC) laboratories.

The basic measurement errors, those of temperature, thickness and power, are listed in Table 1. These uncertainties may be compared with the uncertainty in repeating measurements on the same specimen. The agreement is satisfactory since the repeatability precision determined for both 450 and 600 mm square guarded hot plate (GHP) on the two sigma level was less than about 0.2%.

Having established the basic measurement error, the other errors of the GHP method, i.e., unbalance error, E_g, and specimen edge heat loss error, E_e, can be analyzed.

Unbalance Error

The lateral heat flow across the gap and through the specimen

393

Table 1. Estimated Standard Errors, in Per Cent, and in Thermal
          Resistance Determined with GHP Apparatus on 25 mm Thick
          Specimen. Hot and Cold Temperatures:
          (T_c = 13 ±2°C, T_h = 35 ±2°C)

Source of Error	Repeatability	Uncertainty of the Tests	
		Standard	Differential
Power on main heater			
voltage divider	–	0.06	0.06
potentiometer	0.02	0.15	0.15
standard resistors	–	0.03	0.03
potentiometer	0.002	0.06	0.06
Temperature difference			
cold plate thermocouples	–	0.50	–
potentiometer	0.04	0.04	–
hot plate thermocouples	–	0.30	–
potentiometer	0.02	0.04	0.04
differential measurements	–	–	0.50
Thickness measurements	0.04	0.04	0.04
Metering area (at specified temperature)	–	0.001	0.001
Unbalance error	0.02	0.06	0.06
The uncertainty estimated with two standard deviations	0.13	1.2	1.1

can be expressed as follows[1]:

$$Q_g = \Delta T_g \ (C_1 + C_2 \ \lambda_x) \tag{1}$$

where

C_1 is the total conductance across the gap from the metering
      area to the guard area (W/K). It is the sum of all the
      conductances multiplied by the areas for the heat flow paths
      across the gap, i.e., through the air gap between the face
      plates, through the heater core insulation or other mechani-
      cal connections between main and guard areas, and through
      thermocouple and thermopile wires crossing the gap.

$C_2\lambda_x$ is the lateral heat flow through the specimens per unit of temperature difference (W/K). The heat flow is assumed to be proportional to the lateral conductivity of the specimens, λ_x, (W/mK).

C_2 then is the effective area for heat flow parallel to the plate divided by the effective length of the heat flow path (m^2/m or simply m).

Equation (1) is identical to an equation of Woodside and Wilson[2] for isotropic materials ($\lambda_x = \lambda_y$) and for test conditions where the specimen contribution to the heat flow described by Eq. (1) is constant, and thus the parameter $C_2 \lambda_x$ becomes a constant.

C_1 and partly C_2 can be considered characteristic of the plate, both are proportional to the main heater perimeter and roughly inversely proportional to gap width. C_1 is also influenced by the size, length, and number of thermocouple and thermopile wires crossing the gap; the thickness of the face plates; and the thermal conductance of any material that bridges the gap. The value of C_1 may only be constant if the edge of the main heater face plate is isothermal and the inside edge of the guard ring face plates is isothermal. Since C_1 represents the sum of the heat flow through at least three paths, a ΔT may exist on one path even though the mean ΔT for the whole gap is zero.

The heat flow through the metered area of the specimen normal to the plate is:

$$Q = \Delta T \cdot A \cdot \frac{\lambda_y}{L} \tag{2}$$

where

$\Delta T = T_h - T_c$, the temperature difference between the hot and cold plates (K)

A = the area of the main heater, that is, to the centre of the gap (m^2)

λ_y = thermal conductivity of the specimen in the direction normal to the plate (W/mK)

L = thickness of the specimen (m).

The power, Q', supplied to the main heater differs from Q by the lateral heat flow, Q_g.

$$Q' = Q + Q_g \tag{3}$$

The unbalance error then is:

$$E_g = \frac{Q_g}{Q} = \frac{Q_g}{Q'-Q_g} \tag{4}$$

When Q_g is very small compared with Q', Eq. (4) can be simplified to:

$$E_g = \frac{Q_g}{Q'} \tag{5}$$

In this report, the unbalance error, E_g, is defined as a ratio of heat flow across the gap between metering and guard ring heaters to the power supplied to the metering area. The uncertainty of the unbalance error is ±0.06%. The repeatability of the unbalance error is ±0.02%.

Error Due to Specimen Edge Losses

When the ambient air temperature is maintained at the mean temperature of the specimen, half of the specimen edge temperature is higher and half lower than the ambient; the result is heat exchange, and a distortion of heat flow lines. Any difference between the average heat flux in the metering volume of the specimen and the heat flux on the main heater causes so-called edge loss error.

The error caused by multidimensional heat flow may be reduced in the following ways:

(1) By limiting sample thickness relative to the width of the guard heaters.

(2) By using additional edge insulation.

(3) By limiting unbalance between the main and guard heaters since unbalance creates additional distortion of the heat flow lines.

(4) By maintaining ambient air temperature equal to the specimen mean temperature and, therefore, limiting the difference between heat loss and heat gain at the specimen edges.

The edge loss error depends on specimen thickness, edge insulation, and ambient air temperature. Thus, the accuracy with which it can be determined depends largely on how well all of the other errors can be evaluated, on the testing technique, and on the characteristics of specimens used for the determination. These factors are discussed further in the experimental analysis.

Experimental Analysis of Edge Loss Error

Let us assume that the unbalance errors have already been determined, necessary corrections introduced, and the only error of the GHP apparatus is the edge loss error.

Edge losses are expected to depend on spatial distribution of the ambient temperature (variation between sides or elevation), ambient temperature variation with time, difference between mean specimen and mean ambient air temperatures as well as the level of edge insulation applied in the test.

The effect of edge insulation and ambient air temperature was checked on 25 mm thick polystyrene specimens tested in the 450 mm GHP with a 300 mm metering area. Figure 1 shows the thermal resistance of a 150 mm polystyrene stack composed of six specimens divided with paper septa.

The stack of polystyrene specimens was tested with and without different edge insulations (R_e = 1.7, 2.4, and 2.9 m^{2}K/W).

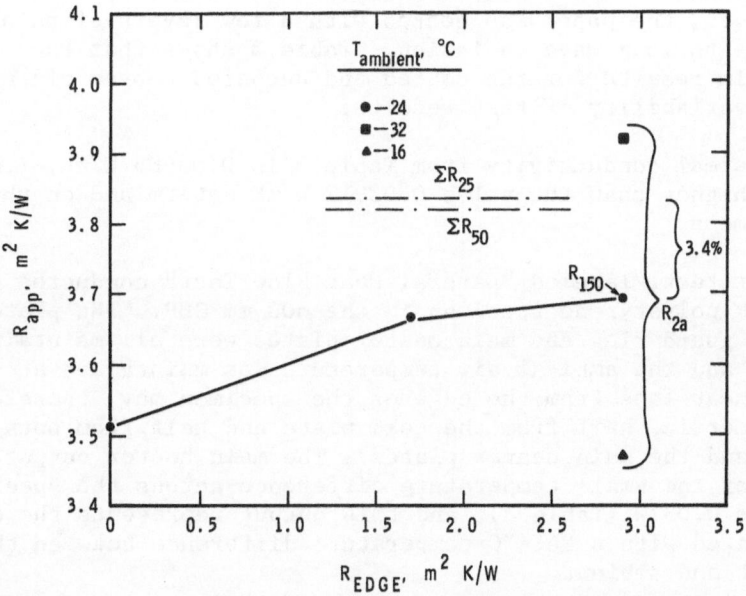

Fig. 1. Apparent thermal resistance of 150 mm thick polystyrene/paper stack, tested on 450 mm GHP as a function of edge insulation. A sum of thermal resistance of 25 and 50 mm layers is shown also.

Figure 1 shows that the lack of edge insulation introduced significant errors for the 150 mm thick polystyrene specimen. The difference between values obtained without edge insulation and with insulation having an R_e = 2.9 m^2K/W tested at ambient temperature equal to mean specimen temperature was 4.6%. Although the presence of some edge insulation (R_e = 1.7 m^2K/W) increased the apparent thermal resistance by 4.3%, the additional increase of edge insulation to R_e = 2.9 m^2K/W was not very effective and resulted in a further increase of only 0.3%. The test results with the ambient temperature at 16°C and 32°C differed from those with an ambient temperature of 24°C by about -7% and +7% respectively. This indicates that for thick specimens ambient temperature must be controlled fairly close to specimen mean temperature.

The apparent resistance (3.694 m^2K/W) obtained on the 149.4 mm specimen with R2.9 edge insulation and with ambient air temperature equal to specimen mean temperature was about 3.3% lower than the sum of the individual R values for the six 25 mm thick layers as reported in Table 2.

Thermal resistance of the same material was determined in the other GHP using other pieces of polystyrene to build up the test specimen for the 600 mm square plate.

In addition to a standard Kraft paper septum, a special septum was used, i.e., the paper was coated with a few layers of paint known to be opaque to long wave radiation. Table 3 shows that the difference in results for the coated and uncoated septum did not exceed the variability of test results.

The thermal conductivity from Table 3 is 0.03985 W/mK, i.e., about 1.8% higher than the value 0.03912 W/mK determined on the 25 mm thick specimens.

Another test, labeled "Lateral Heat Flow Test" conducted on a 150 mm thick polystyrene specimen in the 600 mm GHP. The plates, cold plate, guard ring and main heater plates were all maintained at about 35°C, and the ambient air temperature was maintained at about 10°C. The heat loss from the edge of the specimen may, therefore, be assumed symmetric, half from the cold plate and half from both the guard ring and the main heater plates. The main heater output, corrected for the small temperature difference across the specimen (0.04°C) was 0.057W (Table 4), and this output represents the edge loss associated with a 26.4°C temperature difference between the heater plate and ambient.

In a standard test, the specimen is exposed to a temperature difference that varies from $(T_h - T_c)/2$ to zero or to a mean temperature difference of $(T_h - T_c)/4$. The percentage error indicated by this heat flow can then be expressed as:

Table 2.   Apparent   Thermal Resistance   of Separate or Stacked 25 mm
Polystyrene Specimens at 24°C   Tested With Paper Septum on
450 mm GHP Apparatus

Test Code	Specimens Tested	Thickness mm	R m^2K/W	λ W/mK
60–74	3+4/5+8	50.00	1.272	0.03930
–85	3+4/5+8	50.07	1.272	0.03936
–76	7+10/9+12	48.87	1.247	0.03921
–77	1+2+3+4/5+6+8+11	100.14	2.510	0.03990
–129	1+2+3+4/5+6+8+11	100.12	2.509	0.03991
–127	1+2+3+4/5+6+8+11	100.15	2.527	0.03964
–128	1+2+3+4/5+6+8+11	100.10	2.520	0.03971
–92	4/8	25.04	0.640	0.03913
–93	3/8	25.02	0.641	0.03903
–94	4/5	24.97	0.637	0.03923
–91	3/4	24.97	0.638	0.03913
–96	3/4	24.99	0.637	0.03922
–97	3/4	24.97	0.638	0.03913
–116	1/2	24.89	0.637	0.03910
–117	1/6	25.02	0.640	0.03910
–118	1/11	24.82	0.634	0.03911
–119	2/11	24.84	0.635	0.03912
–120	2/6	25.02	0.640	0.03907
–73	1+2/6+11	49.91	1.272	0.03925
–134	1+3+4/2+5+8	75.03	1.892	0.03965
–68	1+2+3+4+7+10/5+6+8+9+11+12	149.42	3.694	0.04045
–121*	1+2/6+11	49.85	1.234	0.04041
–122*	1+2/6+11	49.84	1.236	0.04031

Averages: at 25 mm λ_a = 0.03912 (average 3/4 λ_a = 0.03916)

   with paper
      septum: at 50 mm λ_a = 0.03931   at 100 mm λ_a = 0.03979
                  at 75 mm λ_a = 0.03965   at 150 mm λ_a = 0.04045
   without paper
      septum: at 50 mm λ_a = 0.0404

Densities of the specimens:

1	15.7 kg/m³	7	15.1 kg/m³
2	15.7 kg/m³	8	15.7 kg/m³
3	15.8 kg/m³	9	15.0 kg/m³
4	15.7 kg/m³	10	14.6 kg/m³
5	15.2 kg/m³	11	15.2 kg/m³
6	15.6 kg/m³	12	14.5 kg/m³

Table 3. Apparent Thermal Resistance of 150 mm Polystyrene/Septum
          Stack Tested on 600 mm GHP at 24°C Mean Temperature

Test Code	Septum	R m^2/W	λ W/mK
360–164	Paper	3.745	0.03984
−180	"	3.757	0.03969
−181	"	3.746	0.03981
	Paper	Mean	0.03978
360–173	Paper and Paint	3.754	0.03980
−174	"	3.731	0.04005
−175	"	3.720	0.04014
−177	"	3.774	0.03956
−178	"	3.763	0.03968
	Paper and Paint	Mean	0.03985 W/mK
	"	Grand Mean	0.03982
	"	St. Deviation	0.00018

Mean thermal conductivity at 25 mm thickness
$$\lambda = 0.03912 \text{ W/mK (from Table 3)}$$

Difference between mean measured on 150 and 25 mm thickness

$$\frac{0.03982 - 0.03912}{0.03912} = 1.8\%$$

Table 4. Lateral Heat Flow Through Polystyrene/Paper Stack With
          150 mm Thickness

Specimen Thickness mm	T_h °C	T_c °C	$\frac{T_c-T_h}{R}2A$ W/m^2	Main Heater, $2Q$ W	Total Flux, Q W	T_{amb} °C	T_h-T_a K
149.5	35.05	35.09	0.02	0.055	0.057	9.5	25.6

$$E_e' = \frac{1}{4} \frac{\Delta T_o}{\Delta T_\ell} \frac{Q_\ell}{Q_o} 100 \qquad (6)$$

where

E_e' = maximum edge loss error in a standard test due to lateral heat flow, %.

ΔT_ℓ = temperature difference between plates and the ambient air, °C.

ΔT_o = temperature difference between main heater and cold plate, $(T_h - T_c)$ °C.

Q_o = heat output on main heater, W/m^2, in a standard test.

Q_ℓ = as above but in a lateral heat flow test.

Substituting data from Tables 4 and 6, one obtains the maximum edge loss error:

$$E_e = \frac{1}{4} \times \frac{22.8}{25.6} \frac{0.057}{1.093} 100 = 1.2\%$$

This would suggest that the 1.8% error indicated by Table 3 consists of about 1.2% edge loss and about 0.6% unbalance error.

Experimental Analysis of Unbalance Errors

Table 5 shows the results of unbalance tests on 450 mm GHP. In the test, the guard ring temperature was controlled from 0.5 to 0.6°C above and below the main heater temperature. The temperature difference was calculated from the differential thermopile output. The values for $C_1 + C_2\lambda$, shown in Table 6, were calculated from the difference in heat flow for the positive and negative test values. Heat flows were corrected to the same temperature difference for the tests with plus and minus guard ring unbalance, ΔT_g. $C_1 + C_2\lambda$ values calculated in this manner are shown in Table 5.

Two conclusions may be drawn from the results shown in Table 5, namely:

(1) Unbalance errors are dependent on the specimen thickness, as shown by an approximately 16% increase in $C_1 + C_2\lambda$ between 25 mm and 100 mm thicknesses.

(2) the unbalance test may be performed on specimens subjected

Table 5. Calculations of $(C_1+C_2\lambda)$ on 450 mm GHP from Differences in Q^1 for Plus and Minus Unbalance Using Polystyrene/Paper Stacks

Specimen		Conditions		Results				
Code	Thick-ness	ΔT	ΔT_g	Q^1 (1)	$Q^1_1-Q^1_2$	$C_1+C_2\lambda$	R_o (2)	λ
3/4	24.99	22	−0.593	6.206				
		22	0.578	6.615	0.409	0.350	0.6374	0.0392
3-4/5-8	50.00	22	−0.520	3.022				
		22	0.505	3.401	0.378	0.369	1.272	0.0393
1-3-4/ 2-5-8	75.03	33	−0.573	3.001				
		33	0.591	3.482	0.481	0.413	1.894	0.0396
1-2-3-4/ 5-6-8-11	100.13	22	−0.574	1.391				
		22	0.570	1.859	0.468	0.409	2.514	0.0398
1-2-3-4- 7-10-12- 9-11-8- 6-5	149.42	22	−0.683	0.802				
		22	0.954	1.477	0.675	0.412	3.772	0.0396

(1) Test heat flow corrected to same ΔT for each pair.
(2) R-value from Q^1 interpolated to $\Delta T_g = 0$.

Table 6. Calculation of $(C_1+C_2\lambda)$ on 600 mm GHP from Difference in Q^1 on Polystyrene/Paper Stack With 150 mm Thickness

Specimen Thickness mm	Conditions		Results				
	ΔT	ΔT_g	Q^1 (1)	$Q^1_1-Q^1_2$	$C_1+C_2\lambda$	R_o (2)	λ
	K	K	W	W	W/K	m^2K/W	k/mK
149.22	22.46	−0.665	1.55631				
	23.09	0.661	0.63270	0.92361	0.697	3.740	0.0399
avg. 22.8			Q_o=1.0931				

(1) Test heat flow corrected to the same ΔT for each pair.
(2) R-value from Q^1 interpolated to $\Delta T_g = 0$.

to a significant distortion of heat flow lines without affecting
$C_1 + C_2\lambda$, as shown by results on 75, 100 and 150 mm thick specimens.
The additional specimen edge loss in tests on the thick specimens did
not appear to alter the mean to indicated gap-temperature-difference
ratio.

The values of $(C_1 + C_2\lambda) = 0.35$ to 0.41 W/K indicate that when
ΔT_g is controlled to ±0.01K the maximum probable error on heat input
is about ±0.004 W. The percentage error related to the polystyrene
specimens used in the evaluation and tested at a 22°C temperature
difference would be 0.06% at 25 mm thickness and about 0.39% at
150 mm thickness.

Table 6 shows the calculation of $(C_1 + C_2\lambda)$ for 600 mm GHP for
the same specimens of polystyrene as those used in 450 mm GHP verification. The number $(C_1 + C_2\lambda)$ being 70% higher indicates more
"thermal bridging" through the gap than in 450 mm GHP. For 0.01°C
unbalance on 150 mm polystyrene/paper stack, the probable error
becomes 0.66%.

COMPUTER ANALYSIS OF GHP ERRORS

A commercial* stress-strain and thermal analysis program, ANSYS,
was used to simulate the heat flux and temperature distribution
through a specimen in a GHP. The simulations were mainly concerned
with edge loss errors and did not analyze the heat transfer across
the air gap.

Two different models were examined:

(i)   three-dimensional model with boundary conditions of
      uniform but different heat flux in main and guard heaters,
      model 1

(ii) three-dimensional model with isothermal plate surfaces,
      model 2.

The calculations were performed on a quarter of the specimen
divided into one thousand parallelpipeds and a quarter of the metal
plates divided into one hundred parallelpipeds. The specimen thermal
conductivity was assumed to be uniform throughout and equal to
0.03912 W/mK. The air gap resistance was selected in such a way that
it produced the same sensitivity to temperature unbalance as had been
found by experiment.

---

*Cybernet computer network was used via Control Data Canada Ltd.

Model 1 was based on uniform but different heat fluxes supplied under the face plate areas of the heater plate. The cold plates were assumed to be isothermal. A trial and error procedure was required to set the heat fluxes. The heat flux applied to the main heater plate was such that it would produce the required temperature gradient in one-dimensional heat flow and the heat flux initially applied to the guard ring heater was about 15% higher. When the program was executed, an unbalance between metering and guard areas was calculated and another value of the guard heater flux tried, until an unbalance of less than 0.02K was obtained.

The first question examined with model 1 was whether there was a significant difference between the integrated temperature difference over the gap and that indicated at a central point on each side, i.e., four points only. For 150 mm polystyrene in the 600 mm GHP with 22K difference across the specimen, a temperature variation of 0.002K was indicated, but the difference between the integrated ΔT_g and ΔT_g indicated at the center of the gap was less than 0.001K. It indicates that the choice of whether the thermocouples are placed at the center or 1/3 of the air gap side is perhaps less important than previously expected.

Table 7 shows R-values measured and calculated with the three-dimensional programs for a 150 mm thick specimen in the 450 mm GHP. Model 2 indicates very small errors, i.e., 0.3%. This model is based on isothermal plate temperatures that are assumed to extend to the center of the gap and, therefore, heat is supplied to all the specimen, including the area at the air gap so that the average heat flux entering the central part of the specimen, q_o, and the average heat flux on the metering area, q, are equal. Thus, the apparent thermal resistance $R_{app} = \dfrac{\Delta T}{q}$ and the thermal resistance of the central part of the specimen $R_O = \dfrac{T}{q_o}$ are equal.

Model 1 does not simulate heat generation in the air gap layer as the heat flux at each of the boundary points is separately formulated in the input data. An input flux of 5.738 w/m^2 was selected since it would give a 22K temperature difference for a thermal resistance of 3.834 m^{2}K/W.

Table 7 shows how the calculated average heat flux, q_o, entering the specimen varies with the level of unbalance. Using this heat flux estimate for calculation of thermal resistance R_O, one obtains a value close to that of model 2 and only slightly affected by the unbalance. The apparent thermal resistance based on heater output divided by the metering area, q, however, shows a strong dependence on the unbalance.

It is not certain that model 1 simulates a real GHP better than model 2, particularly as the apparent thermal resistance exceeds the true R-value. It does indicate, however, a different ratio between the heat flux entering the specimen and the average heat flux on the metering area from that calculated for model 2.

Table 8 shows measurements and calculation for the 600 mm GHP. The thermal conductivity measured on 150 mm is 1.8% greater than that determined on 25 mm thick layers. The error of 1.8% on 600 mm GHP corresponds to the error of 3.3% obtained on 450 mm GHP for these

Table 7. Measured and Calculated R-Values of 150 mm Thick Polystyrene
         Specimen Tested in 450 mm GHP Compared To Measurements on
         25 mm Thick Specimens

Description	Heat Flux on the					Thermal Resistance m^2K/W	
	Main Heater		Meter-ing Area				
	Input	Into Speci-men					
	q_{in} W/m^2	q_o W/m^2	q W/m^2	ΔT K	$R_o=\frac{\Delta T}{q_o}$	$R_{app}=\frac{\Delta T}{q}$	
Measured on 25 mm thick (λ=0.03912) and recalculated for 150 mm thickness	–	–	–	–	–	0.15/0.03912 = 3.834	
Measured on 149.4 mm thick (λ=0.04045) and recalculated for 150 mm thickness	–	–	6.090	22.58	–	0.15/0.04045 = 3.708	
Calculated with model 2 (isothermal plates)	5.868	5.752	5.752	22.00	3.825	3.825	
Calculated with model 1 for ΔT_g = 0.013K	5.738	5.712	5.623	21.79	3.815	3.875	
Calculated with model 1 for ΔT_g = 0.053K	5.738	5.589	5.623	21.29	3.810	3.786	
Calculated with model 1 for ΔT_g = 0.112K	5.738	5.425	5.623	20.62	3.800	3.667	

specimens.   The calculations with both models 1 and 2 indicate
insignificant errors.

Calculations of edge loss errors in a standard test based on the
results of "lateral heat flow test" indicate an error of 1.1%.
A computer simulation using model 2 on the other hand gives 0.9% error,
i.e., good agreement. Thus, both measurements and calculations as
done in the tests labelled "lateral heat flow tests" may be useful
for assessment of the errors in testing "thick" specimens.

The computer calculations for standard testing conditions do not
agree with the measurements. It would appear, therefore, that the
computer program does not adequately simulate the true pattern of
heat flow in a GHP, particularly in the area of the gap.

DISCUSSION

Errors in measurement with GHP apparatus arise from uncertainty
in basic electrical and thickness measurements and from deviations
from one-dimensional heat flow. The errors in basic measurements,
for example, repeatability precision (for a given level of heat
flux), are a characteristic of the equipment used. The extraneous
heat leakage, i.e., unbalance error and edge loss, may also depend on
the nature of the material under test.

Table 8.   Comparison between Errors Measured and Calculated with
           Three-Dimensional Programs for 600 mm GHP

Description	Error, %
Measured at 150 mm λ_{app} = 0.0398 W/mK Measured at 25 mm λ_{app} = 0.0391 W/mK	1.8
Calculated for standard test conditions from measurements, the lateral flow test $T_h=T_c=35°C$, $T_a=9.5$	1.2
Calculated for normal flow, model 2 or model 1 with $\Delta T_g=0$.	Insignificant
Calculated for lateral flow test, model 2	0.9

The materials used in this study were selected to minimize errors caused by material inhomogeneity and testing conditions. Polystyrene with a density of 15 to 16 kg/m^3 was selected as a relatively stable material, unaffected by humidity changes during testing. It does, however, exhibit a significant transmission of long wave radiation making the apparent thermal conductivity dependent on thickness even when tested with the same mean temperature and temperature difference. To reduce changes in the fraction of long wave radiation in the total heat flux, i.e., to eliminate the so called "thickness effect" a Kraft paper septum was used for each 25 mm thick layer (Pelanne[3]). Thus, 25 mm thick layers were tested without paper, 50 mm had one paper septum, and 75 mm had two septa.

The thermal conductivity determined for the polystyrene specimens stacked from 25 to 150 mm with and without septa are shown in Fig. 2. A gradual increase in apparent conductivity with the specimen thickness is evident.

Part of the error commonly attributed to the edge losses could well be an effect of increased heat exchange between metering and guard ring heaters, labelled in this paper as the unbalance errors.

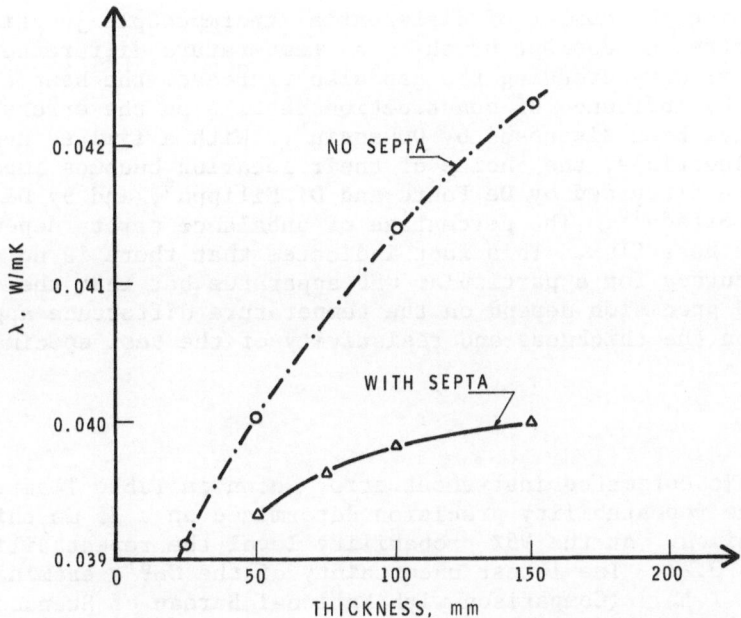

Fig. 2. Apparent thermal conductivity of polystyrene stack with and without paper septa.

Temperature distribution on the plate, particularly in the guard ring area, can produce a variation along the gap. Thus a heat flow across the gap might occur even though the average output from the thermopile is zero. The guard ring heater has to compensate for the heat loss to ambient from the edge of the hot plate and from the edge of the specimen. These losses are proportional to the perimeter length so that when only a single guard ring heater winding is used there is proportionally less heater area to compensate for the heat loss at the corners than at the center of each side. Heat must, therefore, flow laterally and a reduced temperature may be expected in the corners.

The guard ring temperature will be more uniform if two guard ring heater windings are used, a narrow perimeter winding and an inner winding as discussed by Maréchal[4], Rousselle[5] and Bode[6]. The perimeter winding can be used to supply the edge loss without requiring lateral heat transfer in the face plates (Troussart[7]). One suitable technique for operating such an instrument consists of adjusting the heat flux on the inner guard heater to a level equal to or slightly lower than that on the main heater and using the differential thermocouples across the gap to control the heat output on the outer guard ring. Theoretically, the same heat flux should be used on both sides of the gap, but a slightly lower flux on the inner guard ring heater permits more convenient control of the unbalance temperature.

Increasing the number of differential thermocouple junctions permits a better assessment of the mean temperature difference, but the additional wire crossing the gap also increases the heat transfer. The influence of construction details on the errors of the GHP method has been discussed by Doussain[8]. With a limited number of thermopile junctions, the choice of their location becomes important. This has been discussed by De Ponte and Di Filippo[9], and by De Ponte, Marotti and Strada[10]. The percentage of unbalance errors depends on the level of heat flux. This fact indicates that there is no single value of accuracy for a particular GHP apparatus but that the accuracy and precision depend on the temperature difference applied as well as on the thickness and resistivity of the test specimen.

SUMMARY

The basic estimated instrument error shown in Table 1 agrees well with the repeatability precision determined on a 25 mm thick, "ideal" specimen. At the 95% probability level the repeatability precision is 0.2%. The lowest uncertainty of the GHP's examined is estimated at 1.1%. (Comparison with National Bureau of Standards 1000 mm line-heat-source GHP apparatus discussed in these proceedings showed agreement within 0.3% on low density glass fiber specimens of 25 mm thickness.)

Uncertainty of R-value determination depends on thickness and resistivity of the specimen, as well as the temperature difference applied to hot and cold plates. Using the limits of total uncertainty of 2 to 2.5% the specimen thickness is limited to less than 60 mm on 450 mm GHP and to less than 100 mm on 600 mm GHP. Thicker specimens, even with careful conditioning and using edge insulation in excess of 2 m^2K/W, may produce an uncertainty larger than 2.5%.

ACKNOWLEDGMENTS

The authors wish to thank J.G. Theriault for his contribution to the development of the testing methods, and N. Normandin and R.G. Marchand for making the measurements.

They also wish to express their appreciation to D.G. Stephenson, Energy Coordinator, DBR/NRCC, J.G. Hust and M.C. Siu, National Bureau of Standards, and C.M. Pelanne, Johns-Manville R/D, for detailed discussion on the problems of testing thermal insulating materials.

This paper is a contribution from the Division of Building Research, National Research Council of Canada, and is published with the approval of the Director of the Division.

REFERENCES

1. Woodside, W., Analysis of errors due to edge heat loss in guarded hot plates, American Society for Testing Materials Symposium on Thermal Conductivity Measurements, Spec. Tech. Publ. No. 217, 1957, pp. 49-62.
2. Woodside, W., and Wilson, A.G. Unbalance errors in guarded hot plate measurements, Thermal Conductivity Measurements, ASTM Spec. Tech. Publ. No. 217, 1957, pp. 32-46.
3. Pelanne, C.M., Discussion on experiments to separate the effect of thickness from systematic equipment errors in thermal transmission measurements, ASTM Spec. Tech. Publ. No. 718, D.L. McElroy and R.P. Tye Eds., American Society for Testing and Materials, 1980, pp. 322-334.
4. Maréchal, J.C., Mesure de la conductivité thermique par la méthode du champ thermique unidirectionnel, Bulletin RILEM Matériaux et Constructions, Vol. 1, No. 5, 1968, pp. 443-456.
5. Rousselle, J.C., A guarded hot plate apparatus for measuring thermal conductivity from -80 to +100°C. Proceedings of the 7th Conference on Thermal Conductivity, Gaithersburg, Maryland (Nov. 13-16, 1967).
6. Bode, K., Wärmeleitfähigkeitsmessungen mit dem Plattengerät: Einfluss der Schutzringbereite auf die messunsicherheit, Int. Journal Heat and Mass Transfer, Vol. 23, 1980, pp. 961-970.

7. Troussart, L.R., Three-dimensional finite element analysis of the
    guarded hot plate apparatus and its computer implementation.
    Journal of Thermal Insulation, Vol. 4, 1981, p. 225-252.
8. Doussain, R., Influence du mode de construction des plaques
    chauffantes à anneau de garde et de l'isolation latérale sur
    la mesure de la conductivité thermique des matériaux isolants,
    International Institute of Refrigeration, com. II & VI, Liege
    1969, Annexe 1969-7, pp. 289-299.
9. De Ponte, F. and Di Filippo, P., Design criteria for guarded hot
    plate apparatus, Heat Transmission Measurements in Thermal
    Insulations, ASTM Spec. Tech. Publ. No. 544, 1974, pp. 97-117.
10. De Ponte, F., Marotti, M. and Strada, M., Correlation between
    balance and metering area definition in a guarded hot plate
    apparatus, Commission B[1], No. 116, 15th International Congress
    of Refrigeration, Venesia 23-29 Sept. 1979, pp. 1-9.

SESSION   TC-11A

Insulations - II

CHAIRMAN

C. J. Shirtliffe
National Research Council
Ottawa, Canada

COMPARISON OF RESULTS OF MEASUREMENTS MADE ON A LINE-HEAT-SOURCE

AND A DISTRIBUTED-HEAT-SOURCE GUARDED-HOT-PLATE APPARATUS

M.C.I. Siu

National Bureau of Standards
Washington, D.C. 20234

ABSTRACT

Measurements made on a guarded-hot-plate apparatus using line heat sources give results that are in good agreement with those obtained from one using distributed heat sources.

INTRODUCTION

Two variations of the guarded-hot-plate apparatus are in operation at the National Bureau of Standards (NBS). The first was constructed about fifty years ago with heating elements uniformly distributed in the hot plate; for this paper, we refer to this apparatus as NBS-GHP-I. Results obtained with this apparatus have been reported.[1,2] The second variation uses line heat sources at specific locations in the hot plate. Two such line-source apparatuses have been completed at NBS. The plates of both line-source apparatuses are circular; 305 mm in diameter in one case and 1000 mm in the other. Results previously obtained with the 305-mm apparatus have been discussed by Siu and Bulik.[3]

It is of interest to compare observed thermal conductivity, k_{obs}, obtained with the line-heat source and the distributed-heat source guarded-hot-plate apparatus. In this paper, selected results obtained from these two types of guarded-hot-plate apparatuses are presented.

EXPERIMENTAL PROCEDURES

Measurements were made on fibrous glass fiberboard (similar to Standard Reference Material SRM 1450[4]) on the NBS-GHP-I and the 305-

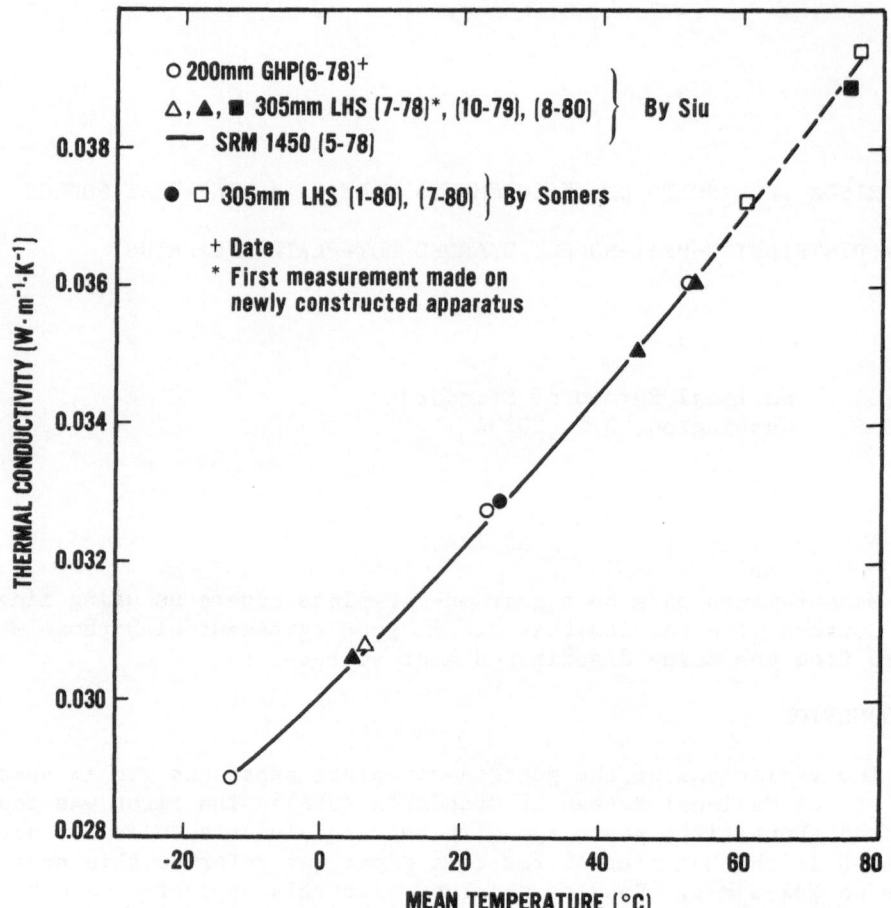

Fig. 1.   Comparative results of the thermal conductivity variation
          with mean temperature. Solid curve represents Lot 1970
          (material similar to SRM 1450) results, Eq. (1); dashes
          represent extrapolation of Eq. (1). Dates that the data
          were taken are given in parentheses.

mm line-source apparatus to compare observed thermal conductivity
and the effect of temperature difference, $\Delta\theta$, between the guard
and the meter sections of the hot plate.

    Another set of measurements was made on 25-mm thick low density
fibrous glass material ($8.5 < \rho < 10.3$ kg/m^3) on NBS-GHP-I and on
the 1000-mm line-source apparatus for the purpose of comparing k_{obs}.

RESULTS AND DISCUSSION

    Figure 1 shows results of measurements of fibrous glass board

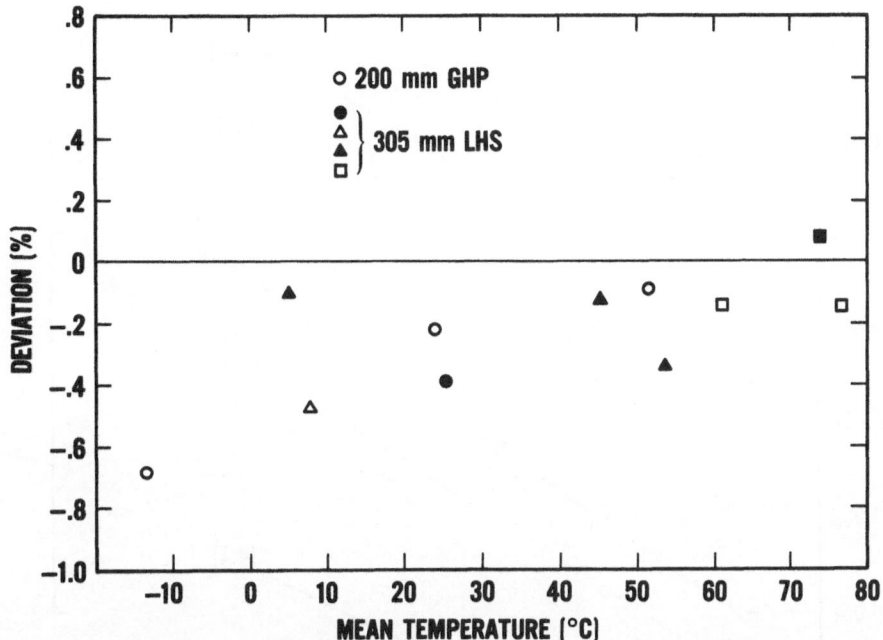

Fig. 2.   Deviations of measured values of thermal conductivity of
          fiberboard (Lot 1970 material) from predicted values, Eq.
          (1).

on NBS-GHP-I and the 305-mm line-source apparatus. The curve shown
in this figure is given by[4]

$$k = 0.02052 + 1.303 \times 10^{-5} \rho + 4.015 \times 10^{-10} T^3 \quad W/m \cdot K \qquad (1)$$

where ρ is the bulk density of the specimen, kg/m^3, (124 kg/m^3 for
the case discussed) and T is the mean specimen temperature in kelvin.
The data shown in this figure and eq. (1) have not been corrected
for thermal expansion effects of the hot plate of the apparatus.
Figure 2 shows deviations of k_{obs} from eq. (1).

Figure 3 shows the effect of gap unbalance, $\Delta\theta$, and the temperature difference across the specimen, θ, for different lots of materials on NBS-GHP-I and the 305-mm line-source apparatus. Figure 4
shows the variation of k_{obs}, as determined on NBS-GHP-I, with $\Delta\theta/\theta$,
for low density fibrous glass ($\rho = 9.80$ kg/m^3). Table 1 gives the
values of the constants k_o and b obtained from a least-squares fit
of the data to the equation

$$k_{gu} = k_o + b \; \Delta\theta/\theta$$

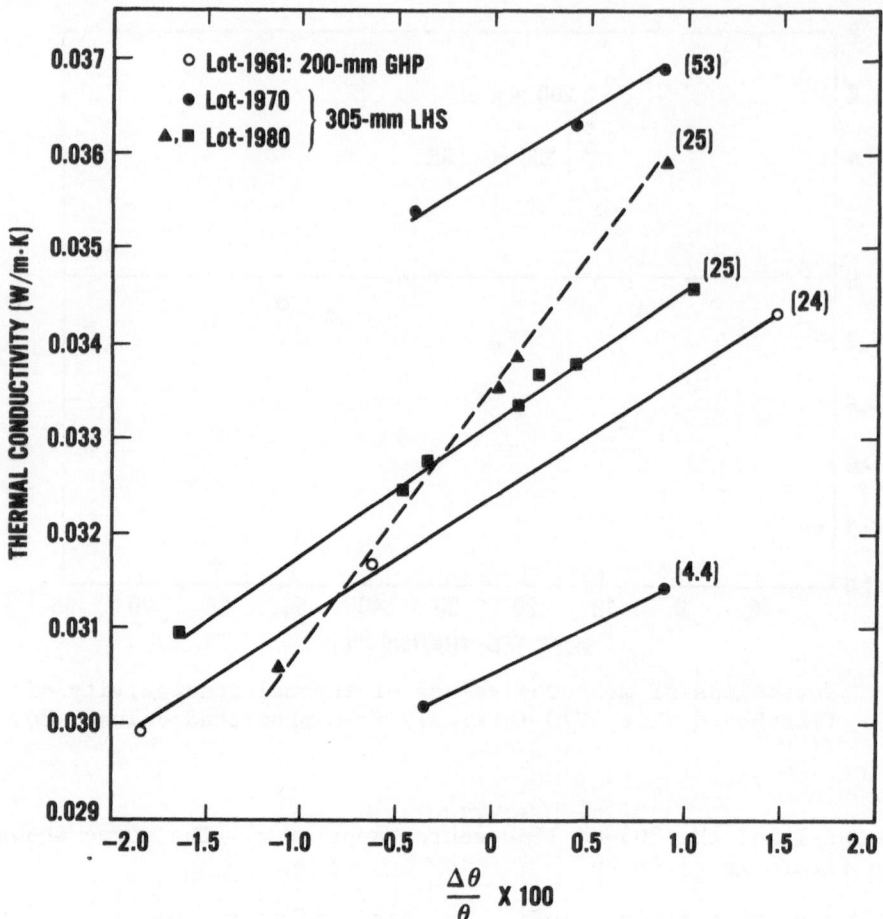

Fig. 3. Variation of the thermal conductivity of fiberboards with
         fractional gap temperature unbalance, $\Delta\theta/\theta$, for several
         different mean temperatures (shown in brackets). Data re-
         presented by triangles correspond to 51-mm thick specimens;
         all other data points correspond to 25-mm thick specimens.

where k_{gu} is k_{obs} under gap unbalance conditions and k_o is k_{obs} at
zero gap unbalance.

    Table 2 gives results of measurements made on low density fi-
brous glass on NBS-GHP-I and the 1000-mm line-heat source apparatus.
Results of the latter apparatus satisfy the equation

$$k(\rho) = 0.03337 + 0.1114\ \rho^{-1}\ .$$

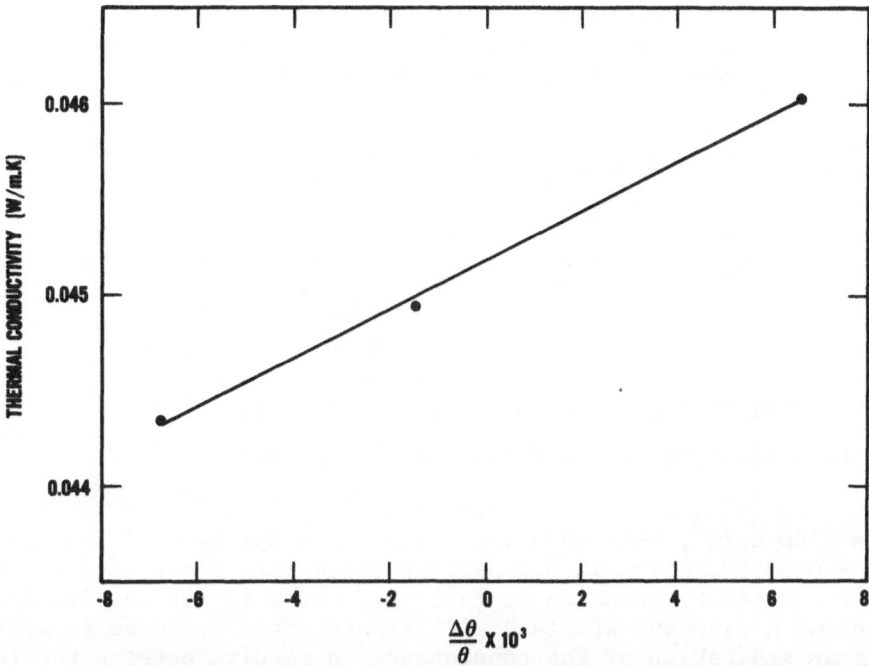

Fig. 4. Variation of the thermal conductivity of low density fibrous glass (25-mm thick; ρ = 9.80 kg/m^3) with fractional gap temperature unbalance, $\Delta\theta/\theta$, as determined on NBS-GHP-I at 24 $^{\circ}$C and a temperature difference of 22 $^{\circ}$C.

Table 1. Values of the coefficients of the Least-Squares Fit Linear Equation

Apparatus	Material	Thickness (mm)	Temperature ($^{\circ}$C)	k_0 (W/m·K)	b (W/m·K)
NBS-GHP-I	Batt*	25.4	23.8	0.04518	0.1271
NBS-GHP-I	Fiberboard (Lot 1961)**	25.4	23.9	0.03238	0.1310
305-mm LHS	Fiberboard (Lot 1980)	25.4	25.1	0.03319	0.1370
305-mm LHS	Fiberboard (Lot 1980)	51.0	25.0	0.03353	0.2786

*Low density fibrous glass material (ρ = 9.80 kg/m^3)

**SRM 1450

Table 2. Thermal conductivity of low density fibrous glass (25 mm
          thick) as determined on NBS-GHP-I and the 1000-mm Line-
          Heat-Source Guarded-Hot-Plate apparatus.

Apparatus	Bulk density* (kg/m^3)	T_{mean} (^oC)	k_{obs} (W/m.K)
1000-mm LHS	9.63	23.88$^+$	0.04528**
1000-mm	10.22	23.88$^+$	0.04461
NBS-GHP-I	9.80	23.8^{++}	0.04518

*Material at the meter section. $^+\Delta T$ = 27.8. $^{++}\Delta T$ = 24.3.
**Last digit is provided for rounding purposes.

For ρ = 9.80 kg/m^3, this equation gives for a specimen of that density k = 0.04507, which is 0.2 percent below that determined on NBS-
GHP-I, k = 0.04518, as shown in Table 2. Owing to the insufficiency
of data for a rigorous statistical analysis, this is to be regarded
only as an indication of the consistency in results between the 1000-
mm line-heat-source apparatus and NBS-GHP-I.

    Figure 1 shows good agreement of results obtained from NBS-GHP-I
and the 305-mm line-source apparatus. Figures 3 and 4 show that the
two types of apparatuses exhibit similar gap temperature unbalance
behavior. The effect of specimen thickness is also shown in Fig. 3
for the 305-mm apparatus. Similarity of performance at gap unbalance
between the two types of apparatuses is displayed in the closeness
in the values of b, Table 1.

CONCLUSION

    Selected measurements made on low (9.80 kg/m^3) and high (124 kg/
m^3) density fibrous glass materials indicate that a line-heat-source
guarded-hot-plate apparatus have similar characteristics and give results that are in good agreement with those obtained from a distributed-heat-source guarded-hot-plate apparatus.

REFERENCES

1. J.L. Fink, J. Res. Natl. Bur. Stand., 5, 973 (1930).
2. H.E. Robinson and T.W. Watson, in ASTM STP 119, American Society
    for Testing and Materials, Philadelphia, Pa., (1951), p. 36.
3. M.C.I. Siu and C. Bulik, Rev. Sci. Instrum.,52(11), 1709 (1981).
4. M.C.I. Siu, in ASTM STP 718, American Society for Testing and
    Materials, Philadelphia, Pa., (1980), p. 343.

DEVELOPMENT OF CALIBRATED TRANSFER SPECIMENS OF THICK, LOW-DENSITY INSULATION

Brian G. Rennex, Robert R. Jones, and David G. Ober

National Bureau of Standards

Gaithersburg, Maryland

Keywords: finite element models; guarded hot plate; low-density thick insulation; Standard Reference Material; thickness effect.

The Need for Low-Density Insulation Calibrated Specimens

The expanded use of low-density insulation material has become an important part of the energy conservation effort in the United States. This is due to the fact that this material can be produced at a relatively low material cost for a given thermal resistance (R value). From the standpoint of fair practice, it is extremely important to industrial and building applications that there be reliable standard specimens for thermal measurements. The purpose of this paper is to discuss the considerations relevant to the production of such standards.

Low-density insulation is commonly sold at thicknesses of about 76, 152 and 305 mm (3, 6 and 12 inches), and the density ranges from 6.4 to 16 kg/m^3 (0.4 to 1.0 lb/ft^3). In the past, the quality control and certification tests of R value usually have been carried out at test thicknesses of 25 or 38 mm (1 or 1.5 inches). These "1-inch" tests provided an R value per unit inch which was multiplied times the thickness of actual use to calculate the R value at a greater thickness. This "linear" extrapolation to greater thickness assumed that the R value per unit inch (and hence

419

its inverse, the thermal conductivity) is constant as a function of
test thickness. This assumption has been increasingly questioned
in recent years.[1-11]

The idea that the thermal conductivity (k value) may not be a
constant as a function of thickness is referred to as the "thickness effect." To understand the physical basis of this effect, we
must look at the classical definition of k and compare it with the
experimentally determined k. Classically, k is defined as a property of a material in which only the conduction heat transfer mode
is operative. That is, k, in one dimension, is equal to the ratio
of the heat flux per unit area passing through the cross-sectional
area of an infinitesimal volume divided by the temperature gradient
across that volume. This volume is, in turn, in an infinite medium
so there can be no interaction with the boundaries.

In the actual apparatus there is interaction between any
volume element of the specimen and the boundaries. The result is
that a measured k value must depend on the apparatus through its
boundary conditions, as well as on the insulation material itself.
This dependence could conceivably lead to the unhappy situation
where each distinct apparatus might measure a distinct k value for
an identical specimen.

In order to achieve a standard k value, regardless of the
measuring apparatus, one can try the following three courses of
action. The first is to minimize the interaction between the measured material volume element and the boundary at the sides. For
example, the apparatus can be made large and guarded to minimize
interactions between a center meter area and the insulation edges.
The second course of action is to model the contribution to the
measured k-value from the material itself as distinct from the
"edge effect" contribution. This means that the "material" contribution to the measured k value should be the same regardless of the
apparatus. This material-adjusted k value would then be appropriate to the definition of a standard thermal conductivity for a
particular specimen. The third course of action is to standardize
any unavoidable boundary effects; for example, the plate emissivities are required to have a value close to 0.9 by the ASTM Test
Method 177 for guarded hot-plate-apparatus.

Now, we have established the context for understanding the
thickness effect. The low-density insulation material has perhaps
as much as a third of its heat transfer via thermal radiation. The
behavior of the coupled radiation and conduction heat transfer near
the apparatus plate boundaries is different from the behavior in
the center of the specimen. The size of the "near-boundary" regions
is expected to be fixed and of the order of the mean free path of
the thermal radiation in the insulation material. As the specimen
becomes thicker, the boundary regions are expected to become less

important, relative to the bulk region, in determining the thermal
resistance of the specimen. Thus, as the thickness is increased,
the thermal resistivity (or conductivity) might be expected to
approach asymptotically a constant value dependent only on the
specimen material.

This information then leads to the question of the appropriate
calibration sample for heat transfer measurement of low-density
insulation. As was mentioned earlier, high-density (128 kg/m^3
[8 lb/ft^3]) calibration specimens have been used in the past. The
radiation component in these materials is much smaller than in the
low-density material, which means that the thickness effect is
expected to be smaller. In addition, the edge effect (or apparatus
systematic error) might be different for the two different specimen
densities. Thus, it is quite possible that a different calibration
curve as a function of thickness might result with high- and
low-density calibration specimens.

The above discussion has given considerations that point out
the need for low-density calibrated transfer specimens until future
research might give a more definite understanding of the physical
processes.

Apparatus Considerations

This section discusses the guarded-hot-plate design
requirements necessary to reduce systematic apparatus errors.
First, the meter area should be large in order that the measured
meter-area heat flux be representative of the heat flow thru the
entire specimen. Next, the guard areas must be large enough to
ensure an approximate one-dimensional heat flow in the meter area.
The gap temperature difference between the meter and guard sections
in the hot plate must be small for the same reason. A one-
dimensional heat flow means that there is no "net" interaction
between the meter-area sample volume and the apparatus edges. The
size constraints lead to the following practical difficulties (1)
the manufacture and support of flat, large plates, (2) the measurement of the average specimen thickness over the meter area, and
(3) increased times.

For the NBS apparatus, circular plates were used, since
cylindrical symmetry makes it easier to model the edge effects.
The diameter of the meter area is 406 mm (16 inches), and the
diameter of the outer guard is 1016 mm (40 inches).

It seemed reasonable in terms of current measurement
capabilities to have as a goal a percentage uncertainty of 1 percent for measured k value of a low-density insulation specimen
with a thickness between 2.54 and 152.4 nm (1 and 6 inches).
Experience in the design and operation of the instrumentation for

this one-meter guarded-hot-plate at NBS indicated that the
measurement of the plate temperatures requires considerable care.
An accuracy goal of within 0.1 percent for the hot- to cold-plate
temperature difference corresponds to a value of 0.03°C (0.05°F)
for a plate temperature difference of 27.8°C (50°F). This kind of
accuracy is most easily achieved with a high quality platinum
resistance thermometer.

The most critical parameters in terms of measurement
uncertainty are (1) the average spacing between the meter area
sections of the plates (at a thickness of 25 mm or 1 inch) and (2)
the control and readout of the above-mentioned gap temperature (at
thicknesses of 152 mm (6 inches) or more.

The design and error analysis of the NBS, 1 meter guarded hot
plate is discussed in reference 10. A detailed description of this
apparatus and its instrumentation capability will be forthcoming in
1982.

Standard Material Considerations

In order to achieve reproducible results on different user
apparatuses, it is necessary to have well-characterized calibration
specimens. Low-density insulation material is intrinsically variable in density across the lateral area and through the thickness
of the sample. This variability poses a potential problem in terms
of the calibration uncertainty, even in the case where the calibrated transfer specimen is actually measured on the NBS apparatus. The reasons are as follows. The thermal conductivity (or the
calibration value) associated with a specimen is a strong function
of the specimen bulk density. At a density of 9.6 kg/m^2, a 3 percent change in density corresponds to a 1 percent change in k-
value. The meter areas and shapes of the NBS and user apparatuses
will in general be different; hence the average densities and the
calibration values will be different.

The meter-area densities were determined by stamp cutting and
weighing the material corresponding to the NBS and user meter
areas. The meter-area density values were determined to within 0.3
percent (or 0.1 percent in the corresponding k-value). The percentage difference between the user and NBS values ranged between 0
and 4 percent, corresponding to a 0 to 1 1/3 percent variance in k-
value. Thus, if an adjustment had not been made, there could have
been as much as a 1 1/3 percent contribution to the calibration
factor uncertainty.

A more detailed NBS report on the characterization of the
low-density calibration transfer specimens will soon be available
in 1982.[12]

Calibrated Transfer Specimen Considerations

The purpose of a low-density calibrated transfer specimen is to enable a user to obtain a value of thermal resistance that does not reflect a significant apparatus systematic error. Another way of saying this is the following. If a specimen R value is measured on several different user apparatuses, all of which have calibrated transfer specimens measured on the same apparatus, then the measured R values should be the same within the measurement uncertainty.

The measured R value depends in a complicated way on the following physical parameters: bulk density, thickness, plate emissivity, mean sample temperature, plate temperature difference, and on material properties such as fiber composition, diameter and orientation, and the optical scattering and absorption coefficients. These many physical parameters have an effect on the interaction between a specimen volume element and the apparatus boundary conditions. Until the orders of magnitude of these effects are determined by research, it was thought to be prudent to have as close a match as possible between the physical properties of the calibrated transfer specimen and the physical properties of the specimens to be tested. For example, if a manufacturer wants to test a product with a density of 8 kg/m^3 (0.5 lb/ ft^3) at a thickness of 76 mm (3 inches), ideally he would want a specimen at the same density and thickness to calibrate his apparatus.

The question that naturally comes up is how to achieve a calibrated transfer specimen with more general applications in order to avoid the expense of a different specimen for each of a number of different products. One solution is to establish calibration curves that bracket the parameter range of interest. For example, suppose a manufacturer tests products over a range of 9.6 to 16 kg/m^3 (0.6 to 1.0 lb/ft^3). Then, if he has calibrated transfer specimens at 8, 12.8, and 17.6 kg/m^3 (0.5, 0.8, and 1.1 lb/ft^3) he can establish a curve at a function of density for the calibration factor. This results in an interpolation error for the calibration factor which is generally significantly smaller than an extrapolation error from a single point.

The goal of a standards program is to provide the most generally applicable standards with the most simple calibration scheme. In order to accomplish this, the significant calibration variables must be determined, and calibration curves must be provided based on these variables.

The initial thick, low-density calibrated transfer specimens to be provided by NBS were chosen with the above considerations in mind. Since the initial predominant need was expressed by the mineral fiber industry, the specimens were selected from a special lot of fiber-glass material produced by the Johns-Manville Corporation. The average density value was 9.6 kg/m^3 (0.6 lb/ft^3). This density value is representative of the insulation products for which the "thickness effect" is expected to occur. A significant portion of the fiberglass batt and blanket products have densities between 8 and 12 kg/m^3 (0.5 and 0.75 lb/ft^3). The calibrated transfer specimen set consists of thicknesses equal to 25, 76 and 152 mm (1, 3 and 6 inches). This allows the user to establish a calibration curve as a function of thickness. Thus, if he wants to measure a 127-mm (5-inch) test specimen, he can interpolate to determine the most appropriate calibration factor. The initial mean temperature is 23.9°C (75°F), and the plate temperature difference is 27.8°C (50°F). Preliminary data indicate that the average k-value at a test thickness of 152 mm (6 inches) is about 3 1/2 percent larger than that at 25 mm (1 inch) at a density of 9.6 kg/m^3 (0.6 lb/ft^3). That is, the "thickness effect" is about 3 1/2 percent.

In the future, it should be possible to provide calibration standards at various values of density or temperature and for various types of material.

Calibrated Transfer Specimens vs. Standard Reference Material

The discussion heretofore has dealt with calibrated transfer specimens. Each of these specimens were actually measured on the apparatus of a standards laboratory. Another approach is the use of Standard Reference Materials (SRM). An SRM specimen is identified as one coming from a lot of "uniform" material which has been statistically characterized and then certified as to a k value mean and uncertainty. That is, the k value and density values are measured and the resulting standard deviations are used to predict the calibration uncertainties of the lot. Specimens issued to a user are not actually measured on the NBS apparatus. The advantage of the SRM is that a larger number of specimens can be provided for the available funds since it is not necessary to perform a time-consuming thermal test on each specimen. The disadvantage is that the calibration uncertainty is larger because it includes a term dependent on the variability of the material within the lot. Let us refer to this as the "lot variability."

In the case of low-density insulation it may be possible to achieve a smaller "lot variability" by optically characterizing the individual specimens. References 11 and 12 discuss the correlation between measured thermal conductivity and visible-light transmittance for this material. A suggested approach is to: (1) purchase a lot of material sufficient to provide several years demand, (2) measure the R value of a statistically representative sampling of the lot as a function of the specimen mean density and the specimen density distribution over the meter area, (3) determine by optical transmission the mean density and density distribution of each specimen meter area appropriate to the user meter area and, finally, (4) provide the SRM specimen with the appropriate value for the calibration uncertainty. The SRM's could, in principle, be provided with calibration curves for a range of temperature, density and thickness values.

The actual details of the Calibrated Transfer Specimen and SRM programs at NBS will depend on the results of research to be carried out in the next few years. The purpose of this paper has been to inform the potential users and others as to the current views and considerations on this important and interesting subject.

BIBLIOGRAPHY

1. J. D. Verschoor and Paul Greebler, "Heat Transfer by Gas Conduction and Radiation in Fibrous Insulations," Transactions, American Society of Mechanical Engineers, Vol. 74, No. 6, p. 961, (1952).
2. B. K. Larkin and S. W. Churchill, J. Am .Inst. Chem. Eng., $\underline{5}$ (4) 467, (1959).
3. R. Viskanta and R. J. Grosh, "Heat Transfer by Simultaneous Conduction and Radiation in an Absorbing Medium," Journal of Heat Transfer, February 1962 p. 63.
4. Charles M. Pelanne, "Experiments on the Separation of Heat Transfer Mechanisms in Low Density Fibrous Insulation," Proceedings, Eighth Thermal Conductivity Conference, C. Y. Ho and R. E. Taylor, Eds, Plenum Press, New York, 1969, pp. 897-911.
5. Claes Bankvall, "Heat Transfer in Fibrous Materials," Journal of Testing and Evaluation, JTEVA, Vol. 1, No. 3, p. 235, (1973).
6. B. Y. Lao and R. E. Skochdopole, "Radiant Heat Transfer in Plastic Foams," 4th S.P.I., International Cellular Plastic Conference, Montreal, Canada, November 15-19, 1976.
7. Charles M. Pelanne, "Experiments to Separate the 'Effect of Thickness' from Systematic Errors in Thermal Transmission Measurements," Thermal Insulation Performance, ASTM STP 718, D. L. McElroy and R. P. Tye, Eds., American Society of Testing and Materials, 1980, pp. 322-334.
8. Marion Hollingsworth, Jr., "Experimental Determination of Thickness Effect in Fibrous Insulations," Ibid, pp. 255-271.
9. Brian G. Rennex, "Thermal Parameters as a Function of Thickness in Low Density Insulation," Journal of Thermal Insulation, Vol. 3, p. 37, (1979). 425
10. M. H. Hahn, H. E. Robinson, D. R. Flynn, "Robinson Line-Heat-Source Guarded Hot Plate Apparatus," Heat Transmission Measurements in Thermal Insulations, ASTM STP 544, American Society for Testing and Materials, 1974, pp. 167-192.

11. C. M. Pelanne, "Thermal and Physical Characteristics of Glass
     Fiber Insulation Produced for the National Bureau of Standards,"
     Johns-Manville Report #436-T-1528, (1981).
12. B. G. Rennex, "Report on the Sample Characterization of the NBS
     Calibration Transfer Specimens," (in preparation).

SESSION   TC-11B

Dielectrics - I

CHAIRMAN

R. Berman
Clarendon Laboratory
Oxford, England

THE THERMAL CONDUCTIVITY OF HELIUM CRYSTALS

CONTAINING NEON IMPURITY

R. Berman and D.M. Livesley

Clarendon Laboratory
Oxford
England

INTRODUCTION

The thermal conductivity of solid helium was first measured by Wilkinson and Wilks.[1] These early measurements did not extend to temperatures below that of the conductivity maximum, but over the range covered the conductivity increased very rapidly with decreasing temperature. Only one other dielectric crystal (Al_2O_3) that had been measured by that time showed a similar dramatic variation in conductivity. Peierls[2] had predicted that a rapid reduction in the frequency of Umklapp processes (U-processes) would give rise to an almost exponential rise in the conductivity of a perfect crystal.

When it became clear[3] that the existence of isotopes of the constituent elements of a crystal could represent sufficient disorder for the effect of U-processes to be masked, even in otherwise perfect crystals, it was evident why helium was almost immune from this influence; the ^3He concentration in ^4He extracted from oil wells is only 0.1 ppm. As it is relatively easy to grow good crystals with different $^3He/^4He$ concentrations, this was one of the systems used to study the isotope effect quantitatively.[4] Helium crystals are extremely compressible, and their density is doubled by growing them at 135 MPa rather than at the minimum pressure of 2.5 MPa. It proved advantageous for interpreting results to make measurements on crystals of different densities.

We have been studying the effect of chemical impurities on the conductivity of helium crystals grown at various pressures. For both isotopic and chemical impurities, a large part of the reduction in conductivity is due to the difference in mass between the

host atoms and the impurities. This mass effect can be calculated
with some certainty so that other contributions to the scattering
of phonons can be estimated from the experimental results. Interesting conclusions could be drawn from the extra scattering observed
in helium crystals containing isotopic impurities, and it was
expected that analysis of experiments with chemical impurities
would also be rewarding.

The late Mark Carritt carried out experiments on hydrogen
impurities in helium, but since the maximum concentrations of hydrogen that can be contained in the starting fluid are small at the
freezing pressures used and the H_2 - ^{4}He mass difference is also
small, the effect on the conductivity was difficult to determine
accurately. The 8 times larger mass difference of ^{20}Ne - ^{4}He
results in a 64 times larger mass effect in phonon scattering for
the same impurity concentration. We present here only our results
for neon impurity.

THE EXPERIMENTS

Crystal Growing

The crystals were grown at constant pressure from helium—neon
gas mixtures. It might be thought that the maximum neon concentration attainable in the gas from which crystals are grown would
correspond to the vapour pressure of neon at the freezing temperature. For example, the freezing temperature of ^{4}He at a pressure
of 100 MPa is 14.3 K. At this temperature the vapour pressure of
neon is $\sim$ 20 Pa, so that the neon concentration would then be only
0.2 ppm. However, when a solid or liquid is subject to a pressure
applied by a different gas, its vapour pressure can be much greater
than the bare vapour pressure measured when the condensed phase is
in contact with only its own vapour. Under the conditions quoted,
the concentration of neon can, in fact, be $\sim$ 1000 ppm, an enhancement of 5000.[5]

Helium—neon mixtures of four known concentrations were made up
and compressed into the 3.2-mm-diameter stainless steel specimen
tube. The crystals were grown by cooling the bottom of the tube
to just below the freezing temperature, and the pressure was kept
constant by supplying the mixture through a capillary at the top.
The temperature at the bottom was decreased slowly to maintain a
steady rate of growth. The time taken to grow a 45-mm-long crystal
was usually $\sim$ 2 h. The crystal was then annealed just below the
freezing temperature for $\sim$ 3 h.

Measurements

The conductivity was measured by the longitudinal flow steady-
state method, and the temperature gradients along two halves of the

specimen tube were determined separately, using gold-iron vs. silver-
normal thermocouples. Arrangements for measuring the conductivity
of the two halves were originally incorporated in the apparatus to
establish whether uniform crystals of pure helium and of isotopic
mixtures were being grown. In the present experiments the results
for the two·halves enabled us to estimate the variation in neon
concentration along the length of the specimen. The overall
accuracy of the measurements was estimated to be ± 7% in general but
± 9% for very high conductivities. About 5% out of this was due to
systematic errors so that the comparative accuracy should be ± 2%
to ± 4%.

The longitudinal conductance of the stainless steel specimen
tube as a function of temperature was measured in a separate experiment. These results were used to correct the total measured
conductivity to give that of the crystal itself. This correction
represented only a small fraction (never more than 20% and usually
very much less) of the overall conductance measured when the tube
contained a crystal.

Results

Figures 1 and 2 show the conductivities of crystals grown at
61 and 136 MPa (600 and 1350 atm) starting from [4]He-Ne mixtures
containing the neon concentrations indicated. Earlier results for
pure [4]He are shown for comparison.[6] The solid lines are fitted to
the averages of the conductivities of the two halves of the specimen
using the Callaway method of analysis,[7] as will be described below.
Results from repeated runs were very consistent, and differed by no
more than the estimated accuracy of ± 2% to ± 4%, so for clarity
only points from one set of measurements for each pressure and
concentration are shown.

ANALYSIS OF RESULTS

It can be seen that although the thermal conductivity is reduced
for each successive concentration at the higher pressure, this is
not so at 61 MPa; the results for 235 and 1140 ppm are very similar
and can be accounted for by assuming that the neon concentration in
the solid is in both cases only about 100 ppm. This apparent limit
is greater than that of about 40 ppm deduced for the gas phase for
this pressure at the freezing temperature by extrapolating the
results of Berman et al.[5] on the solubility of neon in high pressure
[4]He gas. We interpret this as indicating that the maximum equilibrium concentration in the solid phase is greater than that in the
gas phase at the freezing temperature. As the high pressure gas
mixture was cooled towards the freezing temperature in the specimen
tube, it must gradually have shed the excess neon that it could no
longer contain in solution. Whether this formed precipitates in
the gas or on the walls of the tube, it was presumably in such a

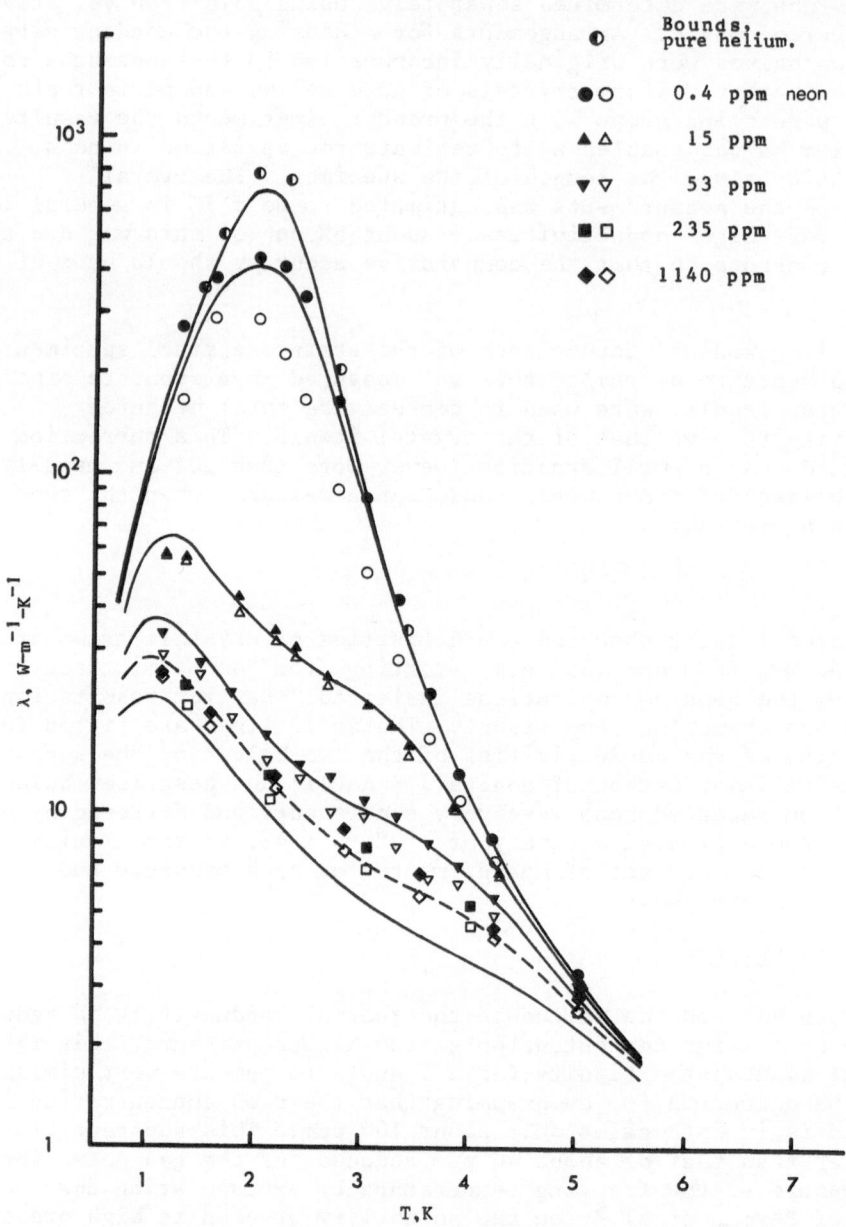

Fig. 1. Thermal conductivity of helium crystals containing neon impurity formed at 61 MPa. Full lines are computed for Ne concentrations 0, 0.4 ppm, 15 ppm and 235 ppm. Open symbols - half of the specimen grown first; closed symbols - half grown last. The broken line is calculated for a concentration of 110 ppm.

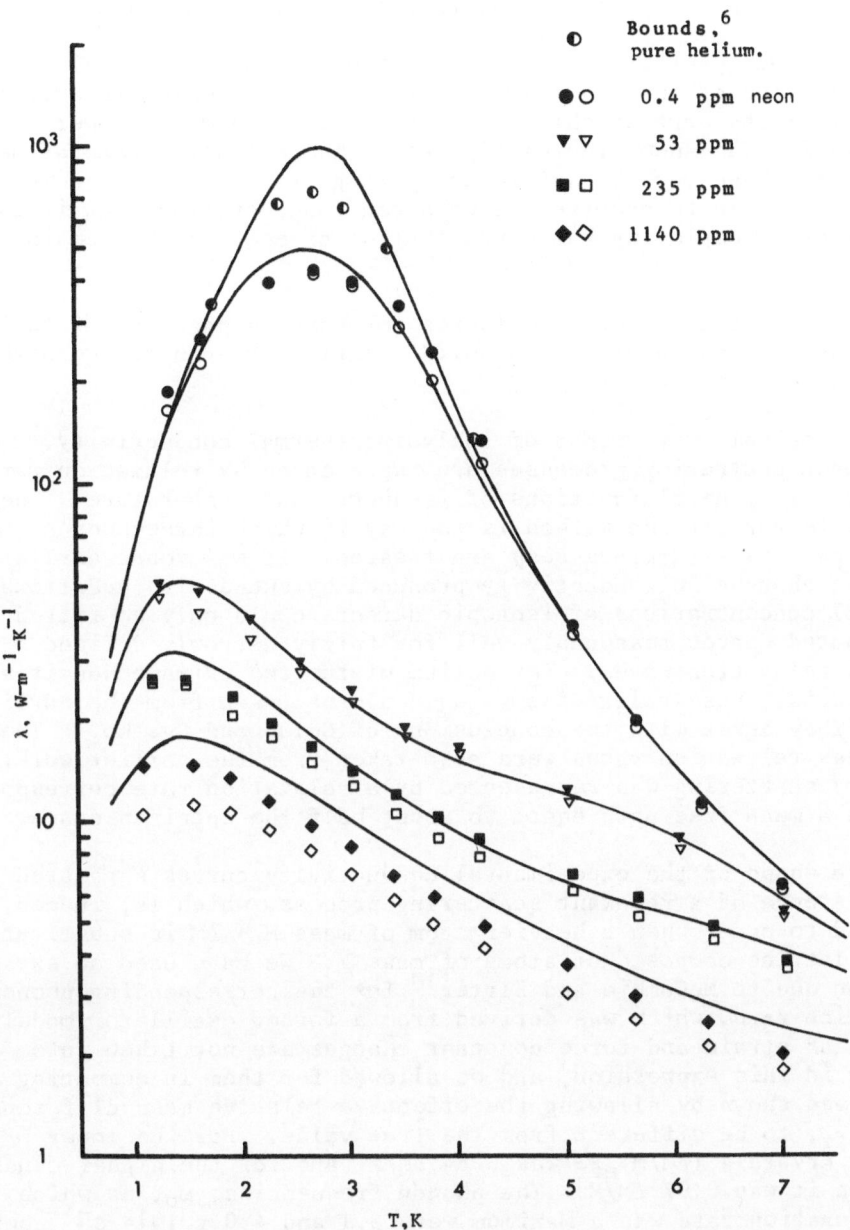

Fig. 2. Thermal conductivity of crystals formed at 136 MPa. Full lines are computed for Ne concentrations 0, 0.4 ppm, 53 ppm, 235 ppm and 1140 ppm.

state that it could be incorporated in the solid as crystallization proceeded. A crystal could thus be produced with a neon concentration greater than existed in solution in the gas from which it grew.

This interpretation agrees with observations on the relative conductivities of the two halves of the crystals, which invariably showed that the part of the crystal grown first had the lower conductivity (as shown in the figures). The diffusion rate of neon in the gas phase must be sufficiently large for the solid grown first to be able to acquire the enhanced concentration that it can hold, thereby depleting the concentration of neon in the remaining gas.

We did not observe any evidence for phase separation of the solid, as is seen in ^{3}He - ^{4}He solid solutions,[8] even at the highest neon concentrations.

In the Callaway method of analysing thermal conductivity,[7] all phonon scattering processes are represented by relaxation rates, which are in general functions of frequency and temperature. The special feature of the method is the way in which three-phonon normal processes (N-processes) are treated. It was found earlier[4] that the changes in conductivity produced by successive additions of small concentrations of isotopic defects could only be fitted by computed curves reasonably well for fairly narrowly defined N-process relaxation rates. For helium of the two extreme densities illustrated, these relaxation rates could be taken from the earlier work. They agree with the conclusions of Golub and Svatko.[9] The U-process relaxation rates were also taken from the earlier work. Boundary scattering was represented by a relaxation rate corresponding to a mean free path equal to about half the specimen diameter.

The shape of the experimental conductivity curves suggested the existence of a resonant scattering process, which is, indeed, expected to occur when a heavier atom of mass $M + \Delta M$ is substituted into a lattice composed of atoms of mass M. We have used an expression due to McCombie and Slater[10] for the corresponding phonon relaxation rate, which was derived from a forced-oscillator model. Effects of strain and force constant changes are not taken into account in this expression, and we allowed for them in computing the curves shown by allowing the effective relative mass difference, $(\Delta M/M)_{eff}$, to be different from the true value. For the lower density crystals $(\Delta M/M)_{eff}$ was 0.84 $\Delta M/M$, and for the higher density crystals it was 0.97 $\Delta M/M$. The phonon frequencies, ω_o, at which the relaxation rate was a maximum were 3.1 and 4.0 x 10^{12} s^{-1}, both of these being about 0.3 times the corresponding Debye maximum frequency. For two intermediate densities $(\Delta M/M)_{eff}$ and ω_o lay between the limits given above.

CONCLUSIONS

Our experiments and analysis show that neon can dissolve in solid helium to form mixed crystals. The maximum neon concentration in the crystals depends on the pressure and corresponding temperature of crystallization and is greater than the maximum concentration that can be held in the fluid under the same conditions. There is a resonant phonon mode associated with the neon impurity, and scattering by it represents a large contribution to the reduction in conductivity caused by the impurity.

Experiments have also been carried out on the Ne – ^{3}He system. The expectations that for crystals with the same Debye characteristic temperature, the mass-difference scattering by neon should be twice as great in ^{3}He and that the resonant mode frequency should be about 15% lower than in ^{4}He were verified.

REFERENCES

1. K.R. Wilkinson and J. Wilks, The thermal conductivity of solid helium, Proc. Phys. Soc. Lond. A64:89 (1951)

2. R. Peierls, Zur kinetischen Theorie der Wärmeleitung in Kristallen, Ann. Phys. 3:1055 (1929).

3. R. Berman, E.L. Foster, and J.M. Ziman, The thermal conductivity of dielectric crystals: the effect of isotopes, Proc. R. Soc. A 237: 344 (1956).

4. R. Berman, C.L. Bounds, and S.J. Rogers, The effect of isotopes on lattice heat conduction, II. solid helium, Proc. R. Soc. A 289:66 (1965).

5. R. Berman, F.A.B. Chaves, D.M. Livesley, and C.D. Swartz, The solubility of solid H_2 and Ne in high-pressure ^{4}He, J. Phys. C (Solid State Physics) 12:L777 (1979).

6. C.L. Bounds, Some thermal properties of solids, D.Phil. thesis, Oxford University, (1968).

7. J. Callaway, Model for lattice thermal conductivity at low temperatures, Phys. Rev. 113:1046 (1959).

8. D.O. Edwards, A.S. McWilliams, and J.G. Daunt, Phase separation in solid ^{3}He – ^{4}He mixtures, as shown by specific heat measurements, Phys. Rev. Lett. 9:195 (1962).

9. A.A. Golub and S.V. Svatko, Thermal conductivity of low-density solid ^{4}He, Fiz. Nizk. Temp. (USSR) 6:957 (1980).

10. C.W. McCombie and J. Slater, The scattering of lattice vibrations by a point defect : I, Proc. Phys. Soc. Lond. 84:499 (1964).

THERMAL CONDUCTIVITY OF ULTRAPURE NaF USING

TWO-FLUID ANHARMONIC PHONON THEORY

Baxter H. Armstrong

IBM Scientific Center
Palo Alto, CA 94304

INTRODUCTION

The two-fluid anharmonic thermal conductivity theory of an earlier paper[1] is extended herein and applied to NaF, with comparison to measurements of Jackson and Walker[2] (JW) on their ultrapure crystal specimen. This theory is based on a division of phonon modes into two groups according to the phonon dispersion curve characteristics. These groups are a resistive category of high-frequency modes in approximate thermal equilibrium and a normal (N-) process dominated lower frequency group of nearly nondispersive modes that deviate importantly from equilibrium. The latter "propagating" group carries the heat, and the former "reservoir" group to which the propagating phonons relax, provides the anharmonic thermal resistance. Dispersive modes above the Umklapp (U-) process combination threshold act as reservoir modes. In addition, nondispersive modes below this threshold act as reservoir modes at low temperatures if the extrinsic scattering rate is strong enough. For anharmonic transition rates, we use average 3-phonon expressions of acoustic absorption theory[3] modified to account for the division of the transition final states into reservoir and propagation modes. The extension beyond Ref. 1 consists of accounting by use of Callaway's method[4] for the effect of "NN processes" as defined in that paper.

The JW thermal conductivity data are fitted by the theoretical formula obtained according to the above ideas using the U-process combination threshold frequency, a maximum propagation mode frequency, and a Grüneisen gamma as the principal adjustable parameters. The values of the frequency parameters inferred from the fit are in good agreement with experimental dispersion curve zone boundary intersections. Good agreement is also achieved between the theoretical curve and the JW data.

THEORY

The Boltzmann Equation and Single Mode Transition Rates

In the two-fluid context, Callaway's approach leads to the linearized Boltzmann equation for the NN region:

$$-\left[\frac{N_q - N(\beta)}{\tau_{NN}}\right] - (N_q - N_q^0)(\tau_{NR}^{-1} + \tau_R^{-1}) = \vec{c}_q \cdot \vec{\nabla}T \frac{dN_q^0}{dT}, \tag{1}$$

where N_q is the occupation number and $\vec{c}_q$ the phonon velocity of the mode with wave vector $\vec{q}$ and angular frequency ω. The equilibrium Bose occupation number is N_q^0 and the shifted Planck distribution $N(\beta)$ is the same as Callaway's except for the limitation to NN processes. This limitation along with determination of β is discussed presently. The transition rates τ_{NR}^{-1} and τ_{NN}^{-1} are computed as follows:[1,3]

$$\tau_{NR}^{-1} = \frac{\gamma_m^2 T\omega \tan^{-1}(2\omega\bar{\tau})}{2\rho v^2} \int_{\omega_R}^{\omega_{RC}} C_{\omega'} d\omega', \tag{2}$$

$$\tau_{NN}^{-1} = \frac{\pi(\gamma_m')^2 T\omega}{4\rho v^2} \int_0^{\omega_R} C_{\omega'} d\omega' \tag{3}$$

where γ_m and γ_m' are mean Grüneisen numbers, $\bar{\tau}$ is the average reservoir relaxation time, ρ is the density, v is the Debye mean sound speed, and ω is in the propagation region.

Equation (2) is the average transition rate for acoustic 3-phonon N-process combination transitions from the propagation region to the reservoir region (phonon splittings are neglected). We assume that U-processes can be included in this rate by adjustment of the Grüneisen number. The integral over the modal specific heat, C_ω, effectively sums over the final states of the transitions, and is therefore taken from the reservoir threshold, ω_R, to the crystal phonon spectrum cutoff, ω_{RC}. Equation (3) is the analogous rate for N-process transitions, termed NN processes, that begin and end in the propagation region. The arctangent function of the original expression for the NN rate is taken as $\pi/2$ because the relaxation time in this case pertains to the propagating modes for which $\omega\bar{\tau} \gg 1$. If the propagation modes are assumed to be nondispersive, Eq. (3) yields

$$\tau_{NN}^{-1}(x) = Q' \, T^5 \, J_4(x_R)x \tag{4}$$

with $Q' \equiv 9\pi(\gamma_m')^2 \, \bar{n} \, s \, R \, k_B/(4\hbar w v^2 T_D^3)$, and J_4 is the $n=4$ member of

the family of Debye integrals $\dfrac{1}{4}\displaystyle\int_0^x \dfrac{t^n dt}{\sinh^2(t/2)}$. The constants appear-

ing in Q' are defined in Ref. 1, save γ_m' which was defined after Eq. (3), above. (All undefined symbols follow the definitions of Ref. 1.)

In the specific heat integral of Eq. (2), the modes lying below ω_{RU} are taken as nondispersive, while those lying above are assumed to be highly dispersive with a frequency-independent density of states. With this assumption, τ_{NR}^{-1} can be evaluated as:[1]

$$\tau_{NR}^{-1}(x) = Q \, T^5 \, x\{[J_4(x_{RU}) - J_4(x_R)]\,\Delta(x_R, x_{RU})$$

$$+ \left[\frac{\bar{n} - \dfrac{1}{3} - \dfrac{2}{3}\left(\dfrac{\omega_{RU}}{\omega_P}\right)^3}{3\bar{n}(\omega_{RC} - \omega_{RU})/\omega_D}\right]\left(\frac{T_D}{T}\right)^2 [J_2(x_{RC}) - J_2(x_{RU})]\} \tag{5}$$

where $Q \equiv \gamma_m^2 \, Q'/(\gamma_m')^2$ and $\Delta(x_R, x_{RU}) = 1$ or 0 according as $x_R \leq x_{RU}$ or $x_R > x_{RU}$, respectively. The first term in Eq. (5) is due to the modes below ω_{RU}. The coefficient of the second term due to the modes above ω_{RU} arises from the somewhat arbitrary choices made in defining the flat density of states and is explained in Ref. 1.

For the extrinsic scattering rate $\tau_R^{-1} \equiv \tau_I^{-1} + \tau_D^{-1} + \tau_b^{-1}$, we use the conventional expressions for impurity (I), dislocation (D), and boundary scattering (b):

$$\tau_I^{-1} = \alpha \, T^4 \, x^4 \tag{6}$$

$$\tau_D^{-1} = T \, \delta \, x \tag{7}$$

with τ_b^{-1} assumed to be independent of T and ω. The NR transitions can be expressed in the single mode relaxation time form because the reservoir (final state) modes are in approximate thermal equilibrium. The extrinsic resistive rates can be expressed in this form because the propagation (initial state) modes are nearly nondispersive.[5]

Equation (1) is identical to Callaway's equation with τ_{NN} identified with his τ_N, and with τ_{NR}^{-1} supplanting his U-process rate.

Therefore, the parameter β, determined by conservation of wave vector flow, becomes

$$\beta = \frac{\displaystyle\int_0^{\omega_R} \tau_c \, \tau_{NN}^{-1} \, G_\omega \, d\omega}{\displaystyle\int_0^{\omega_R} \tau_c \, \tau_{NN}^{-1} \, \tau_{RT}^{-1} \, G_\omega \, d\omega} \tag{8}$$

The upper limit of the integrals is taken as the reservoir threshold ω_R (to be discussed below). The definitions $\tau_c^{-1} \equiv \tau_{NN}^{-1} + \tau_{RT}^{-1}$,

$\tau_{RT}^{-1} = \tau_{NR}^{-1} + \tau_R^{-1}$, $G_\omega \equiv \hbar\omega(\vec{c} \cdot \vec{\nabla}T/T)^2 \, \dfrac{dN_q^0}{dT} \, D(\omega)$, are employed with $D(\omega)$

denoting the polarization average density of states.

Callaway's "effective relaxation time," α_q, can now be written, upon rearrangement of his original result, as

$$\alpha_q = \tau_{RT} + (\beta - \tau_{RT})\tau_c \, \tau_{NN}^{-1} \tag{9}$$

which separates the purely resistive contribution from the NN process correction. Equation (9) with use of the Debye density of states yields the thermal conductivity, κ:

$$\kappa = \frac{k_B(k_B T/\hbar)^3}{8\pi^2 v} \left\{ \int_0^{x_P} \frac{\tau_{RT} \, t^4 \, dt}{\sinh^2(t/2)} + \int_0^{x_R} \frac{(\beta - \tau_{RT})\tau_c \, \tau_{NN}^{-1} \, t^4 dt}{\sinh^2(t/2)} \right\} \tag{10}$$

where $\omega_P = k_B T x_P/\hbar$ is the highest propagation group frequency. The first term of Eq. (10), denoted κ_R, is the result obtained in Ref. 1. The correction term can be recognized as the difference between an inverse and a direct mean of τ_{RT}^{-1}. As such, the Schwartz inequality can be invoked[6] to confirm that it is negative, with its magnitude controlled by the strength of the frequency dependence of τ_{RT}^{-1}. For a frequency-independent τ_{RT}^{-1}, it obviously vanishes, and it remains small as long as τ_{RT}^{-1} is only linear in ω. This correction term vanishes above ω_R because of the definition of the NN region.

It should be noted that the above definitions of NN and NR processes are based on the notion of a mean phonon and an ideal reservoir. In this ideal case, the propagation and reservoir groups would be mutually exclusive with propagating modes lying below the reservoir threshold and reservoir modes lying above (viz., $\omega_R = \omega_P$). This idealization requires modification in the case of real crystals because of the differences between them and the idealization. Our application is therefore not unique and some modes may be counted twice. On the one hand, we will use ω_P as an adjustable parameter to reflect the heat-carrying capacity of the highest (usually longitudinal) nondispersive acoustic phonon branch. On the other hand, thermal resistance due to U-processes sets in for transitions with final phonon in modes at $\omega = \omega_{RU}$ (where ω_{RU} is the U-process combination threshold frequency) and ω_{RU} may be considerably lower than the highest nondispersive longitudinal modes. Therefore, in the absence of a rigorous program, we will cut off the final state integral for $\tau_{NN}^{-1}(\omega)$ and start the final state integral for $\tau_{NR}^{-1}(\omega)$ at ω_R, even though the mode sum over κ will be extended to $\omega_P > \omega_R$ for the first term of κ. The error in this procedure is not likely to exceed the uncertainty in the transition rate expressions, and should be to some extent correctable by adjustment of the precise values of ω_P, ω_{RU}, and the Grüneisen gammas. A more rigorous procedure, involving detailed identification of polarizations in the individual transitions and phonon group definitions is not warranted at the present level of knowledge of the detailed polarization-dependent transition rates.

The Reservoir Modes and Determination of ω_R

We must now specify the reservoir characteristics in order to calculate the foregoing anharmonic transition rates. The high-temperature intrinsic reservoir boundary (defined in Ref. 1) occurs at $\omega_R = \omega_{RU}$, viz., the lowest Brillouin zone boundary (BZB) frequency. The reservoir relaxation time, $\bar{\tau}$, was initially inferred from a fit to the κ data using the approach [Eqs. (43) and (44)] of Ref. 1. It was found that $2\omega\bar{\tau} \gg 1$ for ω over the dominant propagating frequencies (viz., the line widths are narrow as in the case of LiF studied in Ref. 1). Hence, the computation was redone setting $\tan^{-1}(2\omega\tau) = \pi/2$ for reservoir modes, with negligible loss of accuracy.

The low-temperature reservoir, important near the peak of κ, can no longer be estimated as in Refs. 1 and 7. The reservoir definition employed in those references must now be generalized to account for the NN processes. Such a generalization can be based on the small deviation function Φ_q (cf., Eq. 5 of Ref. 1). In the low-temperature reservoir region, extrinsic scattering must dominate in determining Φ_q and this Φ_q must be small compared to what it would be for pure anharmonic scattering. In the propagation region, the situation is reversed. Therefore, the boundary between the two regions can be taken as the frequency at which

$$\Phi_q(\text{ANH}) = \Phi_q(\text{I}) , \tag{11}$$

where $\Phi_q(\text{ANH})$ and $\Phi_q(\text{I})$ are determined by pure anharmonic and pure extrinsic (impurity and isotope) scattering, respectively. Upon cancellation of similar factors this statement reduces to $\alpha_q(\text{ANH}) = \tau_I$ where $\alpha_q(\text{ANH})$ is given by Eq. (9) with $\tau_{RT}^{-1} = \tau_{NR}^{-1}$ (i.e., $\bar{\tau}_I^{-1} = 0$).

The effective relaxation time α_q on the reservoir side of this equality is just τ_I because there will be no shift β or NN scattering by assumption in this region.

Upon neglect of NN processes, the above reservoir definition reduces to $\tau_{NR}(\omega_R) = \tau_I(\omega_R)$, the definition employed in Ref. 1. It should also be noted that the strong frequency variation of τ_I^{-1} relative to τ_{NN}^{-1} and τ_{NR}^{-1} permits the reservoir definition to simulate a threshold even though the functions are all continuous so that a true threshold does not exist.

By use of $\alpha_q = \tau_c(1 + \beta/\tau_{NN})$, the statement (11) can be written as

$$\tau_N(\omega_R)(1 + \beta/\tau_{NN}) = \zeta\,\tau_I \tag{12}$$

where $\tau_N^{-1} \equiv \tau_{NN}^{-1} + \tau_{NR}^{-1}$. The adjustable parameter ζ has been inserted instead of unity as required by Eq. (11) to compensate for the approximate nature of the available transition rates. The magnitude of ζ as determined by the fit to experimental κ data must be close to unity if the proposed method is to be sensible, and this turns out to be the case. If the Debye approximation is made for τ_N^{-1} of Eq. (12), this equation becomes

$$x_R^3 = \zeta\,Q\,T\,J_4(x_D)/[\alpha(1 + \beta/\tau_{NN}(x_R)] \tag{13}$$

where $x_D \equiv \hbar\omega_D/k_B T$, and we have set $\gamma_m' = \gamma_m$ so that $Q' = Q$ [it should be noted that α in Eq. (13) is the impurity scattering strength, not the effective relaxation time]. The Debye approximation used for this result will be reasonably accurate because τ_{NN}^{-1} is defined over nondispersive modes and for a pure crystal at low T, $\tau_{NR}^{-1} \ll \tau_{NN}^{-1}$. Hence, low accuracy on τ_{NR}^{-1} is tolerable. Equation (13), as an implicit equation in x_R, was solved by iteration in the process of fitting Eq. (10) to the JW data. The calculation was started at x_{RU} which was used to determine first values of β and τ_{NN}. The calculation was then repeated to determine successive values of x_R until convergence in these values was obtained to better than one part in 10^3. Figure 1 shows the complete reservoir boundary as inferred for the NaF crystal of JW. The sloping curve at the low temperature end is the boundary determined by the iterative procedure described above for $\zeta = 2.2$, which appeared to give the best average fit to the experimental κ data. The horizontal line is ω_{RU} as inferred from the fit to κ towards higher temperature, as is discussed below.

COMPARISON WITH EXPERIMENT

The thermal conductivity κ was computed according to Eq. (10) using ω_{RU}, ω_p, δ, ζ, τ_b, and a Grüneisen gamma as adjustable parameters (γ_m' was set equal to γ_m). Values for T_D and v were taken as

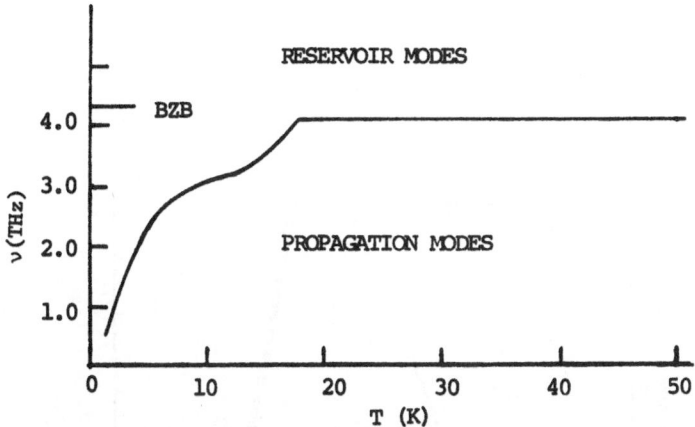

Fig. 1. Inferred reservoir boundary for the Jackson–Walker NaF crystal. The curved section is the low temperature extrinsic
reservoir and the flat section just below the Brillouin
zone boundary is the U–process reservoir.

473 K and 3.687×10^5 cm/s, respectively.[8] This choice of the Debye
mean sound speed causes our τ_b^{-1} to differ slightly from that inferred
by JW. The Rayleigh scattering rate $\alpha = 4.7 \times 10^{-3}$ s^{-1} K^{-4}, established by JW was used, and the spectrum cutoff frequency $\nu_{RC} \equiv \omega_{RC}/2\pi$
= 12.9 THz was obtained from Refs. 9 and 10.

Figure 2 shows the results of the calculation. The lower curve
is κ, and the upper curve is κ without the NN process correction,
viz., κ_R, the first term only of Eq. (10). The inferred values of
the parameters are ν_{RU} = 4.1 THz, ν_p = 10.4 THz, δ = 3000 s^{-1} K^{-1},
γ_m = 1.00, ζ = 2.20, and $\tau_b^{-1} = 8.6 \times 10^5$ s^{-1}. Neutron scattering
data[9] show the lowest BZB intersection at 4.2 – 4.4 THz and the high-
est longitudinal branch intersection is at 9.2 THz. Although the
inferred value of ν_p is 13% higher than this, it should be noted that
the longitudinal branch in the direction with the highest BZB intersection is essentially nondispersive to the zone boundary and joins
to an optical branch with relatively low dispersion. Such optical
branches will act to extend the propagation region. Our inferred δ
lies between the theoretical value δ = 2200 cited by JW and their
inferred value 5200. The Grüneisen gamma inferred is approximately
the same as the low temperature thermal and elastic values given
by Birch, Collins, and White.[11]

The distribution average $\left\langle \tau_{NN}^{-1}(x) \right\rangle$ (as defined by JW) can be

computed from Eq. (4) using our inferred gamma for comparison with

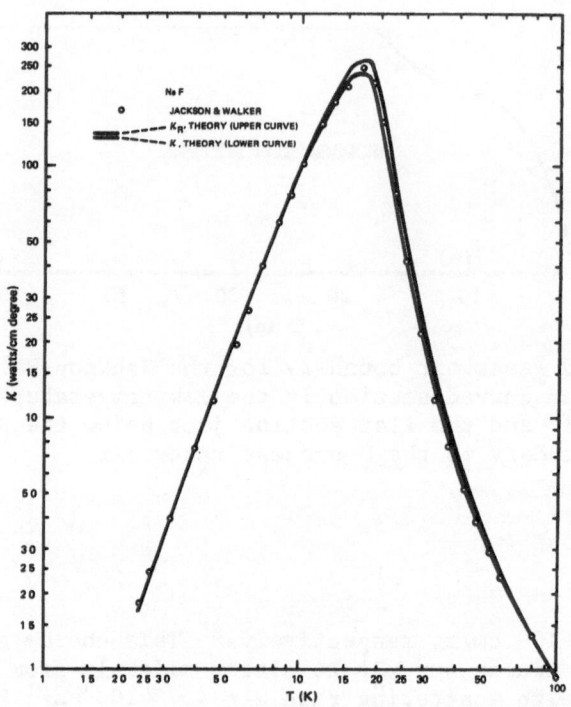

Fig. 2.   Thermal conductivity of the Jackson–Walker NaF crystal.
          The lower curve is the conductivity with the NN process
          correction.   The upper curve is κ_R, the conductivity due
          to direct resistive processes.

the value $60\ T^5$ obtained by JW.   The result is $\left\langle \tau_{NN}^{-1}(x) \right\rangle = 31\ T^5$ in
the low temperature region where τ_{NN}^{-1} is important and $J_4(x_R)$ can be
approximated as $J_4(\infty) = 25.976$.   Thus, reasonable qualitative agreement exists.

ACKNOWLEDGMENTS

    I wish to thank Prof. Howard Jackson for supplying a full-size
graph of his original experimental datum points, and Dr. G. K. White
for calling my attention to the value of the low-temperature Grüneisen
parameters for NaF.

REFERENCES

1. B. H. Armstrong, <u>Phys. Rev.</u> B23:883 (1981).
2. H. E. Jackson and C. T. Walker, <u>Phys. Rev.</u> B3:1428 (1971).
3. H. J. Maris, Interaction of Sound Waves with Thermal Phonons
    in Dielectric Crystals, <u>in</u> "Physical Acoustics," Vol. VIII,
    W. P. Mason and R. N. Thurston, eds., Academic Press, New
    York (1971), p. 279.
4. J. Callaway, <u>Phys. Rev.</u> 113:1046 (1959).
5. P. Carruthers, <u>Rev. Mod. Phys.</u> 33:92 (1961).
6. B. H. Armstrong, Two-Fluid Anharmonic Thermal Conductivity Theory
    and the Peierls-Boltzmann Equation, IBM Scientific Center,
    Palo Alto, CA, to be published.
7. B. H. Armstrong, Two-Fluid Theory of Phonon Scattering in Dielec-
    tric Crystals, in "Phonon Scattering in Condensed Matter,"
    H. J. Maris, ed., Plenum Press, New York (1980), p. 129.
8. O. L. Anderson, Determination and Some Uses of Isotropic Elastic
    Constants of Polycrystalline Aggregates Using Single Crystal
    Data, <u>in</u> "Physical Acoustics," Vol. IIIB, W. P. Mason, ed.,
    Academic Press, New York (1965), p. 43.
9. W. J. L. Buyers, <u>Phys. Rev.</u> 153:923 (1967).
10. A. M. Karo and J. R. Hardy, <u>Phys. Rev.</u> 181:1272 (1969).
11. J. A. Birch, J. G. Collins, and G. K. White, <u>Aust. J. Phys.</u>
    32:463 (1979).

HEAT PULSE THERMAL DIFFUSIVITY MEASUREMENTS ON TRANSPARENT MATERIALS

S.P. Howlett and R. Taylor

Department of Metallurgy, University of Manchester/UMIST
Manchester, United Kingdom

R. Morrell

National Physical Laboratory
Teddington, London, United Kingdom

ABSTRACT

The thermal diffusivity of Herasil fused silica and Pilkington plate glass have been investigated as functions of specimen thickness and surface finish using a heat pulse measurement technique. Similar measurements on synthetic sapphire have been made as a function of specimen thickness only. The specimen faces were sputter coated with tungsten or nickel prior to measurement.

In all materials, increasing the specimen length resulted in an increase in the apparent thermal diffusivity. Rear face polishing reduced the apparent thermal diffusivity in fused silica above ≈800 K, and conversely, for the plate glass specimens, polishing increased the thermal diffusivity value.

Contributions by internal and direct radiative heat transfer to the phonon conduction mechanism are believed to be the important considerations in understanding the changes in apparent thermal diffusivity.

INTRODUCTION

Measurements of thermal diffusivity on transparent materials by the heat pulse technique are sparse. Hirai et al.[1] have measured the thermal diffusivity of C.V.D. (chemical-vapour-deposited) silicon nitride using samples coated with colloidal graphite and Rudkin[2]

has investigated the thermal diffusivity of translucent alumina
with evaporated tungsten coatings also using the 'flash' technique.
Furthermore, heat pulse measurements have been used with varying
degrees of success on several translucent ceramic materials,[3] but
there is a dearth of data on the thermal diffusivity of glasses at
elevated temperatures. Thermal diffusivity measurement by the flash
method has proved to be an increasingly popular method of generating
thermal transport property data. The thermal diffusivity is simply
related to thermal conductivity λ by the following relationship:

$$\lambda = \alpha C_p \rho$$

Hence, when the density, ρ, and specific heat, C_p, are known, thermal
diffusivity measurements provide an attractive method from which
thermal conductivity data may be obtained in a short time and on
relatively thin specimens. The importance of the latter comment
will be discussed later. Although 'flash' thermal diffusivity methods
are potentially attractive, a major problem in their use on diathermous materials is the necessity for an opaque layer to be applied
to the front and rear faces of any specimen. Such coatings provide
faces of high absorbance to promote the transfer of energy from the
incident energy pulse to the specimen, and high emissivity to allow
the rear face history to be monitored without direct optical transmission to the detector from within the specimen. The routine
application of such coatings has proved to be difficult and,
accordingly, has made 'flash' thermal diffusivity measurements on
glasses and other transparent materials difficult. In this work,
various metallic coatings (W, Mo, Ni, Pt) were applied by sputtering.

     The importance of radiative heat transfer in transparent
materials to the overall conductivity has been noted by various
authors.[4,5] Gardon[5] pointed out the importance of specimen thickness in determining the apparent thermal diffusivity of a glass, and
Lucks et al.[6] investigated the importance of surface emissivity on
apparent thermal conductivity.

     In view of these complexities and difficulties, this investi-
gation into the suitability of the flash technique to thermal
diffusivity measurements on transparent materials was undertaken.
Three materials were investigated: fused silica and plate glass of
various thicknesses and surface finishes were studied, and for
comparison, a crystalline material, synthetic sapphire, was studied
as a function of specimen thickness only.

     The materials investigated were chosen to give various degrees
of infrared absorption, i.e., plate glass → fused silica → synthetic
sapphire, a series in which radiative heat transfer should have
varying contributions. The two glasses are considered good radiative
conductors, whereas sapphire is known to have heat transfer
properties dominated by phonon processes.

EXPERIMENTAL DETAILS

Specimen Details

All the specimens, diamond machined to the required size, were 6 mm diameter and thicknesses between 0.5 mm-2.0 mm. To investigate the effect of surface emissivity on thermal diffusivity some specimens were prepared with one polished face. The polishing was performed using successively finer grades of diamond paste on Sturers paper polishing laps, finishing with a 0.25 μm diamond paste. The unpolished specimens had an as-machined finish from 320-grit diamond grinding wheels.

Coating Application

A sputter coating facility was designed and constructed to allow easy replacement of the target cathode. The assembly readily fits most commercial evaporating units with a suitable HT voltage supply and vacuum system. Sample coating was conducted at low argon pressure ($\approx$1.33 Pa) and with a cathode potential of -ve 1.5 kV, for a total time of 1 h. These conditions gave a coating opaque to the incident laser flash of approximately 1 μm thickness. For the various target materials used, the optimum sputtering conditions were determined in an ad hoc manner. The selection of the particular coating to apply to a given substrate was determined by the resilience of the sputtered coating to repeated laser flashes. An unsuitable coating will rupture after a very few laser pulses. Tungsten coatings were found to adhere well to the silica specimens, but were less suitable for the plate glass and sapphire samples. Nickel coatings were found to be suitable. It is believed that the relative thermal expansions of the substrate and coating are of

Table 1.   Specimen Details

Specimen	Thickness (mm)	Surface Finish
Fused silica	0.5	As machined
Fused silica	1.05	As machined
Fused silica	1.05	Rear face polished
Fused silica	2.02	As machined
Fused silica	2.01	Rear face polished
Plate glass	1.92	Rear face polished
Plate glass	1.90	As machined
Plate glass	0.50	As machined
Plate glass	0.51	Rear face polished
Sapphire parallel to 'c' axis	2.14	As machined
Sapphire parallel to 'c' axis	1.06	As machined

primary importance. Fused silica has a low coefficient of thermal
expansion, 0.5 x 10^{-6} K^{-1}, whereas sapphire has an expansion
coefficient of 9.0 x 10^{-6} K^{-1}, which is also a typical value for
plate glass. The thermal expansion coefficients of tungsten and
nickel are 4.5 x 10^{-6} K^{-1} and 13.0 x 10^{-6} K^{-1} respectively. It
appears that for a coating to have a good adherence and, hence, good
rupture resistance to the laser it should have a thermal expansion
somewhat higher than that of the substrate.

Thermal Diffusivity Apparatus

The theory of the 'flash' or heat pulse method has been
adequately detailed in the literature[7],[8] and will not be reviewed
here. In the thermal diffusivity apparatus at the University of
Manchester[9] the heat pulse is supplied by a 100 J solid state ruby
laser (wavelength 0.65 m) of duration $\sim 10^{-3}$ s. The specimen was
heated to the measurement temperature inside a graphite susceptor
using H.F. induction heating. Within the temperature range covered
by this investigation, the temperature change on the rear face was
monitored using a lead-sulphide infrared detector. The amplified
output from the detector was recorded and analyzed using a minicomputer.

During thermal diffusivity measurements, the laser power used
did not exceed 20% of the maximum power output. This restriction
was imposed to prolong the coating life. Also, during measurements
on specimens with one polished face, the polished face was the rear
specimen face in the sample holder.

RESULTS

The temperature dependence of thermal diffusivity for fused
silica as a function of specimen thickness and surface finish is
shown in Figure 1. For the fused silica the effect of increasing
the specimen thickness was to increase the apparent thermal
diffusivity, whereas conversely, polishing the rear face reduced
the observed value of thermal diffusivity. Furthermore, all of the
curves shown in Figure 1 tend towards a constant value at T < 500 K
(the low temperature limit for the PbS detector).

The temperature dependence of the thermal diffusivity data for
plate glass in Figure 2 shows a trend similar to that of the silica
data, with an increasing thickness causing higher values of
diffusivity at any given temperature. However, in direct contrast,
polishing tends to increase the thermal diffusivity for the same
thickness. The observed increase in thermal diffusivity with rear
face polishing is believed to be a true result, since the 2.02 mm
polished specimen was remeasured with the polished surface roughened,

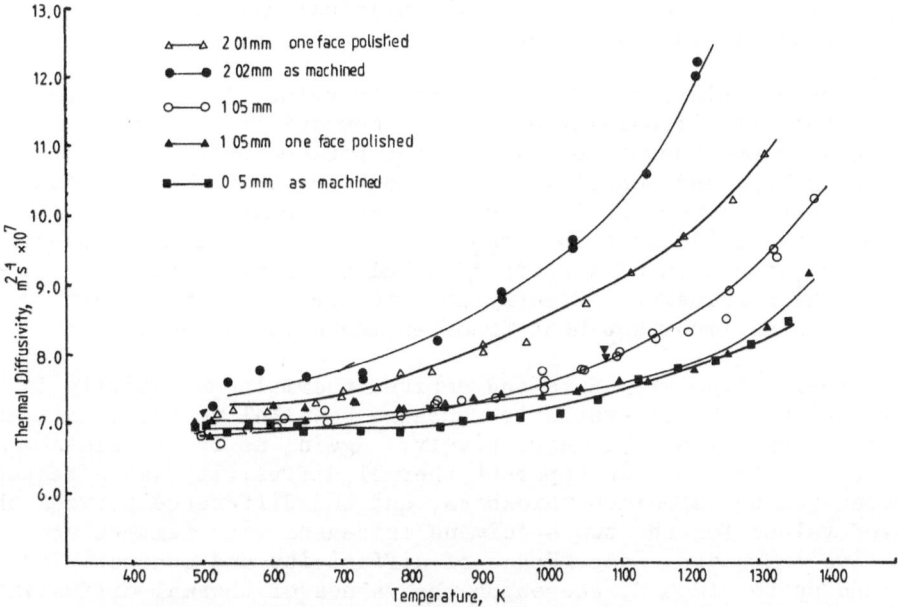

Fig. 1.   Thermal diffusivity of 'Herasil' fused silica as functions
of specimen length and surface finish.

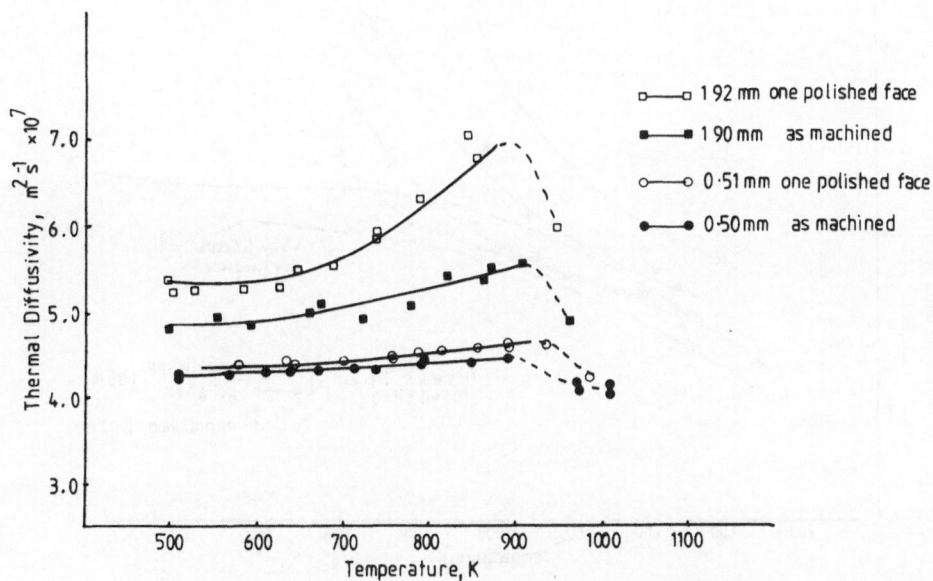

Fig. 2.   Thermal diffusivity of Pilkington   float glass as functions
of specimen length and surface finish.

and good agreement was found with the original data for the
unpolished plate glass specimen.

Figure 3 depicts thermal conductivity values for silica derived
from the thermal diffusivity data. For comparison, the data of
other workers are shown. Good agreement between this work and values
obtained by Wray and Connolly[10] is found below 1000 K. The data of
Wray and Connolly were obtained using a measurement technique which
eliminated radiative contributions to the overall thermal conductivity.
In contrast, the data of Kingery[11] includes contributions by
radiative heat transfer. Clearly in this work, radiative heat
transfer was an important heat transfer mechanism above 1100 K.

Thermal diffusivity data and derived thermal conductivity data
for two thicknesses of synthetic sapphire parallel to the c axis are
shown in Figures 4 and 5, respectively. Again, as in the glass
specimens, an increase in apparent thermal diffusivity was effected
by increasing the specimen thickness, and the difference between the
measured values for the two specimens increased with temperature.
A negative temperature dependence of diffusivity and conductivity
was found up to ≈1200 K, whereupon the values of thermal diffusivity
and, hence, conductivity began to rise with increasing temperature.

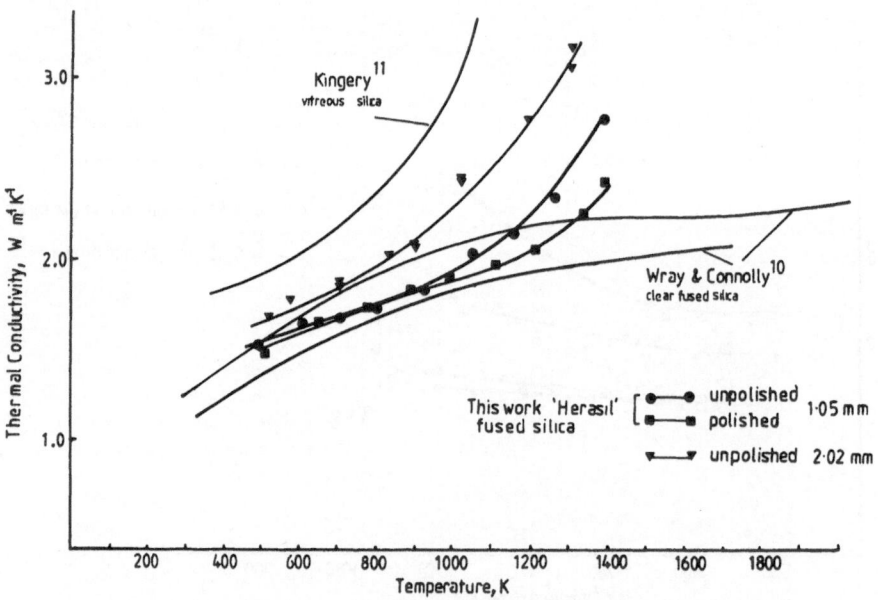

Fig. 3.   Comparison of thermal conductivity data derived from this
          work for silica, with data obtained by other workers.

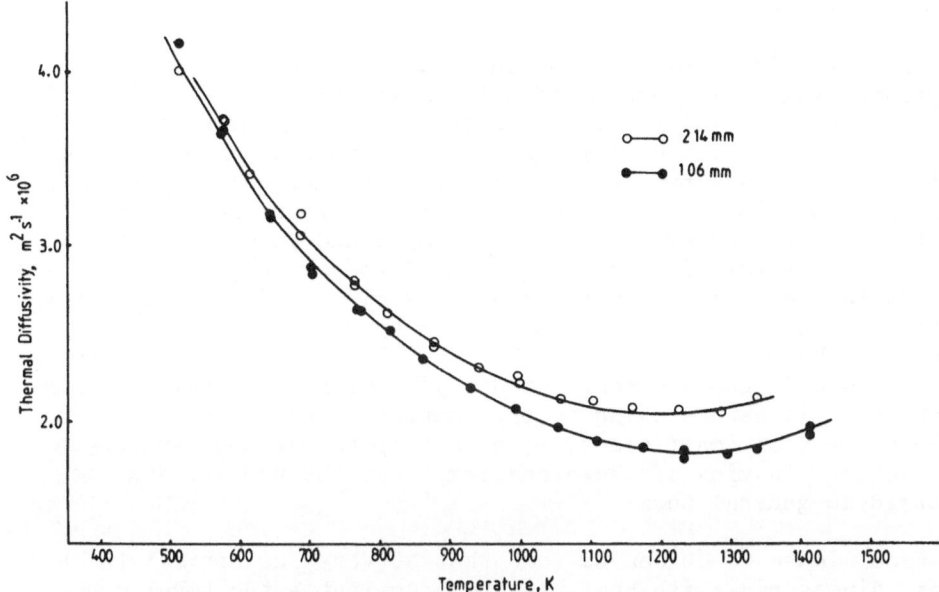

Fig. 4. Thermal diffusivity of synthetic sapphire parallel to the 'c' axis as a function of specimen length.

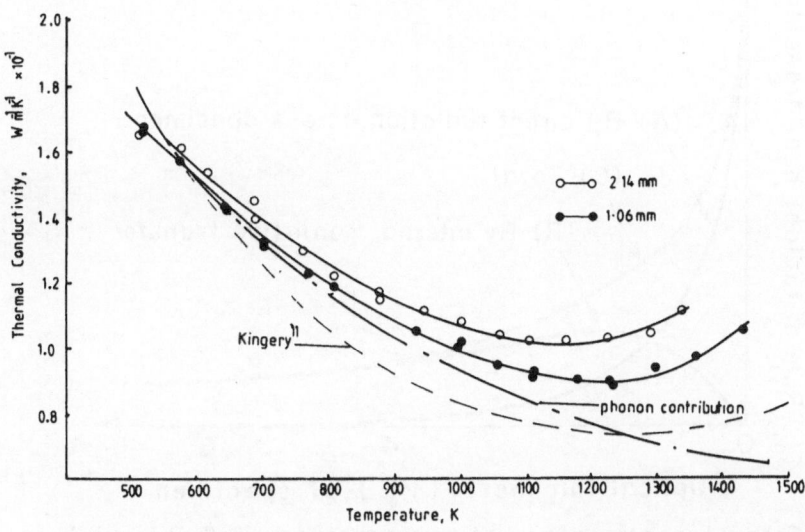

Fig. 5. Derived thermal conductivity data for synthetic sapphire parallel to the 'c' axis.

DISCUSSION OF RESULTS

The deviation from what should be a relatively temperature-independent value of thermal diffusivity for glassy materials[12] is indicative of a radiative heat transfer process becoming increasingly dominant. Gardon[5] reviewed radiative heat transfer in diathermous solids, and several important considerations emerged: Two mechanisms are responsible for radiative heat transfer across the specimen: direct radiative transfer across the sample and internal radiative transport. The relative contributions of the above processes to the total radiative component are shown schematically in Figure 6 as a function of the optical thickness of the specimen, γL, which is defined by the Beer-Lambert law, $I = I_0 e^{-\gamma L}$ (I is the transmitted intensity and I_0 the incident intensity). Such a diagram can only be schematic in nature owing to the complex changes that can occur in the absorption coefficient, γ, with temperature and wavelength variation.[13] In view of these complexities, the present data are discussed in general terms.

From Figure 6, it can be seen that at very low optical thicknesses, direct radiative heat transport dominates the total contribution due to radiation. Considering fused silica where $\gamma(1\text{-}2.5~\mu m)$ is $<0.01 \times 10^{-2}~m^{-1}$ and with a typical specimen thickness of between 0.05×10^{-2} and $0.2 \times 10^{-2}~m$, the optical thickness,

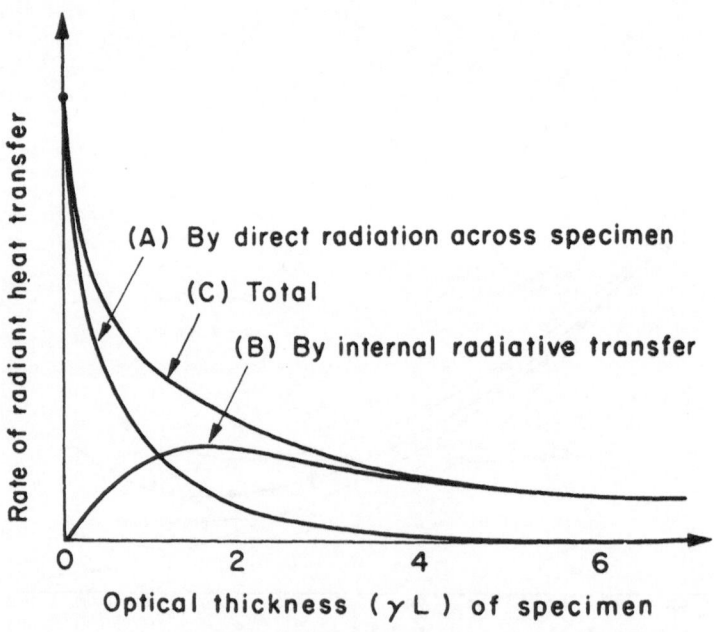

Fig. 6.   Schematic comparison of two mechanisms of radiant heat transfer. (after Gardon[5])

γL, will be very small, $<2 \times 10^{-5}$ m, which corresponds to all radiative transfer being directly transmitted across the specimen. If direct radiation is dominant, polishing (which effectively alters the surface emissivity[14]) will reduce the rate at which the rear face heats up, and since the flash method is dependent on the rear-face heating rate, the diffusivity values will be affected. Figure 1 shows the effect of polishing in reducing thermal diffusivity for silica at a constant thickness; this is clearly indicative of the significance of the direct radiative transfer mechanism.

Although a contribution by direct radiation is clearly important, increasing the specimen length and, hence, the optical thickness, γL, should produce a decrease in the direct radiative component (see Figure 6) and, hence, in the observed conductivity or diffusivity. This was not observed experimentally, and the increase in measured values of diffusivity with increasing specimen length is indicative of an increasingly predominant internal radiative mechanism, even with the relatively thin specimens used in this study. This is believed to be true for all the materials measured, as can be seen from Figures 1, 2 and 4.

The data for plate glass (Figure 2) show a dependence on specimen thickness similar to that of the silica data, i.e., increasing the specimen thickness produced an increase in the measured thermal diffusivity. The effect of thickness on thermal diffusivity was somewhat more pronounced with the plate glass specimens than the silica specimens. This may be attributed to an increase in the internal radiative transfer contribution. Such an explanation would seem likely, since a larger value for γ (0.43×10^{-2} m^{-1}) and a corresponding increase in the optical thickness, γL (to 0.88×10^{-2}), for the 2 mm plate-glass specimen would enhance the contribution due to internal radiative heat transfer. However, rear face polishing on the plate glass specimens produced an increase in apparent thermal diffusivity (see Figure 2). This is the converse of the situation found in the polished silica specimens. The reasons for such behaviour are unclear, but the effects of complex interactions between the phonon and radiative heat transfer processes, together with changes in the spectral absorption and reflectivity of the rear face must be important considerations. Further work on the effects of surface finish is needed to elucidate the predominant operative mechanisms.

Silica and plate glass are recognized as good radiative conductors; in contrast, sapphire is a very good phonon conductor with the thermal resistivity dominated by Umklapp processes at high temperatures. For such a material, there are sufficient precedents to assume:

$$\lambda_{TOTAL} = AT^{-1} + BT^3$$

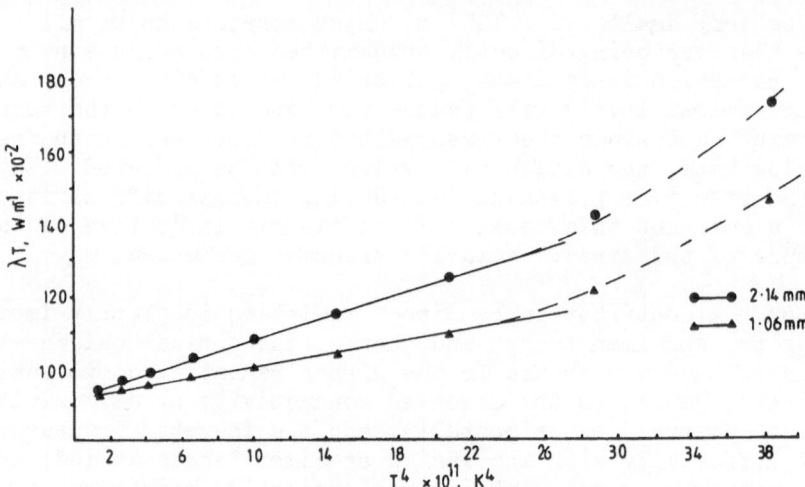

Fig. 7. Plot λT vs T^4 for synthetic sapphire data.

where A and B are constants, AT^{-1} represents the high temperature phonon contribution due to Umklapp scattering (neglecting any point defect scattering), and BT^3, that due to radiative processes. Figure 7 shows the data fitted to the above equation. The fit is good over most of the temperature measurement range; the deviation at higher temperatures could be attributed to the complex variation of the absorption coefficient with temperature or to the experimental effects arising from the instability of the coating. In Figure 5 the contribution due to phonon conduction is shown by the broken line. The phonon contribution has been calculated from the experimental data using $\lambda_{PHONON} = AT^{-1}$. The constant A has been determined by extrapolation of the λT vs T^4 curve to zero temperature. The data shows a good fit to the calculated curve at low temperature and an increasing deviation as the contribution by radiative processes increases. The increase in thermal conductivity and diffusivity is again pointing to a significant contribution by an internal radiative transfer mechanism.

CONCLUSIONS

(1) Thermal diffusivity measurements by the 'flash' technique on diathermous materials are feasible using specimens with opaque, sputtered metallic coatings. The results obtained show good agreement with the thermal conductivity data in the literature. The flash technique has the advantage that thin specimens are used for measurements, which reduces internal radiative transfer and gives values

nearer the 'true' thermal diffusivity.

(2) The temperature dependence of thermal diffusivity for fused silica and plate glass was found to be a function of both specimen thickness and surface finish, with increasing specimen length producing an increase in diffusivity. The role of surface finish on modifying the radiative heat flux, especially on the plate glass specimens, is unclear.

(3) Radiative heat transfer becomes increasingly significant at elevated temperatures in sapphire single crystals.

REFERENCES

1. T. Hirai, S. Hayasai, and K. Niihana, Thermal Diffusivity, Specific Heat and Thermal Conductivity of Chemically Vapour -Deposited Si_3N_4, Ceram.Bull. 57(12): 1126 (1978).
2. R.L. Rudkin, Thermal Diffusivity Measurements on Metals and Ceramics at High Temperatures. U.S. Air Force Rept. ASD-TDR-24 Part II, 1-16, 1963 [AD413 005].
3. S.P. Howlett and R. Taylor, University of Manchester Institute of Science and Technology, unpublished work.
4. W.D. Kingery, Heat-Conductivity Processes in Glass, J.Am.Ceram. Soc. 44(7) 302 (1961).
5. R. Gardon, The Apparent Thermal Conductivity of Diathermous Materials, in: Proceedings of the 2nd Conference on Thermal Conductivity". M. J. Laubitz, ed., NRC Ottawa.
6. C.F. Lucks, H.W. Deem and W.D. Wood, Thermal Properties of Six Glasses and Two Graphites, Ceram. Bull. 39: 313 (1960).
7. W.J. Parker, R.J. Jenkins, C.P. Butler and G.L. Abbot, Flash Method of Determining Thermal Diffusivity, Heat Capacity and Thermal Conductivity, J. Appl. Phys. 32: 1679 (1961).
8. D.A. Watt, Theory of Thermal Diffusivity by Pulse Technique, Brit. J. Appl. Phys. 17: 231 (1966).
9. R. Taylor, Construction of Apparatus for Heat Pulse Thermal Diffusivity Measurements from 300-3000 K. J. Phys. E: Sci. Instrum., 13: 11 (1980).
10. W.L. Wray, T.J. Connolly, Thermal Conductivity of Clear Fused Silica at High Temperatures, J. Appl. Phys. 30 (11) 1702: (1959).
11. W.D. Kingery, Thermal Conductivity: XII, Temperature Dependence of Conductivity for Single Phase Ceramics, J. Am. Ceram. Soc. 44 (7) 302: (1961).
12. C. Kittel, "Introduction to Solid State Physics", 3rd Ed., John Wiley and Sons, New York (1966).
13. L. Genzel, Messing der Ultrarot-Absorption von Glas Zwischen 20° und 1360°C, (Measurement of Infrared Absorption of Glass between 20° and 1360°C), Glastech.Ber., 24 55: (1951).
14. B.J. Keene, K.C. Mills, Emissivities of Slags used in Electro-slag Remelting, Arch.Eisenhuttenw. 52. 311 (1981).

THERMAL CONDUCTIVITY AND DIFFUSIVITY OF CLIMAX STOCK QUARTZ MONZONITE AT HIGH PRESSURE AND TEMPERATURE

W. B. Durham and A. E. Abey

Earth Sciences Division
Lawrence Livermore National Laboratory
University of California
P.O. Box 808
Livermore, California 94550

ABSTRACT

Measurements of thermal conductivity and thermal diffusivity have been made on two samples of Climax Stock quartz monzonite at pressures between 3 and 50 MPa and temperatures between 300 and 523 K. Following those measurements, the apparatus was calibrated with respect to the thermal conductivity measurement using a reference standard of fused silica. Corrected thermal conductivity of the rock indicates a value at room temperature of 2.60 ± 0.25 W/mK at 3 MPa increasing linearly to 2.75 ± 0.25 W/mK at 50 MPa. These values are unchanged (± 0.07 W/mK) by heating under 50-MPa pressure to as high as 473 K. The conductivity under 50-MPa confining pressure falls smoothly from 2.75 ± 0.25 W/mK at 313 K to 2.15 ± 0.25 W/mK at 473 K. Thermal diffusivity at 300 K was found to be $1.2 \pm 0.4 \times 10^{-6}$ m^2/s and shows approximately the same pressure and temperature dependencies as the thermal conductivity.

INTRODUCTION

The desire to measure the thermal properties of rocks at upper crustal conditions of pressure (<100 MPa) and temperature (<500°C) has been stimulated by recent interest in permanent storage of radioactive waste materials in mined underground repositories. An apparatus has been designed and built to make such measurements on large-grained (to approximately 10 mm) polycrystalline rocks (Abey et al., 1981). Earlier papers reported measurements made by this apparatus on rock

salt (Durham et al., 1981) and preliminary results on a quartz monzonite from the Climax Stock, Nevada Test Site (Durham and Abey, 1981).
Those papers also discussed in detail the rationale for making measurements at in situ conditions of pressure and temperature and the
expected thermal response of crystalline rock to changes in pressure
and temperature.

This report presents the latest measurements of thermal
conductivity and diffusivity for Climax Stock quartz monzonite (CSQM)
and presents a recent calibration of the test apparatus. The CSQM
data presented here are the first calibrated data we have reported.
For a detailed description of the apparatus and experimental method,
as the expected thermal response of a model polycrystalline solid containing microfractures, the reader is referred to the earlier papers.

TEST SAMPLES

Two samples of CSQM (referred to hereafter as Run 1 and Run 2)
were tested and a calibration test was made using a sample of fused
silica with known thermal conductivity. The mineralogy of the CSQM
is detailed by Izett (1960). Briefly, the average composition of
core taken from the U-15-A drill hole approximately 300 m horizontally distant from the site of the spent fuel test is 28% by weight
quartz, 25% alkali feldspar, 40% plagioclase, 6% biotite, and 1%
accessory minerals. Grain size in the matrix is 1 to 1.5 mm, but is
marked by quartz phenocrysts (5 to 10% by volume) averaging 4 mm in
diameter and by large orthoclase phenocrysts (5% by volume) averaging
50 mm in length with some as long as 150 mm. The orthoclase phenocrysts are uniformly distributed in the rock. A volume of rock 50 cm^3
or more that does not encounter an orthoclase phenocryst is unusual.
The rock has a connected porosity of approximately 0.54% (Page and
Heard, 1981) and has a density of approximately 2.64 Mg/m^3.

The fused silica reference standard was taken from a single ingot
of General Electric Type 124 clear fused silica. Details of the
chemistry and of the physical properties of the material are available
on product data sheets from the manufacturer. It is a high-purity
silica glass (approximately 54 ppm impurity atoms, 50% of those being
Al) with a low coefficient of thermal linear expansion (0.55 X
10^{-6} K^{-1}). The low expansivity makes it an ideal pressure
calibration standard, given the apparent absence of materials in the
conductivity range 1 to 10 W/mK whose pressure dependencies over the
range 0 to 50 MPa are well-established. Its resistance to thermal
shock makes it unlikely to develop fractures (or microfractures) with
a concurrent extrinsic (crack-related) dependence of any of its
physical properties upon pressure. The intrinsic pressure dependence
in silica glass is negligible over the range 0 to 50 MPa (Bridgman,
1952, for conductivity, 0.38%/100 MPa; Kieffer et al., 1976, for
diffusivity, -0.10%/100 MPa), so it is assumed for the purposes of

calibration that the dependencies of the several thermal properties upon temperature remain unchanged in the pressure range 0 to 50 MPa. The temperature dependence of thermal conductivity at atmospheric pressure is based on independent measurements made on a piece of fused silica cut from the original ingot, immediately adjacent to the center section of our reference standard (Fig. 1).

MEASUREMENT AND CORRECTION PROCEDURES

For all runs, measurement of thermal diffusivity and, for CSQM Run 1, measurement of thermal conductivity were made by the procedure given in Abey et al., 1981. The measurements of thermal conductivity for CSQM Run 2 and for the fused silica reference standard were made by a significantly different technique, which is outlined below. Because the two techniques are different, the calibration run was used to correct only data taken for CSQM Run 2.

The main point of difference between the two techniques is the control of heater powers rather than the control of heater temperatures. The heaters referred to are the three inner sample heaters used to maintain the temperature gradients that allow conductivity to be measured (Fig. 1). The temperatures are those of the controlling thermocouples, also shown in Fig. 1. Of the multitude of heat

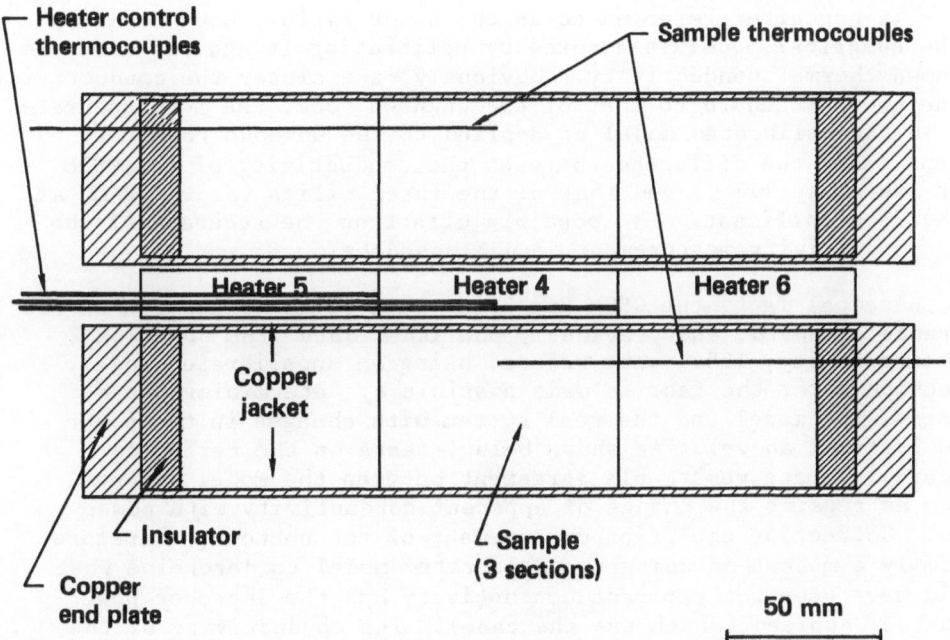

Fig. 1.   Sample assembly cross section.

flow patterns that can be generated in the sample with three independently controllable heaters, any one may be selected by either of
the two methods: heater powers may be set to desired levels (how the
desired levels are decided upon is discussed below), or heater powers
may be set so that thermocouple readings lie at desired levels. In
the ideal situation of perfect instrument precision, no meaningful
distinction can be made between the two methods. It is a characteristic of the experimental configuration, however, that measured
conductivity is considerably more sensitive to thermocouple temperature fluctuation than heater power fluctuation. At 3 W/mK, a temperature fluctuation of $\pm$ 0.1 K at 300 K shifts measured conductivity by approximately $\pm$ 0.3 W/mK. To produce the same shift,
heaters operating at, say, 3 W must fluctuate approximately 0.2 W.
Given standard measurement equipment, therefore, heater control is a
more practical means of achieving satisfactory measurement precision
of thermal conductivity.

    The desired heater power levels are those that produce an
apparent conductivity (a function of the radial temperature gradient
and output power from the central of the three heaters, as given in
Abey et al., 1981) equal to the actual conductivity. These levels are
determined by a numerical simulation of the heat flow in the measurement apparatus. The desired power condition is invariably one in
which the two end heaters operate at the same power, which in turn is
somewhat higher than the power output of the central heater (Fig. 1).
The ratio of the output of one end heater to that of the central
heater is hereafter referred to as the power ratio. Good accuracy
of the numerical model is assured by calibrating it against a sample
of known thermal conductivity. Obviously, the closer the conductivity
of the known standard to that of the unknown rock, the more accurate
will be the calibrated model as applied to the unknown rock. In the
present case, the difference between the conductivity of the CSQM
(near 3 W/mK at 300 K) and that of the fused silica (near 1 W/mK at
300 K) is significant. The possible effect on the accuracy of the
final conductivity measurement is discussed below.

    In actual fact, the CSQM Run 2 was made prior to the run on the
reference standard, and previously published data from CSQM Run 2
(Durham and Abey, 1981) were reduced using an uncalibrated model.
Correction after the fact is made possible by determining the behavior of the model and the real system with changes in the power
ratio (defined above). As shown below, tests on the reference
standard indicate remarkable agreement between the model and real
system as regards the change of apparent conductivity with power
ratio. Correcting the prior measurement of the unknown, therefore,
is simply a matter of using the calibrated model to determine what
should have been the apparent conductivity had the improper power
ratio been applied (which was the case). The conductivity of the
model is then adjusted until the apparent conductivity of the
calibrated model matches the apparent conductivity during the run.
At that point, the modeled conductivity becomes the best estimate of

the real conductivity of the rock. In future situations, with the
calibrated model in hand, the proper ratio will be known and will be
applied during the measurement.

RESULTS

Uncorrected Data

Figure 2 shows uncorrected values of thermal conductivity as a
function of temperature for both CSQM samples. The reduced scatter
between Runs 1 and 2 is the result of the change in measurement procedure discussed in the previous section. Figure 3 gives uncorrected
thermal conductivity as a function of pressure at 313 K (Run 2 data
only) and Fig. 4 shows uncorrected thermal diffusivity as a function
of temperature and pressure for both runs.

Calibration of the Model

Figure 5a illustrates the main features of the calibration
procedure. Apparent conductivities were measured on the fused silica
calibration standard at room temperature at each of four confining
pressures (3, 10, 30, 50 MPa) at several values of power ratio
between 1.7 and 2.9. The results are the points with error bars in
Fig. 5a. The (uncalibrated) model which was used to predict the
power ratios for CSQM Run 2 was applied to the calibration standard, with the results enclosed in the small oval in Fig. 5a. The

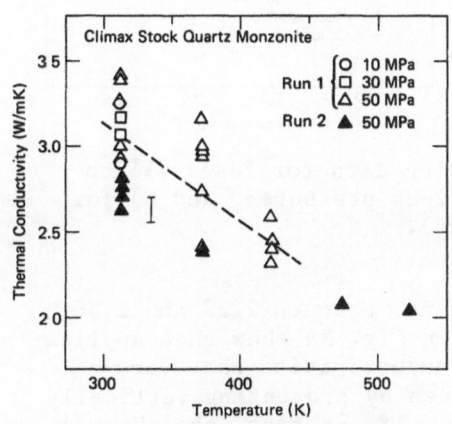

Fig. 2. Thermal conductivity
vs. temperature for
CSQM Runs 1 and 2,
uncorrected. Dashed
line is drawn
freehand through the
data of Run 1 and is
transposed to Fig. 6.

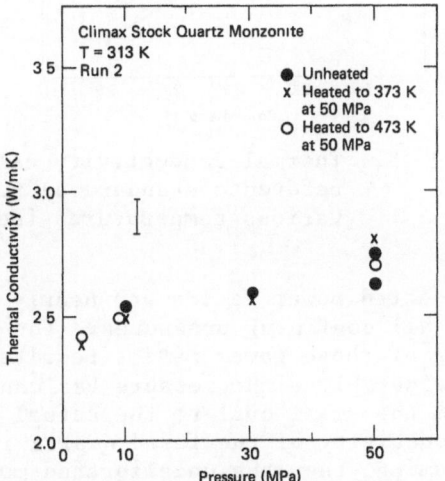

Fig. 3. Thermal conductivity
vs. pressure at 313 K
for CSQM Run 2,
uncorrected.

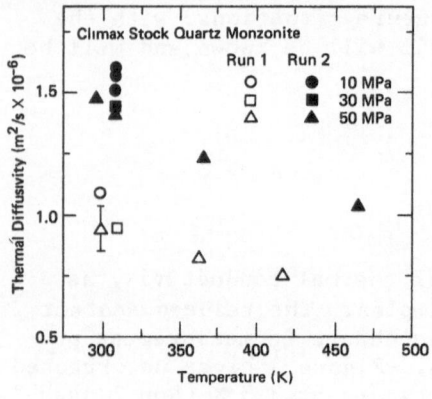

Fig. 4. Thermal diffusivity
vs. temperature and
pressure for CSQM
Runs 1 and 2,
uncorrected.

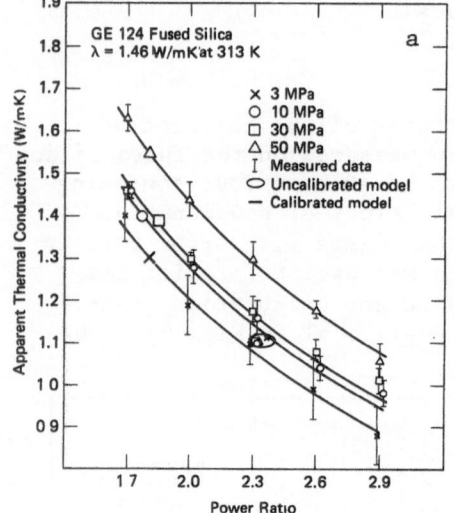

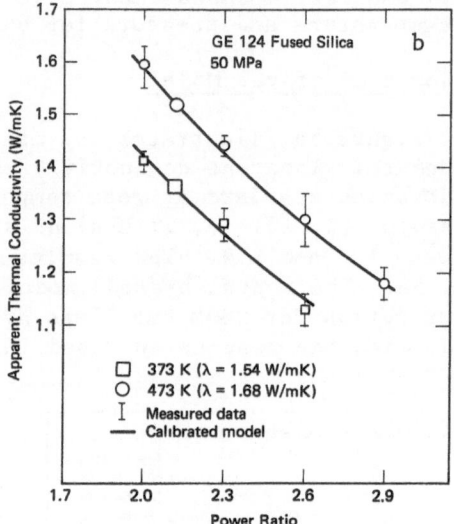

Fig. 5. Thermal conductivity calibration data for fused silica
        reference standard a) for various pressures, and b) for
        various temperatures (see text).

predicted power ratios are nearly constant, between 2.32 and 2.36,
for all confining pressures. The data in Fig. 5a show that application of those power ratios results in conductivities that vary
considerably with pressure (as can be seen by projecting vertically
from the small oval to the actual data). If, in fact, the thermal
conductivity of the fused quartz is constant at 1.46 W/mK as we are
assuming, then the uncalibrated model must be inaccurate.

Figure 5b shows the data for the calibration standard at elevated
temperature and fixed pressure. At the time the data were taken, the
true conductivity was not known, but was assumed to be in the range
1.2 to 1.4 W/mK. Therefore, none of the measured (apparent) values of
conductivity in Fig. 5b reached the true value.

Every attempt was made in the construction of the original model to faithfully copy the physical details of the apparatus. The observation of relatively subtle differences between expected and actual behavior in the fused silica (Fig. 5a) is an indication of the success of the original model. In calibrating the model, two parameters were varied: a) the conductance across the inner jacket-rock interface, infinite in the uncalibrated model, was allowed to decrease with decreasing confining pressure; b) the convective heat transfer from the outer metallic surfaces of the sample assembly, originally perfect, was allowed to decrease with decreasing confining pressure. The adjustment of these two parameters was done manually until the model matched, within an arbitrary error, the observed behavior of the fused silica. The calibrated model in its final form produces the solid lines shown in Figs. 5a and b.

Corrected Data

Following the procedure described above under "Measurement and Correction Procedures," the CSQM Run 2 conductivity values in Figs. 2 and 3 were corrected as shown in Figs. 6 and 7, respectively. Calibration of the thermal diffusivity measurements has not yet been accomplished.

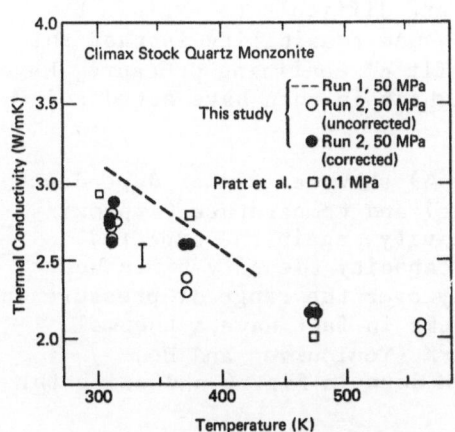

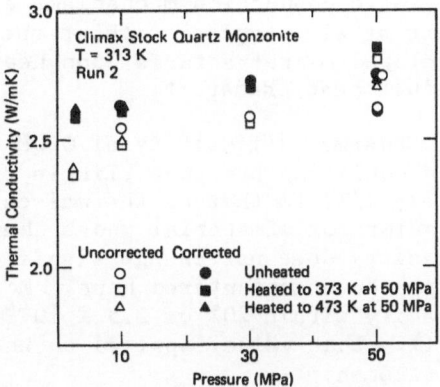

Fig. 6   Thermal conductivity vs. temperature at 50 MPa for CSQM Run 2, corrected and uncorrected. Dashed line is uncorrected data from Run 1, taken from Fig. 2. Error bar indicates precision of data from this study.

Fig. 7. Thermal conductivity vs. pressure at 313 K for CSQM Run 2, corrected (solid symbols). Also shown is the uncorrected data from Fig. 3 (open symbols).

DISCUSSION

The data in Fig. 7 indicate an approximately linear increase in thermal conductivity of CSQM with increasing pressure between 3 and 50 MPa. The best estimate of conductivity at 313 K is 2.60 ± 0.25 W/mK at 3 MPa and 2.75 ± 0.25 W/mK at 50 MPa. The error bands include a measurement precision (reproducibility) of ± 0.07 W/mK (± 1 std. dev.) and an estimated uncertainty in the accuracy of the model of ± 0.20 W/mK. Within the measurement precision, the pressure dependence at 313 K is unchanged by heating under 50-MPa confining pressure to temperatures as high as 473 K, as shown by Fig. 7. The variability in thermal conductivity from sample to sample cannot be determined from our data. Independent measurements of the thermal conductivity of CSQM at 300 K by Pratt et al. (1979) indicate a value of 3.7 W/mK without mention of accuracy. Within a measurement precision of ± 0.2 W/mK (D. Enniss, 1980), Pratt et al. found the thermal conductivity at 300 K to be unchanged by the application of confining pressure over the range 0.1 to 35 MPa.

Under 50-MPa pressure, thermal conductivity changes smoothly with temperature in approximately 1/T fashion (Fig. 6), the usual intrinsic behavior of materials in the phonon conduction range (Touloukian and Ho, 1981). Data at 0.1 MPa from Pratt et al. (1979), also plotted in Fig. 6, show a sharper drop in conductivity with increasing temperature than our own and are difficult to explain by intrinsic conduction mechanisms alone. One possibility is that the Pratt et al. samples, without the benefit of confining pressure, have developed microfractures upon heating which in turn have acted to inhibit heat transport.

Thermal diffusivity of CSQM (Fig. 4) shows a similar dependence upon confining pressure (little or none) and temperature (approximately 1/T) to that of thermal conductivity, again the expected behavior for a material whose thermal capacity (density times heat capacity) does not change significantly over the range of pressure and temperature encountered here. Most rocks in fact have a thermal capacity within 20% of 2.3×10^{-6} Ws/m^{3}K (Touloukian and Ho, 1981). That value, applied to the CSQM data in Fig. 6 and using the relationship

$$\text{Thermal diffusivity} = \frac{\text{thermal conductivity}}{\text{thermal capacity}}$$

produces values intermediate to those of the two runs in Fig. 4. No explanation can be given for the strong disparity between the diffusivity results for Runs 1 and 2 except to say that the precision of the diffusivity measurement has not yet been refined as well as the precision of the conductivity measurement.

CONCLUSIONS

1.   The thermal conductivity of as-cored CSQM at 313 K varies in
approximately linear fashion with increasing pressure from 2.60 ±
0.25 W/mK at 3 MPa to 2.75 ± 0.25 W/mK at 50 MPa. These values
are not altered more than ± 0.07 W/mk by heating under 50-MPa
confining pressure to as high as 473 K. These values are based on
measurements of a single 127-mm-diameter by 203-mm-long volume of
rock. The stated error band includes absolute uncertainty.

2.   Under 50-MPa confining pressure, thermal conductivity of CSQM
varies approximately as 1/T, the expected intrinsic behavior.

3.   The thermal diffusivity of CSQM at 300 K and 50 MPa is 1.25 ±
0.4 X 10^{-6} m^2/s. Within a precision of ± 0.1 X 10^{-6} m^2/s, the
variation of diffusivity with pressure and temperature is proportional
to that of thermal conductivity. The proportionality factor is the
density times the heat capacity.

ACKNOWLEDGMENT

    This work was performed for the National Waste Terminal Storage
Program (NWTS) of the U.S. Department of Energy by the Lawrence
Livermore National Laboratory under contract number W-7405-ENG-48.
The authors gratefully acknowledge the aid of Donald N. Montan in
constructing the computer model.

REFERENCES

Abey, A. E., Durham, W. B., Trimmer, D. A., and Dibley, L., 1981, An
        apparatus for determining the thermal properties of large
        geological samples at pressures to 0.2 GPa and temperatures to
        750 K, UCRL-85784, submitted to Rev. Sci. Inst.
Bridgman, P. W., 1952, "The Physics of High Pressure," G. Bell and
        Sons, London, p. 322.
Durham, W. B. and Abey, A. E., 1981, The effect of pressure and
        temperature on the thermal properties of a salt and a quartz
        monzonite, in: "Proceedings of the 22nd U.S. Symposium on Rock
        Mechanics," Massachusetts Institute of Technology, Cambridge,
        MA, UCRL-85285.
Durham, W. B., Abey, A. E., and Trimmer, D. A., 1981, "Thermal
        Properties of Avery Island Rock Salt to 573 K and 50 MPa
        Confining Pressure," Lawrence Livermore National Laboratory,
        Livermore, CA, UCRL-53128.
Enniss, D., 1980, Terra Tek, Salt Lake City, UT, private
        communication.

Izett, G. A., 1960, "Granite Exploration Hole, Area 15, Nevada Test
    Site, Nye County, Nevada--Interim Report, Part C. Physical
    Properties, Trace Element Memorandum Report 836-C," United
    States Geological Survey, Washington D C, 36 pp.
Kieffer, S. W., Getting, I. C., and Kennedy, G. C., 1976,
    Experimental determination of the pressure dependence of the
    thermal diffusivity of teflon, sodium chloride, quartz, and
    silica, J. Geophys. Res., 81:3018-3024.
Page, L. and Heard, H. C., 1981, Elastic moduli, thermal expansion,
    and inferred permeability of Climax quartz monzonite and
    Sudbury gabbro to 500°C and 55 MPa, in: "Proceedings of the
    22nd U.S. Rock Mechanics Symposium," Massachusetts Institute of
    Technology, Cambridge, MA.
Pratt, H. R., Lingle, R., and Schrauf, T., 1979, "Laboratory Measured
    Material Properties of Quartz Monzonite, Climax Stock, Nevada
    Test Site," Lawrence Livermore National Laboratory, Livermore,
    CA, UCRL-15073.
Touloukian, Y. S. and Ho, C. Y., eds., 1981, "Physical Properties of
    Rocks and Minerals," McGraw-Hill/CINDAS Data Series on Material
    Properties, Volume II-2, McGraw-Hill, New York, p. 413.

SESSION   TC-12A

Insulations - III

CHAIRMAN

R. P. Tye
Energy Materials Testing Laboratory
Biddeford, Maine

AN EXPERIMENTAL AND MATHEMATICAL STUDY OF THE EFFECT OF THICKNESS

IN LOW-DENSITY GLASS-FIBER INSULATION

Mark A. Albers and Charles M. Pelanne

Manville Research and Development Center

Denver, Colorado

ABSTRACT

The "Effect of Thickness" in low-density glass-fiber insulation was determined by three different methods. The results of
each experiment were compared with the assigned values for the NBS
Certified Transfer Specimens. Each of the three methods predicted
the "Effect of Thickness" within 1% of the NBS values. A radiation
absorbing "septa" study was shown to be a useful tool in assessing
apparatus measurement errors.

BACKGROUND

In the past several years the change in thermal conductivity
of low-density insulations when tested at different thicknesses has
been discussed and studied. This change in conductivity has been
termed the "Effect of Thickness." Reports of the magnitude of the
effect varied greatly. Some of those results may have been subject
to considerable equipment-related measurement errors.

Scientists at the National Bureau of Standards assumed the
responsibility of developing "Certified Transfer Specimens" to verify
the test measurements. These specimens, tested in the NBS 1000-mm
Guarded Hot Plate (GHP), will serve to calibrate thermal measurement
equipment with thick, low-density standards.

INTRODUCTION

A series of experimental programs were performed to determine

471

the "Effect of Thickness" (EOT) in a low-density glass-fiber product
by three different methods.   The results of each analysis are compared
with the NBS assigned values for the recently available Certified
Transfer Specimens (CTS).   The CTS are the new low-density (≈ 9.6
kg/m^3, ≈ 0.6 lb/ft^3 glass fiber) reference standards for thermal
resistance measurements at 25.4, 76.2, and 152.4 mm (1, 3, and 6 in.)
thicknesses.

Since Johns-Manville supplied NBS with the glass-fiber material
from which the CTS were developed, additional material was concurrently fabricated from which specimens were selected for this investigation.   Test specimen size was 914 mm x 914 mm (36 x 36 in.).
Thermal testing was performed in a "36-inch heat flow meter" apparatus; the test area was 305 mm x 305 mm (12 x 12 in.).

The Report of This Study is Divided Into Three Areas:

1. Comparison of results of 36-inch heat flow meter tests of
    the CTS with the NBS assigned values.
2. Radiation septa study to assess apparatus errors. Comparison
    of results with NBS values.

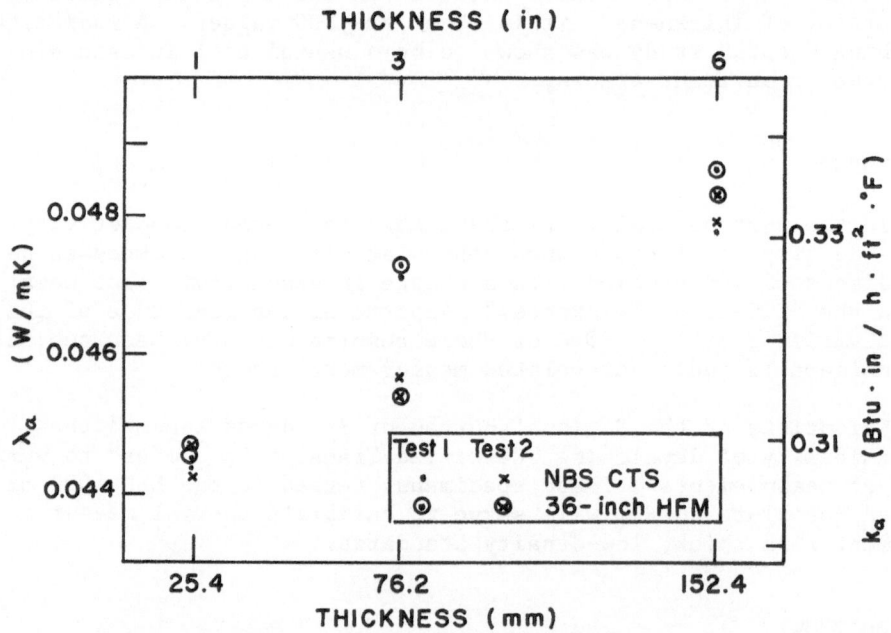

Fig. 1. Apparent thermal conductivity versus thickness - 36-inch
         heat flow tests on the NBS CTS compared with the NBS
         assigned values.

Table 1. Tabulation of Measured and Normalized λ_a Values

CTS	Thickness mm (in)	Density kg/m³ (lb/ft³)	λ_a, W/m·k (BTU·in/h·ft²·°F) NBS assigned	NBS normalized	36-inch HFM	36-inch HFM normalized
6166	25.4 (1)	10.04 (0.627)	0.04428 (0.3070)	0.04488 (0.3112)	0.04434 (0.3074)	0.04494 (0.3116)
6144	25.4 (1)	9.79 (0.611)	0.04425 (0.3068)	0.04449 (0.3085)	0.04439 (0.3078)	0.04464 (0.3095)
1612	76.2 (3)	9.02 (0.563)	0.04713 (0.3268)	0.04604 (0.3192)	0.04721 (0.3273)	0.04611 (0.3197)
1623	76.2 (3)	9.77 (0.610)	0.04552 (0.3156)	0.04579 (0.3175)	0.04545 (0.3151)	0.04572 (0.3170)
2112/2111	152.4 (6)	8.94 (0.558)	0.04783 (0.3316)	0.04650 (0.3224)	0.04862 (0.3371)	0.04581 (0.3276)
2121/2123	152.4 (6)	8.92 (0.557)	0.04790 (0.3321)	0.04654 (0.3227)	0.04822 (0.3343)	0.04683 (0.3247)

Normalized density – 9.61 kg/m³ (0.6 lb/ft³)

3. Mathematical model to predict the EOT from previous 25.4 mm
    (1-in.) specimen thermal testing.

Each of these procedures were used to determine the EOT in low-density
glass-fiber insulations. A comparison was made between each of the
methods and the NBS 1000-mm GHP values.

FULL THICKNESS THERMAL TESTING IN A 914-mm (36-in.) HEAT FLOW METER APPARATUS

Two sets of NBS CTS were tested in a large heat-flow-meter
apparatus (HFM). Each set of specimens consisted of a 25.4, 76.2 and
152.4 mm (1, 3, and 6-in.) thick test specimens. After repeated tests
of each specimen, the average test results were compared with the NBS
assigned values for the CTS. The comparison is shown graphically in
Fig. 1.

Since the densities of the specimens were different, the test
results from the HFM and the NBS values were normalized to 9.61 kg/m^3
(0.6 lb/ft^3) density. The density normalization procedure consists
of determining the percent difference of the test λ from a model
equation prediction at the test density, then incorporating this
difference at the normalized density. The original and normalized
values are shown in Table 1. The results of the HFM tests were
referenced to the NBS 25.4 mm (1-in.) base and plotted, along with the
NBS values, as a percent difference in λ from the 25.4 mm (1 in.)
base. This comparison is shown in Fig. 2.

The HFM test results are therefore within 1% of the NBS CTS
assigned values in predicting the EOT of this glass-fiber material.
Half of this difference can be attributed to equipment measurement
errors, as explained later.

RADIATION SEPTA STUDY

The use of thermal radiation barriers, or "Septa," as a tool to
assess the apparatus errors of measurement in HFMs has been detailed
previously by Pelanne[1] and will only be outlined here.

The transmission of heat by thermal radiation is not reduced as
significantly by increased specimen thicknesses as are the other heat
transfer modes. Therefore, in a low-density insulation at room
temperature, as in other cases when the radiative transfer is significant, the heat transmission cannot be expressed as a true thermal
conductivity. Since the thermal conductivity varies with specimen
thickness, it is termed as apparent thermal conductivity, λ_a.

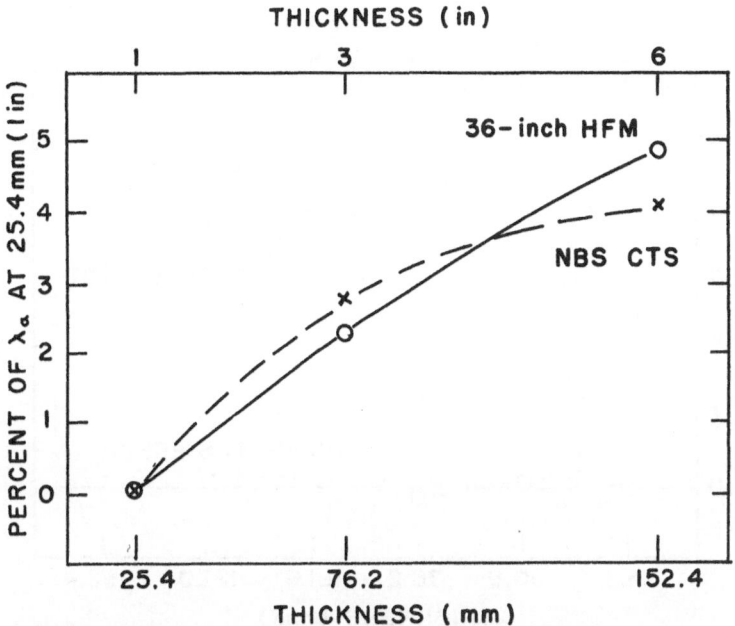

THICKNESS (in)

Fig. 2.   Percent change in apparent thermal conductivity versus thick-
          ness. Test results (avg.) normalized to 9.61 kg/m^3 (0.6
          lb/ft^3) and 25.4-mm base of reference.

     Prior to using the NBS full-thickness CTS, heat flow meters were
calibrated to thinner specimens from NBS GHP tests. These were the
only official references for thermal measurements available in the
United States. The calibration point was usually at 25.4-mm (1-in.)
thickness and all EOT study comparisons were made to that calibration
thickness. Thermal test measurements at thicknesses greater than the
calibration points produce higher λ_a results than those obtained on
the same material measured at the calibration point. This difference,
when compared with the 25.4-mm calibration point, is termed the
MEASURED EOT. Since the test apparatus is not calibrated at these
greater thicknesses, how much of the measured EOT is really specimen
or product EOT and how much is equipment testing error.

     To separate these EOT components, radiation-absorbing barriers
were used at thickness intervals equal to the calibration point thickness, in this case 25.4 mm separation. These septa have the same
emittance as the emittance of the plates in the HFM apparatus
($\varepsilon \simeq 0.95$). The septa, in effect, absorb the radiative energy and
re-emit it, thus acting like a group of HFM plates and simulating a
series of 25.4 mm thick tests. The barriers will, of course, also
stop convection, if present, causing only a negligible change in solid
and air conduction. At the densities of concern here, the effects
attributable to convection have been found to be negligible. This has

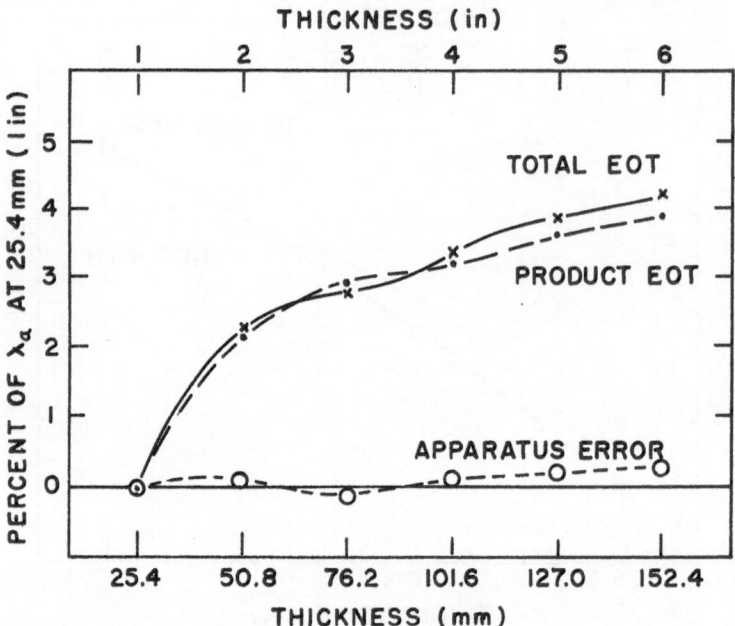

Fig. 3.   Separation of the effect of thickness (EOT) from the
          apparatus errors. Density normalized to 9.61 kg/m^3
          (0.6 lb/ft^3).

been demonstrated by Pelanne by using radiation transparent septa.[1]

Since the septa are not at the same temperatures as the HFM
plates, a small error on the order of 0.1% is introduced when comparing this measured λ_a with other tests. This error is within the
limits of the experiment.

The mathematics of determining the apparatus errors are as
follows: The expected λ_a is determined at any given thickness by
summing the series resistance of its individual 25.4 mm layers. Any
difference between the expected λ_a at a thickness and the measured λ_a,
when using radiation absorbing septa to separate each 25.4 mm layer
from the others, is an apparatus measurement error. Subtracting the
apparatus errors from the measured EOT results in the product EOT.

$$\text{App. Error} = \lambda_{w/septa} - \lambda_{expected}$$

$$= \lambda_{w/septa} - (\text{Calib. thickness}) \times \sum_{i=1}^{n} \frac{1}{\lambda_{a_n}}$$

Product EOT = Measured EOT - Apparatus Error

The results of these calculations are compared on a percentage basis with the average λ_a of the 25.4 mm specimens. The equipment-related measurement errors for some apparatuses can be quite high. The separation of apparatus error on the "36-inch HFM" used in this study is shown in Fig. 3. The test results have been normalized to 9.61 kg/m^3 (0.6 lb/ft^3). It can be seen that there is only a fraction of a percent apparatus error, which can be considered negligible for the most part. The negative error at 76.2 mm (3-in.) thickness may help to explain the low HFM λ_a value measured at 76.2 mm. Also, apparatus error at 152.4 mm (6-in.) thickness helps to explain some of the deviation from the CTS at 152.4 mm.

MATHEMATICAL MODEL PREDICTION OF EOT

Prior to the development of the NBS CTS full-thickness standards, a series of sixty-four 25.4 mm specimen thermal tests were performed on the same material produced for the CTS. With the aid of a simple nonlinear mathematical model of heat transfer in insulation, a prediction of the EOT was made. A plot of the predicted EOT compared with the CTS values and referenced to the same 25.4 mm base is shown in Fig. 4. The prediction of the model based only on 25.4 mm specimen thermal measurements, is within 1% of the NBS CTS values.

The model used is a variation of one that has been used before in one form or another by many authors, including Hamaker, Larkin and Churchill, Viskanta and Grosh, Siegel and Howell, King, Bhattacharyya, Rennex, Pelanne, Striepens and Linford et al.[2-12] A simple breakdown and explanation of the model is shown in the Appendix. The coefficients are calculated for a standard 23.9°C (75°F) mean temperature HFM operating with a 27.8°C (50°F) ΔT and having plate emittances of $\varepsilon = 0.95$. Very simple linear relationships with density have been chosen to represent λ solid and N, the radiation scattering parameter. A much more complex relationship could be developed for N that would include such variables as fiber diameters, fiber orientation and distribution, glass refractive index, and surface characteristics. Nevertheless the model serves well, even in its simplest form.

The usual practice is to perform a nonlinear multiple regression on the measured λ_a and specimen density data to solve for C and N. In actuality, C is relatively small and could be fixed, whereas a linear regression was performed to find N. It is interesting to note that a regression that was performed on multiple thickness data produced an EOT prediction nearly identical to the model based on 25.4 mm (1 in.) thick specimens shown in Fig. 4.

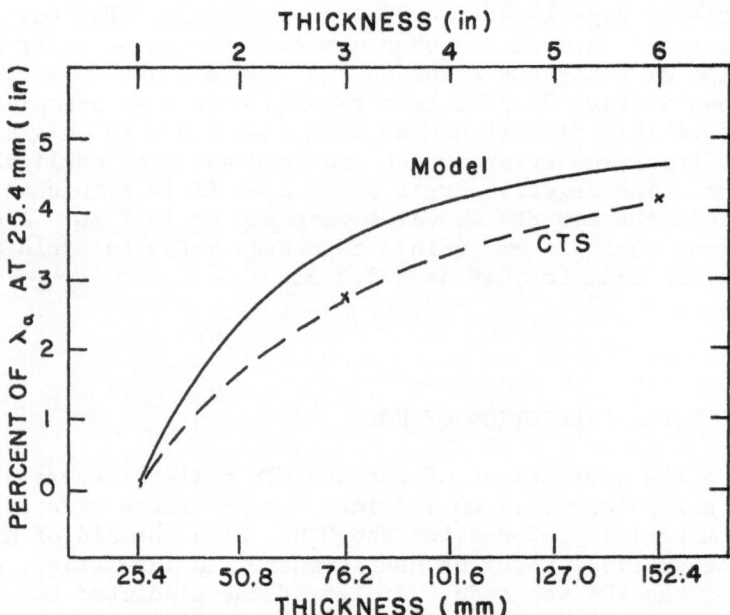

Fig. 4.   Percent change in apparent thermal conductivity versus
          thickness. Model projection of EOT prior to CTS based on
          25.4 mm (1 in.) λ_a measurement only in comparison with CTS
          values normalized to 9.61 kg/m^3 (0.6 lb/ft^3).

CONCLUSIONS

        A summary of the determination of the EOT by the three different
methods is shown in Table 2 and given below:

1.   a.   Density normalization can be used to remove some differences
          in test results and compare all data at a common density.

     b.   The differences between the test results on the 36-inch
          heat flow meter and the NBS assigned values for the CTS are
          relatively small. Since NBS currently estimates their
          uncertainty to be on the order of ± 2%, the differences in
          results are within the experimental error.

2.   a.   A radiation septa study revealed little or no apparatus error
          when testing glass-fiber insulation identical to the CTS
          material in a 914 mm (36 in.) HFM.

     b.   The apparatus error determined in the septa study was sub-
          tracted from the 36-inch HFM test results to produce the
          corrected CTS results. As shown in Table 2, these are within

Table 2. The Percent Effect of Thickness Determined by Three
          Techniques Compared With the NBS Assigned EOT for the
          CTS Material.

	% EOT *	
	76.2 mm (3 in)	152.4 mm (6 in)
CTS		
NBS assigned values	2.7 %	4.1 %
36-inch HFM tests	2.5	5.0
36-inch HFM tests (Corrected for apparatus errors)	2.7	4.6
Septa Study		
Product EOT	2.9	3.9
Apparatus error	−0.2 %	+0.4 %
Math Model Predictions		
Based on 1-in. tests	3.6	4.7
Based on full-thickness tests	3.7	4.7

* All values are normalized to 9.61 kg/m^3 (0.6 lb/ft^3), and
  compared with the NBS 25.4 mm (1-in.) base.

0.5% of the NBS assigned values.

c. As stated above, owing to the uncertainty of CTS values these septa test results are also within the experimental error.

3. a. A model developed from 25.4 mm (1-in.) specimen testing only can predict the EOT in this material, as determined by NBS CTS values, within the experimental uncertainty.

b. The same model developed from full-thickness specimen testing is nearly identical to the 25.4 mm specimen based model prediction.

All of the methods presented here to determine the EOT produced results within 1% of the NBS values. The septa study was shown to be a useful tool in assessing apparatus measurement errors. A septa study should be performed on any such apparatus used for full-thickness thermal testing.

The glass fiber used in these tests was from the same sampling as the NBS CTS and all of the thermal testing was performed on one 36-inch HFM apparatus. This particular apparatus had very little error when testing this material under these conditions, unlike some other equipment which could have considerable error.

REFERENCES

1. C. M. Pelanne, Discussion on experiments to separate the "effect of thickness" from systematic equipment errors, in: "Thermal Transmission Measurements, Thermal Insulation Performance," ASTM STP 718, D. L. McElroy and R. P. Tye, eds., American Society for Testing and Materials, Philadelphia (1980) pp. 322-334.

2. H. C. Hamaker, "Radiation and Heat Conduction in Light-Scattering Material." Philips Research Report 2, Eindhoven, Holland (1974), pp. 55-67, pp. 103-111, pp. 420-425.

3. B. K. Larkin and S. W. Churchill, Heat transfer by radiation through porous insulations, AIChEJ, 5:467-74 (1959).

4. R. Viskanta and R. J. Grosh, Heat transfer by simultaneous conduction and radiation in an absorbing medium, J. Heat Transfer, 84:63-72 (1962).

5. R. Viskanta, Heat transfer by conduction and radiation in absorbing and scattering materials, J. Heat Transfer, 87:143-50 (1965).

6. R. Siegel and J. R. Howell, "Thermal Radiation Heat Trans-
    fer," Second Edition, Hemisphere Publishing Corp.

7. C. R. King, Fibrous insulation heat transfer model, in:
    "Thermal Transmission Measurements of Insulation," ASTM STP
    660, R. P. Tye, ed., American Society for Testing and
    Materials, Philadelphia (1978), pp. 281-292.

8. R. K. Bhattacharyya, Heat transfer model of fibrous insula-
    tions, in: "Thermal Insulation Performance," ASTM STP 718,
    D. L. McElroy and R. P. Tye, eds., American Society for
    Testing and Materials, Philadelphia (1980), pp. 272-286.

9. B. G. Rennex, Thermal parameters as a function of thickness
    for combined radiation and conduction heat transfer in low-
    density insulation, J. Therm. Insul. 3:37 (1979).

10. C. M. Pelanne, "Experiments on the Separation of Heat
    Transfer Mechanisms in Low Density Fibrous Insulation,"
    presented at the Eighth Conference on Thermal Conductivity,
    Purdue University, Lafayette, Indiana, October 7-10, 1968,
    (Plenum Press 1969) pp. 897-911.

11. A. H. Striepens, Heat transfer in refractory fiber insula-
    tions, in: "Thermal Transmission Measurements of Insulation,"
    ASTM STP 660, R. P. Tye, ed., American Society for Testing
    and Materials, (1978), pp. 293-309.

12. R. M. F. Linford, R. J. Schmitt and T. A. Hughes, "Heat
    Transmission Measurements in Thermal Insulation," ASTM STP
    544, American Society for Testing and Materials, Philadelphia
    (1974), pp. 68-84.

APPENDIX

Model for predicting apparent thermal conductivity

$$\lambda_a = \lambda_{air} + \lambda_{solid} + \lambda_{radiation}$$

$$\lambda_a = \lambda_{air} + C\rho + \frac{A}{N + \frac{E}{\ell}}$$

Where: $A = \dfrac{\sigma(T_{hot}^4 - T_{cold}^4)}{\Delta T}$ temperature coefficient,

$A = 165.5 \ W/m^2$

for a 23.9°C (75°F) mean HFM, $\Delta T = 27.8°C$ (50°F)

$E = \dfrac{1}{\varepsilon_{hot}} + \dfrac{1}{\varepsilon_{cold}} - 1$ emittance coefficient,

$E = 1.1053$ for $\varepsilon = 0.95$

$N \simeq N'\rho$ scattering parameter

$\rho = $ density (kg/m^3)

$\ell = $ thickness (m)

$T = $ temperature of hot or cold plate (°K)

$\sigma = $ Stefan-Boltzmann constant

$= 5.6703 \times 10^{-3}$ $(W/m^2 \cdot K^4)$

$\varepsilon = $ emittance of hot or cold plate

Then the model form for a 23.9°C (75°F) mean-heat-flow-meter apparatus would be:

$$\lambda_a = 0.02602 + C\rho + \frac{165.5}{N'\rho + \frac{1.1053}{\ell}}$$

EFFECTIVE THERMAL CONDUCTIVITY OF GLASS-FIBER BOARD AND

BLANKET STANDARD REFERENCE MATERIALS*

D. R. Smith and J. G. Hust

Thermophysical Properties Division
National Bureau of Standards
Boulder, CO 80303

ABSTRACT

Measurements of effective thermal conductivity, λ, have been
performed on a series of specimens of glass-fiber board and glass-
fiber blanket. Measurements of λ were conducted as a function of
temperature from 85 to 360 K, of temperature difference with $\Delta T=10$
to 100 K, of bulk density from 11 to 148 kg/m^3 and for nitrogen,
argon, and helium inter-fiber fill gases at pressures from atmospheric to high vacuum. Results are analyzed and compared with values from the published literature and NBS certification data for
similar material. Polynomial expressions are given for the functional relation between conductivity, temperature, and density for
board and for blanket.

INTRODUCTION

A glass-fiber board insulation has been established as a standard reference material for thermal resistance, SRM 1450a, by the
National Bureau of Standards; it is certified over the temperature
range from 255 to 330 K with atmospheric pressure air as fill gas.[1]
Values of λ for the SRM board and of a lower-density blanket insulation, both 2.54 cm thick, over the temperature range from 85 to
360 K in dry nitrogen gas at pressures ranging from 1.33×10^{-2} Pa
(10^{-4} torr) to 8.4×10^4 Pa (630 torr) are reported. The blanket is

---

*This work was done for the Department of Energy, Oak Ridge
Operation, Oak Ridge, TN 37830, under Interagency Agreement DoE
No. DE-AI05-78OR05965.

under consideration for certification as an insulation SRM of lower
density. The data obtained have been used in the development of a
semiempirical model of heat transfer in fibrous insulation by
Van Poolen, Hust, and Smith which is discussed in a separate paper[2]
of this volume.

APPARATUS AND EXPERIMENTAL PROCEDURE

The instrument used to measure λ was a guarded-hot-plate apparatus constructed in accordance with ASTM Standard Test Method
C 177-76.[3] A complete description of this apparatus can be
obtained from J. G. Hust; a brief description is given below.

The apparatus uses two specimens, matched in density, placed on
opposite faces of the main heater and inner guard. Heat flow
through the specimens is along the vertical direction. The cooler
sides of the specimens face auxiliary heater plates controlled in
temperature. These plates are thermally coupled to lower-tempera-
ture fluid-cooled cold plates, thus permitting variation of both the
temperature gradient and the mean temperature of the specimens. All
heater surfaces are black-anodized aluminum to increase the emissivity. The measured room-temperature emissivity of the plates is
0.83. The assembly of specimens and heaters was surrounded by
exfoliated mica loose-fill thermal insulation to reduce edge heat
losses. Water, refrigerated alcohol, and liquid nitrogen were used
as coolant fluids to vary the mean specimen temperature from 95 K to
350 K. All heater and thermocouple leads were thermally tempered at
an outer guard held at the mean temperature of the specimens and at
copper lugs bolted to the outer surfaces of the aluminum heater
plates.

Thermocouples (36 gauge Type K), used to determine hot- and
cold-face temperatures of the specimens, were laid in shallow
grooves in the surfaces of the heater plates facing both sides of
the specimens. Each thermocouple bead was located in the surface
plane of the heater plate to which it was cemented. The difference
between T_{hot} and T_{cold} is the temperature difference ΔT, through the
specimen.

A 20-pair thermopile (36 gauge Type E) was used as the sensing
element in nulling the inner guard main heater gap temperature difference.

The annular inner guard had an outer diameter of 20.32 cm (8.00
in) and was separated from the enclosed main heater by a gap 1.6-mm
wide. The main heater diameter was 10.16 cm but the metered area,
A, used in calculating the apparent thermal conductivity was taken
to be that enclosed by the midline of the gap between the main and
guard heaters (diameter = 10.32 cm).

The specimen thickness, Δx, was precisely established by the use of three stainless-steel spacer tubes equally placed in notches around the circumference of the specimens. The voids within the tubes were filled with fibrous insulation. Both the metered area, A, and the specimen thickness, Δx, were corrected for thermal expansion.

A digital voltmeter with input impedance of 100 MΩ was used to measure the voltage drop across the heater resistance and also across a precision (0.005%) standard resistor in series with the heater, giving the heater current. A voltage/current calibrator with a stability of 0.05% was used as the heater power supply. The power obtained by the product of heater voltage and current was used as the heat flow term, $\dot{Q}$, in calculating the average λ by the operational form of Fourier's equation: $\dot{Q} = \bar{\lambda} \, A \, \Delta T / \Delta x$.

MEASUREMENTS

Glass-Fiber Board

A total of 158 triplicate λ measurements were conducted on 5 pairs of specimens ranging in density from 106 to 148 kg/m^3. These measurements, some of which were conducted to study measurement uncertainties, were taken in triplicate according to the C 177 specification. Thirteen measurements were performed to determine the detailed temperature dependence (from 85 to 360 K) of a pair of specimens with a density of 123 kg/m^3. The remaining specimens were measured at six temperatures within this interval. One pair of specimens (127 kg/m^3) was used to determine the dependence on gas species and pressure.

Glass-Fiber Blanket

For glass-fiber blanket, 219 triplicate λ measurements were conducted on 5 pairs of specimens ranging in density from 11 to 16 kg/m^3. The pair with a density of 14.75 kg/m^3 was measured at 25 temperatures from 85 to 360 K. The remaining pairs were studied at six temperatures over this range. The pair with a density of 14 kg/m^3 was used for the study of gas species and pressure.

DISCUSSION OF RESULTS

Because of the number of variables that affect the measured λ of these specimens, it is difficult to illustrate these data in a simple manner. A discussion is given first of the temperature dependence of the board and blanket specimens with densities of 123 and 14.75 kg/m^3, respectively. In both cases the fill gas is nitrogen at 8.4 x 10^4 Pa (630 torr) which is the atmospheric pressure at Boulder, Colorado.

A simple power series in temperature was fitted to the data obtained for the average λ of the board with a density of 123 kg/m^3 using a thermal conductivity integral technique.[4] A reduction in systematic deviations between fit and data was obtained in going from a 3-term to a 4-term polynomial, but fits of 5- to 7-term polynomials did not bring about significant further reductions. The data deviate from the 4-term polynomial by up to $\pm 0.7\%$. The temperature dependence of λ of this board is given in the form
$\lambda = \sum\limits_{i=1}^{4} A_i T^i$, where $A_1 = 0.114622$, $A_2 = 1.49249 \times 10^{-4}$.
$A_3 = 1.04399 \times 10^{-6}$, $A_4 = 1.63799 \times 10^{-9}$, λ is in mW/(m·k) and T is in K. Figure 1 shows these data and the fitted curve for λ vs temperature; a deviation plot comparing data and values calculated from the 4-term polynomial is given in figure 2. Previous measurements on glass-fiber board of the same[5] similar[6-11] densities are given in comparison with the present results in figure 3. The earlier work[1,5] on the same lot of specimens and the present results are in agreement to within 1.5% over the range of overlap.

For the glass-fiber blanket specimens of 14.75 kg/m^3 density over a similar temperature range, use of the thermal conductivity integral technique also led to a 4-term polynomial in temperature with $\pm 1\%$ deviations between data and the fitted curve. Thus for the blanket, in the same units as above, $\lambda = \sum\limits_{i=1}^{4} B_i T^i$ where
$B_1 = 0.0986120$, $B_2 = 1.21313 \times 10^{-4}$, $B_3 = -5.62714 \times 10^{-7}$, and
$B_4 = 1.90726 \times 10^{-9}$ over the range from 84 to 360 K. Figure 4 gives the data and fitted curve and figure 5 gives the deviation plot. Figure 6 compares the present results with those of previous investigations[6,11-14] on materials of similar density.

In contrast to the results for the board, for which the effective conductivity is dominated by linear dependence on temperature, the conductivity of the blanket shows a definitely higher power temperature dependence, especially at temperatures greater than 200 K. A direct comparison of the results for the board (123 kg/m^3) and blanket (14.75 kg/m^3) is shown for atmospheric pressure measurements in figure 7. Measurements for high-vacuum conditions, compared in figure 7 with the data for atmospheric gas pressure, show the appreciable role of gas conduction in the total heat transfer process for both the board and the blanket. Below 200 K the board has a greater λ owing to higher fiber packing density, but at temperatures greater than 200 K the blanket has a higher λ owing to the increased radiation component.

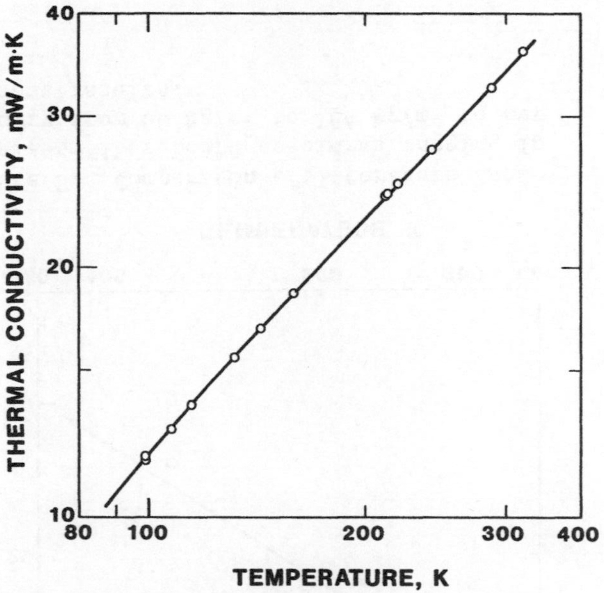

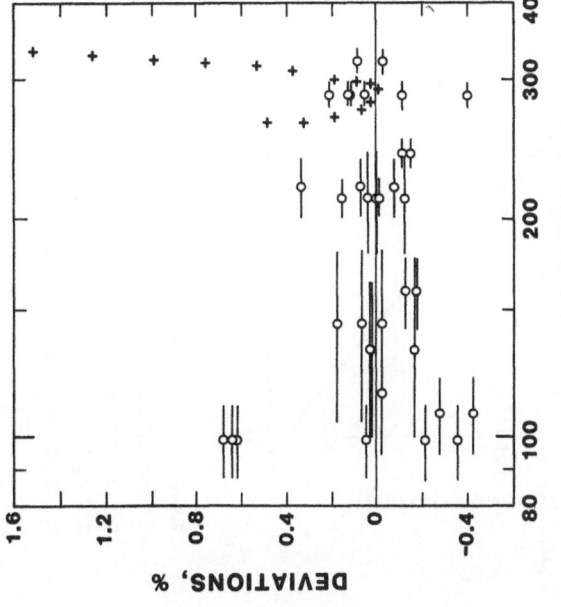

Figure 2. Deviation of observed thermal conductivity integrals (TCI) from calculated values of TCI for SRM 1450a glass-fiber board. Circles show mean temperature, and horizontal bars indicate temperature span of the measurement. Plus signs show differences between the results and NBS certification data.

Figure 1. Thermal conductivity data for SRM 1450a glass-fiber board. Solid line shows results of least squares fit of thermal conductivity integrals; circles are mean λ values plotted at the mean experimental temperatures.

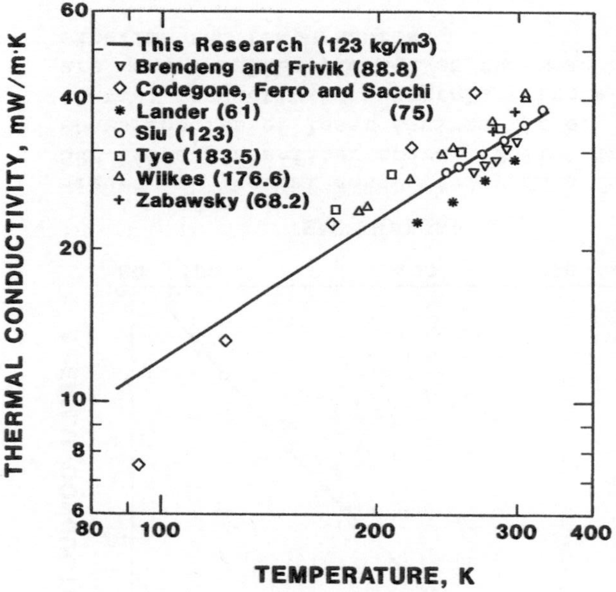

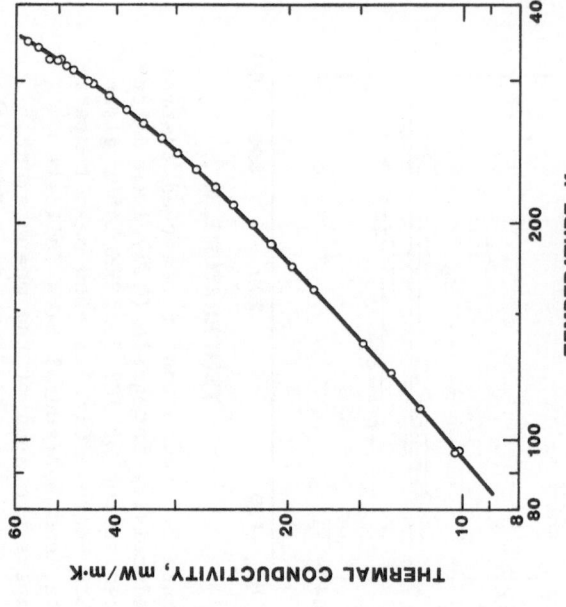

Figure 4. Thermal conductivity of glass–fiber blanket. Solid line shows results of least–squares fit of thermal conductivity integrals; circles are mean λ values plotted at mean experimental temperatures.

Figure 3. Comparison of literature data on glass–fiber board specimens ranging in density from 60 kg/m³ to 184 kg/m³ to our present results.

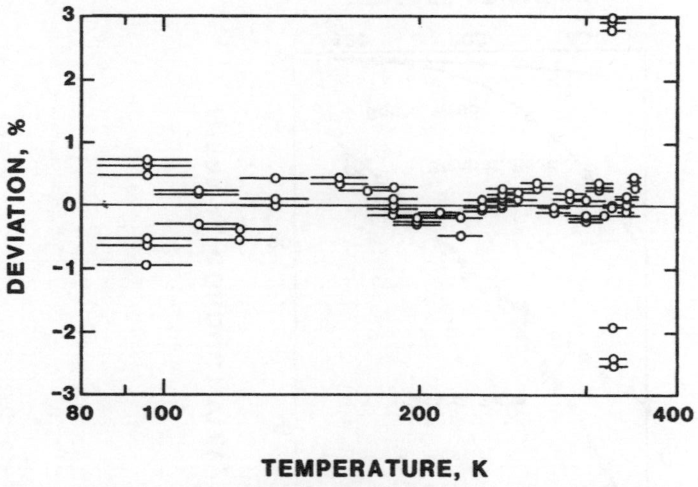

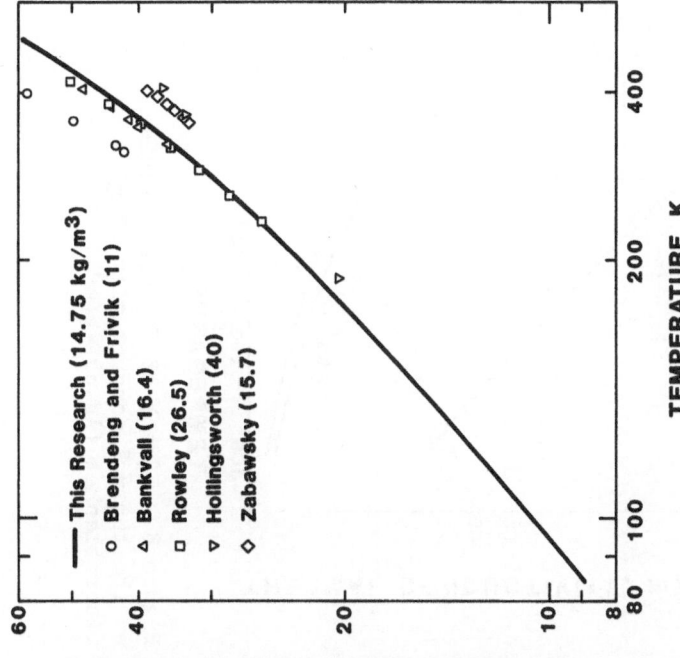

Figure 6. Comparison of literature data on glass-fiber blanket specimens to our present results.

Figure 5. Deviation of observed thermal conductivity integrals (TCI) from calculated values of TCI for glass-fiber blanket. Circles show mean temperature and horizontal bars indicate temperature span of the measurement.

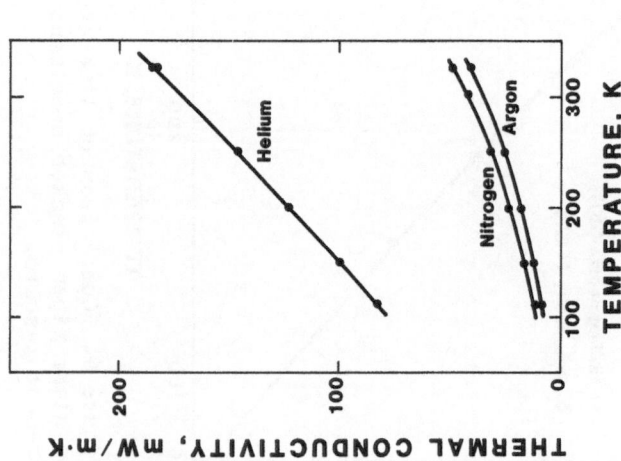

Figure 8. Effect of fill-gas species on the thermal conductivity of glass-fiber blanket (13.65 kg/m^3).

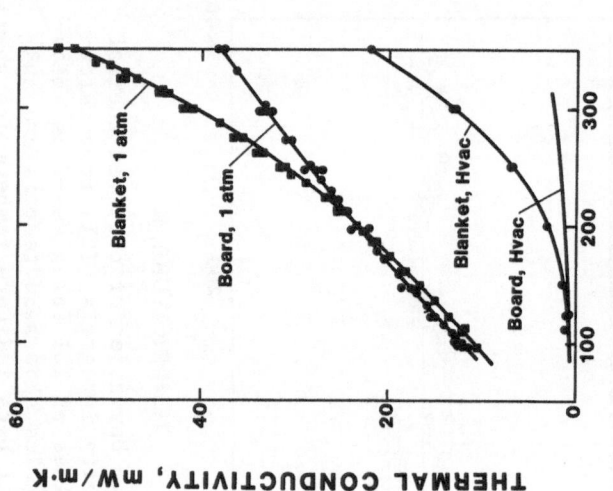

Figure 7. Comparison of the glass-fiber board and blanket thermal conductivities at atmospheric pressure and high vacuum (HVAC) conditions.

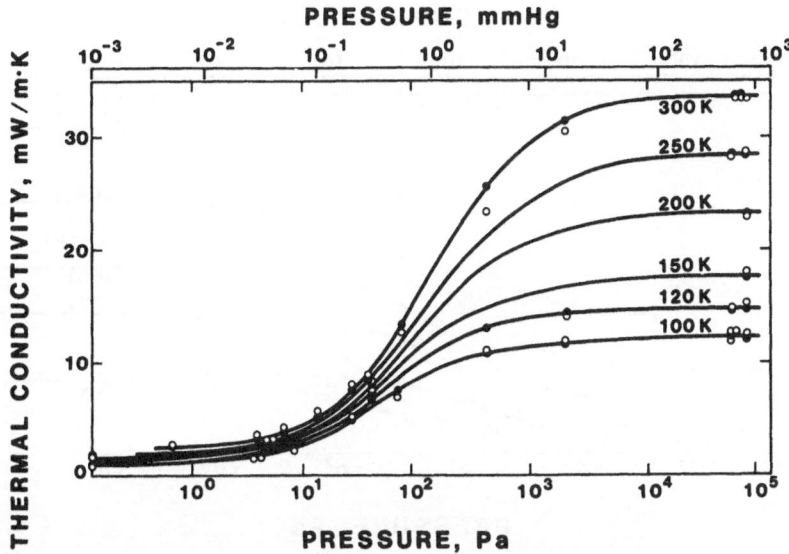

Figure 9. Thermal conductivity of glass-fiber board (127 kg/m^3) as a function of pressure at various temperatures from 100 to 300 K.

The role of gas conduction in the λ is shown in figure 8 for nitrogen, argon, and helium used as the fill gas within a blanket specimen of 13.65 kg/m^3 density. As expected, λ decreases with increasing molecular weight.

Measurements of λ did not show any systematic dependence on the size of the temperature difference for values over the temperature difference range of 10 to 100 K at constant mean temperature for either board or blanket.

A study of the role of gas pressure was carried out for both board and blanket materials, as shown in figures 9 and 10, respectively. The increase of conductivity with temperature at very low pressures is due to the increase in radiative transport and the increase in solid conductivity with temperature. The difference between the low-pressure plateau and the high-pressure plateau is used to estimate the magnitude of the gas conduction component.

The λ difference between the two plateaus, for board and for blanket, is compared with the conductivity of pure nitrogen gas as a function of temperature (fig. 11). Assuming that the solid

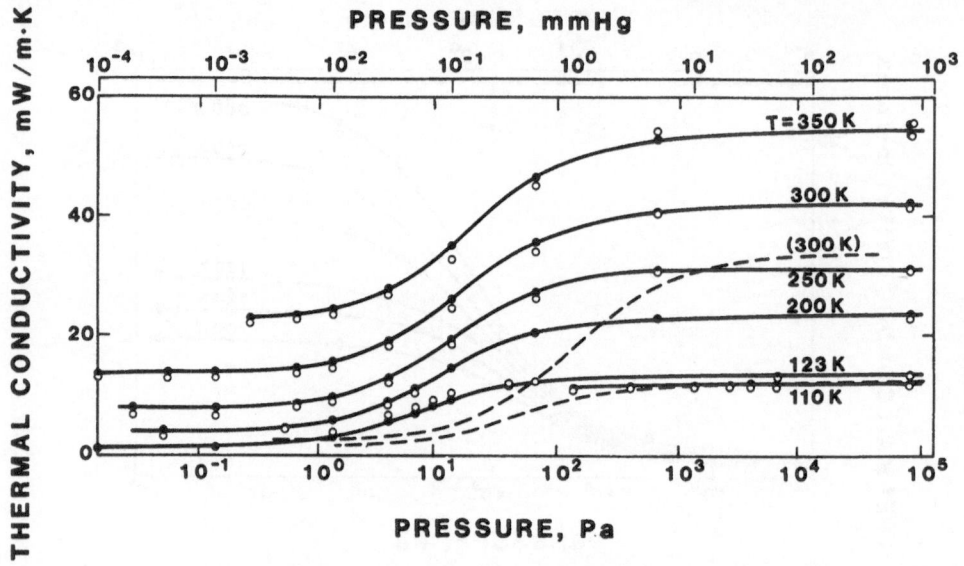

Figure 10. Thermal conductivity of glass-fiber blanket (14 kg/m³)
as a function of pressure at various temperatures from 110 to 350 K.
Dashed lines show values for glass-fiber board at 100 and 300 K.

conductivity and the radiation conductivity are independent of gas
pressure, this difference will yield the gas conductivity term. All
three curves have similar slopes and magnitudes; the difference for
blanket is larger than the conductivity of pure gas, and the difference for board is larger than that of blanket. However, because of
the volume fraction occupied by the fibers, the board and blanket
carry heat via gas conduction through an effective form factor, A/Δx,
larger than the form factor used in measuring pure gas conductivity.
This is the reason that the board and blanket gas conductivity curves
in figure 11 lie above that for pure nitrogen gas. Assuming a planar
model for these insulations, the form factor could be as much as 10%
larger for board and 1% larger for blanket. Since this is not consistent with figure 11, it is likely that the combined conduction and
radiation term increases with increasing gas conduction.

      Measurements of λ for the board and blanket were performed for
specimens having a range of densities. Figures 12 and 13 show conductivity vs density for various mean temperatures. The conductivity of board with relatively high bulk density is expected to increase
with density owing to the greater importance of solid fiber-gas conduction, as is observed. Conductivity of low-density blanket, on the

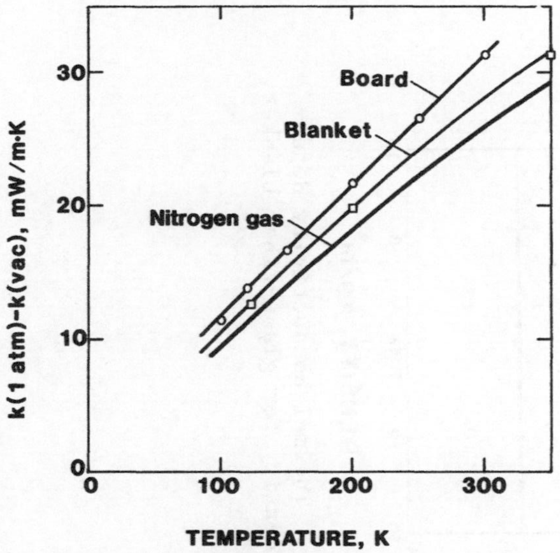

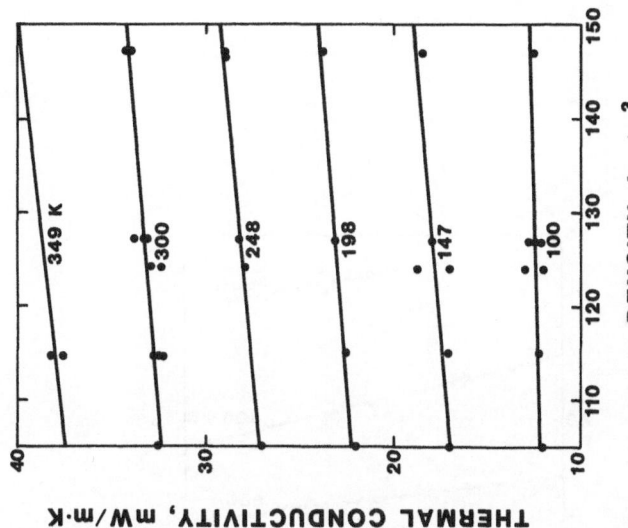

Figure 12. Thermal conductivity dependence on density for glass-fiber board.

Figure 11. Comparison of nitrogen gas conductivity to the difference between the atmospheric pressure and high vacuum plateaus for board and blanket.

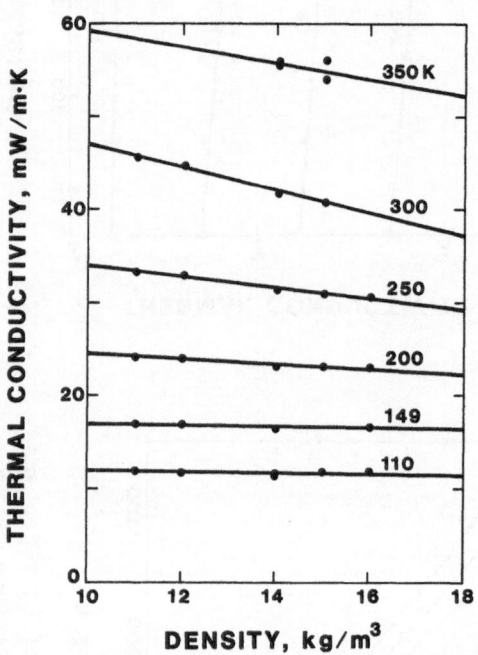

Figure 13. Thermal conductivity dependence on density for glass-fiber blanket.

other hand, is expected to decrease with increasing density because the increased solid fiber content reduces the radiation component by more than it increases the fiber-gas conduction component.

Within the experimental scatter, the slopes of λ vs density for board at various mean temperatures can be represented by a constant, 0.043 mW·m^2/(K·kg). This variation due to density is introduced into the previously given expression for λ vs temperature at 123 kg/m^3 and yields $\lambda = \sum_{i=1}^{4} A_i T^i + 0.043 \, (\rho - 123)$ where the A_i are as given earlier and ρ is the density from 105 to 148 kg/m^3.

For blanket, the slopes of conductivity vs density curves are dependent on the mean specimen temperature. This temperature dependence is represented by a simple power law and the previous expression for λ vs temperature for fiber blanket is adjusted to yield $\lambda = \sum_{i=1}^{4} B_i T^i + 1.6 \times 10^{-9} T^{3.58} \, (14.75 - \rho)$ where the B_i are as given earlier and ρ is density from 11 to 16 kg/m^3.

The λ versus ρ slope given above for board, 0.043, can be compared with the range of slopes reported by Siu[5] for four lots of board, 0.013 to 0.038.

The temperature dependent slopes reported for blanket can not be conveniently compared with any other reported data. Further analyses and comparisons of these and other parameters are made in the paper by Van Poolen, Hust, and Smith.[2]

REFERENCES

1.   M. C. I. Siu, T. W. Watson, and B. A. Peavy, "NBS Certificate, Standard Reference Material 1450a, Thermal Resistance - Fibrous Glass Board," U.S. Department of Commerce, Washington, D.C. (Feb 12, 1979).
2.   L. J. Van Poolen, J. G. Hust, and D. R. Smith, "A Model of Effective Thermal Conductivity for Fibrous Glass Insulation Materials," This volume.
3.   1981 Annual Book of ASTM Standards, Part 18, Thermal and Cryogenic Insulating Materials, American Society for Testing and Materials, Philadelphia.
4.   J. G. Hust and A. Lankford, Comments on the measurement of thermal conductivity and presentation of a thermal conductivity integral method, Int. J. Thermophys., In Press.

5. M. C. I. Siu, Fibrous glass board as a standard reference material for thermal resistance measurement systems, in: "Thermal Insulation Performance," ASTM STP 718, D. L. McElroy and R. P. Tye, eds., American Society for Testing and Materials, Philadelphia (1980), pp. 343-360.

6. E. Brendeng and P. E. Frivik, New development in design of equipment for measuring thermal conductivity and heat flow, in: "Heat Transmission Measurements in Thermal Insulations," STP 544, American Society for Testing and Materials, Philadelphia (1974), pp. 147-166.

7. C. Codegone, V. Ferro, and A. Sacchi, Studies on heat transfer in, refrigeration, in: "Proc. of IIR, Comm. 2, 1966," Pergamon Press Oxford (1966), pp. 23-37.

8. R. M. Lander, Gas is an important factor in the thermal conductivity of most insulating materials, Part II, ASHVE Trans. 27:151-168 (1955).

9. R. P. Tye, Measurements of heat transmission in thermal insulations at cryogenic temperatures using the guarded hot plate method, Prog. Refrig. Sci. Technol., 1:371-378 (1970).

10. G. B. Wilkes, Thermal conductivity, expansion and specific heat of insulators at extremely low temperatures, Refrig. Eng. 52(1):37-42, 68-72 (1946).

11. Z. Zabawsky, An improved guarded hot plate thermal conductivity apparatus with automatic controls, in: "ASTM Symposium on Thermal Conductivity Measurements," STP 217, American Society for Testing and Materials, Philadelphia (1957), pp. 3-17.

12. C. Bankvall, Heat transfer in fibrous materials, J. Test. Eval. 1(3):235-243 (1973).

13. M. Hollingsworth, Jr., An apparatus for thermal conductivity at cryogenic temperatures using a heat flow meter, in: "Thermal Conductivity Measurements of Insulating Materials at Cryogenic Temperatures," ASTM STP 411, American Society for Testing and Materials, Philadelphia (1967), pp. 43-51.

14. F. B. Rowley, R. C. Jordan, and R. M. Lander, Thermal conductivity of insulating materials at low mean temperatures, Refrig. Eng. 50(6):541-544 (1945).

PROFICIENCY TESTING FOR THERMAL INSULATION MATERIALS

IN THE NATIONAL VOLUNTARY LABORATORY ACCREDITATION PROGRAM

Diana Kirkpatrick and Jeffrey Horlick

National Bureau of Standards
Washington, DC

ABSTRACT

The National Voluntary Laboratory Accreditation Program (NVLAP) is administered by the U.S. Department of Commerce to accredit testing laboratories upon request. Accreditation is currently available for laboratories that test carpet, thermal insulation materials, and freshly mixed field concrete. Decisions to accredit laboratories are based on evaluations conducted by the National Bureau of Standards that include questionnaires, on-site assessments, and proficiency testing.

This paper discusses the design and operation of the first two years of the proficiency testing portion of the evaluation of laboratories that test thermal insulation materials.

INTRODUCTION

In 1976 the U.S. Department of Commerce initiated the National Voluntary Laboratory Accreditation Program (NVLAP) to accredit testing laboratories upon request. The task of evaluating laboratories that apply for accreditation rests with the National Bureau of Standards Office of Testing Laboratory Evaluation Technology (OTLET). In general, a three-part approach is used in the evaluation process: questionnaires, on-site assessments, and proficiency testing. Although the capability of a laboratory can be determined through questionnaires and on-site assessments, a good measure of its performance ability can be verified only through proficiency testing.

This paper describes the development of the proficiency testing part of the evaluation process used for assessing laboratories together with the data from thermal resistance and flammability tests obtained during the first two years of operation of the program.

BACKGROUND AND DEVELOPMENT

The use of proficiency testing as part of an evaluation process was examined in depth in the developmental stages of this Laboratory Accreditation Program (LAP) at meetings of NVLAP staff with representatives of industry, ASTM committees, and the testing community. The purpose of the meetings was to establish the elements that should be present in proficiency testing. Participants in these meetings discussed the measurement capability of the "average" laboratory, the advantages and disadvantages of the use of specific materials for thermal conductivity testing, and the types and amounts of information that NVLAP could expect to obtain from various approaches to proficiency testing. These issues were examined in depth by the National Laboratory Accreditation Criteria Committee for Thermal Insulation Materials. This advisory committee was established by the Department of Commerce to assist in the development of criteria for assessing laboratories in the Insulation LAP. Public comment was solicited through publication of the proposed program in the Federal Register.[1] Among the specific issues considered were:

1. The testing areas (e.g., thermal resistance and flammability measurements) that would benefit from proficiency testing;
2. The types of testing systems in need of check measurements;
3. The types of materials most desirable as specimens for each test area and test method, including resistance to problems associated with handling, shipping and storage;
4. The identification of parameters for each test that best characterize the laboratory's performance.

Several guidelines were established:

1. Use proficiency testing only for selected tests in the LAP. Because of major commercial interest, address thermal resistance measurements and flammability tests.
2. Use a variety of materials in successive proficiency testing rounds for each test to allow participating laboratories to become familiar with the characteristics of many materials and to present a more challenging assignment.

3. Use a range of low-density materials for thermal resistance measurements to reveal measurement problems related to radiative thermal components and specimen compressibility.

4. Use a variety of loose-fill materials to reveal problems associated with packing density and specimen preparation techniques.

5. Acquire a collection of precharacterized, reusable proficiency materials for use as test specimens over several years.

6. Develop a statistical package for data analysis for each test method to highlight the critical variables for the test.

INSULATION PROFICIENCY SAMPLE MATERIALS

For reasonably homogeneous materials, proficiency samples can frequently be secured as off-the-shelf items. In some situations, special materials are fabricated to insure uniformity and to enhance a particular characteristic, which truly challenges a laboratory's measurement capability. In NVLAP, the insulation proficiency testing materials have been secured to meet particular specifications. Special runs of material have been purchased from selected manufacturers. Check tests for uniformity and conformance to NVLAP specifications are conducted by NBS personnel or by a qualified outside laboratory prior to shipment to participant laboratories. Low-density fiberglass batts, a variety of loose-fill cellulosic materials, and a rigid foam board are currently used as samples for thermal properties measurements. These materials are selected to determine a laboratory's ability to make thermal resistance measurements, to examine their sample preparation techniques, and to reveal problems in dealing with very compressible materials. For similar reasons, several types of cellulosic loose-fill materials are used for flammability and smoldering combustion tests.

The low-density fiberglass batts are from a NVLAP collection of material which consists of a large number of batts for which there are thermal resistance reference values. Each batt in the collection has a different density and a different resistance value. The densities lie within a range of 0.5 pound per cubic foot to 1.0 pound per cubic foot. Participants in each round of proficiency testing were sent different specimens consisting of batts selected from two narrow subsets of density within this range.

DESIGN OF INSULATION PROFICIENCY TESTS

Each proficiency testing participant receives an information package containing detailed instructions for specimen preparation, conditioning, and mounting. Data sheets for each test with step-by-step directions for conducting the test are included. Cutting instructions are furnished where test specimens must be prepared from a larger sample of material. Although laboratories are generally instructed to conduct each test in accordance with the applicable test method, NVLAP routinely specifies temperature and relative humidity conditions, test density, thickness, and other parameters dictated by the material in question to insure uniformity in procedures and test conditions among the participants. In addition to actual test data, participants are asked to furnish copies of any charts or graphs generated as a part of the test. They also are asked to comment on any unusual events that transpire during a test.

Proficiency testing during the first two years of the Insulation LAP has been used as an evaluation tool to compare the ability of a laboratory to produce accurate results using a given test method with that of other laboratories using the same test method on specimens of the same material within a given time span. In all of the test methods except thermal resistance of fiberglass batts, the data were analyzed to determine consistent performance among laboratories. For the fiberglass batts, laboratory data were compared with predetermined test values.

PROFICIENCY TESTS

(a) Thermal Resistance

When the proficiency testing program was initiated, several versions of certain tests were offered, each version based on the use of a different material. Although this represented an attempt to be responsive to the participant who wished to test a specific material, it created a substantial burden for the laboratory that desired accreditation for all of the materials offered. Currently, samples of one or two materials are provided and there is only one proficiency test requirement for each applicable test method. Proficiency testing is offered twice annually for most test methods. Each offering is identified as a testing round.

(b) Flammability

Proficiency samples were offered for three flammability test

methods in rounds #1 and #3 only. The methods offered are: the
ASTM E84 tunnel test and the Federal Specification HH-I-515D(1)
flooring radiant panel and smoldering combustion tests. Under
the same rationale as that described in the thermal measurements
section, the tunnel test was initially offered for each of the
three materials (batt, board, and loose-fill). However, because
of the limited number of laboratories enrolled for this test, all
three versions were combined, and the proficiency test was
conducted on a single material, a cellulosic loose-fill. This
same material was used for the flooring radiant panel and
smoldering combustion tests.

The loose-fill cellulosic material selected for round #1 was
chosen to produce definite and not borderline results. This
material was used for the the thermal tests also and it presented
a considerable specimen preparation challenge to the
laboratories. The material used for round #3 was an actual
insulation product specially formulated for NVLAP.

RESULTS AND DISCUSSION

(a) Thermal Resistance

The discussion of the data that follows is presented by
material type, with each testing round described separately.
ASTM C177 and ASTM C518 data are grouped under each catagory and
are presented in Tables 1 and 2 and in Figure 1 with the
following statistical parameters included:

MEAN - the average of the test values reported by the
       participating laboratories. Where replicate tests
       are made within a laboratory, the average for each
       laboratory is used to calculate the MEAN. Outliers
       and data reported late are not included.
STD DEV - the standard deviation of the laboratory averages.
LABS IN MEAN - the number of laboratories included in the
               MEAN.
LABS REPORTING - the number of laboratories reporting data.
RANGE - the difference between the highest and lowest
        laboratory average.
% C.V. - the percent coefficienct of variation.
         % C.V. = 100 x (STD DEV/MEAN)

The STD DEV, RANGE, and % C.V. are different ways of
presenting the spread of the laboratory averages.

In round #3, proficiency samples were provided for the first time for ASTM Standard Method C236 Thermal Conductance and Transmittance of Built-Up Sections by Means of the Guarded Hot Box. One inch thick rigid fiberglass boards with foil facing on one side were provided as the test material. Instructions were given for cutting and assembling as necessary to construct a 2-inch thick assembly to serve as the test artifact. Table 3 gives the results for thermal conductance, C, and thermal transmittance, U.

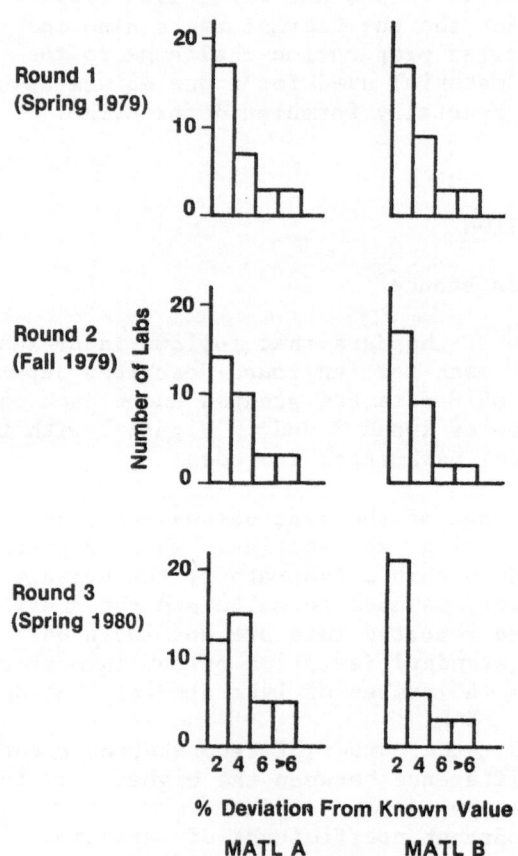

Fig. 1. Proficiency testing results for Batts, ASTM C177 and ASTM C518 Thermal Transmission Properties. Histograms show the number of laboratories with test values within indicated percent of known value.

Table 1. Proficiency Testing Results for Block & Board

Material	Mean	Std Dev	Labs in Mean	Labs Reporting	Range	% C.V.
Round 1, Foam Board #1						
C177	0.2673	0.0067	7	8	0.0160	2.49
C518	0.2639	0.0082	20	20	0.0230	3.09
Round 2, Foam Board #2						
C177	0.2550	0.0055	7	7	0.0100	2.17
C518	0.2541	0.0034	19	20	0.0120	1.33

Table 2. Proficiency Testing Results for Loose Fill

Material	Mean	Std Dev	Labs in Mean	Labs Reporting	Range	% C.V.
Round 1, Cellulosic Noninsulation Material						
C177	0.2797	0.0073	5	5	0.0175	2.62
C518	0.2753	0.0053	10	10	0.0165	1.94
Round 2, Cellulosic Insulation Material						
C177	0.2812	0.0022	5	5	0.0050	0.77
C518	0.2793	0.0042	9	9	0.0130	1.51
Round 2, Mixture Cellulosic Material						
C177	0.2774	0.0038	5	5	0.0090	1.36
C518	0.2752	0.0023	9	9	0.0080	0.83

Thermal conductivity units are Btu $\cdot$ in/h $\cdot$ ft^2 $\cdot$ $^\circ$ F.
Conversion to SI units (watt/meter $\cdot$ degree K) may be achieved by multiplying the thermal conductivity given in the tables by 0.1441.

Table 3. Proficiency Testing Results, Summer 1980, for Thermal Conductance by Guarded Hot Box, ASTM Method C236

	Thermal Conductance, C	Thermal Transmittance, U
Mean	0.1079	0.0933
Std Dev	0.0062	0.0070
Labs in Mean	9	9
Labs Reporting	10	10
Range	0.0177	0.0187
% C.V.	5.8	7.5

C and U are expressed in Btu/h $\cdot$ ft^2 $\cdot$ $^\circ$F

(b) Flammability

The results of the E84 tunnel tests, the smoldering combustion tests, and the flooring radiant panel tests are given in Tables 4, 5, and 6, respectively.

Table 4. Proficiency Testing Results for ASTM Method E84 Tunnel Flame Spread Index, FSI

Material C.V.	Mean*	Std Dev	Labs in Mean	Labs Reporting	Range	%
Round 1, Cellulosic Noninsulation Loose Fill						
	69.3	4.1	7	8	12.2	5.9
Round 3, Cellulosic Loose Fill Insulation Material						
	19.8	3.5	8	8	10.0	17.4

* Average of 3 replicates

Table 5. Proficiency Testing Results for Smoldering Combustion Percent Weight Loss

Material C.V.	Mean*	Std Dev	Labs in Mean	Labs Reporting	Range	%
Round 1, Cellulosic Noninsulation Loose Fill						
	82.5	2.0	12	13	6.0	2.4
Round 3, Cellulosic Loose Fill Insulation Material						
	0.59	0.39	12	12	1.4	66.

* Average of 2 replicates

Table 6. Proficiency Testing Results for Radiant Panel
          Critical Radiant Flux

Material C.V.	Mean*	Std Dev	Labs in Mean	Labs Reporting	Range	%
Round 1, Cellulosic Loose Noninsulation Fill						
	0.12	N/A+	13	13	N/A+	
N/A+						
Round 3, Cellulosic Loose Fill Insulation Material						
	0.20	0.03	13	13	.09	14.

* Average of 3 replicates
+ See text

CONCLUSIONS

     The results of proficiency testing for the first two years
of the Thermal Insulation LAP operation have shown that
reasonably well-behaved results can be generated to assess
laboratory performance by judicious selection of test sample
materials.

     Examination of the data for apparant thermal conductivity
measurements shows no evidence that test values differ as a
function of the use of the test procedures in ASTM C177 or C518.
The relatively small values for the dispersion and the spread of
the data among laboratories confirm that the performance ability
of a laboratory can be verified by comparison with the
performance of a group of similar laboratories. The tabular and
graphical displays of the test data indicate that most
participating laboratories performed well as defined by the NVLAP
performance guidelines.

     The test methods and concepts in the area of flammability
testing of thermal insulation materials are currently in a state
of flux and more than one of the test methods are undergoing
revision. The thermal insulation industry and the standards
community are debating the validity of some of the test methods.
Therefore, no conclusions will be put forward at this time
regarding the performance of the participating laboratories.

     Owing to the limited nature of the proficiency testing data
obtained thus far and since proficiency testing represents only
one element of the total NVLAP evaluation process, the
identification of certain laboratories as occasional outliers

does not necessarily indicate that those laboratories would fail
to receive accreditation. In general, laboratories that produce
outlying results are examined more closely in successive rounds
of proficiency testing to determine if their testing reflects a
consistent pattern of poor performance unworthy of
accreditation. A case in point is a laboratory that reported
proficiency results outside the acceptable range. This
laboratory was asked to perform additional proficiency testing
and has undergone an unannounced on-site examination. As a
result, the laboratory has improved its quality assurance
practices and is now demonstrating the ability to produce
acceptable results within the NVLAP guidelines.

REFERENCES

1. Federal Register, Vol. 42, No. 47, March 10, 1977
    (pp 13326-13336).
2. J. Horlick, "NBS Collaborative Reference Program for Thermal
    Insulation Materials," Thermal Conductivity Report No. 1,
    NBSIR 79-1817, The National Bureau of Standards, Washington,
    DC (Nov. 1979).

FACTORS AFFECTING THE THERMAL PERFORMANCE

OF A PERLITE INSULATED SYSTEM FOR BUILDINGS

R. P. Tye* and S. C. Spinney*

Dynatech R/D Co.
99 Erie Street
Cambridge, MA

ABSTRACT

Loose-fill insulation is becoming more widely used as a new and retrofit insulation for residential and commercial buildings. Perlite is one such material used particularly for insulation of hollow concrete block structures. The thermal performance of such systems, as measured or derived from properties measured under laboratory conditions, may not be realized in practice. The various factors that may influence total performance are discussed and illustrated using the results of thermal performance measurements of typical filled and unfilled concrete block-wall systems. Analytical models for predicting performance are also discussed in relation to the results of the measurements.

*Current address: Energy Materials Testing Laboratory
                   Biddeford, Maine

FACTORS AFFECTING THERMAL PERFORMANCE

Materials used normally for industrial and commercial applications are now being used also as thermal insulation for energy conservation measures in buildings. Perlite loose fill is one such material; its many favorable attributes are given in Table 1.

These positive factors, when combined with its relative ease of handling and installation, have made perlite loose fill attractive for use in hollow block-wall systems.

However, once the thermal insulation material is installed, it becomes a system and in situ, the total system may perform somewhat differently from that predicted from component measurements. This and various other factors that may affect the thermal performance characteristics of the material and systems containing the material are discussed.

General factors that must be considered include both the material and system characteristics, how these were determined, and their relation to in-service performance. In addition, if the thermal performance is being calculated, then the models used for the analyses must be realistic, and the properties used must be accurate and reflect the temperature of the application.

Table 1. Assessed Properties Of Perlite Loose Fill

Density of Use	48-176 kg/m^3	Toxicity	Nontoxic
Heat Capacity	1100 J/kgK	Effects of Age	None
Thermal Resistance for 1m thickness at 24°C	17-26 m^{2}K/W*	Degradation	None
Water Vapor Permeability	High	Corrosiveness	None
Water Absorption	Low	Odor	None
Fire Resistance Properties	Noncombustible. Zero flame spread, smoke developed and fuel contributed.		

*Multiply by 0.144 to convert to Fft^{2}h/BTU for a 1 inch thickness

Thermal performance of thermal insulation materials is generally measured under laboratory conditions using either the guarded-
hot-plate[1] or heat-flow-meter[2] methods. These methods, while being
very accurate, provide results for "ideal" laboratory conditions.
The actual thermal performance can be affected adversely by a number
of factors,[3] such as compaction, moisture absorption, air infiltration, temperature and thickness.[4] Perlite does not absorb moisture,
and it is used at relatively high densities. Therefore, the moisture, air infiltration, and thickness effects are minimal. Furthermore, providing it is not vibrated unduly, it will not compact significantly.

The thermal conductivity of concrete is normally measured by
the guarded-hot-plate method. For concrete block systems, specimens are normally prepared from the faces of one block, and the results are considered to be representative of the particular concrete.
Measurements can be made on conditioned blocks containing equilibrium moisture contents,[5] and these are dried after the tests to
determine the moisture content. Alternatively the specimens are
dried and then measured, and some estimate, such as that proposed
by Valore,[6] is made for the effect of the known moisture content.

For perlite, the major effects relate to the actual system performance rather than effects on the material itself. This is because perlite, if installed correctly in ceilings or attics, does
not introduce holes or voids in the system, and it is not affected
significantly by the environment, moisture and cycling. In an
uninsulated concrete block system, about 25% of the energy transmission is through the concrete webs; but in an insulated system,
about 80% of the energy is transmitted through the webs. Thus, the
thermal conductivity of the concrete is the major factor controlling
the overall block thermal resistance.

The majority of literature information for insulated and non-
insulated concrete block systems has been generated using block density as a guide to thermal transmittance. However, the thermal conductivity of concrete of any one density can vary widely depending
on aggregate and other factors.[5] Using density only to predict
thermal transmittance can often produce erroneous information. The
results of this work indicated that in many cases, the thermal conductivity of concretes can be in considerable error owing to measurement technique. The error increases as the density and hence,
thermal conductivity of the concrete increases. For the range of
densities of typical concretes used for blocks, these errors can be
15 to 30%. Moisture effects increase the thermal conductivity
about 5% for each 1% of moisture contained in the concrete.

The reasons for variations in thermal conductivity at similar densities are the diversity and form of the constituents in the mixes, which are usually proprietary. Thus, to predict thermal performance of masonry constructions, it is essential that information on the thermal conductivity of a specific material is available together with dependence on composition. This information should be published by the concrete industry, since the final values for insulated masonry systems require that it be made available. The present paper describes a study undertaken to illustrate some of the major effects described above.

EXPERIMENTAL MEASUREMENTS

To evaluate the degree of the effects of some of the above factors, a study involving four densities of 2-core 8-in block systems, both unfilled and filled with perlite, was undertaken. The thermal conductivity of the concrete was measured using the guarded-hot-plate method,[1] the thermal conductivity of the perlite was measured using the heat-flow-meter method,[2] on different thicknesses of different densities of products, and the thermal transmittance of the concrete block and insulated block systems were measured using the guarded-hot-box method.[7] The methods and techniques utilized in this study have been well documented,[5] and recognized procedures were followed to minimize errors in each method.

There are additional significant sources of error to be considered when the hot-box method is used on such systems and for this type of comparative study. These considerations include:

a) Avoiding convection through the connecting cores above and
    below the metering section. This can be accomplished by
    adding an impervious plastic barrier between the courses of
    blocks immediately below and above this section and caulking
    the internal joints.

b) Establishing truly steady-state conditions which can take
    several days. Unless sufficient data is taken over regular
    time periods, apparent steady-state conditions can be attained,
    but they may be in error by 10% or more.

c) Eliminating temperature errors due to poor attachment of
    thermocouples to the surface and insufficient numbers of
    sensors. Eighteen temperature sensors were used for each
    surface, they were distributed in a representative manner
    and attached in fine grooves cut into the surfaces of the
    concrete with junctions flush with the surface.

d) Avoiding uncertainties in the comparative results due to the amounts and application of set high conductivity mortar. All face joints were well taped. This taping eliminated air movement through the joints.

RESULTS

The results for the thermal conductivity of the four concretes are given in Table 2. These results show the variation of thermal conductivity with density. An approximate two-fold increase in density produces a factor of five increase in thermal conductivity. In addition, the difference in the results for the 1730 and 1795 kg/m^3 material indicates a definite effect of aggregate and form. Here, a difference of 30% in thermal conductivity was seen for a 4% change in density. The moisture contents were the same to within less than 1%, and thus, had little effect.

Results obtained for perlite indicated that for densities above 96 kg/m^3, there is no thickness effect. For the lower densities, a small effect of less than 2% is present for thicknesses between 1 and 6". A representative value for the perlite at 112 kg/m^3, normally found in filled blocks, is 0.049 W/mK.

The results of the guarded-hot-box studies for the overall heat transmission coefficients of the unfilled and filled block system are shown in Table 3.

Clearly, the effectiveness of adding insulation to the cores is really significant. It is interesting to note that the largest effect is for systems having the lowest concrete density. This is due to the fact that at the lowest densities, a proportionally greater percentage of the energy is passing through the core section. However, as the thermal conductivity of the concrete increases, a larger percentage of the energy passes through the web, thus reducing the effects of adding the insulation.

Table 2. Thermal Conductivity Of Four Concretes at 38°C

Concrete Density kg/m³	Thermal Conductivity W/mK
1170	0.375
1730	1.05
1795	1.46
2070	2.12

Table 3.   Thermal Performance Of Four 8" Two-Core Block Walls

Concrete Density kg/m³	Thermal Transmittance U Value at 24C & 24 kph W/m²K		
	Unfilled	Filled	Reduction, %
1170	1.99	0.85	57
1730	2.55	1.65	35
1795	2.67	1.82	32
2070	3.06	1.99	35

MODELLING OF SYSTEMS

Various mandated minimum requirements for thermal performance of building systems are now in force. It is impractical to measure all buildings or building components, and thus there is a need to predict simply and accurately the thermal performance of insulated systems.

Two relatively simple analytical techniques are in general use to determine the purely resistive components of heat transmission: The "series-parallel" (isothermal planes) and "parallel" techniques. These are given in the ASHRAE Fundamentals Handbook.[8]

Of the two, the series-parallel technique is used more extensively and has been shown to be more accurate,[9] and more representative for the general case. However, this analysis does not include effects of moisture, mass, air infiltration, orientation to the sun, and other factors which influence total performance. The resistive component dominates, and for comparison purposes, it is the one to be recommended as the means to provide most realistic performance values.

The series parallel analysis, together with the appropriate thermal conductivity values, was used for the present walls. The calculated and experimental results are summarized in Table 4.

Table 4. The Measured and Calculated Thermal Transmittance of Four 8-Inch Two Core Concrete Block Walls, Empty and Filled with 112 kg/m³ Perlite Loose Fill Insulation.

Concrete Density kg/m³	Concrete Thermal Conductivity, W/mK	Thermal Transmittance W/m²K							
		Empty		Perlite Filled		Percent Deviation of Calculated Values			
		Calculated	Measured	Calculated	Measured	Empty	Filled		
1170	0.375	1.96	1.99	0.68	0.85	-1	-20		
1730	1.05	2.69	2.55	1.33	1.65	+6	-19		
1795	1.46	2.95	2.68	1.67	1.82	+10	-8		
2070	2.12	3.26	3.01	2.13	1.99	+8	+7		

For the empty two-core concrete block walls, the calculated values average approximately 7% higher than the measured values. For the filled walls, the calculated values show an average 13% deviation, with three calculated values being lower than the measured values and one being higher. This may be considered as very good agreement, considering that the calculations are based on a simplified block with square corners and only a single concrete thermal conductivity determination for each concrete. It is assumed that the density, moisture content, and thermal conductivity of all blocks of one system are the same as that for the one measured experimentally. The analytical technique is thus validated and produces values that, in these cases, are within 20% of measured value. Such values, providing that the concrete thermal conductivity is known, are very good estimates of the actual resistive thermal transmittance and should be considered more valid and rigorous than most previous literature values based only on concrete density[5] and on the other model provided by ASHRAE.

Based on the above factors for a standard 2-core 8 inch block with perlite loose-fill at 112 kg/m^3 having an apparent thermal conductivity of 0.049 W/mK, the series-parallel isothermal planes model can be used to derive the thermal transmittance of filled and unfilled blocks for various thermal conductivities of the block material. Table 5 shows the calculated results for a 7-fold increase in concrete block conductivity from 0.285 to 2.02 W/mK. This range covers that of all concretes currently in use for block constructions. The values confirm the considerable improvement obtained by using loose-fill insulation, especially for the lower block thermal conductivities and, hence, density range.

It should be pointed out that concrete block walls are built using mortar or face shell bonding. The above model can be refined to include the effects of bonding materials. These can be significant if thick applications are used. For example, mortar having a thermal conductivity of approximately 2W/mK can occupy up to 15% of the total area depending on the thickness and distribution used. It can also be refined to allow for the variation in thermal conductivity of perlite or of any other loose-fill insulation suitable for use in such systems. In general, this effect will be less than 7% for a thermal conductivity change from 0.04 to 0.05 W/mK. Finally, the present analysis is valid only for the resistive contributions to heat transmission. It does not include contributions of transient and heat-storage effects or those due to air infiltration. The latter can be significant since the concrete is porous and joints may not be continuous. Thus, actual performance may differ significantly from that based only on measured or derived moisture effects.

Table 5.   Calculated Thermal Transmittance Of
           Concrete Block Wall Systems

Concrete Thermal Conductivity, W/mK	Thermal Transmittance of 8-inch Two-Core Block, W/m^{2}K		Reduction for Fill System, percent
	Unfilled	Filled 112 kg/m^3 Perlite	
0.29	1.77	0.573	67.6
0.58	2.276	0.89	60.9
0.87	2.565	1.17	54.4
1.15	2.785	1.424	48.7
1.44	2.945	1.657	43.9
1.73	3.087	1.867	39.5
2.02	3.212	2.06	35.9

SUMMARY

     The factors influencing the thermal performance of perlite loose-fill systems containing perlite have been discussed. The validity of current analytical procedures has been investigated and verified experimentally to within practical limits for concrete block-wall systems, both unfilled and then filled with perlite loose fill insulation.

ACKNOWLEDGMENTS

     The authors wish to express their appreciation to Mr. R. Milanese and Mr. W. Waugh, the Executive Director and the Chairman of the Technical Committee of the Perlite Institute, for initiating the study and for their helpful comments. They also wish to thank their colleague, Mr. A. O. Desjarlais, for his assistance with some of the experimental and analytical work.

REFERENCES

1. "Steady-State Thermal Transmission Properties by Means Of
    Guarded Hot Plate," Standard Test Method, ASTM C177-76, ASTM
    Stand., Philadelphia (1981).

2. "Steady-State Thermal Transmission Properties by Means Of
    the Heat Flow Meter," Standard Test Method, ASTM C518-76,
    ASTM Stand., Philadelphia (1981).

3. R. P. Tye, E. Ashare, E. C. Guyer, and A. C. Sharon, An
    assessment of thermal insulation materials and systems for
    building applications, ASTM STP 718, pp 9-26, ASTM, Philadelphia
    (1981).

4. T. T. Jones, "Effect of Thickness on the Thermal Conductivity
    of Thermal Insulations." Special Publication 302, National
    Bureau of Standards, Washington, D.C. (1967), p. 737.

5. R. P. Tye and S. C. Spinney, "Thermal Conductivity of Concrete:
    Measurement Problems and Effect of Moisture," of Commission BI
    of Institute International du Froid, Washington, D.C.
    (September 1976).

6. R. C. Valore, Jr., J. Am. Con. Inst. 28(5):502 (1956).

7. ASTM, 236-66 Standard Test Method for "Thermal Conductance and
    Transmittance of Build-Up Sections by Means of the Guarded Hot
    Box." ASTM Stand. Philadelphia (1980).

8. Fundamentals Handbook, Chapter 22, ASHRAE, American Society of
    Heating, Refrigerating and Air-Conditioning Engineers, Atlanta,
    GA (1977).

9. L. S. Shu, A. E. Fiorato, and J. W. Howanski, Heat transmission
    coefficients of concrete block walls with core insulation, ASHRAE
    Special Publication 28, ASHRAE, Atlanta, GA (1981).

SESSION TC-12B

Dielectrics - II

CHAIRMAN

V. V. Mirkovich
Canada Center for Mineral and Energy Technology
Ottawa, Canada

THERMAL CONDUCTIVITY OF SODIUM-CONDUCTING BETA-ALUMINAS

V.V. Mirkovich and T.A. Wheat

Mineral Processing Laboratory
Mineral Sciences Laboratories, CANMET
Energy, Mines & Resources Canada, Ottawa

ABSTRACT

Thermal conductivity of three compositions of ionically conductive beta-aluminas was measured by a comparative method in the temperature range 35° to 650°C. The thermal conductivity was shown to be low and nearly independent of temperature. A pronounced thermal anisotropy was observed.

INTRODUCTION

There has been a progressively increasing interest in the exploitation of various ionically conducting solids since 1967 when the Ford Motor Company announced the development of a high energy-density battery based on the sodium-sulphur cell[1,2]. Since that time, major programs have been under way to exploit this battery commercially in applications such as load levelling and electric vehicles.

In essence, the battery is based on a simple sodium concentration cell in which a potential is developed between a liquid sulphur electrode separated by an ionically conducting solid. Although several materials are presently under active development, by far the greatest effort has been devoted to studying those occurring in the $Na_2O-Al_2O_3$ system in which beta-alumina and the more conductive beta"-alumina (nominally $Na_2O:6Al_2O_3$) occur.

519

A minimum design operating temperature of 300°C is commonly used in a system that operates nominally isothermally; the temperature fluctuations that occur are due to the ohmic heating effects on the charge and discharge cycles. Also reported by the Ford Motor Company, a thermoelectric generator has been developed that exploits the sodium-ion-conducting beta-aluminas in a concentration cell that operates up to 1000°C.

In both of these applications, nonisothermal conditions occur as the device is heated to its operating temperature. In addition, the beta-aluminas crystallize in the hexagonal system and have a perfect basal cleavage that tends to produce a marked degree of anisotropy in the ionic conductivity in which a factor of approximately 1.6 has been found between the fast (// to 001 plane) and slow (⊥ to 001 plane) directions[3]. Because of the difficulty in producing single crystals of uncontaminated beta and beta"-alumina to determine the expected thermal anisotropy in these materials, the present work used a hot-pressing technique to produce oriented specimens of essentially theoretically dense material of each composition.

EXPERIMENTAL PROCEDURE

Synthesis and Fabrication of Materials

The procedure was designed to minimize the extensive ball-milling of reagents that has been used by others and that can give rise to undesirable contamination of the final product. Because the formation of the more desirable beta"-phase is known to be kinetically sluggish and because the use of high sintering temperatures also promotes the loss of Na_2O from the system, a process was developed that resulted in highly reactive material that could be readily fabricated. In essence, high-surface-area Al_2O_3 was first obtained by the calcination of a commercially available $Al(OH)_3$ in air. Subsequently, Na^+, Li^+, and Mg^{2+} ions were adsorbed onto the surface of the calcine by slurrying the Al_2O_3 in an aqueous solution containing one or more of these ions. The dried material was then calcined to remove residual volatiles and to crystallize the beta-/beta"-alumina[4].

Three different compositions were prepared. The binary composition consists of only Na_2O and Al_2O_3. However, because of the difficulty in producing beta"-alumina in a binary composition, stabilizers were used to promote the conversion of the beta" form. The compositions examined are given in Table 1.

Samples of each composition were fabricated into high-density cylinders by hot pressing in graphite dies. To minimize the loss of sodium from the materials during the heating stage and yet

Table 1.   Composition and Density Data of Beta-Alumina

SUBSTANCE		COMPOSITION		DENSITY
		Wt %	Molar	g/cc
BINARY	Na_2O	9.2	6	3.268
	Al_2O_3	90.8	36	
Li_2O- STABILIZED	Li_2O	0.73	1	3.237
	Na_2O	9.13	6	
	Al_2O_3	90.14	36	
MgO- STABILIZED	MgO	1.91	2	3.248
	Na_2O	9.03	6	
	Al_2O_3	89.06	36	

adequately outgas the graphite and carbon wool of the hot press, each sample was heated to between 650° and 750°C under a vacuum of 1 Pa to remove adsorbed water. Thereafter an argon atmosphere was introduced and increased to 100 kPa to suppress the loss of Na_2O; at the same time, the full ram pressure of 25 MPa was applied on the sample, which was then heated to 1450°C. After 15 min, the pressure was released, the temperature lowered to approximately 900°C, and the sample removed from the die at that temperature.

Samples of each composition were examined using x-ray diffraction to determine the beta-/beta" phase ratio and the development of texturing during hot pressing. X-ray diffraction analysis of the pressed materials established the existence of a pronounced orientation in the hot-pressed billet in which the basal (001) plane was found to lie perpendicular to the hot-pressing direction. The degree of texturing was also estimated by measurement of the ionic resistivity in the two principal directions with respect to the pressing direction.

Because hot pressing is known to produce fine-grained microstructures in all materials, fractographs of each material were obtained using a scanning electron microscope to estimate the grain size. A typical fractograph of beta-alumina is shown in Fig. 1. For purposes of comparison, a fractograph of an alumina specimen, AL-300*, used as a thermal conductivity standard in this laboratory, is also shown. It can be seen that there is an order of magnitude difference in the grain size of AL-300 (alpha-alumina) and the beta-aluminas.

---

*Manufactured by the Western Gold and Platinum Corp., Belmont, California. Typical analysis of Al-300: 97.55% Al_2O_3, 1.35% SiO_2 1.05% CaO, 0.03% Fe_2O_3, 0.02% Na_2O.

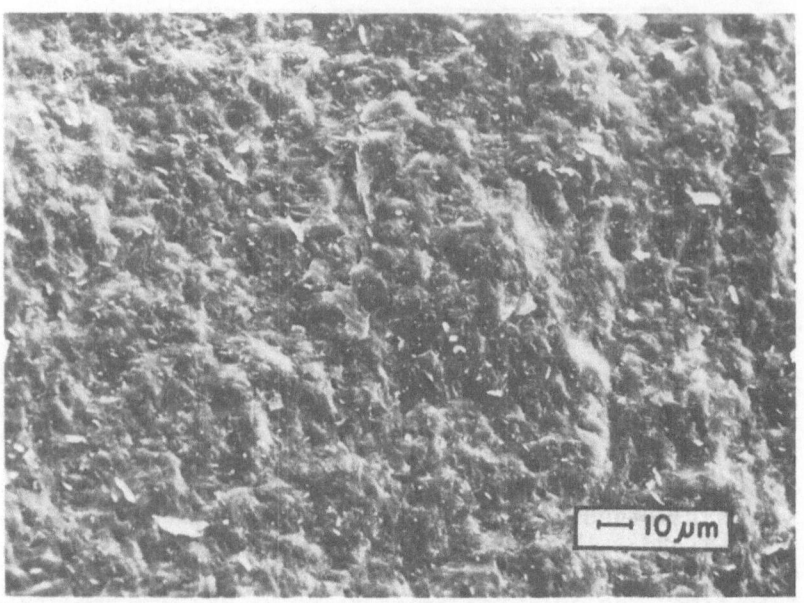

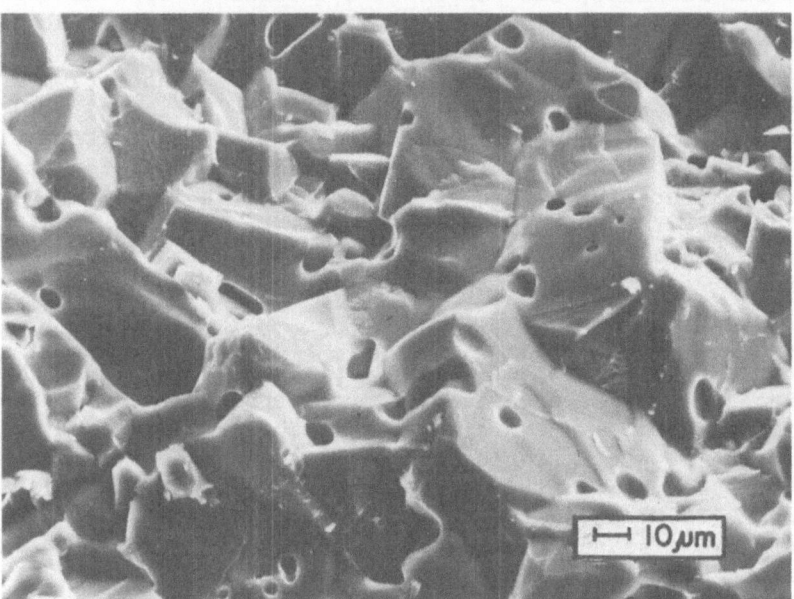

Fig. 1 Scanning electron micrographs of fractured surfaces
of binary beta-alumina (top) and AL-300.

Cylindrical samples of the beta-aluminas for measurement of thermal conductivity were cut so that the longitudinal axis was either parallel or perpendicular to the basal plane of the oriented crystallites.

Determination of Thermal Conductivity

Thermal conductivity was measured by modified comparative method thermal conductivity apparatus described in detail previously[5]. Briefly, the measuring column, consisting of a heat stabilizer and a specimen placed between two standards, is held between the heat source and heat sink. The stabilizer, the standards, and the specimens are cylinders 2.54 cm in diameter by 2.54 cm high. The entire assembly is surrounded by a cylindrical heat guard with five wire heaters wound on the outside to minimize radial heat flux from the heat source and the four cylinders. The modification consists of six stainless-steel cooling jackets: one around each wire heater on the guard and one for the heat sink.

Without the cooling system, the average temperature of the measured sample could not be lowered below 150°C. With the present configuration, the measurements can be made starting at about 35°C.

Pyroceram Code 9606* was used for the standards. The accuracy of measurements is estimated to be ±5%, the same as before the addition of the cooling system.

RESULTS AND DISCUSSION

The thermal conductivity results are shown in Fig. 2. For purposes of comparison, thermal conductivity of AL-300, measured previously,[5] is also included. The beta-alumina curves were fitted to the data by the least-squares method. Each curve represents 10 to 13 measurements. They are expressed by a polynomial of the second degree:

$$\lambda = a_1 + a_2 t + a_3 t^2$$

where λ is thermal conductivity in W/cm°C and t is the temperature in °C. The equation is valid for the temperature range 35 to 650°C. The coefficients a_1, a_2, and a_3 are given, together with standard and average deviations, in Table 2.

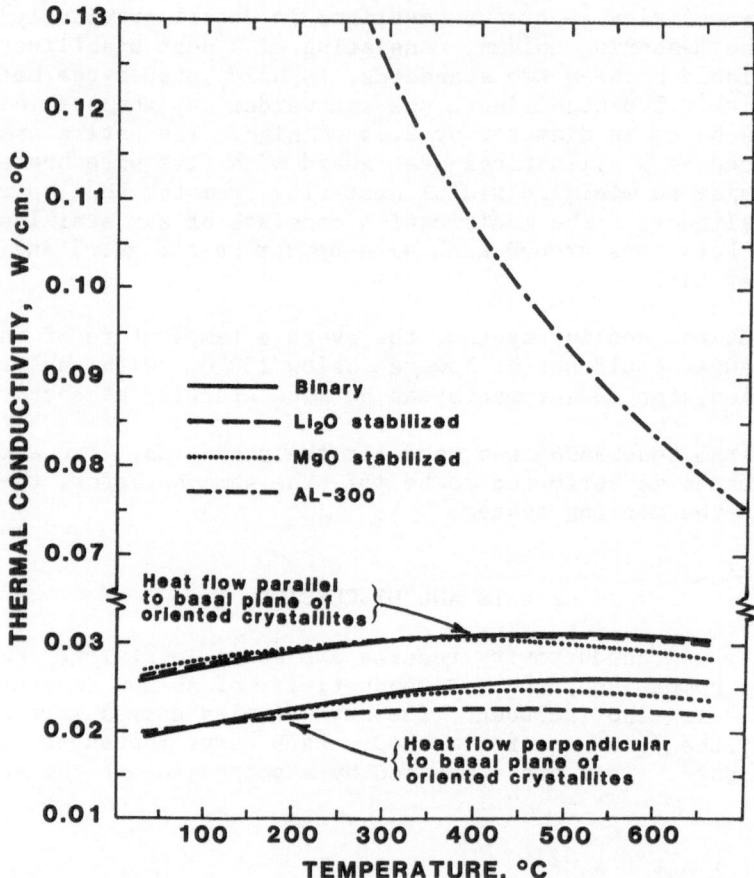

Fig. 2   Thermal conductivity of the beta-aluminas and
         AL-300.

Table 2. Coefficients for the Least-Squares Polynomial
and Standard and Average Deviations

Specimen	Coefficients			Deviations, %	
	$a_1 \times 10^2$	$a_2 \times 10^5$	$a_3 \times 10^8$	Standard	Average
β-Al$_2$O$_3$ $\perp$	1.840	2.672	-2.353	1.6	2.3
β-Al$_2$O$_3$ $\parallel$	2.507	2.378	-2.400	0.7	0.9
Li$_2$O-β-Al$_2$O$_3$ $\perp$	1.966	1.191	-1.264	2.5	2.6
Li$_2$O-β-Al$_2$O$_3$ $\parallel$	2.591	1.773	-1.605	1.0	1.2
MgO-β-Al$_2$O$_3$ $\perp$	1.917	2.039	-1.992	1.7	3.0
MgO-β-Al$_2$O$_3$ $\parallel$	2.632	1.937	-2.500	1.0	1.4

The presence of 9% of Na$_2$O strongly depresses the thermal
conductivity of alumina. At lower temperatures, compared with the
thermal conductivity of a polycrystalline alpha-alumina, this
amounts to more than an order of magnitude. In part, this
substantial lowering of thermal conductivity can also be attributed
to its comparatively fine grain size. The average grain size of
AL-300 is approximately 60 μm, whereas that for the beta-aluminas
is an order of magnitude smaller.

The anisotropic nature of the material is clearly reflected
in the thermal transport properties. The higher values were obtained
in measurements parallel to the basal plane of the oriented
crystallites. However, in contrast to the anisotropy of the ionic
conductivity where a ratio of 1.6:1 was obtained, the thermal
conductivity parallel to the basal plane is only 25% higher than in
the orthogonal direction.

The variation of conductivity with temperature is rather unexpected. Unlike the majority of oxide-based materials for which
thermal conductivity in this temperature range tends to decrease
with temperature for conductivities above 0.01 W/cm°C and generally
increases with temperature below that value, the thermal conductivity
of the beta-aluminas increases with temperature despite having a
conductivity greater than 0.01 W/cm°C. The variation of conductivity
with temperature, however, is small. It reaches a maximum value
between 400° and 500°C.

The effect of compositional change is minimal and unpredictable. Addition of Li_2O and MgO to a binary composition has no measurable effect on thermal conductivity near room temperature. At higher temperature, however, in the case of measurements perpendicular to the basal plane, the addition of either Li_2O and MgO suppresses the thermal conductivity of binary beta-alumina; Li_2O, whose room temperature conductivity is some twenty times lower than that of MgO, has in fact a greater effect. Unfortunately, that is not evident when measurements are made parallel to the basal plane of the oriented crystallites since only minimal changes can be observed.

SUMMARY

The thermal conductivity of beta-aluminas has been shown to be low and essentially independent of temperature. It exhibits minimal change with minor changes in composition. Thermal anisotropy is pronounced but not to the extent expected based on the ionic conductivity data.

REFERENCES

1. N. Weber and J.T. Kummer, Proc. Ann. Power Sources Conf; p 37-91 (1967).

2. N. Weber and J.T. Kummer, Proc. Adv Energy Conv Eng ASME Conf; p 913-16 (1967).

3. G.E. Youngblood and R.S. Gordon, Texture-conductivity relationships in polycrystalline lithia-stabilized β''-alumina, Ceramurgia Int 4(3):93-98; 1978.

4. T.A. Wheat "Synthesis of beta-aluminas from Gibbsite," Division Report ERP/MSL 80-27 (OP), CANMET, Energy, Mines, and Resources, Canada, (1980).

5. V.V. Mirkovich; Comparative method and choice of standards for thermal conductivity determinations, J.Am.Ceram.Soc. 48(8):387-391 (1965).

THERMAL CONDUCTIVITY OF ORIENTATIONALLY-DISORDERED

CRYSTALS UNDER HIGH PRESSURE

P. Andersson, R.G. Ross* and G. Bäckström

Department of Physics
University of Umeå
S-901 87 Umeå, Sweden

ABSTRACT

The transient hot-wire method was used to study the effects of temperature and pressure on the thermal conductivity, λ, of several molecular substances showing varying degrees of disorder in molecular orientation. Molecular solids both in the plastic and normal crystal state were investigated as well as a clathrate hydrate. The measurements were made in the temperature range 100 – 450 K and at pressures up to 2.5 GPa. The conductivity results for the different types of crystal studied are discussed in terms of the simple Debye formula. We discuss the unusual glass-like behaviour of $\lambda(T)$ for the clathrate hydrate in some detail.

INTRODUCTION

Molecular substances exhibit a great variety of behaviour in relation to thermal conductivity, λ. In our laboratory we have undertaken a research program to study the effects of temperature and pressure on λ for several solid organic substances with different degrees of quasi-static disorder in molecular orientation.

We have mainly studied substances that form plastic (or orientationally disordered) crystal phases.[1] In such phases, the molecules can reorient between different directions, but there is positional long-range periodicity. The molecules are often nearly

*Permanent address: School of Mathematics and Physics
                    University of East Anglia
                    Norwich NR4 7TJ, U.K.

spherical in shape, the crystal is deformed easily, and the entropy
of melting is small. A plastic crystal phase occurs at temperatures
near melting. At lower temperatures, a transformation takes place
to a normal crystal phase in which the molecules are fixed in
position and orientation.

We have investigated a simple clathrate hydrate. In such a
phase, individual guest molecules of one substance are encaged by
an ice-like, hydrogen-bonded structure of H_2O molecules, which are
the host molecules. There is virtually no bonding between the host
and guest molecules, and the guests are relatively free to reorient
within their cages, to which they are constrained by high-energy
barriers. This leads to some degree of semidynamic structural disorder. Diffraction experiments have shown that clathrates have a
periodic structure typical of crystalline solids.

EXPERIMENTAL

We used the transient hot-wire method on solids under pressure.
Details of the method and the high pressure arrangements are given
elsewhere.[2,3] The thermal conductivity, λ, and the heat capacity
per unit volume, ρc_p, are measured simultaneously. Here ρ is the
mass density. The temperature and pressure ranges of our experiments
are 100 – 450 K and 1 atm – 2.5 GPa, respectively. The accuracy in
λ is ±3 %, and the precision is about ±1 %. The accuracy in ρc_p is
about ±10 %. In this paper we present and discuss only the thermal
conductivity results.

RESULTS

Temperature Dependence of λ

Typical results for plastic and normal crystal phases are
those of cyclopentane.[4] The phase diagram of this substance is
shown in Figure 1.

The solid phases I and II are both plastic, while phase III
is normal crystalline. Figure 2 shows our results for λ, obtained
at a rather low, constant pressure (0.1 GPa).

A plastic crystal phase shows a small temperature dependence
of λ. The absolute value of λ is small and exceeds that of the
liquid by a few percent. This is in great contrast to the behaviour
in the normal crystal phase. Here λ varies as T^{-1}, as shown for
phase III in Figure 2. At the plastic → normal crystal phase
transition, λ increases strongly.

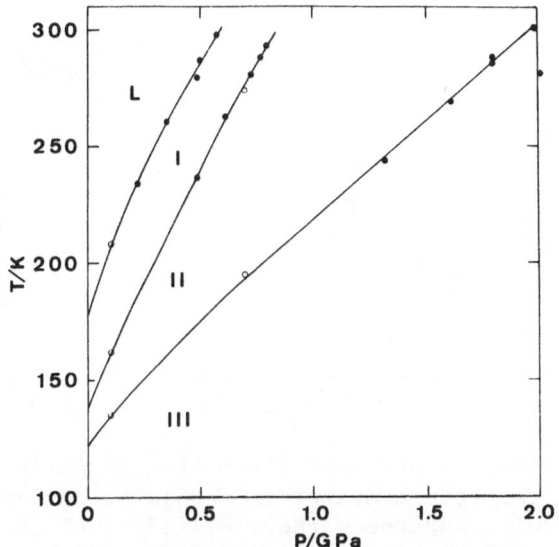

Fig. 1. Phase diagram of cyclopentane.[4]

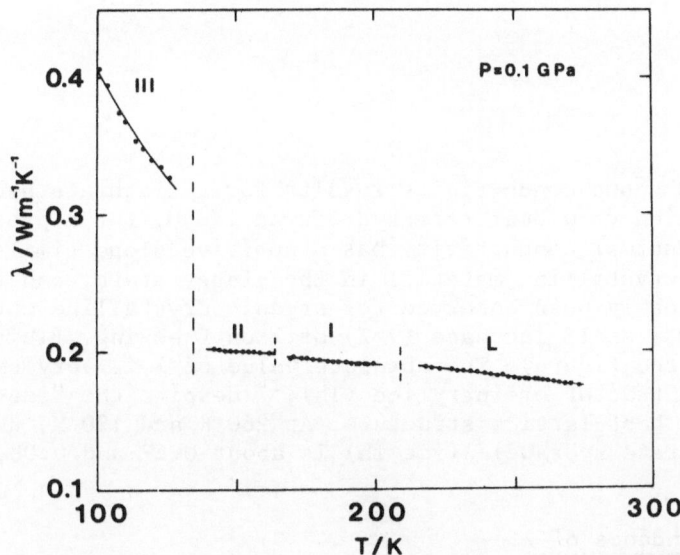

Fig. 2. Isobaric temperature dependence of λ of cyclopentane.[4]

We have invariably found these same features for the plastic phases of all of the other substances we have studied: cyclohexane (I);[5] ammonium fluoride (IV);[6] carbon tetrachloride (Ib);[7] ammonium chloride (I);[8] carbon tetrabromide (I and III);[9] adamantane I);[10] furan (I and IV);[11] and 1,1,1-trichloroethane (Ib).[12]

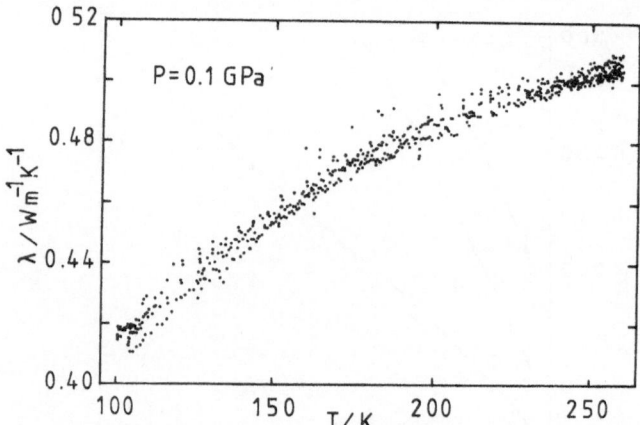

Fig. 3. Isobaric temperature dependence of λ of tetrahydrofuran
·17H₂O clathrate hydrate. It should be noticed that
these results supersede those already published,[13] which
are about 5 % higher due to the presence of a small pro-
portion of ice Ih.

    Figure 3 shows conductivity results for a clathrate hydrate
at a composition very near tetrahydrofuran·17H₂O, for a pressure
of 0.1 GPa. Thermal conductivity has a positive slope similar to
that of a noncrystalline material in the glassy state, and this
has not previously been observed for organic crystalline materials.
There is only a small increase (7 %) of λ on freezing. (This is
not shown in the figure). The absolute value of λ is very small
compared with that of ordinary ice (Ih),[14] despite the "ice-like"
nature of the host-lattice structure. At 260 K and 120 K, the
ratio λ(clathrate hydrate)/λ(ice Ih) is about 0.22 and 0.08,
respectively.

Pressure Dependence of λ

    Figure 4 shows isothermal conductivity results for cyclopen-
tane.[4] The pressure dependences for the plastic phases I and II
are similar to that of the compressed liquid and in sharp contrast
to that of the normal crystal phase. Similar results have been
obtained for other substances. For materials for which compressibility data are known (which are unfortunately few), we can calculate the relative volume dependence of conductivity g =
$-(\delta\ln\lambda/\delta\ln V)_T$. A summary is shown in Table 1.

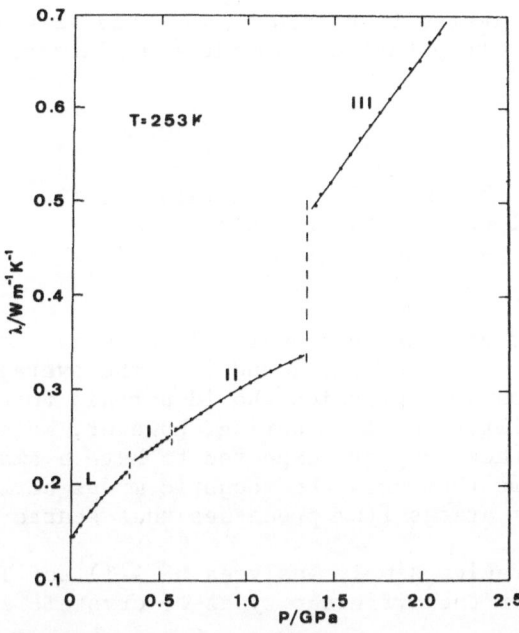

Fig. 4. Isothermal pressure dependence of λ of cyclopentane.[4]

Table 1. Values of Relative Volume Coefficient,
$$g = - (\delta \ln \lambda / \delta \ln V)_T$$

Substance and Phase	g
Adamantane I (plastic)[10]	6.4
Adamantane II[10]	9.8
Hexamethylenetetramine[10]	8.9
Furan I (plastic)[11]	4.0
Clathrate hydrate	0.5

For normal crystals, g is typically about 10, while for
liquids and glasses, g is in the range 1 - 4. Plastic crystals
have intermediate values of g that are close to those for liquids
and glasses.[15]

We also pressurized the clathrate hydrate at 130 K, and λ was
found to increase very little over a range up to 0.8 GPa. The co-

efficient g was estimated to be about 0.5. This value is unusually small, smaller even than that of liquids and glasses.

DISCUSSION

In discussing the thermal conductivity, λ, it is usual to start with the well-known Debye formula

$$\lambda = \frac{1}{3} Cv\ell \tag{1}$$

where C is the heat capacity per unit volume, v is the average phonon velocity or sound velocity, and ℓ is the average phonon mean free path. Such an expression should pertain to each of the lattice modes of vibration. In practice, however, we shall ignore the optic modes, since they are expected to have a small phonon velocity, and assume that only the acoustic modes carry heat. Thermal resistivity arises from processes that scatter phonons.

We shall now review simply analyses of $\lambda(T)$ for T $\geq$ the Debye temperature, θ_D, for the different types of crystals studied.

A normal crystal phase approximates a structurally perfect crystal of monatomic basis. For such a perfect crystal the thermal resistivity arises entirely from three-phonon Umklapp processes. The density of other phonons with which a given phonon can interact is proportional to T, which implies $\ell \propto 1/T$. The temperature dependences of heat capacity and sound velocity are ignored. The resulting theoretical prediction for T $\geq \theta_D$ is thus $\lambda \propto 1/T$. This form of $\lambda(T)$ is found in practice for many real crystals, including molecular crystals in the normal crystal phase.

For a plastic crystal phase the sound velocity is fairly small, and the mean free path is short and limited mainly by structural disorder. Three-phonon Umklapp processes are likely to make only a small contribution to the total thermal resistivity. This picture is consistent with the small absolute value and weak temperature dependence for λ, which are found experimentally for plastic crystal phases.

A clathrate hydrate seems to approximate a glass in relation to $\lambda(T)$. For a glass we take ℓ to be small and temperature independent as a result of structural disorder. The effect of three-phonon Umklapp processes is neglected. The temperature dependence of v is ignored, and C(T) is assumed to be described by a Debye model. The $\lambda(T)$ is then predicted to be similar to C(T), i.e., λ increases modestly with T. Real glasses show approximately the predicted behaviour.

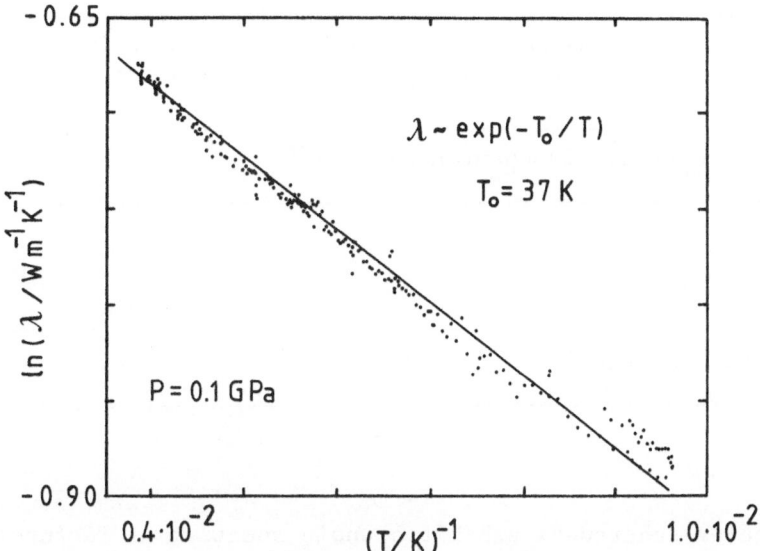

Fig. 5.   Isobaric temperature dependence of λ of tetrahydrofuran
· 17 H_2O clathrate hydrate fitted to equation of form
$\lambda \sim \exp(-T_0/T)$.

To summarize: the thermal conductivity and its temperature
dependence can be described qualitatively in terms of the Debye
formula, equation (1), but a detailed quantitative explanation is
lacking.

We shall now discuss in some detail the spectacular and surprising behaviour of λ for the clathrate hydrate.

We found a relatively poor two-parameter fit to the experimental data, assuming v and ℓ constant and C(T) described by a Debye
model. However, we found empirically that the form $\lambda \sim \exp(-T_0/T)$
provides a good two-parameter fit, and this is shown in Figure 5.
The slope of the fitted line corresponds to a characteristic temperature, T_0, of 37 K. This is nearly the same as the characteristic
temperature (T_e = 36 K) associated with the low-energy excitations
of the tetrahydrofuran (THF) guest molecules, according to far
infrared spectroscopic measurements.[16] It is possible that the
THF molecules cause strong inelastic scattering of phonons, and
that only phonons near the rotation vibration excitation energy,
kT_e, could propagate. The guest molecules will obey Maxwell-
Boltzmann statistics, which might then imply a conductivity
$\lambda \propto \exp(-T_e/T)$. However, a more detailed analysis is evidently
needed to determine whether the agreement between T_0 and T_e is
simply fortuitous.

Table 2. Simplified Summary of the Temperature and
Volume Dependences of Thermal Conductivity
for the Three Types of Crystals Studied

Type of Crystal	Exponent n (in $\lambda \sim T^{-n}$)	$g = -(\delta \ln\lambda/\delta \ln V)_T$
Normal	1	10
Plastic	0.2	5
Clathrate hydrate	−0.3	0.5

CONCLUSIONS

Molecular substances exhibit a whole spectrum of features
in relation to thermal conductivity, and a simplified summary is
shown in Table 2.

The temperature dependence of λ varies from T^{-1} for normal
crystals to $T^{-0.2}$ for plastic crystals and to $T^{+0.3}$ for the
clathrate hydrate tetrahydrofuran·17 H_2O. The volume dependence of
λ varies from $g \approx 10$ for normal crystals to $g \approx 5$ for plastic
crystals and $g \approx 0.5$ for the clathrate hydrate.

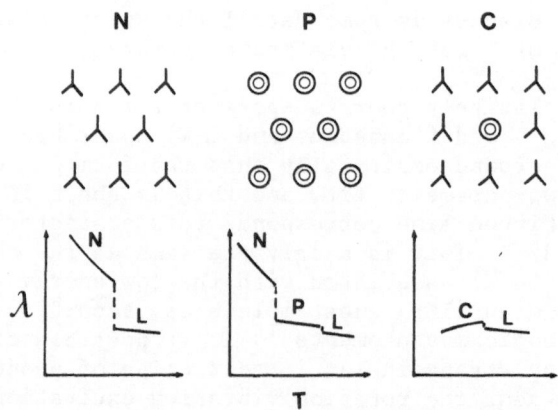

Fig. 6. Schematic diagram of the isobaric temperature dependence
of λ for $T \gtrsim \theta_D$ for the three types of crystals studied.
The types of crystals are represented by simple schematic
drawings and indicated as follows: N, normal; P, plastic;
C, clathrate hydrate; and L, liquid.

Figure 6 is a simplified schematic diagram for λ and its temperature dependence in the high-temperature region (T $\gtrsim$ θ_D) for the three types of crystals studied.

For the normal crystal, λ varies strongly with temperature, as T^{-1}, and the decrease in λ on melting is large, at least 30 %. For the plastic crystal, the temperature dependence of λ is small and similar to that of the liquid. On melting there is a small decrease of λ of 5 - 10 %, while on transforming to the normal crystal there is a large increase of about 25 %. The loss of long-range orientational periodicity is thus much more important for the conduction of heat than the additional loss of long-range translational periodicity that occurs on melting.

In a clathrate hydrate, only the guest molecules are relatively free to reorient, and one would expect such a substance to have a behaviour intermediate between those of a normal and plastic crystal. This does not seem to be the case. For the clathrate hydrate, we observe a decrease of λ on melting similar to that for a plastic crystal phase. Furthermore, the rather few reorienting guest molecules have a very strong effect on the temperature and volume dependences of λ, in fact, stronger than the effect of all molecules reorienting in a plastic crystal phase.

REFERENCES

1. J. Timmermans, Plastic crystals: A historical review, J. Phys. Chem. Solids 18:1 (1961).
2. R.G. Ross, P. Andersson, and G. Bäckström, Thermal conductivity and heat capacity of solid phases of benzene under pressure, Mol. Phys. 38:377 (1979).
3. O. Sandberg, P. Andersson, and G. Bäckström, Glass transitions under pressure detected by heat capacity and thermal conductivity measurements, in: "Proceedings Seventh Symposium on Thermophysical Properties," A. Cezairliyan, ed., ASME, New York (1977), pp. 181-4.
4. P. Andersson, Thermal conductivity and heat capacity of cyclopentane in the range 100-300 K and up to 2.1 GPa, Mol. Phys. 35:587 (1978).
5. P. Andersson, Thermal conductivity and heat capacity of cyclohexane under pressure, J. Phys. Chem. Solids 39:65 (1978).
6. R.G. Ross and O. Sandberg, The thermal conductivity of four solid phases of NH_4F, and a comparison with H_2O, J. Phys. C: Solid State Phys. 11:667 (1978).
7. R.G. Ross and P. Andersson, Thermal conductivity and phase diagram of CCl_4 under pressure, Mol. Phys. 36:39 (1978).
8. R.G. Ross and O. Sandberg, Thermal conductivity and heat capacity of solid phases of NH_4Cl under pressure, J. Phys. C: Solid State Phys. 12:3649 (1979).

9. P. Andersson and R.G. Ross, Thermal resistivity, heat capacity and phase diagram of CBr$_4$ under pressure, Mol. Phys. 39: 1359 (1980).

10. J. Wigren and P. Andersson, Thermal conductivity and heat capacity of adamantane and hexamethylenetetramine under pressure, Mol. Cryst. Liq. Cryst. 59:137 (1980).

11. P. Andersson, Thermal conductivity, heat capacity and phase diagram of furan under pressure, High Temp.-High Pressures 12:655 (1980).

12. R.G. Ross and P. Andersson, Thermal conductivity, heat capacity and phase diagrams of the trichloroethanes under pressure, I. 1,1,1-trichloroethane, Mol. Cryst. Liq. Cryst. 69:145 (1981)

13. R.G. Ross, P. Andersson, and G. Bäckström, Unusual PT dependence of thermal conductivity for a clathrate hydrate, Nature 290:322 (1981).

14. R.G. Ross, P. Andersson, and G. Bäckström, Effects of H and D order on the thermal conductivity of ice phases, J. Chem. Phys. 68:3967 (1978).

15. G.A. Slack, The thermal conductivity of nonmetallic crystals, in: "Solid State Physics," Vol. 34, H. Ehrenreich, F. Seitz, and D. Turnbull, eds., Academic Press, New York (1979), pp. 1-71.

16. J.E. Bertie and S.M. Jacobs, Infrared spectra from 300 to 10 cm^{-1} of structure II clathrate hydrates at 4.3 K, J. Chem. Phys. 69:4105 (1978).

THERMOPOWER AND THERMAL CONDUCTIVITY MEASUREMENTS

ON INTERCALATION COMPOUNDS

J-P. Issi, J. Boxus *, B. Poulaert and J. Heremans [+]

Université Catholique de Louvain - Laboratoire PCES
B - 1348   Louvain-la-Neuve, Belgique

ABSTRACT

The problems associated with the measurement of the thermal
conductivity and thermopower of graphite intercalation compounds
are discussed. Sample holders designed for in-plane and c-axis
measurements are also described.

INTRODUCTION

The electrical properties of graphite intercalation compounds
have been a subject of intensive investigations these last few
years [1]. In addition to the fundamental interest in these quasi
two-dimensional systems, these studies were stimulated by the
promise of realizing synthetic metals with electrical conductivities
as high as those of copper. Increased conductivities are due to the
charge transfer from the intercalate layers to the carbon layers,
where in-plane mobilities are very high. According to the intercalate species, donor or acceptor compounds may be obtained.

The anisotropy of the *electrical resistivity* relative to
pristine graphite ($\sim 10^3$) is usually decreased in donor compounds
and increased in acceptor compounds, where in some cases it may
exceed 10^6. These high anisotropy ratios complicate, to a great
extent, electrical resistivity measurements. When current probes
are applied to a sample, intimate contact cannot be insured over
all the end section of the sample on a microscopic scale whatever
care is taken in realizing these contacts. This is one of the main

* supported by an IRSIA grant
+ Chargé de Recherche FNRS

reasons why four-probe techniques are generally adopted, since when applied to isotropic materials they insure parallel electrical current lines between the emf probes. This is not the case in highly anisotropic solids, like layered materials, where the effective cross sectional area of the sample, in which the current flows along some planes, may be much smaller than the measured one. For this reason contactless ac techniques had to be used for graphite inter-calation compounds [2] instead of the four-probe dc method.

It was only very recently that *thermal conductivity* data were reported on donor [3, 4] and acceptor [4, 5] compounds. We wish to focus primarily here on the specific problems associated with the measure-ment of the thermal conductivity and the thermopower in layered materials. The problems are mainly due to the high anisotropy, small thicknesses, and the mechanical properties of the samples. Also, the graphite intercalation compounds are not stable in air and require particular handling procedures, which are not discussed here. After showing some typical results, we will describe sample holders that we designed to measure the thermal conductivity, the thermopower and the electrical resistivity of graphite intercalation compounds. Since there are about two orders of magnitude difference in the in-plane thermal conductivity and that along the c-axis, we had to consider totally different measuring devices for both cases.

TYPICAL RESULTS

The measurement of the thermal conductivity and thermopower of graphite intercalation compounds revealed interesting features, when compared to pristine graphite. Fig. 1 shows some typical results for the temperature variation of the thermal conductivity of a graphite intercalation compound in-plane (a) and parallel to the c-axis (b). The data are relative to $FeCl_3$ compounds (acceptors) and are compared to that of pure highly oriented pyrolitic graphite (HOPG). The in-plane results (a) indicate that in the lowest temperature range the thermal conductivity of the intercalated samples is entirely electronic, while at higher temperatures both lattice and electronic conduction are effective. The c-axis results (b) on stage-2 $FeCl_3$ compound * are consistent with a lattice contribution over the whole temperature range (cfr refe-rences 4 and 5). This shows that below room temperature the main effect of intercalation is to decrease the lattice thermal con-ductivity relative to pristine graphite and to increase the electronic contribution. Also, contrarily to the case of the electrical conductivity [1], the anisotropy of the thermal conducti-vity (compare Fig. 1a to Fig. 1b) was found to be of the same order as that of pristine graphite in acceptor compounds. As regards the *thermopower*, data were available on donors[6] and acceptors[7,8].

---

* The stage is the number of graphite layers between two intercalate layers

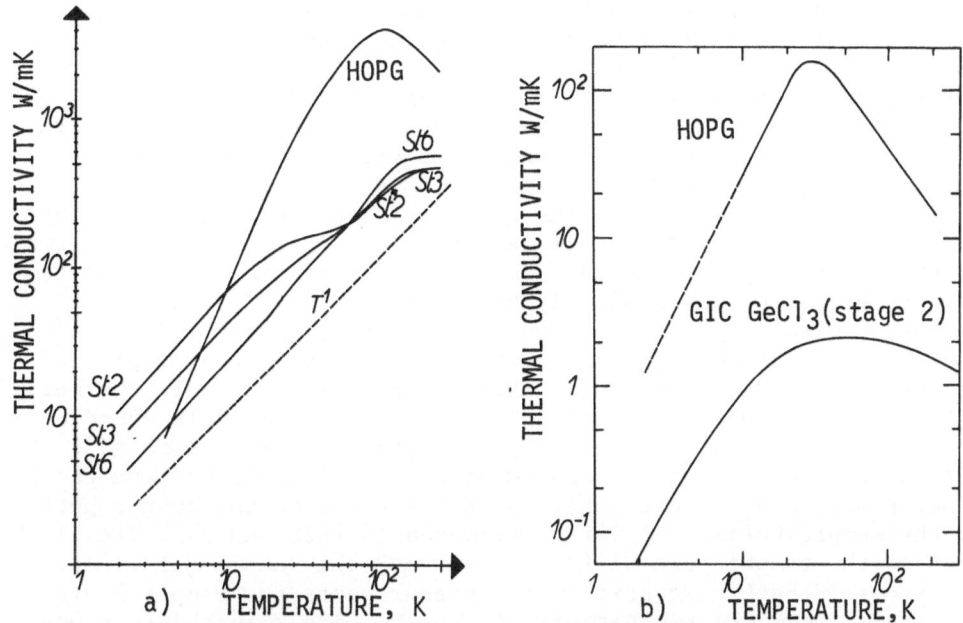

Fig.1 Typical results for the temperature dependence of the in-plane (a) and c-axis (b) thermal conductivity of $FeCl_3$ graphite intercalation compounds.

However a complete study on acceptor compounds [4, 9] as a function of stage and temperature, as well as results on a donor compound as a function of temperature [3], were only reported very recently. The thermopower data essentially revealed that donors and acceptors have negative and positive thermopowers, respectively. The general trend of the temperature dependence of the thermopower is a monotonic increase with temperature followed by a plateau for both donors and acceptors.

IN-PLANE SAMPLE HOLDERS

Around room temperature the intercalated materials have, indeed, a relative high in-plane thermal conductivity which, though lower than that of pristine graphite, is typically of the order of $500 \ W \ m^{-1} K^{-1}$, the room temperature thermal conductivity of copper. In the liquid helium range, the conductivities decrease by one to two orders of magnitude according to the stage and the type of the intercalant [4]. However, contrarily to pristine graphite, large samples of which are readily available, good quality intercalated specimens are thin and brittle. Thicknesses do not usually exceed a fraction of a millimeter (~ 0.5 mm). This raises problems at

low temperatures, where the thermal conductances of the samples may
be low compared with that of the connecting wires. This is particularly true when gold (iron) thermocouples are used as temperature
sensors, since the smallest diameter available commercially is
~ 0.08 mm. So our first step was to check whether the temperature
sensors introduce significant heat leaks and if the thermocouple
junction to sample contact resistance is negligible with respect
to the thermal resistance of the gold (iron) arm of the thermocouple.
To do so, we designed a special sample holder operating as a thermal
potentiometer (Fig.2). In this device, there are two copper posts
that may be heated independently by means of two 400 Ω thin
constantan wires, R_C and R_H. Two thermocouples T_C and T_H measure
each the temperature difference between a region of the sample and
a copper post. One junction of thermocouples T_C and T_H is thermally
anchored to a copper post; the other junction is tightly wound
around the sample and electrically insulated from it. A third
thermocouple, ΔT, measures the temperature difference between the two
copper posts, and a fourth one, T_B, between one of the copper posts
and the sample holder. A fifth thermocouple (not shown in Fig.2)
measures the temperature difference between the sample holder and
the cryogenic bath. To perform the measurements the sample heater
is energized and the two heaters, R_C and R_H, are adjusted to render
the readings of the thermocouples T_C and T_H negligible (less than 1%)
with respect to the temperature difference, ΔT, between the two
copper posts. When thermal equilibrium is reached, there is no heat
flow from the sample through thermocouple wires T_C and T_H, and ΔT
gives the temperature difference between the two thermocouple
junctions anchored to the sample. The sample holder shown in Fig.2,
as well as those shown in Fig.3 and Fig.4, is in good thermal contact,
but electrically insulated from the heat sink of a variable temperature vertical liquid helium cryostat.

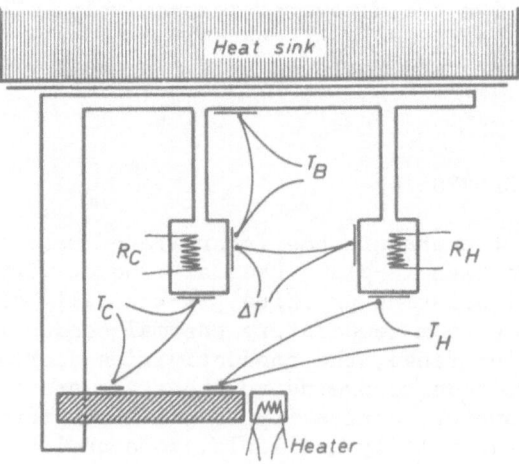

Fig.2   The thermal potentiometer

We have used this method to measure two samples of stage-4
graphite $FeCl_3$ compound [5], which were of very small dimension
(roughly 7 x 2 x 0.2 mm^3). For these samples sizes we found that
the maximum difference observed in the range 2-300 K between measurements performed with and without thermal compensation was less
than 5%. With larger samples, typically 20 x 4 x 0.5 mm^3, that were
subsequently available, we were then confident when using a more
direct measurement technique by means of the sample holder shown
in Fig.3.

Fig.3 represents another four-probe arrangement for in-plane
thermoelectric measurements. One end of the horizontal sample is
attached to the copper sample holder by means of silver ink (type
SC 18 S from Microcircuit Co). To the other end is glued by means
of the same ink, the sample heater, which is supported by a thin-
walled hollow stainless steel cylinder of small diameter, which
maintains the sample strain-free. This supporting tube may be
adjusted horizontally and vertically by means of the screws. A
differential Au (0.03 at % Fe)- Chromel-p thermocouple, ΔT, measures
the temperature difference across the sample. Two thin copper
current leads and two chromel-p emf probes, ΔV, allow the electrical
resistivity to be measured. The emf probes are also used for
thermopower measurements. Note that the length of the gold (iron)
arm of the differential thermocouple, ΔT, in Fig.3 is about 5 times
greater than that of T_C and T_H in Fig.2.

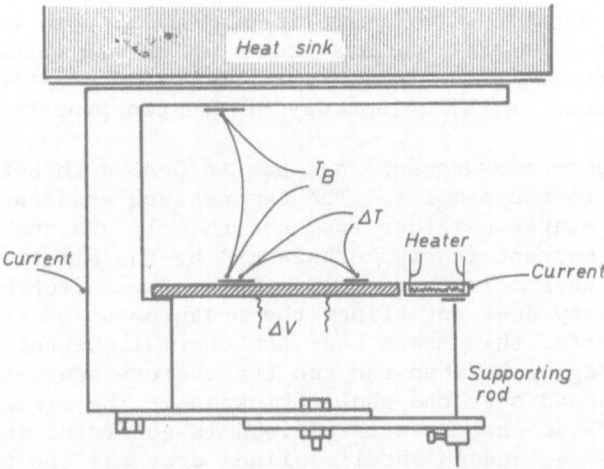

Fig.3 Four-probe arrangement for in-plane thermoelectric
       measurements without thermal compensation.

Because of the very high anisotropy of the *electrical resistivity* of graphite-FeCl$_3$ intercalation compounds, we were unable to measure this property. With the silver ink we used the contact was more uniform over the end section of the sample than with any other conductive glue we have tried. However, the measured room temperature electrical resistivity, ρ, we obtained was still much higher than the one obtained by means of contactless techniques[10]. Also the results were not reproducible when the sample was cycled from room temperature to low temperatures. We ascribe the latter to a rearrangement of the contact area between the graphite planes and the silver ink caused by the differential thermal contraction between the two materials. This view was substantiated by the fact that when the sample was cycled to and fro from 1.5 up to 30 K, reproducible results were obtained provided the sample temperature was maintained below $\sim$ 30 K. In this temperature range thermal contraction is not important. We were thus in a position to measure the temperature variation of the resistance of the fraction of the layers in which the current flows below 30 K and to determine the range where ρ was constant, i.e., the residual range of the resistivity [5].

The situation is different for the *thermal conductivity*. In that case the anisotropy is much smaller than that of the electrical conductivity since it is of the order of 10^2 below room temperature (Fig.1). This is comparable to the anisotropy reported for pristine graphite [11] and it explains why four probe techniques were successfully applied to measure the in-plane thermal conductivity [3][5]. The problems associated with the anisotropy of the thermal conductivity have been discussed in detail by Nye [12] and the particular case of pristine graphite was thoroughly treated by Hooker et al. [11]. The fact that the results of the thermal conductivity were reproducible within experimental error when the samples were cycled in temperature, whereas those of the electrical resistivity measured during the same run were not, is consistent with the large difference in the anisotropy of the two properties.

In *thermopower measurements* one has to deal with both electrical and thermal flow in the samples. The temperature gradient causes a diffusion of the charge carriers from hot to cold and the accompanying electrical current is counterbalanced by the electrical current generated by the thermoelectric field. Since the anisotropy of the thermal conductivity does not affect the measurement of the in-plane thermal conductivity, this means that the thermal current lines are parallel in the region between the two temperature sensors along the sample and are spread over the whole thickness. The electrical current caused by the thermoelectric field is generated internally and will thus also extend in parallel lines over all the thickness of the sample between the emf probes. The feasibility of the in-plane thermoelectric measurements, thus, is tied to that of the thermal and not of the electrical resistivity.

C-AXIS SAMPLE HOLDER

        The situation is entirely different for the c-axis direction.
First, one expects the thermal conductivity to be much lower than
in-plane.   Second, intercalated samples are usually thin.   Third,
one would expect problems associated with cleavage. So the geometry
of the sample (L/A ratio, where L is the active length of the sample
and A its cross section) to be used is completely different, and one
should use flat discs or parallelepipeds; contact resistances are
expected to play an important role. Most of the problems associated
with c-axis measurements on pristine graphite have already been
discussed by Hooker et al. [11] . A great advantage of the anisotropy,
which favors measurements in the c-axis direction, is that isothermal
planes are readily available perpendicular to the axis so that
thermal contact with the heater and heat sink do not have to extend
over the entire basal plane surface to insure isothermal planes.
Also, the temperature sensors and emf probes may be directly applied
on the upper and lower surfaces of the sample (Fig.4). This was
already apparent from the work of Moore et al. [13], and discussed
further by Hooker et al. [11] . Though not essentially different in
principle than the method used by Hooker et al. we have designed a
special sample holder (Fig. 4) for c-axis measurements on intercalated
material. This allows thermal conductivity as well as thermopower
and electrical resistivity measurements to be performed during the
same run. In Fig.4 the heat flow is perpendicular to the surface of
the sample. The differential thermocouple, ΔT, has its junctions
thermally anchored but electrically insulated from the top and
bottom surfaces of the sample.

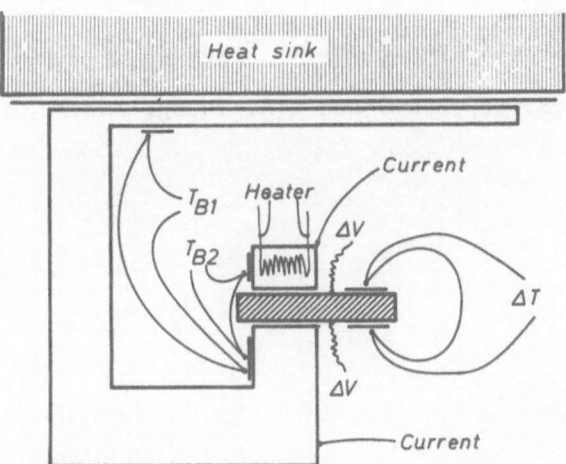

        Fig.4 Sample holder for c-axis measurements of the
               thermoelectric properties.

The thermocouple T_{B2} gives an estimation of the contact resistances when ΔT is known. Combined with the reading of T_{B1}, that of T_{B2} allows the mean temperature of the sample to be estimated. The current and voltage probes are used for the measurement of the c-axis electrical resistivity.

REFERENCES

1. M.S. Dresselhaus and G. Dresselhaus, Advances in Physics, 30, 139 (1981)
2. C. Zeller, A. Dennenstein and G.M.T. Foley, Rev. Sci. Instrum., 50, 602 (1979)
3. J. Heremans, J-P. Issi, I. Zabala-Martinez, M. Shayegan and M.S. Dresselhaus, Physics Letters, 84A, 387 (1981)
4. B. Poulaert, J. Heremans, J-P. Issi, I. Zabala-Martinez, H. Mazurek and M.S. Dresselhaus, Ext. Abst. Bien. Carbon Conf. 15, 92 (1981)
5. J. Boxus, B. Poulaert, J-P. Issi, H. Mazurek and M.S. Dresselhaus, Solid State Communications, 38, 1117 (1981)
6. L.C.F. Blackman, J.F. Mathews and A.R. Ubbelohde, Proc. Roy. Soc., 258A, 339 (1969)
7. A.R. Ubbelohde, Carbon, 6, 177 (1968)
8. F. Maeda, H. Oshima, K. Kawamura and T. Tsuzuku, Extended Abstracts 14th Biennial Carbon Conf., Pennsylvania State University, p.197 (1979)
9. J-P. Issi, J. Boxus, B. Poulaert, H. Mazurek and M.S. Dresselhaus, J. Phys. C. : Solid State Phys., 14, L 307 (1981)
10. J.B. Perrachon, C. Zeller and F.L. Vogel, Extended Abstract Biennial Carbon Conf., 14 304 (1979)
11. C.N. Hooker, A.R. Ubbelohde and D.A. Young, Proc. Roy. Soc. A284, 17 (1965)
12. J.F. Nye, "Physical Properties of Crystals", Oxford University Press (1957)
13. A.W. Moore, A.R. Ubbelohde and D.A. Young, Nature, London, 198, 1192 (1963)

SESSION   TC-13A

Techniques and Apparatus - I

CHAIRMAN

R. E. Taylor
Purdue University
West Lafayette, Indiana

A ROTATING ANNULUS DEVICE FOR PRECISE THERMAL CONDUCTIVITY

MEASUREMENTS ON LIQUIDS; VALUES FOR ETHANOL, JP-4, AND R-113

R. Braun and A. Schaber

Universität-Gesamthochschule Siegen
Siegen, FRG

ABSTRACT

The accurate measurement of the thermal conductivity of liquids requires the elimination of natural convection. The present paper describes the construction and operation of a rotating annulus device with which it is possible to suppress convection reliably. Experimental results for ethanol, JP-4, and R-113 from -20°C to 200°C are presented. These are accurate within ±0.5% to ±3%, depending on the extent of the thermal radiation which in addition may occur. To eliminate the radiation component, an approximate calculation method was employed that considers the spectral absorptivity.

INTRODUCTION

Reliable determination of the thermal conductivity of liquids requires the elimination of natural convection. For this purpose Heckle[1] proposed an investigation of liquids in the annular space between two concentric cylinders rotating about their horizontal axes. If a steady-state heat flux is conducted across the liquid in a radial direction from the inner to the outer wall of the annulus, the direction of the density gradient coincides with the direction of centrifugal acceleration, assuming there is a positive volume expansivity. Consequently convection is suppressed. In this way Heckle was very successful in investigating some liquids in the temperature range 0°C to 30°C. The centrifugal acceleration he used was constant and 30 times larger than the acceleration of gravity. Following the proposal of Heckle a new measuring device has been developed that is described in this paper.

547

MEASURING DEVICE

     With this apparatus it is possible to investigate any liquid --
also gases - under steady-state conditions and to measure thermal
conductivity absolutely in the temperature range -20°C to 250°C and
in the pressure range 0 to 20 MPa. In Fig. 1 a schematic view of the
experimental setup is given. The test fluid is located in the annular
space between two concentric thick-walled copper cylinders. The inner
cylinder contains an electric heating system consisting of three parts:
a main heater and two regulated guard heaters. The heat flux coming
from the heaters is conducted across the liquid to the outer cylinder
and there transferred to a coolant which is thermostatically controlled. As the annular space rotates about its horizontal axis there
will be a very uniform outer heat transfer. The measuring gap, which
has an outer diameter of 60 mm, can be varied from 2 to 7 mm by replacing the internal cylinder. The length of the main heater is
always 200 mm. Figure 1 also shows the solution of the problems caused
by rotation. The electrical connection of the rotating part to the
stationary part of the measuring device is established by mercury
contacts. The very low thermocouple voltages are amplified before
transmission. With the help of expansion bellows on the left side of
the rotating part, the change of volume caused by the temperature
variation can be compensated. This is done hydraulically, so that the
pressure of the test fluid can be controlled optionally during
rotation. Finally, the device has been constructed so that the entire
temperature range can be achieved continuously.

ELIMINATION OF NATURAL CONVECTION

     A driving motor with speed regulation enables the rotation to be
set at any rate. This was employed since preliminary investigations
had shown that a convective flow in the annulus depends on the
rotational rate and therefore can be identified. That needs to be
explained.

     Natural convection that is generated in a horizontal stationary
annulus heated from the inner cylinder with the uniform gravitational
field acting on it is well known. The streamline pattern is demonstrated by Fig. 2. One can estimate that the velocity of convective
motion is small. So turning the annulus and the liquid simultaneously
into a new fixed position only after a longer running-in period, the
streamline pattern that is determined by the direction of gravitational acceleration, g, will appear again. However, if the annulus
is continually turned with an increasing rotational rate, the natural
convection will be more and more suppressed. The convection cells
originally enveloping the whole cavity divide into smaller and
smaller vortexes until finally convection heat transfer completely
vanishes. This can be verified by systematic measurements represented
in Fig. 3. Here the Nusselt number, based on the characteristic

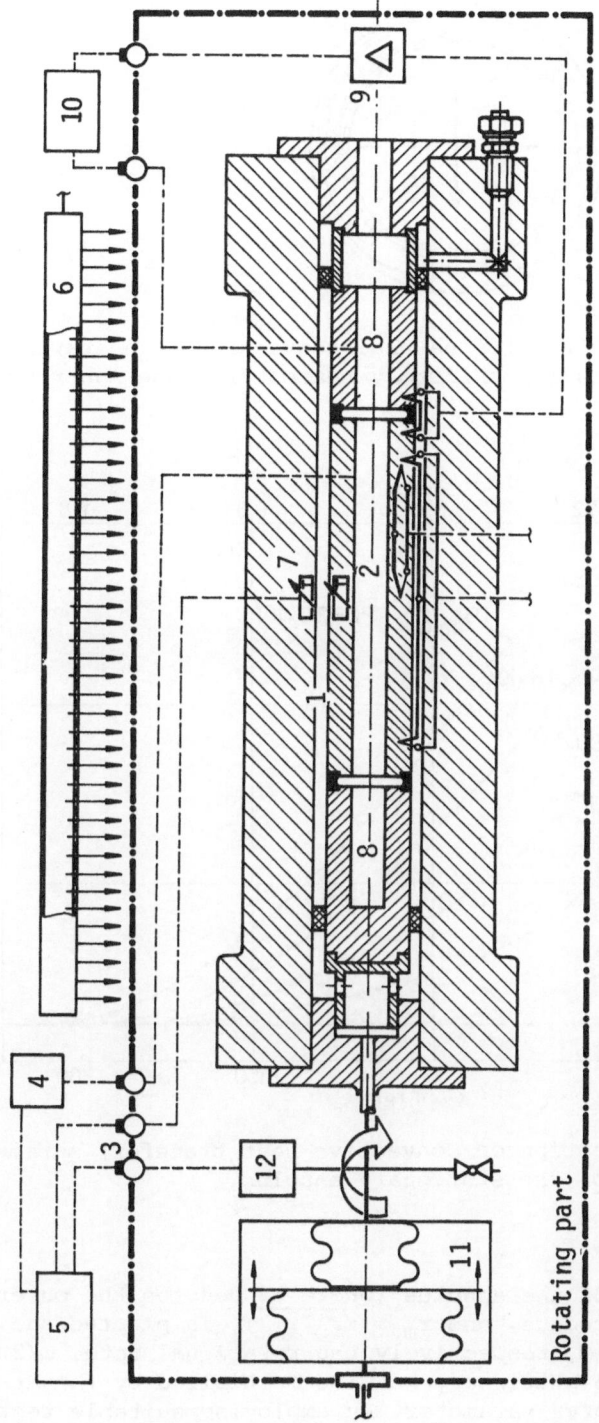

Fig. 1. Schematic view of the experimental setup.
1, measuring gap; 2, main heater; 3, mercury contact; 4, regulated d.c. source; 5, digital voltmeter; 6, thermostatically controlled coolant; 7, four-lead platinum resistance thermometers; 8, guard heater; 9, amplifier (output current independent of load); 10, controller ($\Delta T = 0$ K) and d.c. source; 11, pressure control; 12, pressure transducer.

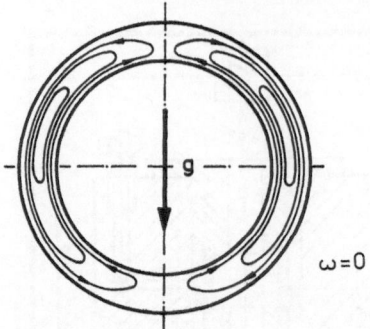

Fig. 2.   Streamline pattern of natural convection in a stationary
         horizontal annulus uniformly heated from the inner cylinder.

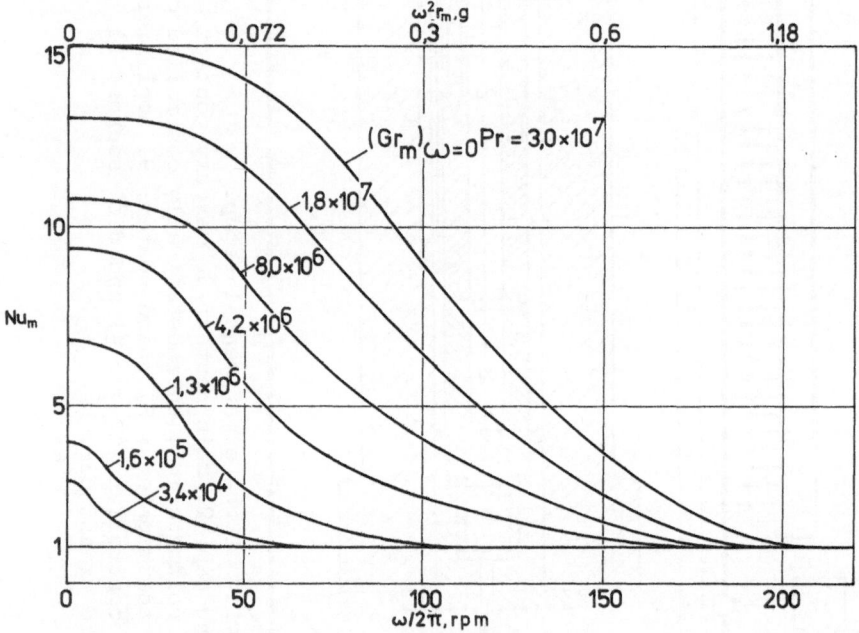

Fig. 3.   Effect of rotation on convective heat transfer; parameter:
         $(Gr_m)_{\omega=0}Pr$ for the stationary annulus.

length, $r_m \ln r_o/r_i$, of the annulus (where r_o denotes the outer
radius, r_i the inner radius, and $r_m = \sqrt{r_o\, r_i}$), is plotted against
the angular velocity, ω, respectively the rotational rate, $\omega/2\pi$. The
Grashof number for the stationary annulus multiplied by the Prandtl
number is used as a curve parameter. By employing suitable test

fluids, the Gr_mPr, which equals the Rayleigh number, can be widely varied. The diagram shows that even with high values of Gr_mPr the Nusselt number goes quickly to unity, where convection does not exist any longer. This may even occur before the centrifugal acceleration has reached the value of the gravitational acceleration.

Figure 3 explains that the high centrifugal acceleration used by Heckle is unnecessary. It can even be dangerous. Of course liquid particles only arrange themselves in a stable stratification, corresponding to the temperature drop, if the bounding surfaces of the annular cavity are sufficiently isothermal. If that is not the case, the acceleration field makes mechanical equilibrium impossible. With the rotational rate increasing, axial pressure gradients come about and with them convective motions. The results of a more detailed investigation are shown in Fig. 4. Here the dependence of measured apparent thermal conductivity, λ_{app}, on the centrifugal acceleration is represented by two curves, one of them concerning ethanol at 30°C, the other ethanol at 100°C. On the left-hand side of the figure one can see that both curves attain a constant value, the "pure" thermal conductivity, before the centrifugal acceleration equals 1 g. The behaviour of λ_{app} observed with further increase in the centrifugal acceleration is shown on the right-hand side. As to the 100°C curve, the uniformity of temperature distribution realized in the measuring gap was exceptionally insufficient. The temperature drop in the axial direction amounted to 0.2 K. At first, λ_{app} remained constant. Above

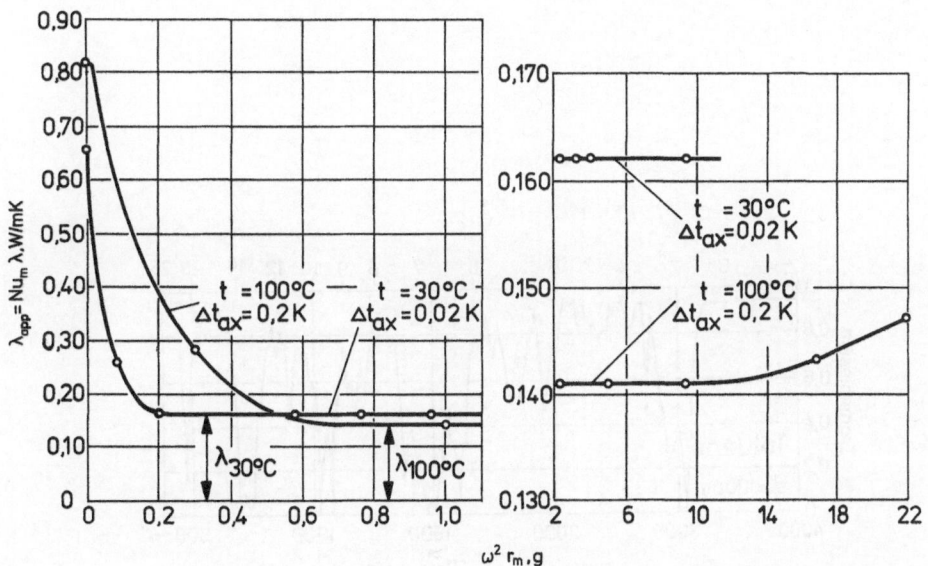

Fig. 4.   Influence of centrifugal acceleration on apparent thermal conductivity, λ_{app}, of ethanol.

13 g, however, the values clearly increased; here convection re-
appeared. The causes were: the isotherms did not run parallel to the
axis and the centrifugal acceleration was too high. From the experiments, therefore, the following important results can be deduced:
- During the measuring process unnecessarily high centrifugal
  acceleration must be avoided.
- Since the convection depends on the rotation, the presence of
  convection in the liquid can be determined by the simple method
  of varying the rotational rate without having to change another
  experimental parameter.

ELIMINATION OF THERMAL RADIATION

        In addition to heat conduction, thermal radiation may occur.
This applies especially to higher temperatures and to liquids that
are semitransparent in the infrared region. In this case, a radiation
component, λ_r, dependent on the layer thickness, s, of the absorbing
and emitting liquid must be eliminated when determining the actual
thermal conductivity, λ. To eliminate λ_r, an approximate calculation
method was used that considers the spectral absorptivity of the liquid.
The method is based on a mathematical model proposed by Kohler.[2] This
model has already been graphically employed by Schödel and Grigull[3]
yielding good results. The numerical investigations were carried out
starting from the simplifying assumptions that the temperature profile in the liquid layer is linear, the cylindrical layer can be
replaced by a plane layer, the bounding surfaces are mirror-like

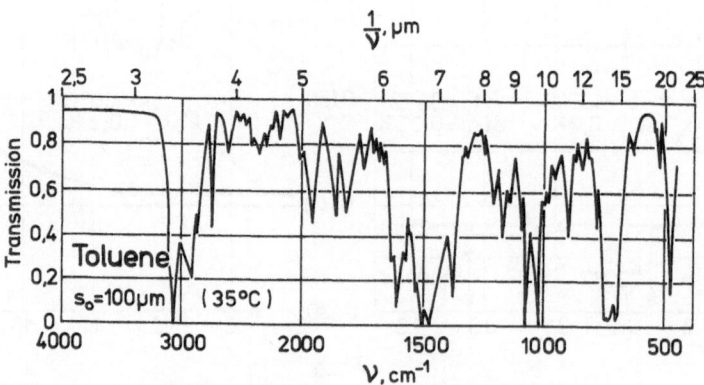

Fig. 5. Infrared absorption spectrum of toluene, fractional trans-
         mission versus wave number.

reflecting and gray emitting with very small emissivities, and the refractive index of the liquid is constant. The radiation component is thus determined by the equation[4]

$$\lambda_r = \frac{4}{3} c_1 c_2 n^2 \int_{\nu=0}^{\infty} \frac{y}{k} \frac{\nu^4 \exp(\nu c_2/\bar{T})}{\bar{T}^2 \{\exp(\nu c_2/\bar{T}) - 1\}^2} \, d\nu,$$

where ν is the wave number, $k(\nu)$ is the spectral absorption coefficient, c_1 and c_2 denote the constants in Planck's spectral energy distribution, n denotes the refractive index, and $\bar{T}$ is the average temperature in the layer. A complicated function of the optical thickness, $\tau = ks$, and the emissivity, ε, is represented by y, which has been calculated and tabulated by Poltz and Jugel[5] for a series of values of the arguments τ and ε. Numerical studies demonstrated that the accuracy of the quantitative data of the infrared absorption spectrum must satisfy extended requirements. Because adequate spectra were not available, they had to be determined, e.g., the one shown in Fig. 5. A spectrophotometer was used to record the transmission spectrum for each liquid on several samples of different thickness. By this means the errors in measurement could be corrected, because the absorption coefficient remains constant. Furthermore, it appeared that to completely eliminate the thermal radiation, the transmission spectra should not have any values larger than 80%. Since common spectroscopy is intent on qualitative statements, such recorded spectra of suitable samples are unusual.

The spectrum of toluene shown in Fig. 5 was used to eliminate thermal radiation from effective conductivities, λ_{eff}, that were measured of several toluene layers of different thickness, s. Figure 6 represents the results. The squares and the curve connecting them reproduce λ_{eff} measurements made by Geller et al.[6] with a stationary vertical annulus. In order to verify the absence of convection all tests have been performed at several different temperature drops across the layer. That is why the increase in λ_{eff} with s can be only due to the rising contribution of thermal radiation. The experimental values are considered to have an uncertainty of ±1.5% or less. If the radiative component, λ_r, is subtracted from these data, in each case a constant, "pure" conductivity independent of layer thickness must result. As the results of calculations, denoted by circles show, that could be verified if our own spectrum was employed. With other available spectra,[8,9] elimination was not sufficient. Radiation heat transfer was calculated for 298 K and for elevated temperatures, too. This is shown in Fig. 7. The spectrum recorded at room temperature was previously modified using the relation between temperature and liquid density. Figure 7 demonstrates that the method employed.is appropriate to eliminate radiation heat transfer also at 328 K and 473 K.

EXPERIMENTAL RESULTS

   With the measuring device previously described, several series
of measurements were carried out on 7 mm thick liquid layers, first
on water and on ethanol to test accuracy.

   The thermal conductivity of ethanol measured at saturation
pressure in the temperature range -20°C to 200°C is a linear function
of the temperature, t:

$$\lambda = (0.1708 - 296 \times 10^{-6} \frac{t}{°C}) \frac{W}{m \ K}.$$

The purity of the sample was higher than 99.8%. This was stated by

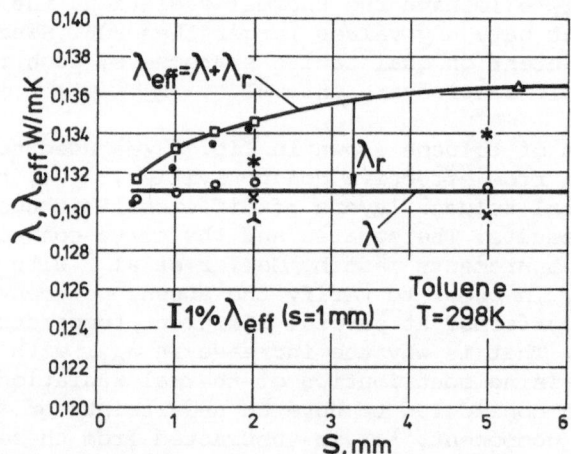

Fig. 6. Determination of the real thermal conductivity, λ, from
         measured effective values, λ_{eff}, by eliminating the
         radiative component, λ_r; T = 298 K.
         λ_{eff} measurements:□ , Geller et al.;[6] ● , Poltz and Jugel;[7]
         Δ, Heckle.[1]
         λ_r calculations: o , with our ir-spectrum; ✳, with DMS--
         ir-spectrum;[8] ⅄, with Merck-ir-spectrum;[9] ✗, of Schödel
         and Grigull.[3]

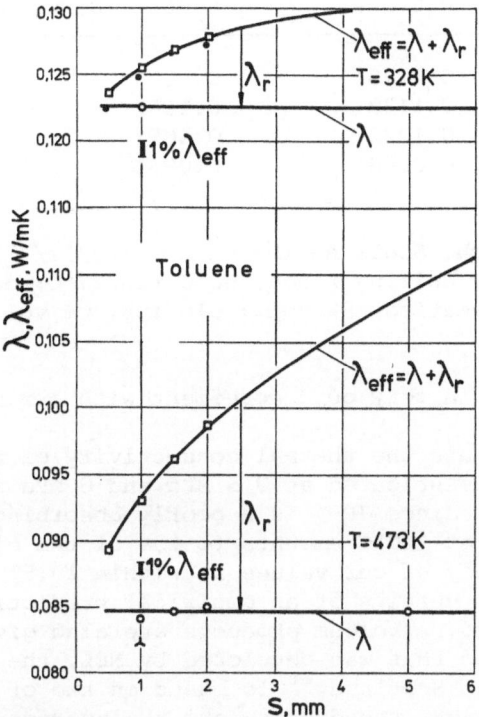

Fig. 7. Determination of the real thermal conductivity, λ, from
measured effective values, λ_{eff}, by eliminating the
radiative component, λ_r; T = 328 K and T = 473 K.
λ_{eff} measurements:□ , Geller et al.;[6] ● , Poltz and Jugel.[7]
λ_r calculations: ⅄, with Merck-ir-spectrum;[9] ○ , with our
ir-spectrum.

a chromatographic analysis. Ethanol is a strongly absorbing liquid.
Nevertheless, the radiation heat flux that had to be eliminated
came to 3% of the total heat flux at 200°C. The values given by the
above equation are accurate within ±0.5% for temperatures up to 50°C,
and within ±1% for temperatures above 150°C. For the temperature
range 10°C to 30°C, precision measurements by Poltz and Jugel[7] are

Table 1.   Thermal Conductivity of Aviation Turbine Fuel JP-4[a]

Temperature °C	λ, W/m K		
	Our Measurements		NEL[10]
	0.5 MPa	6 MPa	Saturation Pressure
25.3	0.1178	0.1209	0.119
60.7	0.1130	0.1162	0.111
100.7	0.1022	0.1051	0.101
140.2	0.0888	0.0931	0.092

[a]Supplier: Deutsche Shell AG (batch no. 118/12/77).
Inspection data: Boiling range, 68°C to 227°C; density (15°C),
762.5 kg/m³ ; aromatics, 9% vol.; olefins, 0% vol.

available. They are in very good agreement with our measured values.

Table 1 represents the thermal conductivity of the aviation
turbine fuel JP-4 investigated at 0.5 MPa and 6 MPa in the temperature
range 25°C to 140°C. Since JP-4 is a poorly absorbing liquid, radiation heat transfer at 140°C amounts to 15% of the effective heat
transfer. The accuracy of our values is within ±0.5% at 25°C and
within ±2% at 140°C. Results of an empirical prediction method
available for refined petroleum products are also given in the table.
The prediction method that was developed by NEL, the National Engineering Laboratory in Scotland,[10] is based on two of the liquid-de-
fining properties, e.g., the density and the average boiling point.
The comparison demonstrates that there is a good agreement between
the experimentally determined values and those predicted by the
NEL-method.

Last, the liquid refrigerant R-113 (purity >99.7%) was investigated. The results, which are related to the saturation pressure
and to the temperature range -15°C to 200°C, are given by the
equation

$$\lambda = (0.0802 - 205 \times 10^{-6} \frac{t}{°C}) \ \frac{W}{m\ K}.$$

The accuracy of the measurements is within ±0.5% to ±3%, depending
on the extent of the thermal radiation. Since R-113 is also a poorly
absorbing liquid, a radiation heat flux of up to 40% of the total
heat flux had to be eliminated. Again, our measurements can be partly
compared with data already available. In the range -30°C to 60°C
these data can be taken from Tauscher[11] and up to 140°C from
Slyusarev.[12] The comparison of all measurements gives fair agreement.

REFERENCES

1. M. Heckle, Konvektionsfreie Messung der Wärmeleitfähigkeit von
    Flüssigkeiten, Chem. Ing. Techn. 41:757-762 (1969).
2. M. Kohler, Einfluß der Strahlung auf den Wärmetransport durch
    eine Flüssigkeitsschicht, Z. Angew. Phys. 18:356-361 (1961).
3. G. Schödel and U. Grigull, Kombinierte Wärmeleitung und Wärme-
    strahlung in Flüssigkeiten, in "Proceedings of the Fourth Inter-
    national Heat Transfer Conference," Vol.3, R 2.2, Paris (1970),
    pp. 1-11.
4. R. Braun, Experimentelle Bestimmung der Wärmeleitfähigkeit von
    Flüssigkeiten mit einem konvektionsfreien Verfahren unter Be-
    rücksichtigung der thermischen Strahlung, Dissertation,
    Universität-Gesamthochschule Siegen, FRG (1980).
5. H. Poltz and R. Jugel, The thermal conductivity of liquids-IV.
    Temperature dependence of thermal conductivity, Int. J. Heat
    Mass Transfer 8:609-620 (1965).
6. V.Z. Geller, I.A. Paramonov, and V.V. Slyusarev, Experimental
    study concerning the contribution of the radiative component to
    the effective thermal conductivity of toluene, J. Eng. Phys.
    1974, 733-737.
7. H. Poltz and R. Jugel, Wärmeleitfähigkeit von Alkoholen, Wärme
    Stoffübertrag., 1:197-201 (1968).
8. B. Schrader and W. Meier, "DMS Raman/IR Atlas," Verlag Chemie,
    Weinheim (1974).
9. E. Merck, "Uvasole, Lösungsmittel und Substanzen für die
    Spektroskopie," E. Merck, Darmstadt (1977).
10. D.T. Jamieson, J.B. Irving, and J.S. Tudhope, "Prediction of
    the Thermal Conductivity of Petroleum and Petroleum Products,"
    National Engineering Laboratory Report 603, Glasgow (1975).
11. W. Tauscher, Measurements of the thermal conductivity of liquid
    refrigerants by an unsteady-state hot wire method, ASHRAE J.,
    11:97-104 (1969).
12. V.V. Slyusarev, Experimental study of thermal conductivity of
    freons and toluene (in Russian), Teplofiz. Svoistva Veshchestv.
    Mater., 11:11-19 (1977).

AUTOMATED THERMAL CONDUCTIVITY MEASUREMENTS

Virginia L. Morris and Chuck Haverlah

McDonnell-Douglas Astronautics Company
5301 Bolsa Ave.
Huntington Beach, CA 92647

ABSTRACT

    Thermal conductivity is a material property of great importance
where any transfer of heat is involved. In recent years, because
of the increase in new materials technology and the increasing
temperature ranges of operation, there has been a corresponding
increase of interest in the measurement of this property. A standard
method for measuring the thermal conductivity of rigid and semi-
rigid materials from ambient to 1000°C in the range of 0.01 to
200 W·m-1·°C-1 uses a cut-bar thermal comparator, an instrument
that employs a secondary technique whereby the thermal conductivity
of the materials of interest is evaluated versus two standards from
the National Bureau of Standards. The use of a commercially available thermal comparator is an expensive technique that requires
considerable operator attention. The accuracy of the method is
operator dependent and data handling is arduous.

    In an effort to reduce the cost and time required for thermal
conductivity measurements and to improve the accuracy of the method,
a Hewlett Packard 9825A minicomputer and a Hewlett Packard 6940B
multiprogrammer were interfaced to a Dynatech TCFCM-N20-R cut-bar
thermal comparator. All hardware items are commercially available
at a moderate cost. The software that was developed allows for
instrument control; data acquisition, reduction, and evaluation;
and error analysis of results. The software package also contains
features designed to protect the thermal comparator from damage
from thermal runaway, thus preventing expensive sample material
damage.

INTRODUCTION

Thermal conductivity, k, is measured at temperature equilibrium by periodic temperature measurements across a sample through which a known heat flux is passing. When measurements at several successive time intervals are the same, it is possible to calculate k. The thermal comparator method, which consists of sandwiching a test specimen between two materials of known k, heating the entire stack to equilibrium, and calculating the specimen conductivity using the assumption of one-dimensional heat flow, is recommended for materials that have a thermal conductivity that lies between 0.01 to 200 W·m-1·°C-1. This includes plastics, glass, ceramics, and some metals and alloys.

The thermal comparator system requires long periods of time for the sample to establish equilibrium at each desired temperature; or alternatively, considerable operator skill and attention to approach the desired temperature in the proper manner to minimize overshoot and cycling of the furnace. This paper describes a system that has been developed to turn the tedious process over to a minicomputer requiring only minimal operator attention.

PRINCIPLE OF MEASUREMENT

The physics of heat conduction is concerned with the description of the mechanisms, on a molecular scale, by which thermal energy is conducted through a material. The principle of one-dimensional steady state heat flux is employed in the cut-bar thermal comparator method by comparing the measured heat flux of the sample to the heat flux of two standard materials of known thermal conductivity. This method uses two reference standards that are placed on either side of the sample material. The sample and standards, which are 5.08-cm-diameter discs, are in turn held between a heater and a liquid-cooled heat sink. For high temperature operation, a cold surface (auxiliary) heater is placed between the lower reference standard and the heat sink.

To avoid radial heat loss from the sample stack, the surrounding temperature must be controlled to match the profile in the test stack to produce one-dimensional heat flow down the test section. The heaters in the guard tube are adjusted to match its temperature profiles with the test section.

Under steady-state conditions, values of the one-dimensional heat flow through the sample and the two reference standards are equal:

$$Q/A_{sample} = Q/A_{top\ standard} = Q/A_{bottom\ standard} \qquad (1)$$

This one-dimensional steady-state flux of heat that is transferred through a given sample is proportional to the temperature gradient and to k, the thermal conductivity:

$$Q/A = -k\frac{dt}{dx} \qquad (2)$$

where Q is the heat transferred per unit time through unit area A, taken perpendicular to the direction where the temperature changes most rapidly (x); and dT/dx is the rate of change of temperature with unit distance. Substituting Equation 2 into Equation 1 produces the following relationship:

$$k_s\frac{dT}{dx} = k_{TS}\frac{dT}{dx} = k_{BS}\frac{dT}{dx} \qquad (3)$$

Equation 3 can then be rearranged to solve for k_s:

$$k_s = 0.5\left(\frac{dx}{dT}\right)_s\left[k_{TS}\left(\frac{dT}{dx}\right)_{TS} + k_{BS}\left(\frac{dT}{dx}\right)_{BS}\right] \qquad (4)$$

where k_s, k_{TS}, and k_{BS} are, respectively, the thermal conductivities of the sample, the top reference standard, and the bottom reference standard (W·m-1·°C-1); Δx is the sample thickness (cm); ΔT is the temperature differential across the stack (°C); and TS and BS, respectively, are used to designate the top and bottom reference standards.

INSTRUMENTATION

The MDAC-HB thermal conductivity system consists of a Dynatech Model TCFCM N20-R thermal comparator, a Hewlett Packard (HP) 9825A minicomputer, and an HP 6940B multiprogrammer. The coolant for the heat sink and the guard shroud is 20°C water, which is circulated through an Aminco 190 1 constant temperature bath. The operating temperatures of the four heat sources (the main, auxiliary, lower- and upper-guard heaters) are monitored on a Soltec Model 3316 recorder.

The HP computer system consists of the microprocessor and memory, a tape drive, a thermal printer, and a real time clock. The test system has been configured for an 8-channel low-level analog-to-digital converter, a 4-channel digital-to-analog converter, and 12 bits of digital input and output. The magnetic tape is used for program and data storage.

The testing system has four temperature controllers (Power Element Trigger types) for powering the heaters. The set point

of these controllers can be adjusted locally from the front panel
or remotely from one of the digital-to-analog channels of the HP
system. A thermocouple located near the heater being powered
by a temperature controller is used for feedback to that controller.

Output from the six monitoring thermocouples in the sample
stack is fed into the multiprogrammer, which converts the thermocouple output to binary temperature data for the microprocessor
and prints the temperature at 1 min. intervals. The microprocessor compares the temperature readings with the previous
measurement to determine when the change is small enough ($\pm0.25°C$)
for the system to be essentially at temperature equilibrium. When
temperature equilibrium is achieved, the microprocessor calculates
the ratio of the heat flux of the top and bottom standard. When
this ratio is equal to 1.00 ± 0.05 (or operator specified accuracy
limit), the microprocessor will then calculate the thermal conductivity, send it to the printer, and increase the test temperature
one operator-specified increment. This process is repeated until
the maximum temperature desired is reached. When the last
temperature point is read, the microprocessor returns the controllers
to a set point below room temperature and signals the run is
complete.

All thermocouples used in the thermal comparator are Type K
(Chromel-Alumel). The temperatures of the six monitoring thermocouples are calculated from their millivolt outputs by Equation 5.[1]

$$T = -0.54 + 24.2486E + 0.141221E^2 - 0.0141736E^3 \\ + 4.83584 \times 10^{-4} E^4 - 7.01963 \times 10^{-6} E^5 + 3.90673 \quad (5) \\ \times 10^{-8} E^6$$

After the temperatures are calculated from the thermocouple data,
the k's of the top and bottom reference standards are calculated
from a polynomial derived from data obtained from the Thermophysical Properties Research Center (TPRC).[2]

Because some sample materials are expensive and, on occasion,
only one sample is available for analysis, additional software
was generated to automatically abort the test if a system error
is detected. Conditions that will initiate an aborted test are
(1) a system imbalance that will not allow equilibrium conditions
to be met within 90 min., (2) a broken thermocouple during the
test run, (3) data not being recorded on tape, and (4) the printer
running out of paper.

The thermal comparator system is equipped with a vacuum bell
jar to allow tests to be run under a variety of environments, e.g.,
nitrogen, vacuum, or air.

ACCURACY

The accuracy of the results is determined by several factors. Use of the computer system has reduced errors in converting milli-volts to temperature to less than 0.1 percent for temperature differences and 0.5°C in actual temperature. The polynomial form used for calculating the thermal conductivities of the reference standards reproduces the values recommended by TPRC to within 0.1 percent.

The major errors involved will be caused by improper choice of reference materials, poor surface preparation of the samples, and uncalibrated thermocouples. Off-the-shelf thermocouples will provide accuracy to within 1° of actual temperature with excellent repeatability.

Lack of surface smoothness and flatness produces errors that change with temperature. This type of error can be reduced to less than 1 percent by smoothing the sample measuring faces to a 32 rms finish and 0.005 cm. parallelism. On very thin samples the depth of the grooves cut in the sample faces (0.025 cm.) can produce an indeterminate thickness up to 10 percent. Consequently, it is necessary to have samples which are at least 0.64 cm. thick.

The greatest error in most experiments is caused by sample inhomogeneity. Samples based on layered materials vary widely, depending on the sampling procedure used. In this respect, a thicker sample reduces the effect of inhomogeneity.

Another possible source of error in determining k is the ability of a material to expand as a function of temperature. The test stack is maintained under a constant, light loading force. If the material expands significantly, it could be compressed and, depending on the degree of compression, the conductive property of the material could be changed.

VERIFICATION OF RESULTS

Confirmational analyses were conducted to verify the validity of a "quasi-equilibrium" approach to thermal conductivity measure-ments. Verification consisted of measuring the thermal conductivity of a known National Bureau of Standards (NBS) standard, Pyrex 7740, versus two Pyrex 7740 standards as references.

Further confirmational analyses consisted of measuring the thermal conductivities of several elastomeric materials using the cut-bar comparator method and obtaining correlational results, using the guarded hot plate method, from different organizations.

Pyrex 7740

Three Pyrex 7740 standards were prepared and the thermal conductivity of one standard was evaluated versus the known conductivities of the other two standards. The standard in the sample
measuring position was rotated to a reference position and the
reference standard was then measured. This rotation continued
until all three standards were evaluated. These results, plotted
versus the published NBS thermal conductivity values, are shown
in Figure 1. The measured thermal conductivities were an average
2.2 percent higher than the NBS reported results with a correlation
coefficient, calculated using Equation 6, of 0.97 and a covariance
of 2.36 x 10^{-3}. The covariance was calculated using Equation 7.

$$\gamma_{xy} = \frac{S_{xy}}{S_x\, S_y} \tag{6}$$

$$S_{xy} = \frac{1}{n-1}\left(\Sigma\, X_i Y_i - 1/n\, \Sigma\, X_i \Sigma Y_i\right) \tag{7}$$

where S_x is the standard deviation of x, S_y is the standard
deviation of y, x,y are the grouped data points, and n is the
number of x,y pairs.

The reproducibility between the three separate Pyrex runs
was ± 1.1 percent. An additional check on the agreement of the
measured values versus the NBS values is a comparison of the line
slopes and intercepts. The average of the three line slopes for
the measured values agreed to within 0.16 ppt and the intercepts
agreed to within 0.4 percent. The determinational coefficient
for the plot of the measured values versus temperature is 0.954
and for the plot of the NBS values versus temperature, 0.999.

Elastomer Analysis

The cut-bar comparator has been used primarily for the analysis
of more rigid materials, e.g., ceramic, metal, and plastic.
Elastomeric materials used for aerospace insulation on flight
vehicles have, in the past, typically been analyzed on guarded hot
plate instruments that require 20 to 23 cm. diameter samples. The
expense of the insulating materials and the lack of sufficient
sample material to construct such a large specimen prompted an
analysis of the applicability of the cut-bar comparator method to
elastomeric materials.

Duplicate specimens of VAMAC 25 and VAMAC 32, where VAMAC is
an ethylene-acrylic copolymer produced by DuPont, were prepared and
analyzed using the comparator method and the guarded hot plate

method. Dynatech Corporation, ran the analyses on the guarded hot plate.[3] The two materials differ in the types and amounts of processing aids used in formulation and the amount of carbon black filler used. The correlation of results for VAMAC 25, whose density is 1.31 g/cm^3, shows a deviation of the line slopes of the k versus temperature plots to be 0.22 ppt with an 18 percent deviation in the y intercepts (Figure 2). The covariance between the guarded hot plate and the cut-bar comparator values was calculated to be 2.5 x 10^{-4} and the correlation coefficient was 1.00.

Figure 3 shows the correlation of results for VAMAC 32, whose density is 1.12 g/cm^3. The covariance between the thermal conductivities analyzed by the guarded hot plate method and the cut-bar comparator method was 2.86 x 10^{-4} and the correlation coefficient was 1.00. The deviation in the slopes of the line plots of thermal conductivity versus temperature is 0.13 ppt and the intercepts have an 8 percent difference.

The next series of analyses were conducted on a RTV 560 methyl-phenyl silicone elastomer produced by General Electric Corporation. The guarded hot plate analysis on this material was completed by SAMSO (Space and Missile Service Organization).[4] The density of this material is 1.43 g/cm.[3] In this case, the covariance was reported as 2.43 x 10^{-4}, and the correlation coefficient was 0.87. The line plot deviation is 0.16 ppt, and the y intercepts differed by 8 percent. The results of their analysis are shown in Figure 4.

CONCLUSIONS

The computerization of the cut-bar thermal comparator has significantly reduced the cost of the analysis of this material property. Subsequently, data can be generated in a more efficient time frame, thus expediting programs that require this data. The accuracy of the technique has been increased by removing from the operator the decision, at each data point, of determining if the material has reached steady-state thermal equilibrium.

However, it should be noted that computerization of these types of instruments does not remove from the operator the responsibility of configuring the test conditions for optimized analytical parameters. Operator specified inputs consist of: (1) optimizing the temperature delta across the sample; (2) establishing the heat flow accuracy limits; (3) establishing criteria for the length of time the sample must meet steady-state thermal equilibrium condition; (4) determining the temperature requirements the monitoring thermocouples must meet, e.g., ± 0.25°C, to establish thermal equilibrium; and (5) setting the size of the temperature increment for each step. The operator should evaluate parameters

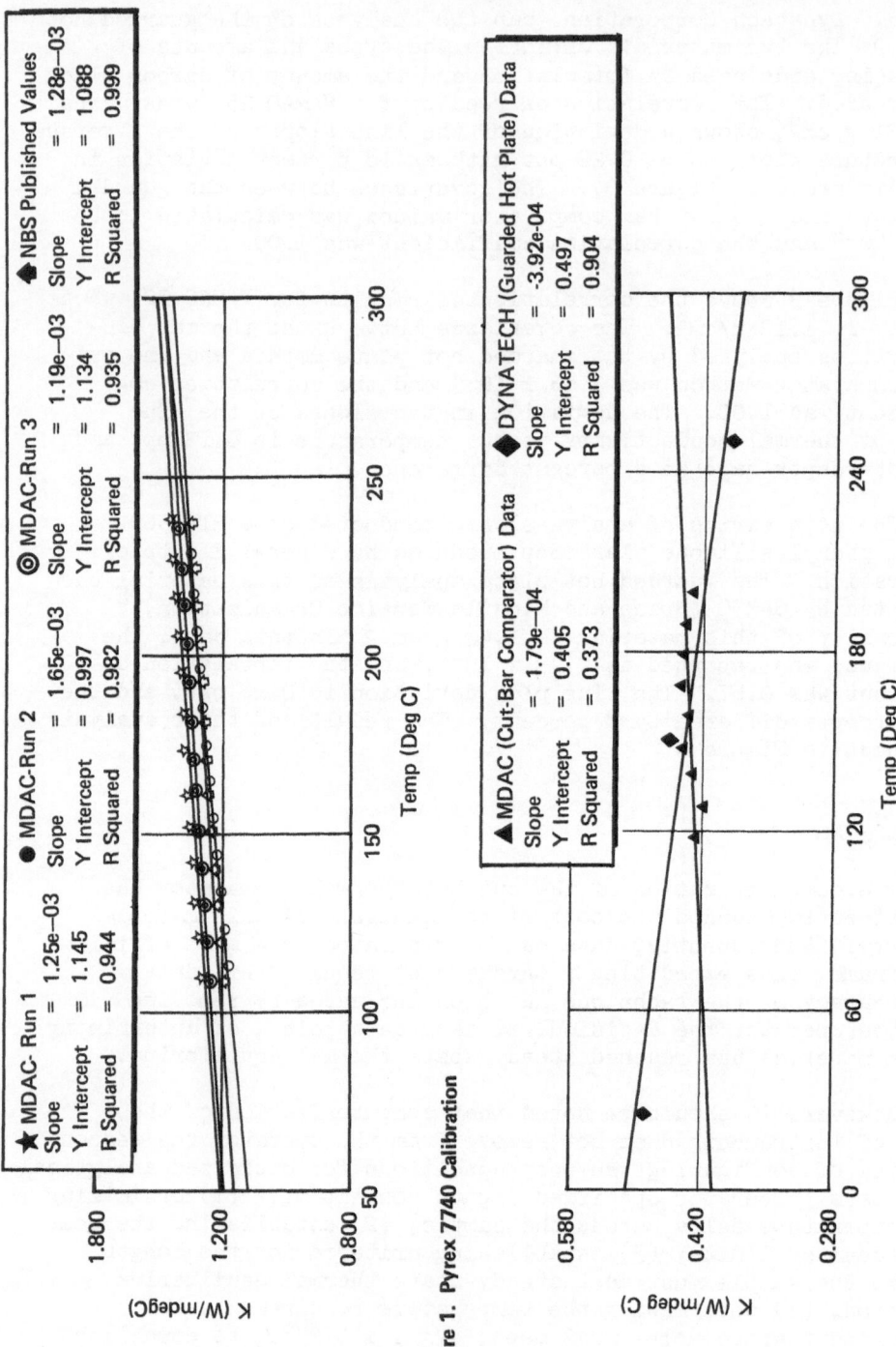

Figure 1. Pyrex 7740 Calibration

Figure 2. VAMAC 25

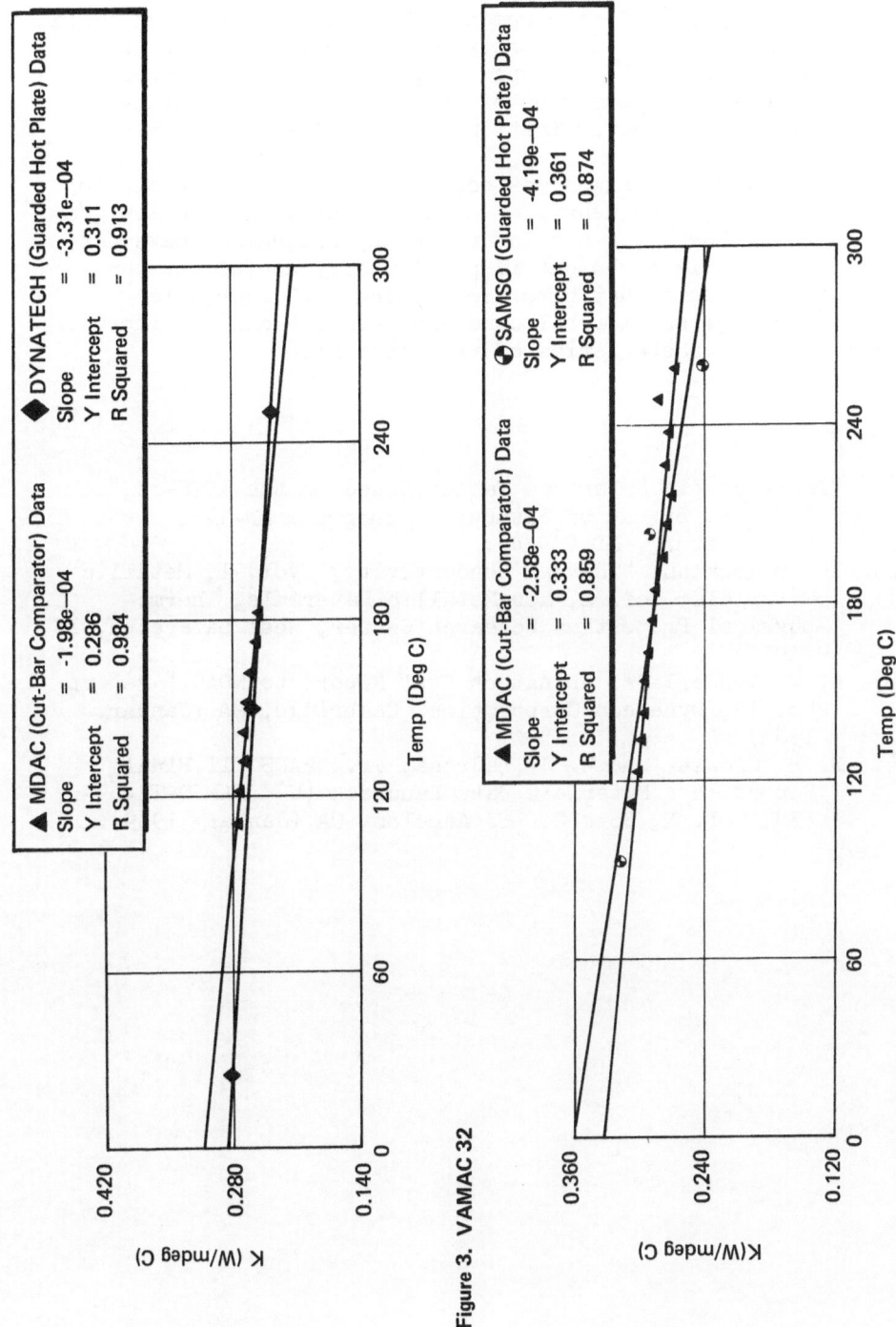

Figure 3. VAMAC 32

Figure 4. RTV 560 Methylphenyl Silicone

that are printed at the end of each temperature increment to
ascertain manually if all requirements for steady-state conditions
are as expected. Mechanical aspects of the test for which the
operator is responsible are proper preparation of the standard
reference materials, preparation of the sample, correct stack
assembly, and accurate determination of Δx, the sample thickness.

The comparator technique has been shown to be applicable to
elastomeric systems with densities of the order of 1.1 to 1.9 g/cm.^{3}
As of this date, the upper and lower density boundaries have not
been evaluated. This technique is particularly useful for the
analyses of elastomers that have been filled with inorganic
materials (e.g., glass microballoons and carbon black) to improve
the performance characteristics of the material.

REFERENCES

1. "Thermocouple Reference Tables Based on the IPTS-68,"
    National Bureau of Standards Monograph MN-125,
    Gaithersburg, MD (1974).
2. Y. Touloukian, "Thermal Conductivity," Vol. 1, Metallic
    Materials; Vol. 2, Non-Metallic Materials, Thermo-
    physical Properties Research Center, West Lafayette, IN
    (1970).
3. A. O. Desjarlais, "Dynatech Test Report to MDAC," Report
    No. 13, Dynatech Corporation, Cambridge, MA (January
    1980).
4. D. P. Crowley and W. J. Ladroga, Jr., RADS III Final
    Report in a Materials Data Handbook (U), SAMSO-TR-69-
    158, Vol. X, Book 2, Los Angeles, CA (January 1969).

AN EVACUATED, LOAD-BEARING POWDER INSULATION FOR A HIGH TEMPERATURE NA/S BATTERY

Harald Reiss

Brown, Boveri & Cie AG
Central Research Laboratory
D-6900 Heidelberg, FRG

ABSTRACT

The development of load-bearing thermal insulations is of increasing importance for the thermal management of high temperature batteries. In order to maintain the energy density as high as possible and to extend the idle periods of the battery to more than 12 to 24 hrs, the Na/S battery requires a thermal insulation with a thermal conductivity less than 10 mW/(m·K). Thermal insulations with such low thermal conductivities can be achieved in vacuum only. The solution to the problem is investigated in three steps: a) selection of an appropriate opacifier, b) allowance for higher residual gas pressures compared with those necessary for evacuated foil insulations, c) preparation of load-bearing insulation plates.

Experimental data are given for the total thermal conductivity as well as for the solid conduction and radiation parts of a mixture of TiO_2 and Fe_3O_4 at different mixture ratios and porosities, and of load-bearing insulation plates prepared from a mixture of fumed silica, TiO_2, Fe_3O_4 and glass fibers, in vacuum and under mechanical load. These insulations can be operated at temperatures above 300 °C at residual gas pressures between 10 and 1000 Pa, which is favourable for the long term stability of the heat shield preventing the battery from cooling. The insulation plates even if highly compacted are shown to be of low density and of high porosity, which is essential for a reduction of the solid conduction part of the total thermal conductivity.

INTRODUCTION

The following report deals with the development of
the thermal insulation for a high temperature Na/S battery. The Na/S battery is designed for application in
electric vehicles. The battery consists of 480 Na/S cells
with an energy density of about 120 Wh/kg, which is roughly
four times higher than the energy density of a conventional lead and sulphuric acid accumulator. The mean temperature of the Na/S battery cells must be maintained above
290°C, since the Na_2S_x polysulfides can be reduced to Na
and S in their fluid state only, and the Na-ion conductivity of the solid ceramic electrolyte decreases at lower
temperatures. Since the idle periods of the electric vehicle can extend to 12 to 24 h, an efficient thermal insulation is required that should avoid external heating
from the stored electric energy. Owing to severe restrictions on the space available in the vehicle and to maintain the energy and power densities (with respect to
volume and weight) as high as possible, the thickness of
the thermal insulation should not exceed 3.5 cm. The results of a computer simulation for the time dependence
of the mean cell temperature for the Na/S battery in its
present state of design restrict the thermal conductivity
of the insulation to values of $\lambda < 8$ mW/(m·K), which can
be achieved only in evacuated devices[1]. As a consequence,
the residual gas pressure, p, of an evacuated insulation
must be kept low during the whole lifetime of the battery
(5 years expected). It would be desirable to operate
within the range $100 \leq p \leq 1000$ Pa.

It is well known that lowest thermal conductivities
(λ below 1 mW/(m·K) at p < 1 Pa) can be achieved with
evacuated foil insulations also at high temperatures. A
measurement of the total heat flow, $\dot{Q}/A$, at $T = 320^{\circ}$C
through a multilayer insulation of 42 Al foils with glass
silk spacers in vacuum is reported in ref.2. Data are given
as a function of the residual gas pressure for air, krypton and xenon. At high gas pressures (p > 10^4 Pa), the
observed $\dot{Q}/A$ of about 160 W/m^2 for the xenon filling corresponds to $\lambda \sim 11$ mW/(m·K), which is approximately (with
some additional convection) the λ for the gas (static).
Since this value is very large compared with the result
of the computer simulation, the gas pressure would have
to be reduced below 1 Pa. For low gas pressures, however,
multilayer insulations are, in a flat plate design, not
self-supporting.

Evacuated powders can replace evacuated foils if

they fulfill three conditions: a) restrict IR-radiation transport to small amounts (as the foils do in the multi-layer concept), b) allow higher residual gas pressures, and c) withstand atmospheric pressure and prevent deformation of the evacuated insulation vessel. Experimental data will be presented with respect to these three conditions.

SELECTION OF AN OPACIFIER

For the radiation transfer through a gray, absorbing, emitting, and isotropically scattering medium between two parallel, gray, diffuse radiating, isothermal plates with temperatures T_1 and T_2, the net radiation flux, $\dot{Q}_{Rad}$, normal to the surfaces can be written as a solution of the equation of transfer as

$$\dot{Q}_{Rad} = \frac{\sigma \cdot \psi \cdot (T_1^4 - T_2^4)}{\{1 + \psi \cdot (1/\varepsilon_1 + 1/\varepsilon_2 - 2)\}} \tag{1}$$

where σ denotes the radiation constant and ε_1 and ε_2 the emission coefficients of the plates. The function ψ contains exponential integral functions and a dimensionless temperature distribution (for details see ref. 3). For large values of the optical density, τ_o, ψ is approximated by $(4/3)/(1.42 + \tau_o)$. The τ_o values of the powders studied here amount to values ≥ 100 (see below). With these τ_o values and $\varepsilon_1 = \varepsilon_2 \sim 0.6$ for an unpolished stainless steel wall of the insulation vessel, the term containing ψ and the ε in the denominator of eq. (1) can be neglected (the physical meaning of neglecting this term is that the wall radiation is soon absorbed in an optically thick medium). Equation (1) can be rewritten as

$$\dot{Q}_{Rad} = \frac{4 \cdot \sigma \cdot (T_1^4 - T_2^4)}{3 \cdot \tau_o} \tag{2}$$

The gray approximation is applicable since the optical constants of the mixtures of opacifiers applied here depend only weakly on the wavelength (at least in that region where the maximum of the radiation occurs). The assumption of isotropic scattering seems valid in view of the small values of the scattering parameter[4].

It will be supposed in the following that the medium has a solid conduction component, λ_{SC}, of the total λ, that is independent of the temperature. For an optimiza-

tion of the extinction properties of the powder insulation and a reduction of λ_{SC}, it is necessary that the
different heat conduction modes can be studied separately.
Therefore, it would be desirable if the total heat flow,
$\dot{Q}$, could be written as the algebraic sum

$$\dot{Q} = \dot{Q}_{SC} + \dot{Q}_{Gas} + \dot{Q}_{Rad} \tag{3}$$

where in vacuum $\dot{Q}_{Gas} \sim 0$. The energy balance requires for
the stationary state of the evacuated insulation that

$$\nabla \cdot (\dot{Q}_{SC} + \dot{Q}_{Rad}) = 0 \tag{4}$$

From a solution of the equation of transfer and eq.
(4) it follows that $\dot{Q}$ can be written with $\dot{Q}_{SC}$ and $\dot{Q}_{Rad}$
exactly separated as in eq. (3) only if the medium is
purely scattering, that is, if the albedo $\omega = S/E = 1$,
$S = E$ and $A = (1 - \omega) \cdot E = 0$, where S, A, and E denote
the scattering, absorption, and total extinction parameters, respectively, or, in an approximation, if the total
extinction, E, is very large. For small E and $\omega < 1$, the
local solid conduction interaction interferes with the
long-range radiation transport process, and $\dot{Q}_{SC}$ is no
longer separable from Q_{Rad} (see ref. 5).

For large A compared with S, the separation of $\dot{Q}_{SC}$
and $\dot{Q}_{Rad}$ in eq. (3) remains valid if the total extinction of the powders is high enough that the conduction/
radiation parameter

$$\tilde{N} = \frac{\lambda_{SC}}{\lambda_{Rad}} = \frac{3 \cdot \lambda_{SC} \cdot E}{4 \cdot \sigma \cdot T^3} \tag{5}$$

is large. In this case, the heat flow is essentially that
of conduction, and the temperature distribution is linear
within the medium, which would be also observed if $\omega = 1$.

It is therefore desirable to secure a high scattering parameter S. As a consequence, opacifiers should be
selected as follows:

a) Use a mixture of opacifiers that is composed of
   absorbing and scattering components

b) Select powders with a high real part and a small
   imaginary part of the complex index of refraction
   and with the lowest thermal conductivity of the
   solid material

c) At least one component of the mixture should have a particle diameter that is of the order of the wavelength Λ of the incident radiation so that the Mie extinction function shows one of the well known resonances[6]. For spherical particles and a refractive index n = 2 (real part), $d_{opt} = \Lambda/2$ defines an optimum particle diameter (see ref. 7): For T = 600 K, $\Lambda_{max} = 5$ µm, $d_{opt} = 2$ to 3 µm.

d) The smaller the particle diameter, the smaller is λ_{SC} and, at fixed residual gas pressure, λ_{Gas}.

e) The smaller λ_{SC}, the higher must be the total extinction, E, so that the conduction/radiation parameter, N, is large, and eqs. (2) and (3) are applicable.

The powders applied here have been selected with respect to these conditions. With $\lambda_{SC} = 5$ mW/(m·K), $T \leq 600$ K and $E \sim 75$ cm^{-1}, we have $N \sim 3$.

With $\dot{Q}_{Gas} = 0$, eq. (3) can be rewritten

$$\dot{Q} = \frac{\lambda_{SC} \cdot (T_1 - T_2)}{L} + \frac{4 \cdot \sigma \cdot (T_1^4 - T_2^4)}{3 \cdot \tau_o} \tag{6}$$

and

$$\lambda = \lambda_{SC} + \frac{4 \cdot \sigma}{3 \cdot E} \cdot (T_1^2 + T_2^2) \cdot (T_1 + T_2) \tag{7}$$

$$= \lambda_{SC} + \frac{4 \cdot \sigma}{3 \cdot E} \cdot T^{*3} \tag{8}$$

where L is the total thickness of the insulation, and T^{*3} denotes a radiation temperature. If λ is plotted versus T^{*3}, the solid conduction part, λ_{SC}, and the total extinction, $E = \tau_o/L$, can be extracted from the intercept and the slope of the function $\lambda = \alpha + \beta \cdot T^{*3}$ (see Fig. 1). α and β should be as small as possible. $\beta = 0$ indicates vanishing radiation flow.

Fig. 1 shows experimental data of λ as a function of T^{*3} for the opacifier TiO_2 and mixtures of TiO_2 with other opacifiers (the curves are least-squares fits to the data). The measurements were performed with the same experimental device as that used for the multilayer insulation[2]. These data are therefore obtained without an external

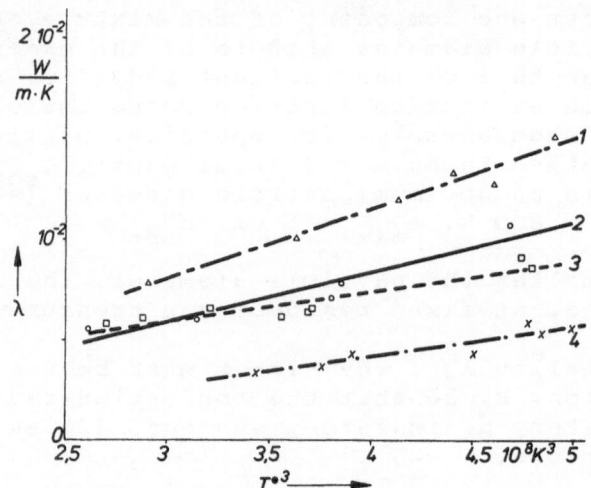

Fig. 1   Thermal conductivity, λ, of evacuated ($p < 100$ Pa) low density powders as a function of the radiation temperature T^{*3}. The interval $2.5 \cdot 10^8 \le T^{*3} \le 5 \cdot 10^8$ K^3 corresponds to $200 \le T_1 \le 380°C$ using $T_2 \equiv 30°C$. Curve 1: TiO_2; curve 2: $Al_2O_3 + ZrO_2$, (1:1 weight parts); curve 3: $TiO_2 + Cr_2O_3$, (1:1 weight parts); curve 4: $TiO_2 + Fe_3O_4$, (2:1 weight parts). For the porosity, π, powder density, ρ, and particle diameter, d, see ref. 2, Fig. 2.

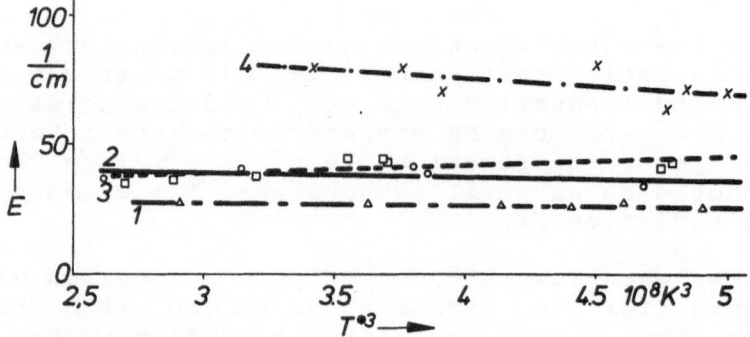

Fig. 2   Total extinction coefficients, E, of the evacuated low density powders of Fig. 1, calculated using eq. (8).

load and at low powder densities. The total extinction coefficients, E, that can be extracted according to eq. (8) are between 25 (curve 1, Fig. 2) and 75 cm^{-1} (curve 4). The E of curve 4 leads, with L = 3.5 cm, to τ_o > 100.

The relative amount of the absorption and scattering parameters, A and S, can be determined by optical methods. For this purpose, we omit the remission term in the equation of transfer and use the method of discrete ordinates (see ref. 8) to transform the equation of transfer into a system of differential equations. As has been shown in ref. 9 using the Markov formula for the substitution of the scattering integral, expressions for the transmission and reflection coefficients can be derived that are of higher accuracy than those obtainable from the Schuster-Schwarzschild approximation. A measurement of reflexion coefficients, R_∞, using samples of infinite optical thickness, τ_o, leads to a simple determination of the albedo ω (see Fig. 3) if the emission term in the source function is experimentally compensated. The experimental method will be described in ref. 10.

Preliminary experimental results for ω at temperatures of about 600 K show for TiO_2 and Fe_3O_4 $\omega \sim 0.1$. This result is not explained by the small particle diameters of TiO_2 (~ 0.03 µm) and Fe_3O_4 (~ 0.2 µm) because we observe, by X-ray fluorescence of $K\alpha$ levels of Ti and Fe in mixtures of these powders, clouds of Fe_3O_4 with diameters between 2 and 6 µm, that is, in the region of the optimum particle diameter (for details, see ref. 10). In order to increase ω, we need at least one additional opacifier.

Application of Fe_3O_4 as opacifier has the advantage that this powder is oxidized to Fe_2O_3 if it is heated. This process consumes the residual O_2 pressure in the evacuated insulation (there will be 2 - 3 kg of the fine Fe_3O_4 powder available in the insulation vessel of $\sim$ 90 liters total volume).

RESIDUAL GAS PRESSURE

Fig. 4 shows experimental results for the total heat flow, $\dot{Q}/A$, for different powders with particle diameters ranging from 2000 to 3000 µm to 0.03 µm as a function of the residual gas pressure, p, of xenon. The curve indicated with 0.03 µm corresponds to the low density mixture of two weight parts TiO_2 with one weight part Fe_3O_4 as mentioned in Figs. 1 and 2 (curve 4). The data were obtained with the same experimental device as that used for

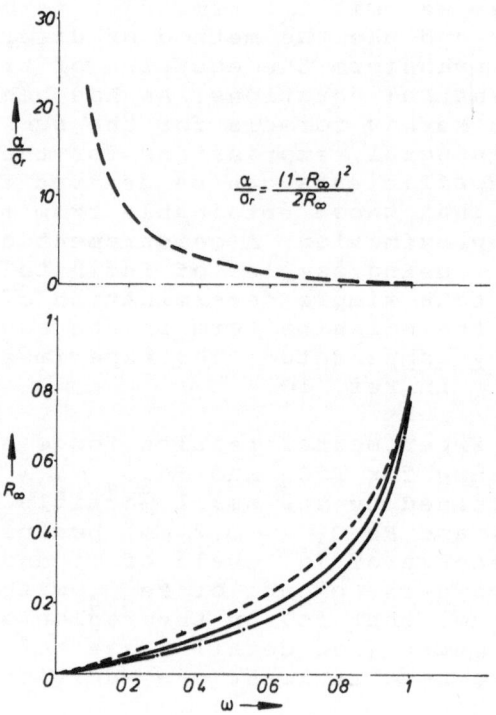

Fig. 3   Coefficients of backscattering, α/σ_r, and reflection coefficients, R_∞, of layers with infinite
optical thickness, τ_0, calculated using different
approximate solutions of the equation of transfer.
In the lower diagram, dashed lines denote results
according to the two-flux model (Schuster-Schwarzschild), full lines and dashed-dotted lines: Discrete ordinate model using Markov integration
formula with 3 zeros for diffuse or parallel incident radiation, respectively (see refs. 9, 10).

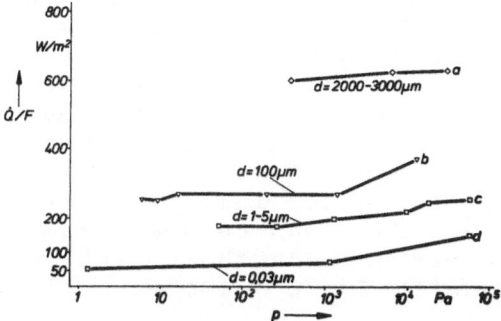

Fig. 4   Total heat flow, $\dot{Q}/\tilde{A}$, through evacuated powders
          as a function of the residual gas pressure, p
          (xenon). T_1 = 320 °C, T_2 = $T_{ambient}$, thickness
          of insulation is 2 cm. Data are given for Al_2O_3
          (mean particle diameter, d = 2000 to 3000 µm
          and 100 µm, respectively), ZrO_2 (1 to 5 µm), and
          the mixture of TiO_2 with Fe_3O_4 (2:1 weight ratio,
          where d = 0.03 µm is that of TiO_2).

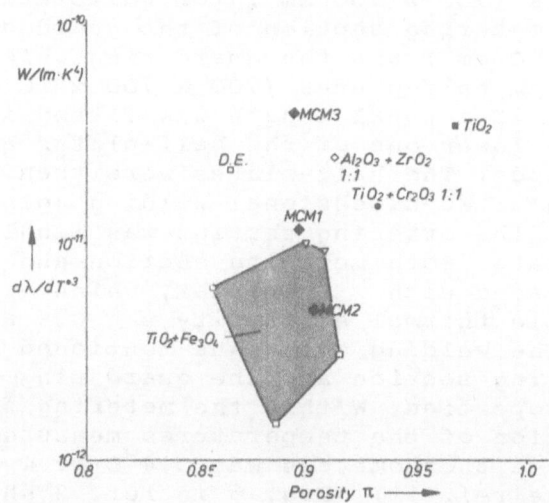

Fig. 5   Coefficients β = d λ/dT^{*3} in the relation λ = α +
          $\beta \cdot T^{*3}$ as a function of the porosity, π. The shaded
          area denotes the range of the β of least-squares
          fits obtained for the experimental λ of TiO_2 +
          Fe_3O_4 mixtures (weight ratios 1:1, 2:1 and 9:1).
          DE and MCM denote Diatomaceous Earth and pressed
          plates prepared from a multicomponent mixture
          (see Fig. 9), respectively. Residual gas pressure,
          p $\leq$ 10 Pa.

the study of the multilayer insulation [2]. With the mixture of TiO_2 and Fe_3O_4, a total heat flow of 50 to 70 W/m^2 between $100 \leq p \leq 1000$ Pa and with a thickness of 2 cm of the insulation is achieved (hot side wall at T = $320°C$, natural convection at the cold side wall). The low heat flow is favoured by large Knudsen numbers, Kn, which can be estimated for p = 100 Pa to Kn > 100 (see ref. 2). For these Kn, λ_{Gas} is rather insensitive to a change of the type of the gas. This can be seen from Fig. 1 in ref. 2 where the differences between the curves for krypton and xenon vanish gradually for p < 100 Pa.

INFLUENCE OF POWDER DENSITY ON λ

The third condition mentioned in the introduction calls for powders of increased powder density, ρ. Only a compacted powder can withstand the external load from the atmospheric pressure on the insulation vessel. For experimental studies of the dependence of λ on ρ, a flat vacuum chamber with a heat flow meter for time-saving tests of a variety of powder mixtures, and a large guarded hot plate device (700 x 700 mm^2) for calibration purposes were used. The metering section of the guarded hot plate device (500 x 500 mm^2) and the guard ring were prepared from two aluminum half-plates (700 x 700 x 10 mm^3). The heating coil of 20 m total length was fitted into grooves machined in the lower one of the half-plates and covered with aluminum rods. The half-plates were then explosion welded yielding a two-dimensional welding joint over the entire surface. The metering section was finally cut out of the whole plate. Both metering section and guard ring were plasma sprayed with TiO_2 powder, which gives the surfaces a stable thermal emissivity $\epsilon > 0.9$ also at high temperatures. The welding technique mentioned before provides the metering section and the guard ring with excellent thermal properties. Within the metering section, the standard deviation of the temperatures measured with 9 Pt-100 resistance thermometers was $0.4°C$ at T = $320°C$ (for details see ref. 11). Fig. 5 in ref. 2 shows a section of the hot plate with the welding joint.

Experimental data of λ for the mixture of TiO_2 with Fe_3O_4 using different mixture ratios and different densities were studied as a function of the radiation temperature, T^*. Fig. 5 shows the slope $\beta = d\lambda/dT^{*3}$ for the curves of these mixtures and for those of other powders for comparison (since, however, the mixtures of TiO_2 with Fe_3O_4 are of very low transparency, very small β result which can be extracted with limited accuracy only. There-

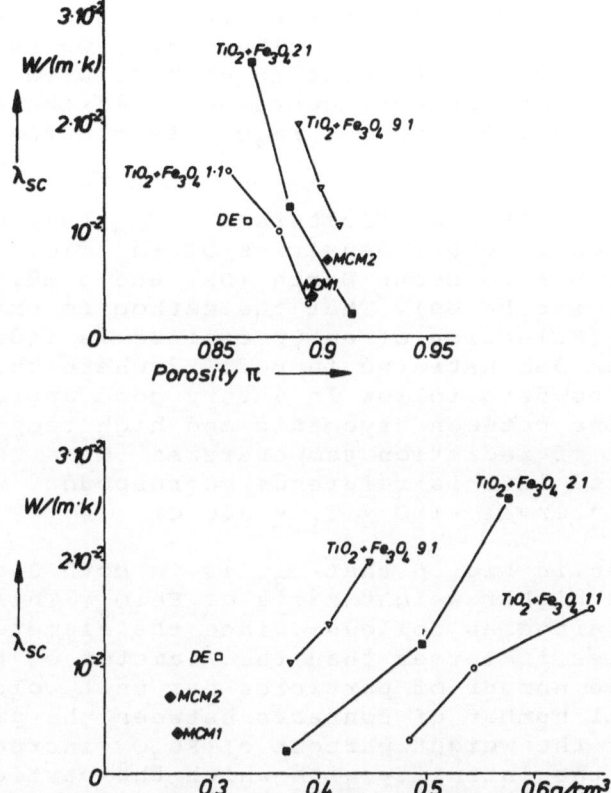

Fig. 6   Solid conduction part, $\alpha = \lambda_{SC}$, in the relation
$\lambda = \alpha + \beta \cdot T^3$ as a function of the porosity, π,
and powder density, ρ, for the mixture of TiO_2
+ Fe_3O_4 and, for comparison, Diatomaceous Earth
(DE) and pressed plates prepared from a multicomponent mixture (MCM) (see Fig. 9). The solid
lines are used to guide the eye. Residual gas
pressure, $p \leq 10$ Pa.

fore, only the region within which these values of β were
obtained is given in Fig. 5). All mixtures of TiO_2 and
Fe_3O_4 yield, for a porosity $\pi > 0.9$, the lowest β and are
therefore more efficient for extinction of IR-radiation
than the other given mixtures or pure materials. Measurements using ZrO_2 as a third component in the mixture of
TiO_2 with Fe_3O_4 are in preparation. For ZrO_2 particles
with $n \sim 2$ and a mean diameter between 1 and 2 μm, we
measured at $T = 600$ K $\omega \sim 0.9$. Therefore, we should observe large ω also in the mixture of TiO_2 with Fe_3O_4 and
ZrO_2 and, as a consequence, decreased β if the absorption
A, which is mainly due to the Fe_3O_4, is constant (see ref.
10).

Fig. 6 shows the coefficients $\alpha = \lambda_{SC}$ that were obtained for higher powder densities of the $TiO_2 + Fe_3O_4$
mixtures, for Diatomaceous Earth (DE) and a multicomponent
mixture (MCM, see below). That the method to extract λ_{SC}
described in "Selection of an Opacifier" is indeed applicable has been demonstrated in ref. 12 where the λ values
of different powders follow in a very good approximation
a straight line between cryogenic and high temperatures
(the interval of radiation temperatures $10^7 \leq T^{*3} \leq 5 \cdot 10^8$
K^3 investigated in this reference corresponds, with $T_2 \equiv$
$30\,^\circ C$, to the interval $-180 \leq T_1 \leq 380\,^\circ C$).

It is seen in Fig. 6 that λ_{SC} is in both diagrams
lower for the higher weight parts of Fe_3O_4. This result
could be explained as follows: Since the diameter of the
Fe_3O_4 particles is larger than the diameter of the TiO_2
particles, the number of particles per unit volume and
thus the total number of contacts between the particles
is reduced if the weight percent of Fe_3O_4 increases. If,
in addition, the intensity with which the particles contact each other is increased owing to an external load
(that is, at a higher powder density, ρ), the λ_{SC} should
increase with increasing ρ more slowly for those mixtures
where the weight percent of Fe_3O_4 is high. This is confirmed by Fig. 6: The slope $d\lambda_{SC}/d\rho$ is smaller for the
1:1 mixture than for the 2:1 or 9:1 mixtures of TiO_2 with
Fe_3O_4, where the TiO_2 weight percentage dominates. On the
other hand, this explanation is in contradiction to the
well known observation that λ_{SC} of single component powders decreases at low powder densities with decreasing
particle diameter. Further work is necessary to clarify
this point.

To withstand the external load without a loss of
volume of more than 10%, the powder density of the TiO_2
$+ Fe_3O_4$ mixture must be increased, as can be extracted

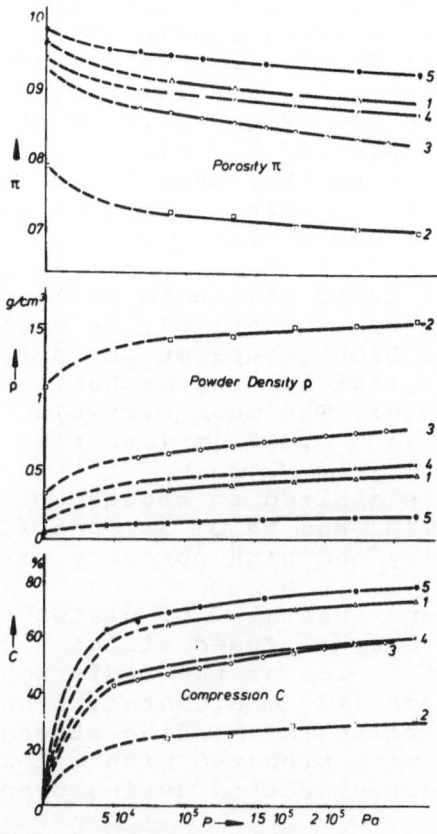

Fig. 7   Porosity, π, powder density, ρ, and compression
         (loss of original volume), C, or pure opacifiers
         TiO$_2$ and Fe$_3$O$_4$, mixtures of opacifiers, and
         pure, fumed silica. Solid lines are least squares
         fits to the data. Curve 1: TiO$_2$; curve 2: Fe$_3$O$_4$;
         curve 3: mixture TiO$_2$ + Fe$_3$O$_4$ (1:1 weight ratio);
         curve 4: as before, 9:1 weight ratio; curve 5:
         fumed silica. Thickness of original volume: 8 cm.
         Measurements were performed in the non-evacuated
         state

from Fig. 7, to $\rho > 0.4$ g/cm^3 for the 9:1 mixture. For
this ρ, λ_{sc} assumes values of more than 10 mW/(m·K) (see
Fig. 6). A powder insulation consisting of the mixed
opacifiers TiO_2 + Fe_3O_4 yields, as a consequence, small
values of λ only if the powder is, by external means,
protected from load; e.g., a low density mixture of TiO_2
+ Fe_3O_4 used for the thermal insulation of pipings for
the transport of hot liquids or gases (ref. 13). For the
preparation of load-bearing insulations, however, it is
necessary to replace the larger part of this mixture by
another powder component of a skeleton or fiber-like
geometric particle structure, because the cubic TiO_2 and
Fe_3O_4 particles favour a large number of contacts per
grain (see Figs. 8 a and 8 b).

The tendency of fumed silica to build up chains, net-
works and skeletons can tentatively be explained by hydrogen bonding between SiOH groups on the surface of the
SiO_2 particles (the real bonding mechanism is not yet
completely understood). The mean particle diameter of
the SiO_2 particles is $\sim$ 0.007 µm (see Fig. 8 c).

Fumed silica was applied to reduce λ_{sc} of the thermal
insulation. Using TiO_2 and Fe_3O_4 as opacifiers, pressed
plates of low density and high porosity were prepared.

Fig. 9 shows experimental λ of plates prepared from
the multicomponent mixture fumed silica + TiO_2 + Fe_3O_4 +
glass fibers as a function of the radiation temperature,
T^*. The glass fibers (13 µm diameter, length 1 to 5 mm)
were applied to increase the bending strength of the
plates. The plates were prepared with a load of 6.5·10^5 Pa
(the powders were not evacuated during preparation of the
plates).

The experimental results in Figs. 1 and 6 favour the
higher weight percent of Fe_3O_4 for the extinction of radiation and for small λ_{sc}. As a consequence, application
of only Fe_3O_4 as opacifier in the mixture with fumed
silica and glass fibers (curve 2 in Fig. 9) yields the
smallest β (see Fig. 5, MCM2), whereas application of only
TiO_2 as opacifier (curve 3 in Fig. 9) yields the largest
β (Fig. 5, MCM3). The α of MCM3 (not given in Fig. 6) is
however very small compared with the $\alpha(MCM1) < \alpha(MCM2)$.
The 1:1 weight mixture of TiO_2 + Fe_3O_4 as opacifiers with
fumed silica and glass fibers (MCM1) yields the lowest
total λ (curve 1 in Fig. 9, MCM1 in Figs. 5 and 6) at
$T^* = 4·10^8$ K^3, which corresponds to 320°C.

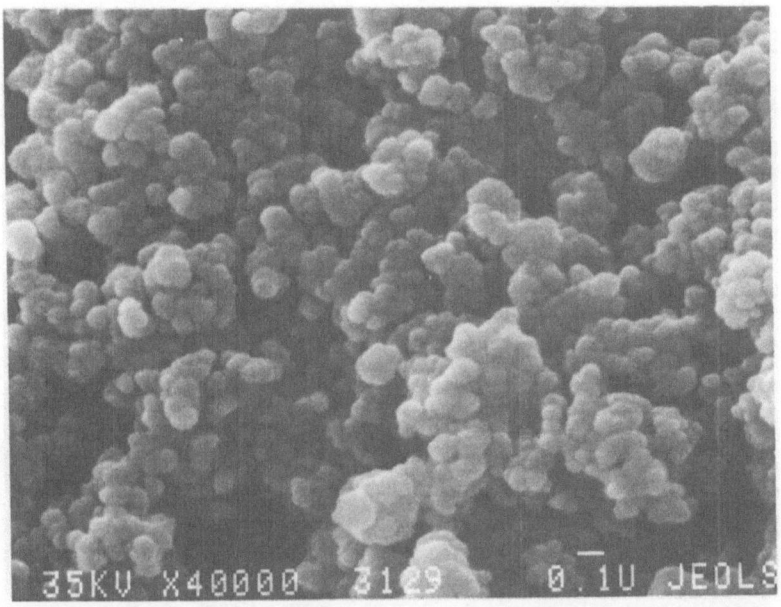

Fig. 8 a Micrograph (Scanning) of TiO_2; magnification
40.000 (courtesy Kontron GmbH, Munich)

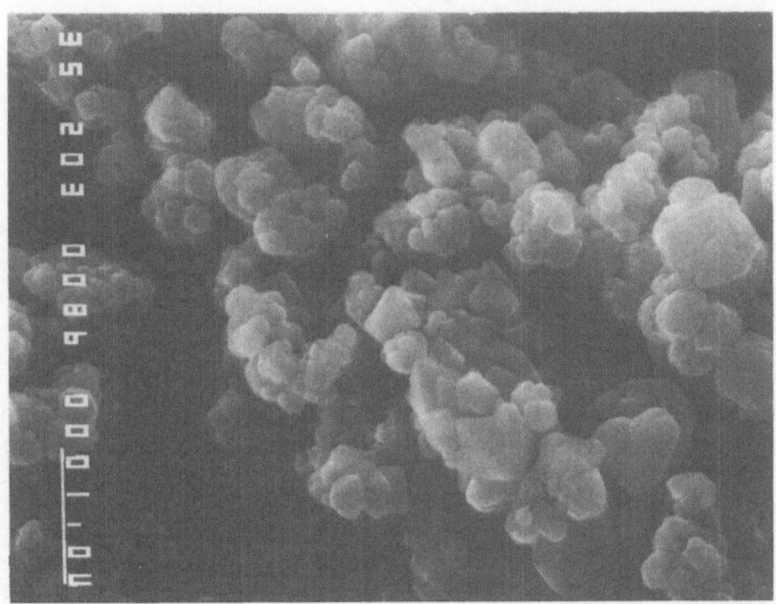

Fig. 8 b Micrograph (Scanning) of Fe_3O_4; magnification
20.000 (courtesy Kontron GmbH, Munich)

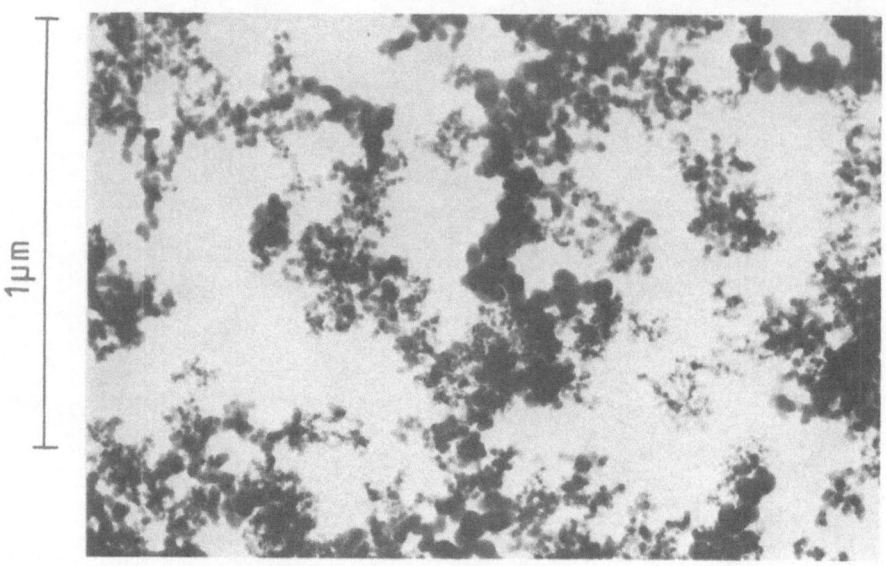

Fig. 8 c   Micrograph (transmission) of fumed silica;
           magnification 65.000 (courtesy Degussa AG,
           Hanau)

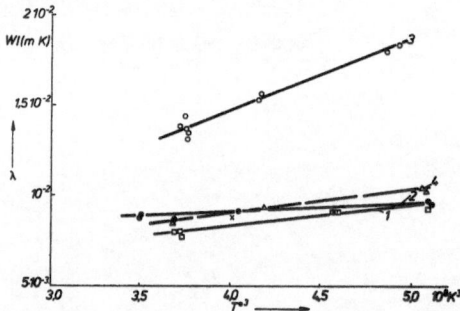

Fig. 9   Thermal conductivity, λ, in vacuum ($p \leq$ 100 Pa)
            of pressed plates prepared from a multicompo-
            nent mixture (MCM) of fumed silica (x) + TiO_2 (y)
            + Fe_3O_4 (z) + glass fibers (10% of total weight)
            at $P = 6.5 \cdot 10^5$ Pa. Curves are least-squares fits
            to the data. Curve 1: x:y:z = 10:1:1; curve 2:
            10:0:2; curve 3: 10:2:0; curve 4: "Minileit"
            (ref. 14), for comparison.

Although the opacifiers constitute only 20% of the entire mass of the pressed plates, it is seen from these results that the preference of high weight percent of Fe_3O_4 in the mixture with TiO_2 concluded from Fig. 1 is confirmed by the sequence of the $\beta(MCM2) < \beta(MCM1) < \beta(MCM3)$. This is easily explained by the well-known high IR transparency of SiO_2-aerogel and fumed silica at wavelengths below 10 μm. In addition, fumed silica obviously exhibits a very small λ_{sc}, which is difficult to determine at high temperatures. As a consequence, an optimization of the extinction and solid conduction properties of the mixture $TiO_2 + Fe_3O_4$ with fumed silica can be performed separately with respect to $\alpha = \lambda_{sc}$ and $\beta = d\lambda/dT$[*3].

Fig. 7 shows that at $P = 2 \cdot 10^5$ Pa the porosity, π, of fumed silica (curve 5) is ≥ 0.93 whereas for TiO_2 $\pi \leq 0.9$. The relative solid part of SiO_2 is thus more than 30% smaller. The compression, C, is however only 7% larger for fumed silica than for TiO_2. This explains the low λ_{sc} of the plates.

The achieved λ of curves 1 and 2 in Fig. 9 are close to the required value 8 mW/(m·K) at high temperatures, at a residual gas pressure of $\sim$ 100 Pa and at high load. The next studies will include measurements of λ for mixtures of fumed silica with opacifiers of still higher total extinction, E, and at reduced load P. The reduction of P to 10^5 to $2 \cdot 10^5$ Pa should further decrease λ_{sc}.

In summary, it has been shown that the conditions a, b and c in the first section can be fulfilled so that an evacuated powder insulation is a way for solving the thermal problems of the Na/S battery.

REFERENCES

1. A.W.Pratt, Heat Transmission in Low Conductivity Materials, in "Thermal Conductivity", R.P.Tye, ed., Vol.1, Academic Press, Inc., London (1969), p. 334

2. H.Reiss, An Evacuated Powder Insulation for a High Temperature Na/S-Battery, AIAA 16th Thermophysics Conference, Palo Alto (1981) AIAA-81-1107

3. R.Siegel and J.R.Howell, "Thermal Radiation Heat Transfer", McGraw Hill Kogakusha Ltd., Tokyo (1972), p. 688

4.   R.Siegel and J.R. Howell, loc. cit., p. 661

5.   R.Viskanta, Heat Transfer by Conduction and Radia-
     tion in Absorbing and Scattering Materials,
     J.Heat Transf. (Febr. 1965) 143

6.   G.N. Plass, Mie Scattering and Absorption Cross
     Sections for Absorbing Particles, Appl. Opt.
     5:279 (1966)

7.   H.H. Weber, Über das optische Verhalten von kugeli-
     gen isotropen Teilchen in verschiedenen Medien,
     Kolloidz. 174:66 (1961)

8.   S.Chandrasekhar, "Radiative Transfer", Oxford Univ.
     Press, London (1950), p. 54

9.   M.G. Kaganer, Untersuchung der Ausbreitung von Licht
     im streuenden Medium durch die Methode der dis-
     kreten Ordinaten, Opt. Spektroskop. 26:443
     (1969), Transl. by G. Wahl, Brown, Boveri & Cie
     AG., Heidelberg, FRG

10. R.Caps, Diplomarbeit, Phys. Inst. Univ. Würzburg
     (1982)

11. D.Büttner, Diplomarbeit, Phys. Inst. Univ. Würz-
     burg (1982)

12. G.L. Serebryanyi, L.B. Zarudnyi and S.N. Shorin,
     Measurement of the Heat Conductivity Coefficient
     of Vacuum Powder-Insulations at High Temperatures,
     Teplofiz. Vysokikh Temp. 6:547 (1968), Transl.
     Plenum Publ. Corp., UDC 536.212.3

13. H.Reiss, Brown, Boveri & Cie AG, Heidelberg, Pat.
     Appl. P 3038142.1 (Oct. 9, 1980)

14. Grünzweig & Hartmann AG, Ludwigshafen, FRG

APPLICATION OF PARAMETER ESTIMATION TECHNIQUES

TO THERMAL CONDUCTIVITY PROBE DATA REDUCTION*

J. A. Koski and D. F. McVey

Sandia National Laboratories
Albuquerque, New Mexico 87185

ABSTRACT

Parameter estimation techniques are applied to determine
thermal conductivity from line heat source probe data. A data
reduction method is described that makes use of experimental
temperature data at both early and late times, does not require
full development of the classical linear temperature increase
versus log-time asymptotic behavior, is not as sensitive to data
noise as other techniques, and permits evaluation of the experiment
design and data assumptions through statistical analysis of final
residuals (difference between the analytical solution and experimental data).

The approach consists of obtaining a nonlinear least-squares fit
between an analytical solution containing the parameters of interest
(conductivity, contact resistance, and, in some cases, density-
specific heat product) and experimental data. The method estimates
the parameters of the analytical solution from experimental data.
The analytical solutions are obtained from the literature.

Examination of the sensitivity coefficients (derivatives of the
parameters with respect to the measured variable) demonstrates the
adequacy of thermal conductivity as a fit parameter. Application
of the method to analysis of laboratory and in situ field data are
presented for illustrative purposes.

---

*This work performed at Sandia National Laboratories supported by
the U.S. Department of Energy under contract DE-AC04-76-DP00789.

INTRODUCTION

Data analysis with the use of parameter estimation theory consists of estimating constants appearing in mathematical models (analytical solutions) of a process or experiment. The desired constants (parameters) serve as the variables in a curve fitting process and are thus determined when the fit is optimized in some sense. Since the mathematical models are often nonlinear in the curve fit parameters, nonlinear least-square methods, such as the Gauss—Newton method, are commonly applied to obtain the solution. The general subject of parameter estimation theory is surveyed in a text by Beck and Arnold.[1]

For the line source thermal conductivity probe,[2] the parameter to be estimated is thermal conductivity. Under certain fairly restrictive conditions, thermal conductivity may be obtained easily from the slope of the probe-temperature-rise versus log-time data. In many cases, other parameters, such as contact resistance between material and probe and the density-specific heat product of the probe and test material, enter into the conductivity determination and severely limit the applicability of the method. The parameter estimation approach with an analytical model permits the the determination of conductivity including influences of other parameters, such as probe and material effects.

Among the advantages of the approach are the following:

- Since both early and late time data may be included in the curve fit process, full development of the asymptotic log-linear region of the temperature rise versus time curve is not required. Overall run times are thus minimized. This can be useful when run times are long, as in the case of low conductivity materials, or where near-field temperature increases must be limited.

- Contact resistance can be estimated simultaneously with the conductivity, simplifying data reduction and experimental design.

- The least-squares curve fit procedures permit treatment of noisy data.

- Experimental data and the theoretical model may be directly compared, so that deviations can be easily detected and analyzed.

- Temperature residuals (the difference between measured and predicted temperatures at various points in time) can be used for statistical tests, including the definition of approximate confidence regions for the parameters used for the curve fit.

- Analysis of the curve fitting process provides information regarding the relative importance of the various experimental parameters and can indicate potential problems or bad data.

ANALYTICAL MODELS

The choice of an analytical model for use with parameter estimation methods is controlled by conflicting needs for rapid computer evaluation and for proper representation of problem complexities. Solutions that represent an adequate compromise are reported by Jaeger[3]. In these solutions, Jaeger assumes radial conduction from a long probe in an infinite medium and includes probe contact resistance and thermal mass effects. Infinite probe conductivity, i.e., no temperature gradients inside the probe, is also assumed. Analytical solutions for the temperature increase in the probe are presented in three different forms:*

Early Time Solution:

For h = 0

$$\frac{kT}{q} = \frac{\omega}{2\pi} \left\{ \tau - \frac{4}{3} \frac{\omega}{\pi^{1/2}} \tau^{3/2} + O[\tau^2] \right\} \qquad (1)$$

For h ≠ 0

$$\frac{kT}{q} = \frac{\omega}{2\pi} \left\{ \tau - \frac{\omega\tau^2}{2h} + O[\tau^{5/2}] \right\} \qquad (2)$$

Long Time Solution:

$$\frac{kT}{q} = \frac{1}{4\pi} \left\{ 2h + \ln\left(\frac{4\tau}{C}\right) - \left(\frac{(4h - \omega)}{2\omega\,\tau}\right) + \left(\frac{\omega - 2}{2\omega\tau}\right) \ln\left(\frac{4\tau}{C}\right) + \ldots \right\} \qquad (3)$$

Integral Solution:

$$\frac{kT}{q} = \frac{2\omega^2}{\pi^3} \int_0^\infty \frac{\{1 - \exp(-\tau u^2)\} du}{u^3 f(u)} \qquad (4)$$

where

$$f(u) = \left[uJ_0'(u) - (\omega - hu^2)J_1(u) \right]^2 + \left[uY_0(u) - (\omega - hu^2)Y_1(u) \right]^2$$

In these solutions, the variables are defined as follows:

q = heat input per unit length of the probe (W/m)
k = thermal conductivity of material (W/m-K)
T = temperature increase (K)

ω = thermal capacity ratio, $\dfrac{2(\rho c_p)_{material}}{(\rho c_p)_{probe}}$ (dimensionless)

---

*Equations (1), (2) and (3) were obtained by Jaeger from Equation (4) by expanding (4) as a series and integrating term by term.

ρc_p = density-specific heat product (J/K-m^3)

τ = Fourier modulus, $\frac{\alpha t}{a^2}$ (dimensionless)

α = thermal diffusivity of material (m^2/s)
t = time (s)
a = probe radius (m)
C = 1.7811 = exp γ, and γ = 0.5772...is Euler's constant

h = contact resistance parameter, $\frac{Rk}{a}$ (dimensionless)

R = probe surface contact resistance (m^2-K/W)
u = variable of integration
J_0, J_1 = Bessel functions of the first kind, orders 0 and 1
Y_0, Y_1 = Bessel functions of the second kind, orders 0 and 1

Figure 1 illustrates the significant gap in the range of validity for the early and long time solutions. The integral equation, which may be evaluated with the aid of a computer, is the only form of the solution valid at all times. As acknowledged by Beck, Jaeger and Newstead[4], the gap between the early and long time solutions can occur in the region of experimental interest.

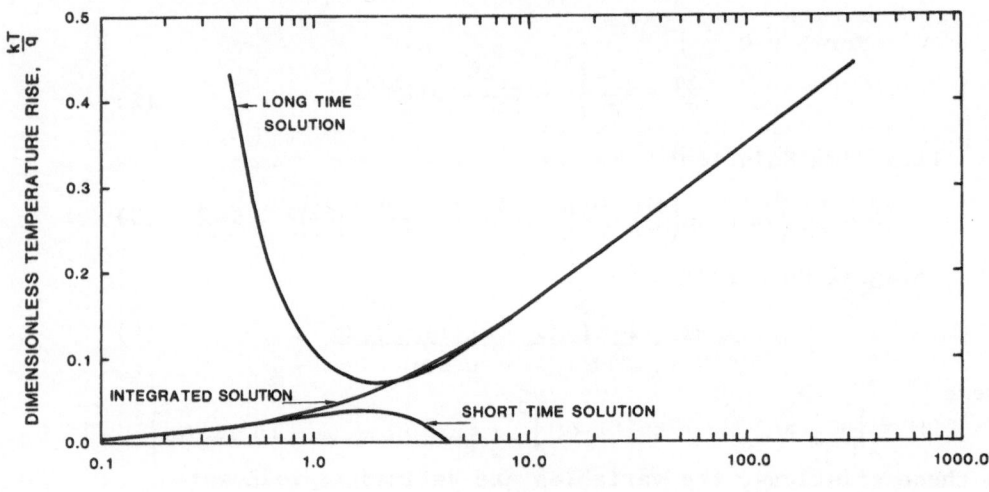

Fig. 1   Comparison of Jaeger's[3] solutions for simulated run with UO$_2$ powder; k = 0.5 W/m-K. Values for other parameters are ω = 1.40, h = 0.0, α = 2.37 x 10^{-7} m^2/s, and a = 0.794 mm.

Blackwell[5] and DeVries and Peck[6] present other analytical solutions that include both probe thermal mass and contact resistance effects as well as probe thermal conductivity. Placed in the proper form, these solutions would also be good candidates for data reduction algorithms. Another alternative, the use of discrete numerical models, is feasible, but careful programming would probably be required to prevent excessive computer execution times.

For the present work, the analytical solutions given in Equations (3) and (4) will be used.

IMPLEMENTATION STRATEGY

Data reduction using the proposed parameter estimation approach is accomplished in the following steps:

1. Acquire temperature versus time data from the conductivity probe in digital form using a data logger or data acquisition minicomputer.

2. Select 20 to 40 representative data points equally spaced over the time or log-time interval of interest.

3. Fit the applicable analytical solution of Equations (1) through (4) to the data using nonlinear least-squares methods. Problem parameters are varied to achieve the fit.

4. The desired parameters are obtained from the best least-squares fit of the analytical solutions to the data.

Other problem parameters, such as heat input to the probe, probe radius, and the density-specific heat product of the probe, are assumed known during the least-squares fit.

The above strategy has been implemented in two different ways with two different objectives in mind. In one version the process was used with a large time-sharing computer to analyze data previously obtained in a field experiment. In this version, Equation (4) is used over the entire range of times. The other version was designed to permit rapid data reduction with the use of a relatively large on-line minicomputer system in the laboratory. Equations (3) and (4) are both used, with Equation (4) covering times before Equation (3) becomes a valid approximation. The transition point between solutions (3) and (4) is a function of h and ω and is defined by empirical functions provided by Koski[7].

The large computer version implements the Gauss-Newton[1] method to perform the curve fit. The sum of squares function to be minimized is

$$S = \left[Y - \eta(\beta) \right]^T W \left[Y - \eta(\beta) \right]$$

(5)

where   Y = a vector containing the n measurements

β = a parameter vector containing the curve fit variables, such as conductivity

$\eta(\beta)$ = a vector containing the n corresponding predictions from the model for the present value of the parameter vector, β

W = an n x n symmetric weighting matrix that is proportional to the measurement errors and equals the identity matrix for the ordinary least squares case

and superscript T refers to the matrix transpose.

For this case Beck and Arnold give effective iterative methods for minimizing the sum-of-squares residuals. The linearized Gauss-Newton method for ordinary least squares estimation for the (k +1)st iteration is

$$b^{(k+1)} = b^{(k)} + P^{(k)} \left[X^{T(k)} (Y - \eta^{(k)}) \right] \qquad (6)$$

$$P^{-1(k)} = X^{T(k)} X^{(k)} \qquad (7)$$

where   b = column vector containing the k parameters being estimated at iterations k and k + 1

$X^k = \frac{\partial \eta}{\partial \beta}$; n x p matrix of sensitivity coefficients for p parameters evaluated at n times or data points.

Equations (6) and (7) represent the ordinary-least-squares form of the procedures implemented in the large-computer code.

The nonlinear least-squares routine (LMSTR) used with this mini-computer version of the program is based on the Levenberg-Marquardt approach and is part of the MINPACK[8] library. The program is implemented on a Hewlett Packard 1000 minicomputer with run times on the order of a few minutes.

SENSITIVITY ANALYSIS

During the nonlinear least-squares curve fitting process, the sensitivity coefficients,

$$x_i = \frac{\partial \eta}{\partial \beta_i} , \qquad (8)$$

must be calculated. The function η is the problem dependent variable, temperature in this case, computed from Equations (3) or (4). The β_i are the fit parameters such as conductivity. When the sensitivity coefficients are normalized so that

$$x_i = \beta_i \frac{\partial \eta}{\partial \beta_i} \qquad (9)$$

and plotted, the sensitivity of the dependent variable to the various parameters can be examined. In this case, X_i represents the change in temperature produced by a unit fractional change in the parameter β_i.

Figure 2 shows the normalized sensitivity coefficients for a simulated run with UO_2 particles in an inert atmosphere near 1300 K. The assumed value for the contact resistance, h = 2.0, is a relatively high value. Note that the thermal conductivity sensitivity coefficient is comparable in magnitude to the other coefficients, especially at later times. This ensures that the probe data are adequately sensitive to the parameter of interest, thermal conductivity.

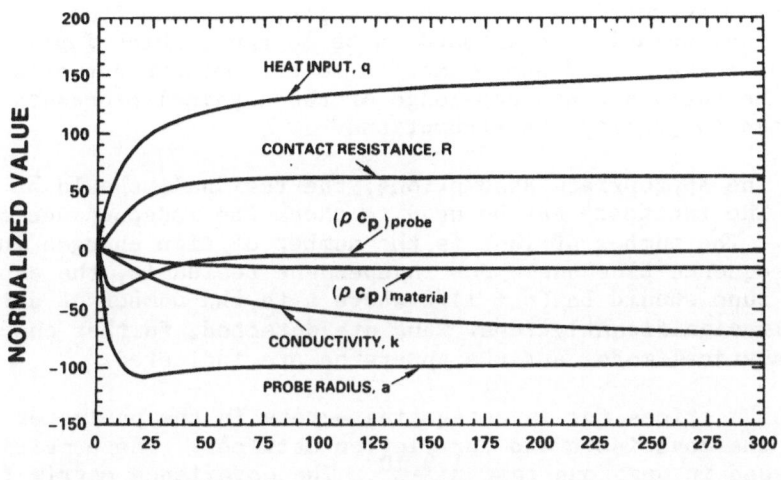

TIME, s

Fig. 2   Sensitivity coefficients for simulated run with UO_2 powder in an inert gas. For the run, k = 0.5 W/m-K, $(\rho c_p)_{material}$ = 2.11 MJ/m^3-K, $(\rho c_p)_{probe}$ = 3.97 MJ/m^3-K, a = 0.794 mm, R = 3.12 mK-m/W (h=2), ω = 1.06.

At later times, the sensitivity coefficients approach a constant value. This indicates that for these times, the sensitivity coefficients are linearly dependent, and unless early time data are also present, parameter estimation techniques cannot independently determine the parameters. Notice that in Equation (7) the p x p matrix $(X^T X)$ must be inverted, and if the terms are linearly dependent, the inversion cannot be accomplished. Only one parameter can be determined accurately from late time data. One manifestation of late time linear dependence is that early time data are required if a measurement of $(\rho c_p)_{material}$ is attempted to determine the material diffusivity, $\alpha = k/(\rho c_p)_{material}$. In general, the region of linearly dependent sensitivity coefficients corresponds to the straight line region of the log-linear time versus temperature rise plot.

ASSESSMENT OF ERRORS

One of the major benefits of the parameter estimation procedure
is the calculation of residuals, the differences between what is
measured and what is predicted. Analysis of residuals permits
assessment of whether the assumptions made in development of the
mathematical model or in design of the experiment appear to be
violated.

Whereas the ordinary least-squares curve fitting process itself
does not require any assumptions regarding errors in the variables,
statistical analyses based on the curve fit residuals do require
several major assumptions (see, for example, Beck and Arnold[1], p. 228).
The analytical solution is assumed to be correct, which implies that
the residuals are a good approximation to the measurement errors.
Tests on the residuals and knowledge of the physical processes involved
must be used to justify the assumptions.

With the appropriate assumptions, the residuals should be independent. The run test[1] may be used to check the independence of the
residuals. The number of runs is the number of sign changes in the
residual sequence plus one. For independent residuals, the expected
number of runs should be $(n + 1)/2$ where n is the number of data
points. If significantly fewer runs are detected, further checking
of the analytical model and the apparatus are indicated.

Useful matrices for investigating errors in the parameter estimates are the covariance and correlation matrices. These matrices
are discussed in numerous texts[1,9,10]. The covariance matrix for the
nonlinear ordinary least-squares case is approximated by the relation

$$\text{cov}(b_{LS}) \approx s^2 \, P \tag{10}$$

where P is the matrix defined in Equations (6) and (7), and s^2 is
the variance of the residuals approximated by the relation

$$s^2 \approx S_{min}/(n - p) \tag{11}$$

where S_{min} is the minimum sum of squares, Equation (5), as determined
in the estimation procedure, with n = number of data points and p =
number of parameters. The standard error in k and h for a two parameter fit may be computed from the P matrix as

$$\hat{\sigma}_1 = \sqrt{s^2 \, P_{11}}$$

$$\hat{\sigma}_2 = \sqrt{s^2 \, P_{22}} \tag{12}$$

where $\hat{\sigma}_1$ and $\hat{\sigma}_2$ are the standards errors for k and h, respectively.

The correlation matrix is derived from the covariance matrix
with the elements of the correlation matrix defined as, r_{ij}

$$r_{ij} = \frac{P_{ij}}{\left(P_{ii} \, P_{jj}\right)^{1/2}} \, .$$

(13)

In general, the statistical assumptions discussed by Beck and Arnold must hold for these matrices to give valid results. When the analytical model is nonlinear in the curve-fit parameters, these matrices are approximations derived from the linearized expressions used in the fitting process. The covariance matrix is useful in determining the standard deviation of the fitted results, and the closely related correlation matrix is useful in indicating strong correlations between curve fit parameters that may cause computational problems. The correlation matrix also provides insights into the interrelationships among the parameters of the mathematical model.

FIELD TEST DATA RESULTS

A typical set of field data obtained in a dome salt formation and analyzed with this method[11] is used as an illustrative example. A plot of the data is presented in Figure 3. The data were taken with a 2.54 cm diameter probe with a heated length to diameter ratio of 37 installed in a drilled hole about 8 to 16 mm oversize. The gap between the probe and the hole was filled with silicone oil to reduce contact resistance.

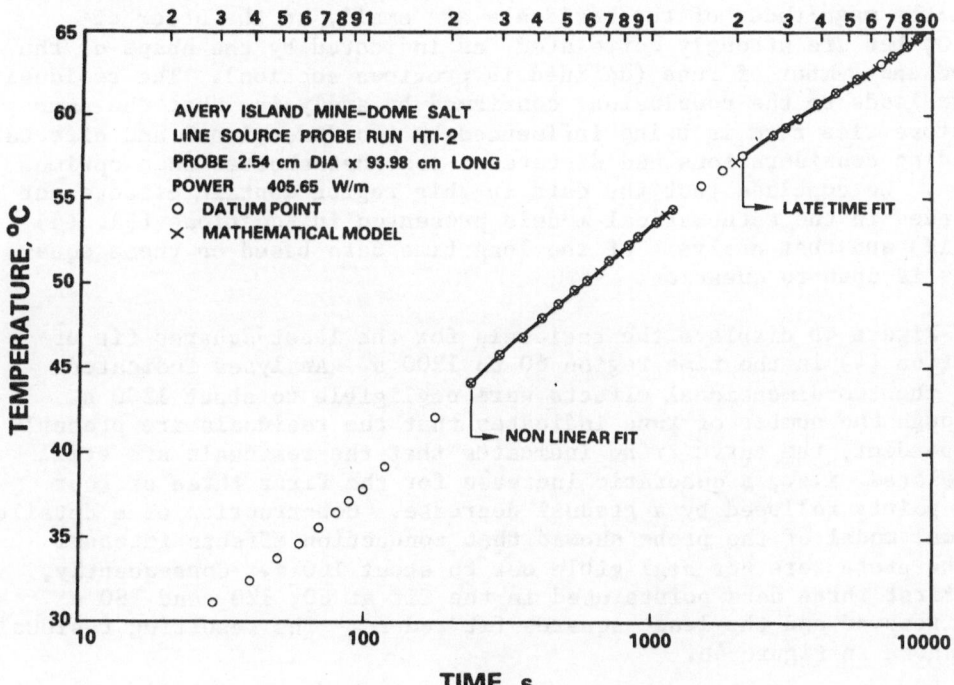

Fig. 3 Rock salt data and curve fits.

Initially, an attempt was made to reduce the data with Jaeger's classic large time linearization [Equation (3)]. At large τ, this equation may be simplified to:

$$k = \frac{q}{4\pi \; (T_2 - T_1)} \; \ln\!\left(\frac{t_2}{t_1}\right) \tag{14}$$

where t_1 and t_2 are times that correspond to temperatures T_1 and T_2.

Although the late time portion of the curve appears to be a straight line, there is a slight curve downward induced by two-dimensional end effects (see Blackwell[12]). As a result, the conductivity was found to increase by as much as 20% as the point (t_1, T_1) was moved toward the end point. Using a simple linear curve fit to obtain the slope also gave varying values for conductivity.

The large computer, nonlinear, ordinary-least-squares technique described above was implemented to fit Equation (4) to the data. On the scale of Figure 3, any noncongruence of the final fit between the mathematical model (X and line) and the data (Θ) is indiscernible to the eye. Notice that this is true both for the separate curve fits at early and late times. As previously discussed, the curve fits are best interpreted by examination of the residuals.

Figure 4a shows residuals for the least-square fits of Equation (4) between 2,040 s and 10,440 s, the supposed linear region. Notice that the magnitudes of the residuals are small, on the order of 0.1°C, but are strongly correlated, as indicated by the shape of the curve and number of runs (defined in previous section). The residual curve leads to the conclusion, confirmed by analysis, that the temperature rise rate is being influenced by two-dimensional end effects. Fielding considerations had dictated a somewhat shorter than optimum probe. We conclude that the data in this region contain effects not included in the mathematical models presented in Equations (3), (4) or (13) and that analysis of the long time data based on these equations is open to question.

Figure 4b displays the residuals for the least-squares fit of Equation (4) in the time region 60 to 1200 s. Analyses indicated that the two-dimensional effects were negligible to about 1200 s. Although the number of runs indicates that the residuals are probably independent, the curve trend indicates that the residuals are still correlated, i.e., a quadratic increase for the first three or four data points followed by a gradual decrease. Construction of a detailed thermal model of the probe showed that conduction effects internal to the probe were not negligible out to about 180 s. Consequently, the first three data points used in the fit at 60, 120, and 180 s were removed and the least squares fit redone. The resulting residuals are shown in Figure 4c.

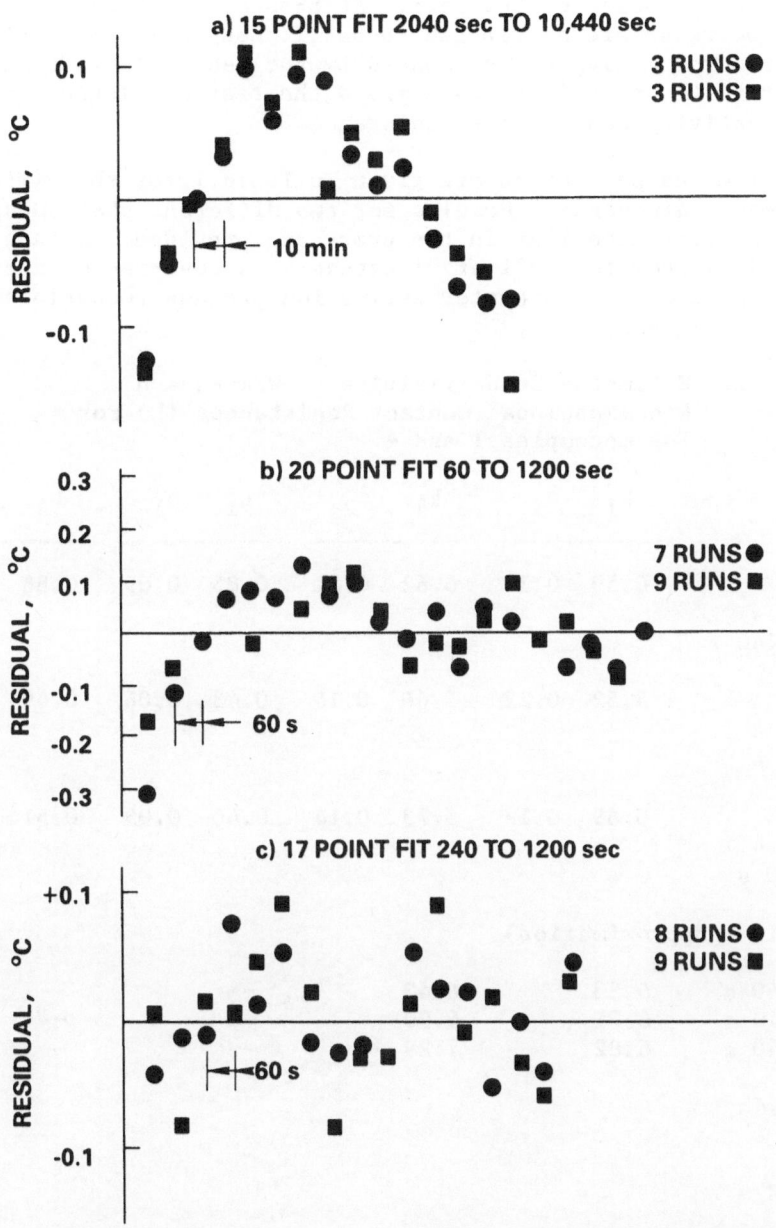

Fig. 5 Residuals for run HT 1-2, <u>in situ</u> dome salt; Thermocouple 1;
Thermocouple 4.

The residuals in Figure 4c are more nearly the desired random scatter when compared with those of Figures 4a and 4b, even though there appears to be a small oscillatory function combined with a linear decrease imposed on the data. At present, both the oscillation and linear decrease are attributed to small changes in the probe input power. The analysis represented by the residuals in Figure 4c indicates that the associated data yield the best available estimate of the conductivity and contact resistance.

The estimated parameters are given in Table 1 for the different data treatments discussed. Results for two different thermocouples located at similar locations in the probe are provided. Standard errors computed from the ordinary–least–squares covariance matrix obtained as part of the parameter estimation process [Equations (12)], are also given.

Table 1. Estimated Conductivities (k W/m–K) and Nondimensional Contact Resistances (h) For Thermocouples 1 and 4

	k_1	$\hat{\sigma}_1$	k_4	$\hat{\sigma}_2$	h_1	$\hat{\sigma}_1$	h_4	$\hat{\sigma}_4$
Long Time Fit of Eq. 4 2040 to 10440 s	6.59	0.19	6.62	0.21	0.85	0.09	0.88	0.10
20 Point Fit of Eq. 4 60 to 1200 s	5.52	0.22	5.66	0.16	0.43	0.06	0.49	0.05
17 Point* Fit of Eq. 4 240 to 1200 s	5.69	0.12	5.73	0.14	0.46	0.03	0.51	0.04
(T versus ℓn t approximation)								
2040 – 10440 s	6.33		6.43		--		--	
4440 – 10440 s	6.71		6.90		--		--	
7440 – 10440 s	6.82		7.29		--		--	

*Recommended

CONCLUSIONS

Our experience to date indicates that parameter estimation is a powerful technique for interpretation of line source thermal conductivity data. This technique removes the requirement that the linear temperature versus log time asymptote be achieved. For large probes, typical of those necessary for <u>in situ</u> measurements, this significantly reduces run time and total temperature increase and permits use of smaller length to diameter ratios. In addition, by using more detailed analytical models or employing finite difference or finite element models in the estimation technique, data can be extracted from experiments in which many of the classic restrictions are relaxed or removed.

Detailed analysis of the residuals can give clear indications of problems with either the model or experiment that may not be given by other methods. In addition, the covariance matrix and sum-of-squares function developed during the fitting process may be used to determine the standard deviation and confidence regions for multiple parameter experiments.

ACKNOWLEDGMENTS

The authors would like to thank D. D. Boozer and C. E. Hickox for their helpful review comments and suggestions regarding this paper. The work of W. D. Drotning in converting the minicomputer version of the computer code to the HP 1000 computer is also gratefully acknowledged.

REFERENCES

1. J. V. Beck and K. J. Arnold, "Parameter Estimation in Engineering and Science", John Wiley and Sons, New York (1977).

2. A. W. Pratt, Heat transmission in low conductivity materials, <u>in</u>: "Thermal Conductivity", Vol. 1, R. P. Tye, ed., Academic Press, New York (1969).

3. J. C. Jaeger, Conduction of heat in an infinite region bounded internally by a circular cylinder of a perfect conductor, <u>Aust. J. Phys.</u>, 9:167 (1956).

4. A. Beck, J. C. Jaeger and G. Newstead, The measurement of thermal conductivity of rocks by observations in boreholes, <u>Aust. J. Phys.</u>, 9:286 (1956).

5. J. H. Blackwell, A transient-flow method for determination of thermal constants of insulating materials in bulk - Part I - Theory, <u>J. Appl. Phys.</u>, 25:137 (1954).

6. D. A. deVries and A. J. Peck, On the cylindrical probe method of measuring thermal conductivity with special reference to soils, <u>Aust. J. Phys.</u>, 11:255 (1958).

7. J. A. Koski, "Analysis of thermal conductivity probes for high
    temperature applications", Sandia National Laboratories,
    Albuquerque, NM, SAND81-0106 (April 1981). (Available from
    NTIS*)

8. J. J. Moré, B. S. Garbow, and K. E. Hillstrom, User guide to
    MINPACK-1, Argonne National Laboratory, Argonne, Illinois,
    ANL-80-74 (August 1980). (Available from NTIS)

9. G. E. P. Box and G. M. Jenkins, "Time Series Analysis, Fore-
    casting and Control", Holden-Day, San Francisco (1970).

10. A. Papoulis, "Probability, Random Variables, and Stochastic
    Processes", McGraw-Hill Book Company, New York (1965).

11. D. F. McVey, Analysis of line source thermal conductivity data
    taken in in situ Avery Island dome salt, Sandia National
    Laboratories, Albuquerque, NM, SAND81-1232 (August 1981).
    (Available from NTIS)

12. J. H. Blackwell, The axial flow error in the thermal conduc-
    tivity probe, Can. J. Phys., 34:412-417 (1956).

*National Technical Information Service, U.S. Department of
Commerce, 5285 Port Royal Road, Springfield, VA, 22161.

A METHOD TO MEASURE THE THERMAL DIFFUSIVITY OF LOW-CONDUCTIVITY MATERIALS WITH A CONVENTIONAL TWIN-PLATE DEVICE FOR MEASUREMENT OF THE THERMAL CONDUCTIVITY

Bu-xuan Wang, Ze-pei Ren, and Zhao-hong Fang

Thermal Engineering Department
Tsinghua University
Beijing, People's Republic of China

ABSTRACT

In this paper, a new method is presented for determining the thermal diffusivities of low-conductivity materials with a conventional guarded twin-plate device for measuring thermal conductivities of the insulating and building materials. The possible errors of measuring results are analyzed in detail. A group of tests have been made for the optical glass K-9. The thermal diffusivity measured has an assessed error of $\pm 7\%$, which is compared with plane heat source method, and both results agree rather well.

INTRODUCTION

The guarded hot plate device has long been recognized as the standard apparatus for measuring the thermal conductivities, λ, of insulating and building materials with high accuracy. In cooperation with others, a guarded twin-plate device of the type DRP-I had been designed and established in the Heat Transfer Laboratory at Tsinghua University. The temperature of the guarded ring hot plate follows automatically the tracks of the main hot plate temperature; the cold plate is cooled with circulating water provided by a thermostatic bath, of which the temperature fluctuation is not greater than $0.1^{\circ}C$. This apparatus is suitable for measuring materials with thermal conductivities ranging from 0.03 up to 1.2 $W \cdot m^{-1} \cdot K^{-1}$, and the maximum overall error of the measuring results is estimated to be within $\pm 4\%$.

However, by the steady-state method, it can be used only to determine the thermal conductivities of the testing material. To ob-

tain the thermal diffusivity, a , of the material, it is still necessary to know its specific heat, c, and density, ρ. For most of the
building and thermal insulating materials, it is not easy to determine their specific heats exactly. We have developed a novel method
to determine the thermal diffusivity of low-conductivity materials,
so as to extend the function of the guarded twin-plate device.

THEORETICAL BASIS

After the steady state in the testing sample is established, the
thermal conductivity, λ, of the testing material may be determined
conventionally. Then, the electric current supplied to the heater is
suddenly cut-off at the instant taking as $\tau = 0$. Thus starts a transient process. The temperature of sample plate decreases gradually as
a result of the heat dissipation to the coolant until a new special
steady state is attained; i.e., a uniform temperature distribution is
finally reached and is equal to temperature of coolant, t_f, throughout the sample and the hot plate. This transient process has been
theoretically analyzed in detail in a previous paper, with the consideration of the influence of heat capacity of the hot plate on the
temperature response in the sample plate. The following assumptions
are made:
1. The heat flux in the sample plate is one dimensional;
2. The thermophysical properties of the sample can be taken as constants, referred to the mean values involved in the test, since the
   actual temperature range is usually controlled within 15-40°C;
3. The temperature of the hot plate has a uniform distribution at any
   time, τ , and is equal to hot surface temperature of the sample,
   $t_{x=0,\tau}$;
4. The temperature of the sample's cold surface remains unchanged;
   i.e., $t_{x=\delta,\tau} = t_{x=\delta, \tau=0}$.

As the result, the temperature response obtained on the sample's hot
surface is:

$$\Theta_1 = \sum_{n=1}^{\infty} \frac{2}{\mu_n^2 (K_c^2 \mu_n^2 + K_c + 1)} \, e^{-\mu_n^2 Fo} \tag{1}$$

where $\Theta_1 = \dfrac{t_{x=0,\tau} - t_{x=\delta,\tau=0}}{t_{x=0,\tau=0} - t_{x=\delta,\tau=0}}$, the normalized temperature excess of

the sample's hot surface;

$Fo = \dfrac{a\tau}{\delta^2}$, the Fourier modulus;

$K_c = C_0/C = (\rho_0 c_0 \delta_0)/(\rho c \delta)$, the ratio of the heat capacity of

the hot plate to that of the sample plate;

μ_n, the nth root of the eigen equation

$$\text{ctg } \mu = K_c \mu. \tag{2}$$

Note that, since the heat capacity, $C(=\rho c \delta)$, of the sample per unit area is $c \rho \delta = \lambda \delta / a$, the dimensionless variables K_c and Fo are both the function of the unknown thermal diffusivity, a. To express equation (1) as an explicit function of the thermal diffusivity, we introduce a new dimensionless variable:

$$R = \frac{Fo}{K_c} = \frac{\lambda \tau}{c_o \delta}. \tag{3}$$

Equation (1) is thence rewritten as:

$$Fo = f(\textcircled{H}_1, R). \tag{4}$$

This function has been calculated out and is plotted in Fig. 1 and summarized in a condensed Table I for $\textcircled{H}_1 = 0.5$.

It is certainly clear, as the transient process goes on, the assumption that the temperature of the sample's cold surface remained unchanged is not satisfied. A more realistic mathematical model adopting a convectional boundary condition has been used and solved numerically by the finite difference method. The convective boundary condition introduces a Biot modulus: $Bi = \alpha \delta / \lambda$. It is reasonable to assume the convective heat transfer coefficient, α, to be con-

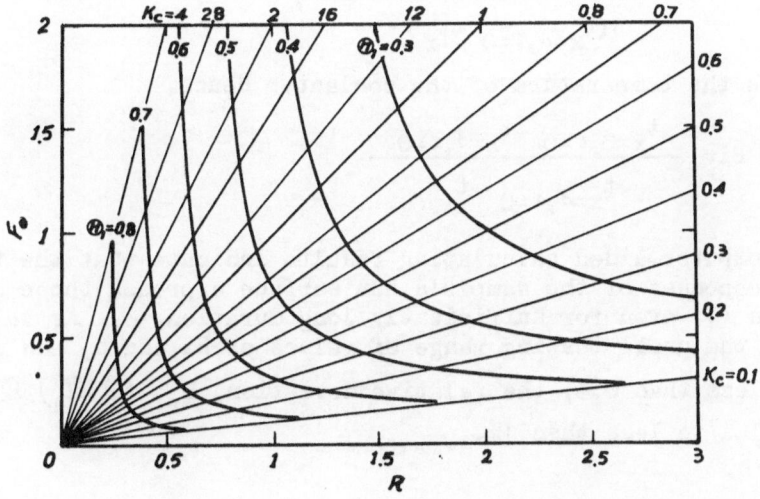

Fig. 1 The function $Fo = f(\textcircled{H}_1, R)$

Table I Condensed table of the function
$Fo = f(\textcircled{B}_1, R)$ for $\textcircled{B}_1 = 0.5$

R	0	2	4	6	8
1.0	0.7241	0.6918	0.6632	0.6378	0.6149
1.1	0.5943	0.5756	0.5585	0.5430	0.5287
1.2	0.5156	0.5034	0.4921	0.4816	0.4718
1.3	0.4626	0.4541	0.4460	0.4385	0.4314
1.4	0.4247	0.4183	0.4123	0.4066	0.4012
1.5	0.3961	0.3912	0.3866	0.3821	0.3779
1.6	0.3739	0.3700	0.3663	0.3627	0.3593
1.7	0.3560	0.3529	0.3499	0.3470	0.3442
1.8	0.3414	0.3388	0.3363	0.3339	0.3315
1.9	0.3293	0.3271	0.3249	0.3229	0.3209
2.0	0.3190	0.3171	0.3153	0.3136	0.3119
2.1	0.3102	0.3086	0.3070	0.3055	0.3040
2.2	0.3025	0.3011	0.2997	0.2984	0.2971

*For example, when R=1.48, Fo=0.4012 from the table. stant, since this convective thermal resistance is much less than the conductive thermal resistance of the sample plate. On this basis, the value of α can be easily determined from the continuity of the heat flux in the steady state as

$$\alpha = \frac{\lambda(t_{x=0,\tau=0} - t_{x=\delta,\tau=0})}{\delta(t_{x=\delta,\tau=0} - t_f)} \; ; \tag{5}$$

where t_f is the temperature of the coolant. Hence,

$$Bi = \frac{t_{x=0,\tau=0} - t_{x=\delta,\tau=0}}{t_{x=\delta,\tau=0} - t_f} \; . \tag{6}$$

The computer-aided calculating results indicate that the temperature responses of the sample's hot surface approach those given by equation (1) even for sufficiently long duration, τ. As shown in Fig. 2, in the usual testing range of values of Bi and K_c and when $\textcircled{B}_1$ is not less than 0.5, the relative deviation, $E_r = (\textcircled{B}_1 - \textcircled{B}_1^*)/\textcircled{B}_1$, can be controlled to less than 1%.

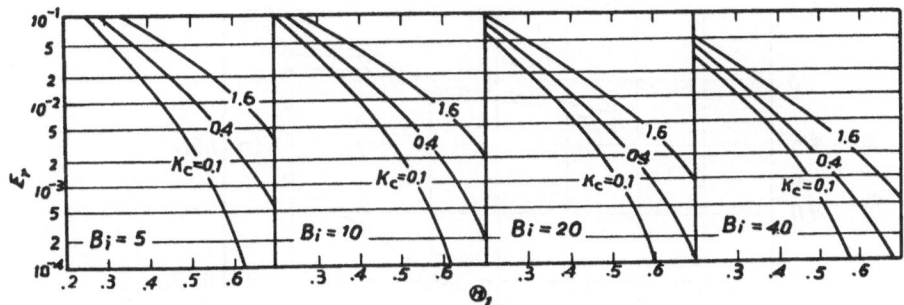

Fig. 2 The relative deviation, E_r

TEST PROGRAM

The testing material was optical glass K-9, whose composition by weight is 66.4% SiO_2, 8.1% B_2O_3, 17.8% K_2O, 4.9% BaO, 1.8% CaO, and 1.0% As_2O_3. The sample plate was 200 mm in diameter and 26.7 mm thick, and both sides of the sample plate are well smoothed. The hot surface of the samples were smeared with a film of grease to reduce the thermal contact resistance.

After a steady state was established in the sample plate, the thermal conductivity of the sample was determined by the conventional steady-state method. Then the current through the heater was suddenly cut-off, and this moment was taken as $\tau = 0$, the starting point of the transient process. The time duration was recorded for the normalized temperature excess of the sample's hot surface decreased to a prescribed value of Θ_1 as 0.5 in our tests. Thermometers were installed at the entrance and exit of the water cooler, and the average of the readings was taken as the temperature of the water coolant of the cold plate of the guarded twin-plate device.

From the measured data, the thermal diffusivity of the glass specimen can be evaluated by two methods, i.e., either by a function table, such as Table I, or by use of a computer.

We calculated the dimensionless variable R from the measured data. From the known value of R, we found the corresponding Fo from Table I or from Fig. 1. The thermal diffusivity of the testing material was thence evaluated from

$$a = \frac{\delta^2}{\tau} \; Fo. \tag{7}$$

Measured data on optical glass K-9 are shown in Fig.3.

ESTIMATION OF ERRORS

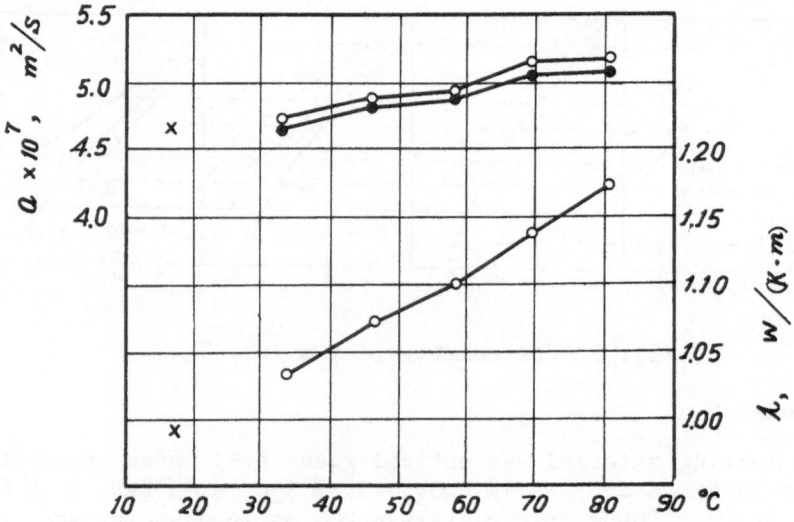

Fig. 3 Measured results for optical glass K-9

Thermal Diffusivity, a :
    o—evaluated for constant cold surface temperature condition
    ●—evaluated for convective boundary condition by computer
    x—measured by plane heat source method[2]
Thermal Conductivity, λ :
    o—measured by guarded twin-plate device
    x—measured by plane heat source method[2]

Estimation of error was based on an analysis of equation (1),
(4) and (7) and the errors of pertinent variables. Systematic and
random errors involved in the calculated thermal diffusivity were assessed, and in view of transmission of errors, the overall probable
error, which is defined as "experimental accuracy", was determined.

By the steady-state method, the thermal conductivity of the sample was measured with an assessed accuracy of ±4%. The heat capacity
of the hot plate was carefully determined with an assessed accuracy
of ±2%. The relative errors of the thickness and the time duration
were less than 0.1%. Combining the uncertainties of the foregoing
variables, the probable error of the dimensionless variable R was assessed to be ±4.5%.

The error of the dimensionless variable $\oplus_1$ comes from the errors
of temperature measurement and is more complicated. When the thermal
diffusivity is evaluated from Table I or Fig. 1, the discrepancy of
boundary conditions brings some error into $\oplus_1$ too, and this error can
be found from Fig. 2. Summarizing all these factors, we have found
that the relative error of $\oplus_1$ does not exceed ±1.5%.

Having discussed the errors of R and $\oplus_1$, it is necessary to analyze the error transmission in order to estimate the error of

From equation (7), $\quad \left| \dfrac{\Delta a}{a} \right| = \left| \dfrac{\Delta Fo}{Fo} \right| + \left| \dfrac{\Delta \tau}{\tau} \right| + 2 \left| \dfrac{\Delta \delta}{\delta} \right|.$ (8)

From equation (4), $\quad \dfrac{\Delta Fo}{Fo} = \dfrac{\oplus_1}{Fo} \left(\dfrac{\partial Fo}{\partial \oplus_1} \right)_R \dfrac{\Delta \oplus_1}{\oplus_1} + \dfrac{R}{Fo} \left(\dfrac{\partial Fo}{\partial R} \right)_{\oplus_1} \dfrac{\Delta R}{R}$

$$= \varepsilon_{\oplus_1} \dfrac{\Delta \oplus_1}{\oplus_1} + \varepsilon_R \dfrac{\Delta R}{R}.$$ (9)

$\varepsilon_{\oplus_1}$ and ε_R are transitive factors of error. From equation (1), we can derive:

$$\frac{1}{Fo} \left(\frac{Fo}{\oplus_1} \right)_R = \left\{ \sum_{n=1}^{\infty} \frac{4 e^{-\mu_n^2 Fo}}{(K_c^2 \mu_n^2 + K_c + 1)^2} \left[K_c Fo - K_c^2 + \frac{K_c}{2\mu_n^2} + \frac{K_c^2}{K_c^2 \mu_n^2 + K_c + 1} - \frac{Fo}{2}(K_c^2 \mu_n^2 + K_c + 1) \right] \right\}^{-1}$$ (10)

The values of $\varepsilon_{\oplus_1}$ calculated from equation (10) and those of ε_R by means of numerical differentiation are plotted in Fig. 4 and Fig.5 By use of these two diagrams, the error of the measured values of thermal diffusivity can be estimated conveniently.

It is clear, a lower value of $\oplus_1$ chosen in the experiment reduces the influences of the errors of $\oplus_1$ and R, whereas the relative error of $\oplus_1$ itself grows rapidly with a decrease of $\oplus_1$, so the value of $\oplus_1$ should be chosen in the range 0.6 to 0.4 on the basis of error estimation. Moreover, it is obvious from Figs. 4 and 5 that the value K_c should be as low as possible to determine the most accurate value of thermal diffusivity.

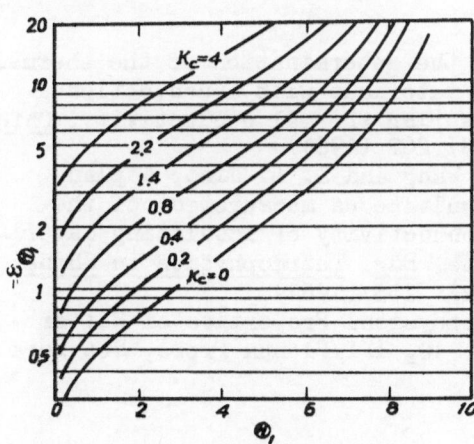

Fig. 4 The transitive factor of error $\varepsilon_{\oplus_1}$

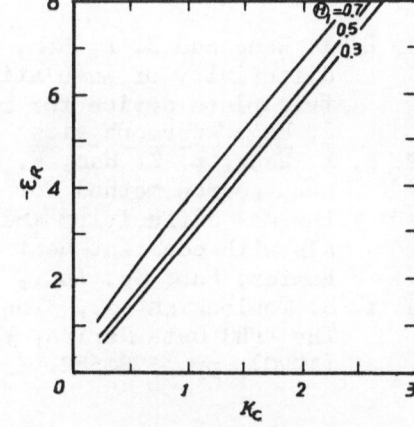

Fig. 5 The transitive factor of error ε_R

For the optical glass K-9, K_c is about 0.35 and $\oplus_1$ is controlled to be 0.5, and the measurement results for thermal diffusivity
have accuracy of $\pm7\%$ if a constant cold surface temperature is assumed or $\pm6\%$ if computer calculation with the convective boundary
condition is adopted.

CONCLUDING COMMENTS

Theoretical analysis and experimental practice have demonstrated the method reported here is feasible for some materials with large
enough volumetric specific heat. This method extends the function
of the conventional guarded hot plate apparatus and is applied easily, so it may be of interest.

The thermal diffusivity data of optical glass K-9 determined by
this method agree rather well with those obtained by the plane heat
source method, of which the maximum error of the measuring results
is estimated within $\pm4\%^2$. The resulting error of these data is believed to be within $\pm10\%$ when K_c is not greater than 0.5 and is roughly the same as the accuracies of other methods reported for determining thermal diffusivities of building materials.

The error analysis shows that K_c severely influences the precision of this method. Because the heat capacity of the 8-mm-thick
duralmin hot plate in our device is quite large compared with those
of samples of common insulating materials, this method brings considerable error into the measurements of the thermal diffusivities of
insulating materials, and it is therefore not preferred for insulating
materials. To improve the accuracy of this method and extend its limits of application, the hot plate should be made of a material with
low volumetric specific heat and made as thin as possible.

REFERENCES:

1. B. X. Wang and Z. P. Ren, On the determination of the thermal
   diffusivity of insulating materials with a conventional
   twin-plate device for measuring thermal conductivity, China.
   J. Eng. Thermophysics 2(3):262 (1981).
2. B. X. Wang, L. Z. Han, W. C. Wang and Z. L. Jiao, A plane
   heat source method for simultaneous measurement of the
   thermal diffusivity and conductivity of insulating materials with constant heat rate, Eng. Thermophysics in China,
   Rumford Pub. Co. Inc., 1(2): 155 (1981).
3. Y. S. Touloukian ed., "Thermophysical Properties of Matter"
   The TPRC Data Series, Vol. 10, IFI/Plenum Press, New York
   (1970), pp. 578-582.

SESSION TC-13B

Composites, Aggregates, and Geological Materials - I

CHAIRMAN

A. E. Beck
University of Western Ontario
London, Canada

THERMOPHYSICAL PROPERTIES OF GEOLOGIC MATERIALS

R. E. Taylor and H. Groot

Thermophysical Properties Research Laboratory
Purdue University
West Lafayette, Indiana

ABSTRACT

The thermal diffusivity, specific heat, and density of certain oil shales and also of a sample from Mt. St. Helens were measured. From these data thermal conductivity values were calculated. Values for the oil shales were determined both before and after removal of hydrocarbons. The shales became hygroscopic after evolution of the hydrocarbons. There was a strong correlation between room temperature diffusivity and density for the oil shales. Thermal conductivity values decreased with increasing temperature for the oil shales to 400°C and then decreased abruptly as oil was removed. After the oil was removed, the specific heat values for all of the shales became nearly identical. The conductivity of the Mt. St. Helens sample increased with increasing temperature indicating a significant radiation heat transfer component even near room temperature.

INTRODUCTION

The thermal conductivity of two types of geologic materials, namely, a dacite dome sample from Mt. St. Helens and certain varieties of oil shale, was determined. Measurements were made of the thermal diffusivity (α), specific heat (C_p), and density (ρ); thermal conductivity (λ) values were calculated as the product of these three quantities, i.e., $\lambda = \alpha C_p \rho$.

The flash method, in which the front face of a small disc-shaped sample is subjected to a short laser burst and the resulting rear-face temperature rise is recorded, was used to measure thermal

611

diffusivity. A highly developed apparatus exists at the Thermophysical Properties Research Laboratory, and we have been involved in an
extensive program to evaluate the technique and broaden its uses.
The apparatus consists of a Korad K2 laser, a high vacuum system including a sample holding assembly, a spring-loaded thermocouple and
an IR detector, appropriate biasing circuits, amplifiers, A-D con-
verters, crystal clocks, and a minicomputer-based digital data
acquisition system capable of accurately taking data in the 40 μs
and longer time domain. The computer controls the experiment, col-
lects the data, calculates the results, and compares the raw data
with the theoretical model.

A Perkin-Elmer Model DSC-2 Differential Scanning Calorimeter
was used to measure specific heat. A reference and a sample holder
were equipped with heaters and temperature sensors that detect fluctuations of the sample holder. A closed-loop electronic system
provided differential power to the heaters to compensate for fluctuations. This differential power was read out in millijoules per
second and is equivalent to the rate of energy absorption or evolution of the sample. By comparing this rate with the rate measured
during the heating of a known mass of sapphire, the specific heat
was calculated. The experiments were performed under computer control, and the specific heat was automatically calculated at equal
temperature intervals.

RESULTS

The bulk density values for the flash diffusivity samples before
and after the diffusivity measurements are given in Table 1. The
weight loss was insignificant for the Mt. St. Helens sample but was
considerable for the oil shale samples as oil was evolved during
heating.

Table 1. Bulk Density Values

Sample No.	Thick (cm)	Width (cm)	Width (cm)	Mass (g)	Density (g cm^{-3})	Density[†] (g cm^{-3})
JDF-6-29-80	0.2824	1.284	1.273	0.6905	1.4956	1.4917
3B-4	0.1748	1.269*	---	0.4207	1.9027	1.5120
H5-14	0.1753	1.269*	---	0.3764	1.6974	1.1842
P2-2	0.1633	1.269*	---	0.3310	1.6018	1.0627
P2-32	0.1773	1.270*	---	0.4866	2.1666	1.8845
P3-10	0.1735	1.267*	---	0.4194	1.9161	1.4967

[†] Measured after heating.
* Sample diameter.

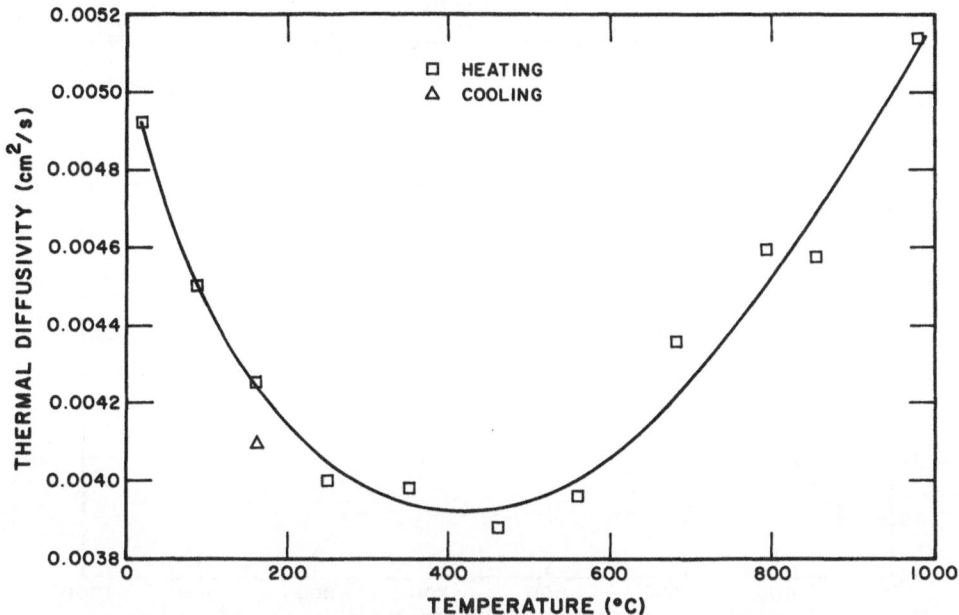

Fig. 1.   Thermal diffusivity of Mt. St. Helens sample.

Thermal diffusivity results for the Mt. St. Helens sample are plotted in Fig. 1.   The diffusivity decreases with increasing temperature from about room temperature to 500°C.   However, at temperatures above 500°C, the diffusivity increases with increasing temperature.   This increase is probably due to a radiation component. Similar behavior has been noted in somewhat porous ceramics with low diffusivity values.   A diffusivity value obtained during the cooling cycle agreed well with that obtained during the heat cycle, showing the material properties were not altered during the tests.

Thermal diffusivity results for the oil shales are plotted in Fig. 2.   The results from room temperature to 400°C were obtained relatively quickly.   Then the samples were maintained at about 400°C for one-half hour and the diffusivity remeasured.   Next the sample temperatures were raised to 500°C where the last measurement was taken.   The diffusivity values decreased with increasing temperature up to 400°C.   At 400°C, the diffusivity values for the different materials behaved differently, but all the samples had a similar diffusivity value at 500°C.   There was a strong correlation between room temperature diffusivity and density.   Sample P2-32 had the highest density and diffusivity, samples 3B-4 and P3-10 had nearly the same density and diffusivity values, sample H5-14 had the next to lowest density and diffusivity values, and sample P2-2 had the lowest density and diffusivity value.   The diffusivity values of the shale after removal of the oil were quite low (about $0.002 \text{ cm}^2 \text{ s}^{-1}$)

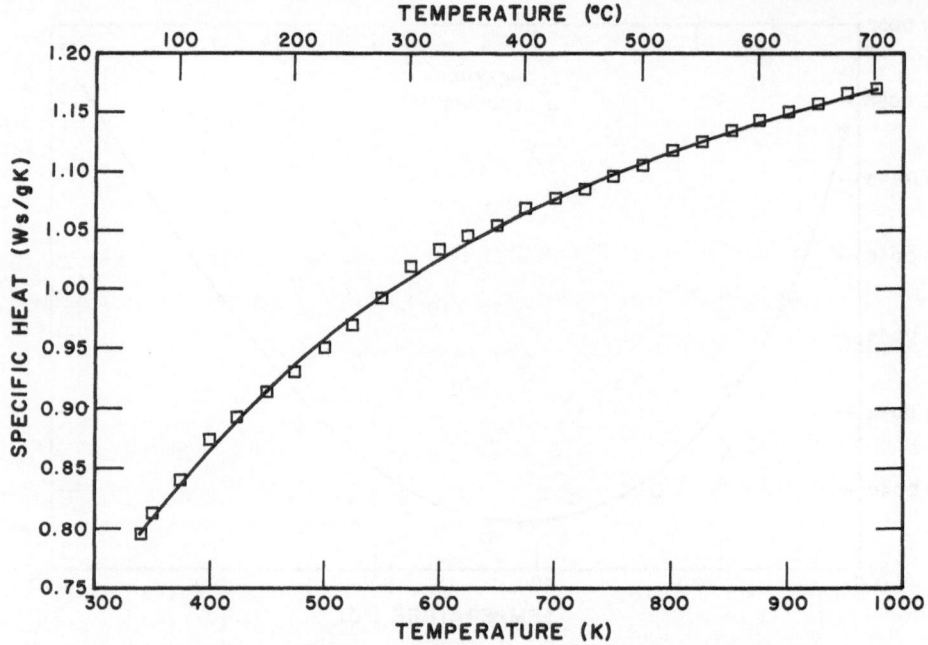

Fig. 2.   Thermal diffusivity of oil shales.

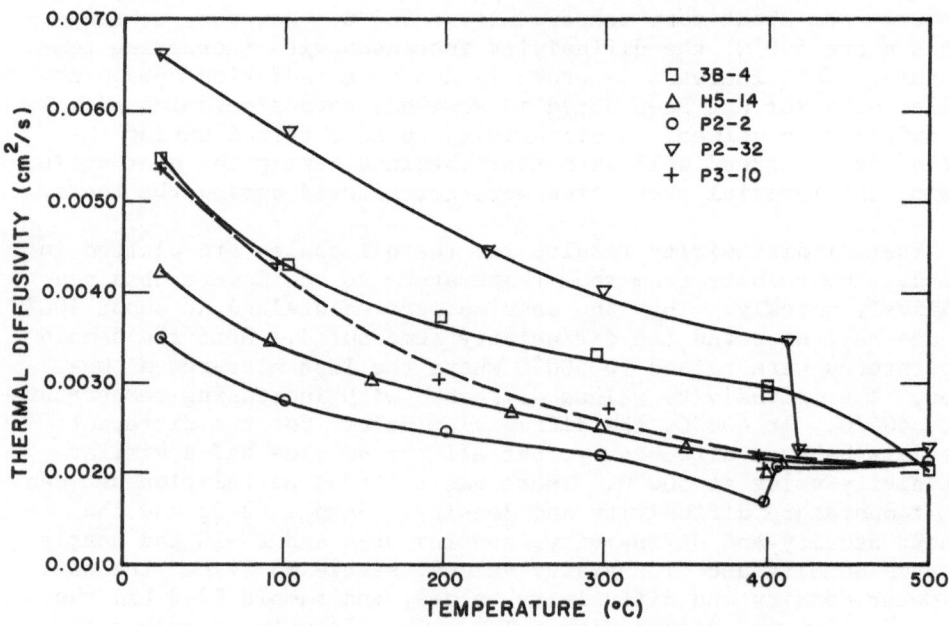

Fig. 3.   Specific heat of Mt. St. Helens sample.

and are comparable to those of air. Thus, if the diffusivity measurements had been made in vacuum, the diffusivity results would be
lower than those reported here.

    The specific heat results for the Mt. St. Helens sample are
included in Fig. 3. These specific heat values follow the usual
behavior for this type of material. Thus the values can be extrapolated safely beyond the range of the measurements, especially since
the diffusivity results exhibited no anomalies.

    The specific heat results up to 300°C for the as-received oil
shales (and those given one or two quick heatings to 300°C) are
plotted in Fig. 4. Presumably a light hydrocarbon was evolved
during this initial heating (see, for example, the differences in
results for P2-2 Runs 1 and 2). For 3B-4 and P2-32, the specific
heat values obtained during the first heating are not shown; these
values are of marginal interest to the present study, since the
stable values of the specific heat are desired. The results obtained
on Run 2 correspond to the stable condition, and the results for subsequent reruns are close to these values. The specific heat results
obtained for the second and third runs were within a few percent of
each other. The results for the second run are shown in Fig. 4,
since the first run data invariably exhibited a peak near 150°C.

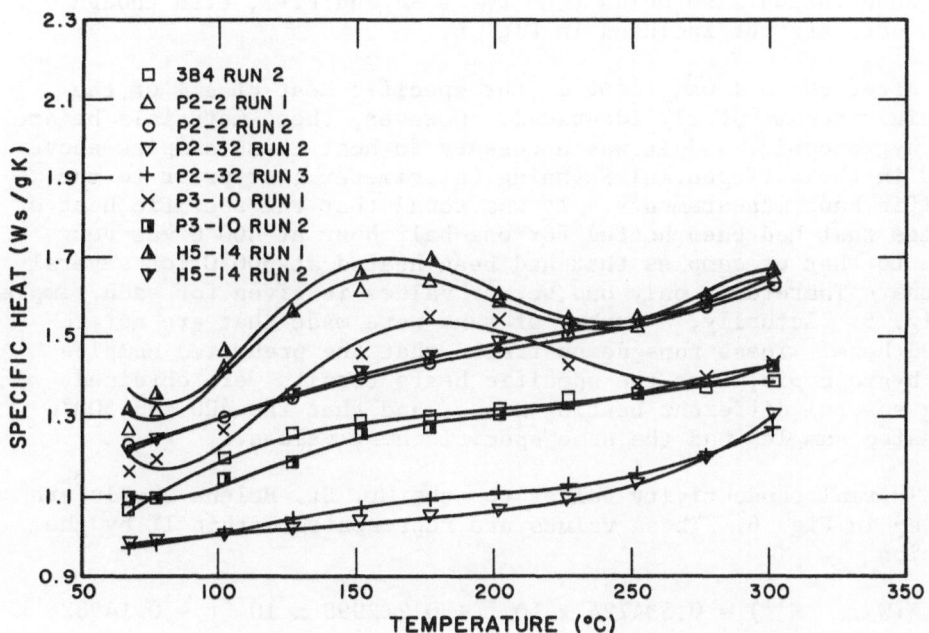

Fig. 4. Specific heat of oil shales to 300°C.

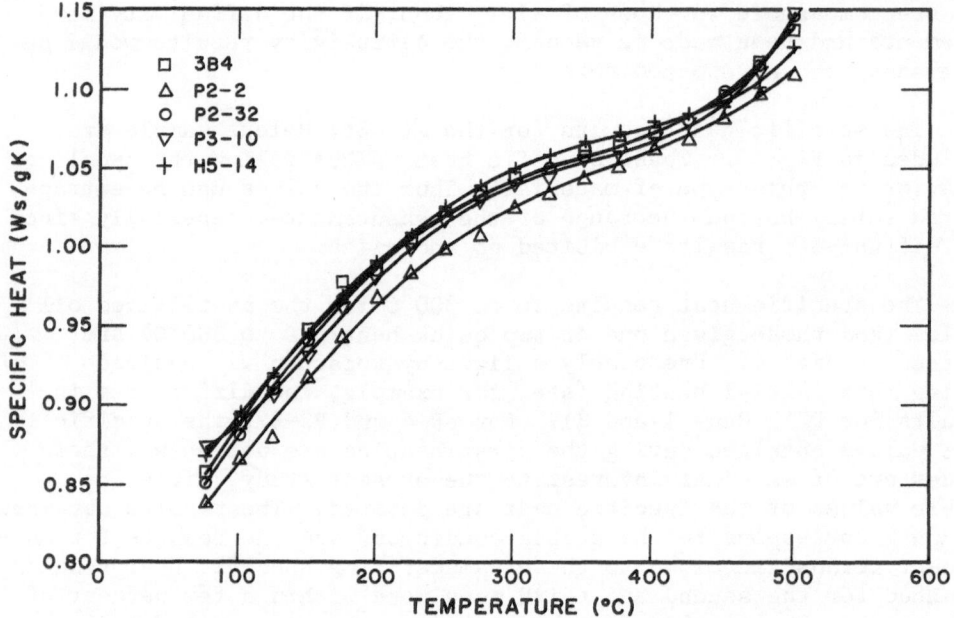

Fig. 5.   Specific heat of oil shales (preheated).

This observation also holds true for 3B-4 and P2-3, even though
these data are not included in Fig. 4.

After the oil was removed, the specific heat of all of the
materials became nearly identical. However, these materials became
very hygroscopic, and it was necessary to heat these samples above
110°C in the Differential Scanning Calorimeter just prior to the
specific heat measurements. It was found that the specific heat of
samples that had been heated for one-half hour at 400°C was very
close to that of samples that had been heated at 500°C for several
minutes. Therefore, only one set of values is given for each sample
in Fig. 5. Actually, a number of runs were made that are not reported here. These runs demonstrated that the preheated samples
were hygroscopic, that the specific heats results were obtained
using several different heating rates, and that the 400 and 500°C
preheated samples had the same specific heat values.

Thermal conductivity values for the Mt. St. Helens sample are
plotted in Fig. 6. These values are represented within 1% by the
equation

$$\lambda (W\,cm^{-1}\,K^{-1}) = 0.534724 \times 10^{-2} + 0.212998 \times 10^{-5}t - 0.149825 \times$$

$$\times 10^{-8}t^2 + 0.369771 \times 10^{-11}t^3 \qquad 23 \le t \le 1000°C.$$

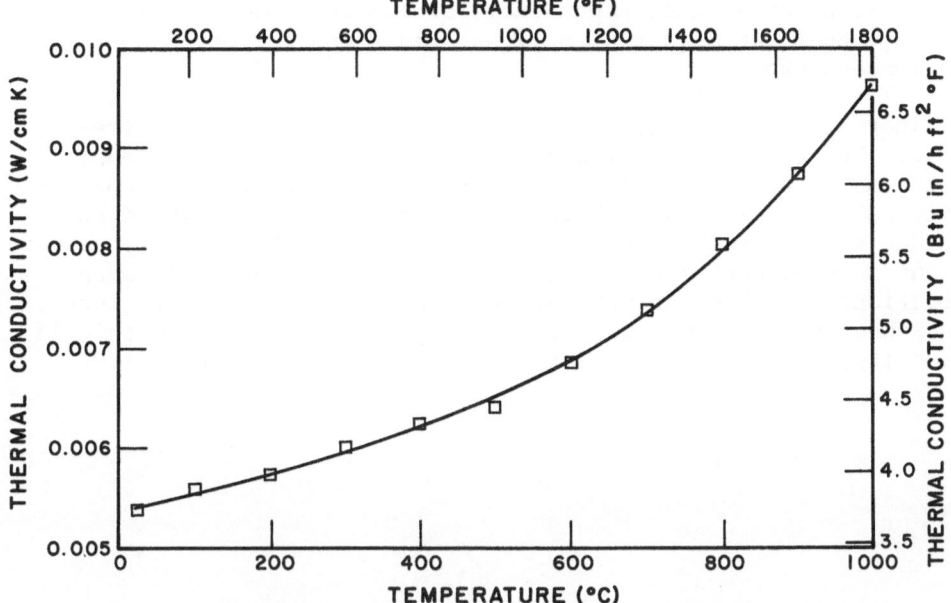

Fig. 6.   Thermal conductivity of Mt. St. Helens sample.

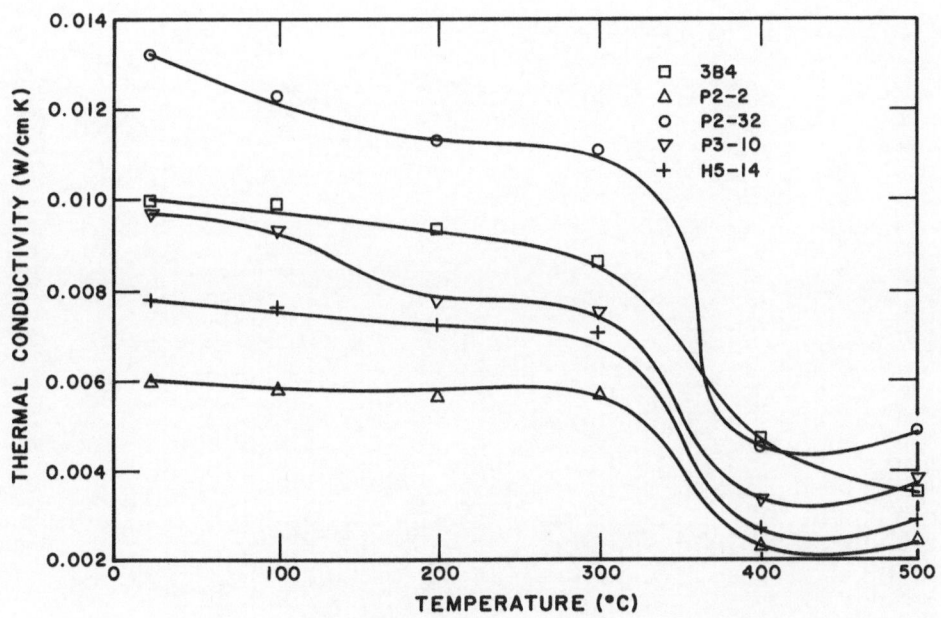

Fig. 7.   Thermal conductivity of oil shales.

The positive temperature dependency for the thermal conductivity
indicates that there is a significant radiation component above
room temperature.

Thermal conductivity values for the oil shales are plotted in
Fig. 7. These results show a strong correlation between density
and thermal conductivity at room temperature. The thermal conductivity values decreased with increasing temperature to 400°C and
then decreased abruptly as the oil was removed. The density values
for the samples heated above 400°C were based on weight loss considerations only and, therefore, are subject to some uncertainty.
Other than this, the measured quantities are accurate to within 1%
on density, 3% on specific heat, and 3% on thermal diffusivity.

INDUCED CONVECTION DURING CYLINDRICAL PROBE CONDUCTIVITY MEASUREMENTS ON PERMEABLE MEDIA

by S. P. Fodemesi* and A. E. Beck

University of Western Ontario, London, Canada, N6A 5B7

ABSTRACT

In oceanic heat flow measurements, an important measured parameter is the thermal conductivity of the sea bottom sediments. The sediments are usually porous with as much as 70% void space. Since the temperature gradients used in the laboratory experiments are several orders of magnitude greater than the natural gradients, there is a real risk that erroneous values are obtained as a result of heat transfer by induced convection.

In this paper we present results from a program of investigation using the transient needle probe thermal conductivity technique on fluid saturated permeable media with a glass bead matrix. To correlate the convection effects on the temperature sensor in the heater probe with convection behaviour in the medium, eight additional radially located sensors were used; all were scanned frequently with a data acquisition system, from the start of the experiment through a few hours of experimental time.

With typical conditions encountered in oceanic heat flow work, induced convection may commence as early as 60 s from the start of the experiment; as expected, the convection effects are worse when the needle probe is oriented horizontally than when it is oriented vertically (gradients orthogonal to the gravitational field), and a correlation is found between permeability and the time of onset and the extent of convective effects. Errors in conductivity as large as 40% have been found.

Empirical techniques are suggested for detecting and correcting for thermal convection using probe sensor data alone.

---

*Now at Esso Resources Canada Limited, Calgary, Canada, T2P 0H6

INTRODUCTION

   In the field of terrestrial heat flow, the depths over which measurements are made and the radius of the area over which the results are applicable are so small compared with the radius of the earth that linear steady state heat flow conditions are assumed to be valid.   The usual technique is, therefore, to measure temperature gradients and thermal conductivity and from their product to obtain a heat flow value.   Depending on the location, corrections often have to be applied (Beck, 1977a) for such things as lateral inhomogeneity or surface temperature changes, or else a large value for the experimental error must be assigned. One potential problem that has not been studied very closely is that of systematic experimental errors in the determination of thermal conductivities of permeable samples, particularly as they might apply to oceanic sediments.

   In a typical oceanic heat flow measurement, a probe is plunged into the sediments to a depth of a few meters and the temperature gradient is measured by means of temperature sensors mounted along the probe either internally or on outriggers. If the probe is hollow and a core is obtained, conductivity measurements are often made on board ship by means of the "needle" probe technique; more recently, in situ techniques of measuring thermal conductivity have been devised (Sass et al., 1981). In both techniques, a transient method is used and in this paper we discuss potential problems that we feel might arise if insufficient attention is given to the probe geometry and proper experimental conditions.

   Most oceanic sediments are extremely porous and permeable; porosities of 50% are not at all uncommon. The grains of the sediment are usually of higher thermal conductivity than the seawater in the pores, so the value of conductivity is very significantly affected by the void fluid. In fact, a number of people do not bother to measure thermal conductivity and simply estimate it on the basis of some figures given by Ratcliffe (1960). Ratcliffe points out that at a porosity of 35-40%, in order to prevent the granular particles from settling to the bottom of his sample, he had to introduce a gel to keep them uniformly suspended in the medium. This technique would also tend to inhibit convection induced by the method of measurement and might explain the sudden change in character from rapidly rising apparent thermal conductivity to one of decreasing conductivity.

   We have designed experiments to investigate the effects of induced convection and of contact resistance between the probe and the sediment.

EXPERIMENTAL DESIGN

Convection occurs most readily when the buoyancy forces and gravitational forces are parallel; when the temperature gradient is orthogonal to the gravitational field then, under similar conditions to the previous case, the onset of convection is significantly inhibited. Therefore, the first objective was to design equipment in such a way that measurements could be made first with the axis of symmetry vertical and then rotated 90° without disturbing the internal arrangements, so that the axis of symmetry was horizontal.

Three probes, about 15 cm long and of differing diameters, but all falling within the range 2.9 to 3.9 mm which is typical of "needle" probes used in oceanic heat flow measurements, were constructed with the thermistor bead temperature sensor inside a hypodermic needle about which was a closely wound heater coil.

To control the matrix conductivity, glass beads from a single manufacturer were chosen with diameters ranging from 0.1 to 3 mm; the use of beads of different diameters gives media of approximately the same porosity, $40\% \pm 1\%$, but very different permeabilities; intrinsic permeabilities varied from 9.1×10^{-8} to 1.8×10^{-5} cm^2. To vary the medium conductivity while keeping the matrix conductivity unchanged, three different fluids were used to fill the voids: air, water, and alcohol.

To check the results obtained by the transient techniques and in an attempt to devise an empirical rule for estimating whether or not convection had occurred during an experiment, eight other thermistors were embedded in the medium at radii of r_1 (= 0.5 cm) and r_2 (= 1 cm) on orthogonal diameters; the temperature sensors were standard commercial equipment of the hypodermic needle style.

In all cases, the experiment was of sufficient duration to reach the quasi-steady-state. Values of thermal conductivity were obtained from the logarithmic asymptote section (short times) and from the quasi-steady-state (long times) sections of the temperature-time curve.

TYPICAL EXPERIMENT

A typical experiment consisted of initiating heating and then measuring temperatures at the needle probe and the distributed temperature sensors with a data acquisition system. Because the system had a limited capacity for data, the sampling rates had to be varied during the experiment. The rates and time spans of acquisition were chosen to minimize errors due to insufficient data at times of rapid temperature rise but at the same time to use the system to capacity for the longest experiment. Sampling rates were 3/s for the first 150 s, 1/s for the next 15 min, 1 every 5 s for the next 20 min and then 1 every 10 s for the remainder of the experiment. An experiment was normally run for up to 2 h. This enabled the apparent thermal conductivity, λ, to be determined by two methods.

First, the transient cylindrical symmetry radial heat flow equation was used:

$$(1) \quad \frac{4\pi\lambda\theta(t)}{Q} = 2h^* - \frac{4h^* - A^*}{2A^*T^*} + [(1 - \frac{A^* - 2}{2A^*T^*})\ln(D)]\ln(t)$$

where T^*, h^*, and A^* are dimensionless parameters related to time, the contact resistance, and thermal capacity respectively (Beck, 1977b); $\theta(t)$ is the temperature rise at time t after the heat is first supplied at the rate of Q per unit time per unit length; $D = 4\alpha^2/1.7811 a^2$, α is the diffusivity of the medium, and a is the radius of the probe. For sufficiently large times, higher order terms, $O(t)$, and the contact resistance can be ignored and λ can be obtained from

$$(2) \quad \lambda = (Q/4\pi)[\ln (t_2/t_1)]/(\theta_2 - \theta_1)$$

where θ_1 and θ_2 are the temperatures at the needle probe at times t_1 and t_2. It follows from this that if the thermal conductivity is plotted against time, when the experiment is in the logarithmic asymptote region, the value of λ should be independent of time; on the other hand, if the equation is incorrect, either because some of the terms that have been ignored are in fact significant or because other forms of heat transfer occur, then the plot of apparent thermal conductivity will be dependant on time.

By running the experiment for a long period of time, but not so long as to violate significantly the assumption of no axial flow from the central region, the temperature at $r_1 = 0.5$ cm rises at approximately the same rate as the temperature at $r_2 = 1$ cm; this implies that most of the heat crossing the r_1 cylinder is leaving via the r_2 cylinder, that very little of the heat being supplied by the heater is being used to heat the material between r_1 and r_2, and the experiment is now in the quasi-steady-state region. If this is the case, then the steady-state cylindrical symmetry heat flow equation (Carslaw and Jaeger, 1959), equation 3, can be used to check the values of thermal conductivity obtained from the transient case.

$$(3) \quad \lambda = [Q/2\pi (\phi_1 - \phi_2)]\ln R^*$$

where ϕ_1 and ϕ_2 are temperatures at r_1 and r_2 at any given time, and $R^* = r_2/r_1$.

It is interesting to note that, in fact, the steady state linear heat flow equation could also be applied, because if r_1 and r_2 are sufficiently close together the heat transfer can be approximated by the linear heat flow case. This is not important for thermal conductivity calculations, but when considering thermal convection effects, the approximation leads to considerable simplification because of the large body of literature concerning convection between two planes at different temperatures. Since only a small fraction of the source energy is being used to warm the material between r_1 and r_2, the heat per unit area flowing across the intermediate cylinder of radius $(r_1 + r_2)/2$ is therefore

$$(4) \qquad Q' = Q/\pi(r_1 + r_2)$$

so for the linear steady state case we have

$$(5) \qquad \lambda = Q'/(d\theta/dr)$$

where $d\theta/dr = (\phi_1 - \phi_2)/(r_2 - r_1)$, thus

$$(6) \qquad \lambda = [Q/\pi(\phi_1 - \phi_2)][(R^* - 1)/(R^* + 1)]$$

If r_1 and r_2 are sufficiently close that the radial and linear steady-state equations give the same results within a stated limit, then equations 3 and 6 can be equated and tested for errors thus

$$(7) \qquad (R^* - 1)/(R^* + 1) = 0.5 \ln R^*$$

The percentage error in equation 7 relative to, say, the linear case can be calculated from

$$(8) \qquad E = 100 \left[1 - \frac{0.5 \ln R^*}{(R^* - 1)/(R^* + 1)}\right]$$

Typical results indicate errors of 1.5%, 4.2%, and 9.8% for values of R^* of 1.5, 2, and 3 respectively. Since in our case R^* is approximately 2, use of the linear heat flow equation would introduce an error of less than 5%. Therefore, under appropriate conditions we can apply empirically derived relationships found in the literature that assume linear steady-state heat flow.

The values of conductivity found by equations 2 and 3 were compared with those obtained on the divided bar (steady state) apparatus (Beck, 1957) and by computation using a model based upon work of Brailsford and Major (1964) for a two component random mixture. Due to an obvious typographical error, the equation given in Brailsford and Major is incorrect and should read

$$(9) \qquad \lambda = [\lambda_m/4R'][\text{TERM} + (\text{TERM}^2 - 8R')^{1/2}]$$

where $R' = \lambda_m/\lambda_f$, λ_m is the conductivity of the continuous (matrix) phase, λ_f is the conductivity of the fluid phase, TERM $= (2R'-1) - 3P^*(R'-1)$, and P^* is the porosity of the medium.

SOME TYPICAL RESULTS

Because we were trying to determine the range of experimental conditions under which one can be certain that negligible convection occurred, the power inputs were varied. Since this leads to problems of easy intercomparison if temperature versus time is plotted, we have, in the figures that follow, normalized the temperature rise and temperature difference.

Convection is obviously a function of temperature gradient, permeability of the medium, and viscosity of the fluid. The onset of convection will also be determined by the relative effect of buoyancy and gravitational force (i.e., whether the experiment is run with the axis horizontal or vertical). We found that convection could not be induced under any reasonable set of circumstances when the matrix consisted of either 0.1 mm or 0.2 mm diameter glass beads, since the permeability was far too low. Discussion will therefore be limited to the cases where the matrix consisted of the larger beads. We show some typical curves in Figures 1 and 2.

Both figures represent the results from experiments on a matrix of 1 mm diameter glass beads using 0.13 W/cm power input. For Figure 1 the fluid was air, whereas for Figure 2 the fluid was water. In both figures, the a,b,c data were obtained with the axis of symmetry vertical and the d,e,f data with the axis of symmetry horizontal.

The data are presented exactly as obtained, that is, there has been no smoothing. They are clearly noisy in places because of the high sampling rate for temperature measurements; in particular, there are marked changes in character when a change occurs in the sampling rate. However, the noise is very high frequency and can be readily smoothed by one of a number of ways; we chose to fit ninth-order polynomials which are the smooth lines shown in Figures 3 and 4. In Figure 1, the following points should be noted: There is clearly no significant convection, since in 1(a) the curves for each set of four thermistors at a given radius are coincident. Figure 1(c) shows the time plot of

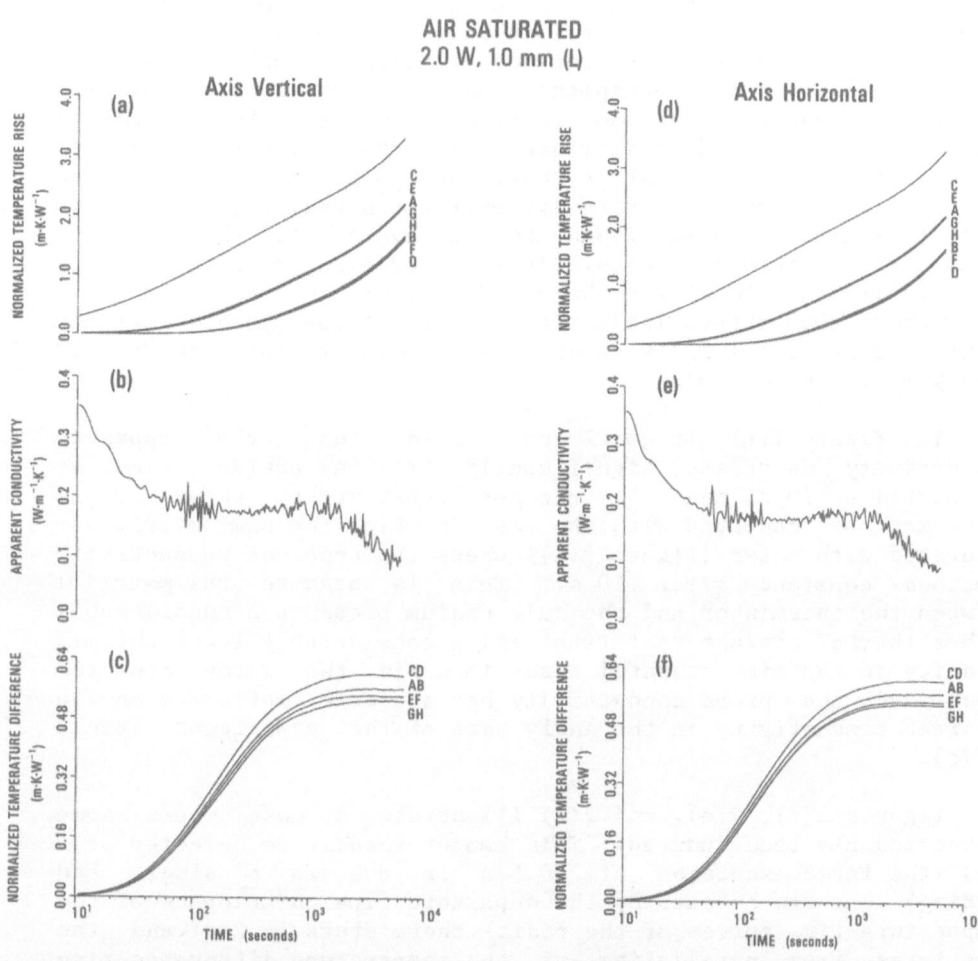

FIG. 1. Typical results for dry (air-saturated) medium with no convection effects; 1 mm diameter beads with 2 W total power input to 3.9 mm diameter, 15 cm long probe.

(a) Normalized temperature rise vs. ln t for probe, 0.5 and 1 cm radius sensors; axis vertical.
(b) Apparent conductivity, using equation 2, vs. ln t; axis vertical.
(c) Normalized temperature differences between pairs of sensors on a radial in the medium; axis vertical.
(d), (e), and (f) similar to (a), (b), and (c) but with axis horizontal.

temperature differences between pairs of thermistors on each of the four radials; in the absence of convection these temperature difference plots also should be precisely equal to one another and appear as one line. However, it must be recognized that we are dealing with a 0.5 cm separation between sensors, and because of the nature of the equipment and the construction of the temperature sensors, it is not possible to centre the thermistor beads precisely at these distances. Therefore, the separation of these four lines is probably a result of positioning errors and does not necessarily indicate differences in radial gradient. In fact, this separation was found to be systematic throughout a given series of experiments. It is interesting to note that the plots in 1(c) are tending to become parallel to the time axis after about 2000 s thus indicating that the temperature difference between radii r_1 and r_2 is constant and therefore that the "quasi" steady state regime holds.

In Figure 1(b) it will be noted that the apparent conductivity decreases significantly from the earliest time, as calculated at 10 s, and only becomes constant at about 100 s. This may be compared with the case in which the same matrix was saturated with water [Figure 2(b)] where the apparent conductivity remained constant after 10 s. This is because the material between the thermistor and the bulk medium presents a considerably higher thermal contact resistance and a considerably lower thermal capacity in the air-saturated case than in the water-saturated case, and the probe conductivity has a greater influence on the apparent conductivity in the early part of the experiment (Beck, 1977c).

Figures 2(d), 2(e), and 2(f) illustrate a case where some convection has been induced. This cannot readily be detected from 2(e) (the large excursion at 10.5 s is due to a single bad reading) but can be seen by the departure from coincidence of the temperature-time curves of the radial thermistors in 2(d) and the departures from parallelism of the temperature difference-time curves in 2(f).

In all cases, the fall off of apparent thermal conductivity with time [Figures 1(b), 1(e), 2(b), 2(e)] after about 1000 s is because the experiment had been running for such a long time that axial heat losses became significant.

EMPIRICAL TECHNIQUES

To this point, we have discussed recognizing the onset of thermal convection qualitatively by observing the graphical results of each experiment. We could readily identify induced

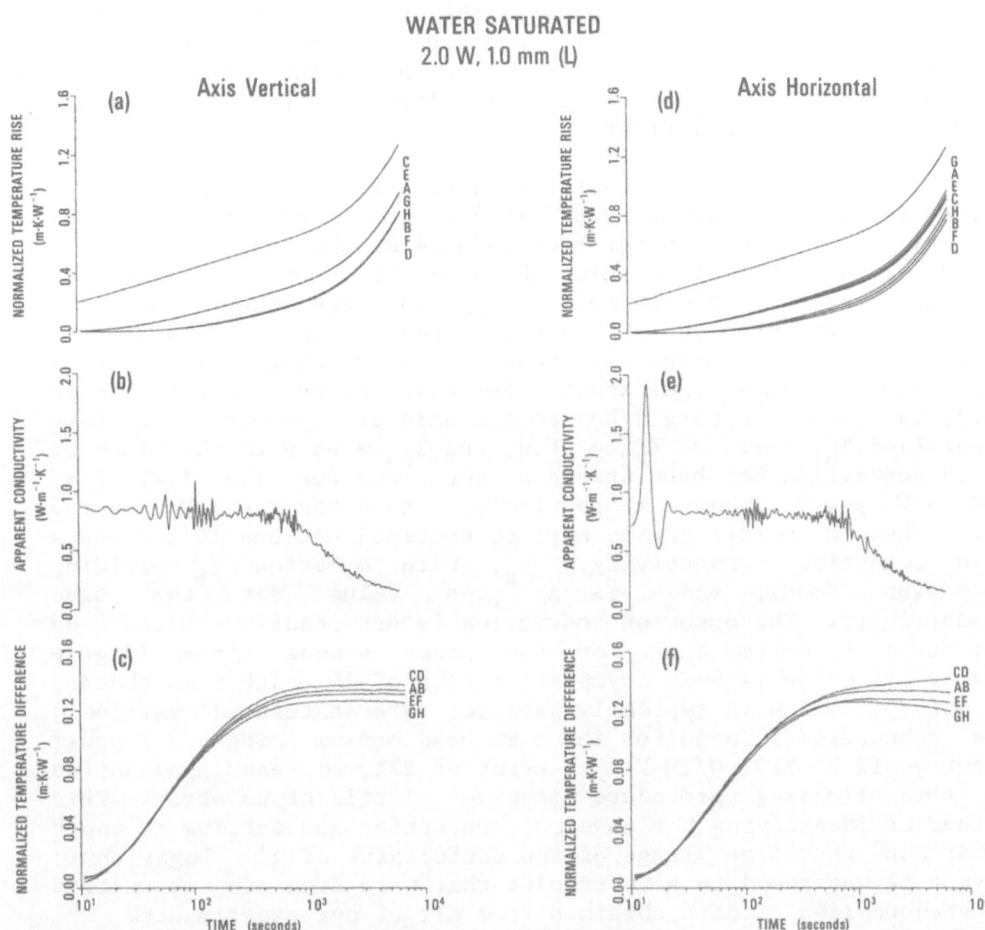

FIG. 2. Typical results for wet (water–saturated) medium, illustrating mild convection effects becoming apparent when axis is horizontal 2(e); 1 mm diameter beads with 2 W total power to 3.9 mm diameter, 15 cm long probe.

(a) Normalized temperature rise vs. ln t for probe, 0.5 and 1 cm radius sensors; axis vertical.
(b) Apparent conductivity, using equation 2, vs. ln t; axis vertical.
(c) Normalized temperature differences between pairs of sensors on a radial in the medium; axis vertical.
(d), (e), and (f) similar to (a), (b), and (c) but with axis horizontal.

convection from the temperature difference-time curves of the four radial sensors. However, in practice, transient thermal conductivity measurements involve the use of an axisymmetric single probe sensor that, obviously, does not allow the measurement of temperature gradients in the medium. Therefore, we will now discuss two methods for detecting the onset of convection from probe sensor data alone.

First, we have successfully used mathematically smoothed apparent thermal conductivity (λ_a) - time curves to locate the time at which convection has been induced at the surface of the probe sensor, thereby defining the upper limit of the logarithmic asymptote region. This method simply involves identifying the time, T_t, at which the λ_a curve departs from an approximate straight line and continues on this path of departure before axial heat loss becomes significant. For example, referring to Figure 3(c), the case involving a horizontal axis of symmetry, we have identified T_{h1} = 112 s, T_{h2} = 58 s, and T_{h3} = 46 s as the times at which convection has been sensed at the probe for the 1 W, 2 W, and 3 W power inputs respectively; thus for the 3 W case the logarithmic asymptote region must be confined between 10 and 46 s when computing conductivity, λ_a, with equation 2, yielding λ_a = 0.86 W/(m-K), which is a good value for the true conductivity. The onset of convection is not readily noticed from the temperature-time curve of the probe sensor alone [Figure 2(a)]. If a logarithmic asymptote region of 15 - 400 s is chosen, an interval which is typically used for water-saturated matrices, the conductivity found for the 3 mm bead medium using a 3 W power input would be 1.09 W/(m-K), an error of 27%, whereas application of this limiting procedure involves little or no error. This method of identifying the onset of convection and setting an upper limit on the time range of the useful part of the logarithmic asymptote was found to give results that were accurate to within 4% when applied to data obtained from all of our experiments.

In cases where convection effects are slight, changes in λ_a become very subtle in the range 15 - 400 s and the location of T_t is difficult, but in these cases (i.e., T_{h1}, T_{v1}, and T_{v2}) use of the standard time range for the logarithmic asymptote gives rise to errors of less than 3%, which are, in effect, insignificant. As a general rule, we have found that if convection is going to be significant, the effects will be observed before T_t = 120 s; there is, therefore, little point in searching for a T_t after this time.

Second, a more quantitative technique has been developed based on the critical Rayleigh number, R_c, (Wooding, 1957) where

(10) $$R_c = k'A'gB'd^2/\alpha_e v'$$

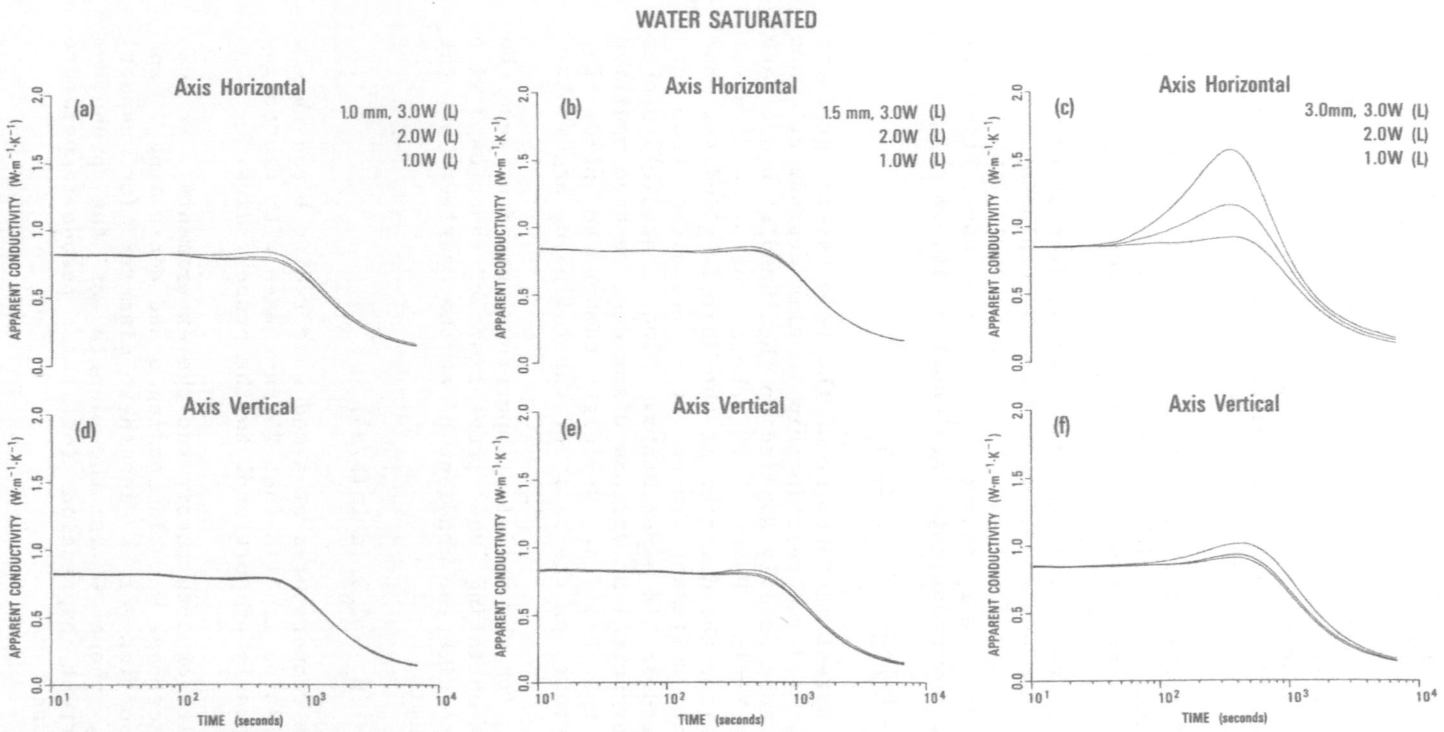

FIG. 3. Results for a typical series of experiments using three medium permeabilities (bead diameters of 1, 1.5, and 3 mm for (a) and (d), (b) and (e), (c) and (f) respectively), three power inputs (1, 2, and 3 W), and two orientations (horizontal and vertical). Where the curves are separable, the upper curve corresponds to the highest power input. Convection effects are clearly negligible at the lower powers whatever the medium permeability, but are becoming quite significant for higher permeabilities when the axis is horizontal; even with the axis vertical, convection is becoming noticeable. 3(a) and 3(d) contain the smoothed curves obtained from the raw data shown in 2(e) and 2(b) respectively.

where k' is the permeability of the medium, A' and v' are, respectively, the coefficient of thermal expansion and the kinematic viscosity of the saturating fluid, g is the gravitational acceleration, B' is the vertical temperature gradient between a layer of thickness d, and α_e is the "effective" thermal diffusivity (Elder, 1965) obtained by dividing the thermal conductivity of the fluid-saturated medium (λ_s) by the thermal capacity, $(\rho C)_f$, of the saturating fluid, i.e., $\alpha_e = \lambda_s/(\rho C)_f$.

Application of equation 10 is limited to two-dimensional flow through a porous medium in which linear steady-state heat flow is assumed and a direct measurement of the vertical temperature gradient is required. More recently, Genceli (1980) has modified the critical Rayleigh number, on the basis of studies of transient convection from heated horizontal cylinders, by replacing B' with q/λ_s (where q is the heat flux) and the characteristic length, d, with the square root of $\alpha_e T_t$ (where T_t is the onset time of convection). If we now substitute into equation 10, we obtain

$$(11) \qquad R_c = k'A'gqT_t/[\alpha_e v'(\rho C)_f]$$

In this case, direct measurement of the temperature gradient is not necessary, and the relationship is time dependent; thus equation 11 can be more readily applied to the "needle" probe case for convection analysis. However, the above expression does not take into consideration the diameter of the heated cylinders, and Genceli found that different values of R_c resulted from each diameter of wire used in his experiments. Since "needle" probes are generally constructed of various diameters, we have modified the expression for the critical Rayleigh number to allow for various probe diameters, on the basis of the following arguments.

The expression for the radial temperature gradient can be obtained by differentiating the progenitor of equation 3 with respect to radius to give the gradient B' at the surface of the probe of radius a as

$$(12) \qquad B' = (\emptyset_a - \emptyset_b)/[a \ln(b/a)]$$

where $\emptyset_a$ and $\emptyset_b$ are temperatures at a and some radius b and $b - a$ is small. This relation implies that B' increases with decreasing a; thus B' is inversely proportional to the probe radius.

From the results of both theory and the experiments, it can be shown that B' increases with increasing q and decreasing λ_s and d. With the introduction of a length-to-diameter (or aspect) ratio, $\bar{A} = L/2a$, where L is the length of the probe, a dimensionally correct expression for the pseudo-temperature gradient B'', given as

(13) $B'' = q\hat{A}/\lambda_s d^2$ where $B'' \propto B'$

can be used to replace B' in cases where magnitude is not important. Substituting equation 13 into equation 10 yields

(14) $\Phi = k'A'gq\hat{A}/[(\rho C)_f \alpha_e^2 v']$

where all temperature-dependent variables in the convection parameter Φ are to be evaluated at the initial temperature of the medium (i.e., before any heating takes place).

Analyses using Φ and the corresponding conductivity ratio $R_1 = \lambda_a/\lambda_t$ (where λ_a is the measured conductivity using the standard time interval for the logarithmic asymptote and λ_t is the determined true thermal conductivity) were successfully applied to cases where the heater axis was horizontal; however, good results were also obtained when the same analyses were applied to cases where the heater axis was vertical. Figure 4 is a graph of R_1 versus Φ plotted on double logarithmic scales; results from all experiments are shown.

Linear regression analyses were performed on data containing a convection component in the heat transfer process (i.e., $R_1 > 1 \pm 3\%$) and the slopes were determined to be

(15) $M_h = \exp(0.234 \pm 3\%)$ and $M_v = \exp(0.134 \pm 8\%)$

where M_h represents the slope of the line in the horizontal case and M_v is that for the vertical case. The critical values of Φ for each of these cases were calculated to be

(16) $\Phi_{hc} = 930 \pm 4\%$ and $\Phi_{vc} = 1490 \pm 10\%$

Therefore, if values of Φ exceed Φ_{hc} in the horizontal cases (or Φ_{vc} in the vertical case), it should be assumed that the onset of convection has occurred during the experiment, and the degree of convection is proportional to $\Phi > \Phi_{hc}$ or $\Phi > \Phi_{vc}$.

From regression analyses, we have formulated correction factors C_h and C_v for the horizontal and vertical cases, respectively:

(17) $C_h = \exp[-0.234 \ln(\Phi_h/930)]$

(18) $C_v = \exp[-0.134 \ln(\Phi_v/1490)]$

which are valid in the range $1 < R_1 < 1.5$. If, for example, in the horizontal case, $\Phi_h > 930$, then equation 17 would give the correct value of conductivity to within a few percent.

Although equation 14 was designed for application to the
transient "needle" probe technique, we have deliberately removed
the time-dependent term so that it is not necessary to identify T_t
from a λ_a - time graph. This results in a more practical method
that can be used to predict if convection currents will be induced
for a given set of parameters before the experiment is run, and to
correct λ_a when it is in error.

The prediction method requires a prior knowledge of α_e and
hence λ_s. Since this is assumed to be unknown, we must use a
two-component model, such as that given by Brailsford and Major
(1964) to estimate the conductivity, λ_s. Then the variables can
be calculated and inserted into equation 14; if the critical
value of Φ is exceeded, convection will occur during the
experiment. If convection is not desired, as in the case of
"needle" probe conductivity measurements, the heat flux, q, must
be lowered until the value of Φ is less than the critical value
for the appropriate case. A more detailed account of the
applications of Φ to needle probe data can be found in Fodemesi
(1979).

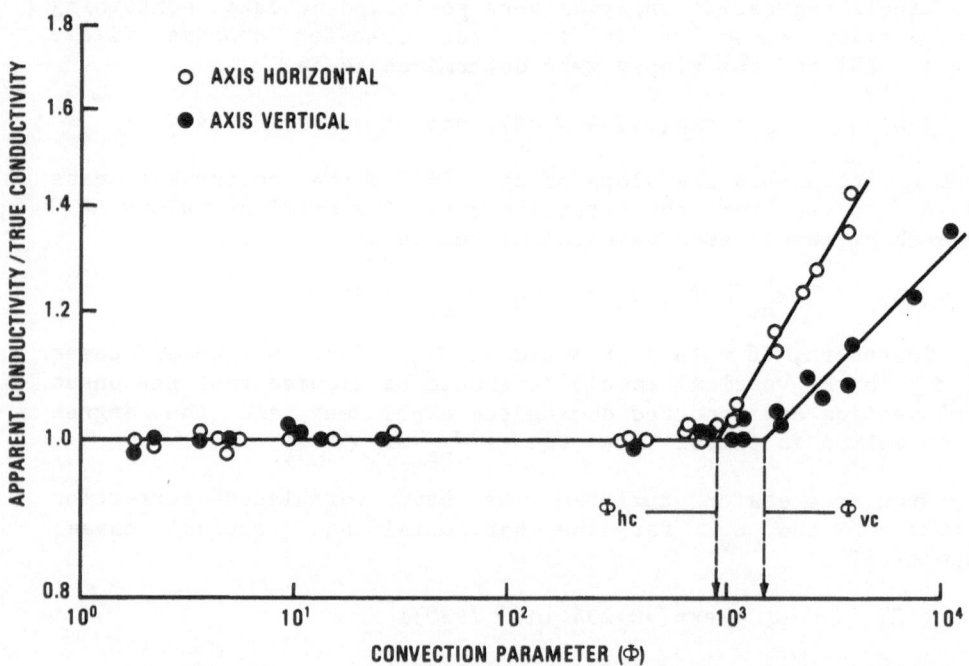

FIG. 4. Experimentally determined heat transfer characteristic
for free convection from a heated axial source in a permeable
medium, showing critical values (hc - axis horizontal, vc - axis
vertical) above which convection becomes increasingly important.

CONCLUSIONS

Induced thermal convection gives rise to an increased rate of heat transfer and hence to a value of thermal conductivity that is too high; the error in the apparent conductivity increases with increasing convection. Two empirical techniques have been developed which can be used to detect and correct convective effects induced during line source conductivity measurements:

(1) A graph of time-dependent variations in apparent conductivity, in addition to being useful for detecting contact resistance and boundary effects, is essential for locating the time of onset of significant convection at the line source and for subsequently removing these effects from the computations of the transient thermal conductivity.

(2) A convection parameter, Φ, offers a quantitative method of correcting for the presence of convection within the specimen. An important property of the convection parameter is that, although it functions as a modified Rayleigh number for porous media, a direct measurement of the temperature gradient is not a prerequisite for its computation. If the permeability and effective thermal diffusivity of the medium are known approximately, the convection parameter can be used to predict whether or not significant convection will be induced at any given power input to the line source. The applications of this method can be extended to other geothermal or engineering problems involving a cylindrical source of heat surrounded by a permeable medium.

Even if convection effects are insignificant, the technique of plotting apparent conductivity against time during the transient experiment enables one to obtain results more rapidly than is usual, and is also effective in signalling when the boundary condition of radial heat flow at the central plane is being violated.

ACKNOWLEDGEMENTS

This work was carried out with the aid of an operating grant from the Natural Sciences and Engineering Research Council of Canada.

REFERENCES

Beck, A. E., 1957. A steady state method for the rapid measurement of the thermal conductivity of rocks. J. Sci. Instr. 34:186-189.

Beck, A. E., 1977a. Climatically perturbed temperature gradients
    and their efffect on regional and continental heat flow
    means. Tectonophysics 41:17-39.

Beck, A. E., 1977b. Geothermal measurements in five small lakes
    of northwest Ontario: Discussion. Can. J. Earth Sci.
    14:332-334.

Beck, A. E., 1977c. A potential systematic error when measuring
    the thermal conductivity of porous rocks saturated with a
    low-conductivity fluid. Tectonophysics 41:9-16.

Brailsford, A. D. and Major, K. G., 1964. The thermal
    conductivity of aggregates of several phases, including
    porous materials. Br. J. Appl. Phys. 15:313-319.

Carslaw, H. S. and Jaeger, J. C., 1959. "Conduction of Heat in
    Solids," 2nd ed., Oxford University Press, Oxford.

Elder, J. W., 1965. Physical processes in geothermal areas, in:
    "Terrestrial Heat Flow." W. H. K. Lee ed., Monograph No. 8,
    American Geophysical Union, Washington, pp 211-239.

Fodemesi, S. P., 1979. The removal of convective effects induced
    during needle probe thermal conductivity measurements, M.Sc.
    Thesis. University of Western Ontario, London, Canada.

Genceli, O. F., 1980. The onset of manifest convection from
    suddenly heated horizontal cylinders.
    Warme und Stoffubertragung 13:163-169.

Ratcliffe, E. H., 1960. The thermal conductivity of ocean
    sediments. J. Geophys. Res. 65:1535-1541.

Sass, J. H., Kennelly, J. P., Wendt, W. E., Moses, T. H., and
    Ziagos, J. P., 1981. In situ determination of heat flow in
    unconsolidated sediments. Geophysics 46:76-83.

Wooding, R. A., 1957. Steady state free thermal convection of a
    liquid in a saturated permeable medium. J. Fluid Mech.
    2:273-285.

MEASUREMENTS ON THERMAL CONDUCTIVITY AND THERMAL DIFFUSIVITY OF ALBERTA OIL SANDS

Nobuhiro Seki,   K. C. Cheng[*] and Shoichiro Fukusako

Dept. of Mechanical Eng.   *Dept. of Mechanical Eng.
Hokkaido University   The University of Alberta
Sapporo, JAPAN   Edmonton, Alberta, CANADA

INTRODUCTION

The increasing demand for and the scarcity of hydrocarbon producible by conventional methods are attracting much attention to the deposits of highly viscous hydrocarbon material such as oil sands and oil shale. It is well known that the Alberta oil sands deposits contain one of the world's major hydrocarbon reserves and the proved reserves of crude bitumen in place amount to as much as the total known conventional world's reserves. Two most prospective in situ recovery techniques developed to date, steam injection and in situ combustion, mainly depend on the reservoir-preheating of the oil sands to increase the mobility of the oil in the formation. Thus, the thermophysical properties of the oil sands are needed for a wide range of temperatures. At the present time, it appears that little information has been published in the open literature on the thermophysical properties of bituminous sands. Although information on porosity, permeability, and oil viscosity of the oil sands reservoir is available, the published information on thermal conductivity and thermal diffusivity appears to be rather limited[1-3].

The objective of the present study is, thus, to obtain values of the thermal conductivity and the thermal diffusivity for the Alberta oil sands over a wide range of temperatures. Alberta oil sands is a four-component mixture of mineral, liquid water, gas, and highly viscous bituminous oil. The most important feature of the Alberta oil sands is the nature of the moisture of the sand particles. Each sand particle is surrounded by a water film and the voids are filled with water, oil, and a gas in varying proportions. For a typical sands grain of 100 µm diameter, the water film thickness is on the order of 2 µm and the bitumen film thickness is 5-6 µm.

It was found that the thermal conductivity values range from about 1.07 W/mK at 15°C to about 0.35 W/mK at 480°C, and the thermal diffusivity values from about 5.28×10^{-1} mm^2/s at 15°C to about 1.97 $\times 10^{-1}$ mm^2/s at 480°C. In order to carry out the above measurement, one 0.22 m^3 container of average-grade oil sands was transported by air from the Oil Sands Research Centre, Alberta Research Council (ARC), Edmonton, Alberta, Canada. It should, therefore, be pointed out that the oil-sands sample received represents a disturbed one. The relevance of the thermophysical properties obtained using the disturbed oil-sands sample in the laboratory to the same properties of the in situ oil sands may not be clear and the possible difference should be recognized. However, it appears that the present study may yield important experimental data, forming the basis for further studies in the future.

DESIGN OF TEST EQUIPMENT

Conventional steady-state measurements of the oil sands sample suffer, in general, from two major disadvantages. First, the desired accuracy of the data requires several hours to establish the corresponding steady-state situation for an individual test. Taking the number of samples and the range of temperature into account, the time required for running these tests appears prohibitive. The second disadvantage is directly associated with the long time period necessary for approaching a steady-state situation. Since the moisture content and the phase changes may have a strong effect on the thermal properties of the oil sands, a uniform distribution of the moisture within the oil-sands sample should be maintained. Unfortunately, it appears that the duration of a steady-state test may encourage moisture migration to the surface kept at lower temperature, which may introduce errors.

For this reason, two transient methods were adopted that require substantially shorter time periods for generating the data. A number of different methods are described in the literature for simultaneously determining the thermal properties, such as thermal conductivity, thermal diffusivity, and specific heat of the samples, by transient heating[4-13].

There are two major requirements for the test apparatus. First of all, possible unaccounted heat losses from the apparatus must be minimized to insure high accuracy of the test data. Second, the ease of construction of the test apparatus and the accessibility for changing the oil-sands samples are also of great importance. Both requirements suggest the application of cylindrical geometry with the oil-sands sample contained in the circular space (an extended Krisher's method originally developed by Saito et al.[11]). As details of Krisher's method are described elsewhere[4,5], the interpretation of the method will not be repeated here.

From the results of the extended Krisher's method, which are in detail interpreted in Ref.11, one obtains the thermal conductivity, the heat capacity, and the thermal diffusivity as follows.

$$\lambda = (q_0/4S\Delta T)r_0^2 \qquad (1) \qquad \rho c = (q_0/S)/(\partial T_c/\partial t) \qquad (2)$$

$$a = \lambda/\rho c \qquad (3)$$

where q_0 is the heat-flow rate per unit length of the cylinder, and other symbols are defined in the Nomenclature at the end of the paper.

EXPERIMENTAL APPARATUS AND PROCEDURES

The design of the test apparatus shown in Fig. 1 was guided by the following requirements: 1) Guard heaters must be employed at both ends of the main heater section to insure solely radial heat fluxes. 2) The time required for running a test must be small by minimizing t_I. 3) Additional guard heaters must be used to minimize heat loss from the main heater to the radial surroundings.

The stated requirements, in conjunction with practical consider-ations for the construction, determined the dimensions as well as the heater power requirements of the test apparatus. The oil-sands sample was placed in the space of copper circular tube, which has a length of 240 mm, a diameter of 32 mm and a thickness of 1.5 mm. The heating wires were wound both on the main and two guard-heating portions of the tube, thus providing uniform heating. The junctions of calibrated iron-constantan thermocouples were embedded in the copper tube for measuring temperatures and for controlling the guard heaters. An additional iron-constantan thermocouple probe was in-stalled along the axis of the tube such that the junction of the thermocouple was located on the central plane of the copper tube.

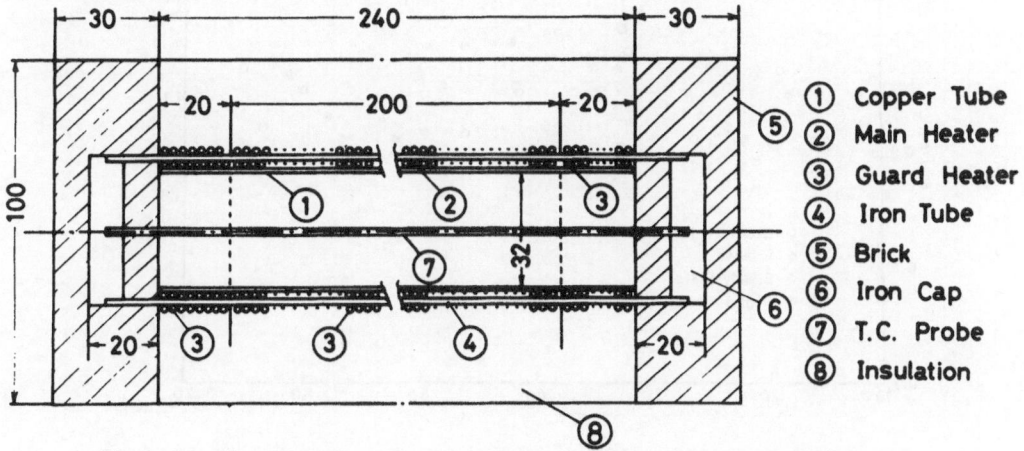

Fig. 1 Sketch of test apparatus (dimension in mm)

By compacting the oil sands into the space of copper tube, the density of the oil-sands sample could be controlled within the desired limits. For the present program, $\rho = 1.95$ Kg/m^3 (which may correspond to the mean density of the undisturbed oil sands) was adopted. After the oil-sands sample was placed in the circular tube, the apparatus was inserted into a thermostatically controlled room built specially for this purpose to establish the desired uniform temperature, T_{in}. During temperature equalization, readings of the various thermocouples attached to the inner surface of the copper tube as well as installed at the central section along the axis of the tube, as previously mentioned, were continuously monitored with a YOKOGAWA data system. After assuring that there were no appreciable temperature differences between the various thermocouple readings, the actual measurement was begun by turning on the main as well as the guard heaters.

RESULTS AND DISCUSSION

For a check of the reliability and accuracy of the test equipment, initial tests have been primarily concerned with the thermal property of the oil-sands sample, which permit comparisons both with accurate steady-state[14] and probe-method[15,16] measurements of the thermal conductivity. For this comparison two sets of test equipment, one similar to that in Ref. 14 and one to that used by Sakate and Nakoto,[16] were constructed. Details of the apparatus for the probe-method are described in Ref. 17.

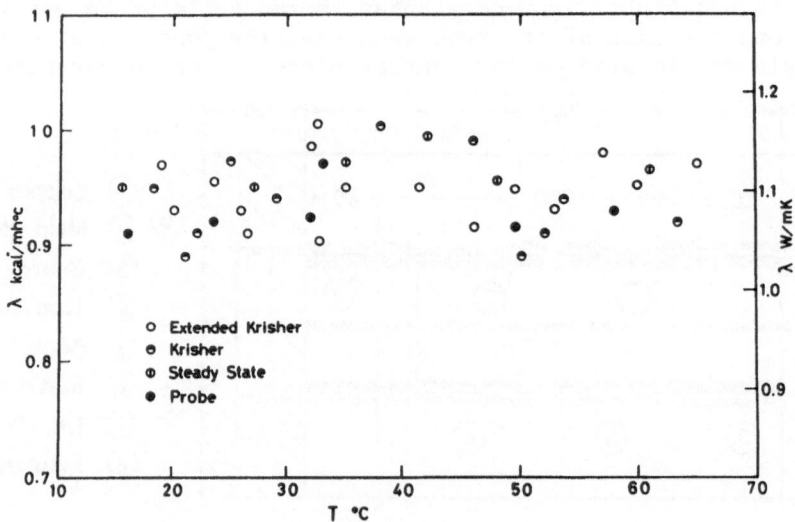

Fig. 2 Comparison of steady-state and transient measurements

Alberta oil sands were transported from Alberta, Canada. From
the typical sample analysis by ARC, Oil Sands Research Centre, details of the composition are: 8.8 t% bitumen, 7.0 t% water, 83.1 t%
soils, and 1.1 t% discrepancy. Under the procedure described in the
previous paragraph, temperature readings were continuously taken at
the axis and the inner surface of the tube containing the oil-sands
sample.

The heat capacities of the copper tube containing the oil-sands
sample and of the heater coil require the following correction for
the evaluation of the thermal data of a sample. The net heat input
to the oil-sands sample cannot be equated with the heat dissipated in
the heater coil because a fraction of this heat is used for raising
both the copper tube and the heater coil itself to the desired temperature level. By taking the heat capacity of copper tube and heater
coil into account, the necessary correction can be calculated. As
the result, this maximum correction in the present program accounts
to about 8.4% of the total heat input.

Figure 2 shows the results of several runs for the various
temperatures from about 15°C to about 65°C using the steady-state
(ASTM C-177) and transient (probe-method) as well as the present
transient methods. It appears that the agreement between the four
methods may be good and within the limits of the accuracy of these
experiments. A random error analysis for the instruments used in the
measurements showed that the accuracy of the thermal conductivities
was within about ± 3% for the steady-state as well as for the transient tests in the range of temperature covered.

Figure 3 shows the effect of temperature on the thermal conductivity of the oil sands. The results show that thermal conductivity depends rather little on temperature in the range 15°C to
70°C. However, it appears that the thermal conductivity may become
a peculiar function of temperature in the range 85° to 120°C. Though
this phenomenon has not been fully understood, it is suspected that
the water film surrounding the sand particles of the oil sands may
play an important role. During the actual measurements, the following behaviors are generally observed. The test begins at the temperature prescribed, and after the initial transient period, it reaches
a quasi-steady state. At this point, axis and inner tube-surface
temperature curves are almost parallel and the temperature increases
almost linearly with time. When the temperature reaches about 100°C,
the temperature curves begin to show small deviations from the ideal
behavior mentioned above. The reason may be mainly due to the fact
that the phase change from liquid to vapor takes place. After the
phase change is completed, it is observed that the system again
becomes quasi-steady state. It should, therefore, be noted that the
results in Fig. 5 for the phase-change region are estimated under the
situation pointed out above. At temperature higher than about 130°C,
it appears that the thermal conductivity may gradually decrease with

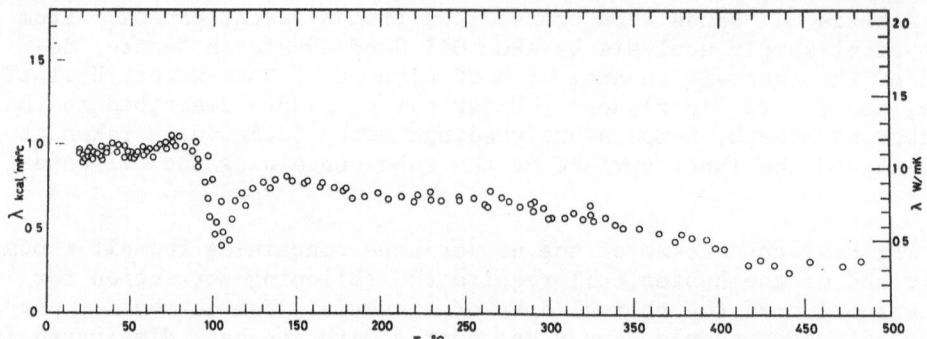

Fig. 3   Thermal conductivity results of oil sands as a function of
         temperature

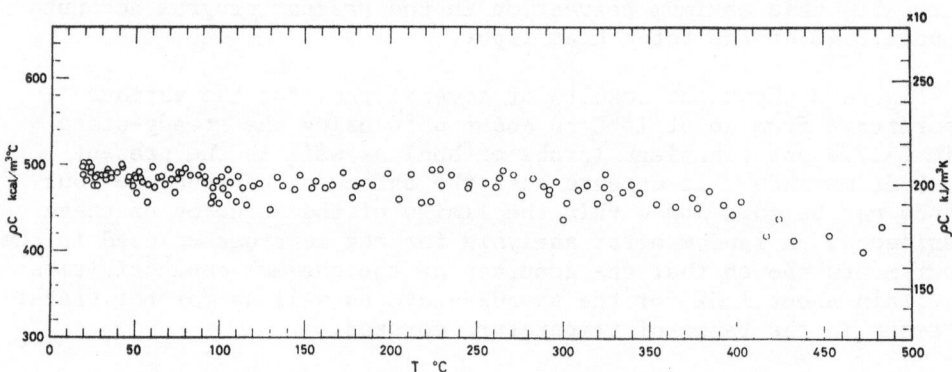

Fig. 4   Heat capacity results of oil sands as a function of
         temperature

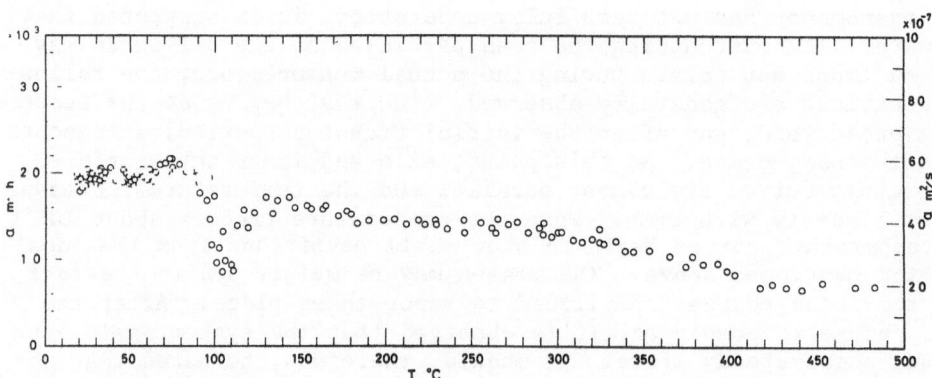

Fig. 5   Thermal diffusivity results of oil sands as a function of
         temperature

an increase in temperature. This phenomenon has not been fully understood either. However, it is thought that it may be due to diffusion of water vapor, which occurs when the field temperature exceeds 100°C, and also to the release of a mixture of nitrogen, carbon dioxide, methane, and other light hydrocarbons upon heating.

From Fig. 4, which shows the effect of temperature on the heat capacity of oil sands, it appears that the behavior of heat capacity may show that for other common soils[11].

When ρc is known, thermal diffusivity, a, can be easily estimated through the relation of $a = \lambda/\rho c$. In Fig. 5 is shown the relation of thermal diffusivity vs. temperature for oil sands. It appears that the behavior of thermal diffusivity may be very similar to that of thermal conductivity.

CONCLUSIONS

The present paper reports the results of the thermal conductivities and the thermal diffusivities of the Alberta oil sands at temperatures ranging from about 15°C to about 480°C. It should, however, be recognized that the oil-sands sample adopted represents a disturbed one. The relevance of the thermophysical properties obtained using the disturbed oil-sands sample in the laboratory to the same properties of the in situ oil sands is not clear, and the possible difference is pointed out. However, it is felt that the present investigation is yielding important data, forming the basis for further studies in the future.

ACKNOWLEDGMENT

The authors are very much indebted to Mr. T. Sugiyama for his help in these experiments.

NOMENCLATURE

a = thermal diffusivity
c = specific heat of sample
L = length of tube
q_0 = heat-flow rate per unit length of tube
r = radius
r_0 = inner radius of tube
S = inner sectional area of circular copper tube
t = time

t_I = time of initial transient period
T = temperature
T_{in} = initial temperature
T_c = temperature of tube axis
T_s = temperature of inner tube surface
$\Delta T = T_s - T_c$
ρ = density of sample
λ = thermal conductivity

REFERENCES

1. K. A. Clark, Some physical properties of a sample of Alberta
    bituminous sand, Can. J. Res., 22F:174 (1944).
2. M. A. Carrigy, The physical and chemical nature of a typical
    tar sand: bulk properties and behaviour, in: "Proceedings,
    Seven World Petroleum Congress," 3:573 (1967).
3. W. H. Somerton, J. A. Keese, and S. L. Chu, Thermal behavior
    of unconsolidated oil sands, Soc. Pet. Eng. J., 413 (1974)
4. O. Krisher, and H. Esdorn, Die wärmeübertragung in feuchten,
    porigen stoffen vershiedener struktur, Forshung, Bd.22:1
    (1956).
5. O. Krisher, and H. Esdorn, Einfaches kurzzeitverfahren zur
    gleichzentigen bestimmung de wärmeleitzahl, der wärme-
    kapazitat und der wärmeeindringzahl fester stoffe, Forsh.-
    Heft, 450:28 (1955).
6. D. Rosenthal, and A. Ambrosio, New method of determining
    thermal diffusivity of solids at various temperatures,
    Trans. Am. Soc. Mech. Eng., 73:971 (1951).
7. L. P. Filippov, Method of simultaneous measurement of heat con-
    ductivity, heat capacity, and thermal diffusivity of solid
    and liquid metal and high temperatures, Int. J. Heat Mass
    Transfer, 9:681 (1966).
8. H. Chang, M. Altman, and R. Sharma, The determination of thermal
    diffusivities of thermal energy storage materials, Trans.
    Am. Soc. Mech. Eng. J. Eng. Power, 89:404 (1967).
9. E. K. Halman, R. W. Gerish, Thermal diffusivity measurements
    from a step function change in flux into a double layer
    infinite slab, Int. J. Heat Mass Transfer, 12:1529 (1969).
10. K. Katayama, and Okada, Transient comparison methods of
    simultaneous measurement of thermal properties, Trans. Jpn.
    Soc. Mech. Eng., 12:1439 (1969).
11. H. Saito, N. Seki, and S. Sakatsume, An investigation of heat
    transfer coefficients of some modeling sands, Trans. Am.
    Found. Soc., 84:243 (1976).
12. R. M. Adbel-Washed, E. Pfender, and E. R. G. Eckert, A transi-
    ent method for measuring thermal properties of soil, Wärme
    Stoffübertrag., 11.1 (1978).
13. J. Mostaghimi, and E. Pfender, Measurement of thermal con-
    ductivities of solids, Wärme Stoffübertrag., 13:3 (1980).
14. ASTM C-177 Standard, and JIS A-1413 Standard.
15. ASTM D-2326 Standard.
16. K. Sakate, and T. Nakoto, Quick measurement of thermal con-
    ductivity, Appl. Phys., 23:314 (1954).
17. T. Sugiyama, Measurements of thermophysical properties of
    Alberta oil sands, B.S. Thesis, Department of Mechanical
    Engineering, Hokkaido University, (1979).

THE APPLICATION OF THERMAL CONDUCTIVITY AND DIFFUSIVITY

MEASUREMENTS IN THE STUDY OF THE MICROSTRUCTURE AND

PHASE TRANSITION OF CERAMIC MATERIALS

Tung-gen Xi, Ben-min Wang, Qi-tao Chen,
He-lin Ni, Zhi-wen Yin, and Tung-sheng Yen

Shanghai Institute of Ceramics
Chinese Academy of Sciences
Shanghai, China

ABSTRACT

In this paper, the thermal conductivities and diffusivities
of three types of new ceramic materials and one ferroelectric
single crystal were measured by using a computerized laser flash
technique, established in our institute. From the measurement data
thus obtained, we further investigated the microstructural changes
or the phase transitions of these materials and tried to examine
with the theories of heat conduction. All of these research results
derived from thermal conductivity and diffusivity measurements were
in good agreement with the experimental results obtained from
structure analysis, high temperature strength tests, and the phase
transitions determined by the electric method. Therefore, thermal
conductivity and diffusivity measurement can be considered as a
new tool or criterion for materials research.

INTRODUCTION

Because the mean free path of the heat conduction carrier,
phonon, is very sensitive to the structural change in dielectric
materials, ferroelectric phase transition has been studied by
Mante and Volger[1] and Nettleton[2] using the thermal conductivity
measurement method. Besides, the microstructural changes, such as
the effects of the formation of micro cracks and the crystallization of mica-glass ceramics, have been examined by Siebeneck et
al.[3] and Siebeneck and Hasselman[4], utilizing the thermal diffusi-

643

vity measurement method. However, the thermal diffusivity studies
of ferroelectric phase transition and the thermal conductivity
studies of the crystallization of the glassy phase at the grain
boundary have not yet been reported, Consequently, in this paper,
these phases were examined by using a computerized laser flash
technique. From the thermal diffusivity temperature curves, it can
be observed that there are discontinuities followed by sharp
rises for lithium tantalate single crystals and doped lead meta
niobate ferroelectric ceramics at their phase transitions from the
ferroelectric state into the paraelectric state. Their Curie temperatures determined from the discontinuity on the thermal diffusivity curves, were found to be in good agreement with the values
measured by electrical methods. Besides, the results also showed
that the thermal diffusivity of poled $PbNb_2O_6$ ceramics was different from those of unpoled samples. This difference is probably
related to the ordered or disordered domain structure orientation
in different samples. The thermal conductivity of the grain boundary phase of hot-pressed Si_3N_4 increased strongly after heat
treatments and its temperature dependence was also changed. This
shows that the glassy phase at the grain boundary has been crystal-
lized. These results were also confirmed by experimental results
from the structure analysis method.

EXPERIMENTAL APPARATUS

The computerized laser thermal diffusivity measurement apparatus, based on the flash method[5], had been established in our
institute[6] (Fig. 1).

Fig. 1. A Computerized Thermal Diffusivity Measurement Apparatus

This apparatus is composed of a laser source with neodymium glass, a vacuum carbon tube furnace, an infrared detector system with lead sulphide, and a computer. In order to correct accuracy, the thermal diffusivity and conductivity of a standard specimen, such as α-Al_2O_3 and Armco iron, were measured by this apparatus. The mean deviations from the mean have been computed to be 2.5%[6]. Data obtained from the standard specimen were compared with recommended values in TPRC[7,8] and found to be in good agreement. The test samples were machined into small disks, 10 mm in diameter and about 1.5 mm in thickness. This apparatus features automatic and high-speed measurement and can be operated in the temperature range 250 to 1800°C. The time interval between laser emission and display of thermal diffusivity on the typewriter is about five seconds.

THERMAL DIFFUSIVITY STUDIES OF FERROELECTRIC PHASE TRANSITION

LiTaO$_3$ Single Crystal

LiTaO$_3$ single crystal samples were grown by the pulling method, with highly pure Li_2CO_3 and Ta_2O_5 as raw materials. The mole ratio of Li_2O and Ta_2O_5 is 0.95. Samples were poled along the Z-axis and cut along the X, Y, and Z planes, respectively. The test samples were prepared and machined into small disk by the crystal growth group of our institute. To avoid the laser light (λ = 1.06 μm) penetrating the sample, the surfaces of the samples were covered with absorbing films.

The temperature dependence of thermal diffusivity of LiTaO$_3$ single crystals is shown in Fig. 2 and Table 1.

Table 1. Thermal Diffusivity, α, of LiTaO$_3$ Single Crystals Along Different Axes

T(°C)	250	300	400	500	600±2 Curie Temp. (°C)	630
α_x (cm^2/s)	0.0072	0.0068	0.0062	0.0057	0.0047	0.0060
α_z (cm^2/s)	0.0086	0.0081	0.0073	0.0067	0.0059	0.0070

In analogy with ordinary gas kinetics, the thermal diffusivity, α, is related to the mean free path, ℓ, of the phonons and phonon velocity, V (nearly equal to the velocity of acoustic waves in the samples), that is,

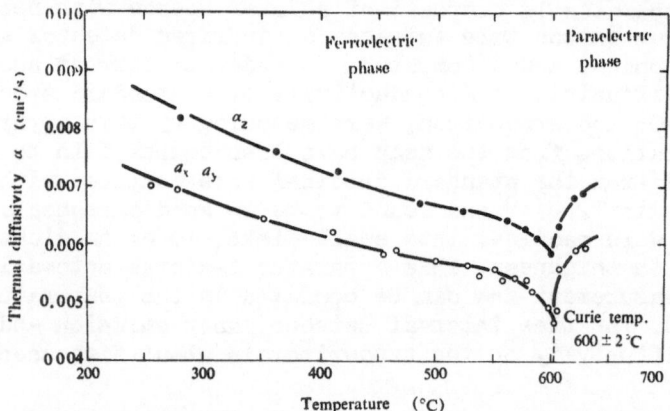

Fig. 2.   Thermal Diffusivity Curves of LiTaO$_3$ Single Crystals
Along Different Axes
(α_z, Parallel to Z-axis. α_x and α_y, Perpendicular to
Z-axis)

$$\alpha = v \cdot l/3 \qquad\qquad (1)$$

The velocity of acoustic waves of poled LiTaO$_3$ single crystals
along the X and Z axes determined by Smith and Welsh[9] was $V_x =$
4.22×10^5 cm/s and $V_z = 6.16 \times 10^5$ cm/s, respectively. From equation
(1), since we already know V_x, V_z and α_x, α_z, we can obtain the
mean free paths, l_x and l_z, along the X and Z axes, listed in Table
2.

Table 2.   Mean Free Path of Phonons in LiTaO$_3$ Single Crystal
Along X and Z Axes

T(°C)	250	300	400	500	600±2 Curie temp. (°C)	630
l_x (10^{-8}cm)	5.12	4.83	4.41	4.05	3.34	4.27
l_z	4.19	3.94	3.56	3.26	2.87	3.40

The results from Fig. 2 and Tables 1 and 2 show that on the
thermal diffusivity-temperature curve of LiTaO$_3$, there is a sudden
change at 600 ± 2°C. This temperature is coincident with the Curie

temperature measured by Fujino et al.[10], and it is also coincident
with the data determined by the crystal property testing group of
our institute using the electric method. At the Curie temperature,
the poled $LiTaO_3$ single crystal changes from the ferroelectric
state into the paraelectric state. Correspondingly, its point group
changes from $C_{3v}(3m)$ without a center of symmetry to $D_{3d}(\bar{3}m)$ with
a center of symmetry. The development of the symmetrical structure
decreases the scatter of phonons and increases the mean free path,
ℓ, of phonons and therefore the thermal diffusivity above the Curie
temperature also increases. The thermal diffusivities of $LiTaO_3$
single crystal along the X-axis and Y-axis are similar, but much
lower than those along the Z-axis. Obviously, the anisotropy of
$LiTaO_3$ thermal diffusivity is due to the different values of V and
ℓ along the X-axis and Z-axis. Since $\ell_z < \ell_x$, the probability of
phonon collisions along Z-axis is larger than that along X-axis.

Doped $PbNb_2O_6$ Ceramics

The $PbNb_2O_6$ ceramic samples, doped with Sr and Ba, were sintered at about 1300°C and suddenly cooled, to maintain their high
temperature structure. Figure 3 shows the thermal diffusivity and
the dielectric constant curves of poled and unpoled $PbNb_2O_6$ ceramics.

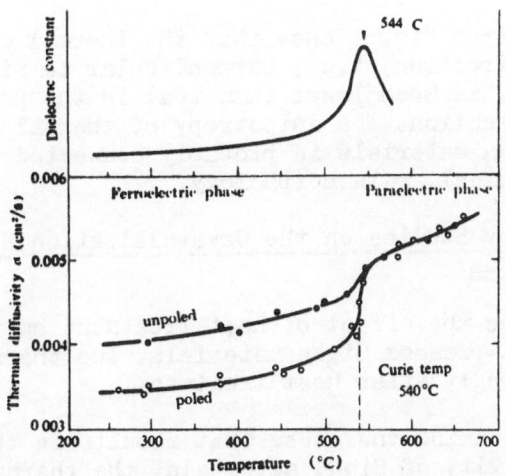

Fig. 3. Thermal Diffusivity and Dielectric Constant Curves of
        Doped $PbNb_2O_6$ Ceramics

The results from Fig. 3 show that there is also a sudden rise
in thermal diffusivity of doped $PbNb_2O_6$ ceramic at about $540^\circ C$.
The Curie temperatures determined from the discontinuity in the
thermal diffusivity curves were found to be in good agreement with
the values measured by the dielectric constant curve. This means
$PbNb_2O_6$ ceramic changes from the ferroelectric state into the para-
electric state at about $540^\circ C$. Besides, the results also show that
the thermal diffusivities of poled $PbNb_2O_6$ ceramics are different
from those of unpoled samples. This difference which is very con-
spicuous below the Curie temperature practically disappears in a
para-electric region beyond the Curie temperature. The most likely
cause should, therefore, be sought in the ordered domain structure
and enhanced phonon scattering from domain walls.

As mentioned above, the thermal diffusivity measurements can
be considered as a new method of studying the ferroelectric phase
transitions.

THERMAL CONDUCTIVITY STUDIES OF THE MICROSTRUCTURE OF HOT-PRESSED Si_3N_4 MATERIALS

Specimens were hot-pressed from a powder mixture of pure Si_3N_4
and sintering aids, i.e. 2%(wt%) Al_2O_3 and 5%(wt%) MgO, under a
pressure of 30 MN/m^2, using a BN-flashed graphite die at $1750^\circ C$ for
2 hours in N_2 atmosphere.

The Effect of Grain Orientations on the Thermal Diffusivity

The thermal diffusivities of hot-pressed Si_3N_4 materials
along different directions are shown in Fig. 4.

The results from Fig. 4 show that the thermal diffusivity in
the hot-pressing direction, i.e., perpendicular to fibrous and
cylindrical grains, is much lower than that in the perpendicular
to hot-pressing direction. The anisotropy of thermal diffusivity
of hot-pressed Si_3N_4 materials is probably connected with their
fibrous and cylindrical grain morphology.

Thermal Conductivity Studies on the Crystallization Behavior of Grain Boundary Phases

Figure 5 shows the effect of heat treatment on the thermal
conductivity of hot-pressed Si_3N_4 materials. The thermal conductivity increased strongly after heat treatment.

In order to examine the cause that results in the increase
of thermal conductivity of Si_3N_4 materials, the thermal conducti-
vities λ_G of the main crystalline phases (α-Si_3N_4, β-Si_3N_4) and
the thermal conductivities λ_B of the grain boundary phases before

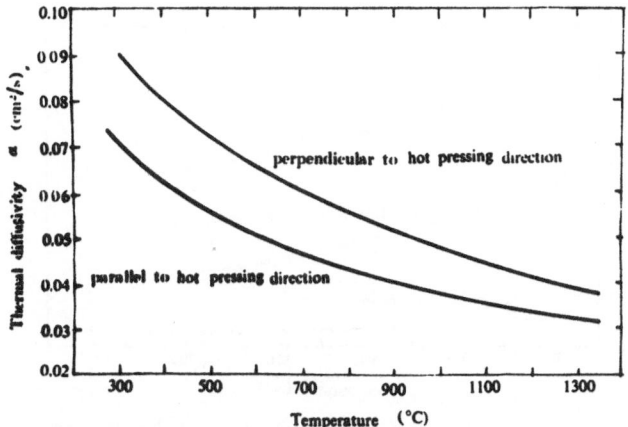

Fig. 4.   Thermal Diffusivity of Hot-Pressed Si_3N_4 Materials
          Along Different Directions

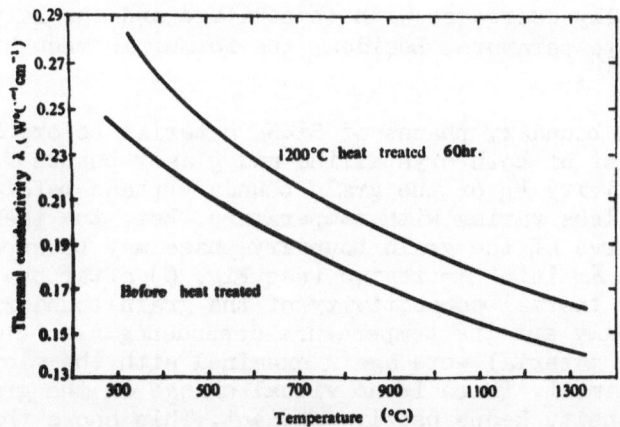

Fig. 5.   Effect of Heat Treatment on the Thermal Conductivity of
          Hot-Pressed Si_3N_4 Materials

and after heat treatment were calculated approximately from our
experimental results, using an empirical formula[11] (see appendix)
for multiphase systems. The calculated results are shown in Fig. 6.

The thermal conductivity behaviors of the crystalline phase

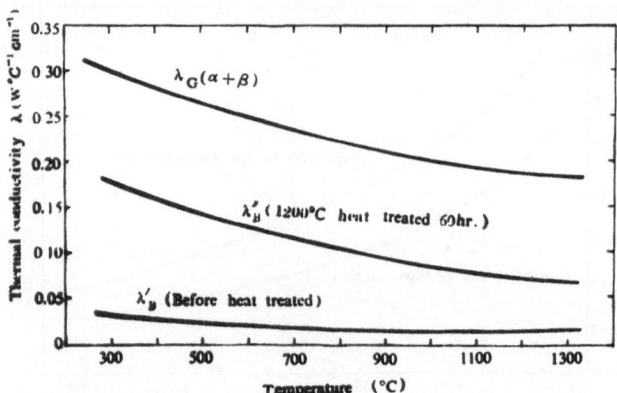

Fig. 6. The Thermal Conductivity λ_G of the Main Crystalline Grain
Phase and the Thermal Conductivity λ_B of the Grain
Boundary Phase of Hot-Pressed Si_3N_4 Materials

and the glassy phase of dielectric materials are quite different;
the former usually decreases with temperature and the latter
increases with temperature. Besides, the former is much higher
than the latter[12].

The grain boundary phases of Si_3N_4 material before heat
treatment consist of both crystalline and glassy phases[13]. The
thermal conductivity λ_B' of the grain boundary phase before heat
treatment much less varies with temperature. But, the thermal
conductivity curve of the grain boundary phase may be expected to
change from the λ_B' into λ_B'' regime (see Fig. 6) after heat treatment, i.e., the thermal conductivity of the grain boundary phase
increased strongly and its temperature dependence also changed. The
grains of Si_3N_4 material were again examined with the microscope
after heat treatment. There is no visual change in the grain size,
and the bulk density keeps nearly constant. This shows that the
glassy phase at the grain boundary had been devitrified after heat
treatment. These results were also confirmed by experimental results obtained from X-ray microanalysis, high resolution electron
microscopy[14], and strength tests at high temperatures.

Hot-pressed Si_3N_4 materials are being considered as a desirable structural material for high temperature applications where
excellent thermal conductivity and high temperature strength are
required. Since heat treatment will promote the devitrification of
grain boundary phases, it has the great advantage of improving
thermal conductivity and strength of Si_3N_4 materials.

All of these results show that thermal conductivity or

diffusivity measurements can be considered as a new tool or criterion for materials research.

APPENDIX

Most ceramic materials are composed of mixtures of one or more solid phases. If the continuous phase has a thermal conductivity, λ_c, and the dispersed phase has a thermal conductivity, λ_d, the resultant thermal conductivity of the mixture is given by

$$\lambda_m = \frac{\lambda_c \left[1 + 2 V_d \left(1 - \lambda_c/\lambda_d \right) \Big/ \left(2 \lambda_c/\lambda_d + 1 \right) \right]}{\left[1 - V_d \left(1 - \lambda_c/\lambda_d \right) \Big/ \left(\lambda_c/\lambda_d + 1 \right) \right]}$$

where V_d is the volume fraction of the dispersed phase. When $\lambda_c \gg \lambda_d$, then the resultant conductivity is $\lambda_m \simeq \lambda_c (1 - V_d/1 + V_d)$. In contrast, if $\lambda_d \gg \lambda_c$, then $\lambda_m \simeq \lambda_c (1 + 2V_d)/(1 - V_d)$.

REFERENCES

1. A. J. H. Mante and J. Volger, Physica, 52:577 (1971).
2. R. E. Nettleton, Ferroelectrics, 1:87 (1970).
3. H. J. Siebeneck, K. Chyung, D. P. H. Hasselman and G. E. Youngblood, J. Am. Ceram. Soc., 60:375 (1977).
4. H. J. Siebeneck and D. P. H. Hasselman, AD-A 029660 (1975).
5. Tung-gen Xi, Sin-yu Zhou, Zong-jie Li, He-lin Ni and Zong-yi Gu, J. Eng. Thermophys. (in Chinese), 1:147 (1980).
6. W. J. Parker, R. J. Jenkins, C. P. Butler, and G. L. Abbott, J. Appl. Phys., 32:1679 (1961).
7. Y. S. Touloukian, R. W. Powell, C. Y. Ho, and P. G. Klemens, "Thermal Conductivity, Thermophysical Properties of Matter", Vol. 1, IFI/Plenum, New York, (1970).
8. Y. S. Touloukian, R. W. Powell, C. Y. Ho, and P. G. Klemens, "Thermal Conductivity, Thermophysical Properties of Matter", Vol. 2, IFI/Plenum, New York, (1970).
9. R. T. Smith, and F. S. Welsh, J. Appl. Phys., 42:2219 (1971).
10. Y. Fujino, H. Tsuya, and K. Sugibuchi, Ferroelectrics, 2:113 (1971).
11. W. D. Kingery, J. Am. Ceram. Soc., 42:617 (1959).
12. Tung-gen Xi, "Thermophysical Properties of Inorganic Materials", Shanghai Science and Technology Press, Shanghai, (1982).
13. R. E. Loehman, and D. J. Rowcliffe, J. Am. Ceram. Soc., 63:144 (1980).
14. D. A. Jefferson, Shu-lin Wen, J. M. Thoms, Yi-tao Chen, Feng-ying Wu, and Tung-sheng Yen, Research Report, University of Cambridge, (1981).

CHAIRMAN

L. L. Sparks
National Bureau of Standards
Boulder, Colorado

THERMAL CONDUCTIVITY OF CONCRETE MORTAR*

L. L. Sparks

Thermophysical Properties Division
National Bureau of Standards
Boulder, CO   80303

INTRODUCTION

The low-temperature thermal conductivity (λ) of concrete is
becoming an increasingly important property because the material is
used to contain and transport cryogenic fuels. A very limited amount
of low-temperature data are available for use by designers of facilities such as ground based storage tanks, sea going tankers, barges,
and pipelines. The variability of existing data is appreciable and
is due to both the large number of variables that affect λ and to the
experimental methods used to determine λ. Perhaps the most difficult
part of finding a meaningful λ for moist concrete is that, even for a
given composition or mix design, the moisture content and distribution
change with time. The thermal gradient necessary in the measurement
of λ also establishes a moisture gradient in the specimen: moisture
migrates toward the cooler part of the specimen.

The conditioning and testing procedure used on the current specimens was intended to minimize the moisture migration problem and to
allow observation of the effect of variable moisture content on λ.
The moisture content of the specimens during the experiment was estimated from mass determinations just before installation in the apparatus, immediately after the system was opened, and after oven drying.
After installation, the temperature of the saturated specimens was
lowered to cryogenic temperatures, where the evaporable moisture was
in the solid phase. A series of tests was made with the highest test

---

*This work was done for the Department of Commerce, Maritime Adminis-
tration, Department of Commerce Building, Washington, DC 20235, under
Program 193000, Project 12-410-54-425.

temperatures below 273 K. The specimens were then warmed to ambient temperature for an extended period to allow for migration and some drying. This conditioning was followed by a second set of λ measurements. The third and final specimen condition was achieved by holding the specimens at approximately 340 K and reduced pressure for several days. This procedure was followed by a third set of λ determinations.

APPARATUS

The apparatus used to make the λ measurements is commonly known as a guarded hot plate and is described in the American Society for Testing and Materials Standard C177 (ASTM, 1978). This method allows an absolute determination of λ and is considered to be the most accurate method available for many materials. The particular apparatus used to make the λ measurements reported here is described in detail by Smith, Hust, and Van Poolen (1981).

On the basis of the percentage errors given by Smith and the magnitude of the parameters for the concrete experiments, uncertainties for concrete are estimated to be 0.004 W/m·K random and 0.010 W/m·K systematic near room temperature; near 80 K the random component is 0.003 W/m·K and the systematic uncertainty is 0.013 W/m·K.

MATERIAL

Commercially important portland cements are available in a variety of compositions, but all are composed primarily of reactive calcium silicates or calcium aluminates. When mixed with water, these materials form insoluble hydration products and make up a class known as hydraulic cements. The thermal and mechanical properties of mortars or concretes made using hydraulic portland cement are influenced by cement composition, cement content, type, size, and amount of aggregate, water to cement ratio, age and aging environment.

The material used here is a mortar consisting of type I portland cement with (water mass)/(cement mass) equal to 0.5 and (aggregate mass)/(cement mass) equal to 3.38. An air entraining agent, neutralized resin, was included in the mix. The aggregate was primarily granite and quartz graded as shown in table 1. Computed air content of the plastic mortar was 17.7 percent and the density was 1.9 g/cm^3 (ASTM C138).

The average density of the aged specimens in the saturated-surface-dry (SSD) condition was 1.869 g/cm^3 and the oven-dry density was 1.686 g/cm^3. Details of the mixing and aging of the specimens is reported by Sparks (1981).

Table 1.   Sieve Analysis of the Clear Creek Sand
Used in NBS Mortar Specimens

Sieve		Percent	Cumulative
Number	Size (mm)	Passing	Percent retained
4	4.75	99.4	0.6
8	2.36	80.5	19.5
16	1.18	59.4	40.6
30	0.60	28.6	71.4
50	0.30	6.6	93.4
100	0.15	0	100

EXPERIMENTAL RESULTS

The thermal conductivity results represent the mortar with three
different moisture contents. The initial average moisture content,
as estimated above, was 10.4 percent of the dry specimen weight.
Figure 1 (◊ and *) shows the data for the mortar in this condition.
The average specimen temperature was above 269 K for approximately
6.5 h while taking the data shown in this figure. Some redistribution of the moisture could be expected with the migration toward the
low-temperature surface. The temperature of the data points shown in
figure 1 does not, in general, reflect the order in which they were
taken. The excursion to temperatures higher than 273 K made no
detectable change in the thermal conductivity at lower temperatures.

Experimentally, the temperature range between 225 and 235 K was
very difficult for the specimens with 10.4 percent evaporable moisture. Equilibrium conditions were tenuous compared with those outside this range. The results in this range (*) are significantly
less precise than the data taken at higher and lower temperatures.
The sequence of observations in this range, in order of decreasing
conductivity, was run 13, run 22, run 14, and run 9; these data were
taken over a period of 3.2 days.

The second set of data (●) shown in this figure was determined
after the specimens had been conditioned at temperatures higher than
273 K for approximately 20 h. This treatment was intended to alter
the moisture content and distribution enough to make an observable,
but not extreme, change in λ. The conductivity decreased by 2.5 percent when the gradient conditions at 317 K were held for 15 h (+).

Subsequent to these tests, the temperature of the specimens was
maintained at 325 ≲ T ≲ 330 K for 72 h while the system pressure was
reduced to less than 26 kPa. The system was purged with dry nitrogen
gas several times during this period. This procedure was intended to
produce a severe loss of moisture from the specimens. The thermal

conductivity data shown in figure 1 (▲) were taken with the specimens in this condition.

Immediately following the third series of measurements, the specimens were removed from the system and weighed. The moisture content was found to be 4 percent of the dry-specimen weight.

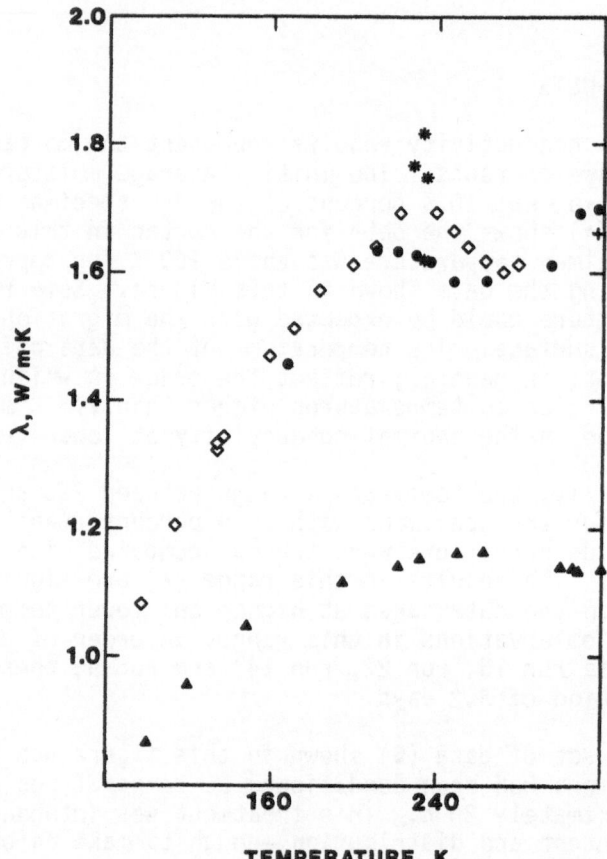

Figure 1. Thermal conductivity as a function of temperature for a concrete mortar with three different moisture contents: ◇, 10.4 percent moisture; , slightly less than 10.4 percent moisture; ▲, 4 percent moisture. Data points of reduced precision are indicated by ∗. The conductivity of the slightly dried specimen decreased by 2.5 percent (+) when the gradient conditions at 317 K were held for 15 h.

DISCUSSION

The current data are compared with values found in the literature in figures 2 and 3 for moist and dry cementitious materials,
respectively. Although the composition of the specimens used for
these curves differs significantly, the conductivity of the dry specimens (fig. 3) follows the general rule that conductivity increases
with increasing density (Tye and Spinney, 1976). This rule is not
strictly applicable to moist materials, as is seen in figure 2.

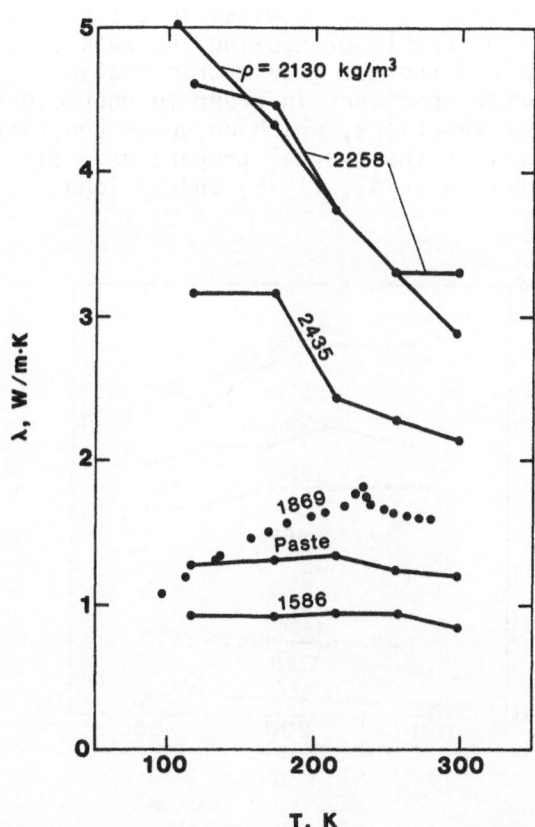

Figure 2. Thermal conductivity as a function of temperature for
various cementitious materials in moist conditions: ρ = 2130 kg/m^3,
sandstone concrete; ρ = 2258 kg/m^3, Elgin sand and gravel concrete;
ρ = 2435 kg/m^3, marble concrete; portland cement paste; ρ = 1586
kg/m^3, expanded shale concrete (Lentz and Monfore, 1965, 1966); ····,
NBS mortar.

A survey of the literature on λ of concrete at low temperatures indicates that the peak in λ of the saturated specimens at 232 K (fig. 1 - ◇, *) has not been previously observed. This may be due to the coarse grid of points in past experiments. The 10 percent increase in λ appears much like that seen in the specific heat of a material undergoing a lambda transition. Since λ is related to specific heat by

$$\lambda = \rho\alpha C_p \tag{1}$$

where α is diffusivity and ρ is density, several differential scanning calorimeter (DSC) determinations of C_p were made on materials similar to the λ specimens. Our limited investigation did not reveal a singular transition capable of causing the peak at 232 K. Further tests utilizing DSC and thermal gravimetric analysis (TGA) will be made on the NBS mortar specimens in order to characterize the complex, interdependent reactions, which occur as the temperature is lowered and which affect the thermal properties. Similar techniques were used by Stockhausen et al. (1979) and by Tognon (1968).

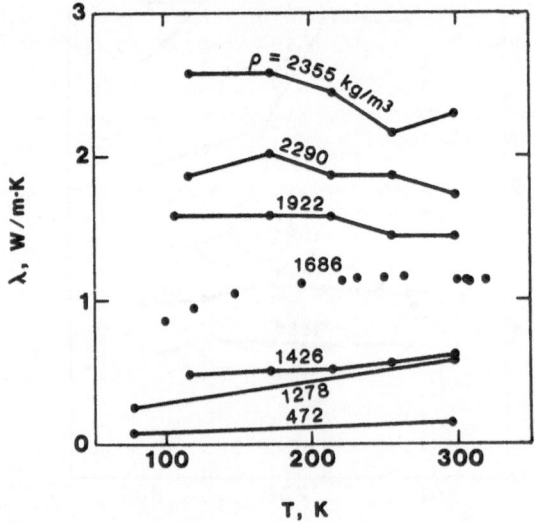

Figure 3. Thermal conductivity as a function of temperature for various cementitious materials in relatively dry conditions: ρ = 2355 kg/m^3, Elgin sand and gravel concrete; ρ = 2290 kg/m^3, marble concrete; ρ = 1922 kg/m^3, sandstone concrete; ρ = 1426 kg/m^3, expanded shale concrete (Lentz and Monfore, 1965, 1966); ρ = 1278, 472 kg/m^3, cellular concrete (Richard, 1977); ••••, NBS mortar.

Stockhausen observed a strong transition at 230 K in a hardened
cement paste, and Tognon quantitatively described the freezing process in a moist cement paste. Any model describing thermal transport
in concrete must account for the relative amount of solid and liquid
phases present as a function of temperature.

The evaporable moisture, water containing various dissolved
salts, is distributed throughout the hydrated material in pores and
capillaries of widely varying sizes. The freezing process occurs
over a wide range of temperatures. Freezing first takes place in the
larger pores and is accompanied by short range migration of moisture
from small pores. This process begins at about 269 K and continues
until roughly 200 K at which temperature most evaporable moisture is
in the solid phase (Tognon, 1968).

Anomalies or inflections in other low-temperature properties at
about 230 K and 269 K have been observed. Tognon (1968), Goto and
Miura (1979), and Rostasy, Schneider, and Wiedemann (1979) found sharp
changes in ΔL/L of moist concrete in this temperature range. Tognon
also found rapid changes in the flexural and compressive strengths in
this temperature range. Although the consensus is that the observed
behavior is due to the presence and effect of evaporable moisture in
the cementitious materials, no detailed model has yet been developed.

Changes observed in ΔL/L of a given specimen as it is thermally
cycled to low temperature are attributed to progressive alteration of
the pore structure. These changes cause hysteresis during a single
cooling-warming cycle and magnitude shifts for repeated cycles. The
internal pore damage was studied by Rostasy, Weiss, and Wiedemann
(1980), who found that the total pore volume was not greatly affected
by thermal cycling to cryogenic temperatures, but that the distribution of pore radii increased significantly in the 50 - 1500 nm sizes.
This redistribution to larger pores would be expected to change the
ice-liquid composition at a given temperature and change the concomi-
tant thermal conductivity. This effect cannot be unambiguously separated from the mild moisture loss occurring between the first two sets
of λ data (fig. 1 - ◇, ●). The inflection points for the second (less
moist) set of data are shifted to lower temperatures and are substantially broadened. This may be the result of moisture loss, alteration of the pore radii, or both. The 2.9 percent decrease in conductivity at 165 K, where all moisture is in the solid state, indicates
that a net moisture decrease did occur between the first and second
sets of data. Further testing involving multiple thermal cycles in
the temperature range where only negligible moisture loss occurs is
necessary for quantitative determination of the effect of thermal
cycling on λ.

The third set of data, taken after aging at high temperatures and
reduced pressure, indicates a much lower conductivity (fig. 1 - ▲).

This is the expected, qualitative effect of a large reduction in moisture content. This large change in moisture content results in both a reduction in magnitude of λ and a different characteristic λ versus T shape (fig. 1). In comparing the data from the "moderately wet" (●) and "dry" (▲) specimens, one finds the double inflection curve between 180 and 300 K replaced by a curve with a single inflection. The conductivity in the range above 310 K is reduced in magnitude for the dry specimen relative to the moist specimen conductivity by about 33 percent but retains the slight positive slope. At 165 K the reduction is 28 percent.

CONCLUSION

The thermal conductivity of porous, moist materials can be measured utilizing guarded-hot-plate systems, provided the test temperatures are low enough to cause partial freezing of the moisture in the pores. The magnitude of λ for the NBS mortar specimens is moisture dependent and is reasonable for a material of this type and density.

Two details were found that are not seen in the very limited literature data for low temperature λ: (1) The moist specimens clearly show inflections at temperatures near 270 K and 232 K. The slope change at 270 K is thought to be due to the initial freezing of moisture in the larger pores. The increasing conductivity as temperature decreases in the range 270 K to 232 K probably reflects the continuous freezing process in pores of decreasing radius. Other properties, particularly thermal strain, also exhibit inflections, discontinuities, or both at these temperatures. (2) The conductivity for both moist and dry mortar decreases at a faster rate than previously observed when the temperature is below 232 K. There is insufficient data available at the present time to develop a model for this behavior.

REFERENCES

ASTM 1978, Standard C138-77, Unit weight, yield, and air content
                                    (gravimetric) of concrete, Part 14;
              Standard C177-76, Steady-state thermal transmission properties by means of the guarded hot plate, Part 18;
       in: "Annual Book of ASTM Standards." American Society for Testing Materials, Philadelphia, PA.
Goto, Y. and Miura, T., 1979, Deterioration of concrete subjected to repetitions of very low temperatures, in: "Proceedings of the Japanese Concrete Institute," Japan Concrete Institute, Tokyo.

Lentz, A. E. and Monfore, G. E., 1965, Thermal conductivity of con-
    crete at very low temperatures, J. Portland Cement Assoc. Res.
    Dev. Labs. 7:39-46.
Lentz, A. E. and Monfore, G. E., 1966, Thermal conductivities of
    portland cement paste, aggregate and concrete down to very low
    temperatures, J. Portland Cement Assoc. Res. Dev. Labs.
    8:27-33.
Richard, T. G., 1977, Low temperature behavior of cellular concrete,
    ACI J. 74:173-178.
Rostasy, F. S., Schneider, U., and Wiedemann, G., 1979, Behavior of
    mortar and concrete at extremely low temperatures, Cem. Concr.
    Res. 9:365-376.
Rostasy, F. S., Weiss, R., and Wiedemann, G., 1980, Changes of pore
    structure of cement mortars due to temperature, Cem. Concr.
    Res. 10:157-164.
Smith, D. R., Hust, J. G., and Van Poolen, L. J., 1981, A guarded-hot-
    plate apparatus for measuring effective thermal conductivity
    of insulations between 80 K and 360 K, NBSIR 81-1657, National
    Bureau of Standards, Boulder, Colorado.
Sparks, L. L., 1981, Thermal conductivity of a concrete mortar from
    95 K to 320 K, NBSIR 81-1651, National Bureau of Standards,
    Boulder, Colorado.
Stockhausen, N., Dorner, H., Zech, B., and Setzer, M. J., 1979,
    Untersuchung von gefriervorgangen in zementstein mit hilfe der
    DTA (Freezing phenomena in hardened cement paste were investi-
    gated by DTA), Cem. Concr. Res. 9:783-794.
Tognon, G., 1968, "Supplementary Paper III - 24: Behavior of Mortars
    and Concretes in the Temperature Range from +20°C to -196°C,"
    The Cement Association of Japan, Tokyo.
Tye, R. P. and Spinney, S. C., 1976, Thermal conductivity of concrete:
    measurement problems and effect of moisture, Bull. Inst. Int.
    Froid, Annexe 1976-2:119-127.

MEASUREMENT OF SOIL THERMAL CONDUCTIVITY USING

AN INFRARED SCANNING TECHNIQUE

C. K. Hsieh

Mechanical Engineering Department
University of Florida
Gainesville, Florida 32611

X. A. Wang

Shanghai Mechanical Engineering Institute
Shanghai, People's Republic of China

INTRODUCTION

In the experimental determination of the thermal conductivity of granular substances, such as powders or insulators, the conventional methods have been (i) the plate (or disk) method and (ii) the concentric sphere (or cylinder) method[1]. In both methods the major source of errors stems from the lateral heat transfer, which cannot be resolved without the use of large test specimens or guard heaters. These conventional methods may not be convenient to use. Thus, a simple test method is given here that utilizes an infrared scanning technique to measure the thermal conductivity. As will be shown later, the method has promise of being a non-destructive test method for property measurements.

METHODOLOGY

The system under investigation is depicted in Fig. 1. The test specimen (soil) occupies the semi-infinite space that is heated over a circular region of radius r_h on its surface. For the ease of computation, this semi-infinite space is modeled as a hemisphere of large radius such that an adiabatic condition can be imposed on the spherical surface. The heat input to the system must be dissipated from the exposed surface. This gives rise to a temperature distribution on this surface, which can be used to

665

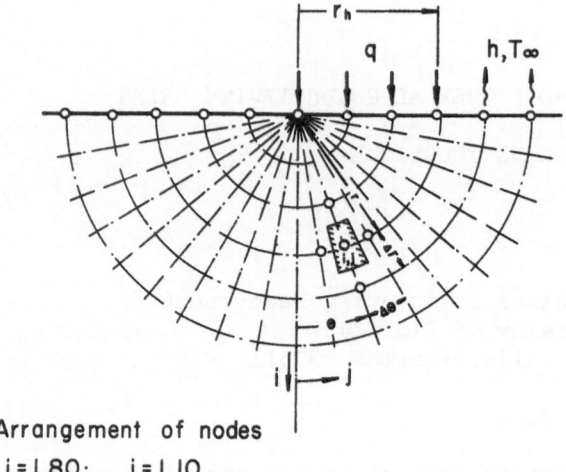

Arrangement of nodes

i = 1,80;   j = 1,10

Fig. 1. Nodal points lay-out in numerical solution

determine the thermal conductivity by temperature matching.

ANALYSIS

Because of the boundary conditions imposed on the system, the problem at hand can not be solved by using an exact solution. A numerical solution must be employed, and the nodal points lay-out for such a problem is shown in Fig. 1. To simplify the analysis, the temperature distribution is assumed to be axisymmetric. There is no heat source or sink inside the system, and the system is in a steady state.

The numerical solution of such a problem is quite straight-forward; no further discussion is necessary. However, a short note is in order to describe how free convection is treated in the analysis. As is well known, the empirical equations documented in the literature are mostly developed for finite surfaces, whereas in the present study, the heat dissipation from concentric rings is important. It is necessary to derive the convective coefficient (h) in terms of the ring radii (r) and temperatures (T). This can be carried out by using the empirical equation[2]

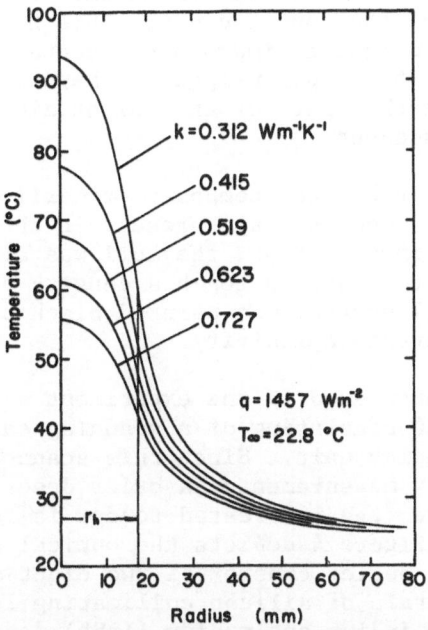

Fig. 2.   A set of surface temperature distribution
          curves for different k values.

$$h = 1.32 \ (\Delta T/1.8r)^{1/4} \qquad\qquad\qquad\qquad (1)$$

to relate the increase in the natural convection as a result of the
increase of the disk radius. It follows that

$$h = 1.1396 \ (\Delta T)^{1/4}(r_2^{7/4} - r_1^{7/4}) \ / \ (r_2^2 - r_1^2) \qquad\qquad (2)$$

This equation was used to derive the nodal equations for boundary
nodes, which were used, in turn, with other nodal equations to
develop a computer program for temperature predictions. A set of
surface temperature data calculated is shown in Fig. 2.

EXPERIMENT

Apparatus

     A schematic diagram showing the construction of the heater is
given in Fig. 3. It consists of two foil heaters (MINCO Products,
75Ω each), with one used as the main heater, the other as the guard
heater. Both heaters were sandwiched between two copper disks of
0.84 mm thickness. Four sheets of asbestos were used as spacers
to separate the heaters. The heat flow direction inside the heater

assembly was monitored with two sets of thermocouples instrumented
on the asbestos side of the copper disks. This heater assembly was
loaded using a copper weight to improve the contact between the
heater and the soil. The copper weight was beveled at an angle of
10° at the top so that this part of the weight did not block the
view of the infrared scanner.

    The soil used for tests was composed primarily of sand, with
little bonding agents, such as clay, present in the sample. Prior
to loading the sample in a sandbox, the soil was allowed to dry in
a laboratory room for two days to reach a density of 1.4 g/cm^3.
After loading, the soil surface was painted black using 3M Nextel
Velvet to raise the surface emissivity.

    The infrared scanner used in the experiment was an AGA
Thermovision System 680 consisting of a scanning camera and an
electronic picture display unit. Since this scanner has not been
widely used in property measurements, a brief description of its
operation is given here (the interested reader is referred to
Ref. 3 for details). Figure 4 depicts the optical system in the
scanner. The scanning optics consist of two eight-sided rotating
prisms, a chopper, a train of silicon collimating lens, and a
liquid-nitrogen-cooled indium antimonide (InSb) detector. The
prisms scan the field of view at a frequency of 1600 lines/s with
100 elements/line, with a resultant picture frequency of 16
pictures/s. The electronics in the display monitor are synchro-
nized with the positions of the prisms in the scanning optics in

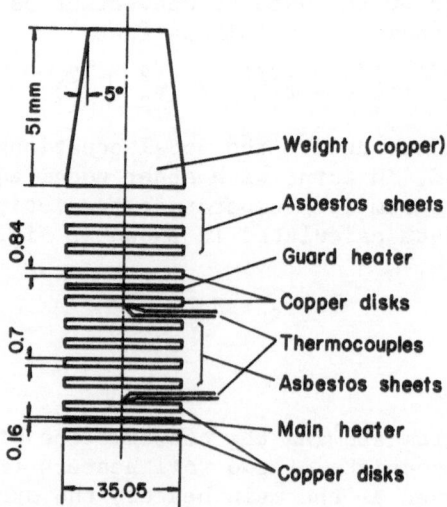

Fig. 3.   A schematic diagram showing construction
          of the heater assembly.

such a way that each point in the optical field of view is transformed into a corresponding point on the monitor screen. The
intensity of the modulated beam in the monitor is a function of the
received infrared radiation. In a normal viewing mode, a real-
time infrared picture is displayed in black-and-white, with the
tone of the warm parts light gray and the cold, dark gray.

To enhance the usefulness of the scanning system suitable for
the present work, two additional pieces of equipment were used with
the system. Since the signal intensity of the original black-and-
white image is hard to discriminate, a ten-color monitor was used
to display the image in color. In addition, a profile adapter was
used to display the scanned signal in an analog mode. This adapter
was particularly useful, since the temperature distribution along
any scan line could be chosen and used for analysis.

Calibration

The calibration of the infrared scanner has been detailed in
Refs. 4-6. Briefly, a checkerboard-like aluminum plate was used
to calibrate the image spatial distortion. This checkerboard has
milled grooves on it, and these grooves were painted black using
the 3M black paint mentioned earlier. Because of the difference
in emissivity, these painted grooves showed up in the infrared
image and were useful to determine the spatial distortion in the
optics.

The scale divisions on the scope of the profile adapter were

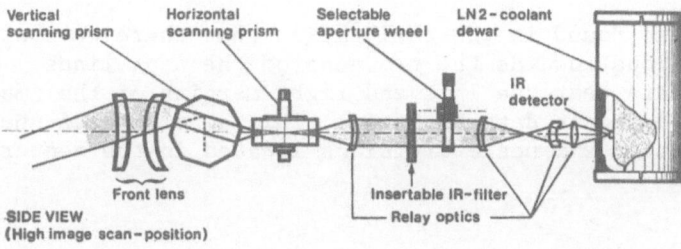

SIDE VIEW
(High image scan-position)

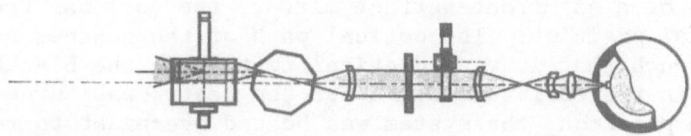

TOP VIEW
(High/right image scan-position)

Fig. 4. Optics of the infrared scanner

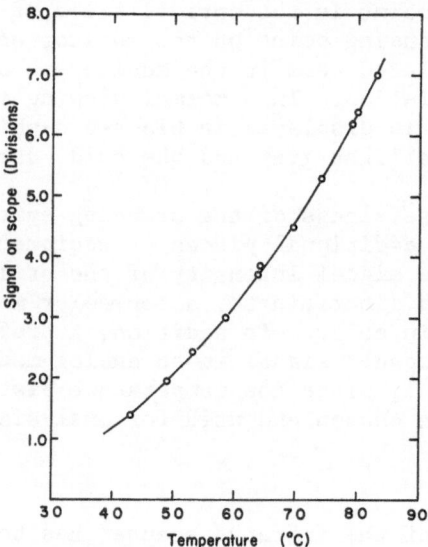

Fig. 5.   Temperature calibration curve.

calibrated using a blackbody cavity (Barnes Engineering, Model
11-101T).   The difference in the emissivity of the blackbody
(ε = 0.994) and the 3M black paint (ε = 0.975) was accounted for
in the calibration by multiplying the blackbody temperature setting
by a factor of 1.0048 to obtain the equivalent paint temperature.
A calibration curve thus constructed is shown in Fig. 5.

It was also found in the calibration that there were slight
signal distortions towards the two ends of the scan lines.   For
this reason, data near the left and right margins of the image
field were rejected in data analysis.   This also reduced the useful data to within six scale divisions located in the center field
of the image.

Testing

The infrared scanner was mounted in a horizontal position.
With the help of a 45° front-surface mirror, the soil was tested
in a horizontal position.   The optical path of the scanned surface
was adjusted such that it was identical to that of the blackbody
used earlier in the calibration.   After the heater was turned on
and placed in position, the system was heated overnight to reach
a steady state.

Prior to data collection, the temperature on the soil surface
was scanned to check temperature symmetry.   This was done by

Fig. 6.   Isotherm contour for testing of temperature symmetry.

operating the scanner in the isotherm mode and viewing the isotherm
contour in the image display. A photograph taken from the color
monitor screen is shown in Fig. 6, where the isotherm (in saturated
white) exhibits the desired circular pattern.

     Next, the profile adapter was adjusted to position the scan
line to pass through the center of the heater. The analog output
of such a scan is shown in Fig. 7, where the smooth curve repre-

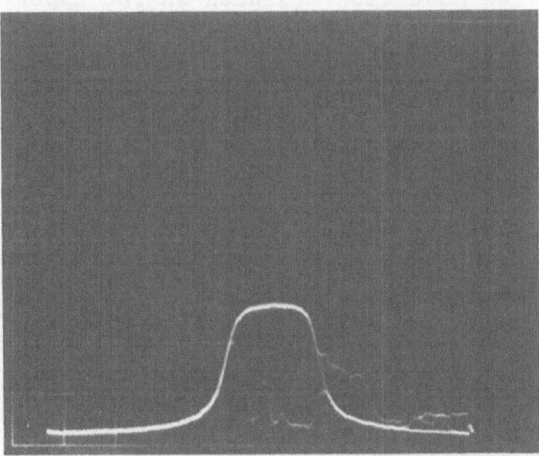

Fig. 7.   Scan of energy radiated from the soil surface.

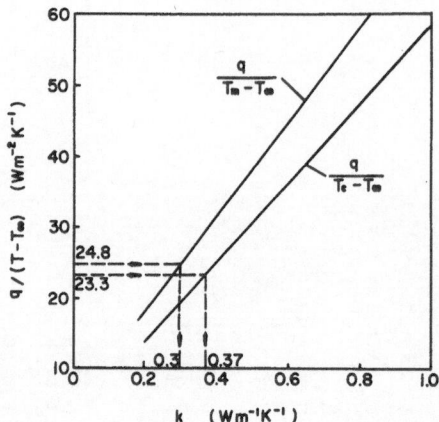

Fig. 8.   Chart to determine range of k for temperature matching.

sents the energy radiated by the soil surface. To scan the soil
surface covered by the heater, a time exposure of 1/2 s was used
in photography, and the heater was removed when the shutter was
open. This caused the scanner to scan the top of the heater and
the hand prior to the removal of the heater. The irregular curves
shown in the picture are the result of the scan of these objects.

RESULTS AND DISCUSSION

      To facilitate determination of the range of the thermal conduc-
tivity to be used in temperature matching, a chart was constructed
for the purpose. The heat flux measured in the experiment was used
in the computer input to evaluate the surface temperature for three
values of the thermal conductivity. Then, a plot of $q/(T - T_\infty)$
versus k can be made, where q is the heat flux, k is the thermal
conductivity. As shown in Fig. 8, the temperature used in the plot
can be either the center temperature, T_c, or the mean temperature,
T_m, of the soil surface that is covered by the heater. Physically,
$q/(T - T_\infty)$ is equivalent to the reciprocal of the soil resistance
between the heater and the ambient. The proportionality of
$q/(T - T_\infty)$ versus k is, thus, not unexpected. It should be noted
that the use of both T_c and T_m in the plot is governed by the
fact that the contact resistance between the heater and the soil
surface may cause some uncertainties in the soil temperature
within the heater radius. Since this figure is used primarily to
determine the range of k useful for temperature matching, the
use of T_m and T_c enables such a range to be found. The final
determination of k still relies on the matching of temperatures
on the exposed surface, as shown in Fig. 9, where the thermal

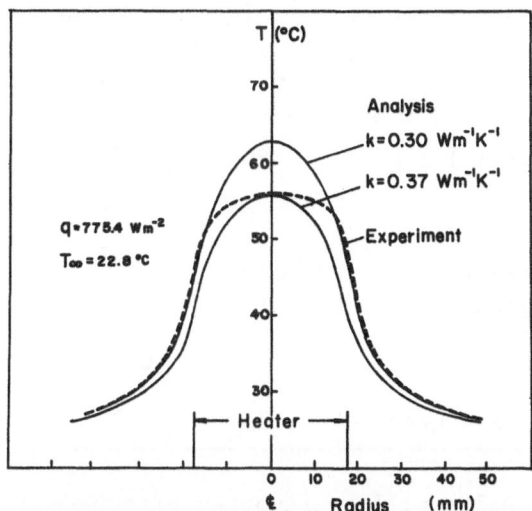

Fig. 9. Temperature matching to determine thermal conductivity.

conductivity was found to be 0.3 W/m K.

The thermal conductivity value determined in the present study appears to differ considerably from those documented in the literature (see Table 1). The lack of detailed specifications of the test samples makes direct comparison difficult. Since a separate measurement using conventional test methods was not attempted in the present study, it is difficult to assess the true accuracy of the method used.

A close examination of the present method reveals that the major source of error is probably the contact resistance between the heater and the soil. This can be verified by comparison of the predicted and the measured temperatures located underneath the heater (see Fig. 9). Because of the softness of the soil sample, increase of load by using a heavier copper weight seems to be impractical. On the other hand, the final determination of the thermal conductivity relies on matching the predicted and the measured temperatures. In this connection, the final error is a function of the accuracy of the heat input, the temperature calibration uncertainties, the numerical error in the prediction of temperatures, and, finally, the accuracy with which the empirical equation (2) can be used to evaluate the convection loss. Of these cited, the last is probably the most important. With all these uncertainties, it is estimated that the error of the method can be as high as 10%. This error is expected to diminish if a hard sample is used for tests. A follow-up study using glass plates for test samples is being

Table 1. Comparison of Soil Thermal Conductivity

Sample Description	T(°C)	k(W/mK)	References
Mineral, dry, density: 1.5 g/cc	20	1.09	1
Organic, dry	20	0.0335	1
Earth, sand (8% moisture), density: 1.5 g/cc		1.05	7
Sand, dry		0.582	7
Sand, dry, density: 1.4 g/cc	23	0.30	Present Study

carried out; its results will be reported elsewhere.

It is probably significant to note that the method presented in this paper can be considered as a nondestructive test method for conductivity measurements. The method can be easily adapted to use in the field. In this application, the original temperature distribution inside the soil must be accounted for in temperature predictions. It is expected that the matching of temperatures for this instance will be more difficult.

ACKNOWLEDGMENT

The partial support from National Science Foundation Grants (Grant No. ENG 78-10982, CME-7919834) is gratefully acknowledged.

REFERENCES

1. Y. S. Touloukian and C. Y. Ho, "Thermophysical Properties of Matter, Volume 2, Thermal Conductivity, Non-metallic Solids," IFI/Plenum, New York (1970).
2. J. P. Holman, "Heat Transfer" (Fifth Edition), McGraw-Hill, New York (1981).
3. W. H. Meyfarth, "Eng. Dig.," 18 (1) (1969).
4. C. K. Hsieh and K. C. Su, Design, construction, and analysis of a continuous-temperature infrared calibrator for temperature measurement using an infrared scanner, Rev. Sci. Instrum., 50(7): 888-896 (1979).
5. C. K. Hsieh, M. C. K. Yang, E. A. Farber, and A. Jorolan, A feasibility study to test structure integrity by infrared scanning technique, "Thermal Conductivity," Vol. 14, Plenum Press, New York, 521-530 (1975).

6. C. K. Hsieh and W. A. Ellingson, The feasibility of using
        infrared scanning to test flaws in ceramic materials,
        "Thermal Conductivity," Vol. 15, Plenum Press, New
        York, 11-22 (1978).
7. F. Kreith and W. Z. Black, "Basic Heat Transfer," Harper
        and Row, New York, 512-513 (1980).

THERMAL DIFFUSIVITY OF SILICON CARBIDE-SILICON COMPOSITES

M. Srinivasan
Carborundum Company
Niagara Falls, New York

L. D. Bentsen and D. P. H. Hasselman
Department of Materials Engineering
Virginia Polytechnic Institute and State University
Blacksburg, Virginia

ABSTRACT

The room temperature thermal diffusivity of silicon carbide-silicon composites was measured by the laser-flash method. The experimental results indicate that the thermal diffusivity and conductivity of these composites decrease strongly with increasing level of impurity content in both the silicon and the silicon carbide. This suggests that the heat transfer properties of these composites can be tailored by controlling impurity levels.

INTRODUCTION

Increased efficiency of energy-conversion systems can be achieved by increasing their operating temperature levels. For this reason, a demand has been created for engineering materials with mechanical properties and corrosion and oxidation resistance well in excess of those exhibited by superalloys. Materials such as the refractory nitrides and carbides, in view of their chemical and structural stability at high temperature, appear to be excellent candidate materials for such applications as the all-ceramic turbine engine or internal combustion engine with retrofitted ceramic components.

The in-service performance of these materials depends not only on their chemical and mechanical properties, but also on their thermal properties, which include the coefficient of thermal

677

expansion and the thermal conductivity and diffusivity. Low values
of the thermal conductivity lead to improved energy-conversion ef-
ficiencies, since heat losses are kept to a minimum. Owing to their
brittle nature, however, the above materials are highly susceptible
to catastrophic failure due to thermal stresses of high magnitude
that result from nonlinear transient or steady-state temperature
distributions. To minimize the possibility of this mode of failure,
the optimum candidate materials should have values of the coefficient
of thermal expansion as low as possible in combination with values
of the thermal conductivity and thermal diffusivity as high as possible. This latter requirement is incompatible with the requirement
of low thermal conductivity for high operating efficiency. Clearly,
appropriate trade-offs must be made. The ability to make such
trade-offs, in turn, requires quantitative information for the values
of the thermal conductivity and thermal diffusivity and a detailed
understanding of the materials and intrinsic variables that affect
their values. In particular, such effects at higher temperatures
are of considerable technical significance.

    Heat transport through a material occurs primarily by phonon,
photon, or electron transport. The heat flux that results from
these mechanisms depends strongly on the associated specific heat,
the temperature, and the existence of structural and chemical
imperfections, such as vacancies, dislocations, grain boundaries,
foreign atoms, optical discontinuities, and any other factors that
contribute to phonon, photon, and electron scattering.[1,2] At the
microstructural level, the conduction of heat is affected by the
presence of pores, inclusions and cracks, grain boundary phases,
and preferred crystallographic orientation.[2,3-9]

    Many of the variables that affect the conduction of heat de-
pend on processing history, which suggests that a measure of control can be exerted over the value of the thermal conductivity or
thermal diffusivity of a material required for a given application.
The purpose of this paper is to present experimental data for some
of the variables that affect the heat conduction properties of
silicon carbide-silicon composites that have demonstrated potential
for ceramic turbine applications.

MATERIALS AND EXPERIMENTAL PROCEDURES

    The silicon carbide-silicon composites were made by a warm
molding process using an industrial grade silicon and silicon carbide powder made by the Acheson process. For two different series
of samples, 400- and 1000-mesh-size silicon carbide powders were
used. A single sample was also prepared from 1200-mesh silicon
carbide powder. The silicon content of the composites was calculated from the measured density of the composites and values of the

densities of silicon and silicon carbide of 2.33 and 3.21 g/cm^3, respectively. For the composites made with the silicon carbide particle sizes of 400, 1000, and 1200 mesh, the final sizes of the silicon inclusions within the silicon carbide matrix were approximately 15, 5, and 4 μm, respectively. Typical photomicrographs are shown in Fig. 1.

For purposes of interpretation of the data, additional samples included in the study consisted of dense, sintered α-silicon carbide and a representative sample of the silicon metal used for the preparation of the silicon carbide-silicon composites. The principal impurities in this silicon metal, as determined by spectrochemical means, are listed in Table 1. An additional sample consisted of zone-refined silicon of at least 99.99% purity.

The heat conduction properties of the silicon carbide-silicon composites were determined by measurements of the thermal diffusivity by the laser-flash method,[10] using a glass-neodymium laser. The specimens of appropriate size and geometry were carbon-coated to prevent direct transmission of the laser beam. For measurements of the thermal diffusivity above room temperature, the specimens were held in a graphite resistance furnace containing a nitrogen atmosphere. At room temperature and up to about 500°C, the transient temperature of the specimen was monitored with a liquid-nitrogen-cooled indium antimonide infrared detector. Above 500°C the specimen temperature was monitored with a silicon photodiode.

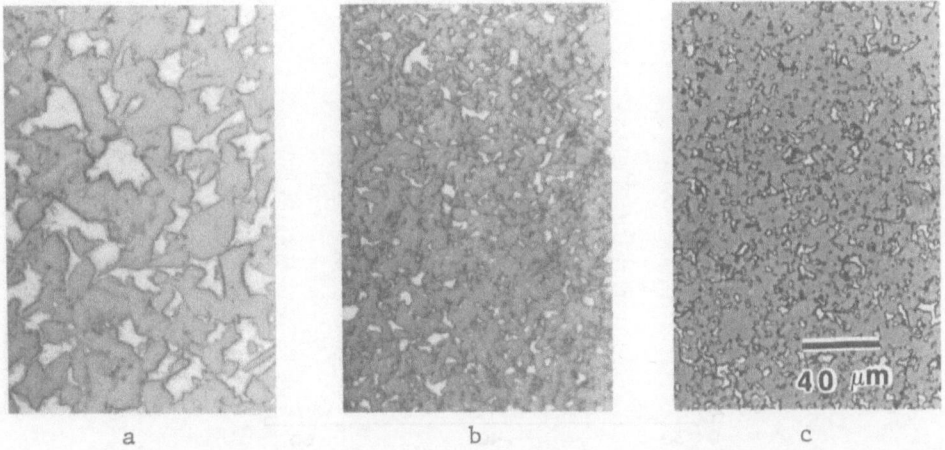

Fig. 1. Optical micrographs of silicon carbide-silicon composites with average silicon particle sizes of a: 15 μm; b: 5 μm, and c: 4 μm.

TABLE 1.   Principal Impurities in Industrial Grade Silicon Used
for Preparation of Silicon Carbide-Silicon Composites

Element	Atom Percent	Element	Atom Percent
Boron	0.6	Titanium	0.004
Magnesium	0.001	Vanadium	0.001
Manganese	0.01	Copper	0.1
Iron	0.1	Chromium	0.002
Nickel	0.004	Calcium	0.06
Aluminum	0.01		

EXPERIMENTAL RESULTS AND DISCUSSION

Figures 2 and 3 show the dependence of the thermal diffusivity
on silicon content at room temperature for the two series of silicon carbide-silicon composites with silicon inclusion sizes of 15

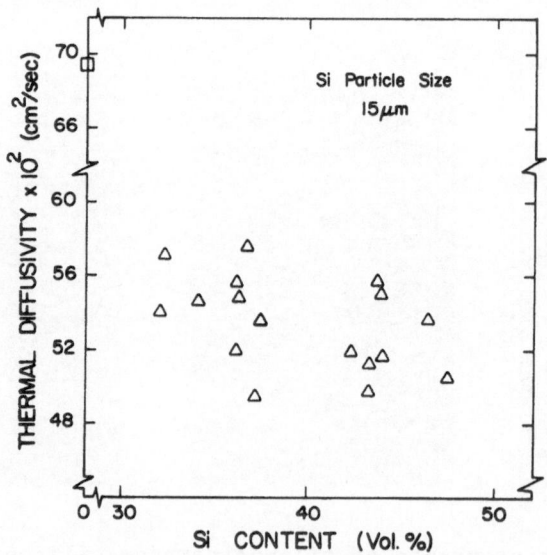

Fig. 2.   Thermal diffusivity at room temperature of silicon
carbide with 15 µm silicon inclusions as a function
of silicon content.

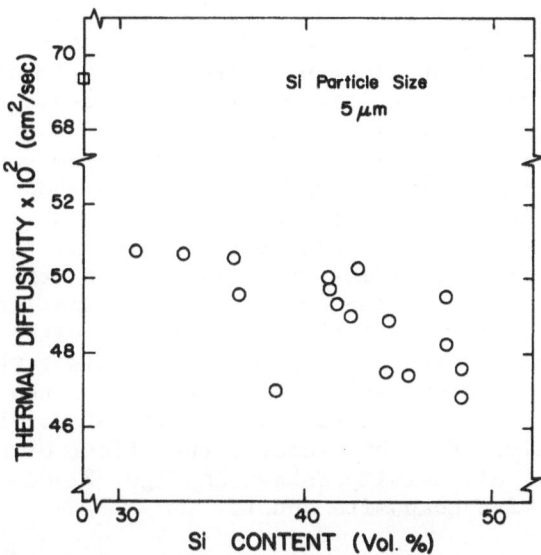

Fig. 3.   Thermal diffusivity of silicon carbide with 5 μm silicon
          inclusions as a function of silicon content (∿ 25°C).

and 5 μm, respectively. Table 2 lists the data for the thermal diffusivity at room temperature for the single specimens of single-
phase α-silicon carbide, the silicon used for the preparation of the
composites, the zone-refined silicon, and the silicon carbide-silicon
containing 20.82 vol.% of 4 μm silicon inclusions. For the latter
material, the thermal diffusivity was found to be independent of
orientation.

TABLE 2.   Values for the Thermal Diffusivity at Room Temperature
           for Individual Samples of Silicon, Silicon Carbide, and
           Silicon Carbide-Silicon

Materials	Thermal Diffusivity (cm^2s^{-1})
Single-Phase Silicon Carbide	0.694
Industrial Silicon	0.181
Zone-refined Silicon	0.668
Silicon Carbide-Silicon (8 specimens)	0.774 ± 8%*

*Coefficient of Variation.

Upon comparison with the value for the thermal diffusivity of the single-phase α-silicon carbide and the inverse dependence of the thermal diffusivity on silicon content, the data shown in Figs. 2 and 3 indicate that the addition of the silicon-dispersed phase to the silicon carbide matrix results in a significant decrease in the thermal diffusivity. This is expected from the general theory of the thermal conductivity of composites, which shows that a dispersed phase with low thermal conductivity within a matrix of higher thermal conductivity leads to a decrease in overall thermal conductivity of the composite, regardless of the nature of the distribution of the dispersed phase.[3-6] The same conclusion applies to the thermal diffusivity of composites for components with comparable values of the specific heat per unit volume. This latter condition holds for the present samples with values for the thermal diffusivity of the silicon and silicon carbide of 0.181 and 0.694 cm^2s^{-1}, respectively. For this reason, the effect of silicon content on the thermal diffusivity (shown in Figs. 2 and 3) is in general agreement with composite theory.

Comparison of the data shown in Figs. 2 and 3 shows that for a given silicon content, the thermal diffusivity is a function of the silicon particle size. Such an effect of inclusion size is not predicted by the theory of the thermal conductivity of composites, unless such a change in inclusion size is accompanied also by a corresponding change in the geometry or orientation of the inclusions. The photomicrographs in Fig. 1 show no conclusive evidence for such a change in geometry or orientation. Also, such an effect of orientation should result in an anisotropy of the thermal diffusivity, which was not observed. If, however, the hypothesis that a change in silicon inclusion size is accompanied by a change in particle geometry or orientation is correct, such differences, at least qualitatively, should be observed in the elastic properties of the composites as well.[11] As shown in Fig. 4, no such effect is indicated by the data for Young's modulus[12] for all the specimens with values of thermal diffusivity shown in Figs. 2 and 3. This represents positive evidence that the apparent effect of the silicon inclusion size on thermal diffusivity cannot be attributed to corresponding changes in inclusion geometry or orientation or both.

For another explanation for this apparent effect of inclusion size, it should be noted that, generally, theories for the thermal conductivity of composites do not consider the effect of an interfacial resistance to heat flow. One possible source of such an interfacial barrier to heat flow is the difference in Young's modulus of the silicon and silicon carbide of a factor of three. The interface between the silicon and silicon carbide in these composites represents an elastic discontinuity, which could contribute to increased phonon scattering. At room temperature such a discontinuity may affect the contribution to heat flow by low

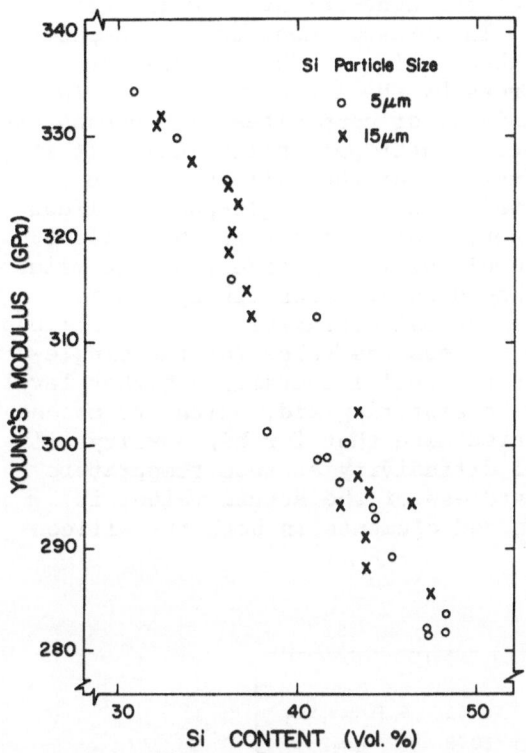

Fig. 4. Young's modulus
of elasticity at room
temperature for specimens
of Fig. 2 and 3.

frequency phonons primarily. Such an effect is expected to be more
pronounced for the smaller silicon inclusions than for the larger
ones and may decrease the thermal diffusivity at room temperature
by perhaps a fraction of a percent, but it does not explain the
average of 5 to 6% difference between the two sets of data shown in
Figs. 2 and 3.

The most likely explanation, albeit only qualitative at this
time, is that the changes in silicon inclusion size are accompanied
by corresponding changes in the thermal diffusivity of the silicon
and possibly the silicon carbide as well. Such changes in the
thermal diffusivity can result from changes in impurity content.
Proof for the validity of this hypothesis is offered by the large
difference in the value of the thermal diffusivity of 0.668 cm^2s^{-1}
for the zone-refined silicon and the corresponding value of 0.181
cm^2s^{-1} for the silicon used for preparation of the composite
specimens.

The value of 0.774 cm^2s^{-1} for the thermal diffusivity of the
sample with 20.82 vol.% silicon with a particle size of 4 μm also
is relevant to the above discussion. This value of thermal diffusivity is higher than the corresponding values of the sintered single-

phase α–silicon carbide as well as the zone-refined and industrial
grade silicon in the composites. This result contradicts composite
theory, which states that the highest value of the thermal conductivity of a composite is governed by the law of mixtures. The
same holds for the thermal diffusivity of composites with components
of comparable values for the specific heat per unit volume. If it
is assumed that the thermal diffusivity of the silicon carbide in
this sample is identical to the value for the single-phase silicon
carbide, then the value of the thermal diffusivity of the silicon,
inferred from composite theory, would be far in excess of the value
for the zone-refined silicon. This does not seem likely. A more
plausible explanation is that the thermal diffusivity of the silicon
carbide in this composite sample exceeds the value for the single-
phase silicon carbide. This latter material contains a higher level
of foreign elements introduced as a sintering aid, which are absent
in the composites. It is speculated here that for high-purity silicon carbide, the value of thermal diffusivity at room temperature
may well approach 1 cm^{2}s^{-1}. Regardless of the actual value, it
seems likely that impurities or added elements in both the silicon

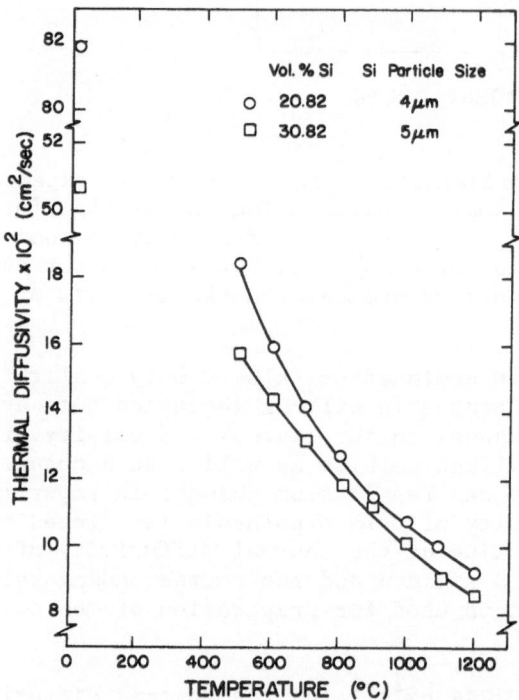

Fig. 5.   Temperature dependence of the thermal diffusivity of two
          silicon carbide-silicon composites for two values of si-
          licon content.

and silicon carbide can affect the heat conduction behavior of silicon carbide-silicon composites significantly.

Figure 5 shows the temperature dependence of the thermal diffusivity of the sample with 20.82 vol.% of 4 μm silicon inclusions and for one of the samples of Fig. 3. These data indicate that the relative difference in the thermal diffusivity decreases with increasing temperature.

Figure 6 compares the temperature dependence of the thermal diffusivity of two of the samples of Figs. 2 and 3 with comparable silicon content. These data are of interest in that they indicate that above 700°C the values for the thermal diffusivity are almost identical. It is suggested here that this effect may constitute evidence for an electronic contribution to the heat transfer by thermally activated holes and electrons, intrinsic as well as extrinsic due to the presence of the various impurities. In parti-

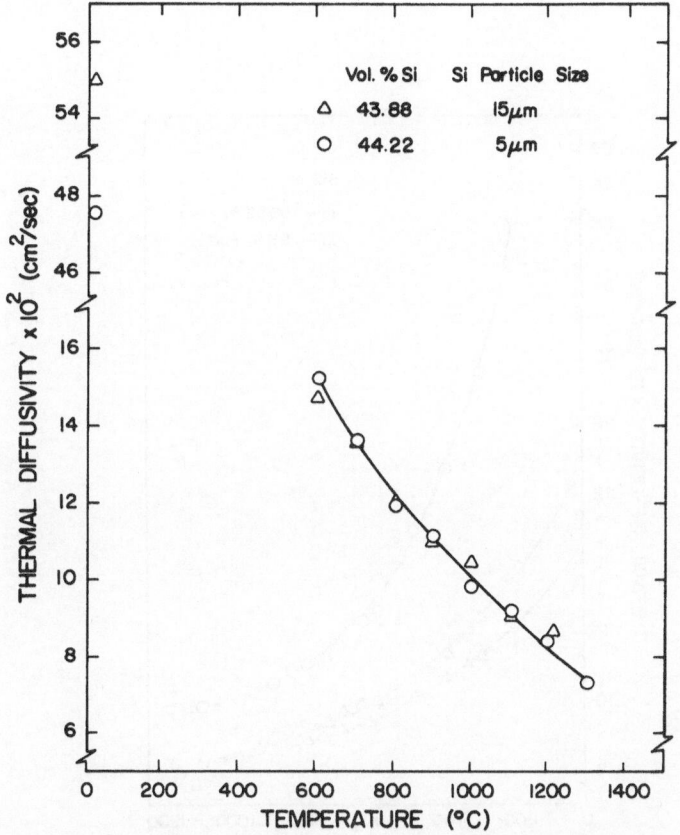

Fig. 6.   Temperature dependence of thermal diffusivity of silicon carbide-silicon composites with comparable silicon contents and two different particle sizes.

cular, this would be the case for the sample with the lower value
of the thermal diffusivity at room temperature, presumably because
of its higher impurity content. Indeed, if this hypothesis is
correct, impurities in silicon or silicon carbide will lower the
thermal diffusivity at low temperature, but may increase it again
at higher temperatures if these impurities are of the type that
affect the electrical conductivity. The higher boron content in
the silicon in the present samples could especially be very
effective in this respect.

Figure 7 shows the temperature dependence of the thermal diffusivity of the samples of zone-refined and industrial silicon.
For these samples also, the relative difference in thermal diffusivity decreases with increasing temperature. Of interest is
the irreversible increase in the thermal diffusivity of the industrial silicon following heating and cooling. This effect could
arise from the formation of precipitates of impurities originally

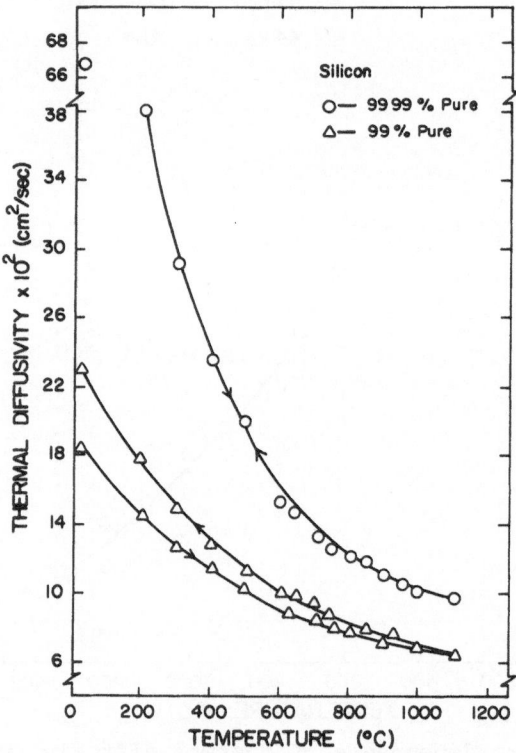

Fig. 7.   Temperature dependence of the thermal diffusivity of an
          industrial grade and purified silicon.

in solid solution. No evidence was obtained in support of this hypothesis.

The experimental results obtained in this study indicate that the heat conduction properties of silicon carbide-silicon composites can be affected by a number of variables. The evidence presented indicates that impurities are particularly effective in this respect. Of practical significance is that such an effect would allow some degree of control over the desired thermal conductivity or thermal diffusivity of such composites. High impurity contents (in the absence of other adverse effects, of course) would be desirable to maximize thermal insulating properties. On the other hand, components or structures subject to thermal stress failure should be as pure as possible in order to obtain high thermal conductivity.

ACKNOWLEDGMENTS

The silicon carbide-silicon specimens investigated in this study were made by Carborundum Company, Niagara Falls, New York. The measurement of the thermal diffusivity and manuscript preparation were carried out at Virginia Polytechnic Institute and State University as part of a research program supported by the Office of Naval Research under contract N00014-78-C-0431.

REFERENCES

1. R. Berman, "Thermal Conductivity in Solids," Clarendon Press, Oxford (1976).
2. W. D. Kingery, H. K. Bowen, and D. R. Uhlman, "Introduction to Ceramics," 2nd Ed., John Wiley & Sons, New York (1975).
3. A. E. Powers, "Conductivity in Aggregates," Knolls Atomic Power Laboratory, Report KAPL-2145, (March 1961).
4. H. Fricke, Phys. Rev. 24:575 (1924).
5. L. Rayleigh, Philos. Mag. 34:481 (1892).
6. J. C. Maxwell, "A Treatise on Electricity and Magnetism," 3rd Ed., Oxford University Press, Oxford (1904).
7. D. P. H. Hasselman, J. Compos. Mater., 12:403 (1978).
8. G. Ziegler and D. P. H. Hasselman, J. Mater. Sci. 16:495 (1981).
9. G. Ziegler, L. D. Bentsen and D. P. H. Hasselman, Comm. Am. Ceram. Soc. 64(2):C-35 (1981).
10. W. J. Parker, R. J. Jenkins, C. P. Butler, and G. L. Abbott, J. Appl. Phys. 32:1679 (1961).
11. D. P. H. Hasselman, J. Gebauer, and J. A. Manson, J. Am. Ceram. Soc. 55(12):58-91 (1972).
12. M. Srinivasan, M. Kasprzyk, Am. Ceram. Soc. Bull, 58:887 (1979).

THERMAL AND ELECTRICAL CONDUCTIVITY

OF METALS EMBEDDED WITH CERAMIC GRANULES

Wolfgang Neumann

Department of Metallurgy
Austrian Research Centre Seibersdorf
A-1082 Vienna, Austria

INTRODUCTION

In some cases with the development of new composites there is a need of a sufficient characterization of their thermal conductivity. Sometimes the prediction of the conductivity data from the conductivity of the components or from other measurements, e. g., electrical conductivity, is preferred to direct measurement.

A lot of studies have been performed in this field already, even some dealing with heterogeneous materials, but a formula or model that fits the results in general has not been found yet. Therefore, the different kinds of composites require special investigations. Their results may contribute to an extension or confirmation of existing models.

The present work should be seen in this connection. It deals with conductivity measurements of nearly spherical granules embedded in a metal matrix. The aim of the activity is the prediction of thermal conductivity of these composites, which developed as an alternative solidified high level waste form, either from thermal conductivity of the pure components or from electric resistivity measurements.

Experiments

A comparative method with longitudinal heat flow in steady-state conditions was used for the determination of the thermal conductivity as function of temperature. The apparatus based on this very well-known method is even commercially available and is described in detail in many papers.[1,2]

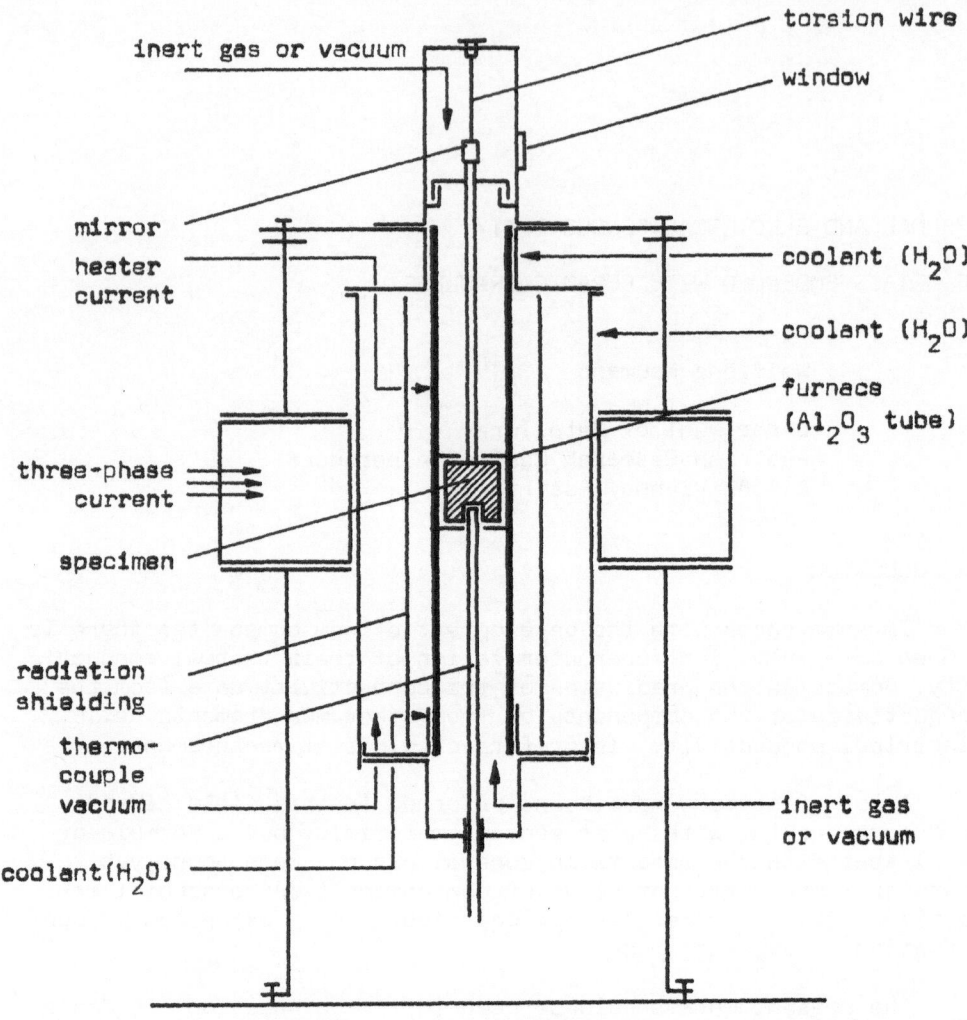

Fig. 1.   Sketch of the rotating field apparatus

Determination of the electrical conductivity was carried out in a rotating field apparatus (Fig. 1).

The measurement with this device is based on induction of eddy currents in a rotating magnetic field. The suitable magnetic field is supplied by a stator of a polyphase induction motor. The specimen located in the centre of the stator would rotate if the suspension rod did not cause a torsional moment in the opposite direction.

Theoretical considerations[3] result a linear relation between deflecting angle and conductivity:

$$\alpha = \frac{\pi}{4} \frac{1}{k} w1R^4H^2\sigma$$

σ .... electrical resistivity
w .... frequency
l .... height)
                of specimen
R .... radius)
H .... magnetic induction
k .... directive force

From this formula one can easily determine the electrical conductivity by comparison with a specimen of well-known conductivity and of the same dimensions.

$$\frac{\alpha_1}{i_1^2} : \frac{\alpha_2}{i_2^2} = \sigma_1 : \sigma_2$$

$i^2 \sim H^2$ rotating field generating electric current

This method, actually developed for measuring liquid metals, is also advantageous in the present work because no electrodes are necessary. Therefore, one does not need to take into account any contact resistance. In addition, one can go to higher temperatures without any change in the experimental arrangement.

The metal-granules composites were prepared by a vacuum-casting technique. The ceramic granules were porous $AlPO_4$ beads with an average diameter of about 2 mm. They were poured in a mould, and the space between them was filled with metal by vacuum-pressure casting. Specimens produced in that way were:

- Al-12% Si alloy with 50% granules
- Pb-1%   Bi with 45% granules
- Al-12% Si with 60% pores, produced by dissolving NaCl
  beads after embedding in the metal matrix.

Specimens dimensions for

- thermal conductivity measurements: diameter 25 mm, height
  25 mm. (For some experiments the diameter of the Al-Si-
  granules composite was stepwise reduced to 22 mm, 19 mm,
  and 16 mm to observe a possible influence of the ratio
  granules diameter/specimen diameter.)
- electrical conductivity measurements: diamter 14 mm, height
  21.8 mm.

RESULTS

The thermal and electrical conductivity data are graphically shown in Fig. 2 and Fig. 3, respectively.

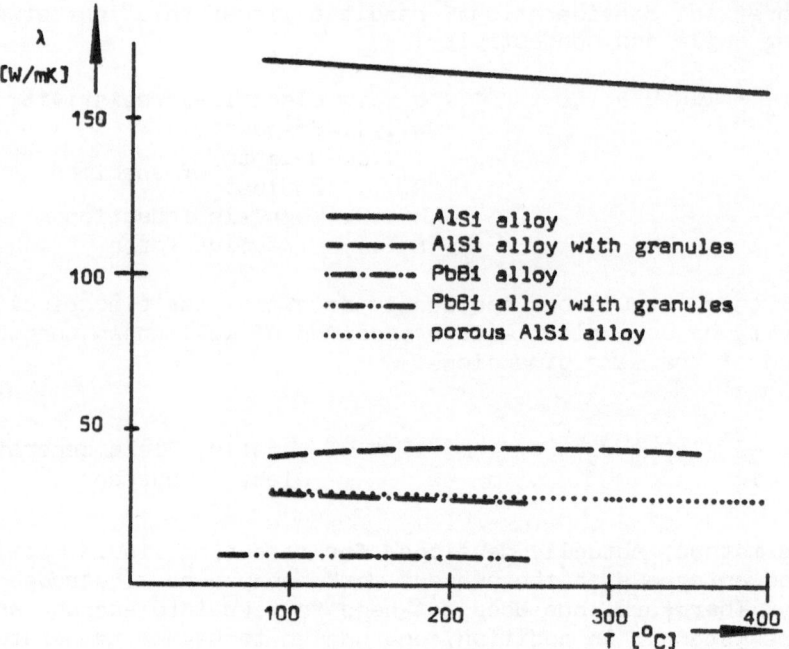

Fig. 2.   Thermal conductivity.

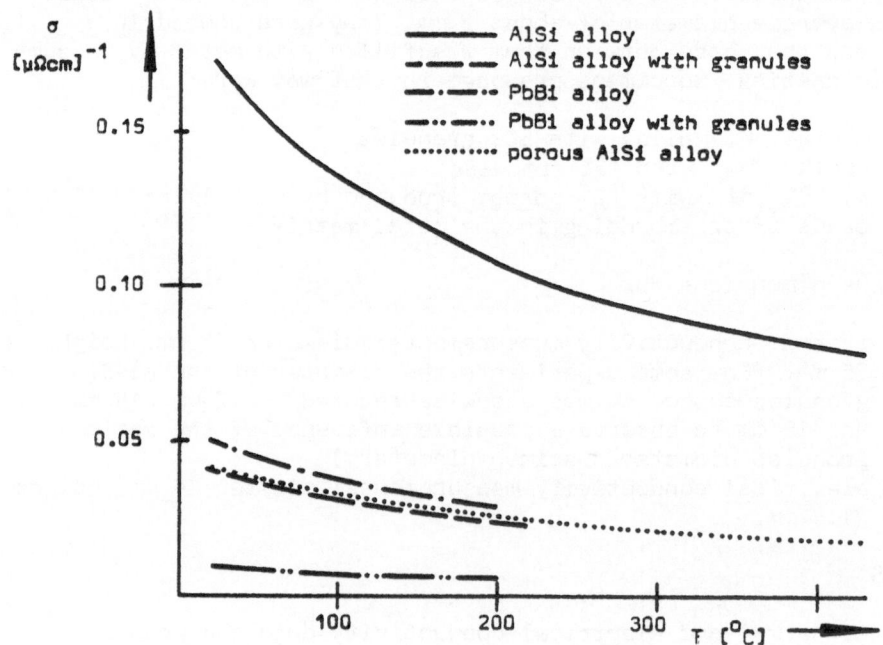

Fig. 3.   Electrical conductivity.

In these diagrams, the results are plotted without any correction related to specimen dimensions or granules content. Nevertheless, the difference between these two graphs is obvious. First, the significant stronger influence of the temperature on the electrical conductivity than on the thermal, in accordance with the Wiedemann-Franz-Lorenz law. Second, the surprisingly low electrical conductivity of the Al-Si-granules composite.

For a quantitative comparison, it is necessary that the results are related to composites with the same content of dispersed phase. Therefore, the thermal conductivity data experimentally obtained were adapted to 60% granules content by calculation using the Bruggemann's equation. Previous investigations have shown that this model fits best the slope of thermal conductivity as function of high granules content.

$$\frac{\lambda - \lambda_s}{\lambda_m - \lambda_s} \left(\frac{\lambda_m}{\lambda} \right)^{1/3} = 1 - f$$

λ ..... effective thermal conductivity of the composite
λ_s .... thermal conductivity of the granules
λ_m .... thermal conductivity of the matrix
f ..... ratio of dispersed phase

The results of the calculation are listed in Table 1. The calculated ratio, ρ, thermal conductivity of composite (for $f = 0.60$) divided by thermal conductivity of the matrix material, is almost the same value (~ 0.18) independent if pores exist or granules are embedded.

In addition, a possible specimen size effect should be considered by comparing the results qualitatively. There may be a little influence of specimen size of composites caused by the granules/specimen diameter ratio, as shown in experiments with different specimen diameters (16, 19, 22, 25 mm) (Fig. 4).

The electrical conductivity data are summarized in Table 2.

The ratio, λ divided by σT, was determined and is also listed in Table 2. In addition, the correlation of λ vs σ was calculated using the equation
$$\lambda/T = L\sigma + C$$
which results the following Lorenz function

Al-Si alloy	L = 3.75
Al-Si granules composites	L = 3.14
porous Al-Si	L = 2.50
Pb-Bi	L = 2.42
Pb-Bi granules composites	L = 3.24

Table 1.   Thermal Conductivity Data

Al-Si-granules composite   Al-Si

T [K]	λ_{exp} (50% gran.) [W/mK]	λ_{calc}* (60% gran.) [W/mK]	λ_{exp} [W/mK]	$\lambda_{exp}/\lambda^0_{exp}$	$\lambda_{calc}/\lambda^0_{calc}$
323	41.4	29.8			
373	42.7	30.7	168	0.254	0.182
423	43.4	31.2	165	0.265	0.189
473	43.6	31.4	163	0.267	0.193

porous Al-Si

T [K]	λ_{exp} (60% pores) [W/mK]	λ_{exp} [W/mK]	$\lambda_{exp}/\lambda^0_{exp}$
523	28.4	162	0.175
573	27.9	161	0.173
623	27.4	159	0.172
673	26.8	157	0.171
723	26.3	155	0.170

Pb-Bi-granules composite   Pb-Bi

T [K]	λ_{exp} (45% gran.) [W/mK]	λ_{calc}** (60% gran.) [W/mK]	λ^0_{exp} [W/mK]	$\lambda_{exp}/\lambda^0_{exp}$	$\lambda_{calc}/\lambda^0_{calc}$
323	8.9	5.5			
373	8.8	5.5	29.6	0.297	0.186
423	8.7	5.4	29.0	0.300	0.186
473	8.6	5.3	28.0	0.307	0.190

*$\overline{\lambda_{calc}}$(60% gran.) = 0.72 × λ_{exp}(50% gran.) } according to

**λ_{calc}(60% gran.) = 0.62 × λ_{exp}(45% gran.) } Bruggemann's law

Table 2. Electrical Conductivity Data

Al-Si-granules composites   Al-Si

T [K]	σ_{exp} (50% gran.) $[\mu\Omega cm]^{-1}$	Q^* (V^2/K^2) $\times 10^{-8}$	σ^o_{exp} $[\mu\Omega cm]^{-1}$	Q^* (V^2/K^2) $\times 10^{-8}$	$\sigma_{exp}/\sigma^o_{exp}$
293	0.039		0.175		0.223
323	0.035	3.65	0.156		0.224
373	0.030	3.80	0.138	3.26	0.220
423	0.026	3.96	0.122	3.19	0.215
473	0.024	3.83	0.110	3.13	0.218

porous Al-Si

T [K]	σ_{exp} (60% gran.) $[\mu\Omega cm]^{-1}$	Q^* (V^2/K^2) $\times 10^{-8}$	σ^o_{exp} $[\mu\Omega cm]^{-1}$	Q^* (V^2/K^2) $\times 10^{-8}$	$\sigma_{exp}/\sigma^o_{exp}$
373	0.0303		0.137	3.26	0.220
423	0.0280		0.122	3.19	0.230
473	0.0248		0.110	3.13	0.225
523	0.0222	2.43	0.100		0.222
573	0.0200	2.45	0.091		0.220
623	0.0182	2.63	0.085		0.214
673	0.0166	2.40	0.079		0.210
723	0.0150	2.40	0.075		0.200

Pb-Bi-granules composites

T [K]	σ_{exp} (45% gran.) $[\mu\Omega cm]^{-1}$	Q^* (V^2/K^2) $\times 10^{-8}$	σ^o_{exp} $[\mu\Omega cm]^{-1}$	Q^* (V^2/K^2) $\times 10^{-8}$	$\sigma_{exp}/\sigma^o_{exp}$
293	0.0094		0.0497		0.189
323	0.0085	3.25	0.0430		0.198
373	0.0070	3.37	0.0348	2.27	0.201
423	0.0061	3.29	0.0301	2.29	0.203
473	0.0056	3.25	0.0266	2.22	0.211

$^*Q = \dfrac{\lambda}{\sigma T}$

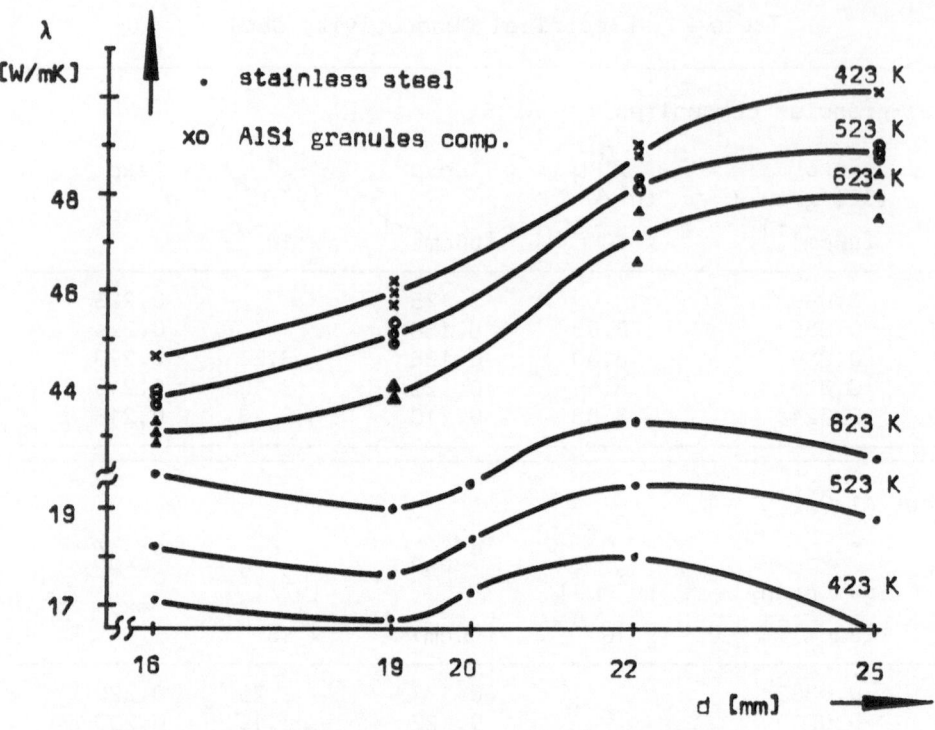

Fig. 4 Influence of specimen diameter

Because of the few experimental values and their achievable
accuracy, the calculated data, C, which should represent the
lattice component of the thermal conductivity, are not very reliable. Besides, the surprisingly high L of Al-Si-Alloy (this may
be caused by the poor accuracy of λ determination of such a good
thermal conductive material) the L values of composites are significantly higher than of the specimens without granules.

It can be concluded from the results that an estimation of
thermal conductivity values are feasible in principle, the accuracy of the predicted data is influenced by some parameters. Besides, the accuracy of the measurement itself, of course, there
are the specimen size and the specimen preparation (e. g., the
possibility of oxide inclusions during casting process) that influence the correlation between λ and σ values. Although it is
recommended to decide case by case which L value is used, the use
of 3.5 leads to a rough approximation of the thermal conductivity
of these composites.

REFERENCES

1.   V. V. Mirkovich, Comparative method and choice of standards
      for thermal conductivity determinations. J. Am. Ceram.
      Soc. 48:387 (1965).
2.   E. Hübner and W. Neumann, "Determination of Thermal Conducti-
      vity by a Comparative Method with Longitudinal Heatflow",
      SGAE 2837-ME 114/77, published by Austrian Research Centre
      Seibersdorf, Vienna.
3.   W. Braunbek, A new electrode-less method to measure electrical
      conductivity, Z. Phys. 73:312 (1932).
4.   W. Neumann, Thermal conductivity measurements of granules-
      metal composites, High Temp.-High Press., to be published.

NONSTATIONARY THERMAL BEHAVIOR OF REINFORCED COMPOSITES: A BETTER

EVALUATION OF WALL ENERGY BALANCE FOR CONVECTIVE CONDITIONS

Agnès M. Luc and Daniel L. Balageas

ONERA, 29 avenue de la Division Leclerc

92320 Châtillon, France.

INTRODUCTION

The increasing use of directional reinforced composite materials as thermal protection in reentry vehicle heat shields shows the necessity of investigating their unsteady thermal behavior. For better understanding of the phenomena, very simple geometric configurations are generally studied : periodically alternated laminae of two or three components or parallel fibers regularly embedded in a matrix.

In the special case of unsteady heat conduction, when the principal heat flow is parallel to the layering, experimental[1,2,3], analytical[4,5], or numerical[2,3,6] studies point out that the thermal diffusivity of such materials, measured by classical methods, depends either on the thermal excitation frequency (stationary wave method) or on the sample thickness (flash method). In the particular case of flash method which is the most universally used, it was demonstrated that for important sample thickness ($e/\omega > 4$), the Parker's diffusivity is nearly to the diffusivity corresponding to the steady effective thermal properties :

$$a_{stat} = k_{stat}/c_{stat} = \left[\tau k_1 + (1-\tau)k_2\right]/\left[\tau c_1 + (1-\tau)c_2\right]$$

For nonstationary heating prediction, the assumptions of a uniform heat flow absorption and equivalent homogeneous medium with thermal diffusivity a_{stat} are always used. Studies[3,6,7] have shown that in transient conditions like those of flash diffusivity measurements when considering small times and/or small thicknesses, the apparent diffusivity is different from a_{stat}. Furthermore, if the assumption of a uniform heat flow absorption is justifiable for flash method (an absorbing layer may be deposited on the front face of the sample), for a convective heat flux that corresponds to the practical use of such materials, this assumption is no longer satisfactory ; there is, indeed, a coupling between the in-depth thermal response of the composite and the prescribed boundary conditions. The net incoming heat flux distribution is no longer uniform on the front face, and the unidirectional treatment may lead to intolerable errors.

699

This paper deals with the incoming heat flux and wall temperature evaluations and presents a method leading to a more accurate evaluation of these parameters while preserving the unidirectional model simplicity.

EQUATIONS OF PROBLEM

The analysis considers a laminated composite as illustrated in Fig. 1 (space period ω) with a thermal contact resistance at the interface between the two constituents that are homogeneous and isotropic. The sample is subjected to a convective heating at the front face. The set of equations to solve is :

$$\rho_\alpha C_\alpha \frac{\partial T_\alpha}{\partial t} = k_\alpha \left[\frac{\partial^2 T_\alpha}{\partial x^2} + \frac{\partial^2 T_\alpha}{\partial y^2}\right] \qquad \alpha = 1,2 \qquad (1)$$

$$x = 0 \qquad h\left[T_r - T_\alpha\right] = - k_\alpha \frac{\partial T_\alpha}{\partial x} \qquad \alpha = 1,2 \qquad (2)$$

$$y = 0 \quad \text{and} \quad y = \frac{\varpi}{2} \qquad k_\alpha \frac{\partial T_\alpha}{\partial y} = 0 \qquad \alpha = 1,2 \qquad (3)$$

$$y = \tau \varpi/2 \quad, \quad k_1 \frac{\partial T_1}{\partial y} = k_2 \frac{\partial T_2}{\partial y} = \frac{T_1 - T_2}{R} \qquad (4)$$

$$t = 0 \quad, \quad T_\alpha = 0 \qquad \alpha = 1,2 \qquad (5)$$

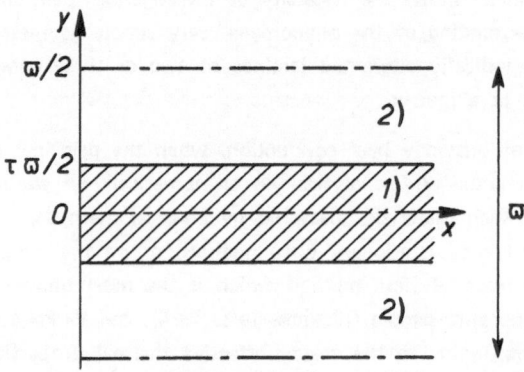

Fig. 1 — Composite geometry.

UNIDIMENSIONAL MODELS

The energy balance over the full composite space period, at the front face is :

$$\frac{1}{\varpi} \int_0^\varpi h\left(T_r - T_{(0,y,t)}\right) dy = \overline{\varphi}(0,t) \qquad (6)$$

Let $\overline{T}(x,t) = 1/\omega \int_0^\omega T(x,y,t)\, dy$ be the arithmetic average temperature, equation (6) becomes :

$$h\left(T_r - \overline{T}(0,t)\right) = \overline{\varphi}(0,t) \qquad (7)$$

This shows that the mean entering heat flux, $\overline{\varphi}$, is exactly evaluated using $\overline{T}$. Therefore, the choice of the assumptions governing the determination of $\overline{T}$ influences directly the evaluation of $\overline{\varphi}$.

Quasi-Stationary Model (QSM) and Uncoupled Component Model (UCM)

The exact solution for a homogeneous material with constant properties subjected to a convective heating is given by :

$$\frac{T(x,t)}{T_r} = erfc\left(\frac{x}{2\sqrt{at}}\right) - exp\left(\frac{hx}{k} + \frac{h^2}{b^2}t\right)erfc\left(\frac{x}{2\sqrt{at}} + \frac{h}{b}\sqrt{t}\right) \quad (8)$$

For the front face :

$$\frac{T(o,t)}{T_r} = 1 - E\left(\frac{h\sqrt{t}}{b}\right) \quad (9)$$

The composite may be treated as a homogeneous equivalent material with thermal properties evaluated from the steady-state values : equivalent thermal conductivity, $k_{stat} = \tau k_1 + (1 - \tau) k_2$, and equivalent volume heat capacity, $c_{stat} = \tau c_1 + (1 - \tau) c_2$. Therefore, equation (9) becomes :

$$\frac{\overline{T}_{stat}(t)}{T_r} = 1 - E\left(\frac{h\sqrt{t}}{b_{stat}}\right) \quad (10)$$

with the effusivity, b_{stat}, defined by : $b_{stat} = \sqrt{k_{stat} \cdot c_{stat}}$. This model is called "quasi-stationary model" (QSM), subscript stat. It is evident that it becomes accurate only when the steady-state is approached, i.e., when $\overline{T}$ approaches T_r, and then after a long time. On the other hand, the first moments, when the thermal phenomenon is highly transient, must be treated with another model. When the wall is cold (small $\overline{T}/T_r$), the composite components behave as if they were uncoupled and independent[6]. Then we must use equation (9) successively for the two components :

reinforcement : $\overline{T}_1(t)/T_r = 1 - E\left(h\sqrt{t}/b_1\right)$, $b_1 = \sqrt{k_1 \cdot C_1}$

matrix : $\overline{T}_2(t)/T_r = 1 - E\left(h\sqrt{t}/b_2\right)$, $b_2 = \sqrt{k_2 \cdot C_2}$

For the entire composite, the average wall temperature is :

$$\overline{T}_o(t) = \tau \overline{T}_1(t) + (1 - \tau) \overline{T}_2(t)$$
$$\frac{\overline{T}_o(t)}{T_r} = 1 - \tau \cdot E\left(\frac{h\sqrt{t}}{b_1}\right) - (1 - \tau) E\left(\frac{h\sqrt{t}}{b_2}\right)$$

Taking into account that this model is appropriate during the first moments, and using in equation (11), for the function E, the first-order approximation of its Taylor series expansion, gives :

$$\overline{T}_o(t)/T_r \simeq \left(2/\sqrt{\pi}\right)h\sqrt{t}\left(\tau/b_1 + (1 - \tau)/b_2\right) = \left(2/\sqrt{\pi}\right)\left(h\sqrt{t}/b_2\right) \quad (11)$$

The composite behaves like a homogeneous material with an effusivity, b_0, given by :

$$\frac{1}{b_o} = \frac{\tau}{b_1} + \frac{1 - \tau}{b_2} \quad (12)$$

Then the average wall temperature is given by :

$$\frac{\overline{T}_o(t)}{T_r} = 1 - E\left(\frac{h\sqrt{t}}{b_o}\right) \quad (13)$$

This model is called "uncoupled component model" (UCM), subscript 0.

Evaluation of the Entering Heat Flux

The entering heat flux may be evaluated using the two models :

$$\overline{\varphi}_{stat}(t) = hT_r \cdot E\left(h\sqrt{t}/b_{stat}\right) \qquad (14)$$

$$\overline{\varphi}_o(t) = hT_r \cdot E\left(h\sqrt{t}/b_o\right) \qquad (15)$$

When using the QSM for a cold wall, the error $\Delta\overline{\varphi}_{stat}/\overline{\varphi}$ may be evaluated identifying $\overline{\varphi}_o$ to the exact entering mean heat flux :

$$\Delta\overline{\varphi}_{stat}/\overline{\varphi} = \left(\overline{\varphi}_{stat} - \overline{\varphi}\right)/\overline{\varphi} \simeq \left(\overline{\varphi}_{stat} - \overline{\varphi}_o\right)/\overline{\varphi}_o$$

This expression gives, for $t \to 0$, a relative error $\Delta\overline{\varphi}_{stat}/\overline{\varphi}$ proportional to $\overline{T}/T_r$.

$$\frac{\Delta\overline{\varphi}_{stat}}{\overline{\varphi}} \simeq \frac{2}{\sqrt{\pi}} \frac{h\sqrt{t}}{b_o}\left(1 - \frac{b_o}{b_{stat}}\right) = \frac{T_o}{T_r}\left(1 - \frac{b_o}{b_{stat}}\right) \simeq \frac{\overline{T}}{T_r}\left(1 - \frac{b_o}{b_{stat}}\right) \qquad (16)$$

When using the UCM for a quasi-adiabatic wall, the error, $\Delta\overline{\varphi}_o/\overline{\varphi}$, may be evaluated assuming $\overline{\varphi}_{stat}$ to be the exact entering mean heat flux :

$$\Delta\overline{\varphi}_o/\overline{\varphi} = \left(\overline{\varphi}_o - \overline{\varphi}\right)/\overline{\varphi} \simeq \left(\overline{\varphi}_o - \overline{\varphi}_{stat}\right)/\overline{\varphi}_{stat}$$

The first-order approximation of the asymptotical expansion of the function E when $t \to \infty$ in the expression of $\Delta\overline{\varphi}_o/\varphi$ gives :

$$\frac{\Delta\overline{\varphi}_o}{\overline{\varphi}} = \left(\frac{b_o - b_{stat}}{\sqrt{\pi}h\sqrt{t}}\right)\bigg/\left(\frac{b_{stat}}{\sqrt{\pi}h\sqrt{t}}\right) = -\left(1 - \frac{b_o}{b_{stat}}\right) \qquad (17)$$

The UCM leads to an error $\Delta\overline{\varphi}_o/\varphi$, for a quasi-adiabatic wall ($\overline{T}/T_r \simeq 1$), which is limited by $-(1 - b_0/b_{stat})$.

For intermediate values of the wall temperature, only numerical calculation gives the accurate values of the incoming heat flux and errors, $\Delta\overline{\varphi}_{stat}/\overline{\varphi}$ and $\Delta\overline{\varphi}_o/\varphi$. These calculations will be developed in the following section.

COMPARISON OF PRESENT METHODS WITH EXACT RESULTS

The governing equations (1-5) are solved numerically by finite differences using an implicit scheme. This allows to evaluate $\overline{T}$, $\overline{\varphi} = h(T_r - \overline{T})$, and an apparent effusivity, $b(t)$, which would give at every time a wall temperature of homogeneous medium identical to the average wall temperature $\overline{T}$:

$$b(t) = h\sqrt{t}\bigg/E^{-1}\left(1 - \overline{T}/T_r\right) \qquad (18)$$

Besides, $\overline{\varphi}$ is compared to the incoming heat flux calculated by QSM and UCM. Values for $\Delta\overline{\varphi}_{stat}/\overline{\varphi}$ and $\Delta\overline{\varphi}_o/\overline{\varphi}$ are then evaluated.

General Evolutions of the Apparent Effusivity and Relative Errors on the Incoming Heat Flux

The material properties used for most of the calculations are those of a silico-phenolic composite. They are given in Table 1. For the first evaluations, the parameters values are : $R = 10^{-4}$ W^{-1} m^2 K, $\tau = 0.33$, $\omega = 0.72$ mm, $h = 1\,000$ Wm^{-2} K^{-1}, and $T_r = 1\,000$ K. Figure 2 presents the evolution of $(b - b_0)/(b_{stat} - b_0)$ versus $\overline{T}/T_r$. The apparent effusivity, b, increases from the UCM value, b_0, for $\overline{T}/T_r$ close to 0 towards the QSM value, b_{stat}, for $\overline{T}/T_r$ close to 1.

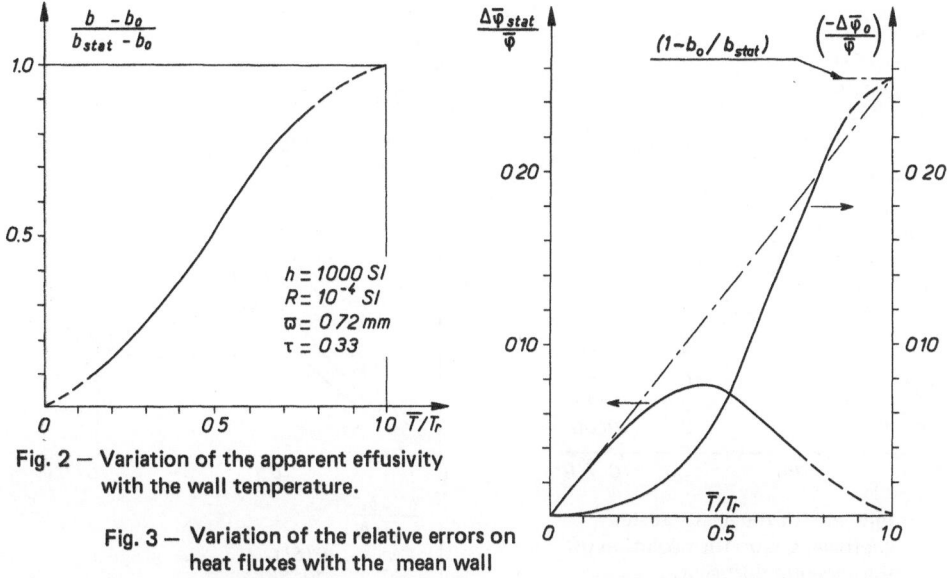

Fig. 2 — Variation of the apparent effusivity with the wall temperature.

Fig. 3 — Variation of the relative errors on heat fluxes with the mean wall temperature.

Figure 3 depicts the evolutions of relative errors, $\Delta\overline{\varphi}_{stat}/\overline{\varphi}$ and $\Delta\overline{\varphi}_0/\varphi$, versus $\overline{T}/T_r$, and shows that they are consistent with the limits calculated in the preceding section.

The two descriptions of relative errors and apparent effusivity above are equivalent and lead to the same conclusions concerning the validity of QSM and UCM.

Influence of Boundary Conditions : Recovery Temperature and Heat Transfer Coefficient

The evolution of b, $\Delta\overline{\varphi}_{stat}/\overline{\varphi}$, and $\Delta\overline{\varphi}_0/\overline{\varphi}$ with $\overline{T}/T_r$ are independent of T_r. This was verified for $100\ K \leqslant T_r \leqslant 10\,000\ K$.

An increase in the heat transfer coefficient value corresponds to an increase of the validity duration of the UCM (see Fig. 4).

Influence of the Composite Coupling Characteristics : Period and Contact Resistance

These parameters act in the same direction : an increase in ω or R corresponds to more accuracy and an increase of the validity duration of UCM (see Fig. 5 and Fig. 6).

Influence of the Thermal Properties of the 2 Components of the Composite

The nature of the constituents influences the values $(1\text{-}b_0/b_{stat})$ that determines : (i) the initial slope of $\Delta\overline{\varphi}_{stat}/\overline{\varphi}$, (ii) the limit of $\Delta\overline{\varphi}_0/\overline{\varphi}$, and (iii) the variation range of the apparent effusivity.

If $c_{12} = c_1/c_2$, and $k_{12} = k_1/k_2$, we obtain :

$$1 - \frac{b_0}{b_{stat}} = 1 - \left[1 - z + \frac{z}{\sqrt{k_{12}.c_{12}}}\right]^{-1} \times \left[1 - z + z\,k_{12}\right]^{-1/2} \times \left[1 - z + z\,c_{12}\right]^{-1/2} \tag{19}$$

Fig. 6 illustrates the evolution of $(1\text{-}b_0/b_{stat})$ with the volume rate, τ, for different values of k_{12} and $c_{12} = 1$. It is obvious that realistic values of these parameters may lead to important errors

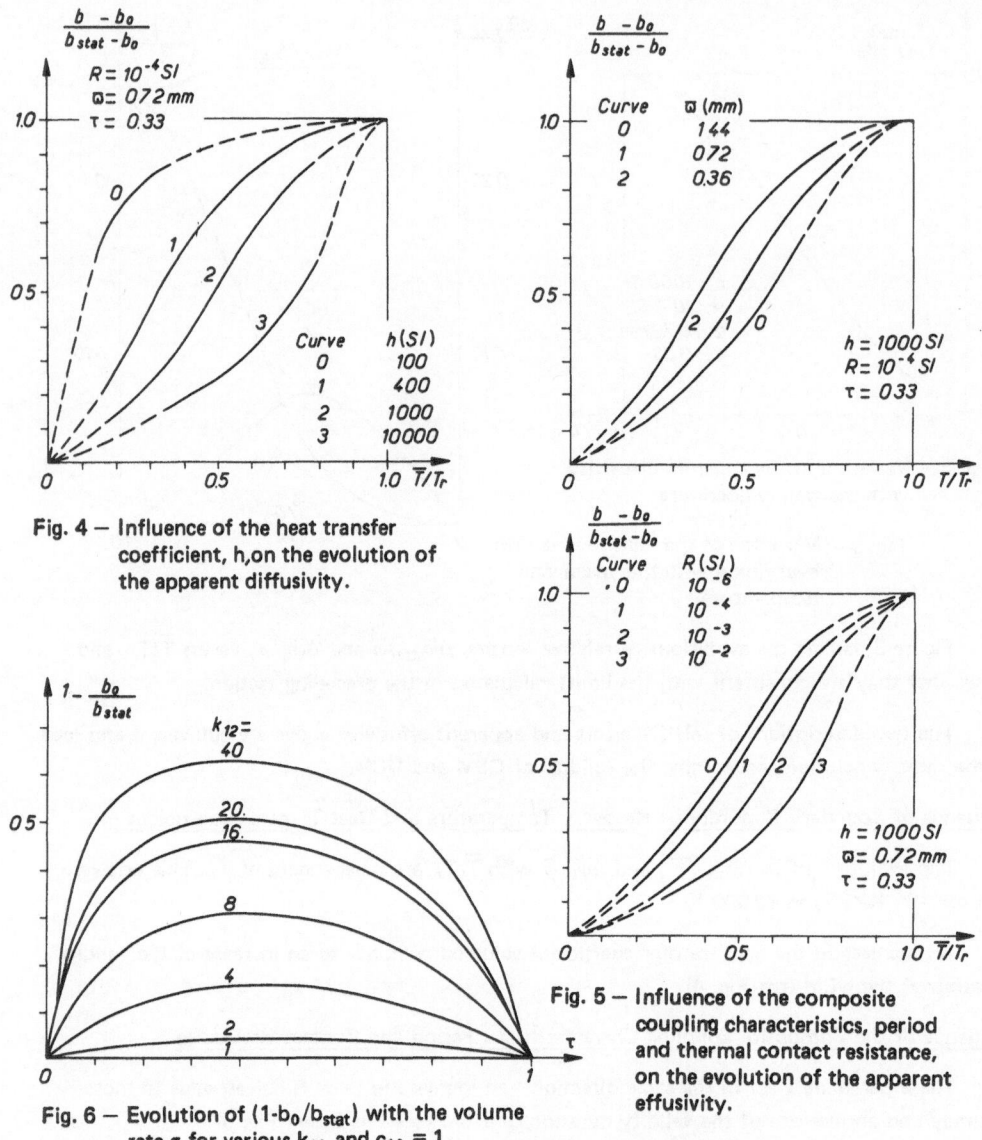

Fig. 4 — Influence of the heat transfer coefficient, h, on the evolution of the apparent diffusivity.

Fig. 6 — Evolution of $(1-b_0/b_{stat})$ with the volume rate τ for various k_{12} and $c_{12} = 1$.

Fig. 5 — Influence of the composite coupling characteristics, period and thermal contact resistance, on the evolution of the apparent effusivity.

Table 1 — Thermal Properties of Composites

Reinforcement Matrix	τ $(1-\tau)$	k_1 k_2	c_1 c_2
Silica Resin	0.33 0.67	1.33 0.21	$1.88\ 10^6$ $1.74\ 10^6$
Silica Resin	0.50 0.50	1.33 0.21	$1.88\ 10^6$ $1.74\ 10^6$
Carbon Resin	0.33 0.67	12.6 0.21	$1.30\ 10^6$ $1.74\ 10^6$

characterized by limit values $(1-b_0/b_{stat})$ reaching several tens of percent.

The evolution of the relative error, $\Delta\bar{\varphi}_{stat}/\bar{\varphi}$, is shown in Fig. 7 for different composites with the same boundary conditions (h,T_r) and coupling characteristics (ω,R). The composite properties used are summarized in table 1.

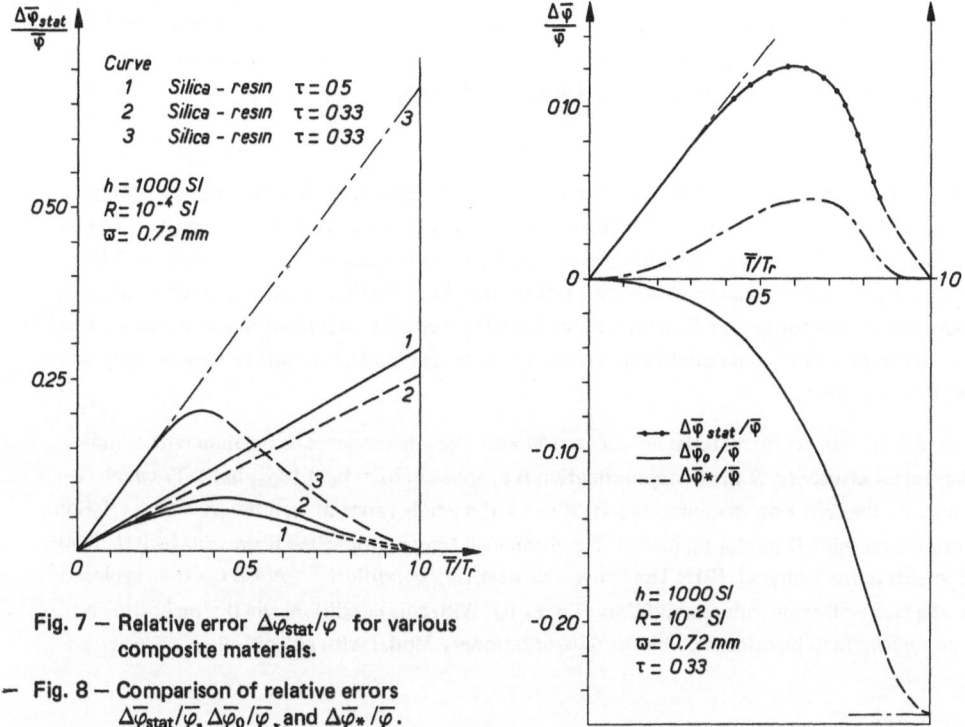

Fig. 7 — Relative error $\Delta\bar{\varphi}_{stat}/\bar{\varphi}$ for various composite materials.

— Fig. 8 — Comparison of relative errors $\Delta\bar{\varphi}_{stat}/\bar{\varphi}$, $\Delta\bar{\varphi}_0/\bar{\varphi}$, and $\Delta\bar{\varphi}_*/\bar{\varphi}$.

BETTER EVALUATION OF HEAT TRANSFER USING A UNIDIMENSIONAL MODEL

A new approach is required, avoiding long computing times and leading to better evaluation than those of QSM and UCM.

Using the above mentioned simulation results, the composite may be treated as a homogeneous medium with an apparent effusivity, b_*, that depends on $\bar{T}/T_r$. Figure 4 and 5 show that a linear approximation

$$\left(b_* - b_0\right)/\left(b_{stat} - b_0\right) = \bar{T}/T_r \tag{20}$$

is in relatively good agreement with the actual laws of apparent effusivity, especially when considering realistic conditions. Then, the mean wall temperature is given by :

$$\frac{\bar{T}_*}{T_r} = 1 - E\left[\frac{h\sqrt{t}}{b_0 + (b_{stat} - b_0)\frac{\bar{T}_*}{T_r}}\right] \tag{21}$$

A numerical solution of eq. (21) gives $\bar{T}_*/T_r$.

The error, $\Delta\overline{\varphi}_*/\overline{\varphi}$, resulting from this model is compared in Fig. 8 to those of UCM and QSM. The present method is more accurate than the other two. In the present case, the error is more than half the QSM error.

With this simple method, the heat flux boundary condition at the front face : $\varphi = A + B\overline{T}$ is replaced by a prescribed wall temperature-time history condition given by eq. (21): $\overline{T}_*/T_r = f(t)$. The heat diffusion in the composite may be solved with this new boundary condition. In particular when the considered depth is great relative to the space period ω, the field temperature may be calculated with the QSM using equivalent diffusivity $a_{stat} = k_{stat}/c_{stat}$.

CONCLUSION

To evaluate the thermal behavior of laminated material (chosen as the most simple configuration of directional reinforced composites) subjected to a convective heat flux, the laminae being parallel to the principal heat flux direction, homogeneous material approximations leading to 1-D solutions may be used. Two such models were studied : the Quasi-Stationary Model which is generally used but is only valid for quasi-adiabatic wall, and the Uncoupled Component Model which is proposed here for quasi-cold wall conditions. Neither of these two models are satisfactory at intermediate wall temperatures.

In fact the composite material behaves *at the wall* like a homogeneous medium with temperature-dependent effusivity. A linear approximation is proposed : $b_* = b_0 + (b_{stat}-b_0)\overline{T}/T_r$ which reduces notably the errors on the entering heat flux for the whole range of wall temperature variation, while preserving the 1-D model simplicity. The mean wall temperature-time history $\overline{T}_*(t)$ is then obtained solving numerically eq. (21). The convective heat flux condition $\overline{\varphi} = A + B\overline{T}$ is then replaced by a prescribed wall temperature condition : $\overline{T} = \overline{T}_*(t)$. With this condition, the thermal diffusion *in the composite* may be calculated by the Quasi-Stationary Model with an equivalent diffusivity : $a_{stat} = k_{stat}/c_{stat}$.

REFERENCES

1. V.Truong, Transient heat transfer in laminated composites - An experimental study, Ph. D. Dissertation, University of Massachusetts, Boston (1975).

2. M.Lafond, Etude du comportement thermocinétique de matériaux composites fibreux soumis à des sollicitations impulsionnelles, Ph.D. Dissertation, University of Paris VI, Paris (1979).

3. M.Lafond-Huot and J. Bransier, Caractérisation thermique de matériaux composites fibreux soumis à un flux de rayonnement impulsionnel, to be published in Letters Heat Mass Trans.

4. A.M. Manaker, Thermal transients in laminated composites, Ph.D. Dissertation, University of Massachussetts, Boston (1973).

5. A.H. Nayfeh, Continuum modeling of low frequency heat conduction in laminated composites with bonds, J. Heat Trans. 102: 312 (1980).

6. A. Luc, Caractérisation des transferts thermiques dans les milieux composites stratifiés en régime transitoire, ONERA Internal Report, ONERA, Châtillon-sous-Bagneux, France (1980).

7. A. Luc and D.L. Balageas, Le problème de la définition et de la mesure de la diffusivité thermique des matériaux composites à renforcement, ONERA TP 1981-30, ONERA, Châtillon-sous-Bagneux, France (1981).

NOMENCLATURE
Symbols

a	thermal diffusivity, $a = k/c$ $(m^2 \, s^{-1})$
b	thermal effusivity, $b = \sqrt{k \cdot c}$ $(J \, m^{-2} \, s^{-1/2} \, K^{-1})$
c	specific heat per unit volume $(J \, m^{-3} \, K^{-1})$
E(u)	function defined by : $E(u) = \exp(u^2) \times \operatorname{erfc}(u)$
e	specimen thickness (m)
h	heat transfer coefficient $(W \, m^{-2} \, K^{-1})$
R	thermal contact resistance per unit area $(W^{-1} \, m^2 \, K)$
t	time (s)
T	temperature (K)
$\overline{T}$	arithmetic mean wall temperature (K)
T_r	recovery temperature
x,y	space coordinate (m)
τ	reinforcement volume fraction
ω	space period of the composite reinforcement (m)
$\overline{\varphi}$	mean incoming heat flux $(W \, m^{-2})$

Indices

o	value calculated with the Uncoupled Component Model (UCM)
1,2	medium 1 (reinforcement), medium 2 (matrix)
stat	value calculated with the Quasi-Stationary Model (QSM)
*	value calculated with the model with temperature-dependent effusivity.

SESSION   TC-14B

Miscellaneous

CHAIRMAN

J. G. Hust
National Bureau of Standards
Boulder, Colorado

THE THERMOPHYSICAL PROPERTIES OF COATED

REINFORCED CARBON-CARBON COMPOSITES

R. P. Tye* and A. O. Desjarlais

Dynatech R/D Company
99 Erie Street
Cambridge, MA

ABSTRACT

Coated reinforced carbon-carbon is the baseline material for
the space shuttle leading edge structural subsystem. The leading
edge structural subsystem is part of the thermal protection system
for the orbiter, which will operate at equilibrium temperatures
greater than 1500 K. The purpose of the material is to limit the
temperature of the primary structure to its maximum design value
during boost and entry.

The material is a reinforced carbon-carbon laminate consisting
of a graphite cloth carbon matrix substrate protected from oxidation by a diffusion bonded silicon carbide coating.

Thermophysical property data are presented and discussed for
several generation materials for utilization in the leading edge
structural subsystem. The specific properties of interest were
thermal conductivity for heat flow, both parallel and perpendicular
to ply direction, and specific heat. The approximate temperature
range of interest was 120 to 2000 K.

---

*Current address: Energy Materials Testing Laboratory, Biddeford,
                   Maine

711

INTRODUCTION

The baseline material for the leading edge structural subsystem (LESS) components for the Space Shuttle/Orbiter Thermal Protection System (TPS) consists of a carbon-coated carbon composite manufactured by Vought Corporation. This material provides an aerodynamic moldline for the orbiter nose and wing leading edge and is mechanically attached to the fuselage and wing primary structure. The overall TPS must limit the temperature of the primary structures to their maximum design values during entry and reentry, and LESS consists of components that are subjected to operating temperatures in excess of 1530 K during these conditions.

The reinforced carbon-carbon (RCC) is fabricated from layers of graphite cloth contained in a carbon matrix formed by pyrolysis. To prevent oxidation at the very high operating temperatures encountered, the outer cloth layers are converted chemically to silicon carbide to form a stable protective coating. Subsequently, enhancement of the oxidation resistance is obtained by impregnating the laminates with tetraethyl-ortho-silicate (TEOS) and then reducing this to silicon oxide. Finally, Type A coating enhancement consisting of a brushed surface sealant to fill in the craze cracks and surface porosity was developed for the production material.

For the design and operation of the systems, reliable values for the thermophysical properties of the baseline material in its various forms developed over a period of time are essential. In addition, the extent to which the material can be fabricated with reproducible properties is an important factor as are the effects of the conditioning due to the reentry cycle of a mission.

Over a period of about four years, Dynatech R/D Company has carried out a number of studies for Vought Corporation on the thermophysical properties of the RCC materials in their various development and manufactured forms over the range 1000 to 2000 K. Thermal conductivity has been measured with the heat flow both parallel to and perpendicular to the ply. Measurements have been made on the uncoated and coated RCC and on the latter in the virgin and conditioned forms in various thicknesses. This paper presents results of the various thermal conductivity and heat capacity measurements undertaken.

EXPERIMENTAL PROGRAMS

Four series of measurements were undertaken. The objectives of each are outlined in Table 1 and summarized as follows:

Table 1. Details of Experimental Program

Program	Material	Condition	Number	Density Range, kg/m³	Property	Temperature Range, K
1.	RCC/uncoated 15 ply nominal 4.5 mm	Virgin	4	1400–1428	λ, Cp	120–1600
	RCC/coated 15 ply nominal 4.5 mm	Virgin	4	1753–1842	λ, Cp	"
	RCC/coated 19 ply nominal 6 mm	Virgin	4	1627–1743	λ, Cp	"
	RCC/coated 35 ply nominal 10.5 mm	Virgin	4	1512–1534	λ, Cp	"
	RCC/coated 19 ply nominal 6 mm	Conditioned†	2	1574–1596	λ, Cp	"
	RCC/coated/TEOS 35 ply nominal 10.5 mm	Virgin	2	1501–1515	λ	"
	RCC/coated/TEOS 35 ply nominal 10.5 mm	Conditioned†	1	1524	λ	"
2.	RCC/coated/TEOS 19 ply nominal 6 mm	Virgin	3	1580–1712	λ, Cp	120–1700
	RCC/coated/TEOS 19 ply nominal 6 mm	Conditioned†	3	1548–1588	λ	"
	RCC/coated/TEOS 35 ply nominal 10.5 mm	Virgin	3	1460–1513	λ, Cp	"
	RCC/coated/TEOS 35 ply nominal 10.5 mm	Conditioned†	3	1472–1526	λ	"
3.	RCC/coated/TEOS 19 ply nominal 6 mm	Virgin	3	1537–1594	λ	300–2000
	RCC/coated/TEOS 19 ply nominal 6 mm	Conditioned†	3	1505–1594	λ	"
4.	RCC/coated/TEOS/Type A enhancement 19 ply nominal 6 mm	Virgin	3	1561–1610	λ*	300–2000
	RCC/coated/TEOS enhanced coating 19 ply nominal 6 mm	Virgin	1	1576	λ*	"

*For heat flow parallel to ply only. All other cases for heat flow parallel and perpendicular to ply.
†Subjected to a radiant heating cycle in a carbon arc facility to simulate entry and reentry conditions.
Nominal coating thickness stated to be 0.64 mm.

1. To determine baseline data for the uncoated, coated,
    and conditioned materials in the various thicknesses
    being manufactured by the processes based upon the
    original experimental developments;

2. To study the effects of the TEOS impregnation and other
    changes in manufacturing processes on the properties;

3. To extend the upper temperature limits of the measure-
    ments on the TEOS impregnated material;

4. To check the effect of the Type A coating enhancements
    on the properties and their reproducibility with the
    previous materials. Thermal conductivity was deter-
    mined for the parallel-to-ply heat flow direction only.

     Specimens approximately 51 mm square and of the appropriate
thicknesses were fabricated from the above plates for thermal conductivity measurements in the heat flow in the perpendicular-to-ply
direction. For the parallel-to-ply direction, sufficient strips
of each material, 12.8 mm wide, were cut from the plates, turned
through 90°, and clamped together to form a specimen 51 mm square
and 12.8 mm thick. The clamps were removed once the specimen was
mounted in the test stack for measurement. Where necessary, the
flat surfaces of a specimen were machined such that the faces were
parallel, but a minimum amount of the coating was removed. For the
parallel-to-ply heat flow direction, the internal faces of the strips
were similarly machined to ensure that they would be in contact when
clamped together.

EXPERIMENTAL DETAILS

Thermal Conductivity

     Measurements were undertaken using the basic flat slab compara-
tive method with longitudinal guarding of the temperature gradient
in the test stack.[1-3] However, because of the limited thicknesses
of the specimens, their range of thermal conductance over the complete range of temperature required, and the possible effects of contamination of thermocouple instrumentation, some special experimental techniques were devised.

     Some preliminary experiments over the approximate range 300 to
600 K with uninstrumented test specimens using a constant applied
load indicated that highly variable contact resistances were present.
Thus, stacking was not possible and instrumented specimens were necessary over the full range.

For measurements over the approximate range 100 to 800 K, fine gage Chromel/Alumel thermocouples in an ungrounded protective nickel alloy sheath were fitted tightly into fine grooves cut in the surface of each specimen. The total thickness of the sheath and the groove dimension was 0.25 mm for the perpendicular heat flow specimens, and 0.5 mm for the parallel heat flow specimens. For the higher temperatures, these thermocouples were replaced with similar protected platinum -10% rhodium/platinum thermocouples. The 0.25 mm sheathed instrumentation used to prevent contamination was the smallest that was commercially available for the overall range of temperature. It was chosen to keep the thickness of the coating at a maximum, so that reliable measurements could still be made on a "representative" specimen. For the thinnest samples used in the first program, the depth of the groove represented approximately 40% of the thickness of the coating, but only 11% of the total specimen. For the majority of the specimens, the depth represented 8% or less of the total.

The three different reference materials used for the investigation were 13-mm-thick Pyroceram 9606, with 0.4-mm grooves cut across each flat parallel surface, and 32-mm-thick Inconel 702 and 33-mm-thick RVD graphite, each with 1-mm diameter holes drilled to the center at positions about 3 mm from each flat and parallel surface. During the course of the various investigations, the thermal conductivity of each of these materials was checked using one of the others as a reference. Those measured results were found to be within ±3% of the accepted value for the different materials.

The general procedure was as follows: Measurements were undertaken in a dry oxygen-free environment at regular temperature intervals from 100 to 800 K. Thermocouples and powder insulation around the test stack were then changed and additional slab insulation added at the bottom of the test stack. Measurements were then made at regular temperatures up to the highest temperatures required. In general, the equilibrium time at each of the higher temperature levels above 1100 K was of the order of 6 h. At least one repeat measurement was made at a lower temperature to investigate the possibility of any change in property due to the heating.

Specific Heat

Measurements were made by a combination of adiabatic calorimetry,[4,5] between 100 and 600K and drop calorimetry,[6,7] from 600 to 1700 K with the test specimens sealed in appropriate nickel alloy containers of known properties. For the former, the specimen was 63-mm square by approximately 13-mm thick and fabricated from a number of smaller pieces cut from the material supplied. For the drop calorimeter, a number of 25-mm or smaller diameter discs were stacked together to form a specimen having a total thickness of the order of 75 mm. Prior to any measurement on the test material, the adiabatic calorimeter was calibrated using a pure copper specimen and the drop calorimeter with a high purity alumina.

RESULTS & DISCUSSION

Thermal Conductivity

The results for the various programs are given in Tables 2 to 4. The values represent an average, together with the standard deviation obtained for each family of specimens. To obtain the average, the individual experimental results for the thermal conductivity, λ, of each specimen were plotted with respect to temperature. Smooth curves were then drawn through these points, and in all cases, none of the experimental points was more than ±5% from an individual curve.

In the majority of cases, none of the points was more than ±3% from the individual curve. Values of λ were then obtained from these curves at regular temperature intervals and averaged, and the standard deviation was determined.

In the first program, a repeat measurement was undertaken on a randomly chosen specimen to investigate the reproducibility of the experimental procedure. This specimen and associated test stack, including a different type of reference material pair, were completely reinstrumented and the experiment repeated. The results are given in Table 5. In addition, after comparing the results of the first and second programs, it was seen that the newer material was some 30%($\perp$) and 50%($\parallel$) higher than the original material. To investigate the possibility of any significant experimental error between the two series of measurements, a random specimen from the first program was retested using the apparatus and techniques of the second program. The results are also contained in Table 5.

In each case, there is excellent agreement between the values for the respective specimens for the repeat measurements, especially allowing for the fact that small changes in value (3 to 4%) were obtained after the first heating on both specimens. These results indicate that the experimental procedures were reproducible and that the measured differences in the materials are real.

It is not possible to compare the present results with those of other investigations since the materials, although uniform in nature, are particularly anisotropic. This is true even for the uncoated material since it would be unlikely that the combination of fiber, cloth, initial resin, impregnation, carbonization, other process

Table 2. Thermal Conductivity and Standard Deviations of RCC Materials in Program 1

		Thermal Conductivity, λ, and Standard Deviation, σ, W/mK at											
		123	223	323	473	573	773	973	1173	1373	1573	1673	473 K
15 ply uncoated ⊥ 4	λ	0.78	1.60	2.45	3.02	3.22	3.35	3.52	3.85	4.02	4.12	4.20	3.10
	σ	0.064	0.14	0.19	0.17	0.15	0.24	0.30	0.39	0.31	0.34	0.36	0.17
15 ply uncoated ∥ 4	λ	1.26	3.10	4.49	5.80	6.20	6.70	7.20	7.72	8.15	8.25	8.38	5.65
	σ	0.10	0.14	0.10	0.08	0.08	0.08	0.29	0.61	0.30	0.21	0.21	0.07
15 ply coated ⊥ 4	λ	1.68	3.05	4.50	5.55	5.92	6.05	6.40	6.70	6.70	6.72	6.75	4.62
	σ	0.13	0.53	0.64	0.53	0.39	0.44	0.41	0.28	0.29	0.29	0.33	0.59
15 ply coated ∥ 4	λ	5.00	8.70	11.50	12.58	12.68	12.85	13.35	13.35	13.35	13.30	13.20	10.98
	σ	0.24	0.43	0.47	0.74	0.79	0.72	0.60	0.52	0.44	0.34	0.34	0.59
19 ply coated ⊥ 4	λ	1.68	2.98	4.37	5.78	5.90	5.92	6.05	6.10	6.15	6.12	6.10	5.62
	σ	0.33	0.68	1.01	0.68	0.67	0.63	0.66	0.61	0.66	0.68	0.65	0.56
19 ply coated ∥ 4	λ	3.15	5.38	7.00	8.22	8.40	8.60	8.70	8.75	8.80	8.78	8.70	8.10
	σ	0.19	0.39	0.32	0.53	0.48	0.29	0.29	0.37	0.38	0.43	0.36	0.51
19 ply coated and conditioned ⊥ 2	λ	1.00	2.00	2.75	3.45	3.60	3.60	3.75	3.95	4.05	4.10	4.20	3.45
	σ	0.14	0.28	0.50	0.78	0.71	0.71	0.64	0.64	0.78	0.85	0.99	0.78
19 ply coated and conditioned ∥ 2	λ	2.80	4.45	6.10	7.85	8.05	8.05	8.20	8.70	8.75	8.80	8.80	7.85
	σ	0.14	0.49	0.85	1.63	1.63	1.34	1.27	0.57	0.35	0.28	0.28	1.48
35 ply coated ⊥ 4	λ	1.42	2.62	3.75	4.60	4.85	5.25	5.52	5.68	5.82	5.90	6.02	4.52
	σ	0.05	0.10	0.17	0.22	0.17	0.17	0.25	0.36	0.38	0.37	0.22	0.25
35 ply coated ∥ 4	λ	2.72	4.82	6.95	8.45	8.75	8.98	9.28	9.40	9.35	9.35	9.35	8.20
	σ	0.33	0.59	1.04	0.89	0.95	1.13	1.13	1.11	1.01	0.97	0.87	0.59
35 ply coated/TEOS ∥ 2	λ	4.8 (173)	7.15(273)	--	9.70	10.05	10.75	11.20	11.70	12.00	12.15	12.20	9.05
	σ	0.56	1.77	--	2.12	2.05	1.77	1.42	0.99	0.89	0.78	0.71	2.19
35 ply coated/TEOS and conditioned ∥ 1	λ	4.6 (173)	6.2 (273)	--	8.6	9.0	9.7	10.4	11.0	11.6	11.8	11.9	8.2

Numbers in parentheses refer to special temperature of test

Table 3. Thermal Conductivity and Standard Deviation of Coated RCC Materials in Program 2.

| | | Thermal Conductivity, λ, and Standard Deviation, σ, W/mK at | | | | | | | | |
		123	273	423	573	873	1173	1423	1623	423 K
19 ply virgin ⊥	λ	1.97	4.03	5.43	6.50	7.03	7.67	7.73	7.63	4.67
	σ	0.06	0.06	0.21	0.30	0.21	0.15	0.15	0.15	0.12
19 ply virgin ∥	λ	5.03	9.37	11.70	12.30	13.17	13.73	13.93	13.53	10.30
	σ	0.15	0.38	0.53	0.95	1.59	1.72	1.36	0.84	0.53
19 ply conditioned ⊥	λ	1.77	3.60	4.47	5.17	5.83	6.47	6.83	6.90	4.07
	σ	0.06	0.10	0.25	0.45	0.47	0.61	0.57	0.60	0.49
19 ply conditioned ∥	λ	3.40	6.73	9.77	10.43	11.57	11.60	11.60	11.37	8.57
	σ	0.26	0.61	0.46	0.45	1.43	1.64	1.59	1.55	0.64
35 ply virgin ⊥	λ	1.43	3.57	5.07	6.00	6.77	7.70	7.93	7.63	4.60
	σ	0.06	0.06	0.06	0.10	0.12	0.10	0.06	0.12	0.20
35 ply virgin ∥	λ	2.63	6.90	10.03	11.90	13.03	13.97	13.93	13.73	9.23
	σ	0.15	0.30	0.25	0.30	0.31	0.21	0.12	0.06	0.15
35 ply conditioned ⊥	λ	1.27	2.77	3.97	4.57	5.47	6.27	6.67	6.93	4.07
	σ	0.06	0.06	0.15	0.15	0.32	0.42	0.40	0.38	0.06
35 ply conditioned ∥	λ	2.17	6.47	9.23	10.37	11.63	12.60	12.83	12.93	8.93
	σ	0.12	0.31	0.75	0.91	1.21	1.44	1.17	1.25	0.70

Table 4. Thermal Conductivity and Standard Deviation of Coated RCC Materials in Programs 3 & 4.

Materials & Population		Thermal Conductivity, λ, and Standard Deviation, σ, W/mK at							
		313	523	823	1123	1373	1673	1923	423 K
19 ply virgin $\perp$ 3	λ	4.13	5.72	6.58	7.18	7.35	7.25	7.18	4.28
	σ	0.38	0.49	0.38	0.45	0.48	0.46	0.43	0.10
19 ply virgin $\parallel$ 3	λ	7.93	9.95	10.93	11.08	11.17	11.08	10.95	8.23
	σ	0.42	0.56	0.73	0.68	0.65	0.73	0.88	0.35
19 ply conditioned $\perp$ 3	λ	3.75	4.83	5.50	6.23	6.53	6.52	6.40	3.08
	σ	0.23	0.31	0.23	0.12	0.20	0.28	0.30	0.28
19 ply conditioned $\parallel$ 3	λ	7.63	9.42	10.17	10.35	10.35	10.20	9.95	7.90
	σ	0.54	0.80	0.95	1.00	0.90	0.92	0.77	0.79
19 ply virgin $\parallel$ 4	λ	8.54	10.02	11.01	11.29	11.32	11.30	11.22	8.36
	σ	0.25	0.10	0.22	0.22	0.20	0.19	0.24	0.27

Table 5. Repeat Measurement of the Thermal Conductivity of Coated RCC Materials.

Description of Retest	Thermal Conductivity, λ, W/mK at					
	473	573	773	973	1173	473 K
Specimen first tested in March 1976	11.2	-	12.0	-	12.4	10.6
Complete reassembly & retest in April 1976	10.2	-	10.8	-	11.7	10.2
Specimen first tested in May 1976	6.3	6.4	6.4	6.6	6.3	6.0
Repeat in May–June 1977 in different apparatus	6.0	6.2	6.4	6.6	6.6	5.9

variables, and final density would be identical. Thus, the general discussion highlights the major factors found to influence the properties.

The thermal conductivity of the uncoated material is in the general range for so-described 2D reinforced carbon-carbon materials. The variability with temperature and the degree of property anisotropy are also comparable to that found for other materials of a similar type and form. The infiltration of silicon carbide to form a protective surface layer is seen to increase the thermal conductivity, as would be expected for this higher density material. However, the different forms of this coating also have a significant effect on this increase and its subsequent change owing to a high temperature conditioning treatment.

The present discussion concentrates on the results for the 19-ply material. However, the comments and observations are generally applicable to the other thicknesses of the material.

An examination of all of the results indicates the following:

1. Since the variability in results obtained for the uncoated materials for both directions was small, it appears that the major variable in the coated specimens was related to the coating adhesion and surface characteristics.

2. The inclusion of a coating increased the thermal conductivity significantly; the increase was dependent on the total thickness. The coating had the greatest effect on the thinnest specimens owing to its greater contribution to the total. It was also dependent on the temperature level. Much larger percentage increases were seen at temperatures below 800 K than at the higher temperatures. Qualitatively, this can be explained by the contribution of the very high thermal conductivity values of the silicon carbide at the lower temperatures. However, the thermal conductivity of silicon carbide reduces sharply as the temperature increases, and its contribution is smaller at high temperatures.

3. The thermal conductivity of all enhanced coated materials was significantly higher than those having the original coating.

4. The conditioning treatment reduced the thermal conductivity especially seen for the materials of the first program. Materials of subsequent programs showed smaller changes between the virgin and conditioned specimens. This indicates that the "improved" coatings developed in the later stages of the development program and for the production material were

much more effective in protecting the substrate and reducing delamination and oxidation. The superior protective properties of these coatings is further illustrated by the relatively small differences in density between the virgin and conditioned specimens compared with those in the first program.

5. Much smaller, but significant reductions in thermal conductivity were obtained on most specimens on cooling. The repeat points were usually lower and with an equivalent or lower variability in the results compared with the values on heating. This could indicate a stabilization of the material or some further degradation. However, it should be pointed out that a specimen was allowed to equilibrate at each of the elevated temperatures for many hours. This would not be truly representative of use conditions, and if degradation was occurring, it was very slow.

6. The overall variability decreased for the materials produced for the latter programs, as illustrated in Table 6. This is due presumably to the development and use of the improved coatings.

7. A high variability was seen for the parallel-to-ply direction due to the test specimen. It was often found necessary to machine off small amounts of the coating to ensure that adjacent surfaces were in contact when clamped together. Variation in surface roughness required the removal of varying percentages of coating, thus changing its contribution from specimen to specimen.

8. Further examination of the results, including those in Table 6, indicates that the anisotropy in thermal conductivity for this material is approximately 2. There is some variability in this value, particularly in comparing the results at temperatures above 1000 K. This variability is due to the larger uncertainty in results for the parallel direction for the reasons listed in 7.

Specific Heat

The results for each sample measured are given in Table 7. The averaged values at regular temperature intervals were obtained in a similar manner as those for thermal conductivity. In all cases, no one experimental point was more than ±3% from the curve drawn through them. The variability in the second series is much lower than that of the first, due presumably to the total specimens being

Table 5. Repeat Measurement of the Thermal Conductivity of
Two Specimens of Coated RCC Materials.

| Temperature, K | Thermal Conductivity, W/mK | | | | | | | |
| | Program 1 | | Program 2 | | Program 3 | | Program 4 | |
	Uncoated*	V	C	V	C	V	C	V
			for ∥ Direction					
500	5.9	8.3	7.9	12.0	10.1	9.8	9.3	9.9
1500	8.2	8.9	8.8	13.9	11.5	11.2	10.4	11.3
			for ⊥ Direction					
500	3.1	5.8	3.5	6.1	4.8	4.9	4.8	–
1500	4.1	6.2	4.1	7.7	6.8	7.3	6.5	–

*Measured on 15-ply material but assumed to be representative of bulk material

V = Virgin; C – Conditioned

Table 7. Specific Heat of Uncoated and Coated RCC Materials

Material		Specific Heat, Cp, and Standard Deviation, σ, J/kgK at											
	Temperature, K	123	223	323	423	473	573	773	973	1173	1373	1573	1673
Uncoated	Cp	420	630	795	900	965	1110	1300	1380	1530	-	-	-
15 ply coated virgin	Cp	420	620	795	-	940	1050	1205	1320	1425	1510	1590	1650
19 ply coated virgin	Cp	420	635	815	-	965	1010	1235	1280	1380	1505	1610	1655
35 ply coated virgin	Cp	460	695	840	-	1005	1110	1275	1380	1465	1550	1655	1715
19 ply coated & TEOS	Cp	420	690	880	-	1035	1160	1310	1385	1485	1560	1650	1675
Average of coated	Cp	430	660	832.5	-	986	1082	1256	1341	1439	1531	1626	1674
4	σ	20	38	36.6	-	42.1	66	46	50.4	46.4	27.8	31.5	29.5
19 ply coated virgin	Cp	410	620	-	895	-	1170	1300	1380	1450	1550	1690	1730
19 ply coated virgin	Cp	400	620	-	900	-	1180	1310	1380	1450	1560	1700	1750
19 ply coated virgin	Cp	400	610	-	890	-	1170	1300	1380	1480	1540	1670	1700
35 ply coated virgin	Cp	380	590	-	870	-	1150	1275	1350	1430	1530	1650	1700
35 ply coated virgin	Cp	380	590	-	870	-	1160	1280	1350	1410	1510	1620	1680
35 ply coated virgin	Cp	390	590	-	880	-	1160	1290	1360	1450	1540	1660	1710
Average of coated	Cp	393	603	-	884	-	1165	1292.5	1367	1438	1538	1665	1712
6	σ	12.1	15.05	-	12.8	-	10.5	13.3	15.1	16	17.2	28.8	24.8

fabricated from the larger and more uniform pieces that were available. All values are consistent with those for a basically carbonaceous material.

At the lower and upper temperature ranges of the measurements,
the values obtained for the only uncoated specimen measured are consistently a little higher than the averaged values for the coated
material in the approximate range 300 to 600 K. The curves cross,
but the differences are small. Silicon carbide has a lower specific
heat than carbon and if allowance is made for the relative contribution of the thickness of the coating; then the resultant curve for
the coated material should be lower than that for the uncoated. Calculated specific heat values for the coated materials using the present results for the uncoated materials combined with those for
silicon carbide taken from TPRC,[8] are in very good agreement with
the experimental results. For example, at 600 K, the calculated
value is 1040 J/kgK compared with 1070 J/kgK as measured. At 1200 K,
the respective values are 1430 and 1425 J/kgK.

SUMMARY AND CONCLUSIONS

Measurements of the thermal conductivity and heat capacity of
various forms of silicon-carbide-coated reinforced carbon-carbon
materials have been evaluated over all or parts of the range 100 to
2000 K. Effects of a conditioning procedure simulating entry and
reentry conditions have also been studied. Increased thermal conductivity accompanied by increasing stability of the property with
conditioning was seen as improved coatings were being developed.
The specific heat values are consistent with those for a carbonaceous
material.

ACKNOWLEDGMENTS

The authors wish to thank M. Macginnis and W. Whitten of Vought
Corporation for their assistance and suggestions during the course
of the various measurement programs. They also wish to acknowledge
the assistance of their colleagues, S. C. Spinney and R. R. Heroux,
with some of the experimental work.

REFERENCES

1. V. V. Mirkovich, J. Am. Ceram. Soc., 48:387 (1965).
2. R. P. Tye and J. R. Hurley, "Proceedings of the Fifth Tempera-
    ture Measurement Society," Vol II (1967).
3. "Thermal Conductivity," Vol I, R. P. Tye, ed., Academic Press,
    London (1969).
4. L. C. Hoagland, F. DeWinter, and A. R. Reti, Trans. Instrum.
    Soc. Am., 6, 111 (1968).
5. R. P. Tye and S. C. Spinney, ASHRAE Trans. 84(1), 675 (1978).
6. D. C. Ginnings and R. J. Coruccini, J. Res. Nat. Bur. Stand.,
    38, 592 (1947).
7. J. B. Conway and R. A. Hein, "Advances in Thermophysical Proper-
    ties at Extreme Temperatures and Pressures," S. Gratch, ed.
    (1965).
8. "Thermophysical Properties of Matter," Vol 5, Y. S. Touloukian,
    E. H. Buyco, and T. Makita, eds., Plenum Press, New York
    (1969).

A TECHNIQUE FOR MEASURING THE APPARENT THERMAL CONDUCTIVITY OF FLAT INSULATIONS*

J. P. Moore, D. L. McElroy, and S. H. Jury

Metals and Ceramics Division
Oak Ridge National Laboratory
Oak Ridge, Tennessee 37830

ABSTRACT

A technique for measuring the apparent thermal conductivity, λ^*, of flat insulations (either batt or loose) is described. This technique consists of a flat Nichrome-screen wire sandwiched between two samples and heated electrically. Results from a limited experimental test are described and compared with results from a pipe insulation tester. Thermal modeling of this system shows that accurate measurements with λ^* down to 0.01 W/m·K and insulation thicknesses below 0.089 m (3-1/2 in.) should be straightforward. Measurement of λ^* for material as thick as 0.305 m (12 in.) should also be possible with a lateral guard scheme to reduce heat flow along the screen wire heater.

INTRODUCTION

Thermal insulation is important for energy conservation in domestic and industrial operations.[1] To assess the economic feasibility of a particular insulation, the thermal conductivity, λ, or a related property, such as the thermal resistance, must be known and well understood.

---

*Research sponsored by the Division of Materials Sciences, U.S. Department of Energy, under contract W-7405-eng-26 with the Union Carbide Corporation.

Unfortunately, λ is extremely difficult to measure accurately especially in the case of poorly conducting thermal insulators,[2] and techniques for measuring λ of insulations tend to be complex and require great effort to obtain accurate data. Typical of these are the guarded hot plate[3] and heat meter.[4] The difficulties encountered with these traditional techniques have been discussed by Pratt.[5]

The λ of pipe insulations is generally easier to measure than that of flat specimens, primarily because the traditional pipe tester has a radial geometry[6] and can be used with very long systems. A wound steel-screen wire was recently used as the central (or core) heater of a pipe tester.[7] The high electrical resistance of the wire screen ensures that a large amount of joule heat can be uniformly generated in the screen, and the low thermal conductivity of the screen minimizes lateral heat flow out of the ends of the system.

The inherent accuracy of this radial system is difficult to realize for blanket or batt insulations. We developed a simplified method for measuring λ of flat insulations as a function of thickness, density, temperature, and emittance of the cooling plates.[8] This paper describes the results of an experimental test on a wall insulation, compares these results with those from the aforementioned pipe tester, and presents detailed thermal analysis calculations that indicate the optimum use of this concept for insulations with thicknesses between 0.0251 and 0.3045 m (about 1 and 12 in.).

Heat transfer through a nonopaque material, such as glass and fibrous insulations, includes a term due to radiation, and this causes the measured λ to be dependent on the emittance of bounding surfaces and on the specimen thickness. For this reason the thermal conductivity discussed herein is called the "apparent λ" and denoted by $\lambda*$.

EXPERIMENT

This concept was tested with the simple apparatus shown in Fig. 1. A sheet of insulation was placed against a metal panel 0.426-m wide by 0.911-m long, and a Nichrome screen 0.33-m wide by 0.76-m long was laid across the insulation and pulled taut at the four corners. An opposing insulation batt was placed on a similar metal panel to form a sandwich. Mounting posts were used between the panels at each corner so that the total thickness of the system could be varied by changing the heights of the posts. A Chromel versus constantan thermocouple was attached to the screen near its center with thermally conducting epoxy. Steel wire of 0.127-mm (0.005-in.) diameter was welded onto the screen so that the voltage

Fig. 1.   Tester for measuring the apparent thermal conductivity of
flat insulation showing the copper buss on each end of the
screen, piano wire at each corner to hold the screen taut,
and a thermocouple attached to the screen.

drop across a known length, ℓ, could be determined when a current
was passed through the screen.

After the sandwich was bolted together, a measured dc current
was passed through the screen, and the heat generated passed through
the two layers of insulation. The thermocouples on the screen and
those mounted on the outside of the metal panel were monitored until
steady state was reached. The thermocouple outputs, the current, I,
and the voltage, E, were measured with standard potentiometric
equipment. The λ* of the specimen was then calculated using the
equation

$$\lambda^* = \frac{P}{4A}\frac{\Delta X}{\Delta T} = \frac{EI}{4\ell w}\frac{\Delta X}{\Delta T} \quad ,$$

where ΔX was the distance between the metal panels, ΔT was the
temperature difference between the Nichrome screen heater and metal
panels, and w was the strip width.

The λ* results at 298 K from this test are shown in Fig. 2 for
three values of insulation density (sample thicknesses between
0.01588 and 0.0363 m). A λ* value obtained on pie-shaped wedges

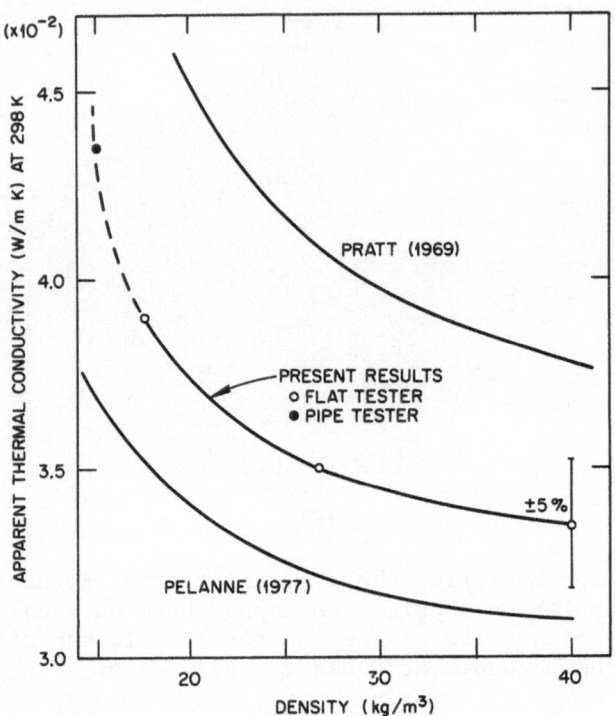

Fig. 2. The apparent thermal conductivity versus density of a
         fibrous insulation at 298 K showing the agreement of the
         flat tester results with those from a pipe tester on the
         same material at a lower density.

with the pipe tester is also shown at a density of 15 kg/m^3
(0.93 lb/ft^3). The data points can be connected by a smooth curve
of the shape normally obtained for a material of this type.[5,9]
These results indicate the feasibility of the concept.

THERMAL ANALYSIS

Scope

     A finite-difference generalized heat-conduction program known
as HEATING5[10] was used to evaluate the temperature profile for
several configurations of the flat screen tester for steady-state
operations.

     A schematic of the problem is shown in Fig. 3. The screen is
located in the x-y plane with the z-axis normal to the flat screen,
to the flat slabs of insulation, and to the bounding surfaces at
z_1 and $-z_1$. The screen extends from y_1 to $-y_1$, and electrical

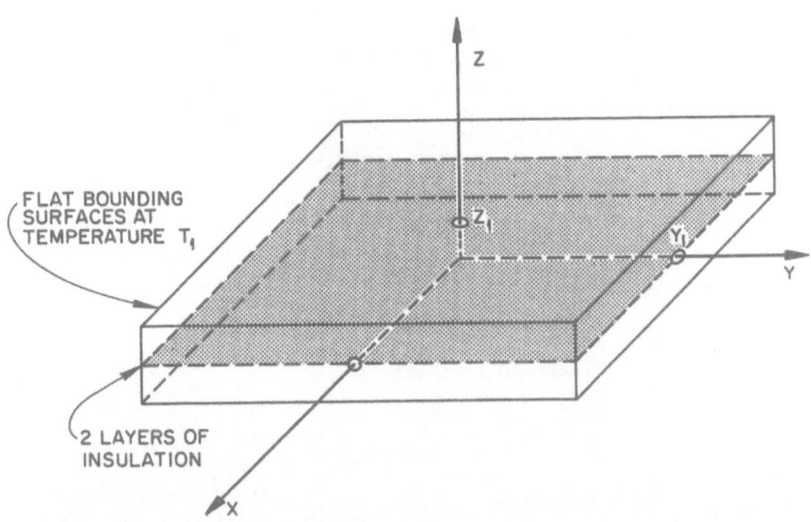

Fig. 3. Three-dimensional schematic of the system for thermal
         modeling showing the screen in the x-y plane with insula-
         tion on each side. A uniform volumetric heat generation
         rate was assumed for this heater.

current to energize the heater flows parallel to the y-axis. Three
configurations for the screen heater are described in Fig. 4, which
shows views of the x-y plane at z = 0. The configuration shown in
Fig. 4(a) was for a screen heater that stopped 76.2 mm (3 in.) from
each side so that the insulation in the 76.2-mm-wide strip restricted
lateral heat flow. In the second configuration, shown in Fig. 4(b),
the screen extended all the way to each side of the insulation. In
the configuration shown in Fig. 4(c), the screen extended to the
outer edges of the insulation, but slots cut in the screen permitted
a higher heat generation rate in the outer strips of the screen.
Presumably this higher heat generation rate could be provided by
having higher current densities in these lateral strips. Configurations 4(b) and 4(c) are essentially identical when $J_1 = J_2 = J_3$,
where J_1 is the current density in the central screen, and J_2 and
and J_3 are the current densities in the lateral guards.

These configurations were evaluated for specimen thicknesses of
25.1 mm (∼1 in.), 88.6 mm (∼3 1/2 in.), 152.1 mm (∼6 in.), 228.3
(∼9 in.), and 304.5 mm (∼12 in.). The screen length was assumed to
be 1.219 m (4 ft) for all calculations involving configurations
4(a) and 4(b). Calculations for configuration 4(c) were done for
an assumed screen length of 1.829 m (6 ft) and a screen assumed to

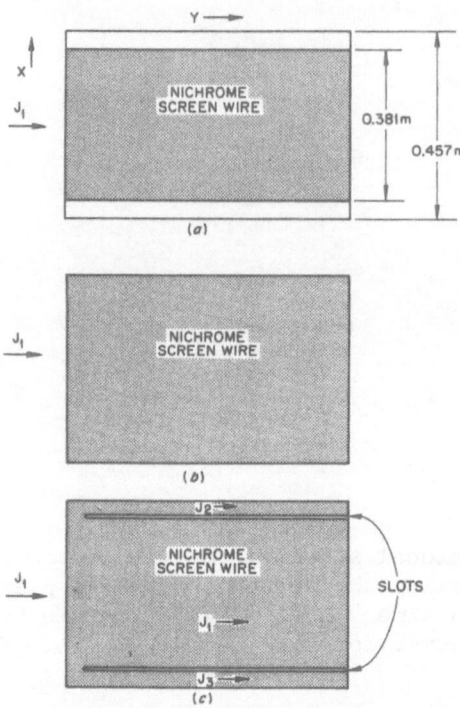

Fig. 4.   Three screen configurations that were examined: (a) screen
           heater terminated 76 mm (3 in.) from lateral edges,
           (b) screen heater assumed to extend to the lateral edge
           of the insulation, (c) slots cut in the screen permitted
           different heat generation rates in the lateral regions of
           the screen. This could be accomplished by current densities
           J_1, J_2, and J_3 through the different sections of the screen.

be infinitely long. Most calculations were for an assumed specimen
λ^* of 0.01 W/m·K, but some were for assumed values of 0.001 and
0.04 W/m·K.

     Since symmetry allowed us to restrict the calculations to the
positive half of each axis, the dimensions used were all one-half
the total package size. The assumption of the thermal conductivity
of the screen in the lateral direction to be 1.6 W/m·K was based on
a consideration of the thermal conductivity of Nichrome and the
volume fraction of the screen actually occupied by the wire.[8] Some
of the calculations were done in three dimensions and some in two
dimensions. An isothermal boundary temperature of 298 K (25°C) was
assumed for the plane heat sink, for the end of the system, and for
the sides. In all calculations a screen power density of 3 kW/m^3
and a total screen thickness of 0.533 mm were assumed.

When lateral heat flow occurs in the screen, the outer edges
of the screen are reduced in temperature, and the size of the isothermal area is diminished. As long as temperatures are measured
in the central isothermal zone, error due to the edge heat loss is
negligible. In some cases the losses are so severe that there is
no isothermal region, and a measurement error exists regardless of
where the screen temperature is measured.

Calculated Errors for Configurations 4(a) and 4(b) and a Specimen λ^* of 0.01 W/m·K

Figure 5 shows the calculated percentage error as a function
of x, where x is the distance between the screen center and the
position at which the screen temperature is measured. For a
specimen thickness of 0.0251 m ($\sim$1 in.), the error is less than 1%
for x less than 0.25 m. At a specimen thickness of 0.0886 m
($\sim$3-1/2 in.), the error is less than 1% for screen temperature
measurement for all values of x up to about 0.1 m. The error
curve for the 0.3045-m ($\sim$12-in.) thick insulation shows that the
error would be about —25%, even when the temperature was measured
at the screen center.

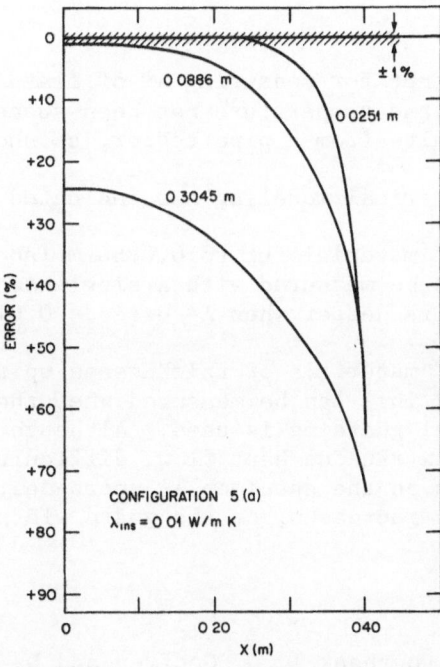

Fig. 5. Percentage errors predicted for configuration 5(a),
λ = 0.01 W/m·K, and system size of 0.914 by 1.219 m
(3 by 4 ft) for three insulation thicknesses.

If the flat screen wire heater passes all the way to the lateral boundaries and has the same heat generation rate [configuration 4(b)], the results are not significantly different from those obtained for configuration 4(a).

Apparently, therefore, the error due to lateral heat flow would not be important for the thinner specimens with thicknesses of 0.0251 and 0.0886 m, but the error could be as high as 25% with a specimen thickness of 0.3046 m. This high error for the thick specimen would only be about 6 to 7% for a specimen with a λ* as high as that described in Fig. 2.

Calculations for Configuration 4(c)

Calculations showed that configuration 4(c) would have less error for thick insulations than either 4(a) or 4(b). This low error could be achieved in thick insulations with a low λ* (0.01 W/m K) only when the power density in the lateral guards was so great that a temperature spike would occur in the guard. This leads to uncertainty about the optimum position to measure temperature for matching purposes. The best way to use these guard strips would be to adjust the power until dt/dx in the screen was zero.

CONCLUSIONS

1. A simple concept for measuring λ* of flat insulations as a function of density and temperature has been successfully tested and compared with results from a pipe tester, as shown in Fig. 3.

2. Extensive numerical modeling has indicated that:

   a. The λ* of materials up to 0.0886-m (about 3-1/2 in.) thick can be measured with a single-section unguarded screen wire heater when λ* exceeds 0.01 W/m·K.

   b. The λ* of materials of thicknesses up to 0.3045 m (about 12 in.) can be measured when the simple concept of lateral guarding is used. Although this guarding reduces extraneous heat flow, difficulty in using it increases as the specimen λ* decreases; the specimen thickness increases; or the guard width decreases.

ACKNOWLEDGMENTS

The authors wish to thank T. G. Godfrey and D. W. Yarbrough for reviewing the manuscript and S. J. Phillips for typing the text.

REFERENCES

1. R. G. Donnelly, V. J. Tennery D. L. McElroy, and J. O. Kolk,
    "Industrial Thermal Insulation: An Assessment,"
    ORNL/TM-5283, Oak Ridge National Laboratory, Oak Ridge,
    Tennessee (March 1976).
2. R. P. Tye, Ed., "Thermal Conductivity," Vol. 1, Academic
    Press, New York (1969).
3. Steady-state thermal transmission properties by means of the
    guarded hot plate, ANSI/ASTM C-177, in: "1980 Annual Book of
    ASTM Standards," Part 18, American Society for Testing and
    Materials, Philadelphia, pp. 20–53 (1980).
4. Steady-state thermal transmission properties by means of the
    heat flow meter, ANSI/ASTM C-518-76, in: "1980 Annual Book
    of ASTM Standards," Part 18, American Society for Testing
    and Materials, Philadelphia, pp. 222–253.
5. A. W. Pratt, Heat transmission in low conductivity materials,
    in: "Thermal Conductivity," Vol. 1, R. P. Tye, ed., Academic
    Press, New York (1969).
6. Thermal conductivity of pipe insulation, in: "1977 Annual Book
    of ASTM Standards," Part 18, American Society for Testing
    and Materials, Philadelphia, pp. 97–102 (1977).
7. S. H. Jury, D. L. McElroy, and J. P. Moore, Pipe insulation
    testers, in: "Symposium Proceedings Advances in Heat
    Transmission Measurement on Thermal Insulation Material
    Systems," American Society for Testing and Materials,
    Philadelphia, (1978).
8. J. P. Moore, D. L. McElroy, and S. H. Jury, "A Technique for
    Measuring the Apparent Thermal Conductivity of Flat Insula-
    tions," ORNL/TM-6494, Oak Ridge National Laboratory, Oak
    Ridge, Tennessee (October 1979).
9. C. M. Pelanne, J. Therm. Insul. 1:48 (1977).
10. W. D. Turner, D. C. Elrod, and I. I. Siman-Tov, "HEATING5 — An
    IBM 360 Heat Conduction Program," ORNL/CSD/TM-15, Oak Ridge
    National Laboratory, Oak Ridge, Tennessee (March 1977).

THERMAL CONDUCTIVITY OF CARBONATE GNEISS

UNDER APPLIED UNIAXIAL PRESSURE

T. Ashworth, T. M. Alexander, and E. Ashworth

Physics and Mining Engineering Departments
South Dakota School of Mines and Technology
Rapid City, South Dakota 57701

ABSTRACT

Results are presented for the thermal conductivity of three pairs
of carbonate gneiss specimens (sedimentary rock torched by igneous
intrusion, Needles Formation, S.D.) at a nominal temperature of 25°C
for applied uniaxial pressures ranging from 0 to 17.2 MPa (0 to 2500
psi). These data represent the preliminary phase of an investigation
of pressure and moisture-content dependence of the thermal conductivity of rocks. Zero load conductivities of 2.1 ± 0.2 W m^{-1}K^{-1}
have been obtained for temperature gradients of 40 to 200°C m^{-1};
some small dependence on temperature gradient was seen in all specimens. Upon pressure cycling, the conductivity at first increases
quickly [(by 8% for a load of 3.5 MPa (500 psi)] and then increases
more slowly and nearly linearly at higher loads [(total increase
12.5% at 17.2 MPa (2500 psi)]. During the unloading part of the
cycle, the conductivity decreases slowly (by less than 4%) as the
pressure less than 3.5 MPa, back to a value close to the initial
value. Thus a very clear hysteresis is seen in this pressure range.
The specimens were used in an "ambient" condition.

INTRODUCTION

A number of currently important problems in the areas of energy
and mining technology would benefit from greater availability of
thermal conductivity data; several references list these applications.[1-4] The work we describe here is part of a developing
program to measure and analyze the thermal conductivity of a range
of composite materials, especially rocks. Here we report measurements on a material of both local and scientific interest. As part
of a continuing, wide-ranging study of the geology of the Black Hills,

variations in and the implications of degree of metamorphism is
being studied. Gneiss is a fairly competent rock with a high degree
of metamorphism. It also contains a considerable variety of well-
dispersed minerals. This type of material is common in mines, so
the data will have some practical value.

SPECIMEN SPECIFICATION AND PREPARATION

Specimens of carbonate gneiss were collected from the Needles
Formation, Black Hills, S.D., 6.23 Km at 146° true from Harney Peak;
a granite outcrop is located about 10 m to the east of the collection
site. Both jointing and apparent bedding planes of the fairly
weathered rock were roughly parallel to the ground.
Specimen composition:

Calcite 50%: 0.02 - 1.00 mm Anhedral to Subhedral

Sphene   7%: 0.1 - 0.4 mm Euhedral to Subhedral

Actinolite 25-27%: 0.2 - 1.5 mm Subhedral
        (locally 5-45%):

Plagioclase 10%: 0.3 mm - 1.5 mm Subhedral

The above analysis was obtained by thin section techniques and
chemical tests.

Cores were cut from the rock using a 15 cm diamond-studded
drill. Cylindrical slabs approximately 2 cm thick were cut and
polished using a centerless grinder. Each surface was smooth and
flat to 0.0005 cm or better. In each case the slabs were cut
parallel to the natural planes in the material. They were then
dried for 72 h at 95°C.

EXPERIMENTAL DETAILS

Thermal conductivity measurements were made in a simple un-
guarded-plate apparatus using dual samples; a schematic drawing of
the stack is shown in figure 1; a description of the system and the
thermistors used for temperature measurement is given in reference
1. At the beginning of each measurement sequence, with no power
dissipation in the heater, the stack is allowed to come to equilibrium. Its temperature is determined and then, still with no
power dissipation, the pressure is cycled. Thus a one-point calibration of all the thermistors is performed and the absence of
significant pressure effects on the thermistors is ascertained.
Correction for edge heat losses and the deviations from nonlinear
heat-flow were made by finite element analysis; the correction was
no larger than 5% of the thermal conductivity.[5] The overall precision has been estimated to be 5% or better. Measurements on a

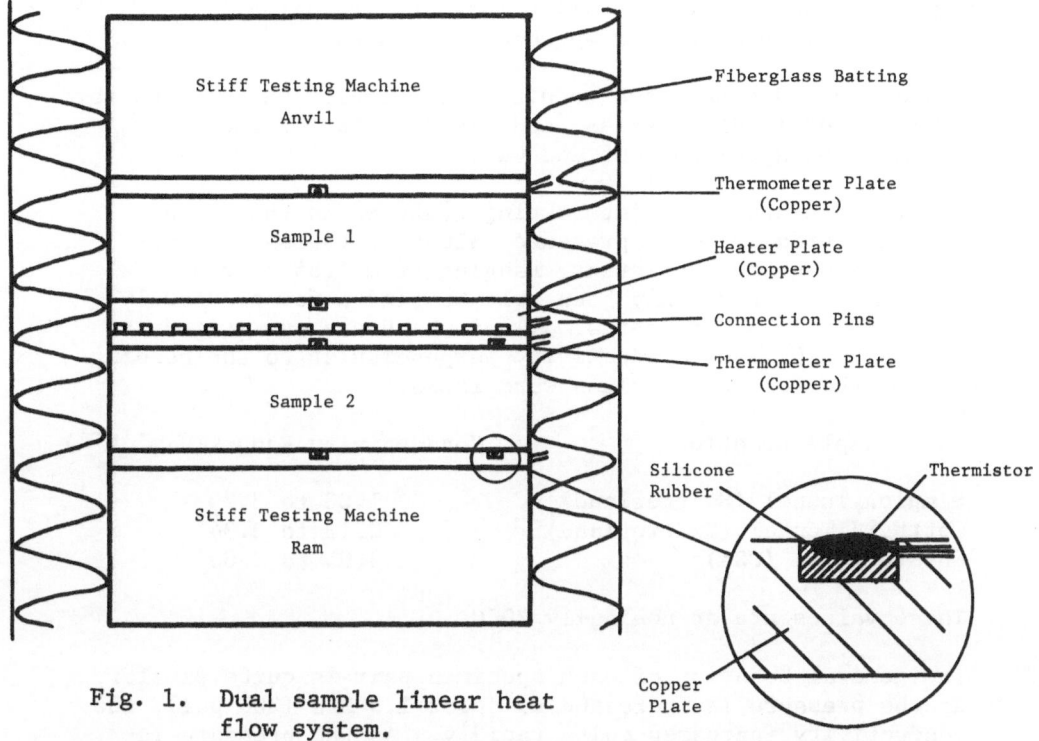

Fig. 1.   Dual sample linear heat
          flow system.

calibrated nylon specimen* gave a value of 0.363 W $m^{-1}K^{-1}$, (at a
pressure of 0.5 MPa and 25°C) within 1.5% of the published value
for this material.[6-9]

Measurements were taken on the gneiss specimens at pressures
ranging from near zero to 17.2 MPa and for power dissipations in
the heater of 2.88 to 9.34 W (which correspond to temperature
gradients of approximately 40 to 200°C m^{-1} across the specimen).

RESULTS

Specimens were prepared in pairs by taking adjacent slices
from a particular core. These matching pairs were used together in
the apparatus, and for the purpose of reducing the data, they were
assumed to have the same conductivity. Thus the data presented in
figure 2 represent the average thermal conductivity, λ, of specimens
CG1 and CG2. Graph 1 in this figure shows λ as a function of
pressure for temperature gradients of about 55 and 100°C m^{-1}; graph
3 shows λ as a function of temperature gradient; and graph 2 shows
λ as a function of pressure for a gradient of about 180°C m^{-1} for the

*Comoco Grand Nylon 66, Supplied by Boulder Plastic, Inc.
 Boulder, Colorado.

loading and unloading cycle. All values are at nominally 25°C. A
similar format is used in the subsequent figures.

Figures 3 and 4 show the results from samples CG3 and CG4 for
first and second loading cycles, and figure 5 is for the first
loading cycle on specimens CG5 and CG6.

There are a number of interesting features in this data:

1)   The (essentially) zero pressure values of thermal conduc-
       tivity are quite consistent, ranging from 1.85 to 2.30
       $W\,m^{-1}K^{-1}$. We have only been able to find a few previously
       reported values for gneiss, all of which are summarized in
       reference 10. For conductivity perpendicular to the bedding
       plane, the following values were found:

Sample Location	Conductivity Range ($W\,m^{-1}K^{-1}$)
Simplon Tunnel (Switzerland)	1.83 to 3.23
Gotthard Tunnel (Switzerland)	2.13 to 3.34
Chester, Vt. (USA)	2.05 to 3.63

       These values are at nominally 20°C.

2)   The general behavior of each specimen pair is quite similar.
       As the pressure is increased to about 1.5 MPa (200 psi), the
       conductivity increases quite rapidly. As the pressure in-
       creases, the rate of increase decreases, and between about 5
       and 17 MPa the λ–P relationship is nearly linear. Defining
       the pressure coefficient of thermal conductivity, p_λ, as
       $(1/\lambda)(d\lambda/dp)$, the coefficient has an average value of above
       0.5 MPa^{-1} at low pressures, and a value of about 0.01 MPa^{-1}.
       From reference 10 we estimate an average value of p_λ of
       0.0035 MPa^{-1} for Bermuda limestone up to a pressure of 68 MPa
       (10,000 psi); this material has one of the largest pressure
       effects of those listed in Clark's compendium.

3)   Upon unloading, conductivities do not, initially, relax back
       to the corresponding values exhibited during the loading
       portion of the curve. Thus a significant hysteresis is
       demonstrated.

4)   Values obtained on the second loading cycle on specimen CG3
       and CG4 are similar to those on the first cycle at the lowest
       and highest pressures, but in between some are up to 4% lower,
       which changes the general character of the pressure dependence
       and the hysteresis. This possible effect has not yet been
       fully explored.

5)   In all specimens, there is a slight increase in apparent con-
       ductivity when measured at larger heat flow (larger temperature

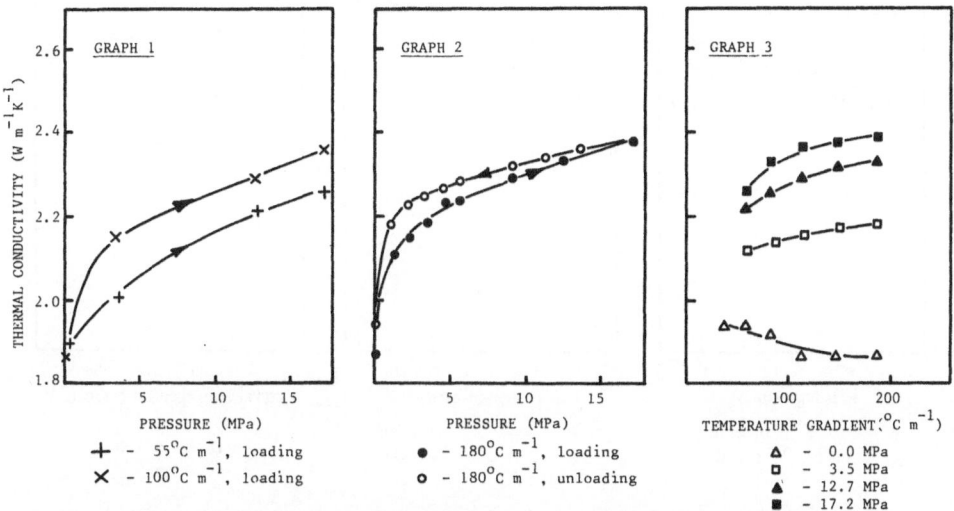

Fig. 2. Thermal Conductivity of Carbonate Gneiss
Specimen CG1 and CG2; First Loading.

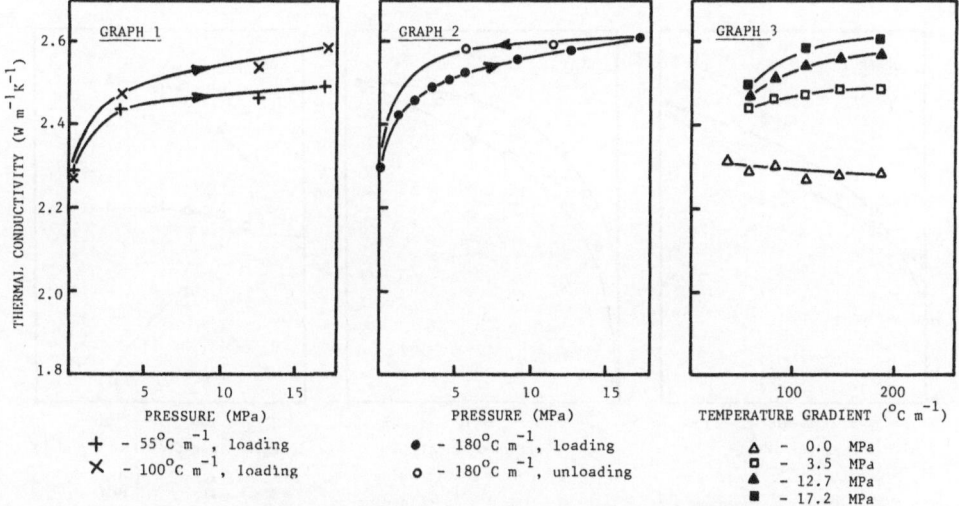

Fig. 3. Thermal Conductivity of Carbonate Gneiss
Specimen CG3 and CG4; First Loading.

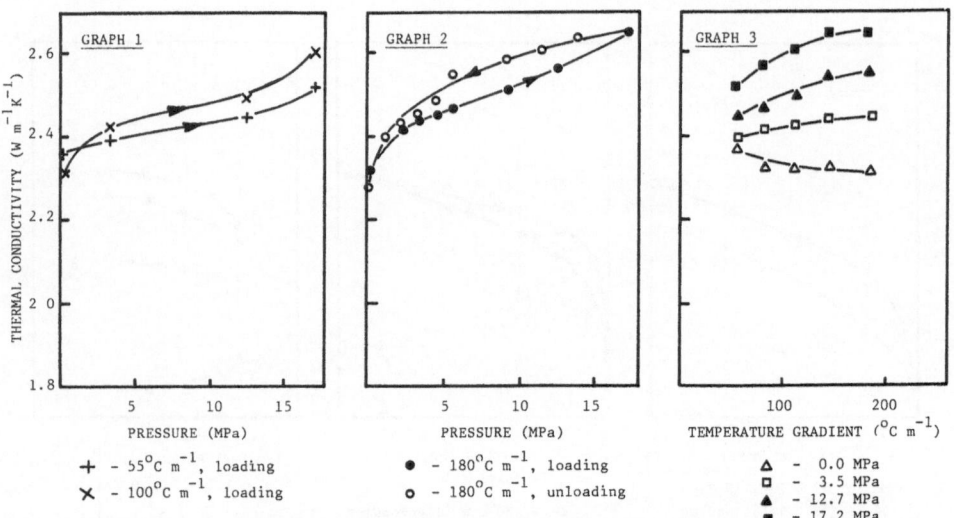

Fig. 4.   Thermal Conductivity of Carbonate Gneiss
           Specimen CG3 and CG4; Second Loading.

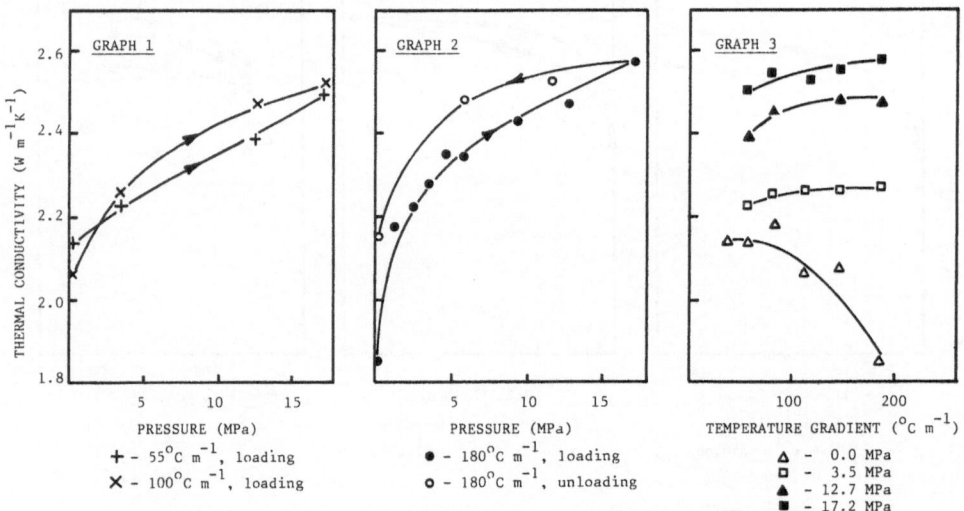

Fig. 5.   Thermal Conductivity of Carbonate Gneiss
           Specimen CG5 and CG6; First Loading.

gradient) for applied pressures greater than 3 MPa. Determination of the dependence of apparent conductivity on temperature gradient is a major objective of our studies. The data presented here cover only the lower temperature gradients, so we have not yet attempted to analyse the effect. Some thermal resistance may exist between the sample and the plates. Each surface is smooth and flat to 0.005 cm or better. The effect of thermometer location and greasing the surfaces is being studied. Variations in the mean temperature should not have contributed to the effect shown in figures 2 through 5, since temperature differences across the sample did not exceed $5^{\circ}C$ (mean temperature variations of less than $3^{\circ}C$) and the temperature coefficient of thermal conductivity for these types of material is generally small and negative ($\sim - 0.001/^{\circ}C$).

6) For (essentially) zero pressure measurements, the conductivity decreases somewhat as the temperature gradient is increased. Thus apparent pressure dependence may be due to lower precision in the low-gradient measurements, but while it was seen in all gneiss samples, it was not seen in measurements on specimens of nylon and granite.

CONCLUSIONS

Thermal conductivity values obtained for carbonate gneiss fall at the lower end of the range of previous measurements; only a $\pm$ 10% variation was seen among our three specimens. The small variation is probably due to sample collection from a limited area, whereas samples from the Swiss sites were probably representative of rock throughout the tunnels. Our values may be slightly low because we used thermometers in copper plates. Although the mounting method does cause the thermometer to make actual contact with the specimen, we may be measuring slightly too large a temperature difference across the specimen, as demonstrated by Tye and Spinney while working with concrete.[11] However, it is also possible that previously reported data were taken with some small pressure across the specimen (for the purpose of holding the system together). In view of the large pressure coefficient found for low pressures, these data should not be directly compared with our near-zero pressure data.

Carbonate gneiss was chosen for study because its geology suggested that it could be an interesting material. This has turned out to be the case. The large pressure dependence of thermal conductivity of the gneiss specimen, which has not been observed with other materials and rocks, clearly indicates that the applied pressure is causing structural modification of the material. While hysteresis is present, indicating some modifications of the internal energy during compression, the effects of the pressure loading are apparently reversed when the applied force is removed. We anticipate that our continued studies, particularly with water content, will indicate more about the specific nature of these changes. The apparent dependence of thermal conductivity on temperature gradient is also receiving further study.

REFERENCES

1. T. Ashworth, R. A. Murdock, and E. Ashworth, Thermal Conductivity Systems for Measurements on Rocks under Applied Stress, "Proc. 16th International Conference on Thermal Conductivity;" in press.
2. M. T. Morgan and G. A. West, Thermal Conductivity of the Rocks in the Bureau of Mines Standard Rock Suite, Report Tm-7052, Oak Ridge National Laboratory, Oak Ridge, Tennessee (1980).
3. J. N. Sweet and J. E. McCreight, Thermal Conductivity of Rocksalt and Other Geologic Materials from the Site of the Proposed Waste Isolation Pilot Plant, Sandia National Laboratory, Report 73-1665, (1980).
4. W. B. Durham, Thermal Conductivity and Diffusivity of Avery Island Salt at Pressure and Temperature, in: "Proc. 16th International Thermal Conductivity Conference."
5. E. Ashworth and T. Ashworth, A Simple Apparatus for Thermal Conductivity Measurements of Rocks and Similar Poor Conducting Materials, in: "Proc. 20th U.S. Symposium on Rock Mechanics," Soc. Petroleum Engineers, Austin (1979), pp. 27-33.
6. T. Ashworth, L. R. Johnson, C. Y. Hsiung, and M. M. Kreitman, Use of the Linear Heat-Flow Method for Poor Conductors and its Application to the Thermal Conductivity of Nylon, Cryogenics 13:34.
7. R. Berman, E. L. Foster, and H. M. Rosenberg, Br. J. Appl. Phys. 6:181 (1955).
8. D. E. Kline, J. Polym. Sci. 50:441 (1961).
9. R. B. Stewart and V. J. Johnson, eds., "A Compendium of the Properties of Materials at Low Temperature," Phase II, 60-56 Part Iv, WADD Technical Report, U.S. Air Force, Dayton, Ohio (1961).
10. S. P. Clark, Jr., Thermal Conductivity, in: "Handbook of Physical Constants," The Geological Society of America, Inc., New York (1966), pp. 461-462, 468-471.
11. R. P. Tye and S. C. Spinney, Thermal Conductivity of Concrete: Measurement Problems and Effect of Moisture, Institute International du Froid, Commission B1 Annex, Washington, DC (1976-2). International Institute of Refrigeration, Paris (1977).

THE THERMAL CONDUCTIVITY OF TWO CLATHRATE HYDRATES

J.G. Cook and M.J. Laubitz

Division of Physics
National Research Council of Canada
Ottawa, Ontario K1A 0R6, Canada

INTRODUCTION

It has recently been discovered that the permafrost regions may contain large amounts of natural gas in the form of natural gas hydrates (Cherksy and Makogan, 1970; Bily and Dick, 1974). The ice structure is not in the conventional hexagonal (Ih) form for these hydrates, but is more open and contains relatively large voids, or cages; these cages contain gas molecules, which interact only weakly with the host lattice. Natural gas hydrates are only one example of what are known as clathrate hydrates, which have been studied for some time (Davidson, 1973); the guest species may be atoms or molecules with diameters between 3.8 and 6.8 Å.

During the fall of 1980 Dr. Don Davidson of the Division of Chemistry of the National Research Council of Canada pointed out to us that no data for the thermal conductivity (κ) of clathrate hydrates were then available, and that such data would be valuable. For example, κ is an important parameter for the various gas recovery schemes now being investigated. Furthermore, if κ of the gas hydrates differs from that of ice (as the recent data of Ross et al. (1981) suggest), a thermal conductivity probe lowered into a drill hole could detect gas hydrates.

It seemed to us that the phenomenon of thermal conduction in these ices would be a very interesting one from a viewpoint of solid state physics, for it has to do with phonon propagation and scattering in a lattice that contains large and loosely held guest molecules. We know of no comparable system that has been studied by solid state physicists. For these various reasons we decided to measure κ of two different clathrate ices. We give our results below.

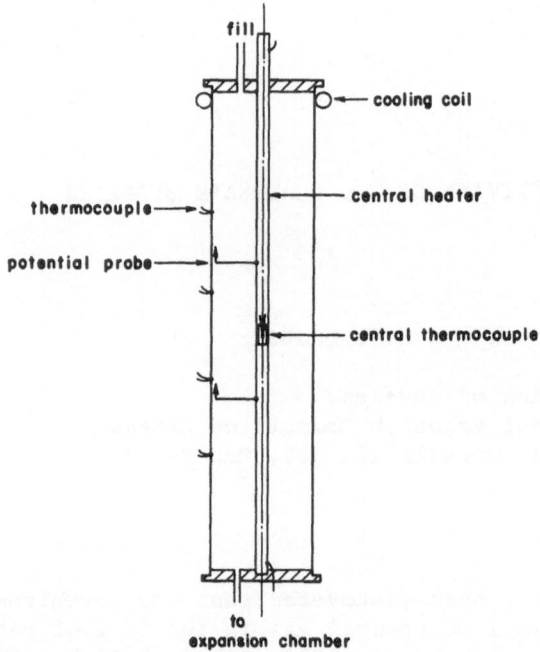

Fig. 1 The experimental chamber as used in the present investigation.

APPARATUS AND SAMPLES

We employed a sample container based on a steady state radial
heat flow geometry. The radial configuration avoids many problems
related to guarding the axial heat flow through a poorly conducting
rod specimen (McElroy and Moore, 1969). The apparatus is sketched
in fig. 1. The outer container is a stainless-steel tube, of 0.09
cm. wall thickness, 7.5 cm. diameter, and 36 cm. length. The inner
tube is also stainless steel, 0.02 cm. wall thickness, and 0.95 cm.
diameter. The end plugs were of plexiglass. Prior to filling, the
container was placed inside a deep dewar. A previously prepared
sample, just above its freezing point, was run into the container
and a rubber expansion volume below it. An antifreeze solution was
then pumped from a regulated bath through the cooling coils shown in
fig. 1 to freeze the sample. At first we found undercooling of the
gas hydrates by as much as 4°C. To prevent this and ensure the
sample froze from the top down, an antifreeze solution at -20°C was
first pumped through; then, after freezing had begun, the temperature
of the fluid was raised to -5°C. The central thermocouple and outside
thermocouples could be used to follow the gradual freezing of the
sample; in each case it progressed from the top down and took approximately two weeks to reach the bottom. Then, the space around the

whole sample was very slowly filled with cold antifreeze to avoid cracking of the sample; during the measurements, a pump was used to rapidly circulate the antifreeze at a suitable temperature around the specimen.

To measure the thermal conductance of the specimen, a known electrical current was passed through the central tube; fine potential probes attached to it permitted a determination of the power conducted away by the sample. A movable and calibrated thermocouple was used to determine the temperature at the inside surface of the heater tube; note that the radial thermal gradient at this surface is zero. The temperature distribution of the outside surface was determined by means of the thermocouples shown in fig. 1, which were of the same lot as the inner one. To avoid the effects of differences among the thermocouples, the thermal conductivity was measured with the so-called low power, high power, low power method (Laubitz and McElroy, 1971). Normally, the inner tube in the high power measurement was 2°C above that of the outer tube.

We consider errors owing to end effects (Peavy, 1963) and to thermal gradients in the container tubes to be well below 1%. Each of these could be corrected for, but even in total they are of little consequence to the present investigation. One systematic error is the contact resistance between the sample and its containing tubes. We made some simple preliminary experiments to determine if pure water and the gas hydrates would stick to stainless steel upon freezing and found that they did; this suggests that the sample-container contact resistance would be relatively small, and we did not attempt to place thermocouples inside the sample itself.

To check on the performance of our system and to provide a basis for comparison for our gas hydrate results, we first made some measurements of κ of ice prepared from distilled and degassed water. We found κ at -8°C to be 20.6 ($\pm$0.6)mW/cmK, which, to within combined experimental uncertainty, agrees with the recent results of Andersson et al. (1980).

The first clathrate hydrate studied, of ideal composition guest $\cdot 17H_2O$, was formed by mixing distilled water and tetrahydrafuran (THF), in the ratio of 4.25:1 by weight. It was found to freeze near 4°C, and is known (Davidson, 1973) to be frozen in what is called a structure II lattice, that has cavities of two different sizes. The THF molecules occupy the larger cages, while some of the small cages may contain N_2 or O_2 previously absorbed by the water.

A hydrate sample of ethylene oxide (EO) of composition $EO \cdot 7.01H_2O$ was prepared by passing EO through deionized water at 10°C, and freezing the solution in the thermal conductivity cell as before. This hydrate is known to form a structure I hydrate

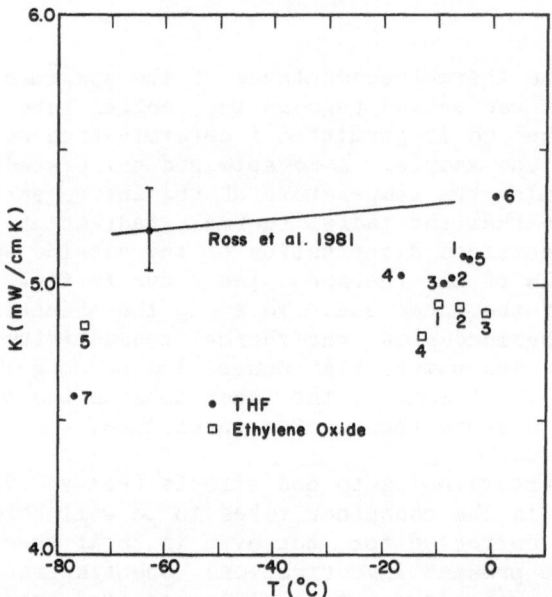

Fig. 2   Thermal conductivity data for THF hydrate, as reported by Ross
         et al. (1981) and as obtained in the present work, and our
         data for EO hydrate.

(Davidson, 1973); it also has cavities of two sizes, the smaller one
of which is similar to that of the type II hydrate. The EO molecules
occupy all the large cages and some of the small ones.

We have also measured the heat capacity and heat of fusion of a
small (5 g) sample of THF hydrate, which was prepared separately
from the solution referred to above. The uncertainty in these data
is of the order of 5%, which is largely due to the sample container
we used for these exploratory measurements, which was ten times
heavier than the sample itself, being designed for specific heat
measurements of gases under pressure (Ancsin, 1982).

RESULTS

Our measured κ values of the two hydrate structures are shown
in fig. 2, together with the recent results for THF hydrate obtained
by Ross et al. (1981). Our low temperature points were measured
using dry ice around the specimen and should be regarded as tentative.
The agreement between our THF data and those of Ross et al. is

within experimental uncertainty; these authors have since discovered (Andersson, 1981) that their sample contained some Ih ice, which increased its κ. When this is corrected for, their data pass through the mean of ours. It can now be said with certainty that κ of THF·17H$_2$O just below its freezing point is only weakly dependent on temperature and a factor of 4.5 below that of Ih ice. It may also be seen that κ of the EO hydrate, which has a quite different structure, is nevertheless very close to that of the THF hydrate.

We found the heat capacity of a second sample of THF hydrate to be 1.78 J/g from −5 to −2°C; a large anomaly in the heat capacity was also observed, centered on −1.5$_5$°C, of a total area of 4.8 J/g. Melting of the sample commenced near 2°C, was complete at 4.2$_4$°C, and required 267 J/g. No other heat capacity data appear to be available. Our value, on a weight basis, is 15% below that of ice, and, on a volumetric basis, 9% lower. The phase diagram of the THF−H$_2$O system (Dyadin et al., 1973) indicates that the peak near −1.5°C is due to a slight excess of H$_2$O in the sample.

DISCUSSION

A number of different mechanisms may be responsible for the relatively low κ of the hydrates studied. One may separate these into two categories, one in which the only role of the guest molecules is to stabilize the hydrate, and another in which the guests play an active role in scattering phonons and reducing κ. In the first category, one may place the following:

a) The heat capacity of the hydrate will be somewhat lower than that of pure ice, as we have observed above.

b) The group velocities of the phonons in the hydrate are likely to be lower than the corresponding velocities in the pure ice. Whalley (1980) has calculated the speed of sound for some hydrates and found values approximately 5% below that of ice, whereas Pandit and King (1981) have observed somewhat larger decreases in propane hydrate. Furthermore, it has been noted elsewhere (compare Slack, 1979) that a lattice with more than one atom per unit cell possesses optical modes of vibration, which will tend to be more localized, to have lower group velocities, and hence to be less effective in conducting heat. This effect may be quite marked in the hydrates under study here, which have a large number of H$_2$O molecules per unit cell.

c) The hydrate lattice is more open than that of pure ice; generally, such a lattice has larger atomic vibrations and, hence, stronger anharmonicity and phonon−phonon scattering. The fact that the hydrate lattice by itself is unstable also suggests that atomic vibrations are relatively large.

Turning to the ways in which the guests may directly reduce κ, we note that they may produce so-called Bjerrum defects (Davidson, 1973) in the lattice, which are basically misfits in the orientation of adjacent host molecules. These defects may be expected to scatter phonons, but their density is difficult to determine. Phonons may conceivably also be scattered by the guests in a more direct sense, in that they may exchange energy and momentum with the various modes of rotation, vibration, and translation, of the guests. It is difficult for us to say which are the more important in the present context.

It is important to note, however, that some of the mechanisms referred to above – such as a reduction in the heat capacity – will decrease κ, but will not alter its temperature dependence. We agree with Ross et al. (1981) that the weak temperature dependence of κ of the hydrates examined here, suggests that the mean-free-path of the phonons has been reduced to the level of the interatomic spacing. Ross et al. suggest that the interaction of the lattice phonons with the frequent "reorientation" of the guests is responsible for the sharp reduction in the phonon mean free path, but, in our opinion, this identification is premature; too little is known at the present time about the coupling between the host lattice phonons and the various motions of the guests. The effectiveness of the other mechanisms mentioned here is also difficult to determine at this time.

CONCLUSION

We have reported κ data for two clathrate hydrates at atmospheric pressure, ethylene-oxide hydrate and THF hydrate. The κ data for these different structures just below 0°C are very similar, and at least a factor of four below that of Ih ice. We are unable to explain this decrease in κ in detail. With Ross et al. (1981) we believe that it is reasonable to expect that κ of methane hydrate, a compound of considerable importance, will also be low, but in the absence of any satisfactory explanation for κ of the two hydrates studied so far, we have set out to measure κ of methane hydrate.

ACKNOWLEDGEMENTS

The assistance of a number of our colleagues at NRC has been explicitly referred to above; we are grateful to them, to Dr. John Ripmeester, who prepared the sample solutions, and to Dr. John Ancsin, who performed the heat capacity measurements. We have also benefitted from many discussions with Dr. Don Davidson and his colleagues.

REFERENCES

Ancsin, J., Melting curves of ice, in: "Proceedings of the 6th Symposium on Temperature", to be published (1982).

Andersson, P., 1981, Umea University, private communication.

Andersson, P., Ross, R.G., and Backstrom, G., 1980, Thermal resistivity of ice Ih near the melting point, J. Phys. C: Solid State Phys. 13:173.

Bily, C., and Dick, J.W.L., 1974, Naturally occurring gas hydrates in the Mackenzie Delta, Bull. Can. Pet. Geol. 22:340.

Chersky, N., and Makogan, Y., 1970, Solid gas—world reserves are enormous, Oil Gas Invest. 10:82.

Davidson, D.W., Clathrate hydrates, in: "Water, a Comprehensive Treatise", Vol. 2, F. Franks, ed., Plenum Press, N.Y. (1973).

Dyadin, Y.A., Kuznetsov, P.N., Yakovlev, I.I., and Pyrinova, A.V., 1973, The system water-tetrahydrofuran in the crystallization region at pressures up to 9 kbar, Dokl. Akad. Nauk SSSR 208:103.

Laubitz, M.J, and McElroy, D.L., 1971, Precise measurement of thermal conductivity at high temperatures (100-1200K), Metrologia 7:1.

McElroy, D.L., and Moore, J.P., Radial heat flow methods for the measurement of the thermal conductivity of solids, in: "Thermal Conductivity", Vol. 1, R.P. Tye, ed., Academic Press, N.Y. (1969).

Pandit, B.I., and King, M.S., 1981, Compressional and shear wave velocities in propane gas hydrates, in: "Proceedings of the Fourth Canadian Permafrost Conference", to be published.

Peavy, B.A., 1963, Steady state heat conduction in cylinders with multiple continuous line heat sources, J. Res. Nat. Bur. Stand. 67C:119.

Ross, R.G., Andersson, P., and Backstrom, G., 1981, Unusual PT dependence of thermal conductivity for a clathrate hydrate, Nature 290:332.

Slack, G.A., 1979, The thermal conductivity of nonmetallic crystals, Solid State Phys. 34:1.

Slack, G.A., 1980, Thermal conductivity of ice, Phys. Rev. B22:3065.

Whalley, E., 1980, Speed of longitudinal sound in clathrate hydrates, J. Geophys. Res. 85:2539.

THERMAL DIFFUSIVITY OF POCO GRAPHITE AND STAINLESS STEEL SRM 735-S

Roy Taylor

Department of Metallurgy
University of Manchester/UMIST
United Kingdom.

ABSTRACT

The thermal diffusivity of POCO graphite AXM-5Q1 has been measured over the temperature range 500-3000 K and the stainless steel SRM 735-S from 500-1200 K. Equations were least square fitted to the data and the results discussed and compared with other available data.

INTRODUCTION

Two materials that are candidate standard reference materials for thermophysical properties are POCO-AXM-5Q1 graphite and a stainless steel SRM 735-S. Specimens of these materials have been disseminated by the National Bureau of Standards to selected laboratories throughout the world for thermophysical property measurement. The POCO graphite has a medium grain size and has been graphitised to 2500°C. The stainless steel was a fully austenitic alloy obtained by homogenising at 1050°C. Although it has been reported that measurements should be possible up to 1200°C,[1] it was felt that it would be prudent to confine measurements to below 1050°C. The thermal diffusivity was measured using the laser flash technique. Thermal diffusivity measurements on another candidate standard reference material, pure iron SRM 735-S, have already been reported.[2]

EXPERIMENTAL DETAILS

Measurements were made using the laser pulse thermal diffusivity apparatus developed at UMIST. Details of this equipment have been

published.[3] The heat pulse is supplied by a solid-state ruby laser
of 100-J maximum output, and the specimen is heated to the measurement temperature inside a graphite susceptor heated by an induction
coil. Radiation from the specimen rear face is collected and
focussed on to a PbS infrared sensor. The lower limit of operation
of this detector is 500 K. The amplified signal from this detector
is sampled and analysed using a MINC PDP-11 microcomputer, which
possesses full data reduction capability. Auxiliary programmes
permit automatic compensation for finite pulse time effects, heat
losses using Cowan's[4] analysis from the voltage (temperature)
amplitude ratio $[\Delta T(10t_{\frac{1}{2}})/\Delta T/t_{\frac{1}{2}}]$ or $\Delta T[5t_{\frac{1}{2}})/\Delta T(t_{\frac{1}{2}})]$, and thermal
expansion of the specimen.

Two specimen rods of POCO graphite, 6.35 mm diameter x 75 mm
long, were received from NBS, Boulder. The average density of these
rods was 1.775 gm cm^{-3}; these were coded AXM-5Q1-102 and AXM-5Q1-401.
From each of these rods, two samples, 3-3.5 mm long, were made. In
the case of the stainless steel, one sample, 1.1 mm long, was
received.

RESULTS AND DISCUSSION

POCO-AXM-5Q1 Graphite

Four samples of this material were measured, two of type 401
and two of type 102 graphite. Some 94 individual measurements were
made over the temperature range 500-3000 K. No measurements were
taken after cooling from 3000 K, so we are unable to determine if
any irreversible change had occurred. In Fig. 1 the diffusivity
data is plotted as a function of temperature. These data are listed
in Table 1. Attempts were made to fit the data using various curve
fitting routines including a polynomial of up to 6th order. The
best fit was given by the equation

$$\alpha = A + BT + C/T \tag{1}$$

where the coefficients are A = -0.06452, B = 2.889 x 10^{-5}, C =
226.85. Although this equation was the best fit of the equations
tried, a visual inspection of figure 1 will show that there is
systematic deviation from the data points. The plotted curve lies
below the data at 1000-2000 K and predicts too high a diffusivity
at temperatures above 2500 K.

It is instructive and informative to compare these data with
other diffusivity data and derived and measured conductivity data.
An excellent resumé of the thermal transport properties of graphite
has been carried out by Minges,[5] and thermophysical property data
for a similar graphite has been obtained under an AGARD cooperative
programme.[6] Our data fit very well with Fitzer's[6] recommended

Table 1.   Thermal Diffusivity and Derived Thermal Conductivity
           for POCO AXM-5Q1 Graphite.

T (K)	α (cm^2 s^{-1})	λ (W cm^{-1} K^{-1})	Specimen number	T (K)	α (cm^2 s^{-1})	λ (W cm^{-1} K^{-1})	Specimen number
481	0.413	0.838	401-1	1266	0.158	0.522	102-1
498	0.401	0.840	401-1	1284	0.156	0.518	102-1
521	0.385	0.837	401-1	1290	0.157	0.521	401-2
530	0.374	0.826	102-1	1315	0.154	0.513	102-1
545	0.368	0.830	401-1	1341	0.150	0.501	102-1
551	0.359	0.817	102-1	1373	0.149	0.499	102-1
565	0.358	0.830	401-1	1392	0.147	0.490	401-2
571	0.352	0.822	102-1	1423	0.144	0.486	102-2
577	0.345	0.813	401-1	1504	0.136	0.461	401-2
584	0.340	0.807	102-1	1554	0.130	0.442	102-2
587	0.322	0.767	102-1	1575	0.132	0.449	401-2
596	0.335	0.809	401-1	1575	0.132	0.449	102-2
615	0.313	0.772	102-1	1615	0.128	0.440	401-2
633	0.296	0.743	102-1	1686	0.125	0.428	401-2
640	0.300	0.760	102-1	1717	0.125	0.432	102-2
653	0.293	0.751	102-1	1788	0.122	0.423	401-2
666	0.286	0.743	102-1	1840	0.111	0.385	401-2
691	0.277	0.736	102-1	1890	0.114	0.396	401-2
715	0.265	0.717	102-1	1890	0.111	0.384	102-2
733	0.259	0.713	102-1	1982	0.112	0.390	102-2
754	0.254	0.707	102-1	1992	0.119	0.414	401-2
778	0.246	0.697	102-1	2094	0.111	0.390	401-2
801	0.240	0.690	102-1	2155	0.107	0.373	401-2
821	0.233	0.677	102-1	2155	0.107	0.376	102-2
832	0.230	0.673	102-1	2165	0.100	0.351	401-2
863	0.223	0.662	102-1	2185	0.106	0.372	401-2
888	0.217	0.653	102-1	2186	0.103	0.363	401-2
917	0.212	0.644	102-1	2247	0.107	0.374	401-2
944	0.205	0.631	102-1	2318	0.106	0.371	102-2
968	0.200	0.621	102-1	2318	0.0985	0.346	102-2
989	0.195	0.610	102-1	2318	0.0982	0.345	102-2
1013	0.193	0.607	102-1	2410	0.0969	0.341	102-2
1042	0.188	0.597	102-1	2410	0.0987	0.347	102-2
1066	0.182	0.581	102-1	2502	0.0880	0.310	102-2
1091	0.181	0.580	102-1	2512	0.0942	0.332	102-2
1108	0.183	0.591	102-1	2512	0.0958	0.337	102-2
1113	0.179	0.574	102-1	2615	0.0949	0.334	102-2
1140	0.174	0.566	102-1	2615	0.0935	0.329	102-2
1165	0.170	0.553	102-1	2705	0.0830	0.292	102-2
1185	0.167	0.547	102-1	2707	0.0883	0.310	102-2
1218	0.166	0.545	102-1	2707	0.0979	0.344	102-2
1218	0.163	0.535	102-1	2810	0.0831	0.292	102-2
1220	0.165	0.543	401-2	2933	0.0862	0.304	102-2
1243	0.160	0.530	102-1	3005	0.0836	0.295	102-2

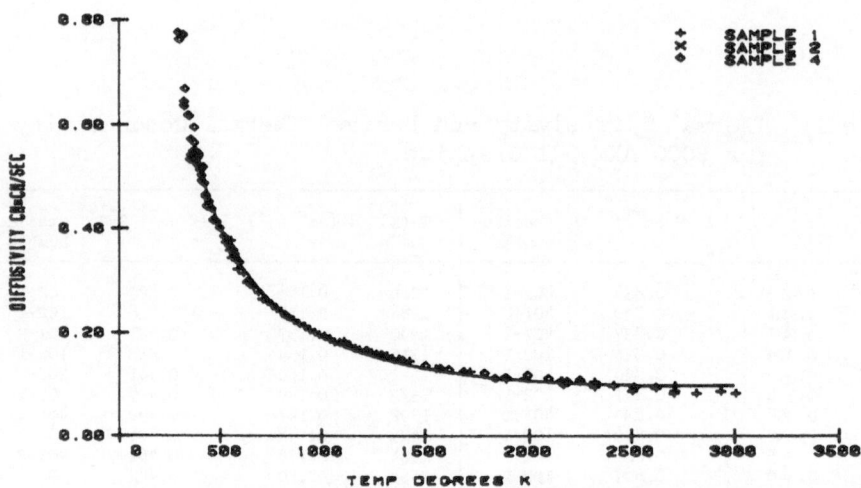

Fig. 1.   Thermal diffusivity of POCO graphite type AXM-5Q1.   Key
          Type 401; + Type 102 specimen 1; x Type 102 specimen 2.

diffusivity curve at low temperatures but are higher by up to 10%
at elevated temperatures.   However, to compare our data with
conductivity data, as Minges has done, it is necessary to have
reliable specific heat data and accurate compensation for the change
in density of the material with temperature.   A specific heat curve
was supplied by AFML[7] based on data obtained by Taylor and Groot[8]
from 300-1000 K and Cezairliyan and Righini[9] from 1500-3000 K.   In
spite of the gap in measurement, good extrapolation is observed
between the two sets of data, which have been fitted by a 5th order
polynomial:

$$Cp(J\ gm^{-1}\ K^{-1}) = -0.6853 + 5.9199 \times 10^{-3}T - 5.5271 \times 10^{-6}\ T^2$$

$$+ 2.6677 \times 10^{-9}T^3 - 6.4429 \times 10^{-13}T^4 + 6.1622 \times 10^{-17}T^5 \qquad (2)$$

Also supplied by AFML was expansion data from 300-3000 K, which may
be fitted by the 2nd order polynomial[7]

$$\Delta L/L = -0.19112 + 0.58767 \times 10^{-3}T + 0.11495 \times 10^{-6}T^2 \qquad (3)$$

Minges[5] shows a density variation with temperature that is linear
and was derived assuming a mean expansion coefficient of 9×10^{-6}
K^{-1}, which is broadly in accord with equation 3.   However, using
equation 3 to compute the volume change with temperature will lead
to a density variation with temperature given by

$$\rho_T = \frac{\rho_{293}}{(0.9943 + 1.7563 \times 10^{-5}T + 3.5387 \times 10^{-9}T^2)} \quad (4)$$

where ρ_{293} is the measured room temperature density, in this case 1.775 gm cm^{-3}.

Using equations 2 and 4, the thermal conductivity has been calculated for all measurements. These derived data are plotted as thermal resistivity (figure 2). The following straight line

$$R = \lambda^{-1} = 0.680 + 9.82 \times 10^{-4}T \quad (5)$$

represents these data quite well from 500 to 1600 K. However, above this temperature there is a systematic deviation from a straight-line plot, observed thermal resistivities being lower than those predicted by equation 5.

Within the auspices of the current programme, we have been able to compare our results with thermal diffusivity data obtained by Maglic et al.[10] over the temperature range 500-1700 K. The data agree to within 1.5 to 3%, depending on whether Maglic et al.'s heat loss corrections were evaluated from the rising or falling part of the curve. However, Maglic et al. do not quote a density figure for their samples, and this will clearly influence the results via the tortuosity and porosity factor.[11]

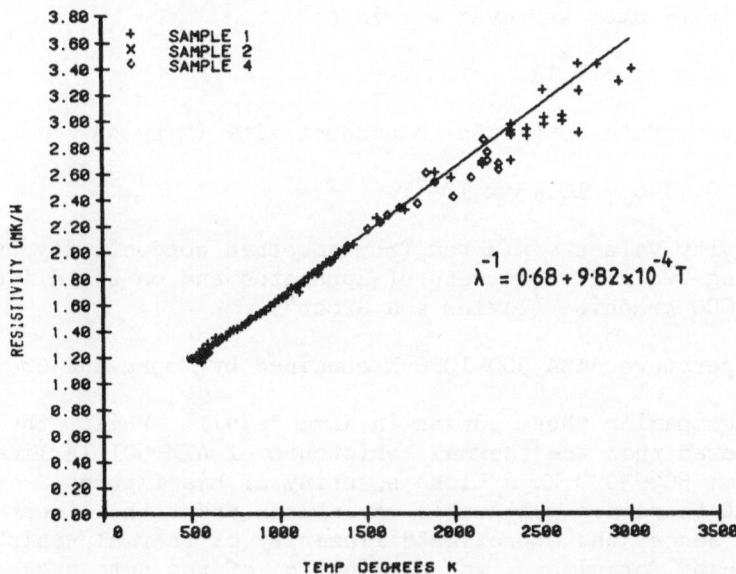

Fig. 2.   Thermal resistivity of POCO graphite fitted to equation 5.
Key   Type 401; + Type 102 specimen 1; x Type 102 specimen 2.

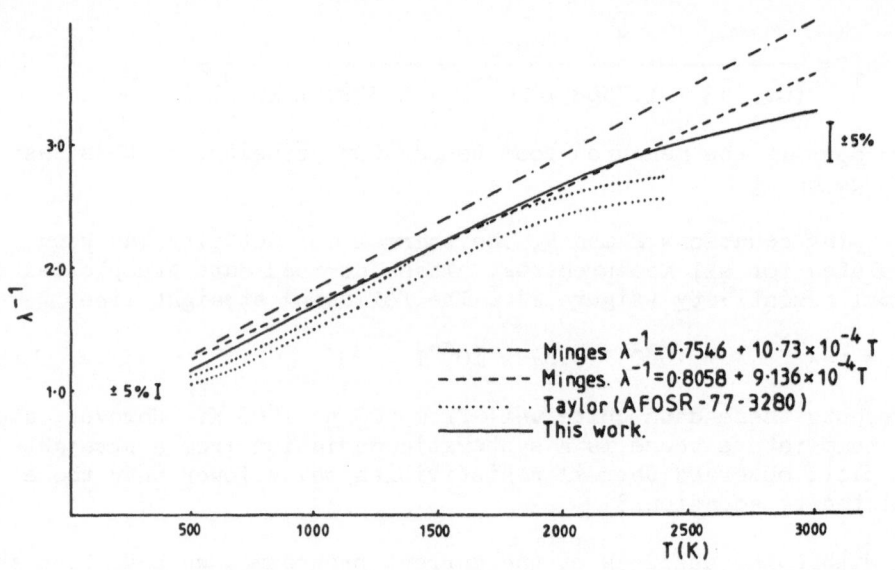

Fig. 3.   Comparison of thermal resistivity curves for POCO graphite.

In Fig. 3 we have evaluated our derived thermal resistivity using the smoothed curve and have compared our data with those obtained by other workers, viz.,

1.   Conductivity data surveyed by Minges[5]

$$\lambda^{-1} = 0.8058 + 9.136 \times 10^{-4}T \qquad\qquad (6)$$

2.   Diffusivity data converted to conductivity (Minges)

$$\lambda^{-1} = 0.7546 + 10.73 \times 10^{-4}T \qquad\qquad (7)$$

3.   Resistivity values evaluated from smoothed conductivity data measured using two different sets of apparatus and on two different samples of POCO graphite (Taylor and Groot[8]).

4.   Low temperature data 500-1000 K obtained by Moore and Graves,[12]

It is worth comparing these curves in some detail. Whilst the survey by Minges showed that the thermal resistance of AXM-5Q1 is linear to about 10% from 300-3000 K, a close scrutiny of his figures 2 and 3 suggests that there are systematic variations about the linear fits determined. Hence, the approximate linearity of thermal resistivity plots is perhaps fortuitous, and examination of the data suggests that at higher temperatures there is a negative deviation from linearity. This is substantiated by our data and also those obtained by Taylor and Groot, which show a substantial systematic deviation

above 1800 K. Taylor and Groot concluded that the thermal conductivity of POCO graphite contained an electronic component ranging
from a few percent at 500 K to at least 15% at 2400 K. Moore and
Graves suggest that this may be as high as 25% at 3000 K, assuming
a minimum value for the Lorenz function of 2.443×10^{-8} $(V/K)^2$. Our
observations lend support to the views that there is an electronic
component contributing to the thermal conductivity, although further
work would be required to quantify the magnitude of this contribution.

However, if we attempt to compare measured conductivity data
with those derived from diffusivity measurements, an interesting
trend is apparent. In the survey by Minges, from which he deduces
a mean grain size of 170 nm and his diffusivity data are converted
using a room temperature density of 1.75 gm cm^{-3}, the thermal
resistivity values from derived diffusivity data are some 2-12%
higher than those from directly measured conductivity data. Within
the current CODATA investigation, the density of material supplied
appears to be somewhat higher, 1.77-1.79 gm cm^{-3}, and the calculated
grain size needed to fit the conductivity curve near the peak is
150 nm.[12] Hence conductivity data should be higher, and this is
borne out by the results obtained by Taylor and Groot[8] and Moore and
Graves,[12] which show a significantly higher thermal conductivity
than the data surveyed by Minges. However whilst the diffusivity
data obtained at UMIST and by Maglic et al.[10] are in very good agreement, it is also apparent from Fig. 6 that conductivity values derived
from diffusivity data are lower at temperatures less than 1200 K
than the conductivity data measured directly by Taylor and Groot
and Moore and Graves. Although there are differences in their
observed behaviour (Moore and Graves observe linear resistivity
from 450-1000 K whereas Taylor and Groot's data show a positive
deviation below 800 K) previous data of these workers have proven to
be reliable, so it may be assumed this difference is real. Various
reasons may be postulated to explain this effect but many of these,
such as possible errors in specific heat or density or both may be
readily discounted, since possible errors in these parameters are
insufficient to explain the magnitude of the difference observed at
low temperatures. This is some 6-15% at 500 K. Two possibilities
are left: firstly, that the differences are real, and secondly,
that there is some fundamental discrepancy between diffusivity and
conductivity measurements on graphite. If the former is the case,
this must raise doubts about the suitability of graphite as a standard reference material.

Diffusivity measurements necessitate smaller samples (2-4 mm
thick) than those generally used for conductivity measurements. It
is unlikely that, if material variability exists, diffusivity
measurements would consistently be carried out on poorer material,
i.e., smaller grain size or lower density. However, the possibility
that damage can occur during specimen preparation cannot be ruled
out entirely since this would have a proportionately greater effect

on smaller samples. This could be either from surface damage or the impaction of detritus within the sample. It would be instructive to have diffusivity measurements carried out on samples machined from material for which well-quantified conductivity data is available. The author has had no experience of any similar effect on samples measured from metallic standard reference materials.

Stainless Steel SRM 735-S

The change in sample length for this material was assumed to be 19×10^{-6} K^{-1}. Since transient times were of the order of 0.03-0.038 s for the 1.08 mm thick sample, these necessitated a correction for finite pulse time, since the laser dissipation time, τ, was 1.3×10^{-3} s. Using the analysis outlined by Clark and Taylor[9] assuming a square wave pulse, this involved calculating diffusivity using

$$\alpha = \frac{0.2613 \, L^2}{1.882 \, t_{\frac{1}{2}} - \tau} \tag{8}$$

this involved a finite pulse time correction of some 2-2.5% for the transients corresponding to this sample. The results are plotted in Fig. 4 and listed in Table 2.

For the data obtained, no detectable difference was obtained during cooling, which proved that no structural change had occurred

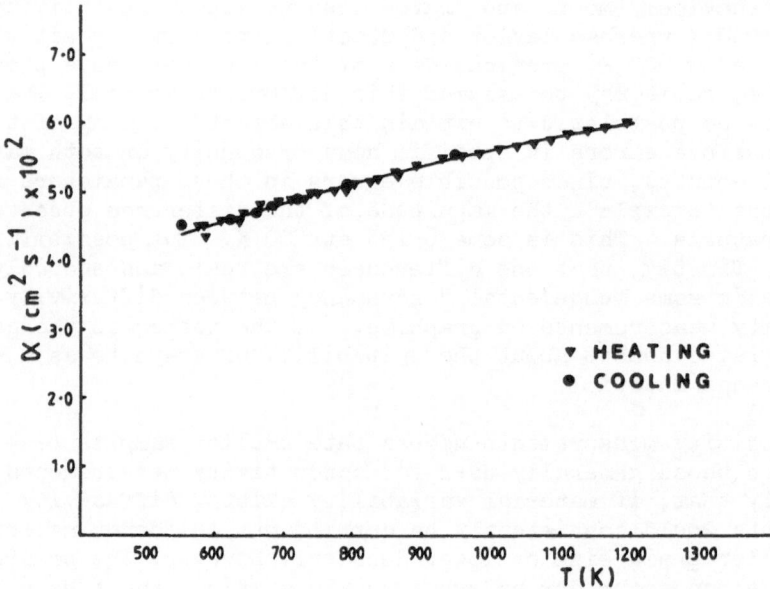

Figure 4.   Thermal diffusivity of SRM-735-S.

Table 2.   Thermal Diffusivity of Stainless Steel SRM 735-S

T (K)	α (cm^2 s^{-1})	T (K)	α (cm^2 s^{-1})
575	4.38	840	5.05
583	4.39	859	5.11
586	4.22	866	5.09
610	4.47	891	5.22
637	4.45	928	5.31
640	4.56	956	5.35
662	4.67	980	5.40
678	4.63	1010	·5.46
693	4.72	1036	5.50
710	4.74	1060	5.53
719	4.74	1084	5.56
728	4.75	1110	5.64
730	4.80	1138	5.70
739	4.84	1162	5.73
744	4.83	1193	5.83
747	4.86	951	5.37
749	4.79	788	4.98
755	4.83	680	4.61
789	4.87	663	4.62
801	4.97	621	4.47
836	4.98	553	4.41

during heating to 1200 K. The data may be fitted to a simple second-order polynomial

$$\alpha = 0.02375 + 4.2083 \times 10^{-5}T - 9.722 \times 10^{-9}T^2 \quad (T \text{ in K}) \quad (9)$$

Data available on a similar material obtained in a collaborative programme have been reported by Fitzer.[6] Although these data showed considerable scatter, the most exhaustive information obtained with the programme was that due to Wheeler,[6] who showed that irreversible changes occurred after heating to >1400 K. Over the temperature range 500-1200 K, our data agree with his results to ±3%. Maglic et al.'s data on the SRM 735-S also agree with data reported here to better than 2% over all the temperature range.

CONCLUSION

Data are reported for thermal diffusivity measurements on POCO graphite AXM-5Q1 and stainless steel SRM 735-S. The results are compared with other available data. Attention is drawn to what appears to be a fundamental difference in diffusivity and conductivity data for the POCO graphite.

REFERENCES

1. K. Kirby, - private communication, National Bureau of Standards,
    Washington D.C.
2. R. Taylor and C.M. Fowler, Thermal diffusivity of pure iron and
    selected iron alloys, in: "Thermal Conductivity XV", V.V.
    Mirkovich, ed., Plenum Press, New York and London (1978).
3. R. Taylor, J. Phys. E. Sci. Instrum. 13: 1193-99 (1980).
4. R. D. Cowan, J. Appl. Phys. 34: 926-7 (1963).
5. M. Minges, Int. J. Heat Mass Transfer 20: 1161-72 (1977).
6. E. Fitzer, "Thermophysical Properties of Solid Materials.
    "Cooperative Measurements on Heat Transport Phenomena of
    Solid Materials at High Temperature", Report 606 (1973)
    AGARD NATO France.
7. L.S. Theibert, AFML Wright-Patterson AFB, Ohio, private
    communication. (1979).
8. R.E. Taylor and H. Groot, "Thermophysical Properties of POCO
    Graphite", AFOSR-77-3280 (1978). Purdue Univ., Indiana.
9. A. Cezairliyan and F. Righini, Rev. Int. Hautes Temp. Refract.
    12, 124-31 (1975).
10. K.D. Maglic, N. Perovic, and Z. Zivotic, Thermal diffusivity
    of four candidate standard reference materials in: "Thermal
    Conductivity 17", Plenum Press, New York (1982).
11. R. Taylor, K.E. Gilchrist and L.J. Poston, Carbon 6: 537-44
    (1968).
12. J.P. Moore and R.S. Graves, Thermal conductivity and electrical
    resistivity of a POCO AXM-5Q1 graphite from 80 to 970 K,
    in:"Thermal Conductivity 17", Plenum Press, New York (1982).

THE DEVELOPMENT OF LOW-DENSITY GLASS-FIBER INSULATION AS THERMAL

TRANSMISSION REFERENCE STANDARDS

Charles M. Pelanne

Manville Corporation
Research and Development Center
Denver, Colorado

ABSTRACT

Tests demonstrating the value of low density glass-fiber insulation materials as reference specimens are reported. The tests conducted over more than fifteen years show excellent reproducibility. Specimen density, apparent thermal conductivity, apparatus, location and operator are variables involved in these tests conducted at 24ºC (75ºF) mean temperature.

The reliability of the apparent thermal conductivity measurements and the demonstrated need to have reference materials with heat-transfer characteristics similar to the materials being measured promoted the development of low-density glass fiber standards reference to the National Bureau of Standards.

INTRODUCTION

Advances in thermal insulation technology, both in measurement techniques and improved understanding of heat flow mechanisms prevailing in insulating materials, have prompted revisions in the conceptual approaches to the determination of the thermal resistive properties of insulations. An important ingredient for this change has been the development of relatively low-density thermal insulations. Figure 1, graphically illustrates the point. In years past, insulations were difficult to make at low densities. Most were made at high densities; their typical performance is shown on the right side of the minimum of this curve. In recent years, manufacturing processes allowed fabrication of lower density materials; their typical performance is shown on the left side of the minimum. The measurement technology

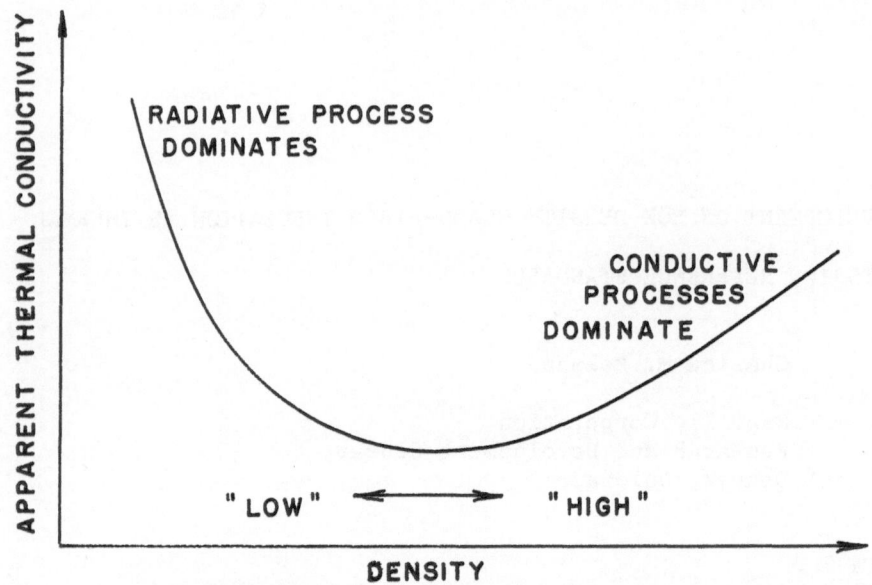

Figure 1.   Schematic diagram of apparent thermal conductivity
            relationship to density of the insulation at a given
            mean temperature.

was geared to meet the requirements of the higher density products.
The mechanisms of heat flow involved in the earlier process approximated the conditions required for the measurement of an intrinsic
property, such as thermal conductivity.

     In materials, such as low-density insulations, that permit heat
flow by various modes, as shown in Figure 2,[1] the influence of radiation can be such that the boundaries have an influence on the results.
The emittance, the temperature, and the distance between the plates
can affect the results. Discrepancies in measurements referenced to
thermal conductivity standards were often attributed to measurement
uncertainties and experimental errors. The influence of the emittance
of the apparatus surface plates on the measurements of 10.6 kg/m^3
glass-fiber insulation is illustrated in Table 1.[2] The effect is more
pronounced on measurements made on lower density insulations, as will
be illustrated later (see Figure 3). Materials themselves, exhibiting
non-uniformities that could not be readily quantified, provided an
explanation for these "errors." These "errors" could have been, in
some instances, attributed to our limited understanding of the heat
transfer process through insulations.

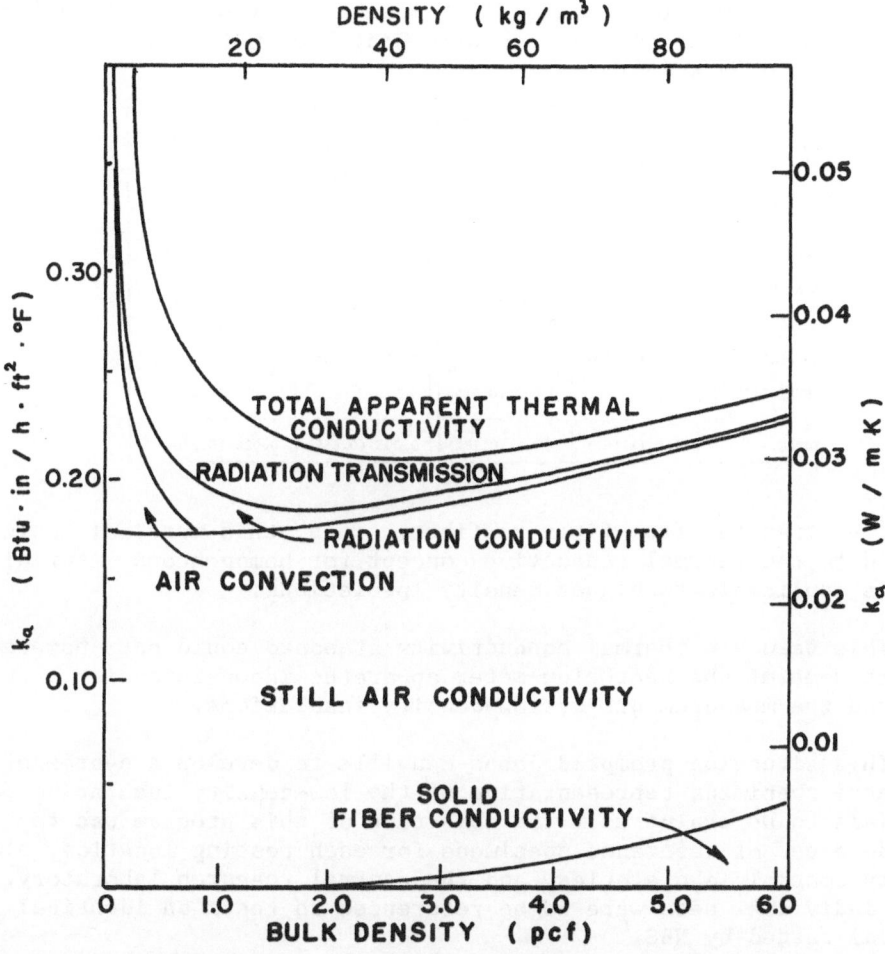

Figure 2.   Breakdown of "thermal conductivity" components versus
            density for a typical glass fiber insulation.   38°C
            (100°F) Hot Face 10°C (50°F) Cold Face 25.4 mm (1 in)
            thickness, black boundaries

    This problem was compounded by the increased use of the heat-flow
meter apparatus, which utilized reliable standard reference materials
for their calibration.   The only reliable reference standard available
in the U.S. for the thermal measurements was the relatively high-density glass-fiber insulation standard (No. 1450)[3] supplied by the
National Bureau of Standards (NBS) in Gaithersburg, Md.   This standard
made of a relatively high-density (100 to 130 kg/m^3) glass-fiber
board, was selected because it had a thermal conductivity relatively
unaffected by small variations in the density.   The heat-transfer
process through this material was essentially through conductive
modes, air conduction, solid conduction, and a minor amount of

Table 1. Performance of Glass Fiber Insulation (10.6 kg/m^3-25.4mm)
          Conditions of Test: Upward Heat Flow-Hot Face 38°C,
          Cold Face 10°C

SOURCE OF HEAT FLOW	BOUNDARY EMITTANCE			
	BLACK SURFACE ε=0.95		GOLD SURFACE ε=0.08	
	W/m·K (Btu in/hr ft² F)			
	NO INSULATION	GLASS FIBER	NO INSULATION	GLASS FIBER
RADIATION	0.152 (1.054)	0.011 (0.077)	0.007 (0.046)	0.004 (0.030)
CONDUCTION AIR	0.026 (0.178)	0.026 (0.177)	0.026 (0.178)	0.026 (0.177)
SOLID	0	0.002 (0.016)	0	0.002 (0.016)
CONVECTON	0.063 (0.435)	0	0.063 (0.435)	0
INTERACTION		0.002 (0.014)		0.007 (0.047)
TOTAL	0.240 (1.667)	0.041 (0.284)	0.095 (0.659)	0.039 (0.270)

radiative transfer from fiber to fiber. Thus, this material could be
related to the thermal conductive concept for homogeneous materials
that is applicable to higher density insulations.

This valuable thermal conductivity standard could not, however,
reflect some of the heat-flow-meter apparatus inconsistencies that
affected the measurements of low-density insulations.

This situation prompted Johns-Manville to develop a system of
reference specimens representative of the low-density insulating
materials to be evaluated. The objective of this program was to
provide a set of reference specimens for each testing location, plant
quality control laboratories, and the central research laboratory.
These individual sets were to be referenced to tests on identical
material tested by NBS.

This paper discusses the background that led to this decision and
the process used to develop these reference specimens.

BACKGROUND

In the early 60's, at the completion of a company round robin on
test apparatus, an unexpected change in the measured thermal conductivity value of the specimens was observed. A check with the same
apparatus, intended to determine if there had been a deterioration of
the specimen, showed a systematic difference between the results
obtained at the start and at the end of the program. These results
are plotted in Figure 3. Upon investigation it was found that the
cold surface plate had been replaced by a new one made of brass. Its
surface had not been painted black as required. A calibration check
of the equipment with the high-density glass-fiber reference standard

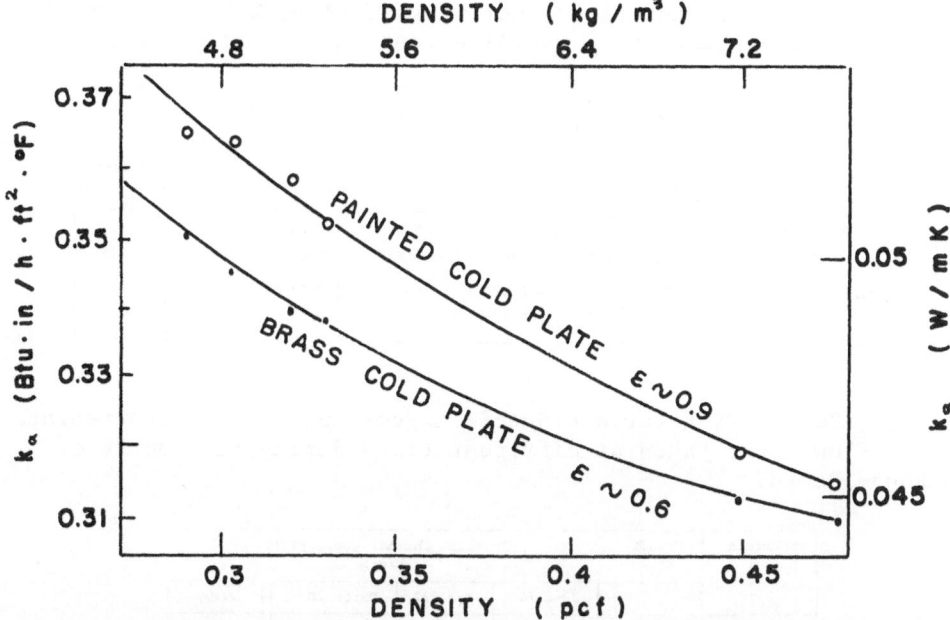

Figure 3.   Apparent thermal conductivity vs. density for 25.4 mm
            (1 in.) specimens. Influence of cold plate emittance on
            the measured values for the same specimens.

failed to reveal the problem. This oversight and the consequent
results led to an in-depth study of the radiative phenomenon in low-
density insulations and to a new approach to the referencing system to
be applied to our measurement equipment.

    As can be observed from the plot of the individual points on the
graph, the results obtained on these specimens of very low-density
(4.6 to 7.7 kg/m^3) retained a consistent relationship to each other
except for the systematic shift, despite a number of repeated measurements at different locations. The reliability of low-density fibrous
insulations in maintaining their apparent thermal conductivity was
further substantiated by repeated measurements on a number of specimens, which were subsequently used to compare the performance of a
number of apparatus located in various company laboratories. The high
degree of reproducibility over time is illustrated by the heat-flow-
meter test results given in Tables 2, 3 and 4. The specimens used for
these measurements were from production materials; no special attention was given to their preparation. The only special care given to
these specimens was to store them in cardboard boxes to protect them.

    The experience gained from these and other measurements on
similar low-density glass-fiber insulations and a careful analysis of
the mechanisms of heat transfer through these insulations led us to

Table 2. Apparent Thermal Conductivity of Glass Fiber Specimens
Measurements by 2 Manville Test Laboratories, spanning
13 years.

Specimen Identification	Density, kg/m³	Apparent Thermal Conductivity (W/m·k) Test Thickness 25.4mm – 23.9°C mean temperature					
		Laboratory A		Laboratory B		$\bar{x}$	σ
		Date	λ_a	Date	λ_a		
6385-5	22.9	Nov. 1969	0.03598	1965	0.03577		
		April 1978	0.03606	June 1978	0.03643	0.03606	.00028
6385-10	47.9	Nov. 1969	0.03182	1965	0.03173		
		April 1978	0.03179	June 1978	0.03202	0.03184	.00013

Table 4. Secondary Reference Specimen Reproducibility Measurements
by Heat Flow Meter taken at different times during the course of
the Round Robin.

Specimen Number	Density kg/m³	Apparent Thermal Conductivity W/m·K			
		June 74	Feb. 77	March 78	Average
0500-4	11.13	0.0448	0.0446	0.0448	0.04473
0500-9	11.09	0.0447	0.0444	0.0448	0.04463
1000-18	16.92	0.0384	0.0382	0.0384	0.03833
1000-24	16.97	0.0386	0.0384	0.0384	0.03846
1500-49	25.49	0.0355	0.0354	0.0356	0.03550
1500-50	25.57	0.0355	0.0354	0.0356	0.03550
2000-14	32.97	0.0342	0.0340	0.0342	0.03413
2000-15	32.84	0.0342	0.0340	0.0343	0.03416

develop an internal calibration system referenced to guarded hot
plates in our Research Laboratory.

Subsequently, it was decided that our measurement system should
be referenced to NBS, in order to validate the results obtained with
all the company heat-flow-meter apparatuses.

THE NBS REFERENCED STANDARDIZATION PROGRAM

In 1974 a program was initiated to reference all heat-flow-meter
apparatuses to measurements made at NBS on low-density glass fiber.
The approach was to take many measurements of a large number of specimens covering the range of densities most commonly measured on the
equipment. From these tests a regression curve would provide the
necessary information to select the representative specimens to be
submitted to NBS for tests. From these results, all individual
specimen values obtained earlier would be adjusted, and thus referenced to NBS measurements. The uncertainty in the density determin-

Table 3. APPARENT THERMAL CONDUCTIVITY (W/m·k) OF GLASS FIBER INSULATION
Mean Temperature 23.9 C – Test Thickness 25.4 mm
Period Covered 1967 – 1978 – 5 Manville Laboratories
– 6 Test Apparatus

Specimen I.D.	Density kg/m³	R/D - A - 1 Unit			R/D-B	R/D-C	QA-D-2 Units		QA-E-2 Units		x̄	σ
		4/67	1/69	3/71	6/78	4/78	5/71	5/71	5/71	5/71		
1-3-MM	10.4	0.04357	0.04373	0.04399 0.04399	0.04398	0.04338 0.04346	0.04360	0.04402	0.04399	0.04399	0.04379	0.00025
2-1-MM	18.9	0.03642	0.03639	0.03635 0.03650	0.03684	0.03642	0.03639	0.03643	0.03649	0.03663	0.03648	0.00014
3-2-NT	25.0	0.03447	0.03463	0.03459 0.03469	0.03503	0.03461	0.03457	0.03456	0.03447	0.03505	0.03466	0.00020
4-1-MM	32.0	0.03288	0.03291	0.03291 0.03296	0.03330	0.03284	0.03277	0.03238	0.03303	0.03317	0.03292	0.00025
5-3-NM	43.7	0.03187	0.03205	0.03179 0.03205	0.03235	0.03198	0.03198	0.03183	0.03216	0.03231	0.03204	0.00019

Test Location – Date of Test

ation applicable to the test area would have no impact on the
data, since each specimen value would relate to the regression
curve and the initial individual measurements.   The hypothetical
correction process is illustrated in Figure 4.

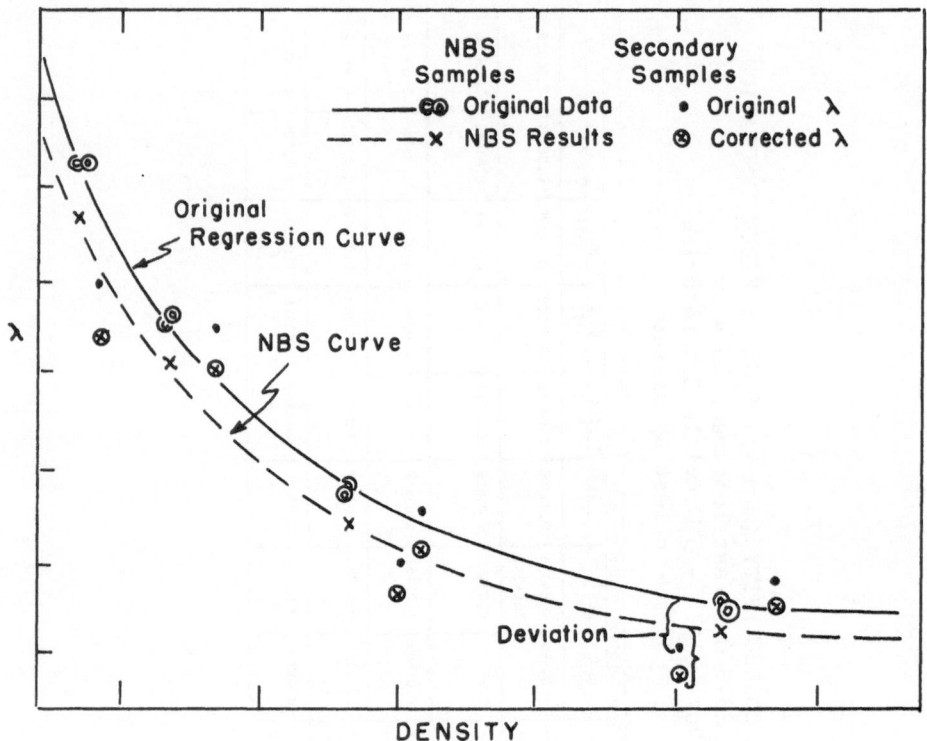

Figure 4.   Hypothetical Correction Process

CALIBRATION PROGRAM

The objectives and limitations of the program were stated as
follows:

Objectives:

1.   To standardize heat-meter "k" measurements at 24°C
     (75°F) mean temperature with 10°C (50°F) cold-face and
     38°C (100°F) hot-face temperatures at all Johns-Manville
     testing locations.

2.   To provide each testing location with secondary reference
     glass-fiber samples of the types normally tested at that
     location.   Such samples calibrated against primary re-
     ference samples that have been tested by NBS in
     Gaithersburg.

3. To provide for further reference primary glass-fiber reference specimens tested at NBS.

4. To establish a checking procedure to insure an accuracy referenced to NBS measurements. A reproducibility of measurements, and limits of confidence over a range of glass-fiber densities representative of products tested.

Limitations: The scope of this program was limited to tests under the following conditions:

1. Low-density glass-fiber insulation

2. Thickness of test samples: 25.4 mm (1 in.)

3. Emittance of hot and cold surface plates: $E \geq 0.95$ (Black)

4. Hot-surface temperature: $38^{\circ}C \pm 1^{\circ}C$ ($100^{\circ}F \pm 2^{\circ}F$)

5. Cold-surface temperature: $10^{\circ}C \pm 1^{\circ}C$ ($50^{\circ}F \pm 2^{\circ}F$)

The tests were to conform to the requirements of ASTM test specification C-518 and Johns-Manville test methods.

Deviation from any of the above requirements would result in only a comparative test not within the scope of this proposed program. Figure 5 schematically represents the process followed during the course of this program.

Initial Testing and Specimen Selection

The specimens, supplied from production materials, approximated this range of densities: 11, 18, 26 and 34 kg/m^3 (0.7, 1.1, 1.6, and 2.1 pcf). The specimens to be tested were first selected from a large population on the basis of density to obtain four groups of specimens. The individual specimens selected were measured four times in random order in our central laboratory heat-flow meter to determine an unreferenced apparent thermal conductivity value. After an extensive data analysis, a number of specimens were selected as most representative. These were submitted to NBS for tests.

NBS Tests

The tests conducted at NBS were performed on the 200 mm (8-in.) guarded hot plate with a 100 mm (4-in.) test area. This apparatus requires two specimens for each test. A total of eight pairs of specimens, two of each density groupings were submitted. In addition to the tests at $24^{\circ}C$ ($75^{\circ}F$) mean temperature, a number of specimens were also tested at $-18^{\circ}C$ ($0^{\circ}F$) and $55^{\circ}C$ ($130^{\circ}F$). The results of these measurements are shown on Figures 6 and 7.

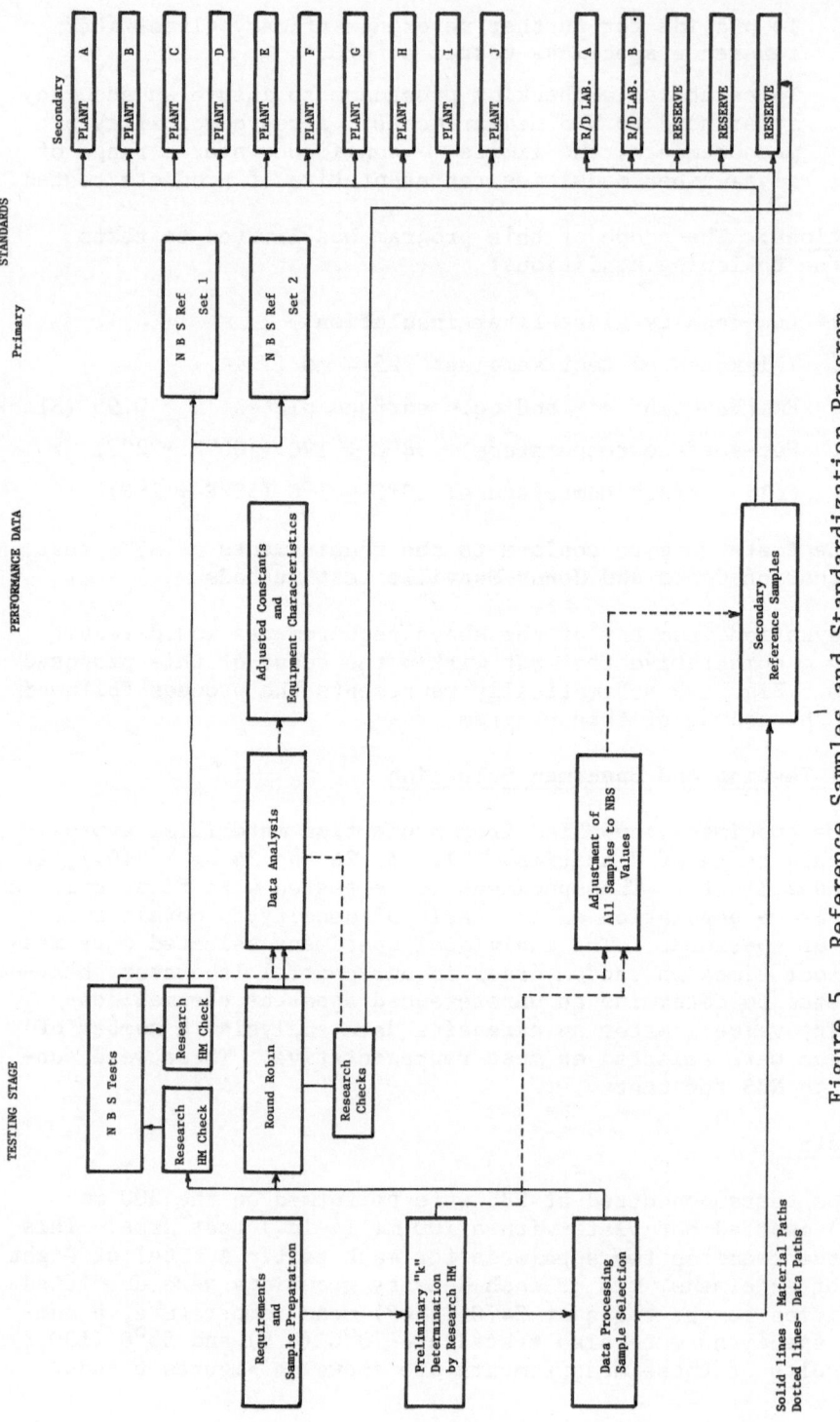

Figure 5. Reference Samples and Standardization Program

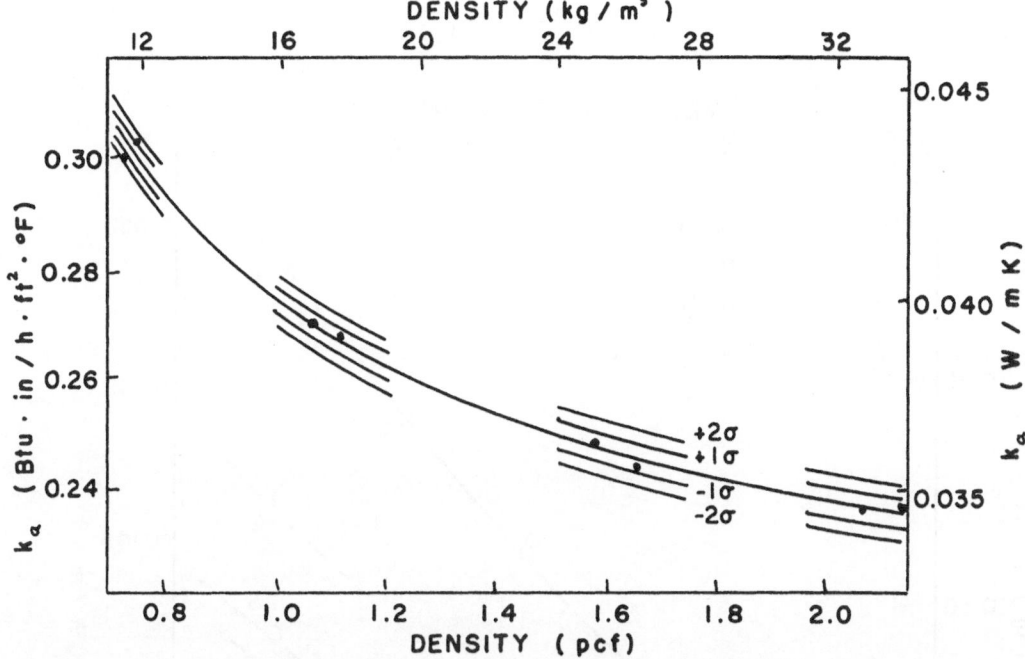

Figure 6. Results of NBS tests on standard reference specimens
24°C (75°F) Mean Temperature

Figure 6 shows the regression curve of the apparent thermal
conductivity at 24°C (75°F) mean temperature versus density for
the total population with the NBS data points. The limits shown
on the graph represent the errors caused by density uncertainty,
apparatus uncertainty, and operator influence for the initial
tests on all the specimens.

Figure 7 shows the measured and extrapolated "k_a" data obtained from the NBS tests versus mean temperature versus specimen density.

These specimens were tested again four times before and after
the NBS tests to insure that no change had occurred. Since then,
they have been checked a number of times; the values have remained
constant within the expected variability of the equipment (σ =
0.00011 (BTU units σ = 0.0008)). These primary reference specimens are retained in the central laboratory and used only for
verification purposes.

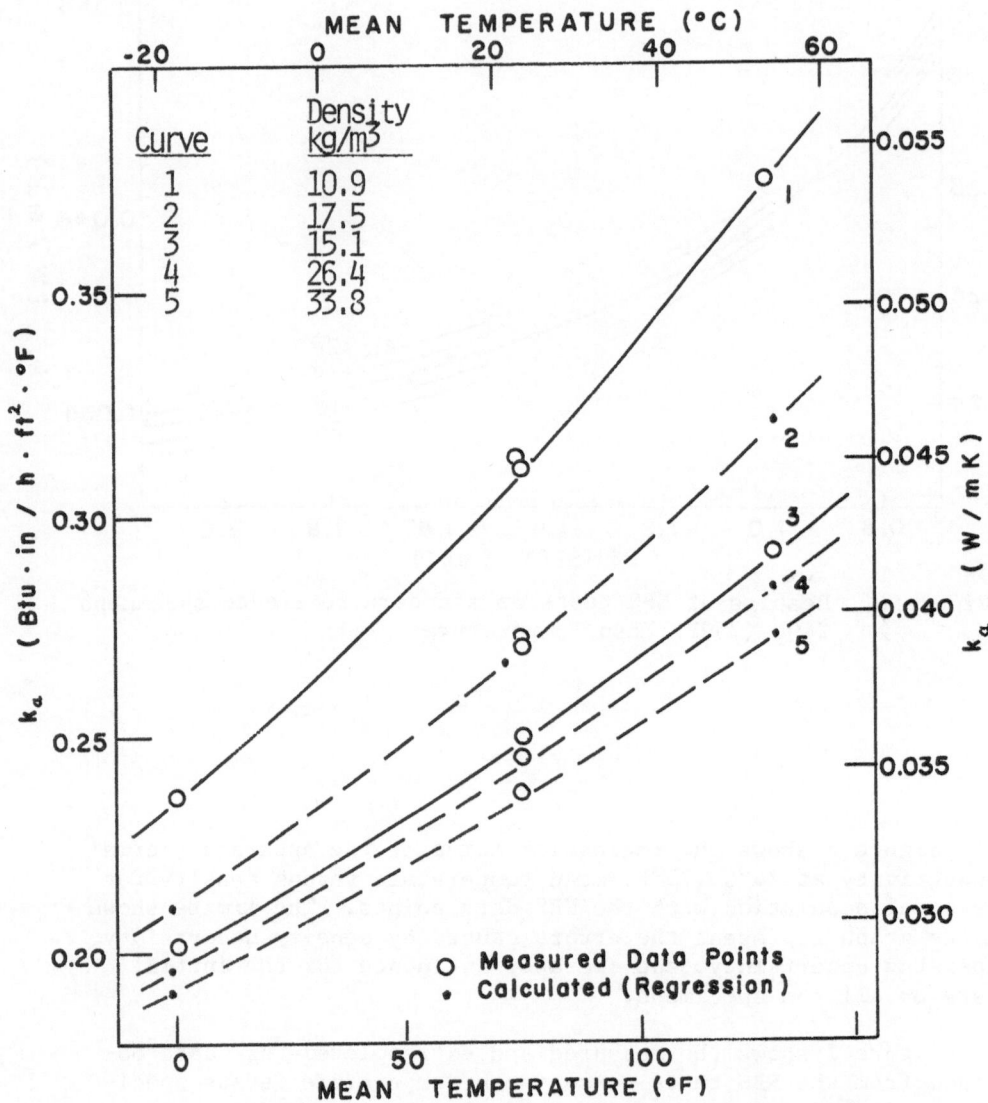

Figure 7.   Measured and extrapolated data from NBS tests at
            25 MM test thickness. Apparent thermal conductivity
            vs. mean temperature glass fiber insulation (rotary)

Internal Round Robin

Concurrently with the tests performed at NBS, a round robin was conducted between the research and the quality control laboratories to establish the characteristics of each apparatus. These specimens were also tested before and after the round robin.

Adjustment of Calibration Constants

The data obtained from the initial round robin, and NBS tests were analyzed. From this analysis, new values were assigned to each specimen, and new calibration constants were provided for each apparatus. All adjustments were rather small, on the order of 1% or less, since the original calibrations were based on the high-density standard (No. 1450) from NBS and measurements by guarded hot plate, on low density glass-fiber insulations.

Secondary-Reference Specimens

To maintain consistent equipment operation, each testing location was provided with two sets of secondary-reference specimens. These specimens were individually packed in a specially designed and identified box. Instructions were provided to each testing location on how to develop the use of day-to-day working reference specimens and to maintain a control chart to keep track of the equipment operation. In case of divergence, the laboratory personnel were instructed to refer to their secondary-reference specimen.

Round Robins

The specimens prepared under this program have proven to be very reliable. The reproducibility of the measurements has been checked at various times during the course of a round robin conducted under the auspices of ASTM Subcommittee C.16.30 on thermal measurements. In no case was their value found to change, see Table 4, when compared to measurements performed on the primary reference specimens tested by NBS. The results of the round robin measurements will be reported after the round robin specimens have been returned for a final evaluation.

SUBSEQUENT APPLICABILITY

The demonstrated value of low-density glass-fiber specimens as reference materials has been in their use in a number of programs at NBS to evaluate thermal measurement equipment. Currently, this type of low-density glass-fiber insulation is used at NBS to provide reference specimens for the evaluation of equipment for tests at thicknesses from 25.4 mm to 152 mm (1 to 6 in.).

CONCLUSIONS

1. Low-density glass fiber materials when carefully selected
    can effectively be used as reference materials. The
    measurements on these materials show that their thermal
    resistance characteristics remain reproducible.

2. Some of the deficiencies in thermal conductivity test
    equipment cannot be detected only with high density in-
    sulations. Low density insulations in which the radia-
    tive heat transfer process in dominant must be used to
    verify the performance of the test equipment.

3. Standard reference materials having the same heat trans-
    mission characteristics as the thermal insulations must
    be used to evaluate the performance of test equipment
    used to measure low density thermal insulations.

REFERENCES

1. C. M. Pelanne, Heat Flow Principles in Thermal Insulations,
    J. Therm. Insul. (1977) 1:48-80.

2. C. M. Pelanne, Experiments on the Separation of Heat
    Transfer Mechanisms in Low-Density Fibrous Insulation,
    in: "Proceedings of the 8th Thermal Conductivity Con-
    ference," Plenum Press, New York (1969), pp 897-910.

3. M. C. I. Siu, Fibrous Glass Board as a Standard Reference
    Material for Thermal Resistance Measurement Systems, in:
    "Thermal Insulation Performance," ASTM STP 718, American
    Society for Testing and Materials, Philadelphia (1980),
    pp 343-360.

A MODEL OF APPARENT THERMAL CONDUCTIVITY

FOR GLASS-FIBER INSULATIONS

L. J. Van Poolen

Calvin College
Grand Rapids, Michigan

J. G. Hust

Thermophysical Properties Division
National Bureau of Standards
Boulder, Colorado

D. R. Smith

Physics Department
South Dakota School of Mines and Technology
Rapid City, South Dakota

ABSTRACT

The heat transfer through glass-fiber insulations is modeled.
The model is based primarily on parallel and independent components
of heat transfer due to: a) gas-fiber-gas and b) fiber conduction,
as well as, c) radiation, and d) convection. The mathematical model,
containing six adjustable parameters, is derived from a physical
model of sequential parallel layers of space (gas filled or evacuated)
and fibers. The total heat transfer, from which one may derive an
apparent thermal conductivity, is a function of the fill gas properties (thermal conductivity, collision diameter, and molecular weight),
insulation properties (solid fiber conductivity, bulk density, fiber
emittance, and fiber diameter), thermodynamic parameters (mean temperature and pressure), and boundary conditions (specimen thickness,
temperature difference, and surface emittance). From data generated
at the National Bureau of Standards, Boulder, Colorado, as well as
published and unpublished data, the parameters in the model were
determined. The functional character of the model using these
parameters is presented and compared qualitatively to available data.

INTRODUCTION

Emphasis on energy conservation and the resulting increased use of insulations have created a need for a better understanding of insulation performance. Efforts continue to be made to develop improved models of heat transfer in insulations.[1-8] Tong and Tien[9] have reviewed many of the models in the literature dealing with thermal radiation transfer in fibrous insulations.

This paper presents a mathematical model for heat transfer in glass-fiber insulations based on a layered physical model. The model consists of n parallel layers of fiber perpendicular to the heat flow direction. A conduction heat path through the solid fiber matrix from one boundary to the other is also assumed. Plausibility arguments are used to introduce coupling, not explicitly derived, between modes of heat transfer.

The fiber layers are separated by a distance d_f and are taken to be of thickness D, the fiber diameter. The heat transfer per unit time per unit area is denoted by q. The two outer boundaries are at temperatures T_1 and T_2 and are separated by a distance Δx.

The total heat flow per unit area, q_t, is:

$$q_t = q_g + q_f + q_{cv} + q_r ,\tag{1}$$

where q_g represents series conduction through the gas-fiber-gas layers. The second term, q_f, represents conduction through fibers from boundary to boundary. The third and fourth terms, q_{cv} and q_r, represent convective and radiative heat transfer, respectively.

The apparent thermal conductivity (averaged with respect to temperature) is defined as:

$$\bar{\lambda} = \frac{q_t \cdot \Delta x}{\Delta T} .\tag{2}$$

In the next section, the terms in eq (1) are developed.

CONDUCTION

Figure 1 shows the resistive analogue of the planar model applied to the fill-gas conductive mechanisms. Resistances to heat flow in the (n+1) gas layers ($1/h_g$) and in the (n) fiber layers ($1/h_f$) are shown where h is a heat transfer coefficient having units of energy/ time/temperature difference/area. The heat flow

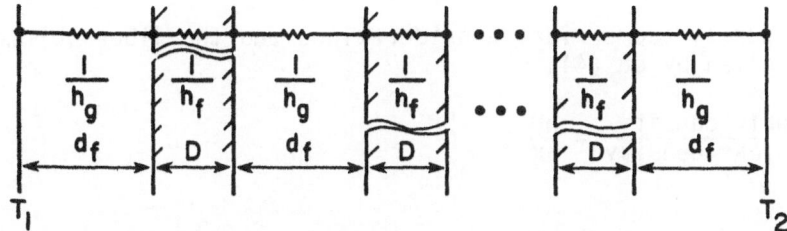

Figure 1. Resistive analog for gas-gas/fiber heat transfer term.

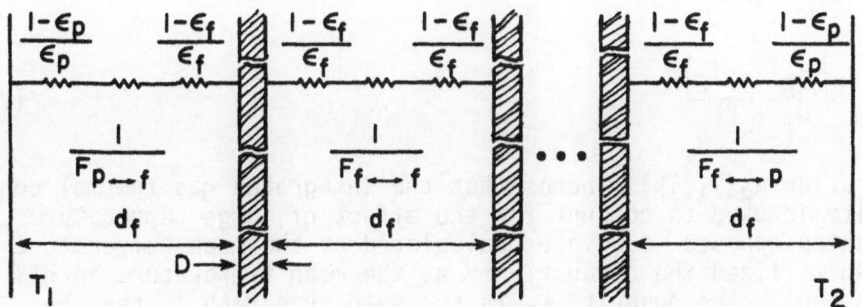

Figure 2. Resistive analog for radiation heat transfer term.

per unit area is the overall ΔT divided by the sum of the resistances, i.e.,

$$q_g = \frac{\Delta T}{\frac{n+1}{h_g} + \frac{n}{h_f}} \cdot \tag{3}$$

From purely geometrical considerations and the fiber volume fraction, given by eq (4),

$$f = \frac{\text{bulk density}}{\text{glass density}} = \frac{nD}{\Delta x} , \tag{4}$$

we obtain

$$q_g = \frac{\Delta T}{\left(\frac{1-f}{d_f h_g} + \frac{f}{Dh_f}\right)\Delta x} , \tag{5}$$

where d_f, the distance between fiber layers, is given by

$$d_f = 4(1-f)G_1 D\Delta x/(\pi f\Delta x + 4G_1 D) . \tag{6}$$

Equation (6) has been altered slightly, from that given in a more detailed analysis by Verschoor and Greebler,[1] to yield a value for d_f of Δx in the limit as $f \to 0$. The parameter, G_1, is introduced to account for uncertainty in the fiber diameter, D. For the insulations and plate spacing considered here, d_f varies from about 0.1 to 1.2 mm. The fiber diameter is 5 μm and Δx is 2.54 cm. Also, from Dushman,[10]

$$h_g = \frac{\lambda_g(T_1,T_2)}{d_f + 2L_g} \cdot \tag{7}$$

The notation $\lambda_g(T_1,T_2)$ denotes that the integrated gas thermal conductivity is used to account for the effect of large temperature differences opposed to a value calculated at the mean temperature (Dushman utilized the conductivity at the mean temperature in his formulation). The quantity L_g is the mean free path in the gas and is defined as:

$$L_g = \frac{K_b \overline{T}}{\sqrt{2} \, \pi d_g^2 P} , \tag{8}$$

where K_b and d_g are the Boltzmann constant and gas collision diameter, respectively. The fill-gas pressure is designated by P, and $\overline{T}$ is taken as $(T_1+T_2)/2$.

Conduction through the solid fibers is taken to be proportional to the conductivity of the fibers, to the volume fraction of the fibers, and inversely to the tortuosity of the fibers. The tortuosity is taken to be proportional to the volume fraction. It has been suggested[3] that the solid conduction term also varies inversely with fiber diameter, D. For glass fibers the conductivity, λ_f, is approximated, over the temperature range $100 \leq T \leq 300$ K, by

$$\lambda_f = 0.053 \ T^{0.56} \ \text{mW} \cdot \text{m}^{-1} \cdot \text{K}^{-1} \ . \tag{9}$$

To allow again for the effect of large ΔT, the integrated equation for the solid conduction term is written as:

$$q_{fo} = \frac{0.053 G_2 f^2}{[1+G_3 D]} \frac{\left(T_1^{1.56} - T_2^{1.56} \right)}{1.56 \ \Delta x} , \tag{10}$$

where G_2 and G_3 are fitted parameters. It is reasonable to argue that the effective tortuosity is also influenced by the degree of radiation and gas conduction between fibers. Thus, as the sum of these heat transfer mechanisms increase, one would expect the fiber conduction term to increase. We have accounted for this coupling in the following manner. Define the ratio, R, as

$$R = \frac{q_r + q_g}{q_{fo}} \ .$$

Then enhance the above value of q_{fo} using R to obtain

$$q_f = (1 + G_4 R/[1+R]) \ q_{fo} \ , \tag{11}$$

where G_4 is a fitted parameter. This enhancement is based on a series model of resistances.

CONVECTION

Fournier and Klarsfeld[11] give criteria that indicate the conditions for convective heat transfer to occur. Such conditions exist for our data near 110 K with the low density material only.

An expression for free convection heat transfer, based on engineering heat transfer principles,[12] is utilized to describe convective heat flow between two horizontal, infinite plates at different temperatures. Utilizing well-known kinetic theory expressions, this heat transfer can be described by:

$$q_{free\ conv} = C_1 \left\{ \left(\frac{P}{d_g}\right)^{2/3} \left(\frac{1}{TM}\right)^{5/6} \Delta T^{1/3} \right\} \Delta T \ , \tag{12}$$

where M is molecular weight. Turbulent flow is assumed and C_1 is 16.0×10^{-30} in SI units.

The presence of fibers diminishes the heat transfer given by eq (12). Allcut,[13] in an analysis of convective heat transfer, presents data indicating a possible exponential decrease of heat flow with insulation bulk density. Utilizing this and accounting for the effect of fiber diameter,

$$q_{cv} = e^{-fG_5/D} \cdot q_{free\ conv} \ , \tag{13}$$

where G_5 is a parameter to be fitted.

RADIATION

Figure 2 indicates the resistive analogue of the layered model as applied to the thermal radiation passing through the insulation.

Following engineering heat transfer theory,[12] the terms containing ε_p or ε_f (emittances of surface and of fiber layers, respectively) represent surface resistances, which, in turn, represent an averaging of multiple reflections and absorptions of photons at these surfaces. The $F_{p \leftrightarrow f}$ and $F_{f \leftrightarrow f}$ are geometric shape factors set equal to unity; i.e., it is assumed that the layers seen in figure 2 are infinite in extent.

The radiative heat flow is derived by summing the resistances corresponding to each pair of surfaces and assuming that q_r is the same throughout:

$$q_r = \frac{\sigma_{sb} d_f G_6}{\Delta x (1-f)} \left\{ \frac{T_1^4 - T_2^4 - T_1^{*4} + T_2^{*4}}{\frac{1}{\varepsilon_f} + \frac{1}{\varepsilon_p} - 1} + \frac{T_1^{*4} - T_2^{*4}}{\frac{2}{\varepsilon_f} - 1} \right\} \ , \tag{14}$$

where σ_{sb} is the Stefan-Boltzmann constant.

The temperatures T_1^* and T_2^* are obtained from

$$T_1^* = T_1 - 0.5 \, d_f \, \Delta T/\Delta x$$

and

$$T_2^* = T_2 + 0.5 \, d_f \, \Delta T/\Delta x \; .$$

It is noted that when $\varepsilon_p = \varepsilon_f$, the terms containing T_1^* and T_2^* cancel. The coefficient G_6 is obtained by fitting the model to data.

TOTAL HEAT TRANSFER

The total heat transfer is then represented by the sum of equations (5), (11), (13), and (14). This sum contains six unknown parameters, $G_1 \ldots G_6$. Currently G_3 for the solid conduction and G_5 for the convection are adjusted for qualitative agreement with data, and then G_1, G_2, G_4, and G_6 are fitted to data by a nonlinear fitting routine. Values of these parameters are listed in table 1. Equation (2) is then used with the above value of q_t to obtain the commonly defined apparent thermal conductivity.

The complete set of heat transfer data by Smith, Hust, and Van Poolen[14,15], obtained in a guarded-hot-plate apparatus at the National Bureau of Standards, Boulder, Colorado, as well as other published and unpublished data were used to determine the parameters in the model.

Table 1. Parameters, G_i, for heat transfer components as defined by eqs. (6), (10), (11), (13), and (14).

i	G_i
1	0.47
2	0.46
3	2×10^4
4	2.6
5	0.008
6	4.8

These parameters are valid for all variables in SI units, i.e., T in K, ρ in kg/m^3, P in Pa, and D and Δx in m.

MODEL BEHAVIOR

To illustrate the characteristics of the model, functional plots
of λ, similar to those found in the literature, were generated.
Figure 3 represents λ versus T for the conditions shown. (For
figures 3, 4, and 5, the boundary conditions are set at ΔT = 25 K,
Δx = 2.54 cm and ε_p = 0.85). The curve for lower density material
exhibits the upward curvature expected due to radiation; the curve
for higher density material exhibits a more nearly linear relationship due to the large influence of gas conduction at atmospheric
pressure and solid conduction at high vacuum conditions.

Figure 4 shows λ versus bulk density. The relatively large
negative slopes at low density represent radiation effects, and the
shallow minimum and slightly positive slopes at higher densities
indicate diminished radiation effects relative to increasing importance of gas and solid conduction mechanisms.

Figure 5 shows λ versus gas pressure. The pressure at the ini-
tiation of the decrease in conductivity, the two orders of magnitude
drop in pressure until the curve flattens, and the total decrease in
λ from atmospheric pressure to high vacuum conditions agree qualitatively with data of Smith et al.,[16] Bankvall,[3] and Pelanne.[4]

SUMMARY

This model for the apparent thermal conductivity of glass-fiber
insulation represents experimental data over a wide variety of conditions. The model was fitted to thermal conductivity data ranging
from approximately 0.5 to 200 mW·m^{-1}·K^{-1}.

A total of over 200 data points were used to fit the parameters
of the model. The vast range of measurement conditions include: a)
air, nitrogen, argon, and helium fill gases, b) fill gas pressures
from atmospheric to high vacuum, c) specimen densities from 11 to
147 kg/m^3, d) temperatures from 107 to 336 K, and e) plate emittances
of 0.04 and 0.83.

At the upper end of the thermal conductivity range, the model
fits to within about five percent of the measured conductivities.
An examination of the systematic differences between specimens
reveals that characterization parameters exist that are neither measured nor contained within the model. It is reasonable to expect
that the resin content and the fiber orientation of the specimens
affect the conductivity.

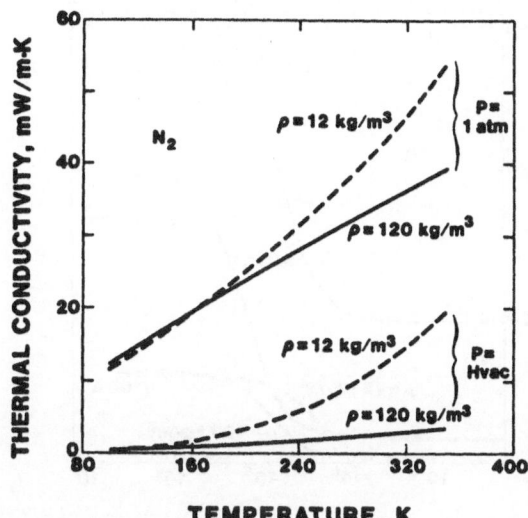

Figure 3.   Calculated curves of λ versus T for various fill-gas
            pressures and densities.

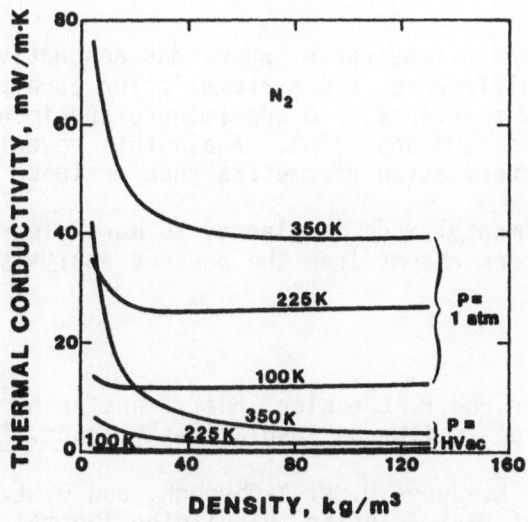

Figure 4.   Calculated curves of λ versus density at various mean
            temperatures and fill-gas pressures.

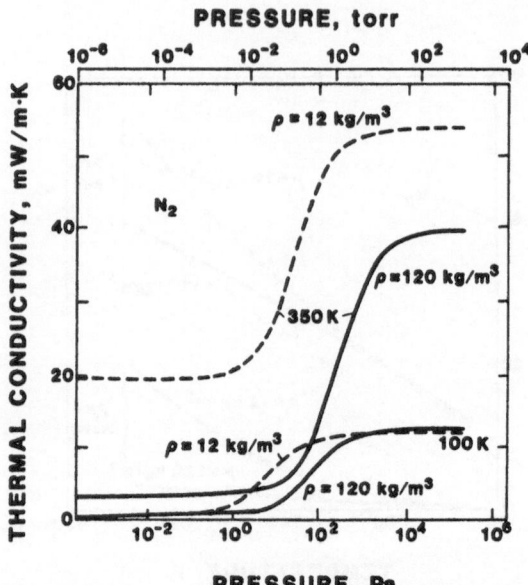

Figure 5.   Calculated curves of λ versus pressure at various
             densities and mean temperatures.

At the lower end of the range, where gas conductivity is small,
the data-to-model differences are systematic for each pair of specimens but are widely scattered from specimen-to-specimen. The differences are as much as +50% and -100%. Again this reveals the presence
of undefined characterization parameters such as those mentioned above.

Further experimental work is planned to more clearly define the
significant parameters absent from the present analysis.

REFERENCES

1.   J. D. Verschoor and P. Greebler, Heat transfer by gas conduction
      and radiation in fibrous insulations", Trans. ASME 74:961
      (1952).
2.   H. A. Fine, S. H. Jury, D. W. Yarbrough, and D. L. McElroy,
      "Analysis of Heat Transfer in Building Thermal Insulation,"
      ORNL/TM-7481, Oak Ridge National Laboratory, Oak Ridge,
      Tennessee (1980).

3.   C. G. Bankvall, Mechanisms of heat transfer in permeable insu-
     lation and their investigation in a special guarded hot plate,
     in: "Heat Transmission Measurements in Thermal Insulations,"
     ASTM STP 544, American Society for Testing and Materials,
     Philadelphia (1974), pp. 34-48.
4.   C. M. Pelanne, Experiments on the separation of heat transfer
     mechanisms in low-density fibrous insulation, in: "Thermal
     Conductivity," C. Y. Ho and R. E. Taylor, eds., Plenum Press,
     New York (1969), pp. 897-911.
5.   J. J. Thigpen and B. E. Short, "The Apparent Thermal Conductivity
     of Fibrous Materials", Paper No. 59-A-293, ASME, New York
     (1959).
6.   C. R. King, Fibrous insulation heat transfer model, in: "Thermal
     Transmission Measurements of Insulation," ASTM STP 660, R. P.
     Tye, ed., American Society for Testing and Materials, Phila-
     delphia, (1978) pp. 281-92.
7.   A. H. Striepens, Heat Transfer in Refractory Fiber Insulations,
     in: "Thermal Transmission Measurements of Insulations," ASTM
     STP 660, R. P. Tye, ed., American Society for Testing and
     Materials, Philadelphia (1978) pp. 293-309.
8.   S. P. Vnukov, V. A. Ryabov, and D. V. Fedoseev, Thermal conduc-
     tivity of glass-fiber systems, J. Eng. Phys. (USSR)
     21(5):1350-4 (1971).
9.   T. W. Tong and C. L. Tien, Analytical models for thermal radia-
     tion in fibrous insulations, J. Therm. Insul. 4:27-44 (1980).
10. S. Dushman, "Scientific Foundations of Vacuum Technique," John
     Wiley & Sons, New York (1949).
11. D. Fournier and S. Klarsfeld, Some recent experimental data on
     glass fiber insulating materials and their use for a reli-
     able design of insulations at low temperatures, in: "Heat
     Transmission Measurements in Thermal Insulations," ASTM STP
     544, American Society for Testing and Materials, Philadel-
     phia (1974), pp. 223-242.
12. F. Kreith, "Principles of Heat Transfer," 3rd ed., Intext,
     New York (1973).
13. E. A. Allcut, An analysis of heat transfer through thermal
     insulating materials," in: "General Discussion on Heat
     Transfer, Institute of Mechanical Engineers, London (1951),
     pp. 232-235, 486-7.
14. D. R. Smith and J. G. Hust, "Effective Thermal Conductivity of
     a Glass Fiberboard Standard Reference Material," NBSIR
     81-1639, National Bureau of Standards, Boulder, Colorado,
     (1981).
15. D. R. Smith, J. G. Hust, and L. J. Van Poolen, "Effective Ther-
     mal Conductivity of a Glass Fiberblanket Standard Reference
     Material," NBSIR 81-1640, National Bureau of Standards,
     Boulder, Colorado (1981).

16. D. R. Smith, J. G. Hust, and L. J. Van Poolen, Effective thermal conductivity of glass-fiber board and blanket standard reference materials, in: "Proceedings of the Seventeenth International Thermal Conductivity Conference," Plenum Press, New York (1982), pp. 483-496.

AUTHOR INDEX

Abey, A.E. 459
Afshar, R. 295
Albers, M.A. 471
Alexander, T. 737
Anderson, A.C. 185
Andersson, P. 527
Andrew, J.F. 209
Armstrong, B.H. 437
Ashworth, E. 737
Ashworth, T. 737

Backstrom, G. 527
Bai, L.-Y. 349
Balageas, D.L. 699
Baroncini, C. 285
Beck, A.E. 619
Bentsen, L.D. 677
Berg, J.I. 329
Berman, R. 105,429
Bogaard, R.H. 45,195
Bomberg, M. 393
Boulant, J. 381
Boxus, J. 537
Brandt, R. 117
Braun, R. 547
Brockerhoff, P. 369

Cabannes, F. 125
Chen, Q.-T. 643
Cheng, K.C. 635
Chi, T.C. 195
Cook, J.G. 745

Deshpande, M.S. 45
Desjarlais, A.O. . . . 711
Di Filippo, P. 285
Dong, L.-K. 349
Durham, W.B. 459

Fang, Z.-H. 601
Fine, H.A. 359
Fodemesi, S.P. 619
Frohn, A. 315
Fukusako, S. 635

Graves, R.S. 153,219
Groot, H. 611

Hardy, N.D. 105
Hasselman, D.P.H. . . . 677
Havill, T.N. 195
Haverlah, C. 559
Heremans, J. 537
Ho, C.Y. 195
Horlick, J. 497
Howlett, S.P. 447
Hsieh, C.K. 665
Hust, J.G. 105,483,777

Isaacs, L.L. 55
Issi, J.P. 537

James, H.M. 195
Jones, R.R. 419
Jury, S.H. 359,727

Kirkpatrick, D.M. . . . 497
Klarsfeld, S. 381
Klemens, P.G. 25,177,209
Koenig, J.R. 137
Koski, J.A. 587
Kuan, C.-N. 265

Langlais, C. 381
Latini, G. 285
Laubitz, M.J. 745
Lee, C.O. 229

Lee, H.J.   229
Lee, S.K.   229
Lin, J.-H.   341
Livesley, D.M. . . .   429
Luc, A.M.   699
Lucks, C.F.   3

MacDonald, W.M. . . .   185
Maglic, K.D.   163
McCurdy, A.K.   63
McElroy, D.L.   219,359,727
McVey, D.F.   587
Minges, M.L.   73
Mirkovich, V.V. . . .   147,519
Moore, J.P.   153,727
Morrell, R.   447
Morris, V.L.   559
Murad, S.   295

Nagasaka, Y.   251,307
Nagashima, A. . . .   251,307
Naziev, Ya.M. . . .   275
Neuer, G.   117
Neumann, W.   689
Ni, H.-L.   643

Ober, D.G.   419
Omotani, T.   251

Pacetti, M.   285
Pears, C.D.   137
Pelanne, C.M. . . .   763
Perovic, N.   163
Poulaert, B.   537
Prabhuram, ...   241

Reiss, H.   569
Ren, Z.P.   601
Rennex, B.G.   419
Roder, H.M.   257
Ross, R.G.   527
Rowley, R.L.   31

Sahota, M.   105
Saksena, M.P.   241
Saxena, S.C.   295
Schaber, A.   547
Seki, N.   635
Shakhverdiyev, A.N. .   275
Siu, M.C.I.   413
Smith, D.R.   483,777
Solvason, K.R.   393
Sparks, L.L.   655
Spinney, S.C.   507
Srinivasan, M.   677

Tainsh, R.J.   105
Taylor, R.   447,753
Taylor, R.E.   611
Tye, R.P.   507,711

Van Poolen, L.J. . . .   777

Wang, B.-M.   643
Wang, B.-X.   601
Wang, W.-Y.   55
Wang, X.-A.   665
Wei, Z.   341
Weaver, F.J.   219
Westerdorf, M.   315
Wheat, T.A.   519
White, G.K.   95
Williams, R.K.   219

Xi, T.-G.   643

Yarbrough, D.W. . . .   265,359
Yen, T.S.   643
Yin, Z.   643

Zhou, B.-L.   341,349
Zivotic, Z.   163

SUBJECT INDEX

Ablation	341,349
Alloys	195,209,219
Alumina	125,447,519,711
Argon	241
Binary Mixtures	31,285
Butanol	285
Calcium Fluoride	63
Carbon Monoxide	295
Carbonate Gneiss	737
Cement	655
Ceramics	643
Composite	45,689,699,711
Concrete	507,655
Contact Resistance	25
Convection	619
Copper	73,195
Crystal	63,429,527
Cyclohexane	275
Cyclopentane	527
Debye Temperature	185
Density	137
Eicosane	265
Electrical Resistivity	95,105,137,153,219,251,689
Ethanol	537
Ethylene Oxide	745
Gases	295,315
Glass	447
Glass Beads	619
Glass Fiber	359,413,419,471,483,727
Graphite	45,55,73,117,137,147,153,163, 711,753
Gruneisen Parameter	177,437
Guarded Hot Plate Apparatus	393,413,419,483,507,559,727

791

Heat Flow Meter   419,507,727
Helium   369,429
Heptadecane   265
Hot Wire Apparatus   527
Hydrate   527,745
Hydrogen   257

Insulation   329,359,369,381,413,419,471,
                                   483,497,569,727
Iron   73,105,163,643
Isotope Effect   429

Knudsen Number   315

Lead   195
Line Heat Source   413,619
Liquids   31,241,251,275,285,307
Lorenz Number   209,753

Matthiessen's Rule   95
Mean Free Path   185,329,419,643
Mercury   241
Metal Powder   25
Methylphenyl Silicone (RTV)   559
Mica   643
Mortar   507,655

Nickel   447
Nitride   677
Nitrided Steel   229
Nitrogen   241,315

Ocean Sediments   619
Octadecane   265
Oil Sand   635
Oil Shale   611

Pentadecane   265
Perlite   507
Phonon Conduction   63,151,185,209,219,359,429,
                                   437,447,677,745
Platinum   95,195
Plutonium Alloys   209
Polystyrene   393
Porosity   125,137
Potassium Nitrate   251
Pyrex   559
Pyroceram   73,711

Quartz   459

Radiation	25,329,359,381,419,447,537
Reactor	369
Rock	125,459,737
Rock Wool	359
Sapphire	447
Seismic Waves	177
Silica	447,459,699
Silicon	63
Silicon Carbide	677,711
Silicon Nitride	125
Sodium Chloride	307
Sodium Fluoride	437
Sodium Nitrate	251
Soil	665
Specific Heat	45,55,185,611,711
Standard Reference Materials	45,55,73,95,105,117,125,137, 147,153,163,419,483,497,753
Steel	73,105,163,219,753
Ternary Mixture	31
Tetradecane	265
Tetrahydrofuran	527,745
Thermal Diffusivity	117,147,153,163,209,229,341, 447,459,635,643,677,699,753
Thermal Expansion	95,163,177,219,251,413,619,677
Thermoelectric Power	193,539
Thoria	125
Toluene	251,275,285,537
Tungsten	73,105,117,447
Umklapp Process	429,437,447
Velocity of Sound	137
Water	275,285,307,315
Zirconia	125